Three-Dimensional Nanoarchitectures

Weilie Zhou · Zhong Lin Wang
Editors

Three-Dimensional Nanoarchitectures

Designing Next-Generation Devices

Springer

Editors
Weilie Zhou
Advanced Materials Research Institute
University of New Orleans
2000 Lakeshore Drive
New Orleans, LA 70148, USA
wzhou@uno.edu

Zhong Lin Wang
School of Materials Science and Engineering
Georgia Institute of Technology
771 Ferst Drive, N.W.
Atlanta, GA 30332-0245, USA
zhong.wang@mse.gatech.edu

ISBN 978-1-4419-9821-7 e-ISBN 978-1-4419-9822-4
DOI 10.1007/978-1-4419-9822-4
Springer New York Dordrecht Heidelberg London

Library of Congress Control Number: 2011930518

Printed on acid-free paper

Springer is part of Springer Science+Business Media (www.springer.com)

Contents

Contributors

Baobao Cao Advanced Materials Research Institute, University of New Orleans, New Orleans, LA 70148, USA, bcao1@uno.edu

Gaoping Cao Research Institute of Chemical Defense, Beijing 100191, China, caogaoping@tom.com

Jiajun Chen Advanced Materials Research Institute, University of New Orleans, New Orleans, LA 70148, USA, jchen2@uno.edu

Xing Cheng Department of Electrical and Computer Engineering, Texas A&M University, College Station, TX 77843-3128, USA, chengx@ece.tamu.edu

Jeong-Hyun Cho Department of Chemical and Biomolecular Engineering, The Johns Hopkins University, Baltimore, MD 21218, USA, itrustyou@gmail.com

Ik Su Chun Micro and Nanotechnology Laboratory, Department of Electrical and Computer Engineering, University of Illinois, Urbana, IL 61801, USA, ichun2@illinois.edu

Yi Cui Department of Materials Science and Engineering, Stanford University, Stanford, CA 94305, USA, yicui@stanford.edu

Eric N. Dattoli Department of Electrical Engineering and Computer Science, University of Michigan, Ann Arbor, MI 48109-2122, USA, edattoli1@gmail.com

Shujiang Ding NanoScience Technology Center, University of Central Florida, Orlando, FL 32826, USA, sjding@mail.ucf.edu

Shanui Fan Department of Electrical Engineering, Stanford University, Stanford, CA 94305, USA, shanhui.fan@stanford.edu

Hong-Jun Gao Institute of Physics, Chinese Academy of Sciences, Beijing 100190, China, hjgao@iphy.ac.cn

Pu-Xian Gao Department of Chemical, Materials and Biomolecular Engineering and Institute of Materials Science, University of Connecticut, Storrs, CT 06269-3136, USA, puxian.gao@ims.uconn.edu

David H. Gracias Department of Chemical and Biomolecular Engineering, The Johns Hopkins University, Baltimore, MD 21218, USA, dgracias@jhu.edu

Hai-Ming Guo Institute of Physics, Chinese Academy of Sciences, Beijing 100190, China, hmguo@iphy.ac.cn

Haiping Hong Department of Material and Metallurgical Engineering, South Dakota School of Mines and Technology, Rapid City, SD 57701, USA, haiping.hong@sdsmt.edu

Yan Hong NanoScience Technology Center, University of Central Florida, Orlando, FL 32826, USA; Department of Mechanical, Materials and Aerospace Engineering, University of Central Florida, Orlando, FL 32826, USA, yanhong@mail.ucf.edu

Mark Horton Department of Material and Metallurgical Engineering, South Dakota School of Mines and Technology, Rapid City, SD 57701, USA, mark.horton@mines.sdsmt.edu

Mainul Hossain School of Electrical Engineering and Computer Science, University of Central Florida, Orlando, FL 32826, USA, mainul@knights.ucf.edu

Ching-Mei Hsu Departments of Materials Science and Engineering and Electrical Engineering, Stanford University, Stanford, CA 94305, USA, chingmei1219@gmail.com

Steve Hu Department of Chemical and Biomolecular Engineering, The Johns Hopkins University, Baltimore, MD 21218, USA, shu5@buffalo.edu

Sangmoo Jeong Department of Electrical Engineering, Stanford University, Stanford, CA 94305, USA, popomoo@stanford.edu

Wenzhi Li Department of Physics, Florida International University, Miami, FL 33199, USA, liwenzhi@fiu.edu

Xiuling Li Micro and Nanotechnology Laboratory, Department of Electrical and Computer Engineering, University of Illinois, Urbana, IL 61801, USA, xiuling@illinois.edu

Yat Li Department of Chemistry and Biochemistry, University of California, Santa Cruz, CA 95064, USA, yatli@ucsc.edu

Jin-Hee Lim Department of Chemistry and Advanced Materials Research Institute, University of New Orleans, New Orleans, LA 70148, USA, jlim@uno.edu

Zhiqun Lin Department of Materials Science and Engineering, Iowa State University, Ames, IA 50011-2300, USA, zqlin@iastate.edu

Gang Liu Department of Chemical, Materials and Biomolecular Engineering, University of Connecticut, Storrs, CT 06269-3136, USA; Institute of Materials

Science, University of Connecticut, Storrs, CT 06269-3136, USA, liugang.zzq@gmail.com

Qi Liu Institute of Physics, Chinese Academy of Sciences, Beijing 100190, China, qliu@semi.ac.cn

Wei Lu Department of Electrical Engineering and Computer Science, University of Michigan, Ann Arbor, MI 48109-2122, USA, wluee@eecs.umich.edu

Zeyu Ma NanoScience Technology Center, University of Central Florida, Orlando, FL 32826, USA; Department of Mechanical, Materials and Aerospace Engineering, University of Central Florida, Orlando, FL 32826, USA, zma@mail.ucf.edu

Xin Miao Micro and Nanotechnology Laboratory, Department of Electrical and Computer Engineering, University of Illinois, Urbana, IL 61801, USA, xmiao3@illinois.edu

Suman Neupane Department of Physics, Florida International University, Miami, FL 33199, USA, suman.neupane@fiu.edu

G.P. Peterson Woodruff School of Mechanical Engineering, Georgia Institute of Technology, Atlanta, GA 303, USA, bud.peterson@gatech.edu

Satish Chandra Rai Advanced Materials Research Institute, University of New Orleans, New Orleans, LA 70148, USA, scrai@my.uno.edu

Jian Shi Department of Materials Science and Engineering, University of Wisconsin-Madison, Madison, WI, USA, jshi7@wisc.edu

Ming Su NanoScience Technology Center, University of Central Florida, Orlando, FL 32826, USA; Department of Mechanical, Materials and Aerospace Engineering, University of Central Florida, Orlando, FL 32826, USA; School of Electrical Engineering and Computer Science, University of Central Florida, Orlando, FL 32826, USA, mingsu@mail.ucf.edu

Limin Tong State Key Laboratory of Modern Optical Instrumentation, Zhejiang University, Hangzhou 310027, China, phytong@zju.edu.cn

Daniel Vennerberg Department of Materials Science and Engineering, Iowa State University, Ames, IA 50011, USA, dcv001@iastate.edu

Jun Wang Department of Materials Science and Engineering, Iowa State University, Ames, IA 50011, USA, junwang@iastate.edu

Kai Wang Advanced Materials Research Institute, University of New Orleans, New Orleans, LA 70148, USA, kwang1@uno.edu

Xudong Wang Department of Materials Science and Engineering, University of Wisconsin-Madison, Madison, WI, USA, xudong@engr.wisc.edu

Ye-Liang Wang Institute of Physics, Chinese Academy of Sciences, Beijing 100190, China, ylwang@iphy.ac.cn

Zhong Lin Wang School of Materials Science and Engineering, Georgia Institute of Technology, Atlanta, GA 30332-0245, USA, zlwang@gatech.edu; zhong.wang@mse.gatech.edu

John B. Wiley Department of Chemistry and Advanced Materials Research Institute, University of New Orleans, New Orleans, LA 70148, USA, jwiley@uno.edu

Xukai Xin Department of Materials Science and Engineering, Iowa State University, Ames, IA 50011, USA, xxin@iastate.edu

Qing Yang State Key Laboratory of Modern Optical Instrumentation, Zhejiang University, Hangzhou 310027, China; School of Materials Science and Engineering, Georgia Institute of Technology, Atlanta, GA 30332, USA, qingyang@zju.edu.cn

Yusheng Yang Research Institute of Chemical Defense, Beijing 100191, China, yangyush@public.bta.cn

Zongfu Yu Department of Applied Physics, Stanford University, Stanford, CA 94305, USA, zfyu@stanford.edu

Hai-Gang Zhang Institute of Physics, Chinese Academy of Sciences, Beijing 100190, China, hgzhang@iphy.ac.cn

Hao Zhang Research Institute of Chemical Defense, Beijing 100191, China, dr.h.zhang@hotmail.com

Minghui Zhang NanoScience Technology Center, University of Central Florida, Orlando, FL 32826, USA, zhangmh@nankai.edu.cn

Weilie Zhou Advanced Materials Research Institute, University of New Orleans, New Orleans, LA 70148, USA, wzhou@uno.edu

Jia Zhu Department of Electrical Engineering, Stanford University, Stanford, CA 94305, USA, jiazhu@stanford.edu

Chapter 1
Building 3D Nanostructured Devices by Self-Assembly

Steve Hu, Jeong-Hyun Cho, and David H. Gracias

1.1 The Pressing Need for 3D Patterned Nanofabrication

Macroscale engineering is 3D and most structures around us are machined and assembled with a variety of materials that are precisely shaped and patterned. However, while several methodologies exist for precise patterning and assembly at the nanoscale, they can only be enabled in an inherently 2D manner; moreover, many are serial and extremely expensive processes [1–4]. Top-down methods such as particle replication in nonwetting templates (PRINT) [5] have enabled the mass production of nanoparticles, but they tend to have shapes consistent with only a single layer and limited patterning. Hence, there is a pressing need to develop parallel and cost-effective methods for patterning and assembly of 3D nanostructured devices (Fig. 1.1). In this chapter, we discuss methods that focus on the creation of precisely patterned molecular or engineered building blocks that are then driven to assemble themselves into 3D structures using a variety of driving forces. Some of these methods leverage already existing e-beam and imprint lithographic infrastructure by engineering 3D structures from building blocks that are exquisitely patterned in 2D and then subsequently assembled in 3D. We review the challenges associated with self-assembling methodologies with a focus on lithographically defined building blocks and discuss future challenges and prospects.

It is important to clarify at the outset what is meant by a 3D nanostructured device. We characterize one as being composed of either (a) a homogeneous or heterogeneous material composition or (b) structural elements or patterns with a 1–100 nm size scale resolution. Here, one can draw an analogy to macrostructured functional devices such as cars, houses, or planes. An example of a structure with a homogeneous composition on the macroscale is a bare cement wall, while that with a heterogeneous macroscale composition is a cement wall that is interlaid with a glass window and a wooden door with handles. On the microscale, a composite

D.H. Gracias (✉)
Department of Chemical and Biomolecular Engineering, The Johns Hopkins University, Baltimore, MD 21218, USA
e-mail: dgracias@jhu.edu

W.L. Zhou, Z.L. Wang (eds.), *Three-Dimensional Nanoarchitectures*,
DOI 10.1007/978-1-4419-9822-4_1,

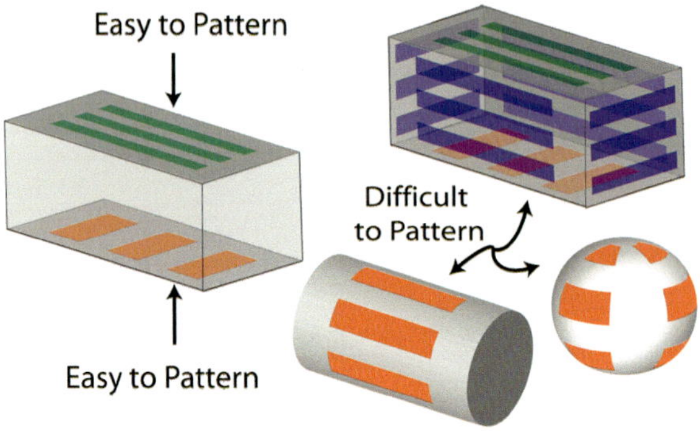

Fig. 1.1 Diagram showing the challenges of fabricating 3D patterned nanostructures. Although it is relatively simple to pattern the *top* and *bottom* of a 3D structure, it is very challenging to pattern perpendicular to the plane or on a curved surface

material such as carbon-fiber reinforced steel has a heterogeneous composition, while glass has a homogeneous composition. This analogy suggests that in general, structures and devices with heterogeneous patterning can provide multifunctionality by leveraging functional traits from each element. Nanoscale engineering seeks to facilitate extreme miniaturization at length scales of 1–100 nm, and there is a need to fabricate 3D structures and devices with both homogeneous and heterogeneous composition and with elements whose sizes range from 1 to 100 nm. One vision for the era of miniaturization was beautifully articulated in Richard Feynman's seminal lecture [6], *There's plenty of room at the bottom*, where he said, "Consider any machine – for example, an automobile – and ask about the problems of making an infinitesimal machine like it." At the present time, human engineering is not even close to creating machines like cars at the nanoscale. In fact, while numerous methods have been developed to grow nanostructures such as spherical or branched nanoparticles, nanorods, nanowires, and nanotubes [7–12], these structures have relatively simple shapes with little or no surface patterning.

Multilayer patterning is essential for the construction of circuits, optical elements, energy harvesting tools, and biomedical devices. Recently, alternative unconventional approaches [13] have been pursued to enable this 3D nanofabrication from the bottom-up. The idea has been put forth that there is a need to build complex structures from precursors using a process that mimics biological assembly. This self-assembly approach focuses on bringing together components with pre-programmed interactions to form organized and functional structures. Self-assembling systems typically consist of components such as atoms, molecules, or larger synthetically structured components. The individual parts can typically interact with each other through a variety of chemical or physical forces. Chemical forces include relatively strong interactions such as those facilitated by covalent and ionic bonds or weak interactions such as hydrogen bonds. Physical forces include interactions driven by electrostatic, magnetic, steric, mechanical stress based, or surface tension forces. These components, when brought together, interact through one of the many forces and self-assemble into organized structures. Typically, components are agitated during self-assembly by a variety of means such as Brownian motion, convection, sonication, or physical shaking. This agitation allows components to explore different orientations and interactions, ultimately enabling stable structures to form within energy minima. Nature is the master of self-assembly, utilizing this approach to create mostly everything from galaxies to humans to seashells and extending all the way down to microorganisms and nanoscale viruses. On closer examination, molecules are self-assembled aggregates of atoms, proteins are self-assembled aggregates of amino acids, and tissues are self-assembled aggregates of cells. It is well known that natural assembly has two distinct self-assembly flavors, namely equilibrium and non-equilibrium self-assemblies. Self-assemblies at equilibrium do not require energy dissipation to maintain their organization; examples include rocks, mountains, or shells. Such assemblies are typically associated with non-living structures. In contrast, non-equilibrium assemblies need to dissipate energy continuously to maintain their organization; examples include microorganisms, humans, and other animals.

Scientists and engineers have long sought to mimic the naturally occurring self-assembly strategy. There are several examples of functional equilibrium self-assembly structures. The stringing together of monomers such as styrene to form synthetic polymers such as polystyrene is a classic example [14]. Synthetic polymers and supramolecules [15] are essentially the chemist's way of engineering self-assemblies using both strong and weak molecular interactions (i.e., chemical bonds). In Section 1.2, we review strategies to enable aggregative self-assembly of 3D nanostructures using molecular linkages and chemical bonds. In Section 1.3, we review strategies for 3D aggregative self-assembly of nanostructures based on physical forces. Section 1.4 focuses on curving, bending, and folding thin films to form 3D curved and polyhedral nanostructures with the possibility of enabling lithographic patterning in all three dimensions.

1.2 Self-Assembly Using Molecular Linkages

Molecular linkages and chemical bonds can be used to direct self-assembly with remarkable specificity and complexity. Through a hierarchy of molecular interactions, naturally occurring molecules are known to form 3D nanoscale functional structures such as viruses [16]. The remarkable experiment by Fraenkel-Conrat and Williams [17] convincingly demonstrated that infectious (active) tobacco mosaic virus (TMV) particles could be self-assembled (reconstituted) from their protein and nucleic acid components. The TMV is an exquisitely patterned nanostructured helix with a helical radius of approximately 4 nm. Presently, the challenge in the creation of synthetic 3D nanostructures composed of molecules lies in understanding how to design the constituent molecular components that will spontaneously form the final structure. Since the number and types of molecular interactions can be large, the rational synthesis of nanostructures using molecules can be arduous. Nevertheless, there have been several elegant demonstrations utilizing molecular linkages such as ligand–receptors, proteins, and nucleic acids. It is also accepted that as compared to self-assembly with physical forces, chemical linkages can provide selective and complex organization [18].

Apart from all molecular assemblies, self-assembly with molecular linkages is also enabled by surface functionalization which serves to modify the surface properties of inorganic nanoparticles so that they may gain some level of specificity in terms of their interactions. The most general form of molecular interaction utilizes ligand–receptor interactions by attaching a ligand molecule to one particle and its complementary receptor molecule to another [19]. Ligand–receptor-based 2D self-assembly has been used in the past for the construction of self-assembled molecular squares [20]. These squares, which are composed of ligand-assisted attached molecules, can incorporate higher order arrangements of 2D structures like diamonds surrounded by smaller squares. Furthermore, functionalization can improve the durability of self-assembled structures since the ligand–receptor attraction is typically enhanced by a conformational fit.

A secondary improvement in functionalization is the ability for the final assembled product to have inherent sensing or actuation capabilities. Since many of the ligands and receptors are key contributors in well-studied biochemical mechanisms, they are capable of sensing. For example, while hydrogen bonding between base pairs in DNA (which will be covered in a later section) can serve as an impetus for self-assembly, aptamer moieties can interact with proteins present in the environment, thereby providing sensing capabilities. Also, many of these ligand–receptor pairs have specific activation and deactivation conditions like temperature or pH, making it possible for the linkages to be made or broken reversibly [18].

It is self-evident that there are a myriad of ligand–receptor combinations that can be chosen to functionalize inorganic particles. For instance, the use of carbon black [21], β-cyclodextrin [22], and methoxysilanes [23] has all been realized for the self-assembly of 3D nanorods and crystals. In this chapter, we will specifically focus on protein (with emphasis on biotin–avidin/biotin–streptavidin) and nucleic acid applications for functionalization since these are the most domineering areas.

1.2.1 Three-Dimensional Self-Assembly Using Protein Linkages

Proteins are ideal for assembling complex nanostructures for many reasons. First, each protein displays a unique conformation in order to ensure specificity, guaranteeing that it will bind strongest when the arrangement and composition of its substrate are correct. Second, proteins can have multiple recognition sites or binding domains, which grants these molecules with an inherent 3D ligand-binding capability. Finally, proteins are easily modified; their structures can be genetically edited [24] to omit or add certain binding subunits.

Several 3D nanostructures composed solely of protein constituents have been demonstrated. The shapes range from nanofibers [25] to nanotubes [26] to protein cages [27] and can be designed at a high level of specificity. Typically, the required proteins are chosen from the Protein Data Bank, which stores the structural information of many characterized proteins. This choice allows for the selection of structural or binding motifs like a terminal α-helix thereby enabling greater control over binding sites. After the systematic selection of these proteins, it is also possible to modify their binding characteristics by varying conditions such as pH [25] or engineering specific regions [28].

Since proteins assemble with high specificity, they can be used in conjunction with nanoparticles through surface functionalization. Functionalization of inorganic particles using proteins can be done using multiple strategies. For example, nanoparticles composed of noble metals such as gold can be first thiolated and subsequently cross-linked with a desired protein's amine group through the use of 1-ethyl-3-(3-dimethylaminopropyl)-carbodiimide/*N*-hydroxysuccinimide (EDC/NHS) reaction scheme [29]. Other charged inorganic nanoparticles like silica can be functionalized by using proteins with charged amino acid residues like lysine, a method which has been applied to form shapes like hollow spheres [30].

One very popular method of functionalizing inorganic molecules is achieved by using the proteins avidin or streptavidin in conjunction with the ligand biotin, a coenzyme. Both avidin and streptavidin are tetrameric proteins which contain four identical binding sites to biotin. The binding strength of biotin to its protein receptor site is extremely strong; it is comparable to that of a covalent bond with a binding constant, K_a, of 1×10^{15} M [31]. With such a high binding strength, the biotin–avidin pair works well in self-assembly, since covalently linked structures are quite stable. Having multiple sites for a ligand can also introduce hierarchical capabilities, where a primary scaffolding molecule can serve to "host" a number of surrounding molecules during assembly.

Let us examine a biotin–streptavidin linkage that has proven successful in 3D self-assembly of inorganic nanostructures. Ferritin is a protein–metal conjugate that is made of a protein shell which encapsulates a ferric oxide core. For biotinylation, ferritin can use its external lysine residues to bind up to 60–70 biotin molecules by a nucleophilic reaction [32]. Once the ferritin has been biotinylated, streptavidin can be added at varying ratios in order to provide adequate streptavidin for cross-linking. However, the streptavidin concentration cannot be too high or it may prevent biotinylated particles from linking to the same streptavidin protein. In one study, this optimal ratio was measured to be approximately one biotinylated ferritin to six streptavidin [32]. To build a 3D nanostructure, biotinylated ferritin can be linked to a primary structure such as a carbon nanotube [33] through hydrophobic interactions to enable the self-assembly of a multilayer nanotube composed of different material layers. An additional concentric layer of a different inorganic metal could also be linked onto this ferritin shell by adding other biotinylated metals.

Another example of controlled self-assembly using biotin–streptavidin can be exemplified through the end-to-end binding of gold nanorods in an experiment performed by Caswell et al. [34]. Biotinylation in this case was achieved using biotin disulfides, which can form disulfide linkages to gold surfaces. Since the nanorods were controllably formed using a cetyltrimethylammonium bromide (CTAB) surfactant, functionalization was hypothesized to only form at the ends of the rods where CTAB was not present. Therefore, the addition of streptavidin would only self-assemble the nanorods end to end instead of connecting them at the sides (Fig. 1.2).

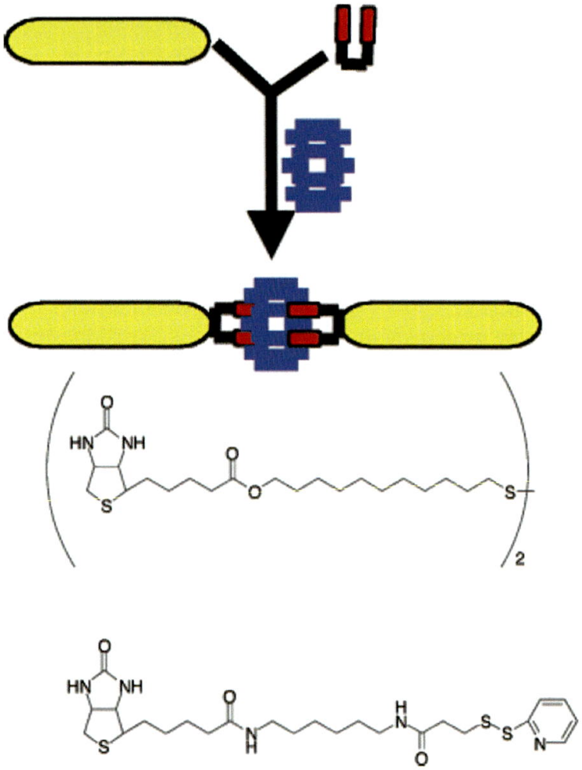

Fig. 1.2 Scheme showing the assembly of gold nanorods (*golden ovals*) by surface functionalization with the biotin disulfide (*red*), and subsequent addition of streptavidin (*blue*) to produce aggregates of nanorods. The chemical structures of the two biotin disulfides are also shown, homemade (*top*) and commercial EZ-Link Biotin-HPDP (*bottom*). Reprinted with permission from [34]. Copyright 2003 American Chemical Society

With the many available methods to biotinylate different inorganic particles, it is possible to create complicated self-assembled structures. In addition, direct control over the region which is biotinylated and the concentration of streptavidin or avidin grants control over the complexity of the assembled structure. Therefore, by using this simple ligand–protein construct, we can expand the possibilities in constructing devices with differing shapes and metal compositions.

1.2.2 Three-Dimensional Self-Assembly with DNA Linkages

Another exciting field enabling the creation of 3D nanostructures is DNA self-assembly. Just like proteins have many benefits for their use in self-assembly, DNA also has significant advantages for its use. DNA is extremely well understood in terms of base pairing and crystallographic structure, making it very predictable and programmable when used as a template for self-assembly [35]. From basic biology, we know that adenine will preferentially bind to thymine, and cytosine will preferentially bind to guanine by hydrogen bonding. Therefore, the assembly of DNA is already quite intuitive from a scientific standpoint.

DNA self-assembly is utilized in two categories for 3D self-assembly: the use of DNA motifs to create origami-like structures made of pure DNA [36] and the use of DNA as a nanoparticle scaffold by functionalization for assembly [37]. The latter portion of DNA technology is very similar to that of protein self-assembly, where the binding originates from specific linkers. Both schemes have brought much potential to enable the creation of devices that can be applied as biosensors, nanophotonic devices, nanoelectric devices [38], and drug delivery vehicles [39].

The folding of 3D DNA structures began with the work of Nadrian Seeman. In 1991, Seeman was able to synthesize a cube solely made from ligated DNA strands by using sticky ends as a linking device [40]. Each vertex of the cube had to be meticulously connected in order to form the final structure, making it a very difficult, low-yield task. Instead, researchers explored other options by starting from stable DNA structures to make 3D nanostructures.

One simple structure is a DNA junction, which is a point where multiple strands of DNA meet via a complementary linkage or through the use of a tris-linker [41]. The junctions allow one to form a DNA folding template, where the structure is already connected and only requires binding intramolecular regions to gain 3D architecture. In one method, a 1669 nt single strand of DNA with multiple junctions was designed with complementary terminal branches. With the assistance of smaller linker sequences, the strand could form crossovers in the center which brought the terminal branches together to form a 3D octahedron [42].

The discovery of the rigidity of triangular-based DNA structures lent credence to the notion that DNA motifs could be used for complicated 3D assembly. The triangular-based structures include trisoligonucleotides (three strands joined by a center tris-linker) and tetrahedrons. The trisoligonucleotides were shown to be able to produce a dodecahedron by utilizing complementary strand sequences as connectors and the center as a vertex [41]. Tetrahedrons are especially viable due to their simple construction protocol; it has been demonstrated that four equal oligonucleotides with complementary regions for each other could be annealed and ligated to self-assemble a DNA tetrahedron with high yield [43]. The tetrahedrons can then be used for advanced functions like protein encapsulation [44] and structural (i.e., bending, twisting) modifications by activating hairpin sequences [39]. In addition, by using a DNA linking strand, it is possible to add a higher level of complexity by assembling multiple tetrahedral DNA motifs [43].

One DNA self-assembly experiment involved the use of DNA "tiles," which are star motifs that can contain multiple symmetric points. A three-point star tile contains a long central strand, three medium strands, three short exterior strands, and three central loops [45]. Each point's end contains a sticky end to be able to latch onto the points of other identical tiles. It was discovered that the length of the central loops allowed for control over the self-assembly of the tiles. Lengthening the loop meant the tile had more curving flexibility and could form 3D shapes more readily through exposing the sticky ends. This, in conjunction with a relatively low DNA tile concentration to increase the chance of connected structures to interact, allowed for the self-assembly of tetrahedral-shaped structures (Fig. 1.3). By adding more DNA tiles and shortening the loop to reduce strand bending, dodecahedrons

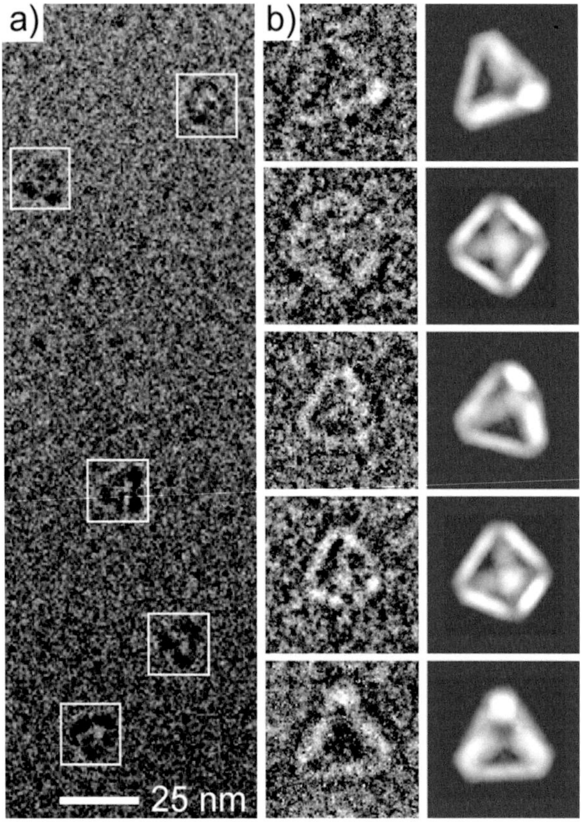

Fig. 1.3 Characterization of the DNA tetrahedron by DLS, AFM, and cryo-EM. (**a**) A representative cryo-EM image. *White boxes* indicate the DNA particles. (**b**) Raw cryo-EM images of individual particles and the corresponding projections of the DNA tetrahedron 3D structure reconstructed from the cryo-EM images. These particles are selected from different image frames to represent views at different orientations. Reprinted with permission from [45]. Copyright 2008 Macmillan Publishers Ltd (Nature Publishing Group)

and buckyballs [45] were formed, possibly arising due to the smaller curvatures of these larger 3D structures. The same phenomenon is observed in five-point stars as well, where longer loops and lower DNA concentrations are found to produce icosahedrons [46].

To simplify DNA folding, a later technique called DNA origami [47] was developed by Paul Rothemund, which involved the use of a long strand of DNA as a large scaffold with hundreds of specific "staple" DNA strands to form crossovers which tighten or bend the scaffold. These staple strands are meticulously programmed to form desired crossovers at repeated turn length intervals through the use of a DNA software system that can pattern the complementary sequences. By using this technology, one can form DNA templates for folding 3D nanostructures with great precision. One experiment displayed the folding of a nanoscale DNA box with a

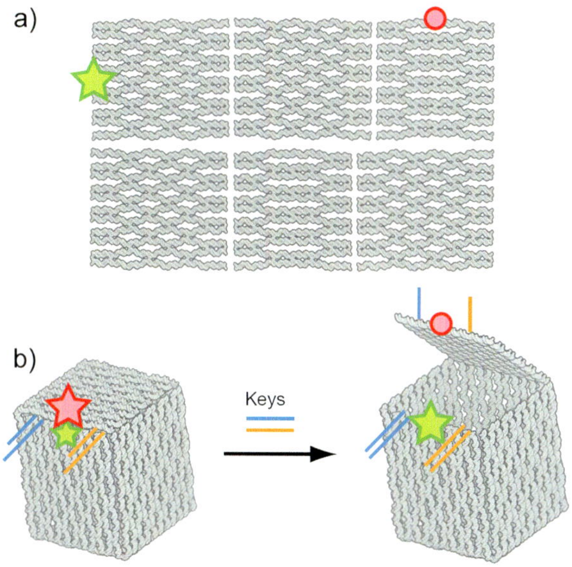

Fig. 1.4 Programmed opening of the box lid. Illustrations of the unlinked faces of the box (**a**) and the controlled opening of the box lid (**b**). The emission from the Cy5 and Cy3 fluorophores is marked with *red* and *green stars*, respectively. Loss of emission from Cy5 is denoted by a *red circle* and the independent lock–key systems are indicated in *blue* and *orange*. Reprinted with permission from [48]. Copyright 2009 Macmillan Publishers Ltd (Nature Publishing Group)

controllable lid [48] through the use of DNA origami by starting with a linked template of six DNA squares. The staple strands assisted in the folding of the squares to form a cubic structure with specific sticky ends to grant complementary oligonucleotide "keys" access to open the box with a conformational change (Fig. 1.4).

DNA, like biotin–streptavidin, can also be used for the self-assembly of inorganic particles through functionalization. There are multiple methods used to tether a strand of DNA to an inorganic nanoparticle. Some involve using charge interactions [49] since DNA is negatively charged, while some metals have cationic properties. These interactions can allow DNA to become a template on which metal nanoparticles may aggregate [50]. Another common functionalizing technique uses gold–thiol linkages to the thiolate DNA [38, 51, 52]. Like the arrangement of protein-linked nanoparticles, DNA-functionalized nanoparticles have high specificity since they will only bind to complementary strands.

Binding specificity was demonstrated through an experiment involving different sized gold particles (diameters of 8 and 31 nm) that were functionalized with complementary oligonucleotides [53]. The particles self-assembled to form a "satellite" structure, which is distinguished by a large central particle surrounded by multiple linked smaller particles (Fig. 1.5). When the oligonucleotides were omitted from these particles, self-assembly did not yield any organized structure [53]. This self-assembly has been applied to not only nanoparticles but also nanostructures like gold nanorods [54]. Added complexity is also possible through the use of biotinylated oligonucleotides that are functionalized onto nanoparticles [55]; streptavidin can be added after hybridization to induce stepwise assembly.

DNA self-assembly is one of the best understood molecular technique today simply because of our ability to manipulate DNA easily. The biological and materials

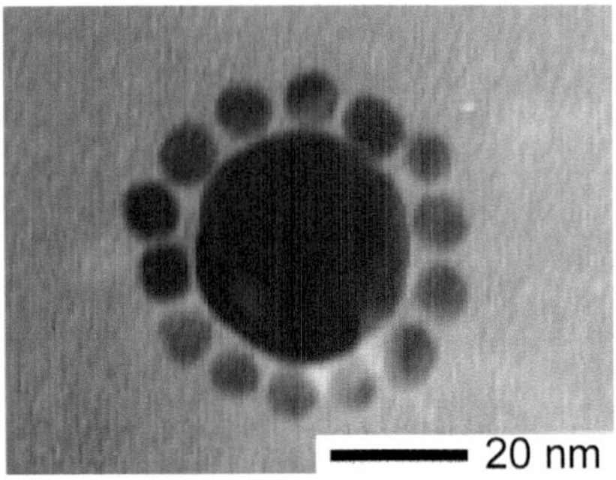

Fig. 1.5 TEM image of the binary nanoparticle network materials supported on holey carbon grids. A nanoparticle satellite structure obtained from the reaction involving 120:1 modified 8 nm particles:modified 31 nm particles and linking oligonucleotides. Reprinted with permission from [53]. Copyright 1998 American Chemical Society

device applications are not trivial, and additional functions can be made with ease. There has been some work to date on RNA self-assembly involving magnesium-mediated RNA and its receptor loop RNA to produce self-assembled gold nanowires [56]. Functionalization occurs through the use of DNA–RNA complementary binding regions secondary to a primary DNA thiol linkage to gold.

1.3 Three-Dimensional Self-Assembly Using Physical Forces

While self-assembly with molecular linkages provides remarkable versatility and specificity in terms of the kinds of interactions that can be engineered, many of these linkages fall apart when the structures are dehydrated or heated. Hence, there is a need to explore the assembly of structures using physical forces. Moreover, devices with magnetic, photonic, or electrical properties often require the incorporation of inorganic components such as metallic or semiconducting materials [57]. There is often a need to organize metallic and inorganic components with nanoscale precision to enable functional 3D nanostructured devices. In this sense, it is already established that self-assembly provides an attractive route to fabricate nanoscale structures [58].

Three-dimensional methods build upon conventional 2D self-assembly methods. The 2D methods typically form arrays or monolayers of nanostructures. Most of these 2D structures are formed by utilizing a physical force to coax nanoparticles into their desired places. For example, the use of surface interactions in Langmuir–Blodgett films at the air–liquid interface can be applied to generate the self-assembly of a monolayer of nanoscale particles [59] and even nanorods [60]. Other methods have involved evaporating thin liquid films over a solid support [61–63]. The 2D self-assembly of an array or monolayer of nanoscale particles by evaporation is driven by capillary forces that pull particles together during removal of the liquid.

After evaporation, the particles are held together by van der Waals or dispersion forces.

To move into the third dimension, one approach focuses on shape control, namely, by synthesizing colloids with a 3D morphology. These inorganic particles are fabricated in solvents with varying pH, concentration, temperature, and voltage; the resulting nanostructures have complex 3D morphologies and resemble flowers, helices, discs, and hollow boxes [64–66]. Another example involves the self-assembly of CdSe quantum dot superlattices by a nucleation-caused crystallization in solution; the lattices are then held together by intermolecular forces [67]. Dipole–dipole interactions are also utilized to create 3D rod-like structures with anisotropic control using a surfactant [68].

The use of physical templates to shape the overall self-assembly has enabled control over both the arrangement and the interactions among constituent particles. Templates can also be structured using electron beam lithography, imprint lithography, particle track etching, or by anodization. Self-assembly of a hexagonal closed packed or face-centered cubic silica colloidal crystal has been demonstrated using a uniform template; small defects in the template reflect small defects in the crystal [69]. By constricting the volume of the template on the nanoscale, one can produce a wide range of structures [70] (Fig. 1.6). Linked templates are also used, which involve a chemically functionalized template to bind to a specific inorganic compound. This helps to produce a hierarchical self-assembly [71], where inorganic molecules can link onto an initial self-assembled structure and then build upon it to form complicated structures such as sieves [72]. The linked template method has been mastered to the extent to which even a nanoscale painting by Picasso [73] has been replicated using colloidal gold!

Van der Waals or dispersion forces are widely used to hold together nanostructured organizations after self-assembly. However, it should be noted that this force is weak, and the nanostructures can often be disrupted by sonication. For example, the cubic assemblies shown in Fig. 1.7 were formed by self-assembly of selectively functionalized hydrophobic units in water. Here, selective functionalization of different faces of silver (Ag) cubes was achieved using hydrophobic (octadecanethiol) and hydrophilic (mercaptohexadecanoic acid) thiols. Selective functionalization results in the formation of linear chains and closed packed crystals; however, the units are held together only by weak van der Waals forces [74]. Similar assemblies were demonstrated using nanowires with amphiphilic segments [75] and were even observed during dissolution of the templates (Fig. 1.8a) [76]. These 3D nanostructures, although self-assembled in impressive 3D geometries, would fall apart upon sonication. Self-assembly with segmented nanowires provides some specificity in the interactions by the inclusion of hydrophobic, hydrophilic [75], or magnetic (Fig. 1.8c–d) segments [77].

Permanent bonding of units after assembly can be achieved using surface tension-based assembly with liquid drops that solidify on cooling or cross-linking. This approach draws inspiration from mesoscale self-assembly approaches [78–81] to permanently bond nanoscale assemblies using adhesives [82] (Fig. 1.8e–f) or solder [83]. These bonded assemblies survive sonication. Additionally, selective patterning

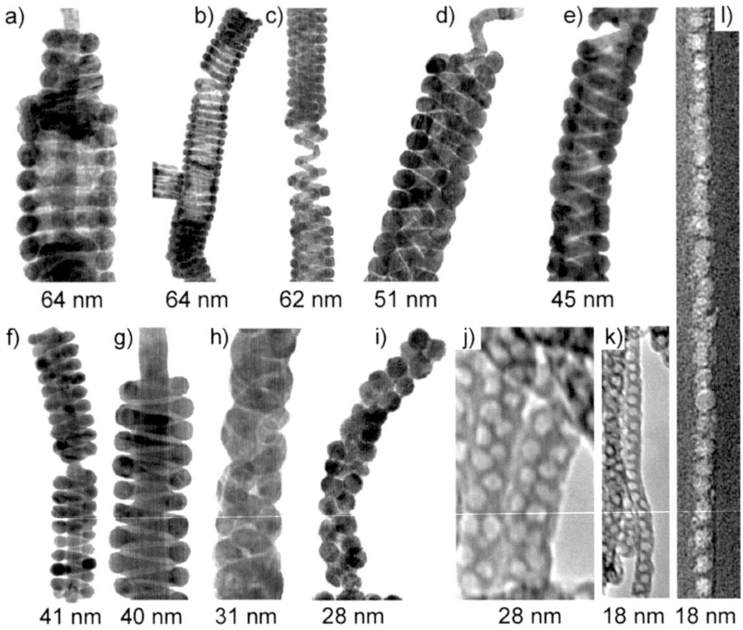

Fig. 1.6 Representative TEM images of mesostructures formed inside alumina nanochannels with differing confinement dimensions. The confining nanochannel diameter is indicated underneath each image. (**a–i**) Silver inverted mesostructures prepared by backfilling the confined mesoporous silica; (**j–k**) free-standing mesoporous silica fibers; (**i**) mesoporous silica embedded inside the alumina nanochannels obtained using a focused ion beam for sample preparation. The structures are (**a**) three-layer stacked *doughnuts*; (**b**) S-helix; (**c**) core–shell D-helix, in which the core and the shell are both S-helix; (**d**) core–shell triple helix, in which the shell is a D-helix and the core is an S-helix; (**e**) D-helix; (**f, g**) S-helix with a straight core channel; (**h**) D-helix; (**i, j**) inverted *peapod* structure with two lines of spherical cages packed along the long axis of the alumina nanochannel; (**k, l**) inverted *peapod* with one line of cages. Reprinted with permission from [70]. Copyright 2007 Macmillan Publishers Ltd (Nature Publishing Group)

of segments of nanowires with self-assembled monolayers enables specific bonding of these segments with hydrophobic monomers that can be subsequently cross-linked (Fig. 1.8f).

1.4 Three-Dimensional Patterned Nanofabrication by Curving and Bending Nanostructures

One attractive strategy to enable patterning in 3D is to leverage precise 2D nanofabrication paradigms to create units that are then rotated or bent into the third dimension. The forces required to achieve this curving or bending can be derived from several mechanisms including thin film stresses, magnetism, pneumatics, swelling, and surface tension [84]. Most of the prior research involving the above mechanisms has been limited to the microscale. Here, we focus on two mechanisms

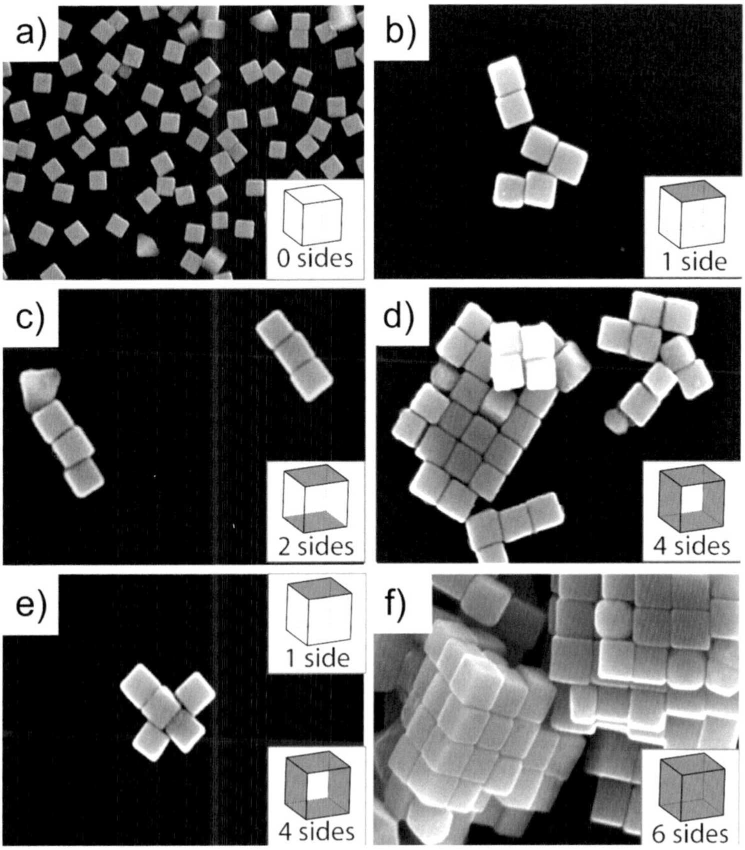

Fig. 1.7 SEM images of Ag nanocubes and the assemblies. Unfunctionalized *cubes* deposited on Si from water are shown in (**a**) for reference. Nanocubes whose faces have been selectively functionalized with hydrophilic and hydrophobic thiolate SAMs and then allowed to assemble in water are shown in (**b–f**). The number of faces on each cube that were rendered hydrophobic is indicated in the *bottom right corner* of each panel, the remaining faces on the cube were rendered hydrophilic. In (**e**), cubes with four hydrophobic sides were mixed with cubes that only had one hydrophobic face at a ratio of 1:4 and then allowed to self-assemble in water. All cubes used in this study had a mean edge length of (97 ± 6) nm (as determined from 123 cubes). Reproduced with permission from [74]. Copyright 2008 Wiley-VCH Verlag GmbH & Co. KGaA

involving thin film stress and surface tension that have enabled nanoscale curvature and bending. We divide the discussion into hingeless (curving) and hinged (bending) structures.

1.4.1 Curving Hingeless Nanostructures Using Stress

The magnitude of stress required to curve nanostructures with diameters less than 100 nm is extremely high (typically greater than a few gigapascals). These high stresses can be generated within strained heteroepitaxial thin films. When an

Fig. 1.8 Summary of nanowire-based 3D assembly. (**a**) A scanning electron microscopy image of gold nanowire bundles observed during dissolution of alumina membranes (with kind permission from Springer Science and Business Media [76], fig. 2a). (**b**) A higher magnification top-view image of a nanowire bundle. The bundles were only weakly held together and broke apart upon sonication (with kind permission from Springer Science and Business Media [76], fig. 2b). (**c**) Scanning electron micrograph (SEM) of multiple bundles of rods. The *light* sections are gold and the *gray* sections are nickel. The aspect ratio of the ferromagnetic sections is ~0.5 (reprinted with permission from [77]. Copyright 2003 American Chemical Society). (**d**) SEM of a single bundle demonstrating the alignment of ferromagnetic sections (reprinted with permission from [77]. Copyright 2003 American Chemical Society). (**e**) SEM images showing two rods held together by the polymerized adhesive. Secondary electron image showing the polymeric adhesive and the rods (reprinted with permission from [82]. Copyright 2004 American Chemical Society). (**f**) A backscattered SEM image of 3D bundles formed using rods composed entirely of Au (reprinted with permission from [82]. Copyright 2004 American Chemical Society)

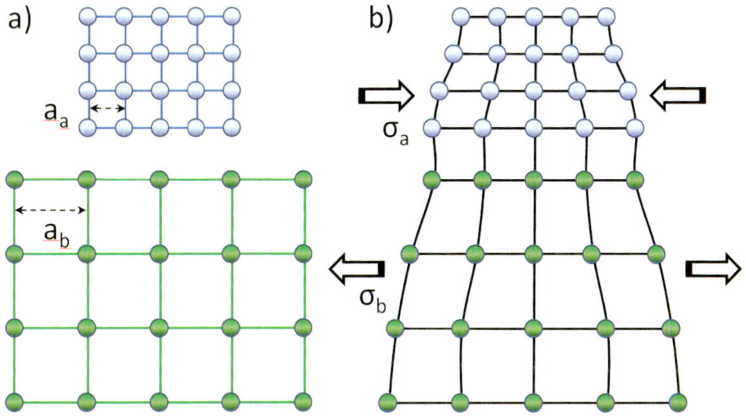

Fig. 1.9 Illustration of a strained heteroepitax. (**a**) Two lattices composed of different materials have different lattice parameters, a_a and a_b; (**b**) after forming a heteroepitax, the first few film layers at the interface strain in order to form a matched lattice and the strain develops compress and tensile stresses

epitaxial film grows on top of a crystalline film with different lattice parameters, the two films result in strained-layer epitaxy, also known as heteroepitaxy (Fig. 1.9) [85–92]. These bilayer films spontaneously curve on release from the underlying substrate on which they are deposited.

One of the well-known heteroepitaxial films is a compound semiconductor bilayer, InAs/GaAs, which is often grown on a sacrificial layer of AlAs [85, 90, 91]. The lattice parameters of InAs and GaAs are 6.06 and 5.65 Å [93], respectively, and the value of the lattice mismatch, $\Delta a/a = (6.06\text{–}5.65)/5.65$, is approximately 7.2%. Because the two different lattice parameters are matched at the interfacial boundary of the two materials, the lattice parameter of InAs progressively decreases and that of GaAs increases away from the interface (Fig. 1.10). The sacrificial layer AlAs can

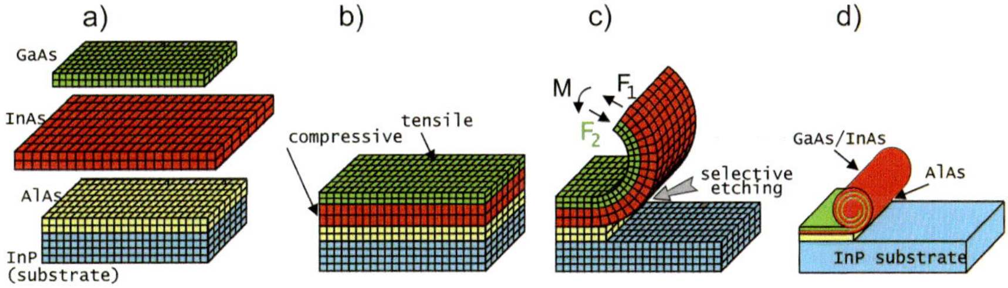

Fig. 1.10 Initial formation stages of free-standing, several monolayer (ML) thick nanotubes (schematically): (**a**) free 2 ML thick InAs and GaAs layers with naturally mismatched lattice constants ($\Delta a/a = 7.2\%$); (**b**) matching of the layers at the interface between them in an InAs/GaAs bifilm MBE grown on an InP substrate; (**c**) bending of the GaAs/InAs bifilm after its partial detachment from the substrate during selective etching of the underlying AlAs sacrificial layer; (**d**) self-rolling of the GaAs/InAs bifilm in a tube scroll during further selective etching. Reproduced with permission from [90]. Copyright 2006 Wiley-VCH Verlag GmbH & Co. KGaA

be etched in an HF-based etchant thereby releasing the bilayer from the substrate. After etching the sacrificial layer, the compressed layer (InAs) stretches and develops an elastic force $F1$. On the other hand, the stretched layer (GaAs) compresses and develops an elastic force $F2$. The directions of $F1$ and $F2$ are opposite and create a non-zero moment (bending moment) of force M. This bending moment is what transforms 2D patterns into 3D curved structures.

The moment of force can be controlled by varying the film thicknesses as well as the material composition to generate different diameters of curvatures [91]. Although this chapter will not cover temperature effects, it has also been found that growth temperature also influences the diameter [91]. Figure 1.11a–d shows SEM (scanning electron microscopy) images of In(Ga)As/GaAs nanotubes with various diameters (35–550 nm). The diameters of the tubes in the figure were tuned based on the two parameters: film thickness and material type. In order to predict the shape of 3D structures, the bilayered nanotube diameter (D) can be modeled using continuum mechanics [94, 95]:

$$D = \frac{d\,[3(1+m)^2 + (1+m\cdot n)\cdot[m^2 + (m\cdot n)^{-1}]]}{3\varepsilon(1+m)^2} \tag{1.1}$$

$$d = d_1 + d_2 \tag{1.2}$$

$$m = d_1/d_2 \tag{1.3}$$

$$n = Y_1/Y_2 \tag{1.4}$$

where ε is the in-plane biaxial strain between the two films, d_1 and d_2 are the thicknesses, and Y_1 and Y_2 are Young's moduli of the first and second layers, respectively. Based on the equations, the diameters of InAs/GaAs nanotubes were calculated and compared with experimental data as shown in Fig. 1.11e. The results show that in order to minimize the diameters, the bilayer thickness d needs to be decreased. Using the thinnest possible bilayer composed of 1 monolayer (ML) of GaAs and 1 ML of InAs, an inside diameter of approximately 2 nm was formed, which is the smallest achievable inside diameter of nanotubes [85].

Although the heteroepitaxial deposition method can create 3D curved structures with extremely small diameters, there are several drawbacks. During film deposition, temperatures as high as 1000°C are often required to generate high stresses. These high deposition temperatures can induce thermal shock on other devices on a substrate and alter device properties. In order to overcome these limitations, extrinsic stress has been utilized by grain coalescence using a tin (Sn) metal [96]. Because tin has a relatively low melting point (232°C) [97], this methodology allows for the creation of controlled stresses at relatively low temperatures. The latter methodology also does not require the specialized thin film deposition equipment needed to generate heteroepitaxial films.

It is well known that nanoscale grains can generate extremely high intrinsic stress through grain coalescence during a film deposition process [98–101]. The grain

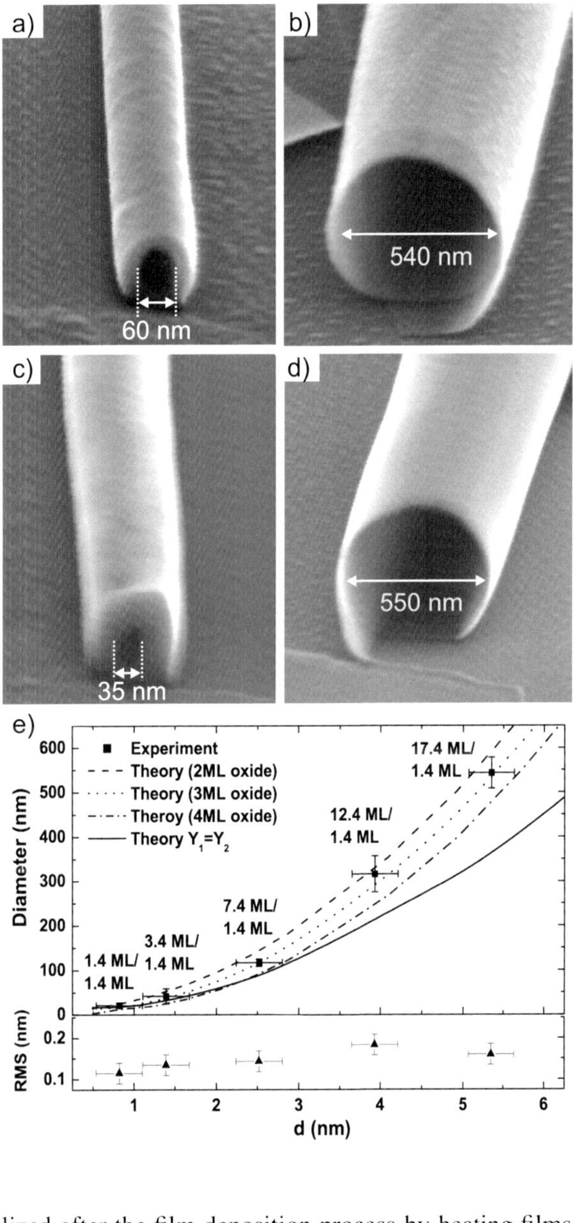

Fig. 1.11 (**a–d**) Four typical tube openings consisting of the following as-grown bilayers: (**a**) 1.4 ML $In_{0.33}Ga_{0.67}As$/6.4 ML GaAs, (**b**) 14.1 ML $In_{0.33}Ga_{0.67}As$/19.1 ML GaAs, (**c**) 1.4 ML InAs/6.4 ML GaAs, and (**d**) 1.4 ML InAs/20.4 ML GaAs. (**e**) InAs/GaAs tube diameter as a function of bilayer thickness. Bilayers were chosen highly asymmetric (see labeling for each data point). Theory (Eq. (1.1)) excellently describes experimental data points. Reproduced with permission from [91]. Copyright 2002 Institute of Physics

coalescence can also be realized after the film deposition process by heating films to create extrinsic stresses [96]. Sn films show a Volmer–Weber or grain growth after thermal evaporation on Si substrates. The reason for this growth mechanism is that the interactions between pairs of Sn adatoms are stronger than interactions between Sn adatoms and the Si surface (Fig. 1.12). In order to induce grain coalescence, exothermic heat is applied to the Sn film with a plasma etcher. Plasma etching of the sacrificial layer Si with a gas mixture composed of tetrafluoromethane (CF_4)

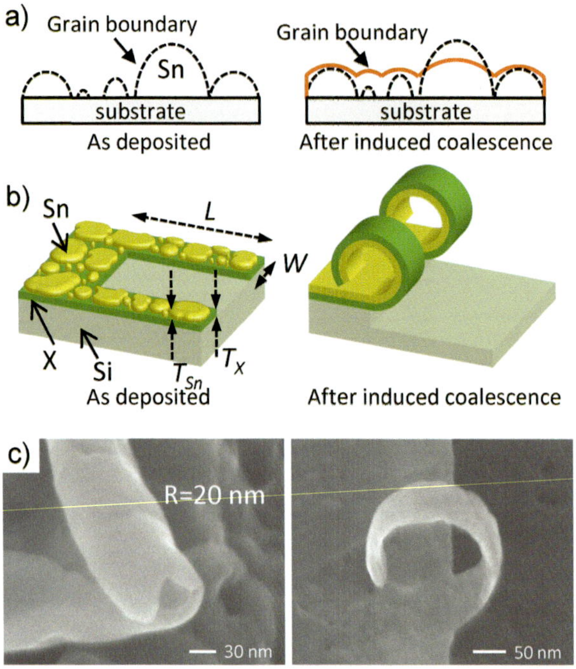

Fig. 1.12 Conceptual sketches and scanning electron microscopy (SEM) images showing the origin of the high extrinsic stress observed within the Sn film that caused Ni/Sn bilayers to curl up with a nanoscale radii of curvature. (**a**) The induction of grain coalescence in Sn films during plasma processing causes a large extrinsic stress. (**b**) When deposited atop a Ni film, the stress within the Sn thin film is large enough to cause the Sn/Ni bilayer to curl up due to grain coalescence. (**c**) SEM image of Ni/Sn bilayers curving into a nanoscale tube with 20 nm radii of curvature. Also shown is a nanoscale ring. Reproduced with permission from [96]. Copyright 2010 Wiley-VCH Verlag GmbH & Co. KGaA

and oxygen (O_2) generates exothermic heat due to the chemical reaction between atomic fluorine (F) and Si [102]. The induced coalescence develops stresses within the Sn film.

It has also been observed that Sn shows a similar grain growth on nickel (Ni) and alumina (Al_2O_3) layers. Due to the grain coalescence of Sn, the stress within the Sn film is large enough to cause Sn/Ni and Sn/Al_2O_3 bilayers to curl up with nanoscale radii of curvature (Fig. 1.12b). Because the metals can be deposited at a low temperature (approximately 25°C) using an evaporator, the layers could be patterned with a photo or an electron beam (e-beam) lithographic process and subsequent lift-off metallization process. After Sn grain coalescence, the smallest diameter D (40 nm) curvature was achieved with a thickness of 5 nm Ni and 5 nm Sn (Fig. 1.12c).

Here, the diameter of nanostructures depends not only on film thickness and material properties but also on the shape of 2D patterns before curving. In order to investigate the geometric factors involved, 2D cantilevers were designed with variable geometries. Experimental results show a direct relationship between

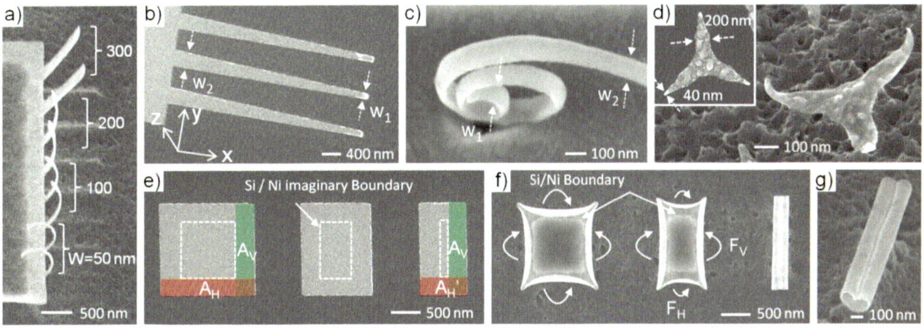

Fig. 1.13 SEM images of the variation of curvature with varying widths showing that nanostructures with both homogeneous and varying radii of curvature can be self-assembled. (**a**) SEM image of the curving of cantilevers with different widths (50, 100, 200, and 300 nm). All cantilevers have the same $L = 1$ μm and thickness Ni 10 nm/Sn 2.5 nm. Cantilevers with the same width show the same radii of curvature, while those of larger widths have larger radii of curvature. This result highlights the reproducibility of the self-assembly process. (**b**) SEM image of cantilevers with varying width along the length of the cantilever, i.e., $W_1 W_2$. (**c**) A cantilever with this varying width curves with a varying radii of curvature due to a varying area moment of inertia, resulting in the formation of a nanospiral. (**d**) Nanoscale *three-fingered talon-shaped* structures before and after coalescence. (**e–f**) *Square* and *rectangular* patterns (Sn 5 nm/Ni 5 nm) before and after coalescence developing different bending forces F_V and F_H. (**g**) *Tilted* zoomed-in image of the nanoscroll shown in (**f**). Reproduced with permission from [96]. Copyright 2010 Wiley-VCH Verlag GmbH & Co. KGaA

cantilever width and radius of curvature (Fig. 1.13a). This behavior was explained by the theory of the area moment of inertia [103]. Because the cantilever beams show biaxial bending, beams with wide widths show larger rolled cross-sectional areas (larger area moment of inertia) as compared to beams with narrow widths. Larger area moment of inertia values show more resistance to curving, resulting in greater diameters. Through varying the width of the 2D structures, homogeneous curvature (tubes, rings, and scrolls) and non-homogeneous curvature (spiral and talons) structures have been fabricated (Fig. 1.13a–d). It has also been observed that square-shaped 2D panels curved equally on all four sides. On the other hand, rectangular-shaped panels curved along the direction of shorter side length because the shorter side length results in a lower area moment of inertia and that area develops less resistance to curving (Fig. 1.13e–g).

This curving strategy can be combined with conventional 2D nanoscale patterning techniques to create simultaneously curved and patterned nanostructures. The e-beam lithographic process can be used to generate sub-5 nm scale patterns on a 2D substrate [104, 105]. Since this assembly process is compatible with e-beam lithography and film deposition near room temperature on a substrate, any desired nanoscale patterns can be created on curved structures (Fig. 1.14). In one demonstration, 2D structures were first patterned through e-beam lithography and lift-off metallization processes. On the 2D structures, pores and the letters JHU and NANOJHU were patterned on Ni structures. On top of the Ni, a Sn thin film was deposited. When grain coalescence was induced by plasma etching of the

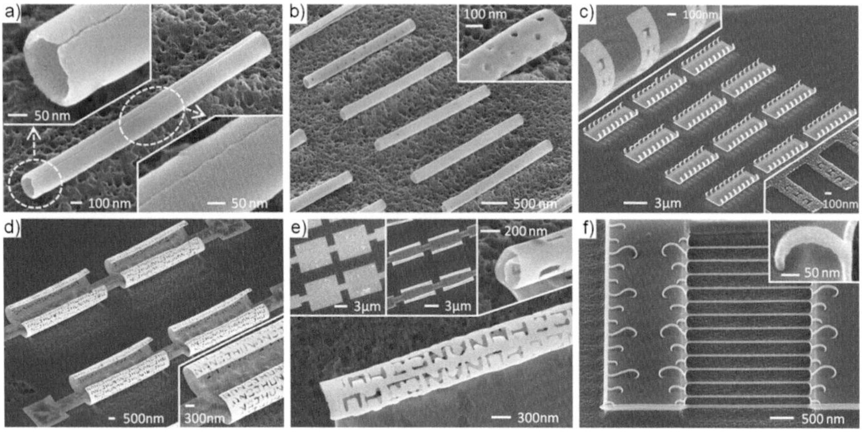

Fig. 1.14 Demonstration of surface patterning (**a–e**), materials versatility (**f**) and parallel nature of the assembly process. SEM images of single rolled nanotubes without (**a**) patterning and (**b**) with patterning of pores. (**c–e**) Nanostructures such as rings and scrolls with the letters JHU and NANOJHU patterns on them. (**f**) Curving nanostructures composed of a dielectric material, namely, alumina (Al_2O_3 6 nm/Sn 5 nm). Reproduced with permission from [96]. Copyright 2010 Wiley-VCH Verlag GmbH & Co. KGaA

Si sacrificial layer, the 2D structures with surface patterns curved and resulted in lithographically patterned nanotubes, nanoscrolls, and nanorings. In addition to metallic compositions, these curved nanostructures could also be fabricated with dielectric materials such as Al_2O_3 (Fig. 1.14f). Because of the material flexibility in this method, this process can be used to fabricate functional electronic and optical devices [106].

1.4.2 Three-Dimensional Nanofabrication by Bending Hinged Panels to Create Patterned Polyhedral Nanoparticles

Self-assembly provides an attractive strategy to create patterned nanoparticles as observed in naturally occurring viral assembly. However, in the absence of sophisticated mechanisms providing specificity in protein assembly of viruses, artificially structured untethered components do not form an organized structure (Fig. 1.15a). One strategy in synthetic self-assembly draws inspiration from macro- and microscale origami wherein folds in paper or hinges within rigid panels are bent to create 3D structures [107–120]. The concept utilized here is to pre-organize precisely patterned panels with a series of hinges; on the microscale, it has been shown that this type of organization [118, 119] in 2D strongly affects self-assembly. Additionally, this concept can be extended to self-assembled structures and materials with thousands of hinges or interconnections [120].

Recently, it was demonstrated that this self-folding approach could be used to construct patterned nanoparticles [121]. Here, e-beam lithographically patterned

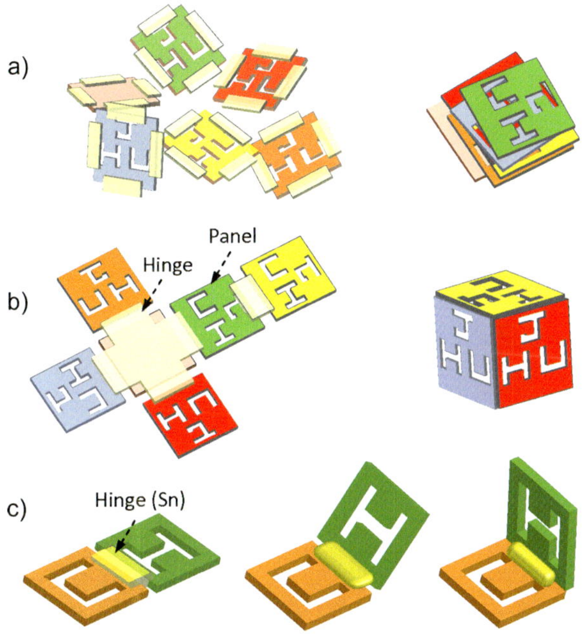

Fig. 1.15 Schematic diagram showing the concept behind the self-assembly process. (**a**) Patterned panels with binding sites that interact without constraints are unlikely to self-assemble into cubes. (**b**) Joining panels to form nets limits the possible interactions and allows them to assemble correctly to form a nanocube. (**c**) Self-assembly is driven by the reflow of tin (Sn) within the hinges of the net; the panel angular orientation needed for self-assembly is derived from the force that is generated when the reflowed hinges minimize their surface area. Reprinted with permission from [121]. Copyright 2009 American Chemical Society

panels were shown to self-assemble by bending Sn hinges into cubic particles. In principle, since the bending angle can be controlled by varying the volume of Sn and the assembly conditions, different polyhedral-shaped particles can be fabricated using this approach. Self-assembly of patterned particles with overall sizes of 100 nm with a resolution of 15 nm in all three dimensions was demonstrated. This assembly approach is limited mainly by the 2D patterning resolution; in principle, particles with smaller sizes and patterns can also be fabricated. Strategies to enable parallel 2D patterning with imprint lithography to create large numbers of particles as well as the incorporation of multilayer patterning to enable the attachment of electronic circuits and optical elements on the surfaces of the polyhedrons are actively being pursued. The primary challenge with the fabrication of large numbers of particles using imprint lithography is that nanoscale multilayer alignment is hard to achieve with sub-100 nm resolution. Typically, this sub-100 nm alignment needs to be achieved by positioning one imprint mask in registry with a pattern on a substrate using piezoelectric positioners; this precise positioning can be challenging to achieve over large substrates due to thermal noise, differences in thermal expansion coefficients, and linear and angular drift or offsets. Hence, advances in mass

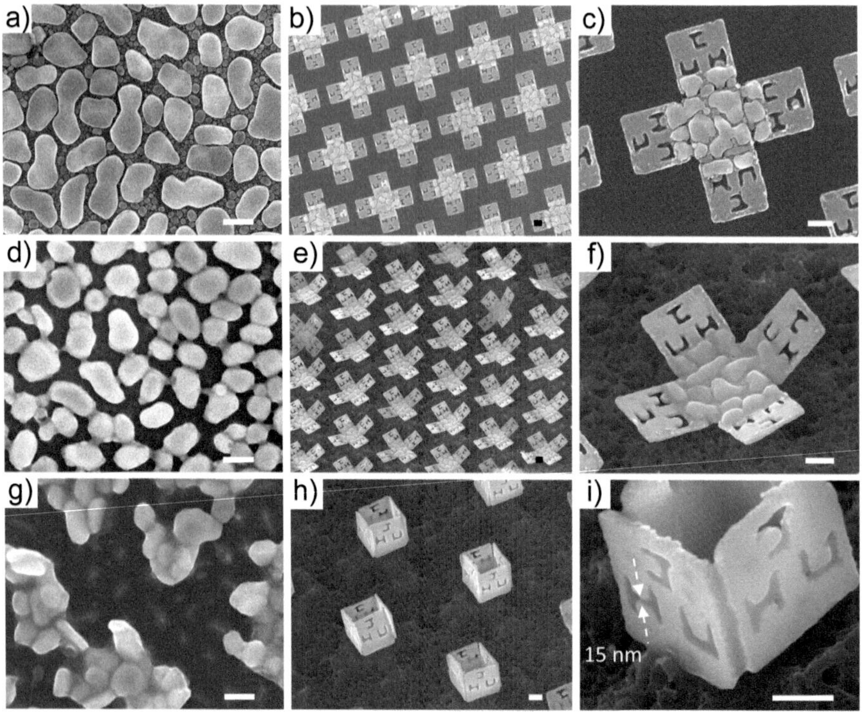

Fig. 1.16 Results of experiments demonstrating that the orientation angle can be controlled by varying the ratio of O_2 to CF_4. SEM images of Sn thin films on a silicon wafer and 500 nm sized 2D nets before and after plasma etching. (**a–c**) Images of a Sn thin film and 2D nets before plasma etching. (**a**) 50 nm thick Sn on a silicon wafer. (**b, c**) Progressively zoomed-in images of Ni panels with Sn hinges. (**d–f**) Images of the Sn film and 2D nets after plasma etching with a 0.2 and 12 sccm flow rate of O_2 and CF_4, respectively. (**d**) The Sn film shows some grain coalescence (of grains less than 50 nm in size) but no significant reflow of large grains. (**e, f**) Progressively zoomed-in images showing that the 2D nets assemble with angles of approximately 45° under these conditions. (**g–i**) SEM images of the Sn film and 2D nets after plasma etching with a 3.6 and 12 sccm flow rate of O_2 and CF_4, respectively. (**g**) The Sn film shows considerable reflow. (**h**) Progressively zoomed-in images showing that the 2D nets assemble with angles of approximately 90° under these conditions. It should be noted that the assembly process is parallel and (**i**) the particles have the letters JHU patterned with line widths as small as 15 nm. Scale bars: 200 nm. Reprinted with permission from [121]. Copyright 2009 American Chemical Society

production of lithographically patterned nanoparticles depend on advances in 2D lithographic multilayer patterning.

1.5 Conclusions

In conclusion, the creation of precisely patterned and nanostructured 3D devices remains a critical challenge that needs to be overcome to enable versatile nanotechnological function. Self-assembly provides an attractive route to create these structures in a highly parallel and cost-effective manner. Nevertheless, rules that govern yield and defect tolerance in self-assembly still need to be uncovered. Nature has had significantly more time to gain experience in evolving components and processes capable of self-assembling with high yield and defect tolerance. The mastery of concepts including balancing positive and negative interactions, strong and weak interactions, hierarchy, switches, and steric factors can enable advanced assembly. Most of the structures described here are static self-assemblies; however, it is envisioned that non-equilibrium and dynamic assembly would enable machine-based function. For example, on the microscale, hinged assembly can be reversibly reconfigured based on chemical cues to create structures such as microgrippers that open and close in response to specific chemicals [122–124]. On the nanoscale, a glimpse of this has been provided by the DNA box (Fig. 1.4b). Hence, there is a need to also study strategies to disassemble these patterned 3D nanostructures to provide reversible self-assembly and reconfigurability thereby enabling advanced nanomachine-based function. If successful, these nanostructures could enable actions like gripping and releasing or opening and closing. It is now understood that biological self-assembly often utilizes weak molecular forces (comparable to the Boltzmann thermal energy) to achieve this reconfigurability. When dynamic functionality is coupled with functionalities derived from patterning, such as the incorporation of electronic circuits, sensors, and binding sites, we will witness a true realization of the era of nanotechnology as laid out by Feynman.

Acknowledgments We wish to acknowledge financial support from the NSF (Grant CMMI-0854881) and the NIH (Grant DP2-OD004346). Any opinions, findings, and conclusions or recommendations expressed in this material are those of the author(s) and do not necessarily reflect the views of the funding agencies.

References

1. R. Chau, J. Kavalieros, B. Roberds, R. Schenker, D. Lionberger, D. Barlage, B. Doyle, R. Arghavani, A. Murthy, G. Dewey, 30 nm physical gate length CMOS transistors with 1.0 ps n-MOS and 1.7 ps p-MOS gate delays. International Electron Devices Meeting Technical Digest 2000, IEEE, San Francisco, CA, pp. 45–48, 2000
2. S.Y. Chou, P.R. Krauss, P.J. Renstrom, Imprint lithography with 25-nanometer resolution. Science **272**, 85 (1996)
3. R.M. Langford, P.M. Nellen, J. Gierak, Y. Fu, Focused ion beam micro- and nanoengineering. MRS Bull. **32**, 417 (2007)
4. S. Roy, Fabrication of micro- and nano-structured materials using mask-less processes. J. Phys. D: Appl. Phys. **40**, R413 (2007)
5. J.P. Rolland, B.W. Maynor, L.E. Euliss, A.E. Exner, G.M. Denison, J.M. DeSimone, Direct fabrication and harvesting of monodisperse, shape-specific nanobiomaterials. J. Am. Chem. Soc. **127**, 10096 (2005)
6. R.P. Feynman, There's plenty of room at the bottom. J. Microelectromech. Syst. **1**, 60 (1992)
7. T.M. Whitney, J.S. Jiang, P.C. Searson, C.L. Chien, Properties of arrays of metallic nanowires. Science **261**, 1316 (1993)
8. M.P. Zach, K.H. Ng, R.M. Penner, Molybdenum nanowires by electrodeposition. Science **290**, 2120 (2000)

9. C.R. Martin, Membrane-based synthesis of nanomaterials. Chem. Mater. **8**, 1739 (1996)
10. J. Hu, T.W. Odom, C.M. Lieber, Chemistry and physics in one-dimension: synthesis and properties of nanowires and nanotubes. Acc. Chem. Res. **32**, 435 (1999)
11. S. Iijima, Carbon nanotubes: past, present, and future. Phys. B: Condens. Matter **323**, 1 (2002)
12. D.J. Milliron, S.M. Hughes, Y. Cui, L. Manna, J.B. Li, L.W. Wang, A.P. Alivisatos, Colloidal nanocrystal heterostructures with linear and branched topology. Nature **430**, 190 (2004)
13. B.D. Gates, Q. Xu, J.C. Love, D.B. Wolfe, G.M. Whitesides, Unconventional nanofabrication. Ann. Rev. Mater. Res. **34**, 339 (2004)
14. H. Ringsdorf, Hermann Staudinger and the future of polymer research jubilees – beloved occasions for cultural piety. Angew. Chem. **43**, 1064 (2004)
15. J.M. Lehn, Perspectives in supramolecular chemistry – from molecular recognition towards molecular information processing and self-organization. Angew. Chem. **29**, 1304 (1990)
16. A. Klug, The tobacco mosaic virus particle: structure and assembly. Phil. Trans. R. Soc. Lond. B **354**, 531 (1999)
17. H. Fraenkel-Conrat, R.C. Williams, Reconstitution of active tobacco mosaic virus from its inactive protein and nucleic acid components. Proc. Natl. Acad. Sci. **41**, 690 (1955)
18. X.Y. Ling, D.N. Reinhoudt, J. Huskens, From supramolecular chemistry to nanotechnology: assembly of 3D nanostructures. Pure Appl. Chem. **81**, 2225 (2009)
19. H. Cölfen, S. Mann, Higher-order organization by mesoscale self-assembly and transformation of hybrid nanostructures. Angew. Chem. **42**, 2350 (2003)
20. D.H. Cao, K. Chen, J. Fan, J. Manna, B. Olenyuk, J.A. Whiteford, P.J. Stang, Supramolecular chemistry and molecular design: self-assembly of molecular squares. Pure Appl. Chem. **69**, 1979 (1997)
21. L.E. Levine, G.G. Long, J. Ilavsky, R.A. Gerhardt, R. Ou, C.A. Parker, Self-assembly of carbon black into nanowires that form a conductive three dimensional micronetwork. Appl. Phys. Lett. **90**, 014101 (2007)
22. X.Y. Ling, I.Y. Phang, D.N. Reinhoudt, G.J. Vancso, J.H. Huskens, Free-standing porous supramolecular assemblies of nanoparticles made using a double-templating strategy. Faraday Discuss. **143**, 117 (2009)
23. K. Mitamura, T. Imae, N. Saito, O. Takai, Fabrication and self-assembly of hydrophobic gold nanorods. J. Phys. Chem. B **111**, 8891 (2007)
24. I. Koltover, S. Sahu, N. Davis, Genetic engineering of the nanoscale structure in polyelectrolyte-lipid self-assembled systems. Angew. Chem. **43**, 4034 (2004)
25. K. Niece, J.D. Hartgerink, J.J.J.M. Donners, S.I. Stupp, Self-assembly combining two bioactive peptide-amphiphile molecules into nanofibers by electrostatic attraction. J. Am. Chem. Soc. **125**, 7146 (2003)
26. E. Gazit, R. Nussinov (eds.), *Methods in Molecular Biology, Vol. 474: Nanostructure Design: Methods and Protocols* (Humana, Totowa, NJ, 2008)
27. J.E. Padilla, C. Colovos, T.O. Yeates, Nanohedra: using symmetry to design self assembling protein cages, layers, crystals, and filaments. Proc. Natl. Acad. Sci. **98**, 2217 (2001)
28. R.A. Mcmillan, C.D. Paavola, J. Howard, S.L. Chan, N.J. Zaluzec, J.D. Trent, Ordered nanoparticle arrays formed on engineered chaperonin protein template. Nat. Mater. **1**, 247 (2002)
29. R. Bhattacharya, C.R. Patra, S.F. Wang, L.C. Lu, M.J. Yaszemski, D. Mukhopadhyay, P. Mukherjee, Assembly of gold nanoparticles in a rod-like fashion using proteins as templates. Adv. Funct. Mater. **16**, 395 (2006)
30. M.S. Wong, J.N. Cha, K. Choi, T.J. Deming, G.D. Stucky, Assembly of nanoparticles into hollow spheres using block copolypeptides. Nano Lett. **2**, 583 (2002)
31. M.S. Ayyagari, R. Pande, S. Kamtekar, H. Gao, K.A. Marx, J. Kumar, S.K. Tripathy, J.A. Akkara, D.L. Kaplan, Molecular assembly of proteins and conjugated polymers: toward development of biosensors. Biotechnol. Bioeng. **45**, 116 (1995)

32. M. Li, K.K.W. Wong, S. Mann, Organization of inorganic nanoparticles using biotin-strepavidin connectors. Chem. Mater. **11**, 23 (1999)
33. M. Li, E. Dujardin, S. Mann, Programmed assembly of multi-layered protein/nanoparticle-carbon nanotube conjugates. Chem. Commun. 4952 (2005)
34. K.K. Caswell, J.N. Wilson, U.H.F. Bunz, C.J. Murphy, Preferential end-to-end assembly of gold nanorods by biotin-strepavidin connectors. J. Am. Chem. Soc. **125**, 13914 (2003)
35. C.A. Mirkin, Programming the assembly of two- and three-dimensional architectures with DNA and nanoscale inorganic building blocks. Inorg. Chem. **39**, 2258 (2000)
36. F.C. Simmel, Three-dimensional nanoconstruction with DNA. Angew. Chem. **47**, 5884 (2008)
37. C.Q. Yi, D.D. Liu, M.S. Yang, Building nanoscale architectures by directed synthesis and self-assembly. Curr. Nanosci. **5**, 75 (2009)
38. R.A. Kiehl, DNA-directed assembly of nanocomponents for nanoelectronics, nanophotonics, and nanosensing. *Proceedings of SPIE 2007* (SPIE, Bellingham, WA, 2007), p. 67680Z
39. R.P. Goodman, M. Heilemann, S. Doose, C.M. Erben, A.N. Kapanidis, A.J. Turberfield, Reconfigurable, braced, three-dimensional DNA nanostructures. Nat. Nanotechnol. **3**, 93 (2008)
40. J. Chen, N.C. Seeman, Synthesis from DNA of a molecule with the connectivity of a cube. Nature **350**, 631 (1991)
41. J. Zimmermann, M.P.J. Cebulla, S. Mönninghoff, G. von Kiedrowski, Self-assembly of a DNA dodecahedron from 20 trisoligonucleotides with C_{3h} linkers. Angew. Chem. **47**, 3626 (2008)
42. W.M. Shih, J.D. Quispe, G.F. Joyce, A 1.7-kilobase single-stranded DNA that folds into a nanoscale octahedron. Nature **427**, 618 (2004)
43. R.P. Goodman, I.A.T. Schaap, C.F. Tardin, C.M. Erben, R.M. Berry, C.F. Schmidt, A.J. Turberfield, Rapid chiral assembly of rigid DNA building blocks for molecular nanofabrication. Science **310**, 1661 (2005)
44. C.M. Erben, R.P. Goodman, A.J. Turberfield, Single-molecule protein encapsulation in a rigid DNA cage. Angew. Chem. **45**, 7414 (2006)
45. Y. He, T. Ye, M. Su, C. Zhang, A.E. Ribbe, W. Jiang, C. Mao, Hierarchical self-assembly of DNA into symmetric supramolecular polyhedra. Nature **452**, 198 (2008)
46. C. Zhang, M. Su, Y. He, X. Zhao, P. Fang, A.E. Ribbe, W. Jiang, C. Mao, Conformational flexibility facilitates self-assembly of complex DNA nanostructures. Proc. Natl. Acad. Sci. **105**, 10665 (2008)
47. P.W.K. Rothemund, Folding DNA to create nanoscale shapes and patterns. Nature **440**, 297 (2006)
48. E.S. Andersen, M. Dong, M.M. Nielsen, K. Jahn, R. Subramani, W. Mamdouh, M.M. Golas, B. Sander, H. Stark, C.L.P. Oliveira, J.S. Pedersen, V. Birkedal, F. Besenbacher, K.V. Gothelf, J. Kjems, Self-assembly of a nanoscale DNA box with a controllable lid. Nature **459**, 73 (2009)
49. E. Dujardin, S. Mann, Bio-inspired materials chemistry. Adv. Eng. Mater. **4**, 461 (2002)
50. A. Kumar, M. Pattarkine, M. Bhadbhade, A.B. Mandale, K.N. Ganesh, S.S. Datar, C.V. Dharmadhikari, M. Sastry, Linear superclusters of colloidal gold particles by electrostatic assembly on DNA templates. Adv. Mater. **13**, 341 (2001)
51. C. Mirkin, R.L. Letsinger, R.C. Mucic, J.J. Storhoff, A DNA-based method for rationally assembling nanoparticles into macroscopic materials. Nature **382**, 607 (1996)
52. J. Zhang, Y. Liu, Y. Ke, H. Yan, Periodic square-like gold nanoparticle arrays templated by self-assembled 2D DNA nanogrids on a surface. Nano Lett. **6**, 248 (2006)
53. R.C. Mucic, J.J. Storhoff, C.A. Mirkin, R.L. Letsinger, DNA-directed synthesis of binary nanoparticle network materials. J. Am. Chem. Soc. **120**, 12674 (1998)
54. E. Dujardin, L. Hsin, C.R.C. Wang, S. Mann, DNA-driven self-assembly of gold nanorods. Chem. Commun. 1264 (2001)

55. S. Park, A.A. Lazarides, C.A. Mirkin, R.L. Letsinger, Directed assembly of periodic materials from protein and oligonucleotide-modified nanoparticle building blocks. Angew. Chem. **40**, 2909 (2001)
56. A.D. Bates, B.P. Callen, J.M. Cooper, R. Cosstick, C. Geary, A. Glidle, L. Jaeger, J.L. Pearson, M. Proupín-Pérez, C. Xu, D.R.S. Cumming, Construction and characterization of a gold nanoparticle wire assembled using Mg^{2+}-dependent RNA–RNA interactions. Nano Lett. **6**, 445 (2006)
57. D. Chowdhury, Combination of self-assembly and nanolithography as an effective nanofabrication methodology for device realization. Curr. Sci. **96**, 923 (2009)
58. G.M. Whitesides, B. Grzybowski, Self-assembly at all scales. Science **295**, 2418 (2002)
59. M. Wen, K.E.H. Qi, L. Li, J. Chen, Y. Chen, Q. Wu, T. Zhang, Langmuir-Blodgett self-assembly and electrochemical catalytic property of FePt magnetic nano-monolayer. J. Nanopart. Res. **9**, 909 (2007)
60. F. Kim, S. Kwan, J. Akana, P. Yang, Langmuir-Blodgett nanorod assembly. J. Am. Chem. Soc. **123**, 4360 (2001)
61. Y. Xia, B. Gates, Y. Yin, Y. Lu, Monodispersed colloidal spheres: old materials with new applications. Adv. Mater. **12**, 693 (2000)
62. R. Micheletto, H. Fukuda, M. Ohtsu, A simple method for the production of a two-dimensional, ordered array of small latex particles. Langmuir **11**, 3333 (1995)
63. C. Petit, A. Taleb, M.P. Pileni, Self-organization of magnetic nanosized cobalt particles. Adv. Mater. **10**, 259 (1998)
64. B.J. Murray, Q. Li, J.T. Newberg, E.J. Menke, J.C. Hemminger, R.M. Penner, Shape-and size-selective electrochemical synthesis of dispersed silver(I) oxide colloids. Nano Lett. **5**, 2319 (2005)
65. I. Sokolov, Y. Kievsky, 3D design of self-assembled nanoporous colloids. Stud. Surf. Sci. Catal. **156**, 433 (2005)
66. Y. Sun, Y. Xia, Shape-controlled synthesis of gold and silver nanoparticles. Science **298**, 2176 (2002)
67. C.B. Murray, C.R. Kagan, M.G. Bawendi, Self-organization of CdSe nanocrystallites into three-dimensional quantum dot superlattices. Science **270**, 1335 (1995)
68. A. Ganguly, T. Ahmad, A.K. Ganguli, Self-assembly of copper succinate nanoparticles to form anisotropic mesostructures. Dalton Trans. 3536 (2009)
69. A. van Blaaderen, R. Ruel, P. Wiltzius, Template-directed colloidal crystallization. Nature **385**, 321 (1997)
70. Y.Y. Wu, G.S. Cheng, K. Katsov, S.W. Sides, J.F. Wang, J. Tang, G.H. Fredrickson, M. Moskovits, G.D. Stucky, Composite mesostructures by nano-confinement. Nat. Mater. **3**, 816 (2004)
71. B.A. Grzybowski, C.E. Wilmer, J. Kim, K.P. Browne, K.J.M. Bishop, Self-assembly: from crystals to cells. Soft Matter **5**, 1110 (2009)
72. J.S. Beck, J.C. Vartuli, W.J. Roth, M.E. Leonowicz, C.T. Kresge, K.D. Schmitt, C.T.W. Chu, D.H. Olson, E.W. Sheppard, S. B. McCullen, J.B. Higgins, J.L. Schlenker, A new family of mesoporous molecular sieves prepared with liquid crystal templates. J. Am. Chem. Soc. **114**, 10834 (1992)
73. S. Liu, R. Maoz, J. Sagiv, Planned nanostructures of colloidal gold via self-assembly on hierarchically assembled organic bilayer template patterns with in-situ generated terminal amino functionality. Nano Lett. **4**, 845 (2004)
74. M. Rycenga, J. M. McLellan, Y. Xia, Controlling the assembly of silver nanocubes through selective functionalization. Adv. Mater. **20**, 2416 (2008)
75. S. Park, J. Lim, S. Chung, C.A. Mirkin, Self-assembly of mesoscopic metal-polymer amphiphiles. Science **303**, 348 (2004)
76. Z. Gu, H. Ye, D.H. Gracias, The bonding of nanowire assemblies using adhesive and solder. J. Miner. Met. Mater. Soc. **57**, 60 (2005)

77. J.C. Love, A.R. Urbach, M.G. Prentiss, G.M. Whitesides, Three-dimensional self-assembly of metallic rods with submicron diameters using magnetic interactions. J. Am. Chem. Soc. **125**, 12696 (2003)
78. J. Tien, T.L. Breen, G.M. Whitesides, Crystallization of millimeter-scale objects with use of capillary forces. J. Am. Chem. Soc. **120**, 12670 (1998)
79. S.R.J. Oliver, N. Bowden, G.M. Whitesides, Self-assembly of hexagonal rod arrays based on capillary forces. J. Colloid Interface Sci. **224**, 425 (2000)
80. T.L. Breen, J. Tien, S.R.J. Oliver, T. Hadzic, G.M. Whitesides, Design and self-assembly of open, regular, 3D mesostructures. Science **284**, 948 (1999)
81. D.H. Gracias, J. Tien, T.L. Breen, C. Hsu, G.M. Whitesides, Forming electrical networks in three dimensions by self-assembly. Science **289**, 1170 (2000)
82. Z. Gu, Y. Chen, D.H. Gracias, Surface tension driven self-assembly of bundles and networks of 200 nm diameter rods using a polymerizable adhesive. Langmuir **20**, 11308 (2004)
83. Z. Gu, H. Ye, D. Smirnova, D.H. Gracias, Reflow and electrical characteristics of nanoscale solder. Small **2**, 225 (2006)
84. T.G. Leong, A. Zarafshar, D.H. Gracias, Three dimensional fabrication at small size scales. Small **6**, 792 (2010)
85. V.Y. Prinz, V.A. Seleznev, A.K. Gutakovsky, A.V. Chehovskiy, V.V. Preobrazhenskii, M.A. Putyato, T.A. Gavrilova, Free-standing and overgrown InGaAs/GaAs nanotubes, nanohelices and their arrays. Phys. E **6**, 828 (2000)
86. O.G. Schmidt, K. Eberl, Nanotechnology: thin solid films roll up into nanotubes. Nature **410**, 168 (2001)
87. O.G. Schmidt, N. Schmarje, C. Deneke, C. Müller, N.Y. Jin-Phillipp, Three-dimensional nano-objects evolving from a two-dimensional layer technology. Adv. Mater. **13**, 756 (2001)
88. A.V. Prinz, V.Y. Prinz, V.A. Seleznev, Semiconductor micro- and nanoneedles for microinjections and ink-jet printing. Microelectron. Eng. **67–68**, 782 (2003)
89. M. Huang, C. Boone, M. Roberts, D.E. Savage, M.G. Lagally, N. Shaji, H. Qin, R. Blick, J.A. Nairn, F. Liu, Nanomechanical architecture of strained bilayer thin films: from design principles to experimental fabrication. Adv. Mater. **17**, 2860 (2005)
90. V.Y. Prinz, Precise, Molecularly thin semiconductor shells: From nanotubes to nanocorrugated quantum systems. Phys. Stat. Sol. B **243**, 3333 (2006)
91. C. Deneke, C. Müller, N.Y. Jin-Phillipp, O.G. Schmidt, Diameter scalability of rolled-up In(Ga)As/GaAs nanotubes. Semicond. Sci. Technol. **17**, 1278 (2002)
92. O.G. Schmidt, C. Deneke, Y.M. Manz, C. Müller, Semiconductor tubes, rods and rings of nanometer and micrometer dimension. Phys. E **13**, 969 (2002)
93. M. Ohring, *Materials Science of Thin Films* (Academic, San Diego, CA, 2002)
94. P.O. Vaccaro, K. Kubota, T. Aida, Strain-driven self-positioning of micromachined structures. Appl. Phys. Lett. **78**, 2852 (2001)
95. Y.C. Tsui, T.W. Clyne, An analytical model for predicting residual stresses in progressively deposited coatings part 1: planar geometry. Thin Solid Films **306**, 23 (1997)
96. J.H. Cho, T. James, D.H. Gracias, Curving nanostructures using extrinsic stress. Adv. Mater. **22**, 2320 (2010)
97. D.R. Lide, *CRC Handbook of Chemistry and Physics* (CRC Press, Boca Raton, FL, 2009)
98. W.D. Nix, B.M. Clemens, Crystallite coalescence: a mechanism for intrinsic tensile stresses in thin films. J. Mater. Res. **14**, 3467 (1999)
99. S.C. Seel, C.V. Thompson, Tensile stress generation during island coalescence for variable island-substrate contact angle. J. Appl. Phys. **93**, 9038 (2003)
100. B.W. Sheldon, K.H.A. Lau, A. Rajamani, Intrinsic stress, island coalescence, and surface roughness during the growth of polycrystalline films. J. Appl. Phys. **90**, 5097 (2001)
101. F. Spaepen, Interfaces and stresses in thin films. Acta Mater. **48**, 31 (2000)
102. A.N. Magunov, Determining the heat of a surface plasmochemical reaction by scanning calorimetry. Instrum. Exp. Tech. **43**, 706 (2000)
103. W.D. Pilkey, *Analysis and Design of Elastic Beams* (Wiley, New York, NY, 2002)

104. S. Yasin, D.G. Hasko, H. Ahmed, Fabrication of <5 nm width lines in poly(methylmethacrylate) resist using a water:isopropyl alcohol developer and ultrasonically-assisted development. Appl. Phys. Lett. **78**, 2760 (2001)
105. K. Yamazaki, H. Namatsu, 5-nm-order electron-beam lithography for nanodevice fabrication. Jpn. J. Appl. Phys. **43**, 3767 (2004)
106. E.J. Smith, Z. Liu, Y. Mei, O.G. Schmidt, Combined surface plasmon and classical waveguiding through metamaterial fiber design. Nano Lett. **10**, 1 (2010)
107. K. Miura, Method of packaging and deployment of large membranes in space. *Proceedings of the 31st Congress of the International Astronautical Federation* (American Institute of Aeronautics and Astronautics, New York, NY, 1980), p. 1
108. I. Stewart, Mathematics: some assembly needed. Nature **448**, 419 (2007)
109. H.J. In, S. Kumar, Y. Shao-Horn, G. Barbastathis, Origami fabrication of nanostructured, three-dimensional devices: electrochemical capacitors with carbon electrodes. Appl. Phys. Lett. **88**, 083104 (2006)
110. S.T. Brittain, O.J.A. Schueller, H.K. Wu, S. Whitesides, G.M. Whitesides, Microorigami: fabrication of small, three-dimensional, metallic structures. J. Phys. Chem. B **105**, 347 (2001)
111. H. Okuzaki, T. Saido, H. Suzuki, Y. Hara, H. Yan, A biomorphic origami actuator fabricated by folding a conducting paper. J. Phys. Conf. Ser. **127**, 012001 (2008)
112. R.R.A. Syms, E.M. Yeatman, Self-assembly of fully three-dimensional microstructures using rotation by surface tension forces. Electron. Lett. **29**, 662 (1993)
113. R.R.A. Syms, E.M. Yeatman, V.M. Bright, G.M. Whitesides, Surface tension-powered self-assembly of microstructures – the state-of-the-art. J. Microelectromech. Syst. **12**, 387 (2003)
114. W.J. Arora, A.J. Nichol, H.I. Smith, G. Barbastathis, Membrane folding to achieve three-dimensional nanostructures: nanopatterned silicon nitride folded with stressed chromium hinges. Appl. Phys. Lett. **88**, 053108 (2006)
115. B. Gimi, T.G. Leong, Z. Gu, M. Yang, D. Artemov, Z.M. Bhujwalla, D.H. Gracias, Self-assembled three dimensional radio frequency (RF) shielded containers for cell encapsulation. Biomed. Microdevices **7**, 341 (2005)
116. T.G. Leong, P.A. Lester, T.L. Koh, E.K. Call, D.H. Gracias, Surface tension driven self-folding polyhedra. Langmuir **23**, 8747 (2007)
117. J.H. Cho, S. Hu, D.H. Gracias, Self-assembly of orthogonal 3-axis sensors. Appl. Phys. Lett. **93**, 043505 (2008)
118. A. Azam, T.G. Leong, A.M. Zarafshar, D.H. Gracias, Compactness determines the success of cube and octahedron self-assembly. PLoS One **4** (2009) e4451
119. D.J. Filipiak, A. Azam, T.G. Leong, D.H. Gracias, Hierarchical self-assembly of complex polyhedral microcontainers. J. Micromech. Microeng. **19**, 075012 (2009)
120. N. Bassik, G.M. Stern, D.H. Gracias, Microassembly based on hands free origami with bidirectional curvature. Appl. Phys. Lett. **95**, 091901 (2009)
121. J.H. Cho, D.H. Gracias, Self-assembly of lithographically patterned nanoparticles. Nano Lett. **9**, 4049 (2009)
122. J.S. Randhawa, T.G. Leong, N. Bassik, B.R. Benson, M.T. Jochmans, D.H. Gracias, Pick-and-place using chemically actuated microgrippers. J. Am. Chem. Soc. **130**, 7238 (2008)
123. T.G. Leong, C.L. Randall, B.R. Benson, N. Bassik, G.M. Stern, D.H. Gracias, Tetherless thermobiochemically actuated microgrippers. Proc. Natl. Acad. Sci. **106**, 703 (2009)
124. J.S. Randhawa, M.D. Keung, P. Tyagi, D.H. Gracias, Reversible actuation of microstructures by surface chemical modification of thin film bilayers. Adv. Mater. **22**, 407 (2010)

Chapter 2
Bio-inspired 3D Nanoarchitectures

Jian Shi and Xudong Wang

2.1 Introduction

Nature designs a variety of peculiar structures and materials for plants and animals with unique functionalities, which have inspired many scientists and researchers to study the structure–property relationship and mimic their structures and functions. The surprisingly precise functionality control of creatures, such as sensing, self-cleaning, self-healing, and flying, has always been igniting human's curiosities and intriguing us to steal ideas from nature [1–5]. In contrast to all manufacturing techniques developed by mankind, nature uses gene, a unique self-assembled way, to design and fabricate structures and tools, to control properties, and, more importantly, to reproduce. The scale of nature products ranges from nanometer to millimeter. Carbon, oxygen, nitrogen, hydrogen, calcium, phosphorus, sulfur, and silicon are favorite elements that nature uses. Biologically functional materials grown by gene-controlled self-assembly have been carefully designed for billions of years in responding to the environment. Longtime evolution has made nature's design quite sophisticated. Hierarchical structures at all size levels are seamlessly integrated together in order to adapt to the tremendous variance of nature. Evolution has already helped nature to produce billions of delicate biological machines on earth. However, for a long period of time, people did not understand the gift of nature in designing and manufacturing. Man-made structures/machines are usually designed by prior estimation of the data and information to fulfill the mechanical, electrical, optical, or biological requirements, which is a static process and far simpler than a biological process. Although gene works as a template in making organisms, the strategy for nature is rather a dynamic process considering the fluctuation of environments. Therefore, nature's selection is more delicate and flexible [6].

X. Wang (✉)
Department of Materials Science and Engineering, University of Wisconsin-Madison, Madison, WI, USA
e-mail: xudong@engr.wisc.edu

W.L. Zhou, Z.L. Wang (eds.), *Three-Dimensional Nanoarchitectures*,
DOI 10.1007/978-1-4419-9822-4_2,

The quick development and usage of nanometer-level characterization tools in biological materials in the past two decades has revealed many surprising discoveries, which correlate the functions of plants and animals to their micro- and nanoscale structures. Bones were typically considered as a well-studied biostructure. However, recently people have started to realize that the mechanical properties of a complex hierarchical bone structure cannot be accurately predicted by the composite rule of mixtures formulae without considering the micro- and nano-structural interactions. For example, plate-shaped crystals of carbonate apatite and various collagens are of the size of a few nanometers, but the mechanical strength and toughness of these natural composite materials are two to three orders of magnitude higher than that of single crystals of pure minerals, which can be partially attributed to the hierarchical architectures [7]. Golden silk spider, *Nephila clavipes*, uses water as the solvent to fabricate spider silk, which also has a special hierarchical structure. Properly controlling the folding and crystallization of main protein constituents and adding auxiliary compounds are reported to be the key steps to form spider silk with super-mechanical properties. It is believed that precisely copying such a spider protocol could help industry to improve the properties of current spinning products [8]. Arzt et al. revealed that the very efficient attaching mechanism of flies, spiders, and geckos is caused by splitting up the contact into finer sub-contacts. They showed that heavier animals exhibit finer adhesion structures, in which the setae of gecko lie in sub-micrometer scale and are believed to produce the strongest adhesion [9]. Self-cleaning mechanism, the so-called Lotus effect, is found in many terrestrial plants and animals. The common characteristic of this functionality is that the hydrophobic property results from the microscopic roughness. This effect has been intensively studied and is promising for the applications as artificial surfaces for cars, facades, and foils [1]. Another famous discovery is the optical property of brittle stars with calcitic microlenses as part of the photoreceptor system. The hierarchical calcite structure acts not only as an optical element but also as a mechanical support [10]. In addition, by investigating the nanostructure of sponge *Euplectella* sp., it has been found that there are seven hierarchical levels such as laminated structures, fiber-reinforced composites, bundled beams, and diagonally reinforced squared grid cells in the sponge skeleton. This configuration is particularly intuitive for structural materials designing and fabrication [11]. Studies of the remarkable features, the evolution history, and the underlying mechanisms of these functional biological species would definitely provide new ideas and insights for creating new designs, improving the controllability and reproducibility of the micro- and nanoscale smart materials and realizing novel functionalities.

Among current manufacturing industry, energy saving and environmental friendly are basic development criteria. Bio-inspired materials and structures are particularly advantageous due to their high adaptability, effectiveness, and efficiency. For example, the bio-inspired materials would obviously play an important role in harvesting and utilizing sunlight energy due to their intrinsic photosynthesis demands. The bio-fabrication process is also considered as pollution free, which is critical in preventing our world from deteriorating. The unique functionalities of biological species would ignite new ideas in creating novel artificial materials or devices with integrated functionalities and time, space, and cost saving. Therefore, the study and utilization of bio-inspired materials and structures would be of great significance in advancing our current technology.

2.2 Historical Perspective

The approach of bio-inspired design has been utilized for thousands of years. It begins with mimicking the motions or functions of animals to improve the capability of human beings, and then the focus gradually shifted to the creation of novel materials and devices. For example, people learned how to fly from birds. Submarines were created by learning from fish. The ancient Chinese ever tried to fabricate artificial silk around 3000 years ago. In general, these lessons from nature were taken by us macroscopically.

The invention of electron microscope opened a completely new small world to us, and new opportunities were brought up when we were able to see the fine features in the nanoworld and understand the mechanisms from the origins. To this end, bio-inspired designing and fabrication become more systematic, efficient, and effective. One example is Velcro, the well-known brand name of fabric hook-and-loop fasteners. In 1941, Mestral used microscope to examine the burr of burdock, which kept sticking to his clothes and his dog's fur when they walked through the burdock and found hundreds of "hooks" on it. He realized this could be man-made if proper materials were chosen and eventually the Velcro was invented. The successful applications of Velcro on weaving, aerospace, and medical treatment demonstrated the excellence of bio-inspired structures. Stepping into the 21st century, many more advanced bio-inspired materials and systems, such as pinecone fabrics inspired by pinecones, interferometric modular display (iMoD) technology inspired by butterfly wings, neuromorphic computer chips inspired by neural networks, have been created. These successful developments are continuously strengthening our confidence in learning and applying bio-inspired materials and systems into our lives. In the rest of this chapter, we will focus on discussing bio-inspired nanophotonics, bio-inspired fabrication, and bio-inspired functionalities.

2.3 Bio-inspired Nanophotonics

2.3.1 Photonic Crystals

A photonic crystal (PC) is a periodically arranged structure composed of alternating dielectric materials with high and low refractive indices, the periodicity of which is controlled in the order of the electromagnetic wavelength, so that the light can "sense" the change of refractive index as it transmits through the structure. The periodicity could be in one, two, or three dimensions, as schematically illustrated in

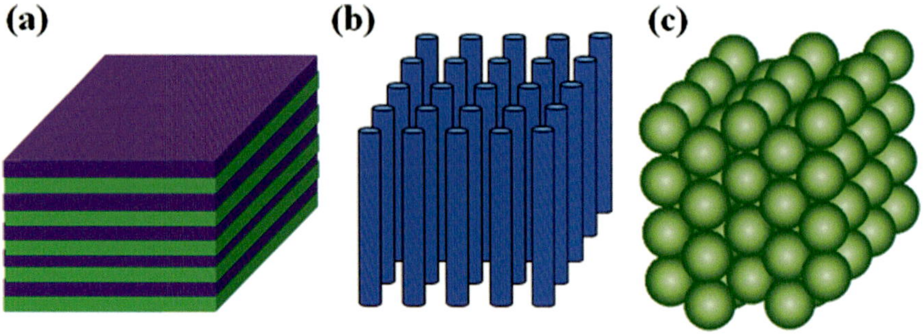

Fig. 2.1 Schematic illustrations of (**a**) 1D, (**b**) 2D, and (**c**) 3D photonic crystals

Fig. 2.1. Most exciting applications of PCs involve the control of light in the visible light regime.

To study the propagation of light in a PC, we start with the Maxwell equations:

$$\nabla \times \vec{E} + \frac{1}{c}\frac{\partial \vec{H}}{\partial t} = 0$$

$$\nabla \times \vec{H} - \frac{1}{c}\frac{\partial \varepsilon \vec{E}}{\partial t} = \frac{4\pi}{c}\vec{j}$$

$$\nabla \times \varepsilon \vec{E} = 4\pi\rho$$

$$\nabla \times \vec{H} = 0$$

where $\vec{E}$ and $\vec{H}$ are the electric and magnetic fields, respectively; ρ and J are the free charges and currents, respectively. In a dielectric medium, the electric, magnetic fields, charges, and currents can be expressed as follows:

$$\vec{E}(r,t) = \vec{E}(r)\exp(i\omega t)$$

$$\vec{H}(r,t) = \vec{H}(r)\exp(i\omega t)$$

$$\vec{j} = \rho = 0$$

At a given ε, a continuous spectrum of $\vec{H}$ as a function of frequency ω could be obtained:

$$\nabla \times \frac{1}{\varepsilon}\nabla \times \vec{H} = \left(\frac{\omega}{c}\right)^2 \vec{H}$$

The operator is found to be a periodic function of position:

$$\vec{H}_{\vec{k}}(\vec{r}) = e^{i(\vec{k}.\vec{r})}\vec{u}_{\vec{k}}(\vec{r})$$

$$\vec{u}_{\vec{k}}(\vec{r}) = \vec{u}_{\vec{k}}(\vec{r} + \vec{R})$$

Assume there is a medium with uniform ε_1, then a linear relationship between frequency ω and wave vector k is obtained in the entire 1D space:

$$\omega = \frac{k}{\sqrt{\varepsilon_1}}$$

When we treat this uniform medium as "artificially" periodic,

$$\varepsilon(x) = \varepsilon(x + a)$$

The above bands would be "folded" by $2\pi/a$

If we add a small "real" periodicity, $\varepsilon_2 = \varepsilon_1 + \Delta\varepsilon$, splitting of degeneracy would happen since states concentrated in high index ε_2 would have lower frequency. This perturbation in index causes the appearance of photonic band gap. Figure 2.2a shows a 3D photonic crystal of silicon inverse opal structure. The corresponding photonic band structure is shown in Fig. 2.2b [12]. According to the band structure, we could see that certain wavelengths (around 1.46 μm) of light are forbidden in such a crystal. Some other unique properties of PCs include energy moves very slowly inside the PCs and the stationary modes act like resonators; light refracts in anomalous directions and negative index of refraction is possible; faster or slower decay for luminescent material could happen and a possible laser action would exist. Self-collimation is another advantageous property of PCs, in which parallel beams could be obtained and no beam spreads at the designed frequencies, while conventional integrated optics requires boundaries to confine beams. In addition, light travels inside PCs without any coupling, cross-talking, space limitation, or special resonating phenomenon, and 100% reflection can be achieved. Based on these unique properties, photonic band gap mirrors and integrated "free-space-like" optics can be realized. PC indeed provides a new and more advanced functionality over conventional optics.

Defects embedded in PCs can trap light, which makes defective PCs working as cavities or waveguides. For example, a PC fiber contains a dielectric core that is located at the center of a periodic dielectric shell. A 2D PC waveguide is fabricated by introducing line defects into a 2D PC slab, in which zero loss transmitting through 90° bending channel with bending radius less than 1 wavelength of the light is possible. Although PCs possess fascinating optical properties, fabrication of such delicate periodic structures in large scale is truly challenging and significantly limits the application of PC-based optics.

Self-assembly, lithography, and nano/micro-manipulation are typical methods for making PCs so far, whereas nature has already shown various success in constructing PCs with well-controlled features. Opal structure is a typical non-biological self-assembled PCs existing in nature. It comprises highly packed homogeneous submicron-sized silica spheres with a face-centered cubic (FCC) arrangement. Some living creatures in nature also have PC structures, which give them beautiful color or

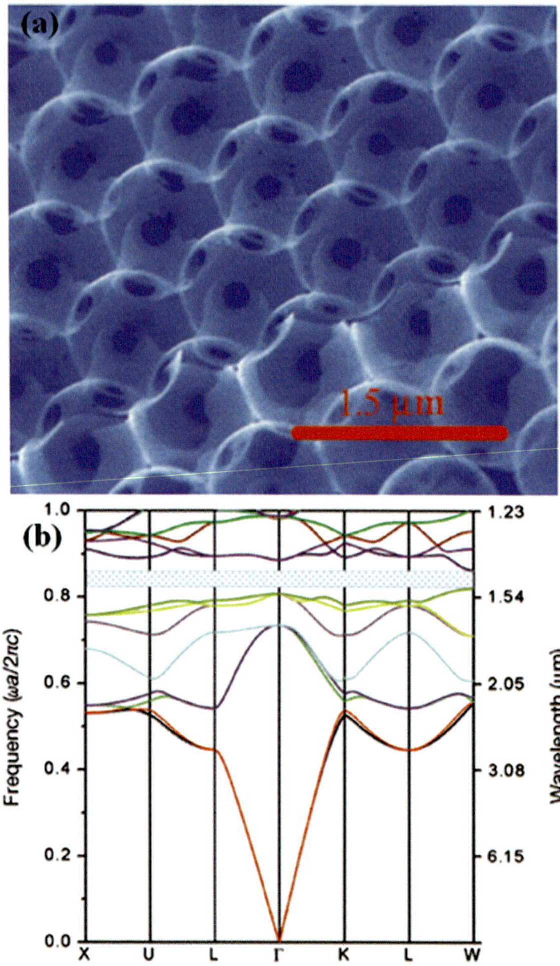

Fig. 2.2 (**a**) SEM image of internal facets of silicon inverse opal. (**b**) Photonic band structure of silicon inverse opal with 88% Si infiltrated into the available opal template voids. Reproduced with permission from [12]. © 1999, American Chemical Society

special optical functionality. For example, the shining color on butterfly wings and some beetles are contributed by 2D or 3D PC structures on their wing or shell scales. Most algae have PC-like skeleton in order to enhance the light sensitivity under water. Spine of sea mouse is composed of 2D PC fibers, which enhance red light reflectance. In general, nature has many kinds of examples as PCs and most could provide us remarkable hints and guidance in advancing our current technologies.

2.3.2 Color Mine in Nature

Before we go into the details of nature PCs, we first briefly discuss the origin of colors. The beautiful colors of living creatures in nature are usually a mixture of physical colors and chemical colors. Pigments are referred as chemical colors, which produce color by selective absorption of light at particular wavelength by chemical bonds or substituent of the pigments. Physical color, also called structural color, is caused by interference effects resulted from the physical structures. Colors from Tyndall effect follow the fact that the intensity of the scattered light depends on the fourth power of the frequency, which explains that blue light is most commonly seen on earth. Grating effect happens when the microstructure of a media is arranged into arrays and reflects light with different wavelengths into different directions due to the interference mechanism. Structural colors are widely found throughout in the organisms of living creatures such as feathers, scales, wings, shells, and insect cuticle [13–17].

Structural color could also be a mixture of two colors. Vukusic et al. reported that the colored regions of the Indonesian male *Papilio palinurus* butterfly are the consequence of a regular 2D array of concavities (~4–6 μm in diameter and ~0.5–3 μm at the greatest depth) and a multilayer structure [16]. The green color sensed by human eyes is the juxtaposition of the yellow region on the flat surface of the concavity and the blue region along the inclined side of the concavity. If this structure is observed beyond a specific angle, the blue color would be enhanced but the green color would disappear. This is because the inclined sides of the concavity produce angle-dependent polarization that intensifies the blue light. This color formation mechanism is called as spatial-averaging color stimulus synthesis, which has also been reported in beetles [18]. In general, structural colors can provide higher visibility than pigment-based colors and is considered as the prerequisite for optical property duplication.

2.3.3 Natural Photonic Crystals

In the last two decades, with the assistance of advanced electron microscopy techniques, PC structures have been widely discovered from many creatures. Their unique and well-controlled submicron structures greatly inspired researchers to create revolutionary techniques to fabricate nanometer-scale photonic structures and systems. In this section, we summarize a few representative natural PC systems that have great potential in bio-inspired PC assembly and application.

2.3.3.1 Spine of Sea Mouse

Spine of sea mouse *Aphrodita* sp. is normally deep red, but spectacular iridescence is seen as strips running along the spine when illuminated from side. Parker et al. found that this special phenomenon was derived from a 2D PC structure [19]. Electron microscope image in Fig. 2.3a shows that the spine is composed of an array of hexagonally close-packed cylinders with a diameter of 400 and 510 nm apart. The walls of the cylinders are made of pure α-chitin with a refractive index of 1.54. Considering the refractive index of seawater that fills in the holes is 1.35, a partial photonic band gap exists along the ΓM direction. The calculated polarization

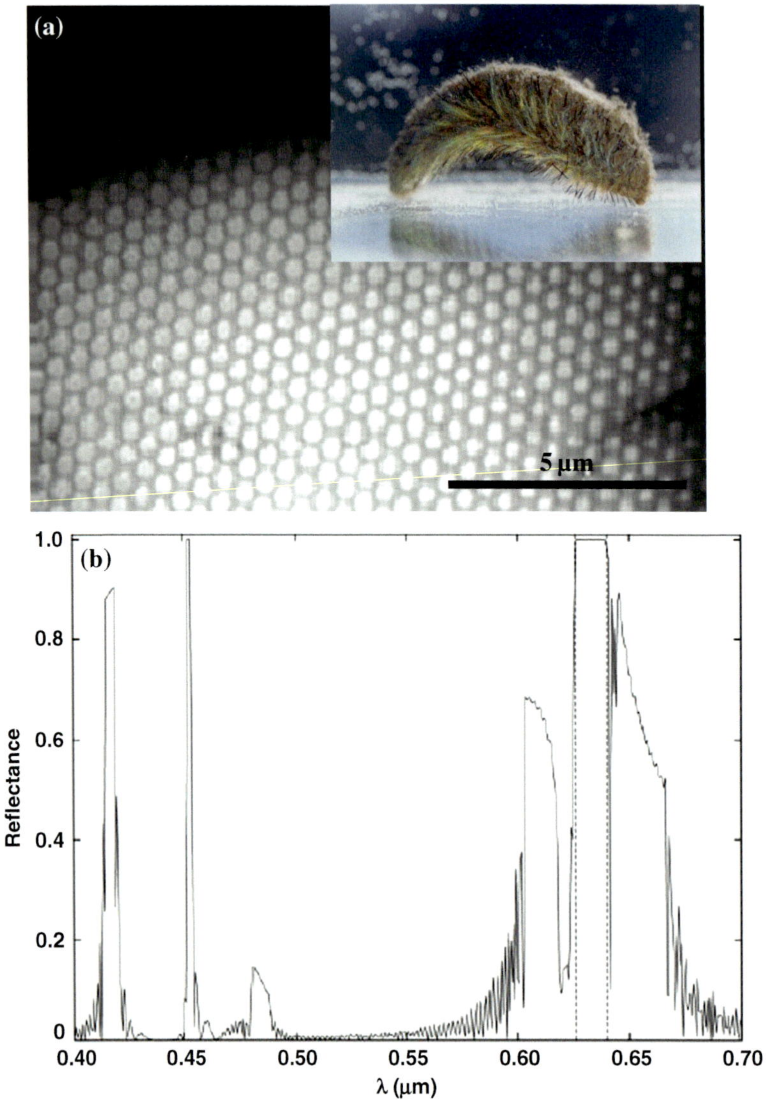

Fig. 2.3 (**a**) SEM image of the cross section of the spine. *Inset* is a photo of a sea mouse. (**b**) Calculated reflectance of a hexagonal cylindrical structure having 500 layers of voids in the chitin of a sea mouse in responding to normal incident light. *Vertical dashed lines* indicate a partial photonic band gap. Reproduced with permission from [19]. © 2001, Nature Publishing Group

reflectance is shown in Fig. 2.3b. Vertical dashed lines indicate the partial band gap in the red light region, which has also been confirmed by experimental measurements. Therefore, enhanced reflectance of red color can be observed in the spines. This narrowband reflectance demonstrates a promising application in optical filters for photonic communications. In addition, the thicker walls at edges of spine are believed to work as mechanical support.

2.3.3.2 Diatom

Diatom is a single-celled marine alga, where *Coscinodiscus granii* is a representative diatom with a central symmetry [20]. Its central feature is a bio-mineralized cell wall (frustules) made from amorphous silica. Each cell has two halves of thecae, which overlap like a petri dish and separate during cell division as shown in Fig. 2.4a. These thecae are distinguished by a valve and girdle bands. Cell walls of the valve and girdle both show very regular arrays of chambers and pores that form specific periodic patterns, which are formed by self-organized phase separation. Basically, the diatom cell can be considered as a "photonic box" with PC walls. The cell diameter of diatoms is around 150 μm, and the refractive index of the silica shell is 1.43. Although the cell size varies tremendously, the size and periodicity of the patterns are almost constant. In detail, the valves of the cells shown in Fig. 2.4b have hexagonally patterned holes with a lattice constant of 900–950 nm and a wall thickness of 700 nm. Girdle in Fig. 2.4c exhibits a square lattice with a lattice constant of 250 nm and a wall thickness of 200–600 nm. Figure 2.4d is the cross section of the valve. Both the valve and girdle are photonic crystal slabs with distinct symmetries and spectral ranges.

In the girdle region, one guided mode along the ΓE or ΓM polarization exists when the thickness of the waveguide (the silica cell wall) is 200 nm and four modes present when thickness increases to 700 nm in air or water, as shown in Fig. 2.4e. Then girdle could guide blue to blue-green light. The decay of evanescent waves in the cell is proportional to the difference in indices between the waveguide and surrounding medium. Photosynthetic units are found existing very close to the cell walls and therefore strongly coupled with the evanescent waves. In the valve region, hexagonal patterns have the grating periodicity larger than the wavelength of visible light and the guided light is in the infrared regime. The valve structure can be regarded as a 2D grating coupler of higher order. Fuhrmann et al. calculated the band structure of the valve utilizing water as the surrounding medium [20]. They found that there are several resonances distributed over the visible range, which make it possible to couple incoming light into the waveguide modes.

2.3.3.3 Butterfly Wings

Most butterflies have beautiful colors on their wings, while only a small portion of them is contributed by periodical submicron structures, which is referred as "physical color." Research has revealed that the irradiant blue color of *Morpho peleides* butterfly wings is resulted from arrays of vertically aligned net-like skeleton structures [13, 21]. The uniform blue color covering the major area of the wings was from millions of ordered scales, as shown in the optical microscope image in Fig. 2.5a. The typical dimension of the scale is ~150 μm in length and ~60 μm in width; 35–40 rows of lamellas align on the scale surface with almost an identical inter-spacing.

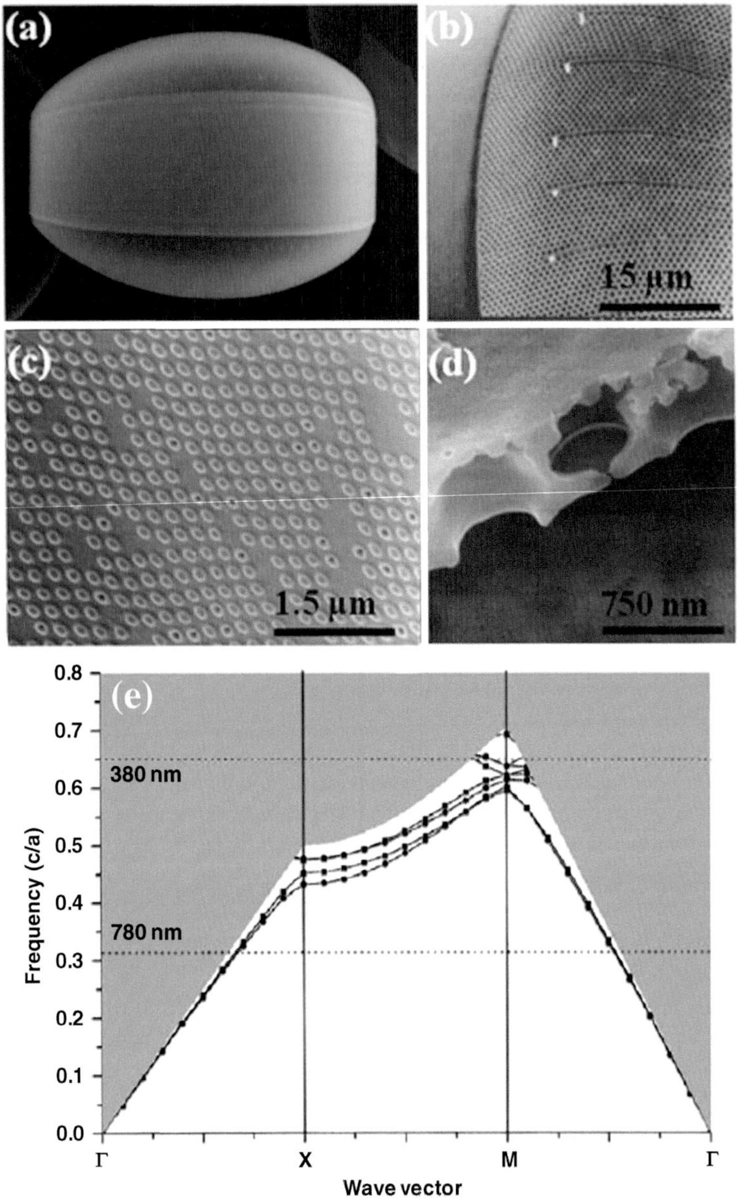

Fig. 2.4 Structure and band structure of diatoms. (**a**) SEM image of the diatom frustules with two overlapping thecae. (**b**) Transition region between the valve and girdle with a hexagonal pattern of areolas. (**c**) Structure of girdle with a square lattice. (**d**) Cross section through the cell wall of the valve. (**e**) Calculated photonic band structure of a thin diatom girdle. Reproduced with permission from [20]. © 2004, Springer

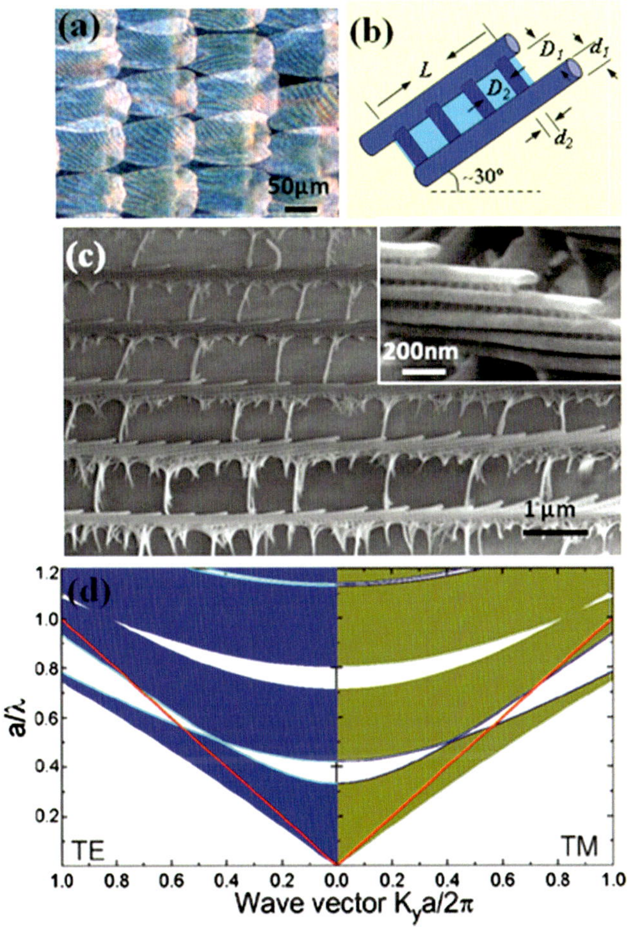

Fig. 2.5 Morphology and band structure of butterfly wing scales. (**a**) An optical microscope image of a butterfly wing scale. (**b**) The schematic of a "unit cell" on the lamellae that is responsible for the scattered colorful light. (**c**) An SEM image taken on the scale surface showing the structure of the lamellae. The *inset* is a higher resolution SEM image showing the periodic nanostructures on a lamella [15]. (**d**) Calculated off-x-axis propagation band structure of the 2D periodic structure of the lamellae using the MPB software. Reproduced with permission from [22]. © 2009, American Institute of Physics

A closer view of these lamellae is shown in Fig. 2.5c. The lamellae are located ~1.6 μm from each other and supported by cross-ribs, which dictate the inter-lamellae spacing and hold the lamellae ~1 μm above the bottom surface of the scales.

The details of the periodic structure on a lamella are shown by a higher magnification SEM image in the inset of Fig. 2.5c, of which one "unit cell" is schematically illustrated in Fig. 2.5b. The cylindrical main ribs are ~1.6 μm in length (L) and ~100 nm in diameter (d_1) with ~60 nm of inter-distance (D_1). Two adjacent ribs are bound together by a row of smaller subribs with a diameter of ~20 nm (d_2), which remain ~50 nm (D_2) away from each other. This configuration constructs a 2D array of 50 × 60 nm rectangular squares surrounded by organic nanometer-scale cylinders (the main and sub-ribs) with a periodicity of ~160 nm ($D_1 + d_1$) along its length direction and ~70 nm ($D_2 + d_2$) along the main ribs. Although the entire lamella is perpendicular to the scale surface, the "unit cells" are largely tilted and the angle between the main ribs and the horizontal plane is only ~30°. This long-range ordered organic structure with a very small periodicity can be considered as a 2D PC slab, and the brilliant blue color was attributed to the existence of the photonic band gap in this structure. Figure 2.5d shows the band structures of both TE and TM modes related to off-x-axis propagation, where the normal direction of the lamella is taken as x-axis [22]. Light is allowed in the shadowed region above the red line while the white region above the red line corresponds to the photonic band gap. The band gap-related reflection peak was found to depend on the incidence beams. A blue shift would occur when the angle between incidence beam and the normal direction of the wing scale increases. Such a delicate structure given by nature contains ~100 nm scale photonic band gap structures with the stop band in the visible light range, which are still very costly to be made by current techniques. Utilizing such structure as a template and coating it uniformly with an inorganic dielectric material would be possible to fabricate PC structures with high reproducibility and low cost.

The structures of waveguides and beam splitters have also been discovered on a *Morpho* butterfly wing scale. The butterfly lamellae structure has a repeating unit of 50 × 60 nm rectangular square surrounded by the ribs. The bigger size ribs were similar to line defects in the 2D matrix of periodic nano-squares. Since the bottoms of these ribs were connected by a big channel, light with a certain wavelength could propagate inside the channel and be confined by the nano-square matrix due to the existence of photonic band gap. Figure 2.6a is an optical microscope image of an Al_2O_3 replica of a wing scale showing bright blue/violet dashed lines, which are the lamellae rows perpendicular to the observing plane (Fig. 2.6b). They suggest that as long as the replica is covered by another selected dielectric material to achieve a good reflection at the structure interface, light would be trapped inside the hollow main ribs with the wavelength corresponding to the photonic band gap of the nano-square matrix so as to realize a 2D PC waveguide array [15]. In addition, the split of lamella could be generally observed when the scale becomes wider, as shown in Fig. 2.6c. This configuration was believed to be able to form a beam-splitting device in the nanometer scale. The light trace of the splitter is shown in Fig. 2.6d. The splitting junction structure and the corresponding light distribution are highlighted by white dashed circles in Fig. 2.6c, d, respectively. At the splitting point, the bright dot first became wider and then split into two separated rows of dots exactly following the lamella structure.

2.3.3.4 Beetles

Pachyrhynchus argus beetle from northeastern Queensland, Australia, has a metallic coloration that can be seen in any direction as shown in the inset of Fig. 2.7a. Parker

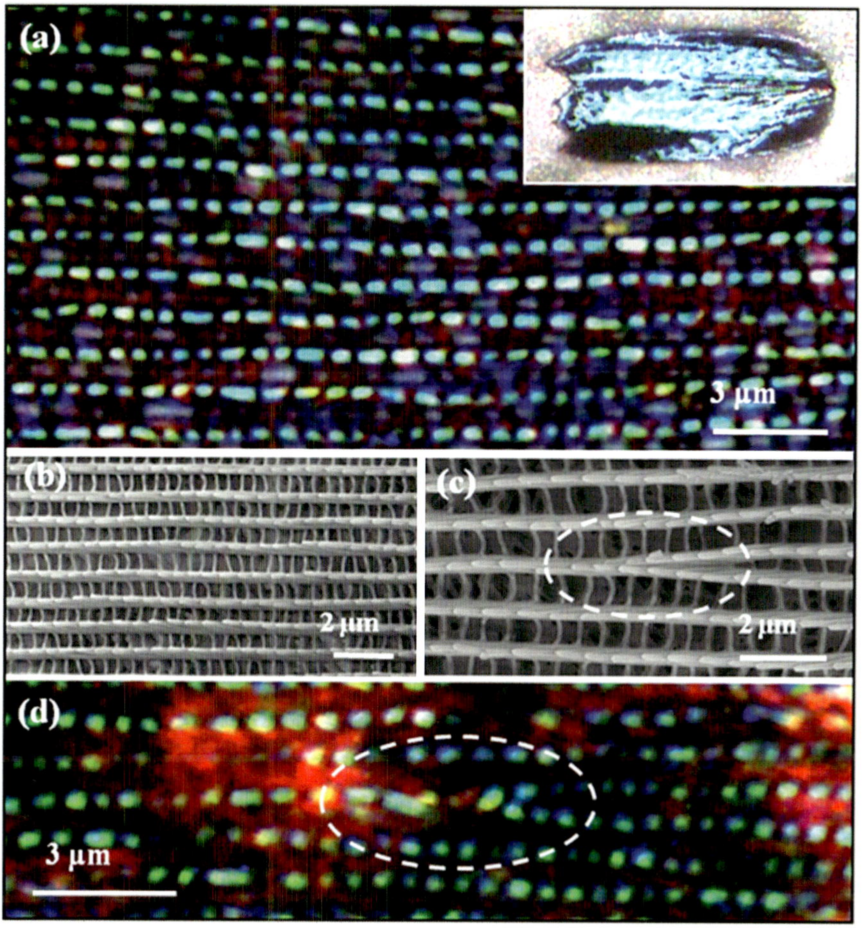

Fig. 2.6 "Waveguide" properties of the alumina replicas of butterfly wing scales. (**a**) A dark field optical microscope image of an alumina replica of the lamella rows; the *inset* is an optical image of an individual scale. (**b**) The SEM image of the replica structure corresponding to optical microscopy image shown in (**a**). (**c**) The SEM image of the bifurcated lamella structure. (**d**) The corresponding dark field optical microscope image showing the potential application as a beam splitter [15]

et al. found that such a brilliant color is resulted from a 3D PC structure analogous to the opal structure [14]. The beetle's scales are flat and consist of two parts: an outer shell and an inner PC structure (Fig. 2.7a). If the inner structure is absent in some scales, the scale would be transparent. When white light is shining onto the scale surface at certain incident angles, the reflectance maximizes at a wavelength of 530 nm. Scanning electron microscopy revealed that the inner structure of the scales is a 3D ordered array of transparent spheres, as shown in Fig. 2.7b. The diameter of these spheres is about 250 nm. This opal-like structure works as the 3D PC and reflects a light in a narrow wavelength range over a wide range of incident angles. To

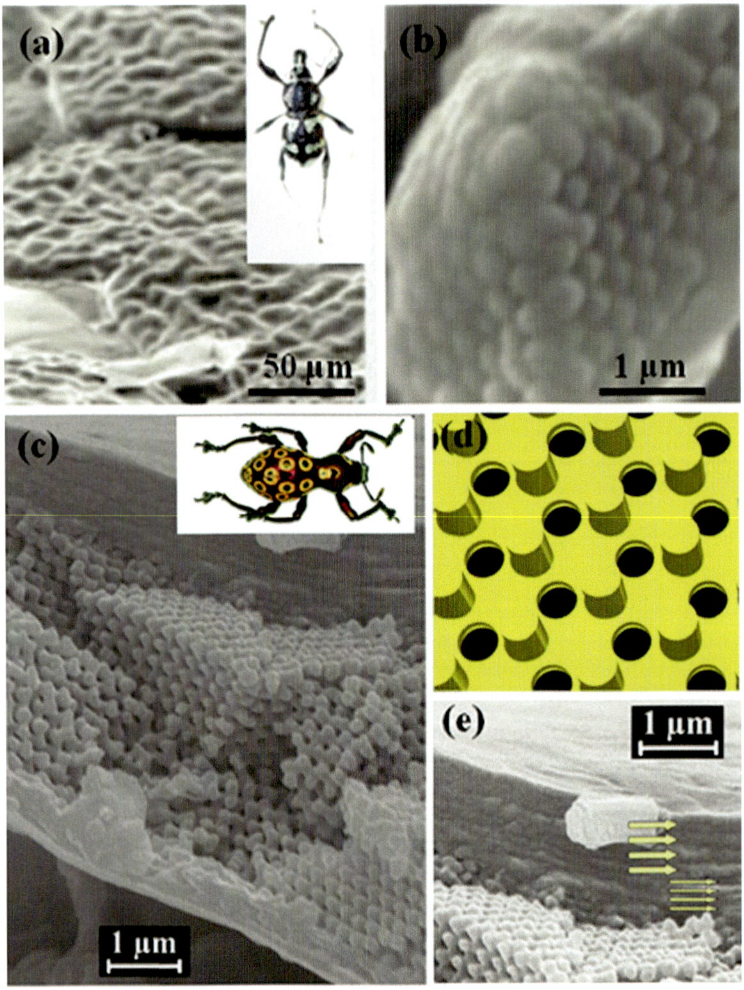

Fig. 2.7 (**a**) SEM image of several overlapped scales on the shell of beetle. *Inset* is the dorsal view of a beetle. (**b**) The fine structure of the scales: 3D ordered array of transparent spheres. Reproduced with permission from [14]. © 2003, Nature Publishing Group. (**c**) SEM image of the scale structure of a *pachyrrhynchus congestus pavonius* weevil. *Inset* is a photo of the weevil. (**d**) Schematic illustration of a chitin layer with three overlapping sublattices. (**e**) SEM image showing the layered cortex containing two distinctive mirrors indicated by two arrays of *arrows*. Reproduced with permission from [23]. © 2007, American Physical Society

calculate the wavelength of maximum reflectance, $\lambda_{\max}$, for an angle of incidence θ, we have

$$\lambda_{\max} = 2d \times 0.816\sqrt{n^2 - \sin^2\theta}$$

where d is the lattice spacing, 0.816 is the spacing between close-packed planes in a unit of sphere diameter, and n represents the average refractive index in the system.

Here n is the sum of refractive index of the submicron spheres and water matrix; 573 nm at $\theta = 20°$ was calculated to be the wavelength of maximum reflectance for this opal-like structure. Experimental measurement matched with this calculated result very well. This fact is quite encouraging since the self-assembled structure from a beetle could serve as an effective and efficient example to fabricate 3D nanophotonic devices.

2.3.3.5 Weevil

Welch et al. revealed that the orange coloration of a weevil, *Pachyrrhynchus congestus pavonius,* is associated with the 3D periodic submicron structure on the scales [23]. The structure is a 1D/3D photonic crystal with an FCC stacking, as shown in Fig. 2.7c. Three coexisting sublattices, as illustrated in Fig. 2.7d, are found in this structure including a sublattice of protrusions acting as spacers, a sublattice of empty sites, and a sublattice of holes. The coherent 3D FCC inverse opal structure is believed to result in the shift of the reflection bands to shorter wavelengths with increasing incidence angles. The inner structure of the scales consists of numerous micro-sized 3D PC grains. The organization of all grains is random, which provides isotropic scattering along all directions, thus the irradiant orange color on the scales is independent of the viewing angle [23]. This complex optical structure is the first example of the generically self-assembled 1D/3D hybrid PC. Figure 2.7e shows two distinct selective sub-layers in the 1D Bragg stack mirrors on top of the weevil scale. They reflect yellowish green color with the wavelength of 528 nm and purplish blue light with the wavelength of 447 nm, respectively. This special 1D/3D structure is promising for camouflage coating due to its broadband and wide angle reflection in the visible and infrared light region.

2.3.4 Other Natural Photonics

Interacting with light is one of the main activities of creatures. In addition to PC-like structure for gaining better control over the light, other features, such as lens and optical fibers, can also often be found in the organisms of some living creatures. Light sensor characterized as natural lens arrays has been discovered in brittle stars. Glass sponge shows a high-tech optical fiber network.

2.3.4.1 Brittle Star

High photosensitivity is found in the *Ophiocoma wendtii*, one type of sea star, which changes color from dark brown during the day to banded gray and black at night as shown in Fig. 2.8a. It also responds quickly to shadows thus escaping from predators. These properties can be attributed to a unique skeletal structure of the dorsal arm plates as shown in Fig. 2.8b. Skeletal elements are composed of a single crystal of oriented calcite, which is shaped into a 3D mesh called stereom. The stereom is

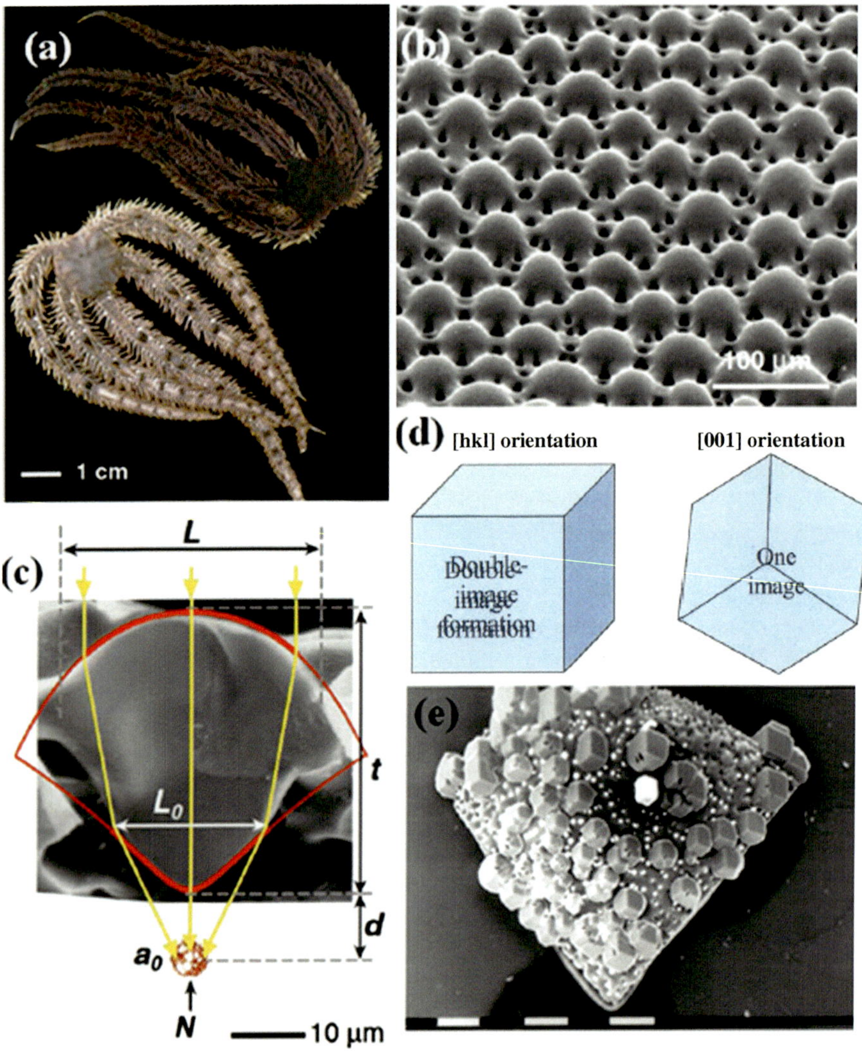

Fig. 2.8 Structure of brittle stars. (**a**) The color change of brittle stars *Ophiocoma wendtii* in the night (*top*) and day (*bottom*). (**b**) SEM of an array of microlenses on the surface of the dorsal arm plate of *O. wendtii*. (**c**) SEM of a cross section of the brittle star lens showing a doublet structure. Calculated profile of an ideal spherical aberration-free lens matches the bio-lens profile very well. (**d**) Schematic illustration of birefringence in calcite crystals. A double image is formed when the crystal is not aligned along the [001] direction, which is the optical axis. (**e**) SEM of the dorsal arm plates embedded with synthetic calcite crystals that grew epitaxially on the surface. Reproduced with permission from [10]. © 2004, Royal Society of Chemistry

~10–15 μm in diameter with a regular double-lens configuration. Electron microscope characterization of these structures revealed that these calcite microstructures have a function in directing and focusing the light on photosensitive tissues [10, 24, 25]. SEM of a cross section of the brittle star lens with doublet structure is shown in Fig. 2.8c. Calculated profile of an ideal spherical aberration-free lens matches this bio-lens profile very well. Aizenberg et al. summarized a number of unique, advantageous properties and design strategies of the brittle star microlens: compact micro-scale porous structure, individually addressed crystalline lenses, wide range transmission adjustment using pigment arrangement, double-lens geometry to reduce the spherical aberration, an inorganic–organic composite that is mechanically strong, calcite crystal orientation along the optical axis of the lens to eliminate the birefringence effect, and angular selectivity [10]. Figure 2.8d schematically illustrates the birefringence in calcite crystals. A double image can be formed when the incident light is not aligned along the [001] direction of the crystal, which is the optical axis. SEM images of the dorsal arm plates showed that the embedded synthetic calcite crystals grew epitaxially on the surface with the [001] direction aligned vertically, as shown in Fig. 2.8e, which indicates that the birefringence effect can be effectively suppressed by this structure.

2.3.4.2 Glass Sponge

Another highly effective and sophisticated optical system designed by nature, in some aspects better than its artificial analogs, is the glass sponge as shown in Fig. 2.9a [11]. Electron microscope revealed that the non-uniform refractive index profile along the spicules' cross section is attributed to the variation of its chemical composition. As shown in the bottom inset of Fig. 2.9a, the core of the glass sponge (1–2 μm in diameter) is Na-doped silica with a high refractive index of 1.45–1.48, which is very close to that of vitreous silica (1.458). The large cylindrical tube (15–25 μm in diameter) surrounding this core has a lower refractive index of 1.425. High organic content and/or low degree of silica condensation is believed to be the result of this low index. The outer part of the spicule shows an increasing refractive index (from 1.433 to 1.438), which corresponds to the well-defined bands of organic molecules. Figure 2.9b shows the measured refractive indices along the diameter of a spicule. Waveguide property is thus introduced by this characteristic "core–cladding" structure of the spicules. Considering that the seawater is the surrounding medium and the cladding diameter is much larger than the core diameter, the spicule is more like a waveguide with a multimode fiber inside, where light is confined over the entire body. Although similar characteristics exist in conventional optical fibers, the glass sponge has many advantages than typical communication optical fibers. The characteristic lamellae that are connected by organic ligand layers in the exterior regions of the spicules significantly improve the mechanical properties of the glass sponge via the crack-arresting mechanism. In addition, the lens-like structures at the end of the spicules are beneficial to the light-collecting efficiency as shown in the top inset of Fig. 2.9a. The dual functionality of the spicule makes it a promising candidate for high-performance optical fibers.

In summary, the fascinating structures, together with their optical properties/phenomena of living creatures, have been developed by nature for millions or even billions of years with unique functionalities, such as communication for reproduction, sensing and signaling danger, coloration/camouflage for hiding

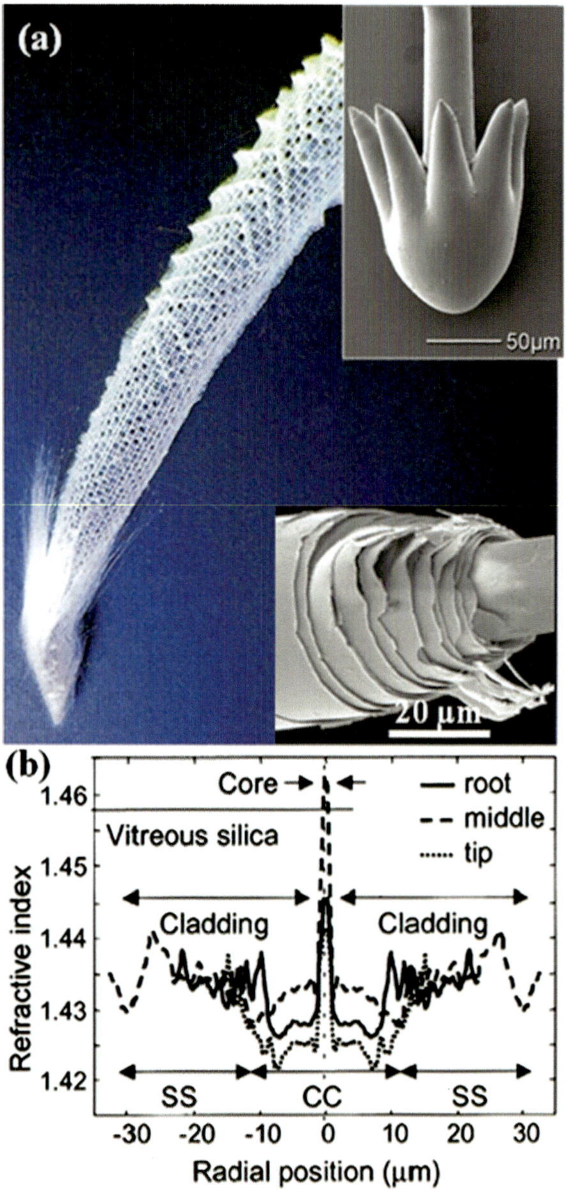

Fig. 2.9 (**a**) Optical image of a glass sponge. *Top inset* is an SEM image of the spicule tip. *Bottom inset* shows the "core–cladding" structure. (**b**) Refractive index value distribution along the cross section of spicule. Reproduced with permission from [11]. © 2004, National Academy of Sciences

from predators, and energy harvesting for food and development. Learning from nature and understanding the mechanism behind would create shortcuts leading to improved synthesis techniques with excellent reproducibility and controllability, advanced micro/nanodevices with superior performance, and even novel applications that cannot be realized by other means.

2.4 Bio-inspired Fabrication of Nanostructures

As discussed in previous sections, nature provides many excellent examples for us to learn and to use. To fabricate those structures would be the first step for us to acquire those gifts from nature. The conventional technology to fabricate micro/nanoscale structure can be generally classified into top-down (typically lithography) and bottom-up (typically self-assembly) processes. Both of them have certain merits and some intrinsic limitations, which can be found in many literatures. For nature, the way to form living devices is genetically directed molecule-level self-assembly, which exhibits extremely high efficiency, reproducibility, controllability, and defect tolerance. To completely mimic such a process would be our ultimate goal in bio-inspired fabrication. However, in the current stage, this strategy is still not realistic. The most common ways for us to take the advantages from the structures and devices designed by nature are either partially mimicking the non-genetically controlled growth process or directly using the fine structures of biological systems as templates to duplicate the structures and functions.

2.4.1 Biomineralization

Biomineralization is a process that living creatures produce minerals inside or around their bodies. Three general biomineralization principles have been suggested to control the crystal formation in biological systems. (i) Biocrystals usually form inside an organic framework that regulates the mineral deposition following a predetermined spatial pattern. (ii) The single crystal structure is transformed from the amorphous form. (iii) The crystallographic orientations are controlled by oriented nucleation on well-defined and chemically modified intracellular sites [24, 25]. Borrowing these biomineralization principles, a crystallization strategy was developed to fabricate large-area single-crystalline calcite with sub-10-μm patterns and well-controlled crystallographic orientations. As shown in Fig. 2.10a, first, a microscale framework was fabricated on a gold-coated glass substrate serving as the deposition template. Second, self-assembled monolayer $HS(CH_2)_nA$ ($A = OH$, CO_2H, SO_3H), which is used to induce the nucleation of the calcium carbonate, was deposited on the Au/glass substrate by the tip of atomic force microscope. The remaining region was passivated with a mixture of alkanethiols of different lengths to suppress the nucleation of calcium carbonate. Then metastable calcium carbonate was filled in the interstices of the framework (Fig. 2.10b). During the following 30 min, calcite crystals nucleated at specific areas where the self-assembled $HS(CH_2)_nA$ monolayer was deposited. Subsequent crystal growth transformed the amorphous calcium carbonate into a crystallized phase. The patterned structure of as-synthesized calcite film is shown in Fig. 2.10c. The crystallinity of the crystal patterns was confirmed by electron diffraction pattern as shown in the inset of Fig. 2.10c. No polycrystalline grains were found in the film. It can be seen that the nucleation stage is one key step of this growth technique, which determines the formation of single-crystal film and the crystal orientation. This large-scale

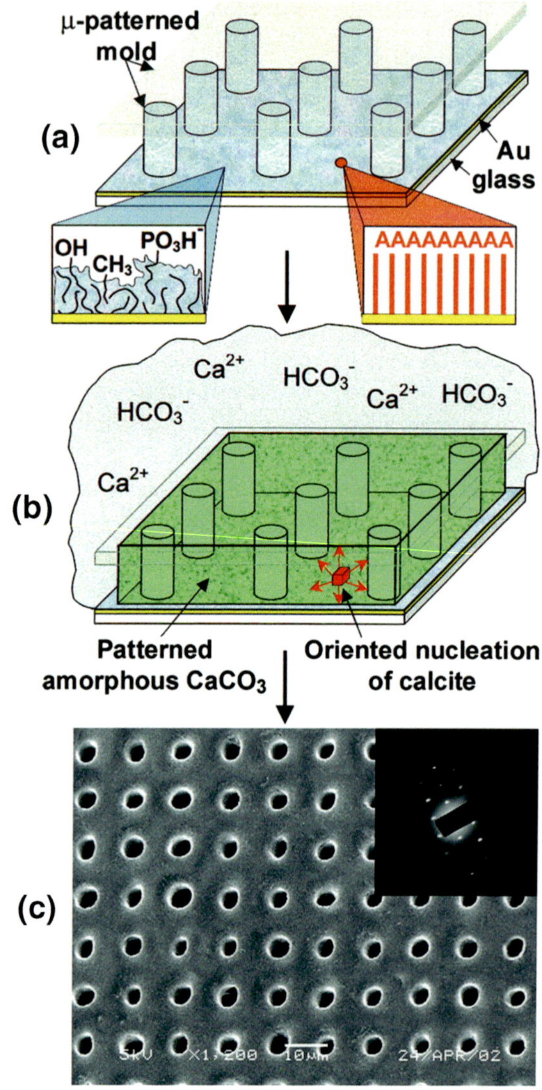

Fig. 2.10 (**a**) Schematic picture of the template with 2D micropatterns for calcite deposition. (**b**) The process of growing single-crystalline calcite on the micropattern. (**c**) SEM of a resulted single-crystalline calcite with micropatterns shown in (**a**). *Inset* is the electron diffraction pattern of (**c**). Reproduced with permission from [24]. © 2003, American Association for the Advancement of Science

bottom-up strategy for synthesizing single-crystalline calcite with controlled orientation and microstructure is an excellent example in applying natural growth methods to the synthesis of micro- or nano-patterned crystalline materials.

2.4.2 Biological Fine Structure Duplication

Although the fascinating fine structures from nature offered many valuable functions, the relatively small selection of materials limits the applications in many man-made devices. For example, nature's favorite building components are mostly organic materials, such as keratin, chitin, and melanin. A few inorganic materials like silica and calcite are also common to see in biological systems. These materials are insulators, offer limited selection of refractive index, and some are sensitive to temperature or other environment conditions. Thus one main effort in bio-inspired fabrication is to duplicate the well-designed biological fine structures with other functional materials while retaining the structure-related functionality. Typically, there are two strategies of duplication: replicating the shape by surface coating and locally replacing the original materials via atom exchange.

2.4.2.1 Replication by Surface Coating

First, we will discuss the structure replication by surface coating. Considering most biological templates are C-, H-, and O-based organic skeletons, such as butterfly wings, a moderate replication condition is required, including non-corrosive environment, proper pH value, and mild temperature. Atomic layer deposition (ALD), in which film growth is a cyclic, multi-step process of alternating surface-limited chemical reactions, has been demonstrated to be a powerful technique for depositing high-quality conformal thin films on various substrates/structures at moderate temperature (<100°C). The ALD process proceeds through saturative steps resulting in a constant thickness increase of conformal films. Taking the deposition of Al_2O_3 as an example, the growth mechanism can be described by the schematics shown in Fig. 2.11. The precursors are $Al(CH_3)_3$ (TMA) and H_2O. In one deposition cycle, TMA vapor is first pulsed into the reaction chamber and reacts with the surface hydroxyl groups (Fig. 2.11a). The chamber is then purged by inert gas to remove the precursor still in gas phase as well as physisorbed reactants (Fig. 2.11b). The second precursor (H_2O in this case) is now pulsed into the chamber and reacts with the dangling methyl groups around Al atoms (Fig. 2.11c). Gaseous by-products (CH_4) and any unreacted precursors are removed during the subsequent purge. By the end of the cycle, Al–O bridges are formed together with a new layer of hydroxyl surface groups waiting for another deposition cycle (Fig. 2.11d). Owing to the self-limited layer by layer deposition process, ALD exhibits excellent thickness controllability and uniformity even on porous surfaces or structures with large aspect ratio. Besides, ALD applies to a wide verity of oxides within a low-temperature deposition window. Therefore, ALD is a promising tool in replicating the biological structures.

Using ALD technique, Huang et al. successfully replicated the fine structure of the wing scale by Al_2O_3, where the thickness of the Al_2O_3 layer could be very well controlled with a resolution less than 1 nm [15]. By increasing the coating thickness from 10 to 40 nm with a 10 nm interval, the reflected color shifted from original blue to green, yellow, orange, and eventually pink. A series of optical microscope images of the wing scales coated with a 10, 20, 30, and 40 nm thick Al_2O_3 layer are shown in Fig. 2.12a. The corresponding reflectance spectra are shown in Fig. 2.12b. This redshift is due to the surface film-enhanced reflection of a particular wavelength that is determined by the thickness of the film and its refraction index. The original butterfly template can also be completely removed by annealing the sample

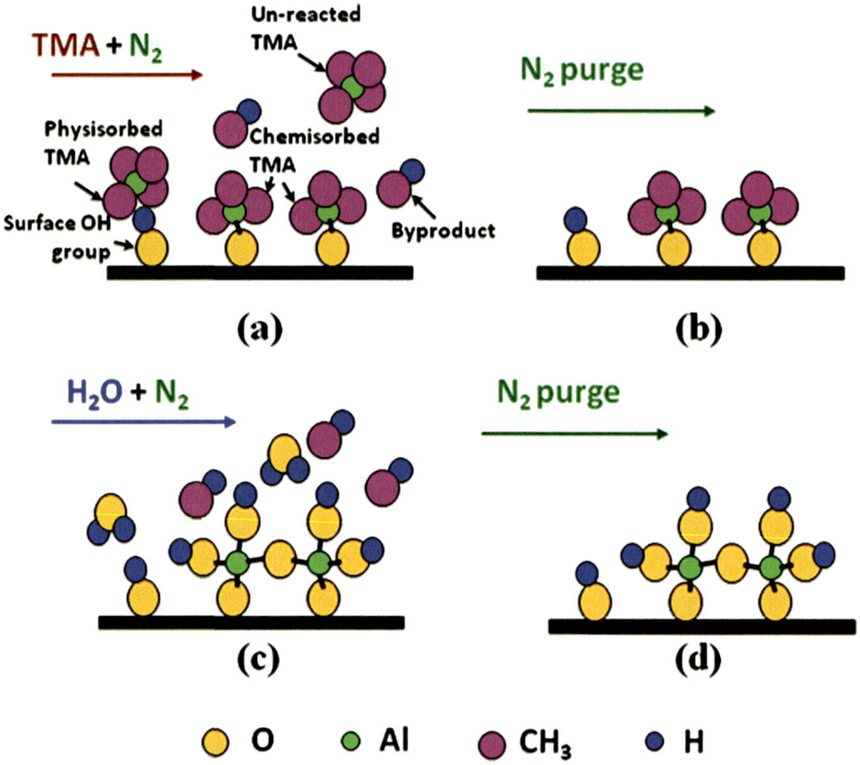

Fig. 2.11 Mechanism of forming Al_2O_3 atomic layer on a flat surface through ALD technique using TMA and H_2O as precursors

to 800°C for 3 h in atmosphere. Meanwhile, the amorphous Al_2O_3 shell was gradually crystallized into a more robust polycrystalline shell structure, with crystal grain sizes smaller than 3 nm. Owing to the excellent uniformity of the alumina film, both the large-scale arrangement of the wing scales and the nanometer-scale periodic structures are perfectly preserved after this vigorously template-removing process. The alumina replica inherits both the nanoscale periodic structure and the optical properties of the butterfly wing template and can be considered as an inverse wing scale structure. The waveguide and beam splitter features were also successfully preserved by the alumina replication, as discussed in Section 2.3.3.

In addition to butterfly wings, ALD also showed success in replicating other biological structures, such as water strider legs for inheriting the hydrophobic property [3] and fly eye for microlens array fabrication [26]. Water striders (*Gerris remigis*) could stand and move on water easily owing to their hydrophobic legs (Fig. 2.13a). The water-repelling capability of the strider legs is attributed to the special hierarchical structure along the leg surface [27]. A high-sensitive balance system revealed that the maximal supporting force of a single leg is 152 dynes, which is 15 times of the total body weight of a strider. SEM shows that these legs have many micro-setae

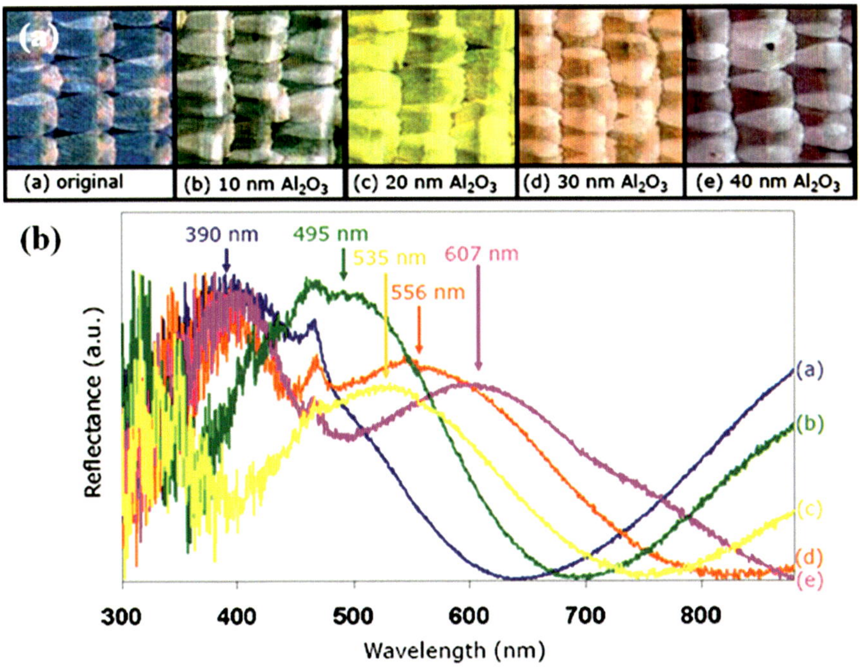

Fig. 2.12 (**a**) Optical microscope images of the butterfly wing scales coated by alumina with different thickness, of which the color changed from original *blue* to *pink*. (**b**) Corresponding reflectance spectra of the alumina-coated butterfly wings [15]

with fine nano-grooves as shown in Fig. 2.13b. These needle-shaped setae have a diameter of few micrometers to a few hundred nanometers and 50 μm in length. Most setae have a 20° incline angle with the leg surface. Such a hierarchical structure is believed to be composed of solid and air by Cassie's law [28]. The air fraction between the leg and the water surface is then calculated to be 96.86%. This structure forms a cushion in the interface, which supports the strider body above the water. This structure was precisely replicated by Al_2O_3 coating using ALD (Fig. 2.13c), and the super-hydrophobic property was also successfully inherited. This research further proved that the non-wetting property of water strider legs is mostly contributed by the fine setae structure. In addition, to simulate this non-wetting property, round copper wire is used to mimic the curvature effect of setae (Fig. 2.13d). By adjusting contact lines at the interface, the system prefers to trap air to form a composite interface regardless of whether the surface is hydrophobic or hydrophilic. This is a solid evidence to verify the Cassie state of non-wetting property of strider legs. This mechanism might provide a practical way to make aquatic machines or systems.

In general, the ALD replication method can be considered as a biological system-friendly technique that precisely duplicates the fine structures down to the nanometer regime. This strategy may create a new pathway toward precise

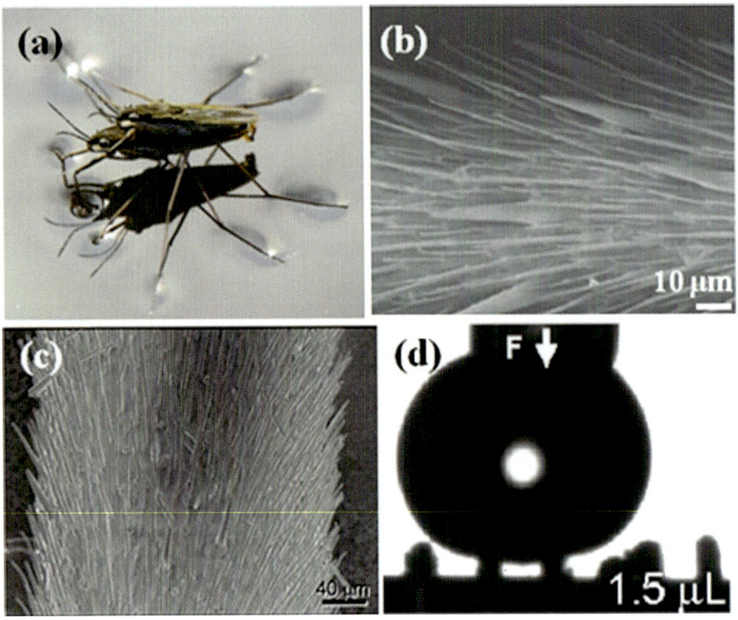

Fig. 2.13 (**a**) Photo of a water strider standing on the water surface (http://en.wikipedia.org/wiki/Gerridae). (**b**) Oriented spindly setae structure on the strider legs. Reproduced with permission from [27]. © 2004, Nature Publishing Group. (**c**) A strider leg coated with alumina by ALD. (**d**) A water droplet on array of alumina-coated wires to mimic the strider leg effect. Reproduced with permission from [3]. © 2008, Institute of Physics

copying natural organisms and creating fundamental building blocks for advanced nanodevices with high reproducibility and lower fabrication cost comparing to traditional lithography techniques.

2.4.2.2 Replication by Atom Exchange

Although precise, ALD technique can only produce a shell configuration. A direct duplication requires a replacement of atoms/molecules in the biological structure with other functional materials. Sandhage et al. developed a low-temperature magnesiothermic reduction method to produce microporous nanocrystalline silicon replicas from diatoms (unicellular algae) through a reaction with magnesium gas and subsequent selective magnesia dissolution [29]. The overall reaction is

$$2\,\mathrm{Mg(g)} + \mathrm{SiO_2(s)} = 2\mathrm{MgO(s)} + \mathrm{Si(s)}$$

After the reaction, MgO was etched away by immersing the structures in a 1 M HCl solution. Figure 2.14a is an SEM image of an *Aulacoseira* diatom frustules. Figure 2.14b shows the MgO/Si composite after reaction of frustules with Mg gas. A silicon-bearing replica product obtained by dissolution of magnesia from MgO/Si

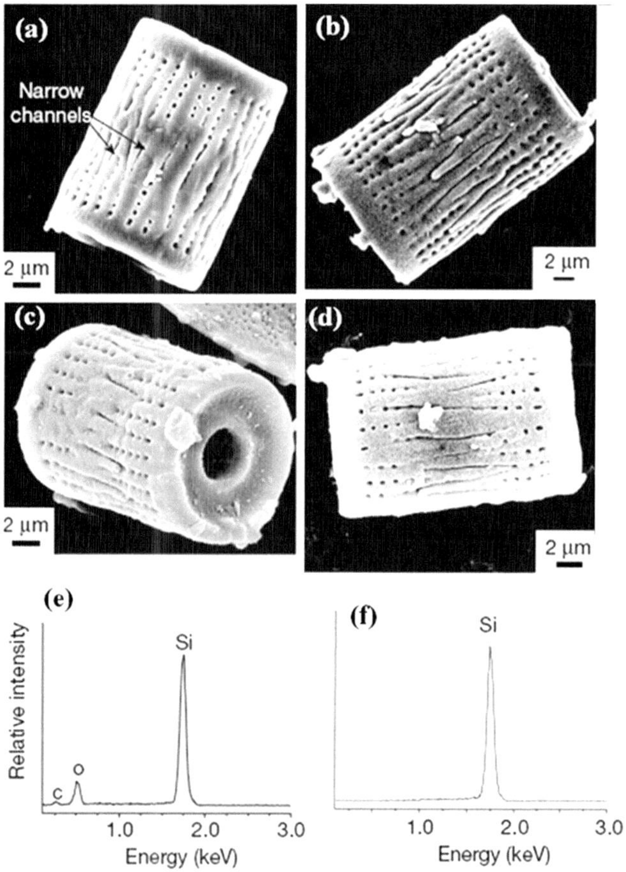

Fig. 2.14 Shape-preserving chemical reduction of silica diatom frustules. (**a**) SEM image of a diatom frustule. (**b**) An MgO/Si composite replica after reacting a frustule with Mg vapor. (**c**) A silicon-bearing replica by selective dissolution of magnesia in an HCl solution. (**d**) Silicon replica after HF treatment. (**e**, **f**) EDS spectra of (**c**) and (**d**). Reproduced with permission from [29].

in an HCl solution is shown in Fig. 2.14c. The final step is to immerse the products into HF solution to remove unreacted SiO_2, as shown in Fig. 2.14d. Energy dispersive X-ray analyses shown in Fig. 2.14e and f were obtained from the structures shown in Fig. 2.14c, d, respectively. The absence of oxygen peak in Fig. 2.14f revealed the necessity of HF treatment. These treatments completely converted insulating SiO_2 diatom frustules into semiconducting Si replica and retained the 3D morphology and nanoscale features. The silicon replicas could be used as sensor, electronic, optical, or biomedical devices. This gas/silica displacement reaction has also been applied in fabricating other non-silica-based oxides like TiO_2 and ZrO_2. In addition, hydrothermal reactions, wet chemical coating, or impregnation methods have been used to transfer the diatom 3D structure into $BaTiO_3$, Zn_2SiO_4, or other materials. Appealing merits of these low-temperature methods are the low cost and the feasibility to large-scale synthesis [30].

2.5 Bio-inspired Functionality

The ultimate goal of the bio-inspire technology is to realize the unique functionality of biological systems using man-made devices. Beyond the fabrication techniques that completely duplicate the fine structure of biological species, learning the designs from nature and then fabricating similar configurations in a different scale so as to duplicate the functionality is considered as another aspect of bio-inspired fabrication. Here we name it as bio-inspired functionality in order to distinguish from the previous section and many successful examples fall into this category. The configuration of Velcro, as we introduced earlier in this chapter, is an excellent example of applying a natural design into man-made systems (http://en.wikipedia.org/wiki/Gerridae). Driven by the question "which marine and freshwater animals don't foul?" Brennan invented the low-friction ship hulls. This design was borrowed from shark scale. The very small scales of sharks, known as placoid scales or dermal denticles, are made up of several small ribs, which keep tiny algal spores from attaching onto the shark's body. Applying this design to ship hulls could eliminate algal spores and/or algae attachment thus significantly reducing the water friction (http://www.robaid.com/bionics/shark-skin-biomimicry-from-hull-coating-to-health-care.htm, http://www.sharklet.com/wpcontent/themes/sharklet/pdfs/archive/SharkskinSolutions.pdf).

iMoD is a major breakthrough of flat-panel display technology, which was inspired by the double-layer scale configuration of butterfly wings (http://www.qualcomm.com/common/documents/white_papers/iMoD_Display_Overview.pdf). An iMod element is a small micro-electro-mechanical system (MEMS) composed of two conductive plates: a thin-film stack on a glass substrate and a reflective membrane suspended below. The thin-film stack and membrane forms an optically resonant cavity, in which the incident light is reflected. A slight phase shift in the optical cavity due to the thin-film-to-membrane distance change would alter the interfering wavelength. The constructively interfered wavelength will be observed as a colored pixel (the on state). When the reflective membrane collapses onto the thin-film stack, ultraviolet light interferes constructively, which is not observable by human eyes, thus shows a black pixel (the off state). This color generation mechanism is much more efficient than conventional color filters and polarizers. The on and off states are controlled by applying a voltage onto the thin-film stack. iMoD has two great advantages compared to conventional LCDs: (1) it requires very little power since it uses ambient light sources and (2) the image of iMoD is quite clear and can be viewed in any direction with a two to three times higher brightness than state-of-the-art technologies.

Inspired by the microlens discovered on brittle stars, man-made microlens arrays have been designed and fabricated. As shown in Fig. 2.15a, Yang et al. demonstrated a simple microfluidic device to simulate the migration of pigment-filled chromatophore cells that control the light transmission in a porous brittle star lens array [31]. When the light-absorbing liquid was pumped into the pores, a reduction of light transmission was detected as shown in Fig. 2.15b. By adjusting the concentration of the liquid, the intensity of the transmitted light could be controlled. In

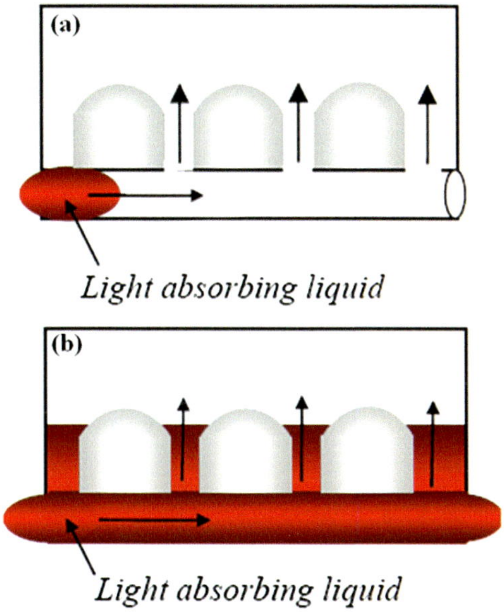

Fig. 2.15 Illustration of the transmission tunability through the lens array by using controlled transport of light-absorbing liquid in the channels between the lenses. The microfluidic assembly and actuation are shown in (**a**) and (**b**). Reproduced with permission from [31]. © 2005, Wiley-VCH Verlag GmbH & Co

addition, changing the liquid species can tune the focal length, numerical aperture, and wavelength selectivity of this lens. This concept would improve current techniques in fabricating optical devices and provide new ideas in designing and making adaptive micro-scale lenses.

The gecko foot is another significant example of creating novel functionality from nature inspiration. Gecko lizards can climb on wall firmly and move swiftly due to its foot's hierarchical structure, which comprises many setae (Fig. 2.16a). A single seta also consists of hundreds of tiny structures called spatulas as shown in Fig. 2.16b [32]. The strong van der Waals force provided by the spatulas is strong enough to support the Gecko's body weight. This function was successfully imitated by using carbon nanotube arrays to simulate the aligned microscopic elastic hairs on gecko foot [33]. The carbon nanotubes could generate a dry adhesive force of $\sim$100 N/cm^2, almost 10 times higher than what a gecko foot can produce. Figure 2.16c shows a book of 1480 g in weight suspended on a glass surface by carbon nanotubes dry adhesion. The arrangement and the structure of the carbon nanotubes are shown in Fig. 2.16d and e. Shear-induced alignment of the entangled top part of carbon nanotubes was believed to result in a significantly large shear adhesion force due to the extremely large contact area of carbon nanotubes on a flat surface. The relatively small normal adhesive force was attributed to the tip-surface contact between carbon nanotubes and walls. Figure 2.16f shows the experimental and predicted total adhesion force as a function of loading angles. This angle-dependent characteristic could be the guideline for achieving the detachment and attachment of carbon nanotube arrays.

In addition to the above examples, temperature adapting fabric, dirt and water-repelling paint, artificial muscles, and many more are all typical functionalities

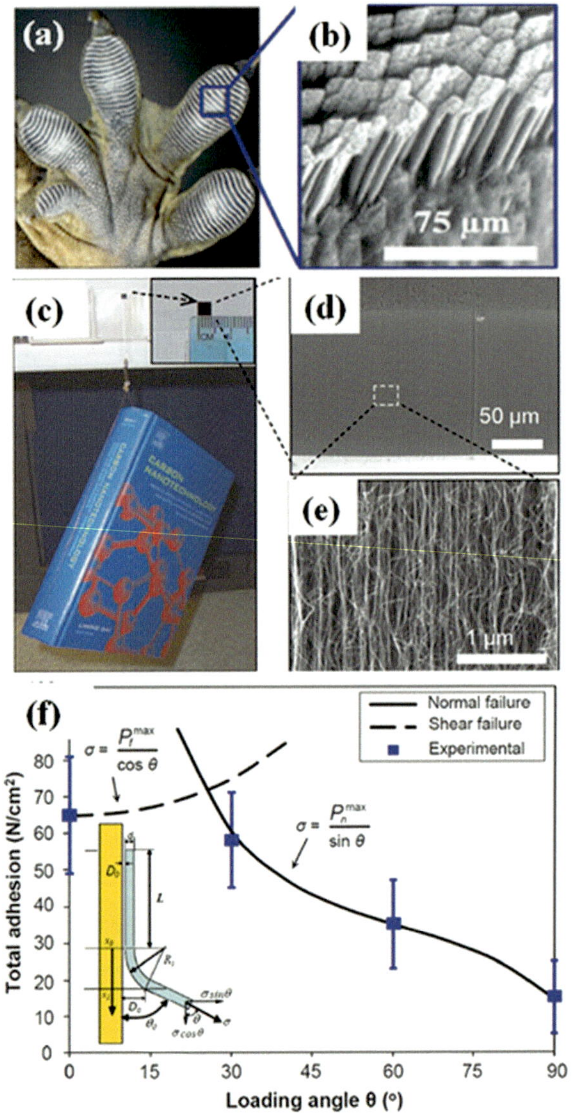

Fig. 2.16 (**a**) Gecko foot showing that the setae are arranged in many lobes along the foot. (**b**) SEM image of gecko setae with numerous smaller spatulas. Reproduced with permission from [32]. © 2007, National Academy of Sciences. (**c**) A book of 1480 g in weight was suspended by the MWNTs pad by mimicking the gecko foot. *Top right inset* shows the dimension of the MWNTs pad. (**d**) and (**e**) The arrangement and structure of MWNT arrays with different magnification. (**f**) Experimental and predicated total adhesion force as a function of loading angle. Reproduced with permission from [33]. © 2008, American Association for the Advancement of Science

inspired by biological systems. With rapid advancing of fabrication and characterization techniques and more in-depth understanding of the biological world, more sophisticated configurations and functions would be realized by human beings.

2.6 Conclusion

Since the beginning of human history, learning from nature has always been one important strategy to advance our technology and civilization. The level of imitation depends on our ability of observation and manufacturing. The rapid development of modern fabrication and characterization technologies in the past two decades enabled us to see features and manipulate objectives down to the atomic level, thus brought us an intriguing new world of nature in the nanometer regime. Numerous extraordinary optical, electrical, and mechanical properties from biological systems were discovered and understood, which greatly encouraged researchers to devote more and more efforts on learning and mimicking the nature's masterpieces for furthering our technology in a more effective and efficient manner.

From the examples discussed in this chapter, we can see that the most important aspect to mimic the nature is extensive and intensive investigation on the evolution history, the structure formation mechanisms, and the origin of functionalities. Interdisciplinary research across nanotechnology, materials science, physics, chemistry, computer science, and biology would play a key role in the development of bio-inspired technology. Although many successes have been shown in this area, there still are a few key questions that are urgently to be addressed in the near future. First, how could a fully controlled self-assembling process be achieved in the fabrication of functional biological features in the nanoscale? Second, how could near 100% reproducibility and repeatability be realized in self-assembly processes like the reproduction of biological systems? Third, how could the level of error tolerance be improved in our electronic and optical systems by learning from the complex nature systems? Last, how could the intelligence functionality be adapted to our man-made systems? At current stage, there may be no specific answers to above questions yet. However, it is still foreseeable that the strength brought together by nanotechnology, biotechnology, and information technology would soon implement more and more crucial nature functionalities to our civilization.

Acknowledgment We thank National Science Foundation under grant No DMR-0905914 and UW-Madison graduate school.

References

1. W. Barthlott, C. Neinhuis, Purity of the sacred lotus, or escape from contamination in biological surfaces. Planta **202**(1), 1–8 (1997)
2. F. Wickson, Narratives of nature and nanotechnology. Nat. Nanotechnol. **3**(6), 313–315 (2008)
3. Y. Ding et al., Modifying the anti-wetting property of butterfly wings and water strider legs by atomic layer deposition coating: Surface materials versus geometry. Nanotechnology **19**(35), 355708–355714 (2008)
4. C. Sanchez, H. Arribart, M.M.G. Guille, Biomimetism and bioinspiration as tools for the design of innovative materials and systems. Nat. Mater. **4**(4), 277–288 (2005)
5. J.P. Youngblood, N.R. Sottos, C. Extrand, Bioinspired materials or self-cleaning and self-healing. MRS Bull. **33**(8), 732–741 (2008)
6. P. Fratzl, Biomimetic materials research: What can we really learn from nature's structural materials? J. R. Soc. Interface **4**(15), 637–642 (2007)
7. J.Y. Rho, L. Kuhn-Spearing, P. Zioupos, Mechanical properties and the hierarchical structure of bone. Med. Eng. Phys. **20**(2), 92–102 (1998)
8. F. Vollrath, D.P. Knight, Liquid crystalline spinning of spider silk. Nature **410**(6828), 541–548 (2001)

9. E. Arzt, S. Gorb, R. Spolenak, From micro to nano contacts in biological attachment devices. Proc. Natl. Acad. Sci. USA **100**(19), 10603–10606 (2003)
10. J. Aizenberg, G. Hendler, Designing efficient microlens arrays: Lessons from Nature. J. Mater. Chem. **14**(14), 2066–2072 (2004)
11. J. Aizenberg et al., Biological glass fibers: Correlation between optical and structural properties. Proc. Natl. Acad. Sci. USA **101**(10), 3358–3363 (2004)
12. A. Blanco et al., Large-scale synthesis of a silicon photonic crystal with a complete three-dimensional bandgap near 1.5 micrometres. Nature **405**(6785), 437–440 (2000)
13. H. Ghiradella, Light and color on the wing – structural colors in butterflies and moths. Appl. Opt. **30**(24), 3492–3500 (1991)
14. A.R. Parker et al., Structural colour – opal analogue discovered in a weevil. Nature **426**(6968), 786–787 (2003)
15. J.Y. Huang, X.D. Wang, Z.L. Wang, Controlled replication of butterfly wings for achieving tunable photonic properties. Nano Lett. **6**(10), 2325–2331 (2006)
16. P. Vukusic, J.R. Sambles, C.R. Lawrence, Structural colour – Colour mixing in wing scales of a butterfly. Nature **404**(6777), 457 (2000)
17. A.E. Seago et al., Gold bugs and beyond: A review of iridescence and structural colour mechanisms in beetles (Coleoptera). J. R. Soc. Interface **6**, S165–S184 (2009)
18. T.D. Schultz, G.D. Bernard, Pointillistic mixing of interference colors in cryptic tiger beetles. Nature **337**(6202), 72–73 (1989)
19. A.R. Parker et al., Photonic engineering – Aphrodite's iridescence. Nature **409**(6816), 36–37 (2001)
20. T. Fuhrmann et al., Diatoms as living photonic crystals. Appl. Phys. B-Lasers Opt. **78**(3–4), 257–260 (2004)
21. M. Srinivasarao, Nano-optics in the biological world: Beetles, butterflies, birds, and moths. Chem. Rev. **99**(7), 1935–1961 (1999)
22. Y. Ding, S. Xu, Z.L. Wang, Structural colors from morpho peleides butterfly wing scales. J. Appl. Phys. **106**(7), 074702 (2009)
23. V. Welch et al., Orange reflection from a three-dimensional photonic crystal in the scales of the weevil pachyrrhynchus congestus pavonius (Curculionidae). Phys. Rev. E **75**(4), 041919 (2007)
24. J. Aizenberg et al., Direct fabrication of large micropatterned single crystals. Science **299**(5610), 1205–1208 (2003)
25. J. Aizenberg et al., Calcitic microlenses as part of the photoreceptor system in brittlestars. Nature **412**(6849), 819–822 (2001)
26. J.Y. Huang, X.D. Wang, Z.L. Wang, Bio-inspired fabrication of antireflection nanostructures by replicating fly eyes. Nanotechnology **19**(2), 025602 (2008)
27. X.F. Gao, L. Jiang, Water-repellent legs of water striders. Nature **432**(7013), 36 (2004)
28. A.B.D. Cassie, S. Baxter, Wettability of porous surfaces. Trans. Faraday Soc. **40**, 546–551 (1944)
29. Z.H. Bao et al., Chemical reduction of three-dimensional silica micro-assemblies into microporous silicon replicas. Nature **446**(7132), 172–175 (2007)
30. U. Kusari et al., Formation of nanostructured, nanocrystalline boron nitride microparticles with diatom-derived 3-D shapes. Chem. Commun. **43**(11), 1177–1179 (2007)
31. S. Yang et al., Functional biomimetic microlens arrays with integrated pores. Adv. Mater. **17**(4), 435–438 (2005)
32. L. Ge et al., Carbon nanotube-based synthetic gecko tapes. Proc. Natl. Acad. Sci. USA **104**(26), 10792–10795 (2007)
33. L.T. Qu et al., Carbon nanotube arrays with strong shear binding-on and easy normal lifting-off. Science **322**(5899), 238–242 (2008)

Chapter 3
Building 3D Micro- and Nanostructures Through Nanoimprint

Xing Cheng

3.1 Introduction to 3D Structure Fabrication Through Nanoimprint

The great success of microelectronic integrated circuits is built on two-dimensional planar processing on silicon wafers. Along with the development of microelectronic fabrication techniques, micro-electro-mechanical systems (MEMS) has become a vast and increasingly mature research area. For MEMS applications in micro-optics and bioengineering, it is often desirable to have three-dimensional (3D) structures. Even in microelectronic circuit fabrication, there are growing interests in 3D multi-layer integration as the miniaturization of solid-state devices becomes more difficult. The growing interests in 3D micro- and nanostructures have led to the invention of many new fabrication techniques, such as those discussed in other chapters of this book.

Nanoimprint lithography is a relatively new patterning technique for soft materials, particularly polymers. Because nanoimprint is inherently a physical molding process, it is the ideal technology for patterning 3D micro- and nanostructures in soft materials. In addition to the dramatic development of the nanoimprint technique in the last 10–15 years, including equipment development and process innovation, there are significant progresses in building 3D polymer structures by nanoimprint. Compared with other 3D fabrication techniques, nanoimprint enjoys advantages such as simple equipment, low cost, and high throughput. Nanoimprint is particularly suitable for large-scale production once the complex 3D template is fabricated. Despite nanoimprint mainly processes polymers, the fabricated 3D polymer structures can be used as sacrificial templates for fabricating corresponding 3D inorganic structures.

This chapter mainly addresses the fabrication of 3D polymer micro- and nanostructures by nanoimprint. It provides a survey of techniques reported in literature

X. Cheng (✉)
Department of Electrical and Computer Engineering, Texas A&M University, College Station, TX 77843-3128, USA
e-mail: chengx@ece.tamu.edu

W.L. Zhou, Z.L. Wang (eds.), *Three-Dimensional Nanoarchitectures*,
DOI 10.1007/978-1-4419-9822-4_3,

in recent years. The applications of those 3D structures in microelectronics, bio-MEMS, micro-optics and photonics, and other areas are not the main focus of this chapter. Only a few representative applications of 3D polymer micro- and nanostructures will be explained after the discussion of their fabrication. Since nanoimprint is by and large a lithographic or patterning technique, it is also used as one of the many steps in creating 3D structures through elaborate layer-by-layer scheme. In those cases, the nanoimprint simply functions as a patterning technique to define the structure that can be used by other micro- and nanofabrication techniques such as reaction-ion etching (RIE) to form 3D structures. For maintaining a concise and focused discussion, this chapter ignores 3D structure fabrications where nanoimprint simply performs the patterning role and emphasizes on the fabrication of 3D structures where nanoimprint itself is the indispensible main process.

3.2 Overview of Nanoimprint Lithography

3.2.1 Fundamentals of Nanoimprint Lithography

Injection molding and hot embossing are standard techniques in industry for manufacturing polymer products from a hard template or die. As the critical feature sizes on the template or die become continuously smaller, interests arises in extending the embossing techniques to mold structures down to deep sub-micrometer level. Nanoimprint lithography, which is based on the principles of embossing, soon attracted widespread attention in both academia and industry since the first successful replication of sub-100 nm features [1, 2].

Early nanoimprint is performed on thermoplastic polymers at elevated temperature and pressure, thus the process is called thermal nanoimprint. The typical processing steps of thermal nanoimprint are shown in Fig. 3.1a. A pre-fabricated hard mold or template is pressed against a polymer film at a temperature above its glass transition temperature (T_g). The polymer flows due to the applied stress and fills up the empty areas on the mold. After cooling down to solidify the polymer, the mold and substrate is separated, leaving a micro- or nanostructure that is complementary to the pattern on the mold in the polymer film. The polymer pattern can then be further transferred into the substrate by dry or wet etching. Many variations of the nanoimprint lithography have been developed since its inception. The most important one is replacing the thermoplastic polymer film with a liquid polymer precursor [3]. The liquid polymer precursor can be cured by UV light through a transparent template, such as fused silica. The processing steps of the UV-curable nanoimprint are shown in Fig. 3.1b. The main advantage of this technique is that no heating is required and the molding can be done with very low pressure, making it highly appealing to microelectronic fabrication. UV-curable nanoimprint can be further performed in a step-and-flash scheme for patterning over large area with high throughput [4]. The step-and-flash imprint lithography (SFIL) has attracted particular attention from microelectronic industry as it is more compatible with current microelectronic fabrication than thermal nanoimprint.

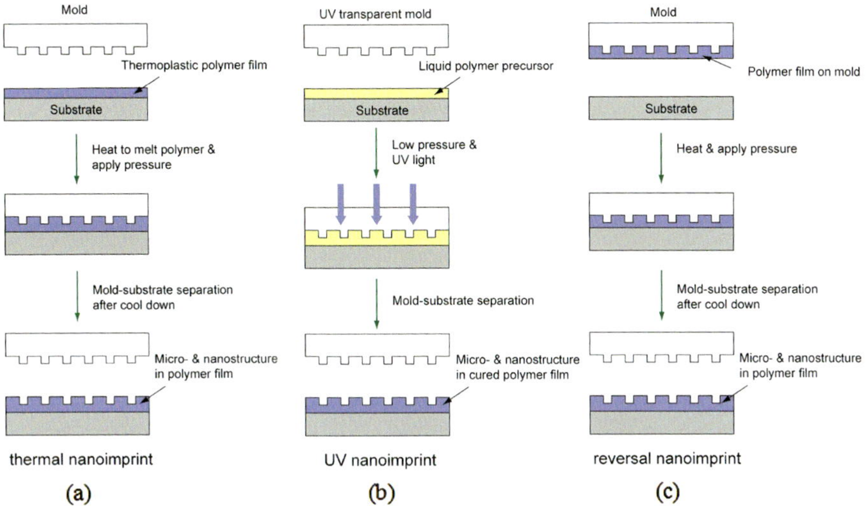

Fig. 3.1 Schematics of nanoimprint lithography and its variations: (**a**) thermal nanoimprint; (**b**) UV-curable nanoimprint; and (**c**) reversal nanoimprint

Among many variations of nanoimprint processes such as combined-nanoimprint-and-photolithography [5] and hybrid-mask-mode technique [6], reversal nanoimprint [7] is particularly important for building 3D polymer structures by nanoimprint in a layer-by-layer scheme. Figure 3.1c shows the processing steps of reversal nanoimprint. The polymer film is first deposited on the mold surface and then transfer bonded to a substrate. This patterning scheme allows sequential addition of structured polymer layers onto pre-existing polymer micro- and nanostructures, leading to 3D polymer structures. Recent development of reversal nanoimprint and transfer bonding, and their applications in 3D polymer structure fabrication, will be discussed in detail. The nanotransfer printing of inorganic semiconductors and metals using PDMS templates can be found in a recent book chapter [8] and will not be discussed here.

In response to the excitement in nanoimprint lithography, many review articles [9–17] and book chapters [18–22] have been published to summarize both the fundamental processes of nanoimprint and the applications of nanoimprint in diverse areas. A good summary of those review literature can be found in Schift's review article [16]. Readers interested in nanoimprint lithography in general can refer to those publications. In the next section, only brief discussions on nanoimprint materials and template fabrication techniques are attempted.

3.2.2 Materials for Nanoimprint Lithography

Since nanoimprint is a mechanical forming technique, it has certain requirements on the mechanical properties of both molds or templates and resist materials. For nanoimprint mold or template, the general requirement is that the material has certain mechanical strength and thermal stability such that it will not deform under nanoimprint conditions. Thermal nanoimprint usually requires hard materials that are compatible with microelectronic processing such as silicon, silicon dioxide, silicon nitride, nickel [23–25], SiC [26], and diamond-like carbon (DLC) [27–29] because those molds or templates are usually fabricated using traditional microelectronic fabrication techniques such as thin film deposition, lithography, and dry etching. The mold needed for building 3D micro- and nanostructures by nanoimprint is usually more complicated. Typical techniques used to fabricate those templates will be discussed in sections below.

Resist materials for thermal nanoimprint should be able to undergo irrecoverable deformation under the applied stress at elevated temperature. Commonly used materials include engineering thermoplastic polymers such as poly(methyl methacrylate) (PMMA), polystyrene (PS) and polycarbonate (PC), and thermally curable precursors such as polyimide [30] or epoxy-based thermal curing formula. SU-8, a very useful resist for MEMS fabrication, can also be mechanically formed by thermal nanoimprint and subsequently cured by UV exposure and baking. A new trend of direct patterning functional thermoplastic polymers such as piezoelectric polyvinylidene fluoride [31, 32] and semiconducting polythiophene [33] is gradually gaining more attention. For UV-curable resist, commonly used precursors include acrylate and vinyl ether formula based on free-radical polymerization [17] and silicone epoxy formula based on cationic polymerization [34]. Due to the need for exposure to UV light, either the mold or the substrate must be transparent to UV light.

Although most mold or template materials used in nanoimprint lithography are inorganic materials due to the requirement of mechanical strength and thermal stability, there are growing interesting in using cured organic materials as soft and flexible mold or template. Cured SU-8 and poly(dimethylsiloxane) (PDMS) are often used as a secondary template from a master template. Fluoropolymer [35] and poly(vinyl alcohol) (PVA) [36] can be thermally imprinted and used as flexible templates. It is even possible to use naturally occurring structures such as cicada wings [37] and plant leaves [38–40] as flexible templates for pattern replication. The flexible templates provide better conformal contact with the substrates, which is particularly useful when patterning non-planar substrates. However, those organic flexible templates require much lower processing pressure and temperature to avoid significant pattern deformation. The resolution of those templates is lower than inorganic templates due to their low modulus.

Because nanoimprint is a contact lithography technique in which the mold is in direct contact with the resist material to be patterned, adhesion between resist material and template can be a serious issue in limiting the process yield. Since adhesion between two interfaces also depends on the separation rate of the interface to create two surfaces [41], demolding after nanoimprint cannot be done very quickly, thus adhesion between mold and resist must be minimized in order to increase the throughput of nanoimprint. Micro- and nanostructures on mold surface will greatly increase the total interfacial area between the mold and the resist, leading to strong adhesion. A general strategy to overcome this issue is to apply mold-releasing agent either on mold surface or in resist formula. Due to the vast differences in surface properties of a wide range of mold and resist materials used in nanoimprint, there is no standard recipe for the choice and the application of mold-releasing agents. A good summary of commonly used anti-sticking strategies used in current nanoimprint process can be found in aforementioned review papers and book chapters.

3.3 Building 3D Nanostructures by Nanoimprint

Three-dimensional micro- and nanostructures fabricated by nanoimprint can be sorted into several different groups: (a) multi-level structures with non-uniform height or depth, (b) hierarchical structures, and (c) multi-layer structures. Schematics of those structures are shown in Fig. 3.2. It should be noticed that the hierarchical structure can be regarded as a special case of the multi-level structure. Due to many practical applications of the hierarchical structures, particularly in microelectronics and micro-optics, it is worth a separate section to discuss their fabrication and application. The most common nanoimprint schemes used to fabricate those 3D structures are direct nanoimprint with pre-fabricated 3D molds and transfer printing or layer stacking using reversal nanoimprint. These two general fabrication schemes are discussed in detail in the following sections. A few other special fabrication techniques reported in literature are also briefly discussed in this section.

3.3.1 Direct Patterning of 3D Structures in One Step

3.3.1.1 Replicating 3D Polymer Structures from 3D Templates

Since nanoimprint is a physical replication process, it is easy to create 3D structures in a polymer layer if the template itself already contains 3D micro- or nanostructures. The replication process would be identical to a conventional thermal or UV nanoimprint process. As a typical example, a three-tiered template made from multiple electron beam lithography (EBL) steps [42] is shown in Fig. 3.3a, and the replicated inverse structure in UV-curable material is shown in Fig. 3.3b.

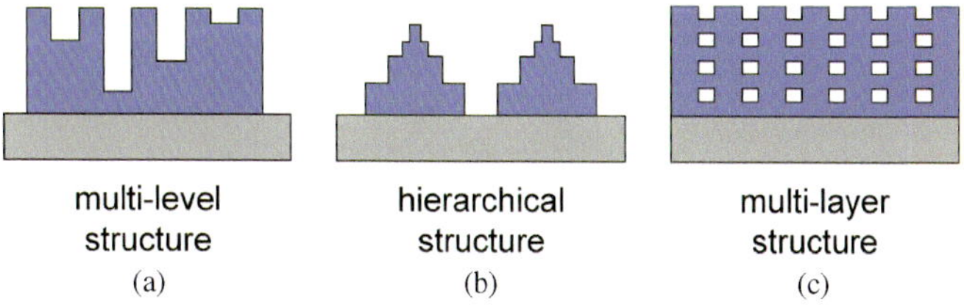

Fig. 3.2 Schematics of 3D micro- and nanostructures that can be fabricated by nanoimprint. (**a**) Structures with non-uniform height; (**b**) hierarchical structures; and (**c**) multi-layer structures

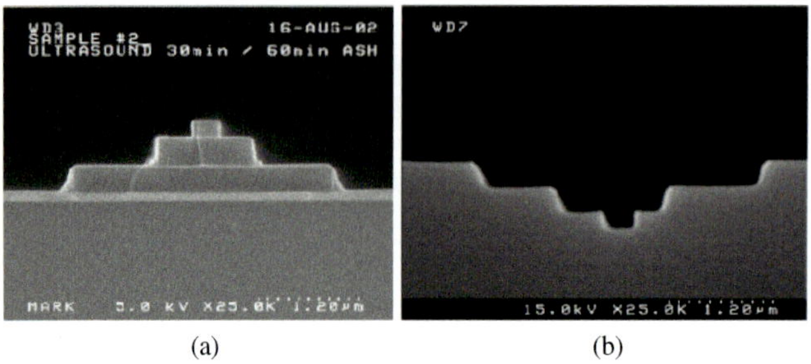

Fig. 3.3 (**a**) Three-tiered template fabricated by multiple lithography steps and (**b**) imprinted structures from three-tiered template. (Reprint from [42])

The key factor in one-step fabrication of 3D nanostructures through nanoimprint is the fabrication of the 3D mold. Generally, multiple steps of EBL and nanoimprint (for nanostructures) or photolithography (for larger structures) followed by reactive-ion etching (RIE) are used in 3D template fabrication [42, 43]. Interference lithography is another very useful technique for patterning periodic structures with sub-micrometer resolution over a large area [44]. Tapered or pyramid structures can be easily obtained using gray-scale lithography [45] or variable dose EBL [46, 47]. Multi-level step profiles can be obtained by controlled exposure in EBL with either variable dose at the same accelerating voltage [48] or variable accelerating voltage at the same dosage [49]. In addition to EBL, focused ion beam (FIB) can also be used to fabricate multi-level template in spin-on glass using variable dose exposure [50]. More intricate 3D structures can also be fabricated by FIB chemical vapor deposition (FIB-CVD) technique [51, 52].

For some special structures, non-conventional techniques can be adopted to achieve simple and efficient fabrication of the 3D templates. Peng and colleagues developed a novel process to fabricate sub-16-nm T-gate structure for high-frequency microwave transistors [53]. Starting from an SiO_2 grating structure (Fig. 3.4a), low-pressure chemical vapor deposition (LPCVD) was employed to conformally deposit a SiN_x coating on the surface of the SiO_2 grating (Fig. 3.4b). The SiN_x coating on top of the SiO_2 was then removed by a selective etching,

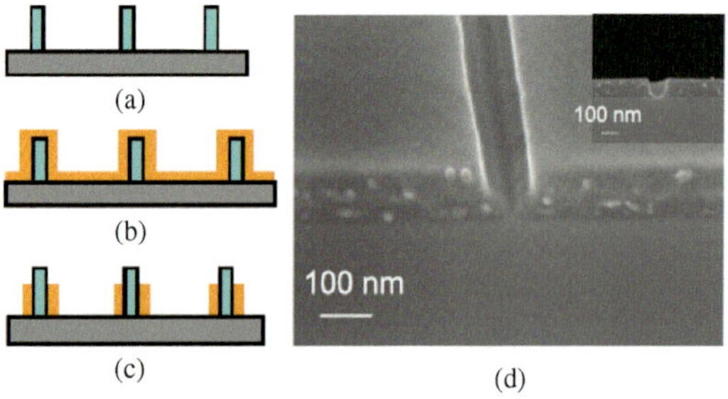

Fig. 3.4 (**a**) Starting SiO_2 grating patterned by nanoimprint lithography; (**b**) conformally deposited SiN_x by LPCVD; (**c**) anisotropic etching of SiN_x to form the T-shaped mold; and (**d**) imprinted resist with complementary T-shaped profile. (Reprint from [53])

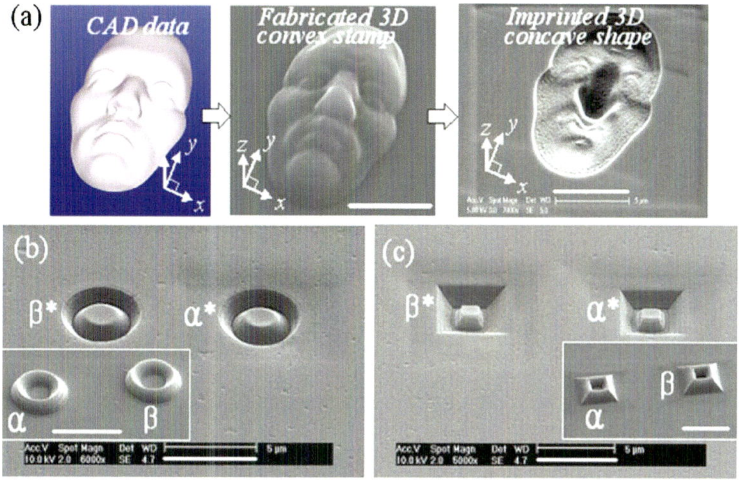

Fig. 3.5 (**a**) From *left* to *right*: CAD data of a face, fabricated 3D template, and imprinted 3D shape; (**b**) SEM images of 3D circular ring pattern on template (insert) and imprinted 3D *circular rings*. Scale bar, 5 μm; (**c**) SEM images of 3D rectangular ring pattern on template (insert) and imprinted 3D *rectangular rings*. Scale bar, 5 μm. (Reprint from [54])

forming a T-shaped mold (Fig. 3.4c). The T-shaped mold can be used to imprint complementary T-shaped profile in resist (Fig. 3.4d), which can be used as a template to form metallic T-shaped gate by either metal evaporation or electroplating.

Two-photon polymerization is an effective technique in the fabrication of arbitrary 3D polymer structures. Park and colleagues [54] fabricated complex 3D templates by depositing DLC on 3D polymer structures generated by two-photon polymerization. The major advantage of this approach is that complex 3D shapes that are difficult to fabricate using other lithography techniques can be easily generated through CAD data. Imprinted 3D shapes of a human face and 3D circular and rectangular rings are demonstrated in Fig. 3.5.

3.3.1.2 Applications of 3D Polymer Structures by One-Step Nanoimprint

In 3D polymer structure fabrication from 3D nanoimprint template, the multiple lithography and etching steps need to be done only once during the fabrication of the template. The subsequent replication of the 3D structures is done in one step. The biggest advantage of this fabrication scheme is the significant reduction of the number of processing steps needed to create polymer 3D structures. Moreover, the alignment of the multi-level structures also needs to be done only once during mold fabrication. The elimination of alignment in replicating the 3D polymer structures not only increases the overall processing speed but also reduces the alignment error and non-uniformity in batch fabrication. Once the alignment is carefully done during mold fabrication, the alignment accuracy and the uniformity of the multi-level structures are guaranteed in all replicated 3D structures. Because of these advantages, many practical applications of 3D polymer fabrication by one-step nanoimprint have been developed.

Dual Damascene Structure for Back-End Processing of Microelectronic Circuit Chips

Microelectronic circuit chip fabrication can be divided into front-end and back-end processing. The front-end processing involves the fabrication of electronic devices in a circuit, such as MOSFETs, while the back-end processing deals with the fabrication of metal interconnects that complete the wiring of the electronic devices in the circuit. In a typical circuit chip, multiple layers of interconnects are needed. The standard structure of metal interconnect is the so-called damascene structure, in which metal wires are embedded in dielectric layers. Metal interconnect wires and their via holes typically form a dual damascene structure. In a standard dual damascene fabrication scheme, a T-shaped opening is created in interlayer dielectrics (ILD), which is filled with metal and then planarized by a chemical–mechanical polishing step. The creation of the T-shaped opening in ILD is accomplished by two lithography processes followed by multiple dry etching steps.

3D printing of T-shaped trench in dielectric layer was recently developed by Molecular Imprints (Austin, TX) to significantly reduce the number of processing steps in dual damascene fabrication [55–57]. With a 3D template, the T-shaped structure can be formed by one-step, UV-curable nanoimprint, as shown in Fig. 3.6a. The two-tiered template (Fig. 3.6b) can be fabricated with two lithography steps.

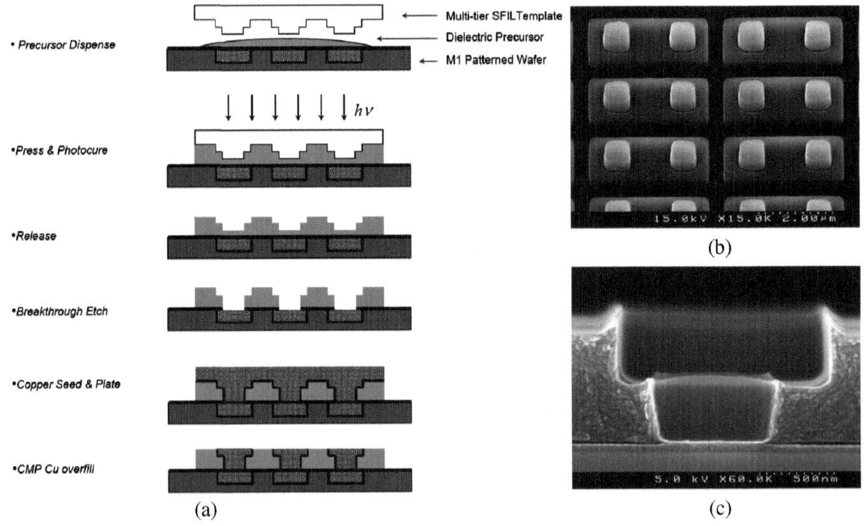

Fig. 3.6 (**a**) Schematics of the dual damascene processing using nanoimprint; (**b**) SEM image of a 3D template for dual damascene processing; and (**c**) a T-shaped trench in ILD patterned by UV nanoimprint from a 3D template. (Reprint from [56])

EBL is used to define the more challenging high-resolution structure, followed by a second lithography step to pattern the less challenging base structure. The details of the fabrication process and other examples of two-tiered templates for dual damascene processing can be found in [58, 59]. An example of replicated T trench in cured UV resist is shown in Fig. 3.6c. The cured low-dielectric-constant UV resist remains in the final structure and serves as the ILD in the damascene structure. Many different types of UV-curable formula based on functionalized T8 silsesquioxane have been developed to provide lower viscosity before curing and better mechanical strength after curing [60]. To complete the dual damascene structure, the T trench in ILD is filled with electroplated copper [61]. After chemical–mechanical polishing of the copper layer, the planarized surface is ready for the fabrication of the next level of interconnect using the same dual damascene process.

The total number of processing steps to accomplish each metal interconnect layer in traditional dual damascene process is about 23. Using 3D nanoimprint lithography, the number of processing steps to form dual damascene structure is reduced to only 8 [55], a 65% reduction of the processing steps. In modern day microprocessor, there are eight or nine interconnect layers. In total, more than 100 processing steps can be eliminated using the 3D UV nanoimprint scheme in microprocessor back-end processing. The significant reduction in processing steps will directly translate into significantly lower cost and higher process yield, making the 3D UV nanoimprint technique economically appealing.

Advanced Optical Components Based on 3D Polymer Structures

Miniaturization of conventional optical components such as mirrors and lenses is an active research area in optics. Microfabricated optical components enable compact and advanced optical systems. Furthermore, novel optical structures can be fabricated to achieve new functionality or superior performance. An example of such advanced optical component is the biologically inspired artificial compound eye. Jeong and colleagues developed a UV-curable molding technique to fabricate the intricate artificial compound eye [62]. The fabrication scheme is shown in Fig. 3.7a–h. PDMS mold with 3D structures was fabricated in two steps. The first step was to pattern the hexagonal micro-lenses in photoresist and then replicate the structure into PDMS (Fig. 3.7a, b). The PDMS thin film is then buckled by partially drawing vacuum underneath the film (Fig. 3.7c, d). A UV-curable epoxy resin was then poured into the buckled PDMS film with micro-lens surface structure and cured with UV exposure (Fig. 3.7e). A second PDMS molding was performed using the cured epoxy template to produce the final PDMS mold (Fig. 3.7f). The 3D compound-eye polymer structure was formed by molding UV-curable SU-8 resin using the 3D PDMS mold (Fig. 3.7g). Finally, the rest of the compound-eye structure, such as polymer cones and waveguide arrays, was accomplished by lens-assisted UV exposure and thermal curing (Fig. 3.7h). The fabricated 3D compound-eye structure is shown in Fig. 3.7i–k.

Tormen and colleagues developed another novel fabrication method for two-tiered compound-eye structure for micro-optical applications [63]. The fabrication

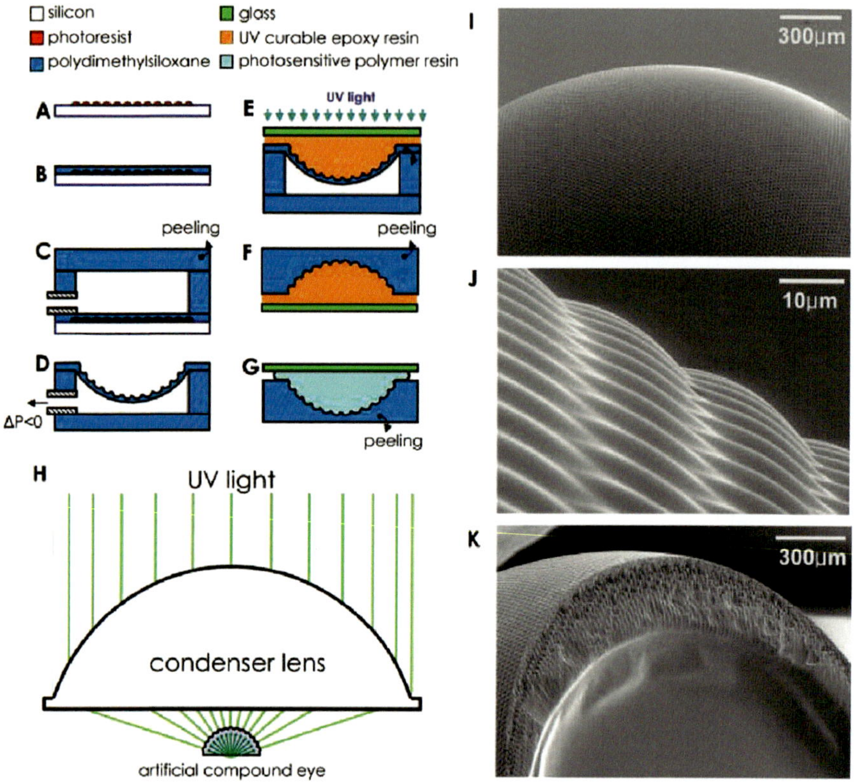

Fig. 3.7 The 3D polymer synthesis of biomimetic artificial compound eyes using (**a**) a honeycomb-packed polymer microlens process, (**b–g**) a reconfigurable microtemplating polymer process, and (H) a self-written waveguide process in photosensitive polymer resin by lens-assisted radial UV exposure. (**a**) Microlens template by a resist reflow method, (**b**) first PDMS molding, (**c**) PDMS bonding, (**d**) PDMS membrane deformation, (**e**) replication with UV-curable polymer resin, (**f**) second PDMS molding, (**g**) photosensitive polymer resin (SU-8) molding, and (**h**) lens-assisted radial UV exposure and thermal cross-linking for self-written waveguides. SEM images of an artificial compound eye. (**i**) Spherical arrangement of 8370 artificial ommatidia on a hemi-spherical polymer dome 2.5 mm in diameter, (**j**) hexagonal microlenses, and (**k**) a cross section with the spherical arrangement of artificial ommatidia consisting of microlenses, polymer cones, and waveguide arrays. (Reprint from [62])

scheme and the replicated 3D SU-8 compound-eye structure are shown in Fig. 3.8. The template fabrication started at drilling a hole and a circular line in Cr thin film deposited on a quartz substrate (Fig. 3.8a). The quartz substrate was then etched in hydrofluoric acid. The isotropic wet etching produces a hemisphere and a hemi-torus in the quartz substrate (Fig. 3.8b). After striping the Cr film, the quartz substrate was further wet etched in hydrofluoric acid till the hemisphere and the hemi-torus touch each other (Fig. 3.8d). FIB was used to mill the hemisphere to form secondary holes, as shown in the SEM image (Fig. 3.8e). The drilled holes can be enlarged by performing hydrofluoric acid wet etching again on the quartz

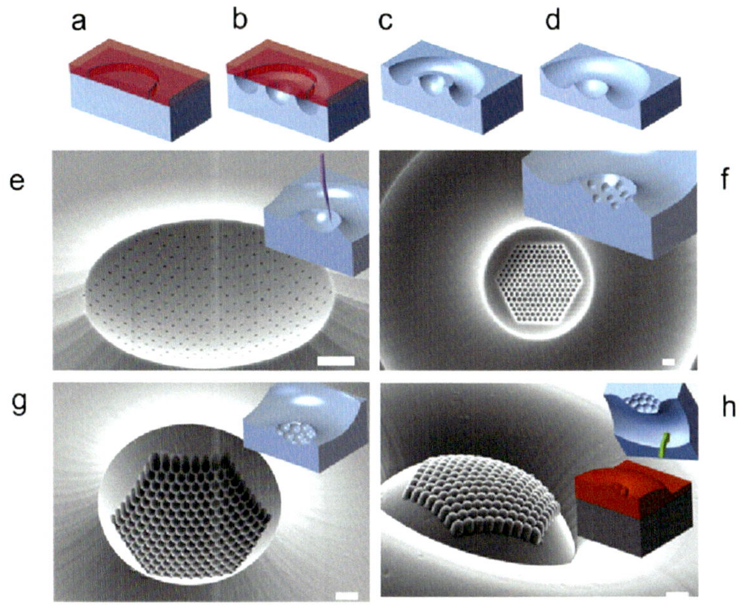

Fig. 3.8 Fabrication steps of 3D compound-eye structure. (**a**) Use FIB to drill hole and line in Cr thin film on a quartz substrate; (**b**) hydrofluoric wet etching of quartz; (**c**) strip Cr film; (**d**) further hydrofluoric wet etching of quartz; (**e**) use FIB to drill holes in quartz hemisphere; (**f**) enlarge FIB-drilled holes by hydrofluoric wet etching; (**g**) further enlarge the holes by hydrofluoric wet etching; and (**h**) embossed SU-8 3D compound-eye structure using the 3D template shown in (**g**). (Reprint from [63])

substrate (Fig. 3.8f, g). The completed 3D template with two-tiered lenses is shown in Fig. 3.8g. This template can be used to mold thermoplastic, thermosetting, or UV-curable polymers. The polymer compound-eye structure can be used as refractive optical elements to generate complex light intensity patterns at different focal planes. Potential advanced optical functions that can be performed by such polymer compound eye include beam shaping, mode conversion, and phase control.

Moth-eye structures have also generated lots of research interests because of its light coupling abilities [64]. The structure can assist light out-coupling for light-emitting diodes [65] and light in-coupling for solar cell panels by providing anti-reflective properties [66]. The surface nanostructures also render the surface superhydrophobic [67], an additional desirable property for outdoor deployment of large-screen displays and solar panels. 3D moth-eye structure, in which smaller lenses sitting on the surface of a much larger concave lens, is also attractive for advanced light control. The fabrication of the 3D moth-eye structure can be done by one-step nanoimprint once the 3D template is made. Kettle and colleagues used FIB milling to produce complex 3D templates [68, 69]. The fabricated fused silica template and the UV-imprinted 3D moth-eye structure are shown in Fig. 3.9a, b, respectively.

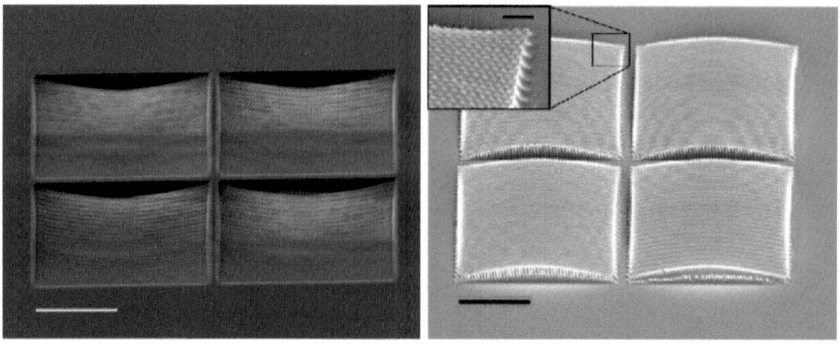

(a) (b)

Fig. 3.9 (**a**) SEM image of a fused silica template fabricated with FIB (scale bar: 4 μm) and (**b**) SEM image of SFIL replication of 3D moth-eye structure (scale bar, 500 nm) (Reprint from [68])

Fabricating 3D diffractive optical structures is another area that increasingly gains attention in integrated optics. For example, grating structures can be placed on top of optical waveguides for controlling the propagation of optical modes. Kim and colleagues [70] used nanoimprint to fabricate rib waveguide with integrated Bragg grating structure in one step from a 3D template. Huang and colleagues [71] fabricated similar waveguide filter in flexible material using molding. Multi-level diffractive optical elements are other examples of optical devices that can greatly benefit from nanoimprint for high-throughput and low-cost replication. Using either gray-scale exposure of resists or multiple lithography steps, multi-level structures can be created for template fabrication, which in turn simplifies the mass production of those diffractive optical elements by one-step nanoimprint [72].

3.3.2 Building 3D Nanostructures by Transfer Bonding and Sequential Layer Stacking

3.3.2.1 Principles of Transfer Bonding and Sequential Layer Stacking

Although building 3D nanostructures in one step using a 3D mold or template is very attractive in terms of simplicity, this technique has limited capability in producing complex 3D structures. To address this limitation, reversal nanoimprint has been extended to transfer bonding a polymer layer onto an existing polymer layer to form 3D structures (Fig. 3.10). A normal nanoimprint step is used to pattern the first polymer layer (Fig. 3.10a). To prepare for reversal nanoimprint, the polymer layer after a second nanoimprint will be transferred onto the mold surface (Fig. 3.10b). This layer will be transfer bonded to the first polymer layer formed in Fig. 3.10a by reversal nanoimprint to form a two-layered polymer structure. The whole process can be repeated many times to form multi-layer complex 3D polymer structures.

The advantages of the transfer bonding and sequential layer stacking technique are multi-fold. First, 3D structures that can be fabricated will no longer be limited

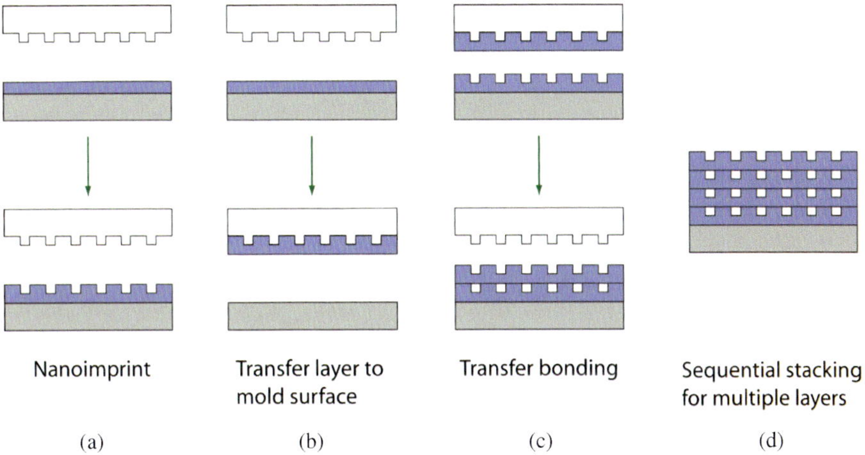

Fig. 3.10 Schematics of building 3D micro- and nanostructures by transfer bonding and sequential layer stacking. (**a**) Normal nanoimprint for patterning the first layer; (**b**) transfer polymer onto mold surface; (**c**) transfer bonding to form two-layered structure; and (**d**) repeat steps (**b**) and (**c**) to achieve multi-layer structure through sequential layer stacking

by those achievable in one-step nanoimprint with a 3D mold. Second, the pattern in each polymer layer, such as the shapes and the dimensions of the patterns, can be varied to suit for the needs of specific applications. Sub-micrometer or nanoscale structures can be used for diffractive optical devices, while structures with tens of micrometer to even millimeter scale can be used for bio-MEMS or tissue engineering. Third, the material in each layer can also be varied provided that good bonding can be achieved between neighboring layers. This allows for the integration of different polymer materials or even inorganic thin films into the whole structure to provide an additional dimension of flexibility in designing the 3D structure to enable new applications. Finally, compared to other 3D polymer structure patterning techniques based on serial writing or tedious layer-by-layer fabrication using multiple lithography and etching steps, the whole processing is simple and has good process throughput.

The most critical steps in building 3D structures by transfer bonding and sequential layer stacking are placing polymer layer onto the mold surface as shown in Fig. 3.10b and bonding polymer layers as shown in Fig. 3.10c. The issues in polymer bonding and strategies to overcome those issues will be discussed later. There are several methods to place the polymer thin film onto the mold surface. The first technique is to directly transfer the polymer thin film to mold surface after nanoimprint as depicted in Fig. 3.10b. For this to work, the polymer–substrate adhesion must be weaker than the mold–polymer adhesion. Since mold surface is usually coated with low-energy surfactant for anti-adhesion for subsequent transfer bonding, it is often necessary to also treat the substrate with a surfactant of intermediate surface energy to reduce polymer–substrate adhesion. Despite the substrate surface energy is higher than that of mold, polymer thin film may still be transferred to mold surface because micro- and nanostructures on mold surface significantly increase the total contact area between the mold and the polymer, hence achieving greater mold–polymer adhesion. If it is difficult to find a suitable mold and substrate surfactant pair to enable polymer film transfer, twist demolding instead of conventional vertical demolding can be employed to successfully transfer polymer film onto mold surface after nanoimprint [73].

Spin coating is the standard technique to prepare polymer thin film on a substrate for nanoimprint, thus it is attractive to use this simple technique to place polymer thin film on mold surface for subsequent transfer bonding. However, directly spin coating polymer solution on mold surface is not trivial. For spin coating to work, the surface tension of the polymer solution must be lower than that of the substrate so that the polymer solution can spread nicely on substrate surface; otherwise polymer solution will bead up and slide off the substrate surface during spin coating, resulting in no film coating. This fact makes it difficult to spin coat polymer solution on mold surface because the mold surface is coated with anti-adhesion surfactant and thus has very low surface energy. A quick solution to this problem is to add fluorinated surfactant additives into polymer solution. This can be combined with using a mold surfactant coating with higher surface energy permitted by anti-adhesion requirement. For example, using OTS or phenyl silane instead of FDTS can lower the difficulty of spin-coating polymer solution on mold surface.

Another factor often needs to be kept in mind when searching for a solvent/mold surfactant combination to enable spin coating of polymer thin film on surfactant-coated surface. It is well known that surface micro- and nanostructures are very effective in modifying surface wetting behavior. Hydrophobic surface with dense micro- and nanostructures tends to become superhydrophobic [74], rendering it nearly impossible to spin coat polymer thin film on mold surface. Two workaround strategies can be used to partially mitigate the problem. The first is to use a solvent with low boiling point, such as tetrahydrofuran or chloroform, which evaporates very quickly during spin coating. The fast evaporation of solvent will leave polymer film on mold surface before polymer solution slides off mold surface. One disadvantage of this technique is that fast evaporation often leads to poor film uniformity. Another way to address the dewetting of polymer solution on mold surface is to increase the viscosity of polymer solution. Viscous polymer solution thin film is meta-stable on low-energy surface due to its slow kinetic response [75]. A side effect of this method is that the final polymer film is often thick.

3.3.2.2 3D Structures Built by Transfer Bonding and Sequential Layer Stacking

The key enabling process for building 3D structures by nanoimprint is the reversal nanoimprint technique developed by Huang and colleagues [7]. The first 3D structure by reversal nanoimprint was achieved by transfer printing polymer layers on mold to the surface of a pre-patterned substrate by Bao and colleagues [76]. Depending on the mechanical properties of the polymer materials and the pattern density in the substrate, two polymer transfer modes occurred (Fig. 3.11). If the

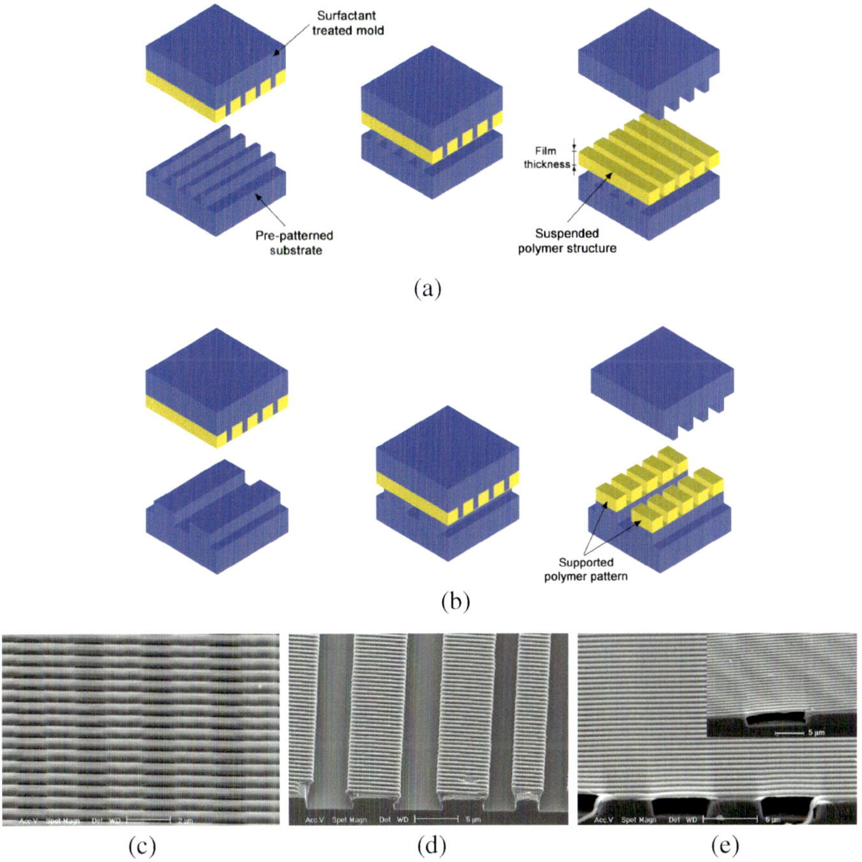

Fig. 3.11 Schematics of nanoimprinting over topography with two different results: (**a**) whole layer transfer; (**b**) layer transfer only on protruded surfaces; (**c**) 700-nm-period PMMA grating transferred onto substrate line patterns with spacing of less than 2 μm; (**d**) 700-nm-period PMMA grating transferred onto substrate line patterns with spacing of 5 μm; and (**e**) 700-nm-period PC grating transferred onto substrate line patterns with spacing of 5–10 μm. (Reprint from [76])

substrate pattern density is very high, the polymer layer will be transferred in whole (Fig. 3.11a). For 700-nm-period PMMA grating, whole layer transfer was achieved on top of a grating with spacing of less than 2 μm (Fig. 3.11c). If the substrate line spacing was increased, only PMMA on line protrusions was transferred due to the weak mechanical strength of the polymer layer (Fig. 3.11b). PMMA is a brittle polymer and its thin film can easily break if the substrate line spacing is greater than 5 μm (Fig. 3.11d). However, for tough polymers such as polycarbonate, whole layer transfer can be achieved over much larger substrate line spacing (Fig. 3.11e). Both transfer modes are useful for creating different 3D structures for specific applications. The substrate pattern size and density and the polymer mechanical property need to be taken into account when designing the 3D structures.

If the substrate pattern is in inorganic materials such as SiO_2, the layer-by-layer stacking technique can have very high process yield by applying larger pressure and higher temperature to ensure intimate polymer–substrate contact for better polymer–substrate adhesion. It is certain that there is an upper limit for both the applied pressure and temperature during transfer bonding. The pressure should not be larger than the threshold of damaging substrate structures, particularly if the substrate structure is on the nanoscale. The temperature at which the transfer bonding is performed should be close to the T_g of the polymer, but not significantly greater than T_g to prevent unwanted substrate pattern filling by significant polymer melt flow during transfer bonding [77].

Building all-polymer 3D structures by transfer bonding is much more difficult because the mechanical properties of the bottom polymer layer are heavily affected by the processing temperature and pressure. For thermoset polymers such as cured SU-8, high processing yield can still be easily achieved because the cured polymer under-layer does not have a softening point and can endure relatively high temperature and pressure, much similar to an inorganic structure. The degree of curing of thermoset polymers can be used as another means to fine-tune their mechanical properties to control the final polymer 3D structures [77]. By controlling the curing degree of each SU-8 layer in conjunction with processing temperature and pressure tuning, multi-level sealed fluidic channels can be successfully fabricated [77–79].

Thermoset polymers enjoy ease of transfer bonding and they are particularly attractive for applications in which the 3D polymer structures are permanent components of the final devices. However, many applications require special polymers that are not thermoset in nature. In addition, polymer 3D structures are often used as a sacrificial template for depositing inorganic materials such as semiconductors, dielectrics, and metals from solution to fabricated 3D inorganic structures. Thermoset polymers are hard to remove after curing. Due to those limitations, the application scope of 3D thermoset polymer structures is limited and it would be greatly desirable to achieve 3D structures in thermoplastic polymers by the transfer bonding and sequential layer stacking. For thermoplastic polymers such as PMMA, PS, and PC, it is much more difficult to build all-polymer 3D structures. Structures in bottom thermoplastic polymer layer can be easily flattened under pressure at elevated temperature. To overcome this issue, Bao and colleagues developed a scheme to build 3D polymer structure in which the polymers used in each layer have progressively lower T_g from bottom to top. In each bonding step, the transfer bonding can be performed at a temperature below the T_g of the polymer layer to be bonded to. In this way, the pattern collapsing during transfer bonding can be minimized. By incorporating an oxygen RIE step to remove the residual layer after each transfer bonding, a 3D interconnected thermoplastic network was achieved as shown in Fig. 3.12.

The need for using polymers of progressively lower T_g has placed a strict constraint on the number of available polymers and the number of total layers for building the 3D polymer structures. It would be advantageous to remove this constraint. The key to achieve multi-layer 3D structures with the same thermoplastic polymer in each layer is to fine-control the process temperature and pressure. As

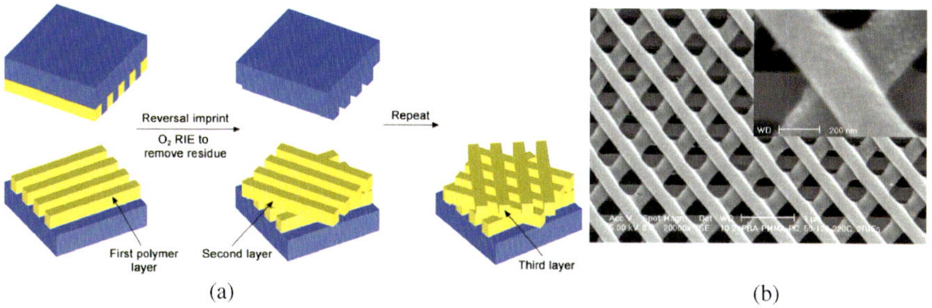

Fig. 3.12 (**a**) Schematics of sequential layer stacking with residue layer removal step and (**b**) SEM image of a 3D thermoplastic polymer structure. Polymers used in layers from bottom to top: PC ($T_g = 150°C$), PMMA ($T_g = 105°C$), and poly(*t*-butyl acrylate) ($T_g = 43°C$). (Reprint from [76])

aforementioned, preserving the structural integrity of the bottom polymer layer requires a lower bonding temperature and pressure. However, achieving good bonding between two polymer layers requires a higher bonding temperature and pressure to ensure good interfacial adhesion. These requirements make it imperative to adopt an intermediate temperature and pressure to achieve desired 3D structures. The pressure needs to be varied depending on the pattern densities. It should be noted that the real pressure at the polymer interface can be significantly greater than the externally applied pressure because the contact area between two patterned polymer layers is significantly lower than that of a uniform film. For most thermoplastic polymers, the best transfer-bonding temperature would be 5–10°C below the T_g of the polymer [73, 80]. Figure 3.13 shows examples of multi-layer PMMA structures bonded at 100°C. It is well known that polymer surface has a lower T_g than that of the bulk [81, 82]. When transfer bonding at a few degrees below T_g, polymer chains at the surface are mobile due to lower T_g at the surface, which ensures good cross-interface polymer chain mixing and entanglement to enhance adhesion. The bulk of the polymer layer is still below T_g to avoid structural collapse during

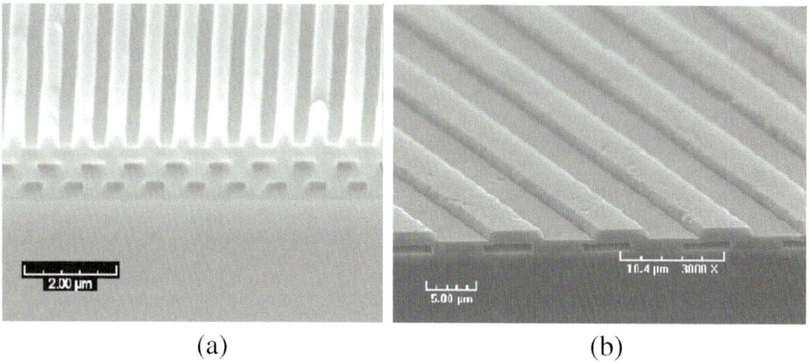

Fig. 3.13 SEM micrographs of 3D multi-layer PMMA structures using direct thermal bonding at 100°C. (**a**) Three-layered, 700-nm-period grating and (**b**) two-layered, 10-µm-period grating. (Reprint from [73])

transfer bonding. When pressure and temperature are optimized and accurately controlled, high-yield bonding can be achieved, allowing the building of multi-layer 3D polymer structures in single thermoplastic polymer.

For nanoimprint lithography, there is always a residue layer left in the patterned structures. The residue layers also exist in each layer of the 3D polymer structures fabricated by transfer-bonding technique and they prevent the interconnection of neighboring layers. For some applications, it is desirable to remove the residue layers, such as in a 3D microfluidic device where fluid flow across different layers is needed. The residue layer can be removed by an RIE step after each transfer bonding. Considering the fact that an over-etching is usually adopted in RIE process to ensure complete residue layer removal, excessive etching of bottom polymer layers can occur. It is possible to remove the residue layer before transfer bonding by performing an RIE on polymer film on mold surface to avoid unwanted etching of bottom polymer structures. However, the over-etch will usually lead to the etching of surfactant coating on mold protrusions, making it necessary to recoat mold surfactant after each transfer bonding. To address the problems introduced by the RIE step, Park and Cheng [83] developed a technique to remove the residue layer by solvent dissolution or dewetting. As shown in Fig. 3.14, the residue layer is exposed to mold surface after transferring the polymer from substrate to mold, which can then be easily removed by soaking the mold in solvent (Fig. 3.14b) or heating the mold to a temperature much higher than the T_g of the polymer (Fig. 3.14c). By carefully controlling the soaking time based on the dissolution rate of polymers in solvents or the heating temperature, the residue layers can be completely removed. The polymer films on mold can then be transfer bonded to patterned polymer bottom layer to form 3D structures. Both techniques have yielded 3D PMMA multi-layer structures without residue layer in each layer (Fig. 3.14c, d). It should be noted that the removal of the residue layer can be strategically employed for certain layers of a complex

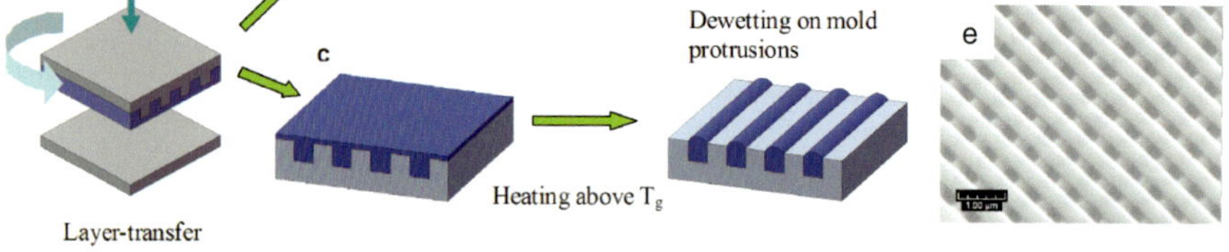

Fig. 3.14 A schematic of (**a**) layer transfer method to place polymer film on mold surface after nanoimprint, (**b**) solvent dissolution method, and (**c**) dewetting method for residue layer removal; SEM micrographs of 3D multi-layer PMMA structures: (**d**) 10-μm-period grating with residue layer removed by the dewetting method and (**e**) 700-nm-period grating with residue layer removed by the solvent dissolution method. (Reprint from [83])

3D polymer structures to introduce connectivity between neighboring layers, leading to highly flexible 3D structure design for advanced microfluidic applications. Moreover, unlike oxygen RIE which can damage functional polymers such as conjugated polymers by breaking chemical bonds, both solvent dissolution and dewetting are benign to functional polymers. This opens up new applications for 3D functional polymer structures to take advantage of their electrical, optical, piezoelectric, magnetic, and biofunctional properties.

Other techniques have also been experimented by various groups to build 3D structures in thermoplastic polymers. Nakajima and colleagues [84] have developed a thermo-differential process in which the upper and lower stages of the nanoimprint apparatus were kept at different temperatures (Fig. 3.15a). The high upper stage temperature melted the polymer, while the low bottom stage temperature kept the bottom polymer layer in solid state, thus achieving both good polymer bonding and bottom layer structural integrity. Three-layered PMMA structures are successfully achieved with this new process (Fig. 3.15b).

The main reason to perform transfer bonding at elevated temperature is for improving adhesion between polymer layers. However, bonding at low temperature is preferred to preserve bottom layer structures. Low-temperature bonding can achieve high process yield if other adhesion promotion techniques are used. Traditional techniques used in polymer joining include plasma surface treatment and using thin adhesive layer or coupling agent [85]. Surface plasma treatment can greatly increase the surface energy of the polymers and create charged species. High polymer surface energy and electrostatic interaction among charged species across the polymer interface are conducive to strong adhesion and often lead to near-perfect polymer bonding. Plasma-activated bonding has been applied to a wide range of polymer materials with excellent results and has become a facile process for building 3D polymer microfluidic devices for bio-MEMS applications [86]. Oxygen plasma treatment is particularly effective for bonding PDMS layers because strong interfacial covalent Si–O–Si bonds are formed. Since PDMS is a thermoset polymer, almost unlimited number of layer can be stacked and bonded at

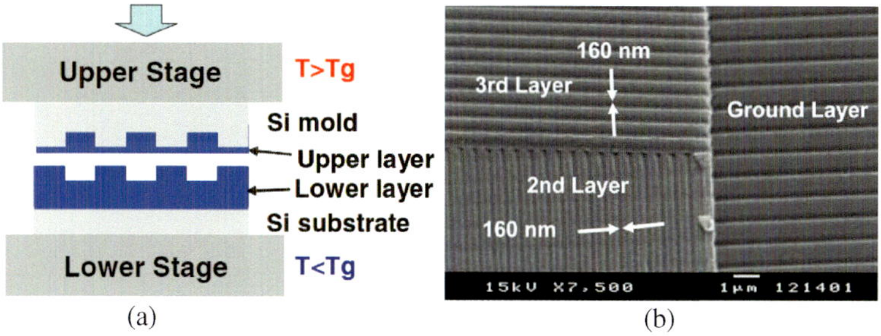

Fig. 3.15 (**a**) A schematic diagram of the thermo-differential process. The upper and lower stages are kept at different temperatures during transfer bonding and (**b**) three-layered PMMA grating fabricated by the thermo-differential process with upper stage at 170°C and lower stage at 30°C. (Reprint from [84])

room temperature after surface plasma treatment. This feature provides unlimited flexibility in designing 3D structures for highly sophisticated micro- and nanofluidic systems for advanced bio-MEMS applications. Kim and colleagues built a particle filtration and separation device based on lateral flow [87] using three-layered PDMS channels (Fig. 3.16). The fabricated device can sort particles into different size ranges that are determined by the dimensions of the fluidic channels in each layer.

When plasma treatment is not available or not desired for the 3D polymer structure, polymer adhesion can be improved by using polymer coupling agents or using a thin adhesive layer. The coupling agents are usually surfactants that can bond well with polymers to be joined. Many coupling agents have been developed in plastic industry and they are commercially available. A summary of coupling agents can be found in Chapter 9 of [85]. A thin adhesive layer can also be used to improve polymer bonding (Fig. 3.17a). For most thermoplastic polymers, SU-8 is a good choice as the adhesive layer because of its low T_g (~64°C). Park and colleagues [73] used thin SU-8 layer to bond PMMA micro- and nanostructures. Because transfer bonding can be performed at a temperature much lower than the T_g of PMMA, the process yield can be very high (Fig. 3.17b). However, the thickness of the adhesive layer needs to be carefully controlled to avoid excessive filling of bottom patterns, particularly if the bottom layer structures are on the nanoscale (Fig. 3.17c–e). Using SU-8 adhesive layer is very attractive because it allows for the integration of dissimilar polymers in one 3D structure. It is even possible to bond inorganic layers, such as dielectrics and metals, with the polymer layers to form more advanced 3D composite structures, thus providing high-level versatility in material choices for building 3D structures.

An interesting variation of the transfer-bonding scheme, which is called combinatorial-mold imprint lithography, is recently developed for the fabrication of 3D polymer structures [88–90]. In combinatorial-mold or duo-mold technique, the 3D polymer structures are formed in one-step nanoimprint with two nanoimprint molds (Fig. 3.18). The polymer 3D structure is detached from one mold after nanoimprint and then transfer bonded onto a flat substrate. A variety of 3D patterns can be easily obtained, including T bars [89] and sealed nanofluidic

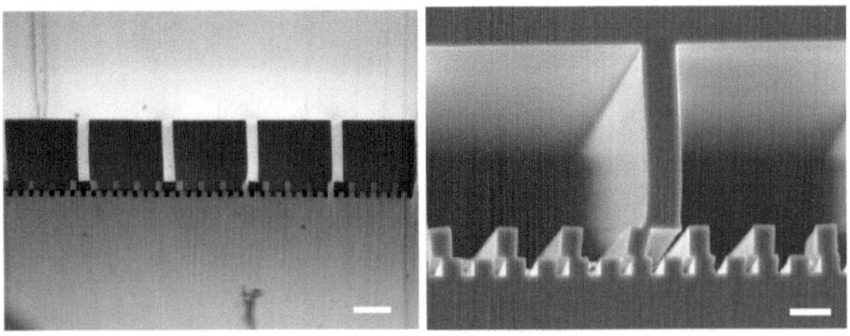

(a) (b)

Fig. 3.16 (**a**) Optical microscope image of the cross section of the fabricated multi-layer particle sorting device. The light area is PDMS and the dark area is microfluidic channel. Scale bar, 50 μm. (**b**) SEM micrograph of the multi-layer particle sorting device. Scale bar, 20 μm. (Reprint from [87])

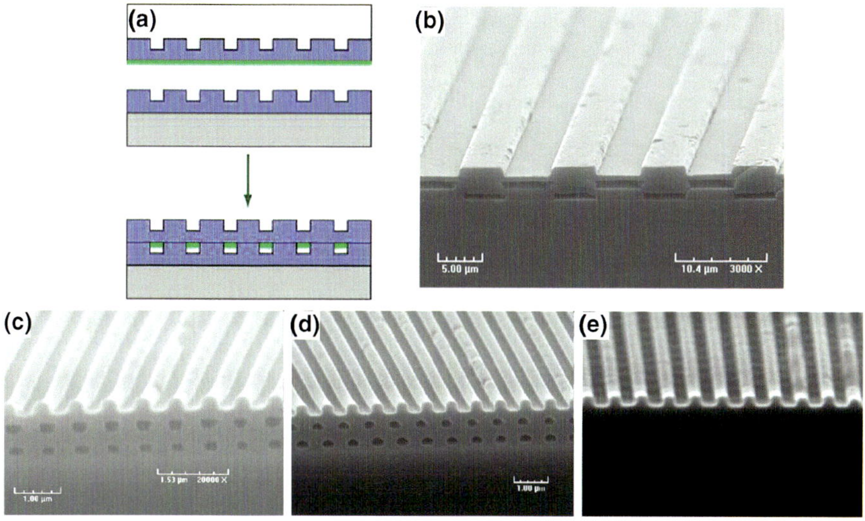

Fig. 3.17 (**a**) Schematic of fabricating multi-layer polymer structures by bonding with thin adhesive layer; (**b**) three-layered, 10-µm PMMA gratings bonded by thin SU-8 adhesive layer; three-layered, 700-nm PMMA grating bonded by SU-8 film spin coated from (**c**) 1%, (**d**) 6%, and (**e**) 10% SU-8 solution. There is progressively more bottom layer filling from (**c**) to (**e**). (Reprint from [73])

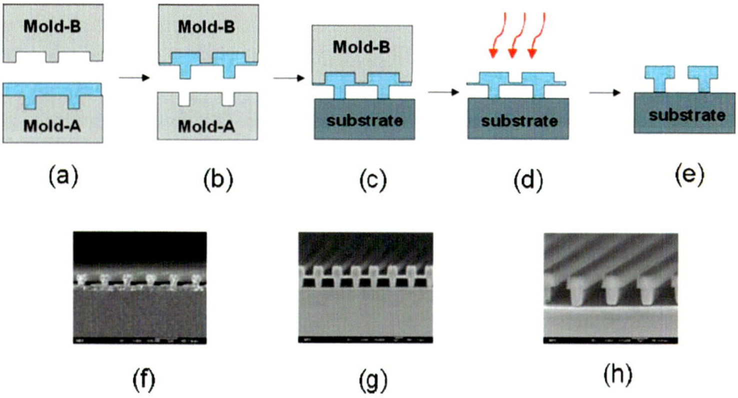

Fig. 3.18 Fabrication of T-bar structure by combinatorial-mold nanoimprint lithography. (**a**) A polymer film or a liquid resin is coated onto mold A; (**b**) mold-to-mold imprinting is carried out at elevated temperature and pressure. The patterned film preferentially adheres to mold B after demolding; (**c**) the patterned film is stamped onto the substrate at elevated temperature and pressure; (**d**) oxygen plasma etching is carried out to remove the suspended residue layer to give (**e**) segregated T-bar structures; (**f**) SEM cross-sectional view of isolated PMMAT bar on an indium tin oxide-coated glass; (**g**) SEM cross-sectional view of the cross-linked PS T-bar (from liquid resin) structure prior to oxygen plasma etching on Si substrate; and (**h**) SEM cross-sectional view of isolated T bars from image (**g**). (Reprint from [89])

channels [88]. Additional RIE steps can be employed to further modify the 3D polymer structure. Other complex structures can also be fabricated if one of the two molds already has 3D multi-level structures patterned by multiple lithography steps. Gadegaard and colleagues [91] used double-side nanoimprint to create complex poly(ε-caprolactone) 3D topography for use as vascular tissue engineering scaffolds.

3.3.2.3 Defect Modes and Process Yield of Transfer Bonding and Sequential Layer Stacking

It is straightforward to see that bonding between sequential layers during transfer bonding is a critical step to achieve 3D polymer structures. The result of the transfer bonding process relies on the competition of two adhesion forces at the mold–polymer and the polymer–polymer interfaces. Successful transfer bonding requires the polymer–polymer adhesion to be stronger than the mold–polymer adhesion. Despite the mold is usually coated with surfactant to reduce adhesion, successful transfer bonding may not be easily achieved because the polymer layer on the mold is to be transferred onto a patterned instead of a flat polymer layer at the bottom. The surface pattern in the bottom polymer layer lowers the total contact area with the polymer layer to be transferred, thus also lowering the polymer–polymer adhesion. In general, it is impossible here to provide a simple guideline for processing parameters due to the large variety of pattern shapes and sizes and polymer materials involved in different applications. Each pattern/material combination yields different levels of adhesion strength at the interfaces, making it necessary to customize processing conditions for each scenario. Figure 3.19 illustrates a few examples that demand special attention. If mold–polymer interface has very large contact area due to high-density (Fig. 3.19a) or high-aspect-ratio (Fig. 3.19b) mold pattern, the mold–polymer adhesion can be very large to prevent a successful transfer bonding. Also, it will be nearly impossible to transfer bond the polymer layer if the mold pattern has an inward slope (Fig. 3.19c). The inward slope on the mold is usually caused by dry or wet etching step during mold fabrication. Care must be taken to ensure that such inward slope is not present in mold structures. Weak polymer–polymer adhesion due to scarce pattern in bottom polymer layer (Fig. 3.19d) often leads to the failure of polymer bonding. Other factors that can disrupt the bonding process include weak mechanical strength of the polymer film (Fig. 3.19e) and the presence of particles (Fig. 3.19f). If the thin polymer layer to be transferred is brittle or has low cohesive strength, large interfacial adhesion forces may rupture the polymer layer to be transferred, resulting in incomplete transfer. For polymers, the strength of higher molecular weight is larger than that of lower molecular weight [80], making the molecular weight of polymers a factor to be considered in process development. Finally, particles will prevent a conformal contact between mold and substrate. Not only is it impossible to transfer bond the polymer layer, but also particles can damage mold.

In addition to the considerations on pattern shapes and sizes, the polymer–polymer bonding is a very complicated issue in that it also strongly depends on

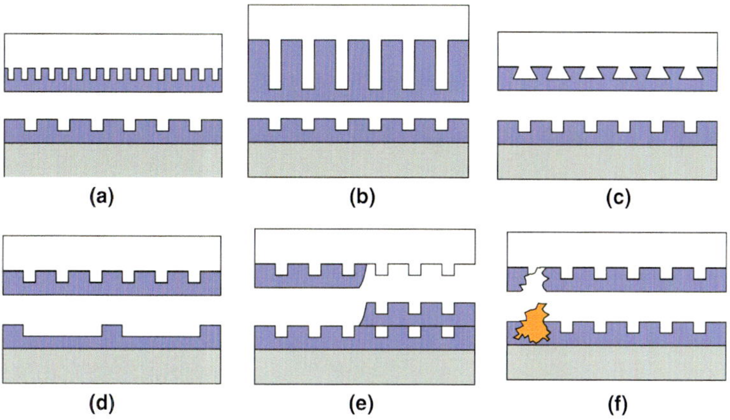

Fig. 3.19 Challenges in building 3D structures by transfer bonding and sequential layer stacking. (**a**) High-density structures on mold; (**b**) high-aspect-ratio structures on mold; (**c**) inward slope of mold structure; (**d**) scarce polymer layer to be bonded; (**e**) weak mechanical strength of the polymer layer to be bonded; and (**f**) presence of particles

the nature of the polymers to be bonded, and case-by-case process optimization is often required for different polymers. Polymer–polymer adhesion strongly depends on inter-chain interaction at the interface. Physical interaction such as chain entanglement across the interface or chemical interaction such as interfacial covalent or hydrogen bonding can yield strong polymer–polymer adhesion. Typical strategies to achieve good polymer–polymer adhesion include plasma treatment of polymer surface immediately before bonding, using a thin adhesion layer or coupling agent, or applying pressure and heat. Plasma treatment can lead to chemical bonding across interface [92, 93]. It is highly effective in improving adhesion for polymers such as PDMA, PMMA, and PC. However, for certain polymers such as functional polymers, plasma etching can degrade or even damage the material. Thus other techniques of adhesion promotion should be used for functional polymers. Thin glue layer such as spin-coated SU-8 from diluted solution [73] or coupling agent such as amine silanes for PMMA can greatly improve interfacial contact and interaction. Bonding polymer layers at elevated temperature can help improve forming cross-interface physical chain entanglement, while increasing bonding pressure can improve interfacial contact. Both can lead to better polymer–polymer adhesion and are often used simultaneously. This approach is effective for thermoset polymers because heating and high pressure will not deform the polymer patterns. However, for thermoplastic polymers, temperature and pressure need to be carefully controlled to avoid the collapsing of the 3D polymer structures. Typically, the temperature should be lower than the T_g of the thermoplastic polymer and the pressure should be lower than the yield strength of the polymer at nanoimprint temperature. Fortunately, available thermal and mechanical properties of polymers [94] and the adhesion engineering between polymer interfaces [85] can offer the guideline for developing optimized processing parameters, such as the temperature

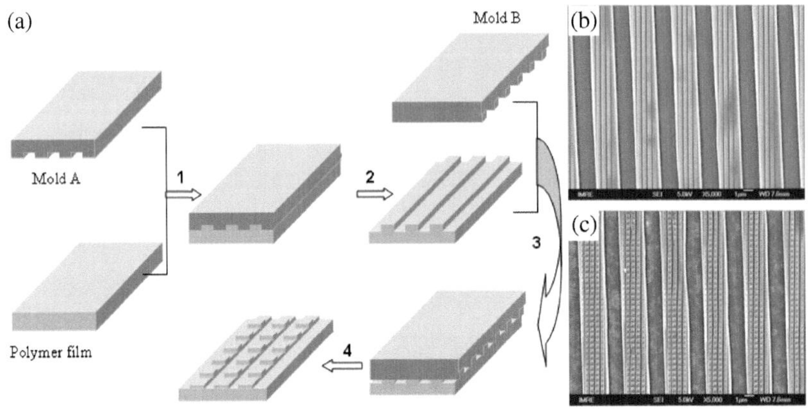

Fig. 3.20 (**a**) Schematic diagram for hierarchical nanoimprinting: a polymer film is imprinted first with mould A at above T_g (step 1); upon demolding, a primary pattern is obtained in the polymer film (step 2); another mould (mould B) is subsequently pressed against the primary pattern at below T_g (step 3) such that a secondary pattern is created (step 4). (**b**) Substrate-supported PC film imprinted first with a 2-μm grating and then with a 250-nm grating in parallel orientation (80°C and 15 bar for 30 s). (**c**) A tertiary 250-nm grating (80°C and 15 bar for 30 s) imprinted perpendicular to the secondary 250-nm grating in sample (**b**). (Reprint from [95])

and pressure for bonding, and the use of adhesion promotion coating, to achieve high-yield fabrication of 3D multi-layer structures for designated applications.

3.3.3 Building 3D Nanostructures by Two Consecutive Nanoimprints

Ordered 3D hierarchical structures can also be obtained by two or more consecutive nanoimprint steps [95] without 3D templates. The first nanoimprint is done at a temperature above the T_g of the polymer to be patterned. The second nanoimprint is done on top of the previously patterned structure at a temperature below polymer T_g (Fig. 3.20a). In this case, the secondary patterns on top of the protrusions of the previously patterned PC structures were formed by purely mechanical forming under high pressure instead of melt flow (Fig. 3.20b). Subsequent mechanical forming can be applied to achieve more complicated 3D hierarchical structures (Fig. 3.20c). This technique can be applied to both freestanding polymer sheets and polymer thin films supported by a substrate.

3.4 Summary and Future Outlook

Embossing is a simple yet effective technique for building 3D structures from 3D templates. With nanoimprint's sub-10-nm resolution, it is convenient to fabricate multi-level and hierarchical 3D micro- and nanostructures using one-step physical molding. There are several advantages of using one-step nanoimprint to fabricate 3D polymer micro- and nanostructures. First, the number of processing steps to build the 3D structures is significantly reduced. Once the 3D template is fabricated, the replication of the 3D structure is accomplished in a single molding step. The reduction in processing steps directly translates into higher throughput, higher processing yield, better uniformity, and lower cost. Second, pattern alignment for multi-level structures, which is often a difficult and time-consuming process, is delegated to the mold fabrication process. Once the 3D mold is fabricated, no pattern alignment is needed for the multi-level 3D structures and the registration of patterns in each level is guaranteed to be uniform in all replicated 3D structures. The key factor in building 3D structures with one-step nanoimprint is the fabrication of the 3D template. Traditional microelectronic fabrication techniques and novel fabrication schemes have been used in 3D template fabrication. Many of the 3D fabrication techniques discussed in other chapters of this book can be employed to fabricate the 3D templates. Notable applications of one-step replication of 3D polymer structures from pre-fabricated 3D templates by nanoimprint or embossing include, but not limited to, dual damascene structure fabrication for microelectronic interconnect processing, T-shaped gate for high-frequency microwave devices, complex-shaped structures for micro-optics such as compound eyes, and multi-level diffractive optical elements.

Although one-step replication of 3D structures by nanoimprint is very attractive, this approach is limited to 3D structures that are friendly to the demolding process after nanoimprint. There are many useful 3D structures, such as 3D photonic crystals and 3D microfluidic devices, which cannot take advantage of this route. To fabricate 3D multi-layer structures, reversal nanoimprint and transfer bonding are very useful. Multi-layer polymer structures can be built by sequential layer-by-layer stacking of polymer structures. The major advantage of this technique is that the material and the structure in each layer can be varied. This provides great flexibility in designing the 3D multi-layer structures to suit target applications. Thus, transfer printing and sequential layer stacking is one of the most versatile techniques for building 3D micro- and nanostructures. Despite that it can be tedious to bond multiple layers to complete the final 3D structure, layer-by-layer stacking is still considered to have higher throughput when compared with other serial scanning techniques to achieve the same 3D structure.

So far the vast majority of reported 3D micro- and nanostructures fabricated by nanoimprint use passive materials. Nanoimprint lithography involves only pressure and heat or UV light in the processing. In most cases, no wet chemicals and highly energetic electron or photon beams are needed in nanoimprint process. These characteristics make nanoimprint widely compatible with functional soft materials. The capability of nondestructively patterning functional soft materials, such as conducting and semiconducting, piezoelectric, and biofunctional polymers, gives nanoimprint a competitive edge to become the technique of choice for fabricating 3D micro- and nanostructures with functional materials. This is expected to open up a whole new horizon of novel applications in electronics, photonics, and bioengineering.

In spite of the many advantages of 3D structure formation by nanoimprint, remaining challenges still need to be addressed to make nanoimprint a facile, practical, and commercially viable solution for 3D fabrication. The key challenge would be improving process yield by engineering surface adhesion at various interfaces. It is easy to realize that the mold–polymer adhesion in one-step 3D nanoimprint is much greater than that in conventional nanoimprint because of the much increased surface area of a 3D structure. For transfer bonding, it is usually the weak bonding between layers due to reduced contact areas that leads to process failure. So reducing mold–polymer adhesion and increasing polymer–polymer adhesion are of paramount importance in improving process reliability. For materials with different mechanical and thermal properties, processing parameters such as nanoimprint temperature and pressure often need to be optimized to achieve best results. Unfortunately, due to the wide range of material properties, there is no simple standard processing parameters that can be established. Much research is still needed in process optimization to achieve high-yield fabrication of 3D micro- and nanostructures through nanoimprint.

References

1. S.Y. Chou, P.R. Krauss, P.J. Renstrom, Science **272**, 85 (1996)
2. S.Y. Chou, P.R. Krauss, W. Zhang, L. Guo, L. Zhuang, J. Vac. Sci. Technol. B **15**, 2897 (1997)
3. J. Haisma, M. Verheijen, K. van den Heuvel, J. van den Berg, J. Vac. Sci. Technol. B **14**, 4124 (1996)
4. M. Colburn, S.C. Johnson, M.D. Stewart, S. Damle, T.C. Bailey, B. Choi, M. Wedlake, T.B. Michaelson, S.V. Sreenivasan, J.G. Ekerdt, C.G. Willson, Proc. SPIE Int. Soc. Opt. Eng. **3676**, 379 (1999)
5. X. Cheng, L. Jay Guo, Microelectron. Eng. **71**, 277 (2004)
6. X. Cheng, L. Jay Guo, Microelectron. Eng. **71**, 288 (2004)
7. X.D. Huang, L.R. Bao, X. Cheng, L.J. Guo, S.W. Pang, A.F. Yee, J. Vac. Sci. Technol. B **20**, 2872 (2002)
8. E. Menard, J.A. Rogers, in *Stamping Techniques for Micro- and Nanofabrication*, ed. by B. Bhushan. Springer Handbook of Nanotechnology (Springer, Berlin, 2010), p. 313
9. L.J. Guo, J. Phys. D: Appl. Phys. **37**, R123 (2004)
10. M. Heckele, W.K. Schomburg, J. Micromech. Microeng. **14**, R1 (2004)
11. D. Resnick, S.V. Sreenivasan, C.G. Willson, Step and flash imprint lithography. Mater. Today **8**, 34 (2005)
12. M.D. Stewart, C.G. Willson, MRS Bull. **30**, 947 (2005)
13. G.L.W. Cross, J. Phys. D: Appl. Phys. **39**, R363 (2006)
14. V.N. Truskett, M.P.C. Watts, Trends Biotechnol. **24**, 312 (2006)
15. L.J. Guo, Adv. Mater. **19**, 495 (2007)
16. H. Schift, J. Vac. Sci. Technol. B **26**, 458 (2008)
17. E.A. Costner, M.W. Lin, W.-L. Jen, C.G. Willson, Annu. Rev. Mater. Res. **39**, 155 (2009)
18. H.-C. Scheer, H. Schulz, T. Hoffmann, C.M. Sotomayor Torres, in *Nanoimprint Techniques*, ed. by N. Hari Singh. Handbook of Thin Films (Academic, Burlington, 2002), p. 1
19. H.D. Rowland, W.P. King, in *Micro- and Nanomanufacturing Via Molding*, ed. by P.J. Hesketh. BioNanoFluidic MEMS (Springer, Berlin, 2008), p. 131
20. M. Worgull, *Hot Embossing Theory and Technology of Microreplication* (William Andrew, Burlington, 2009)

21. H. Schift, A. Kristensen, in *Nanoimprint Lithography – Patterning of Resists Using Molding*, ed. by B. Bhushan. Springer Handbook of Nanotechnology (Springer, Berlin, 2010), p. 271
22. X. Cheng, L.J. Guo, in *Recent Development and Applications of Nanoimprint Lithography*, eds. by G. Cao, Q. Zhang, C.J. Brinker. Annual Review of Nano Research (World Scientific, Hackensack, NJ, 2010), p. 317
23. B. Heidari, I. Maximov, L. Montelius, J. Vac. Sci. Technol. B **18**, 3557 (2000)
24. S. Park, H. Schift, C. Padeste, B. Schnyder, R. Kotz, J. Gobrecht, Microelectron. Eng. **73–74**, 196 (2004)
25. S.-H. Hong, J.-H. Lee, H. Lee, Microelectron. Eng. **84**, 977 (2007)
26. K.A. Lister, S. Thoms, D.S. Macintyre, C.D.W. Wilkinson, J.M.R. Weaver, B.G. Casey, J. Vac. Sci. Technol. B **22**, 3257 (2004)
27. E.S. Lee, J.H. Jeong, K.D. Kim, Y.S. Sim, D.G. Choi, J. Choi, S.H. Park, T.W. Lim, D.Y. Yang, N.G. Cha, J.G. Park, W.R. Lee, J. Nanosci. Nanotechnol. **6**, 3619 (2006)
28. K.I. Nakamatsu, N. Yamada, K. Kanda, Y. Haruyama, S. Matsui, Jpn. J. Appl. Phys. Part 2 **45**, L954 (2006)
29. S. Ramachandran, L. Tao, T.H. Lee, S. Sant, L.J. Overzet, M.J. Goeckner, M.J. Kim, G.S. Lee, W. Hu, J. Vac. Sci. Technol. B **24**, 2993 (2006)
30. B. Cui, Y. Cortot, T. Veres, Microelectron. Eng. **83**, 906 (2006)
31. Z. Hu, G. Baralia, V. Bayot, J.F. Gohy, A.M. Jonas, Nano Lett. **5**, 1738 (2005)
32. S.J. Kang, Y.J. Park, J.Y. Hwang, H.J. Jeong, J.S. Lee, K.J. Kim, H.C. Kim, J. Huh, C. Park, Adv. Mater. **19**, 581 (2007)
33. D. Cui, H. Li, H. Park, X. Cheng, J. Vac. Sci. Technol. B **26**, 2404 (2008)
34. X. Cheng, L.J. Guo, P.F. Fu, Adv. Mater. **17**, 1419 (2005)
35. S.H. Ahn, L.J. Guo, Adv. Mater. **20**, 2044 (2008)
36. C.D. Schaper, A. Miahnahri, J. Vac. Sci. Technol. B **22**, 3323 (2004)
37. G. Zhang, J. Zhang, G. Xie, Z. Liu, H. Shao, Small **2**, 1440 (2006)
38. S.-M. Lee, H.S. Lee, D.S. Kim, T.H. Kwon, Surf. Coat. Technol. **201**, 553 (2006)
39. L. Seung-Mo, K. Tai Hun, Nanotechnology **17**, 3189 (2006)
40. L. Seung-Mo, K. Tai Hun, J. Micromech. Microeng. **17**, 687 (2007)
41. D.A. Dillard, A.V. Pocius (eds.), *The Mechanics of Adhesion*, 1st edn. (Elsevier Science, Amsterdam, 2002)
42. S. Johnson, D.J. Resnick, D. Mancini, K. Nordquist, W.J. Dauksher, K. Gehoski, J.H. Baker, L. Dues, A. Hooper, T.C. Bailey, S.V. Sreenivasan, J.G. Ekerdt, C.G. Willson, Microelectron. Eng. **67–68**, 221 (2003)
43. S. Park, G. Kim, K. Choi, J. Lee, Microelectron. Eng. **87**, 968 (2010)
44. S.R.J. Brueck, Proc. IEEE **93**, 1704 (2005)
45. M. Kurihara, M. Abe, K. Suzuki, K. Yoshida, T. Shimomura, M. Hoga, H. Mohri, N. Hayashi, Microelectron. Eng. **84**, 999 (2007)
46. K. Mohamed, M.M. Alkaisi, R.J. Blaikie, J. Vac. Sci. Technol. B **25**, 2357 (2007)
47. K. Mohamed, M.M. Alkaisi, R.J. Blaikie, J. Vac. Sci. Technol. B **26**, 2500 (2008)
48. M.M. Alkaisi, W. Jayatissa, M. Konijn, Curr. Appl Phys. **4**, 111 (2004)
49. Y. Ishii, J. Taniguchi, Microelectron. Eng. **84**, 912 (2007)
50. J. Taniguchi, K. Koga, Y. Kogo, I. Miyamoto, Microelectron. Eng. **83**, 940 (2006)
51. T. Morita, K. Watanabe, R. Kometani, K. Kanda, Y. Haruyama, T. Kaito, J.I. Fujita, M. Ishida, Y. Ochiai, T. Tajima, S. Matsui, Jpn. J. Appl. Phys. Part 1 **42**, 3874 (2003)
52. S. Matsui, in *Three-Dimensional Nanostructure Fabrication by Focused Ion Beam Chemical Vapor Deposition*, ed. by B. Bhushan. Springer Handbook of Nanotechnology (Springer, Berlin, 2010), p. 211
53. C. Peng, X. Liang, S.Y. Chou, Nanotechnology **20**, (2009)
54. S.-H. Park, T.-W. Lim, D.-Y. Yang, J.-H. Jeong, K.-D. Kim, K.-S. Lee, H.-J. Kong, Appl. Phys. Lett. **88**, 203105 (2006)

55. S. MacDonald, G. Hughes, M. Stewart, F. Palmieri, C.G. Willson, Proc. SPIE Int. Soc. Opt. Eng. **5992**, 59922F (2005)
56. M.D. Stewart, J.T. Wetzel, G.M. Schmid, F. Palmieri, E. Thompson, E.K. Kim, D. Wang, K. Sotodeh, K. Jen, S.C. Johnson, J. Hao, M.D. Dickey, Y. Nishimura, R.M. Laine, D.J. Resnick, C.G. Willson, Proc. SPIE Int. Soc. Opt. Eng. **5751**, 210 (2005)
57. B.H. Chao, F. Palmieri, W.-L. Jen, D.H. McMichael, C.G. Willson, J. Owens, R. Berger, K. Sotoodeh, B. Wilks, J. Pham, R. Carpio, E. LaBelle, J. Wetzel, Proc. SPIE Int. Soc. Opt. Eng. **6921**, 69210C (2008)
58. J. Butschke, M. Irmscher, D. Resnick, H. Sailer, E. Thompson, Proc. SPIE Int. Soc. Opt. Eng. **6607**, (2007)
59. M. Pritschow, J. Butschke, M. Irmscher, H. Sailer, D. Resnick, E. Thompson, Proc. SPIE Int. Soc. Opt. Eng. **6730**, 67300G (2007)
60. G.M. Schmid, M.D. Stewart, J. Wetzel, F. Palmieri, J. Hao, Y. Nishimura, K. Jen, E.K. Kim, D.J. Resnick, J.A. Liddle, C.G. Willson, J. Vac. Sci. Technol. B **24**, 1283 (2006)
61. N. Nagai, H. Ono, K. Sakuma, M. Saito, J. Mizuno, S. Shoji, Jpn. J. Appl. Phys. **48**, 115001 (2009)
62. K.-H. Jeong, J. Kim, L.P. Lee, Science **312**, 557 (2006)
63. M. Tormen, A. Carpentiero, E. Ferrari, D. Cojoc, E.D. Fabrizio, Nanotechnology **18**, 385301 (2007)
64. P.B. Clapham, M.C. Hutley, Nature **244**, 281 (1973)
65. H. Kasugai, Y. Miyake, A. Honshio, S. Mishima, T. Kawashima, K. Iida, M. Iwaya, S. Kamiyama, H. Amano, I. Akasaki, H. Kinoshita, H. Shiomi, Jpn. J. Appl. Phys. **44**, 7414 (2005)
66. A.R. Parker, H.E. Townley, Nat. Nanotechnol. **2**, 347 (2007)
67. A. Kaless, U. Schulz, P. Munzert, N. Kaiser, Surf. Coat. Technol. **200**, 58 (2005)
68. J. Kettle, R.T. Hoyle, S. Dimov, R.M. Perks, Microelectron. Eng. **85**, 853 (2008)
69. J. Kettle, R.T. Hoyle, R.M. Perks, S. Dimov, J. Vac. Sci. Technol. B **26**, 1794 (2008)
70. D.H. Kim, W.J. Chin, S.S. Lee, S.W. Ahn, K.D. Lee, Appl. Phys. Lett. **88**, (2006)
71. C.-S. Huang, E.Y.-B. Pun, W.-C. Wang, J. Opt. Soc. Am. B **26**, 1256 (2009)
72. D.W. Prather, D. Pustai, M.R. LeCompte, S. Shi, Proc. SPIE Int. Soc. Opt. Eng. **4291**, 43 (2001)
73. H. Park, H. Li, X. Cheng, J. Vac. Sci. Technol. B **25**, 2325 (2007)
74. X.J. Feng, L. Jiang, Adv. Mater. **18**, 3063 (2006)
75. P.-G.D. Gennes, F. Brochard-Wyart, D. Quéré, *Capillarity and Wetting Phenomena: Drops, Bubbles, Pearls, Waves* (Springer, New York, NY, 2004)
76. L.R. Bao, X. Cheng, X.D. Huang, L.J. Guo, S.W. Pang, A.F. Yee, J. Vac. Sci. Technol. B **20**, 2881 (2002)
77. W. Hu, B. Yang, C. Peng, S.W. Pang, J. Vac. Sci. Technol. B **24**, 2225 (2006)
78. M. Agirregabiria, F.J. Blanco, J. Berganzo, M.T. Arroyo, A. Fullaondo, K. Mayora, J.M. Ruano-Lopez, Lab Chip **5**, 545 (2005)
79. B. Yang, S.W. Pang, J. Vac. Sci. Technol. B **24**, 2984 (2006)
80. H. Ooe, M. Morimatsu, T. Yoshikawa, H. Kawata, Y. Hirai, J. Vac. Sci. Technol. B **23**, 375 (2005)
81. J.A. Forrest, K. Dalnoki-Veress, J.R. Stevens, J.R. Dutcher, Phys. Rev. Lett. **77**, 2002 (1996)
82. Y. Yang, D.H. Liu, Y.B. Xie, L.J. Lee, D.L. Tomasko, Adv. Mater. **19**, 251 (2007)
83. H. Park, X. Cheng, Nanotechnology **20**, 245308 (2009)
84. M. Nakajima, T. Yoshikawa, K. Sogo, Y. Hirai, Microelectron. Eng. **83**, 876 (2006)
85. M. Chaudhury, A.V. Pocius (eds.), *Surfaces, Chemistry and Applications*, 1st edn. (Elsevier Science, Amsterdam, 2002)
86. P. Kettner, L. Pelzer, T. Glinsner, S. Farrens, D. Lee, J. Phys. Conf. Ser. **34**, 65 (2006)
87. H.C. Kim, J. Park, Y. Cho, H. Park, A. Han, X. Cheng, J. Vac. Sci. Technol. B **27**, 3115 (2009)

88. J.J. Dumond, H.Y. Low, I. Rodriguez, Nanotechnology **17**, 1975 (2006)
89. H.Y. Low, W. Zhao, J. Dumond, Appl. Phys. Lett. **89**, 023109 (2006)
90. K.S. Han, S.H. Hong, J.H. Jeong, H. Lee, Microelectron. Eng. **87**, 610 (2010)
91. N. Gadegaard, K. Seunarine, D.J.A. Smith, D.O. Meredith, C.D.W. Wilkinson, M.O. Riehle, Jpn. J. Appl. Phys. **47**, 7415 (2008)
92. E.M. Liston, L. Martinu, M.R. Wertheimer, J. Adhes. Sci. Technol. **7**, 1091 (1993)
93. C.M. Chan, T.M. Ko, H. Hiraoka, Surf. Sci. Rep. **24**, 1 (1996)
94. J. Brandrup, E.H. Immergut, E.A. Grulke, *Polymer Handbook*, 4th edn. (Wiley, New York, NY, 1999)
95. F. Zhang, H.Y. Low, Nanotechnology **17**, 1884 (2006)

Chapter 4
Electrochemical Growth of Nanostructured Materials

Jin-Hee Lim and John B. Wiley

The fabrication of nanostructures such as nanoparticles, nanorods, nanowires, nanotubes, and core–shell nanowires have received considerable attention because of their potential applications in fields as diverse as electronics, magnetics, optical devices, and catalysts. Among the various methods for fabrication of nanostructures, the template-assisted electrodeposition technique has been extremely prolific in the preparation of a wide range of metallic, semiconductor, and polymeric nanowires because it is a simple and inexpensive technique. Furthermore, with this method it is easy to control the size, shape, and structural properties of nanostructures through the modification of templates and electrodeposition conditions. Commercially available anodic alumina and polycarbonate track etch membranes are commonly used templates. Such templates can, however, have disadvantages over "homemade" membranes due to branched pore structures and randomly distributed pores. Advances in the synthetic techniques in membrane production have led to almost perfectly regular-ordered porous alumina structures with controlled pore diameters (D_p), interpore distances (D_i), and thicknesses. For electrodeposition within a template, one side of the membrane is sputtered or vacuum evaporated with metal thin film, which acts as the working electrode during electrodeposition. Figure 4.1 illustrates a typical electrodeposition system. For preparing nanostructures, the system usually consists of three electrodes: the working electrode (WE), counter electrode (CE), and reference electrode (RE). Nanowires grown in the linear pores of an alumina or polymer membrane grow from the bottom (sputtered metal) surface into the pores since the deposition occurs by electron transfer from the working electrode. This electrodeposition technique offers an effective and versatile approach for the fabrication of a diverse range of materials with a variety of structures. In this chapter, we will work to highlight some of the important nanostructures produced with this approach.

J.B. Wiley (✉)
Department of Chemistry and Advanced Materials Research Institute,
University of New Orleans, New Orleans, LA 70148, USA
e-mail: jwiley@uno.edu

W.L. Zhou, Z.L. Wang (eds.), *Three-Dimensional Nanoarchitectures*,
DOI 10.1007/978-1-4419-9822-4_4,

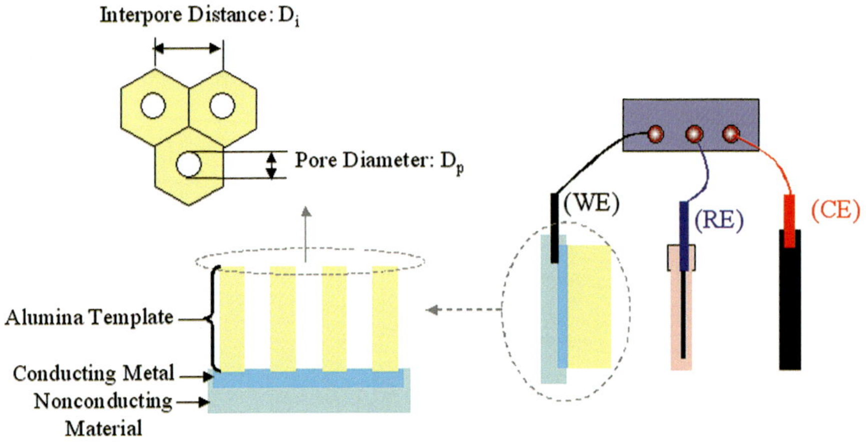

Fig. 4.1 Illustration of the electrodeposition system for synthesis of nanomaterials in a porous template

4.1 Magnetic Nanomaterials

One-dimensional magnetic nanomaterials are of great interest in magnetic recording media, giant magnetoresistance, and spintronics [1]. Most of the previous work in ferromagnetic nanowire arrays has focused on the magnetization reversal mechanism and high-density magnetic recording media [2]. For magnetic recording media applications the wire diameter and the interwire distance should be as small as possible to increase the recording density. Thus, most magnetic studies concentrate on sub-100 nm wires [3]. The development of magnetic recording media is restricted by the superparamagnetic limit. To overcome this limitation either the effective anisotropy of the material or the thickness of the nanostructures can be increased [4]. Magnetic properties of nanomaterials are determined by magnetostatic interaction among nanowires and magnetic anisotropy such as magnetocrystalline and shape anisotropy. The parameters such as length, diameter, and interwire distance of nanowires, which influence the anisotropy and magnetostatic interaction, can be tuned by adjusting templates and electrodeposition conditions.

Fe, Co, and Ni nanowires synthesized in alumina templates with different pore diameters were investigated. Baik et al. [5] prepared Fe nanowires in sulfuric-anodized template (42 nm pore diameter) and oxalic-anodized template (63 nm pore diameter) and measured magnetic properties. The 42 nm diameter Fe nanowires with preferred (110) orientation showed a coercivity of 1654 Oe and a squareness (M_r/M_s, ratio of the remanence (M_r) to magnetic saturation (M_s)) of 0.66 with the magnetic field oriented along the long axis of the nanowires. The coercivity and squareness of the 63 nm diameter Fe nanowires were reduced to 1000 Oe and 0.33, respectively, and exhibit both (110) and (200) orientations, signifying a polycrystalline wire. Increasing pore diameter results in improved crystallinity for the nanowires. Their coercivity and squareness are reduced because of the decreased aspect ratio and hence reduced shape anisotropy.

Similar results were observed in Co nanowires. Huang et al. [6] reported single crystal Co nanowires with diameters of 40, 70, and 110 nm. Corresponding coercivity and squareness are deceased gradually from 40 to 110 nm. The observed effect was attributed to an increase in the magnetostatic interaction field with decreasing distance between the nanowires.

Figure 4.2 shows the coercivity of Ni nanowires as a function of pore widening time (nanowire diameter). Zheng et al. [7] controlled pore diameter by the pore widening process and the coercivity of Ni nanowires with 8–25 nm in diameter was measured. In this measurement, the coercivity was determined with the magnetic field applied along the wire axis. For Ni nanowires deposited in the as-synthesized template, the coercivity was about 480 Oe. The coercivity increased with pore widening time and reached the highest value of 950 Oe. The coercivity then decreased on further increasing the pore diameter. The characteristic coercivity of nanowires was determined by their diameters and interwire magnetostatic interactions.

The size of the crystallites within the Fe, Co, and Ni nanowires, as well as the crystalline structure of the nanowire, depends on the deposition conditions such as the pH of the solutions and the deposition parameters [2, 8]. The crystal structure

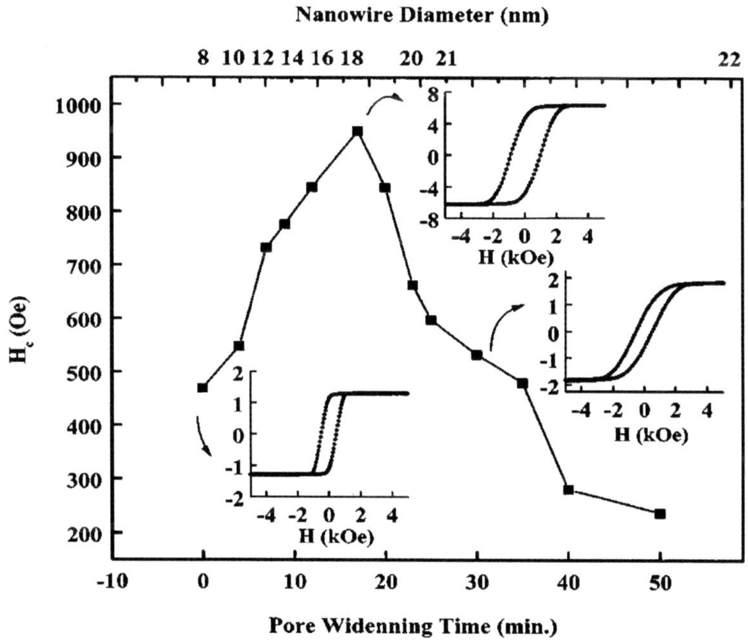

Fig. 4.2 Dependence of the coercivity versus pore-widening time where nanowire diameter is determined by the pore widening time. *Insets* are the hysteresis loops for Ni nanowires with diameters of 8, 18, and 21 nm. Reprinted with permission from [7]. Copyright 2000 by the American Physical Society

of Fe and Ni nanowires are body-centered cubic (bcc) and face-centered cubic (fcc), respectively. However, the structure of Co wires is more complicated. The Co nanowires consist either of hexagonal close-packed (hcp) or fcc structures or a mixture of both [9]. Huang et al. [10] fabricated single crystalline Co nanowire arrays with different orientations. The orientation of nanowires can be controlled by controlling the deposition parameters shown in Table 4.1. Figure 4.3 shows the magnetic properties of Co nanowire arrays with different orientations. When the applied field is parallel to the nanowire axis, the loops are relatively square. It is well known that the ferromagnetic properties of the Co nanowires are strongly dependent on their crystalline orientation. The crystalline orientation of nanowires can be modified by proper selection of the deposition current and pH value.

Table 4.1 Various deposition conditions to control orientations and coercivity of Co nanowires. Reprinted with permission from [10] by American Chemical Society (ACS)

						Hc (Oe)	
Voltage (V)	Current^2 (mA^2)	Pulsed time (ms)	Duty (%)	pH	Orientation	||	⊥
1.5	0.09	5	50	4	hcp[1010]	477	139
2.4	0.64	5	50	4	fcc[111]	1075	149
3.0	2.25	7.5	75	2	fcc[200]	1031	86

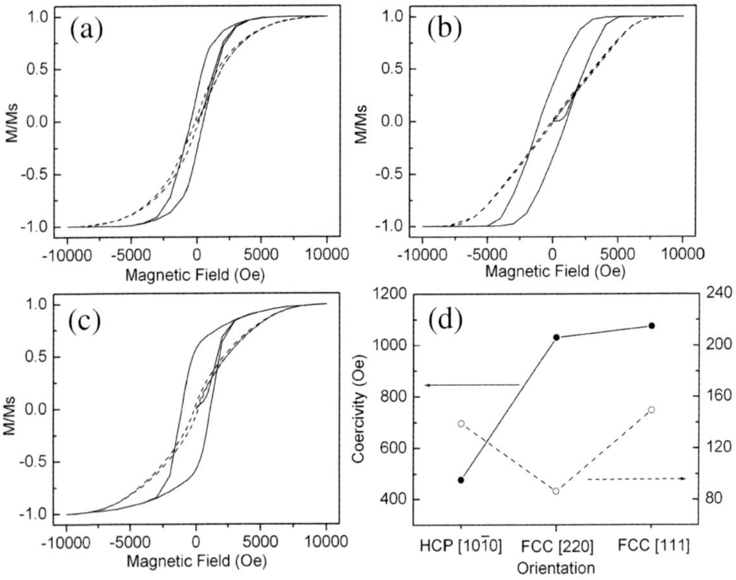

Fig. 4.3 Hysteresis loops at room temperature of Co nanowire arrays with the orientation of (**a**) hvp[10$\bar{1}$0], (**b**) fcc[220], and (**c**) fcc[111], with the applied field parallel (*solid curves*) and perpendicular (*dashed curves*) to the nanowire axes. (**d**) Coercivity versus orientations with the applied field parallel (*solid curves*) and perpendicular (*dashed curves*) to the nanowire axes. Reprinted with permission from [10] by American Chemical Society (ACS)

4.2 Semiconductor Nanostructures

Semiconductor nanowires can be used as chemical gas sensors, light-emitting diodes (LEDs), and transistors because of their large surface-to-volume ratios as well as their tunable optical and electrical properties [11]. Semiconductor nanomaterials including GaN, InN, ZnO, SnO_2, TiO_2, HfO_2, and WO_3 were produced by various physical and chemical methods such as chemical vapor deposition (CVD), thermal evaporation, conventional metal organic chemical vapor deposition (MOCVD), sol–gel approach, and template-based electrochemical method [11–15]. In this chapter, we are only focusing on electrochemically grown nanostructures.

Tin dioxide (SnO_2), an n-type semiconductor with a large band gap ($E_g = 3.6$ eV at 300 K), is an important metal oxide semiconductor owing to its potential applications in the transparent conducting coating of glass, electrochemical modifiers on electrodes, solar cells, and gas sensor devices [12, 16]. Recently, studies on SnO_2 mainly focused on the preparation of SnO_2 films or nanoparticles. However, Zhang et al. [16] first reported uniform SnO_2 nanowire arrays synthesized by electrochemical deposition and thermal oxidization. After that, one-dimensional SnO_2 nanotubes were fabricated using a similar method in 50 nm polycarbonate membranes by Lai et al. [12]. Figure 4.4 shows field emission scanning electron microscopy (FESEM) and transmission electron microscopy (TEM) images of SnO_2 nanotubes. TEM images indicate that the tubular structure is maintained after annealing in air at 400°C. The optoelectronic response of the SnO_2 nanotubes was investigated in air using UV light with a wavelength of 254 nm. Figure 4.5 shows *I*–*V* curves for the SnO_2 nanotubes with UV light on and off, respectively. The electrical conductance of nanotubes increased to 33 times the original value upon UV illumination at 254 nm.

Zinc oxide (ZnO), a wide band gap (3.37 eV) semiconductor with high exciton binding energy (60 meV), is chemically and thermally stable [17, 18]. Many forms of nanosized ZnO such as nanobelts, nanocages, nanocombs, and nanowires are

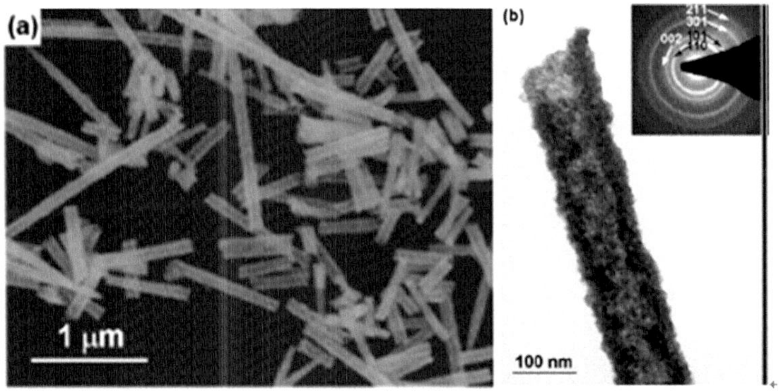

Fig. 4.4 (**a**) FESEM and (**b**) TEM images of SnO_2 nanotubes electrodeposited in 50 nm polycarbonate membranes. Reprinted with permission from [12] by IOP Publishing Ltd

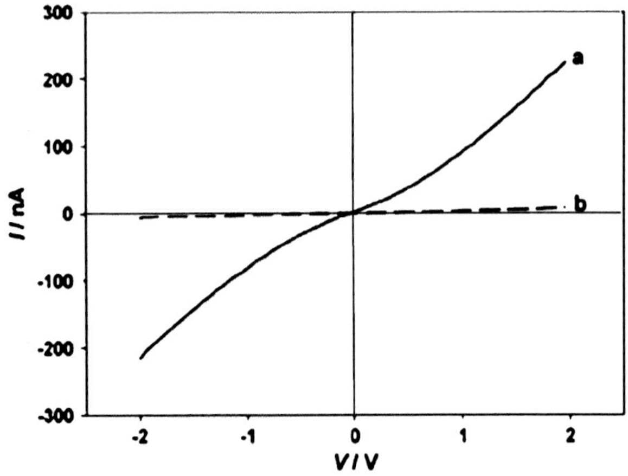

Fig. 4.5 *I–V* curves of annealed SnO_2 nanotube in the (**a**) presence and (**b**) absence of UV illumination at 254 nm. Reprinted with permission from [12] by IOP Publishing Ltd

prepared by Wang [19]. Even though several approaches can be used to prepare a nanostructure with uniform size and morphology, very little work related to the electrochemically synthesized one-dimensional ZnO nanostructure has been reported. Zhang et al. [18] reported direct electrodeposition of ZnO nanotube arrays from aqueous solutions containing $Zn(NO_3)_2{\cdot}6H_2O$ at 85°C. The reaction steps thought to occur in the preparation of ZnO nanotubes are proposed as follows:

$$Zn(NO_3)_2 \rightarrow Zn^{2+} + 2NO_3^-$$

$$NO_3^- + H_2O + 2e^- \rightarrow NO_2^- + 2OH^-$$

$$Zn^{2+} + 2OH^- \rightarrow Zn(OH)_2$$

$$Zn(OH)_2 \rightarrow ZnO + H_2O$$

$Zn(OH)_2$ will form and deposit on the cathode at 85°C.

Generally in template-based electrodeposition it is very hard to fine stable conditions, especially in the fabrication of titanium oxide (TiO_2), hafnium oxide (HfO_2), and tungsten oxide (WO_3) nanomaterials. Utilizing electrodeposition, however, several research groups developed anodization techniques using pure metal films such as Ti, Hf, and W. Among these efforts, synthesis and characterization of anodic TiO_2 are well known. Gong et al. [20] pioneered the synthesis of the first-generation titania nanotube arrays of up to 500 nm length. Recent reports describe the fabrication of TiO_2 layers up to 222 μm in thickness using a variety of polar organic electrolytes that include ethylene glycol, formamide, *N*-methylformamide, and dimethyl sulfoxide in combination with HF, KF, NaF, NH_4F, Bu_4NF, or $BnMe_3NF$ to provide fluoride ions [21, 22]. Highly ordered TiO_2 nanotubes offer a large internal surface area. In addition, TiO_2 absorbs only solar light in the UV region because of a large band gap (=3.1–3.2 eV). It can be used for sensing, heterojunction solar cells, tissue

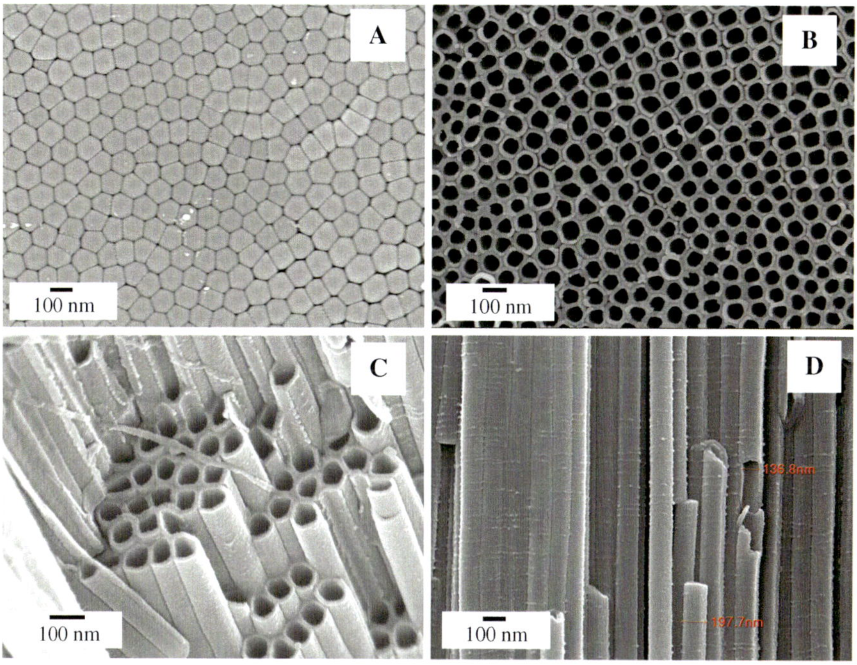

Fig. 4.6 FESEM images of Ti nanotube structures anodized at (**a–c**) 60 V and (**d**) 65 V in an electrolyte comprised of 0.3 wt% NH_4F and 2 vol.% H_2O in ethylene glycol. Reprinted with permission from [22] by American Chemical Society (ACS)

engineering, fuel cell, etc. [23]. FESEM images in Fig. 4.6 show the nanotube arrays anodized in 0.3 wt% NH_4F and 2% H_2O in ethylene glycol for 17 h.

To obtain porous tungsten oxide (WO_3) layers, electrochemical treatment was carried out in 1 M H_2SO_4 + 0.5 wt% NaF by Berger et al. [24]. Synthesized self-organized porous WO_3 layers consist of a regular structure with 70 nm pore and 10 nm wall thickness. As-synthesized sample showed an amorphous structure but it was changed to crystalline monoclinic structure by annealing. Qiu et al. [25] investigated the synthesis of highly ordered HfO_2 nanotubes and their growth mechanism. HfO_2 nanotubes were prepared using thin Hf foil in a solution containing 0.1 M NH_4F with 20 mL DI water and 80 mL ethylene glycol at room temperature. They suggested an HfO_2 nanotube growth mechanism based on volume expansion combined with inhomogeneous surface defects and gas evolution.

4.3 Thermoelectric Nanomaterials

Thermoelectric materials are ranked by the thermoelectric figure of merit, ZT ($=S^2\sigma T/\kappa$), where S is the thermoelectric power or Seebeck coefficient, σ is the electrical conductivity, κ is the thermal conductivity, and T is the absolute temperature [26–28]. It is very difficult to achieve a high ZT value because S, σ, and κ are interdependent on each other. The increased electrical conductivity needed for a high ZT usually leads to an increase in thermal conductivity and a decrease in the Seebeck coefficient. As a result only bulk materials with high atomic weight such as Bi_2Te_3 and its alloys with Sb, Sn, and Pb have been routinely used for the commercial applications [26].

In recent research, theoretical predictions suggest that the thermoelectric properties of nanostructures will be significantly enhanced compared with bulk materials [29, 30]. However, a single nanowire does not carry a high enough current for thermoelectric materials' applications [31]. To collect dense nanowire arrays researchers need to find a new approaches like template-assisted electrodeposition.

Prieto et al. [29] reported ordered $Bi_{88}Sb_{12}$ nanowires with uniform diameters of 40 nm that were prepared by direct electrodeposition under potentiostatic control with potentials more than –0.45 V versus Hg/Hg_2SO_4. Recently, Bi nanowires [32], Bi nanotubes [33], BiSb superlattice nanowires [34], BiSb nanowire junctions [35], and BiSb nanotube–nanowires [36] have been synthesized successfully with electrodeposition techniques. Also Bi_2Te_3 nanowires were prepared by similar methods, as described above [37]. However, most of the nanowires synthesized by template-assisted electrodeposition have a disadvantage in that it is hard to measure thermoelectric properties because the final structure is contained in the polycrystalline structure of a nanoporous template. Further, the preparation of single crystalline nanowires is still a challenge.

4.4 Conducting Polymer Nanostructures

Conducting polymers such as poly(3,4-ethylenedioxythiophene) (PEDOT), polypyrrole (PPY), polyaniline (PANI), polythiophene (PT), and poly(3-methylthiophene) (P3MT) are emerging as promising materials for applications pertinent to displays, energy storage devices, actuators, sensors, etc. [38–41]. Conducting polymers exhibit electrical, electronic, magnetic, and optical properties similar to metals or semiconductors [42–44]. Many of the applications of conducting polymers rely on the redox properties. Conducting polymers can be reversibly reduced and oxidized using chemical or electrochemical methods [45]. The reversible interchange between the redox states gives a possibility to control their properties such as polymer conformation, doping level, conductivity, and color [44].

Lee and Cho [44] have introduced synthesis procedures, mechanism, and applications of PEDOT nanotubes. They showed that fast electrochromics, one of the interesting properties of conducting polymers, responds without sacrificing color contrasts. This electrochromic property can be applied to smart widows, rear-view mirrors, electronic paper, displays, etc. [46]. The PEDOT nanotubes can provide short diffusion pathways for fast redox switching with response times of less than 10 ms. In comparison, PEDOT films usually take 1–2 s to change a color [44].

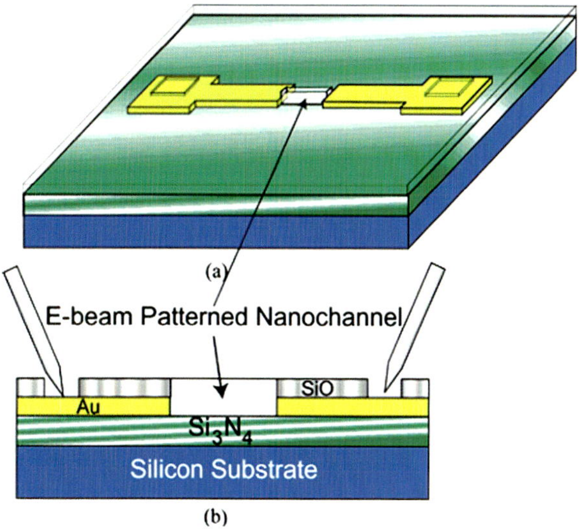

Fig. 4.7 Schematic diagram of (**a**) structure used for the electrochemical wire growth and (**b**) cross-sectional view of the Si substrate, silicon nitride (1 μm), Au contacts, and thermally evaporated SiO. Electrodeposited wire connected between electrodes. Reprinted with permission from [48] by American Chemical Society (ACS)

In the case of PPY and PANI, they are well known as good pH-sensing materials. The conductivity of PPY is directly proportional to the pH [47]. Yun et al. [48] directly electrodeposited PPY nanowires on electrodes and measured their pH-sensing ability. Figure 4.7 illustrates an e-beam-patterned nanochannel, where the PPY nanowires are located. Figure 4.8 shows that when buffered solution is added on top of the polymer wire, the current increased upon the change in resistance. Other efforts include those by Bartlett and co-workers [45] who prepared three-dimensional ordered macroporous conducting polymers from a template of close-packed poly(styrene) latex spheres. Lahav et al. [49] produced freestanding

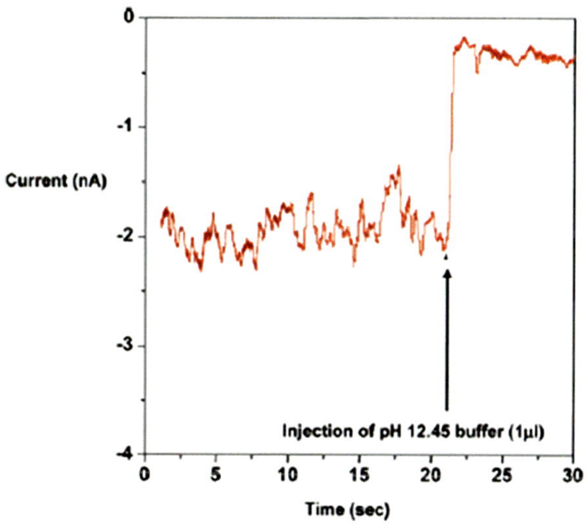

Fig. 4.8 Current versus time for demonstration of pH sensing by adding 1 μL of pH 12.45 buffered solution on top of a single polypyrrole wire (500 nm width). Reprinted with permission from [48] by American Chemical Society (ACS)

composites of PANI nanotubes with gold shells, where both components were of controllable length.

PT and P3MT polymer nanostructures in both undoped and doped states have been studied for organic field effect transistors (OFET), solar cells, sensors, electrochromic devices, and organic light-emitting diodes (OLEDs) [50, 51]. Joo et al. [51] have prepared PT and P3MT nanotubes and nanowires for studying OLED properties. A very large enhancement of the photoluminescence efficiency was observed for a single strand of a light-emitting PT nanotube after coating it with nanometer-scale copper, nickel, or cobalt metal.

4.5 Nanotube and Core–Shell Nanostructures

After carbon nanotubes were discovered in 1991, nanotube structures have attracted much attention because of their fundamental importance as well as their potential applications [52]. Many kinds of metal, semiconductor, and polymer nanotube structures were synthesized in porous membranes by numerous methods including supercritical fluid (SCF) inclusion phase [53], atomic layer deposition (ALD) [54], layer-by-layer deposition [55], sol–gel [56], melt-assisted template wetting [57], or electroless deposition techniques [58]. Although these methods can vary the outer diameters of nanotubes by varying the template used, they can have the disadvantage of being time-consuming processes and can allow for nanotube impurities [59, 60]. Also, it is hard to control wall thickness and length. Thus, electrodeposition methods offer an efficient approach to the production of nanotube structures via templates. Much of success in this area has been achieved on larger pored membranes, for instance, H_3PO_4-anodized AAO or commercial templates with over 150 nm pore diameters. Further the conducting metal at the bottom of templates should be thin enough, not covering the pore channels. Current density and pH of electrolytes also influence the growth of nanotubes [61, 62]. Fukunaka et al. [62] prepared metal nanowire and nanotube arrays by appropriately controlled electrochemical deposition. The shape of Ni nanostructures can be controlled by pH as well as electrode potential. Figure 4.9 shows TEM images of Ni nanostructures. A number of researchers have suggested growth mechanisms for nanotubes but as of yet, no clear consensus has been achieved [52, 62, 63].

One major advantage of these tubular structures is the high surface area through the inner and outer surfaces. In case of magnetic nanotube structures, coercivity is one of the most important properties of magnetic materials for many present and future applications and the understanding of magnetization reversal mechanisms is a permanent challenge for researchers involved in studying the properties of these materials (see Section 4.1) [64]. Furthermore, the synthesis of nanotube arrays opened more opportunities for fabrication of core–shell nanowires. In early studies, the outer surface of solid nanowires was coated with different materials to prepare core–shell nanowires, and thus, solid nanowires only can constitute core in this system. However, core–shell structures can be prepared through deposition within nanotube structures as well [65, 66].

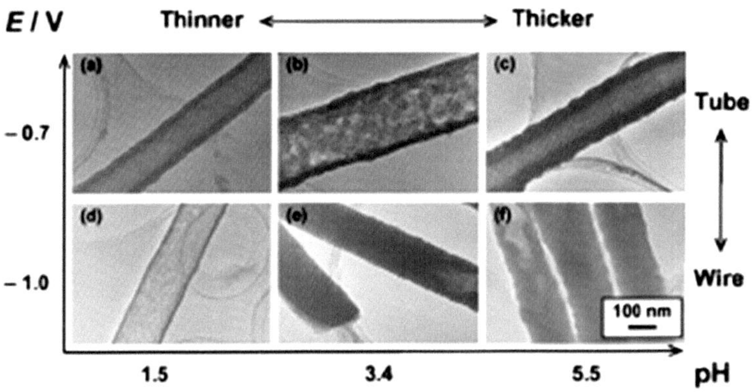

Fig. 4.9 TEM images of Ni nanotubes and nanowires with 200 nm diameter synthesized by controlling the electrodeposition conditions. Reprinted with permission from [62]. Copyright 2006, The Electrochemical Society

Various core–shell nanowires are accessible. The combination of two compounds in one nanowire structure expands their potential application to a variety of areas. For example, core–shell nanowires in which the magnetic core is coated with nonmagnetic materials are used for biomedical applications [67]. The enhancement of the electrical, magnetic, optical, structural, and mechanical properties has occurred in hybrid nanomaterials including inorganic–organic and organic–inorganic nanocomposites [68]. Sensing ability of semiconductor core–shell nanowires exhibited a high response, a good stability, and a short response/recovery time in the detection of low concentrations of various combustible gases [69]. In the case of bimagnetic core–shell nanowires, where both core and shell are ferro- or ferri-magnetic, contacts between magnetic materials lead to effective exchange coupling and therefore cooperative magnetic switching, facilitating the fabrication of nanostructured magnetic materials with tunable properties [70]. Lim et al. [66] described the fabrication of both nanotubes and core–shell nanostructures in alumina templates using a two-step electrodeposition process. They prepared initially ferromagnetic nanotube structures, and during a second electrodeposition process, ferromagnetic cores were synthesized in the nanotubes. Figure 4.10 images show the ferromagnetic nanotubes and core–shell nanowires.

4.6 Porous Au Nanowires

Solid metallic nanoparticles have been exploited for a wide range of applications that take advantage of their high surface-to-volume ratio [71]. Nanorods or nanowires, in contrast to the nanoparticles, exhibit additional degrees of freedom associated with their inherent shape anisotropy and the ability to incorporate different components along their length [72–74]. Additionally, the surface morphology and surface area of nanorods or nanowires can be controlled by chemical methods.

Fig. 4.10 FESEM images of ferromagnetic nanotube and core–shell nanowire arrays: (**a**) Co and (**c**) Ni nanotubes, (**b**) NiCo and (**d**) CoNi core–shell nanowires, which are synthesized using (**a**) Co nanotubes and (**c**) Ni nanotubes, respectively. The interwire distances are about 400 nm in diameters [66]

For example, nanoporous metal nanowires have been prepared by selective chemical or electrochemical dissolution of Au–Ag alloys, known as a 'dealloying' process [75]. These nanoporous gold nanowires are useful for the applications in catalysts [76], sensing [77], surface-enhanced Raman scattering (SERS) [78], localized surface plasmon (LSP) resonance [79], or actuators [80]. Nanoporous Au nanowires can be produced by a template-based approach. Synthesized solid Au/Ag alloys are immersed in concentrated nitric acid, in which only Ag can be soluble. The length of nanostructures was controlled by monitoring of charge and electrodeposition time. Also the surface morphology can be tailored by the ratio of the Au_x/Ag_{1-x} alloy composition. The schematic diagram in Fig. 4.11 shows the approach for the synthesis of nanoporous nanowires.

Fig. 4.11 Schematic diagram of the fabrication of nanoporous Au nanowires. (**a**) Ag sputtering; (**b**) electrodeposition of Ag and Au–Ag alloy nanowires into the alumina membrane; (**c**) chemical dissolution of Ag electrodes and Ag from the Au–Ag alloy nanowires by concentrated nitric acid; and (**d**) alumina removing in NaOH solution

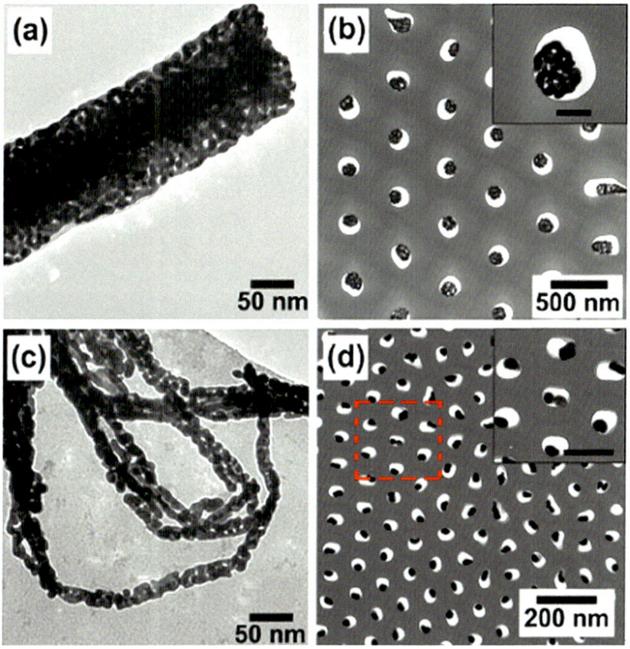

Fig. 4.12 TEM images of (**a**) nanoporous Au nanowires and (**b**) planar view of nanoporous Au nanowire with H_3PO_4-anodized AAO. TEM images of (**c**) nanoporous Au nanowires and (**d**) planar view of nanoporous Au nanowires with $H_2C_2O_4$-anodized AAO. *Insets* of (**b**) and (**d**) are magnified TEM images, *inset scale bar*: 120 nm for (**b**), 100 nm for (**d**). Reprinted with permission from [75] by IOP Publishing Ltd

When the porous Au nanowires are used as catalysts, these wires can react at low temperature, unlike Pt or Pd catalysts, have good thermal stability, and are resistant to oxidation [81]. Recently, Liu et al. [75] reported the fabrication of a composite membrane of nanoporous gold nanowires. They controlled the diameter of nanoporous nanowires using two alumina membranes (AAO), which were synthesized in oxalic and phosphoric acid solutions (Fig. 4.12). When porous nanowires are in the membranes, these structures allow relatively stable handling of nanostructures for further applications and result in effectively rapid flow through catalytic reactions.

For the sensor applications, high surface area can be a significant factor. If the dimension of a metallic conductor is smaller than the mean free path of electrons, the resistance becomes dominated by scattering of electrons at the surface [82]. The mean free path for most metals is on the order of a few tens of nanometers. Searson et al. [82] demonstrated the sensor applications with single nanoporous gold nanowires. First, they made nanoporous nanowires and then dispersed those wires on Si wafers. Subsequently, photolithography was employed for electrode synthesis. (More details on lithographic procedures will be introduced in Section 4.10.) Figure 4.13 shows optical microscope images of a single nanowire device. Figure 4.14 shows the normalized resistance change for the single

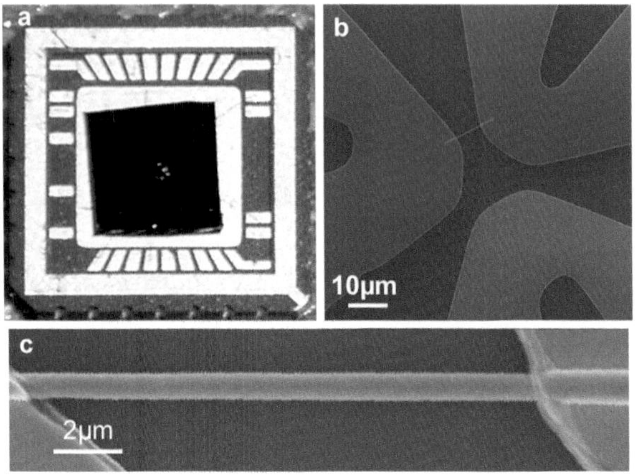

Fig. 4.13 (**a**) Optical microscope image of a chip carrier with a single nanowire device, (**b**) SEM image of a single nanowire patterned by photolithography, and (**c**) magnified image of a nanoporous Au nanowire. Reprinted with permission from [82] by American Chemical Society (ACS)

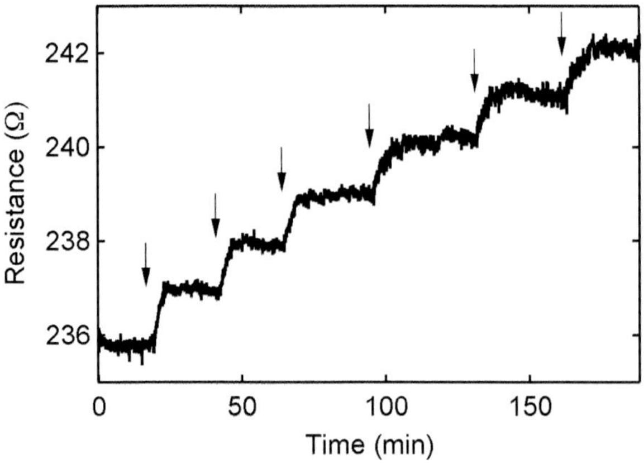

Fig. 4.14 Variation of resistance for a single nanoporous Au nanowire device after the injection of aliquots of 5×10^{-11} mol ODT. The *arrows* indicate the point of injection. Reprinted with permission from [82] by American Chemical Society (ACS)

nanoporous gold wire on introducing 10 μL of 1 mM octadecanethiol (ODT). The initial resistance of the nanowire is 236 Ω, and each aliquot results in a 1 Ω increase in the resistance. This result demonstrated the porous gold nanowires connected with the organic molecules on the surface with dimensions less than the mean free path of electrons. The change in resistance is related to the increase in inelastic scattering of electrons owing to the adsorbates.

4.7 Modification of Nanowires

Many efforts for producing nanomaterials have led to the synthesis of complex nanostructures ranging from spheres and rods to disks, cubes, prisms, tetrapods, and even hollow cubes [83–89]. One of the modified nanostructures is the linear noble metal nanoparticle chain that is desirable for one-dimensional nano-optical devices

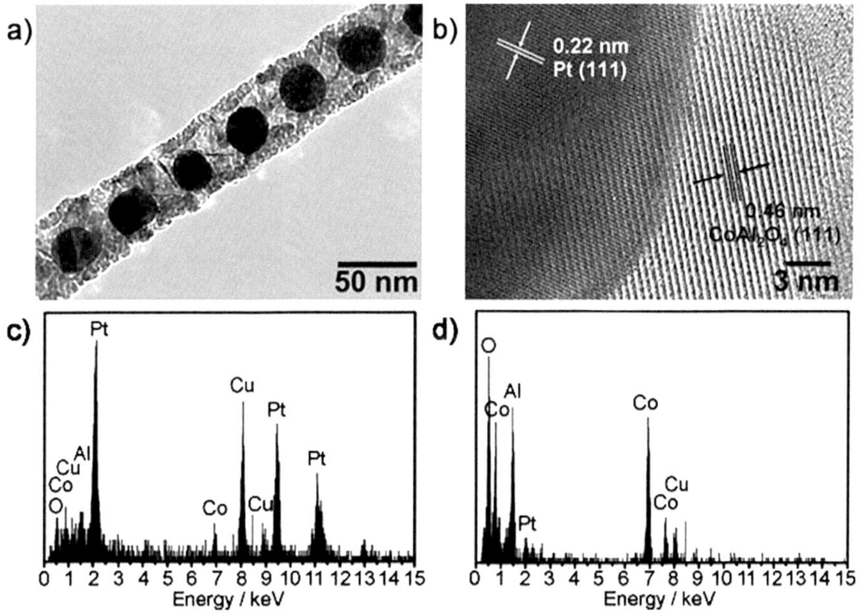

Fig. 4.15 (**a**) TEM and (**b**) HRTEM images of single Pt@$CoAl_2O_4$ nanopeapod. (**c**) and (**d**) EDS spectra collected from the pea and the pod part of the structure, respectively. Reprinted with permission from [92]. Copyright Wiley-VCH Verlag GmbH & Co. KGaA

including plasmonic waveguides and plasmonic printing [90, 91]. The ability of these devices depends on the particle size, shape, and separation in metal chains [92, 93]. Electrochemical-based methods are suitable to meet their requirements. Gösele's group [92] reported on tailor-made inorganic nanopeapods, which consist of well-defined Pt nanoparticles encapsulated in continuous $CoAl_2O_4$ nanoshells; these were synthesized by electrodeposition and subsequent solid-state reaction at high temperature. The size and separation of Co and Pt segments were defined during the electrodeposition process and the Pt particles were formed when Co segments reacted with alumina at high temperature. Figure 4.15 shows TEM images of a single Pt@$CoAl_2O_4$ inorganic nanopeapod prepared using $H_2C_2O_4$-anodized alumina template. Energy dispersive X-ray elemental analyses (EDS) confirmed the elemental composition of the nanopeapods. A second example of modified nanomaterials is the nanoporous nanowire arrays fabricated with nanospheres. Li et al. [94] reported porous Ni and Au nanowires synthesized using sphere-infiltrated membranes. To fabricate porous nanowires, silica or polystyrene nanospheres were placed in a membrane by vacuum filtration. After metal electrodeposition, the spheres were removed to make pore structures as shown in Fig. 4.16. The locations of spheres that were in contact with the channel wall are particularly obvious from the many pores found on the surface of the wires as shown in Fig. 4.16f. The pore size in nanowires and sphere packing rely on the diameter of the spheres and the relative diameters of the spheres to the pores, respectively.

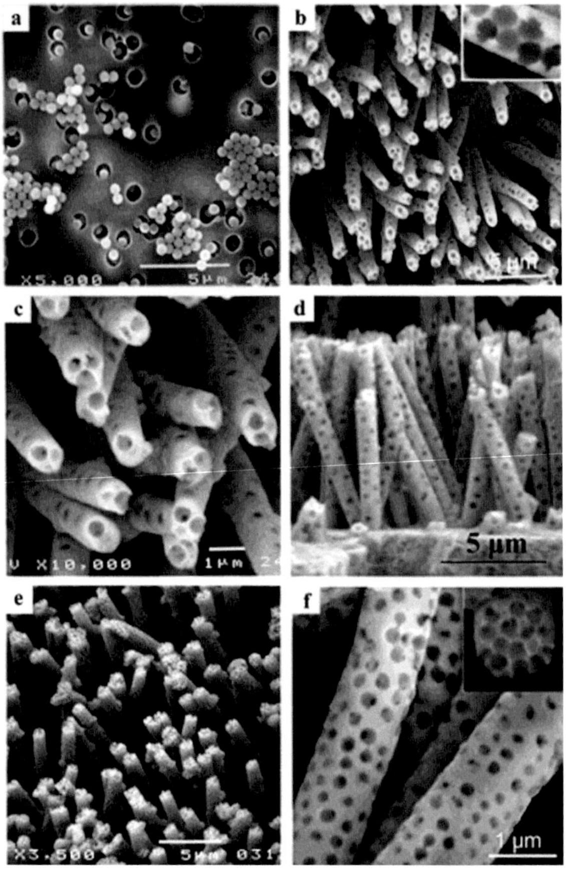

Fig. 4.16 SEM images of (**a**) 500 nm SiO_2 beads filtered polycarbonate membrane and (**b, c**) *top view* and (**d**) *side view* of porous Au wire arrays with 500 nm pores. (**e**) *Top view* and (**f**) *side view* of porous Au wires with 300 nm pores. *Inset in* (**b**): cross section along the length of a wire. *Inset in* (**f**): cross section along width of 300 nm pore wire. Reprinted with permission from [94] by American Chemical Society (ACS)

In addition modified nanowires can be synthesized using modified membranes such as Y-branched [95], triangular and rectangular shapes [96], and mild–hard-anodized alumina templates [97, 98]. During the deposition process, nanostructures start nucleating on the conducting metal surface and then grow along the pore channels of membrane. Therefore, morphology of porous templates is very important to control the shape of nanowires and replicate complex nanostructures.

4.8 Functionalization of Nanowires

One-dimensional nanomaterials with a broad range of chemical compositions have been prepared by electrochemical synthesis techniques. The surface of fabricated nanowires can be functionalized with specific chemical components to extend their application area. Ni nanowires functionalized with antibodies can be utilized for cell separation, which is essential for many cell-based applications in biochemistry,

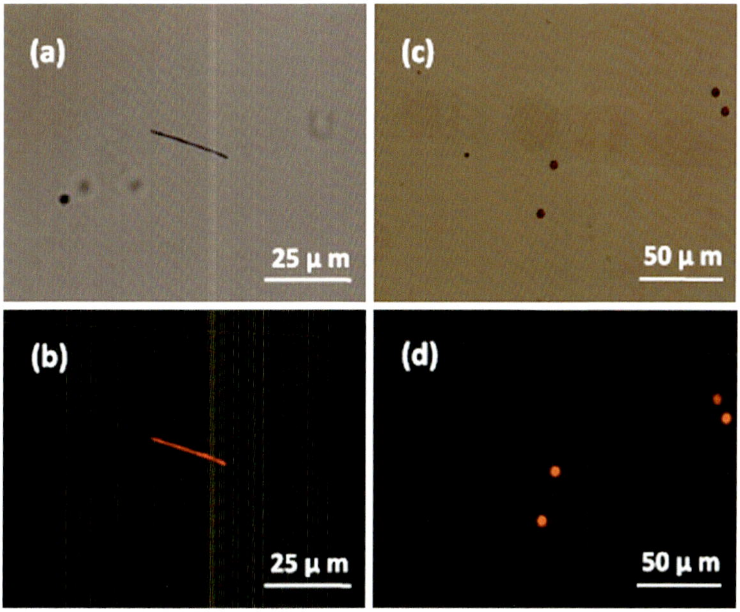

Fig. 4.17 Bright-field images of (**a**) Ni nanowire and (**c**) beads functionalized with CD31 antibody, which is an Alex Fluor 594 (*red*) labeled secondary antibody. Fluorescent images of functionalized (**b**) Ni nanowire and (**d**) beads. Reprinted with permission from [99] by IOP Publishing Ltd

immunology, cell and molecular biology, and clinical research. For cell separation, Gao et al. [99] demonstrated that Ni nanowires functionalized with antibodies can attach to specific cells in a mixed cell population via antibody–antigen interactions and then be separated by an external magnetic field. Figure 4.17 shows images of nanowires and beads functionalized with CD31 antibody. Uniform red fluorescence indicated the immobilization of CD31 antibodies on the surfaces of nanowires and beads.

Further efforts have involved conducting polymers utilized as electrochemical biosensors. These polymeric materials are advantageous because of their electronic conductivity, environmental stability, and easy processing. Hernández et al. [100] reported on the synthesis of biologically functionalized polypyrrole nanowire arrays prepared by a template method. Subsequent studies by Ramanathan et al. [43] introduced a new strategy where polymer nanowires were electrodeposited within a channel between two electrodes on the surface of a silicon wafer.

Complex structures can be realized in this approach. Intricate materials with varied chemical affinities toward different functional groups can be formed in multisegmented single nanowires through electrodeposition. Each of these segments can be functionalized to facilitate positive and negative interactions based on molecular, electrostatic, magnetic, or capillary forces [101]. The functionalization of each segment depends on the differential reactivity of the materials – for example, alkyl thiols bind strongly to gold, histidine binds to nickel, and isocyanides bind to

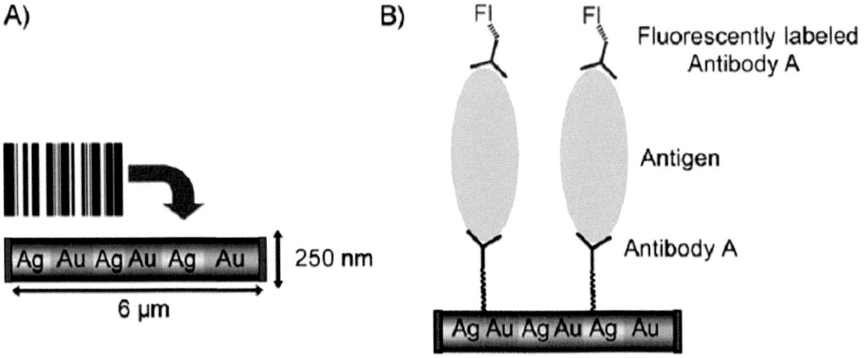

Fig. 4.18 (**a**) Similarity between a conventional barcode and Ag–Au multisegment nanowire. Ni segments (50 nm) are deposited at both ends (not drawn to scale). (**b**) Schematic of the sandwich immunoassay performed on a nanowire. Reprinted with permission from [104]. Copyright Wiley-VCH Verlag GmbH & Co. KGaA

platinum [102]. Functionalized multisegment nanowires with antibodies or oligonucleotides are extremely useful for sandwich immunoassays of proteins or for the detection of DNA hybridization [103]. Tok et al. [104] have demonstrated the feasibility of using antibody-coated multisegmented metallic nanowires for rapid and sensitive single and multiplex immunoassays as shown in Fig. 4.18. The incorporation of a ferromagnetic Ni segments, deposited at both ends on the wire, enables the nanoparticles to be manipulated with magnetic fields. Fluorescence image processing illuminates the degree of binding between the antibody-coated nanowires and a fluorophore-tagged antigen target.

4.9 Nanostructure Arrays on Substrates

Many kinds of metal and polymer nanowires have been synthesized by template-assisted electrodeposition techniques. However, the handling of nanostructures in porous membranes can be very difficult owing to the fragility of membranes and therefore additional processes are needed to make resilient nanodevices. Another approach is to make nanostructures on stable substrates. In the case of alumina templates, this involves the retention of the aluminum metal base after the anodization process. This, however, also has several disadvantages including unstable deposition conditions, and the use of a bulk Al substrates precludes most photonic and electronic applications [105].

To solve these problems, direct fabrication of nanostructures on a thick, solid substrate, which does not deform easily, is particularly important. One of the methods is the usage of alumina membrane on a substrate such as silicon, glass, MgO, or ITO [106–109]. This approach needs first coating of thin Au, Ti, or Pt adhesion layers used as an electrical contact for both anodization and electrodeposition, and as a stop point for anodization [105, 106, 108]. Al thin films can be deposited by several techniques, for example, sputtering, thermal evaporation, or electron beam evaporation. As discussed above, porous membranes can be prepared by anodization of Al. When an Al thin layer is completely anodized, a visible change in color is observed and the current drops quickly. The barrier layer that remains under the porous layer can then be removed, by either immersing the sample in H_3PO_4 solution for several minutes [109, 110], by applying a reversed-bias voltage after complete anodization [106], or by reactive ion etching (RIE) [111]. Several kinds of nanowires such as Ni [105], Au [106], Pd [112], CoPt [108], and CdS [113] were produced on a desired substrate. Jung et al. [107] prepared long-range-ordered AAO templates on a Ti-coated silicon wafer and used this membrane to make Ni nanodots. Figure 4.19 shows SEM images of a nanoporous alumina template on Ti-coated silicon wafer. The templates used for synthesis of nickel nanodots are presented in Fig. 4.19c. The average diameter and the interparticle spacing of 81 and 108 nm, as shown in Fig. 4.20, correspond well to the SEM images in Fig. 4.19.

4.10 Patterning of Nanowires

A large amount of research in nanoscience involves the development of synthetic techniques and the discovery of unique physical and chemical properties of nanosized materials. These efforts move to patterning of nanomaterials for miniaturization of devices such as sensors, displays, data storage, electronic, and biomedical devices. Small features can result in increasingly higher speeds and decreasing energy consumption per computing function [114, 115]. In this section, we are introducing the lithography techniques combined with electrodeposition methods.

For the past several decades, photolithography has been the primary route for semiconductor device fabrication due to its simplicity and high throughput [114]. Photons transfer patterns of a photomask to a photosensitive thin film (photoresist) on a substrate. The solubility of the photoresist is altered during an exposure of photons and the exposed surface is then dissolved in a developer solution [114]. When a template is used as a substrate for lithography, the templates can be patterned and make patterned nanostructure arrays through an electrodeposition. Jee et al. [116] reported patterned alumina templates using a photolithography. Initially they prepared alumina templates by a well-known anodization process in oxalic acid. After the formation of patterns on a template through lithography, patterned alumina templates appeared by wet chemical etching. Patterned alumina templates can be used for electrochemical nanowire growth. Li et al. [117] reported the patterned metal nanowire arrays using this method. Selectively opened pore channels were produced on the template during UV exposure and developing processes, and the opened pores allowed the nanowire growth in the channels by electrodeposition after coating of the open pore surface with conducting metal. The patterned metal nanowires with defined structures were observed in the desired location after removing of templates as shown in Fig. 4.21.

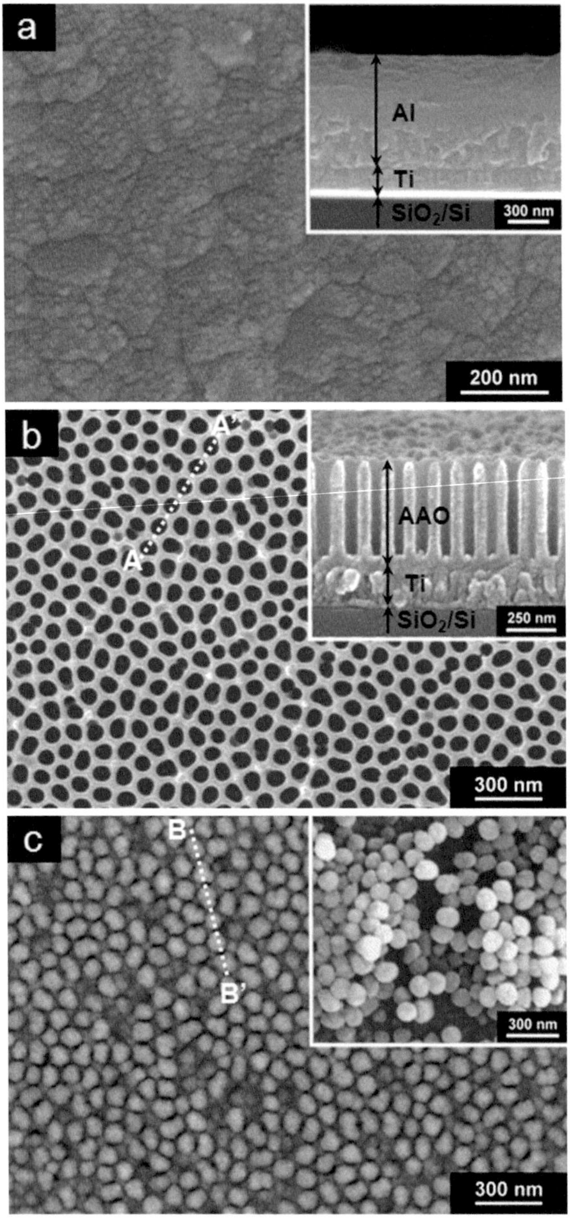

Fig. 4.19 Top surface and cross section (*inset*) view of (**a**) Al and Ti layers on the Si wafer and (**b**) thin AAO template on the substrate after pore widening. (**c**) Top surface view of synthesized Ni nanodot arrays after removal of the template. *Inset* is the collected Ni nanodots. Reprinted with permission from [107] by The Korean Chemical Society

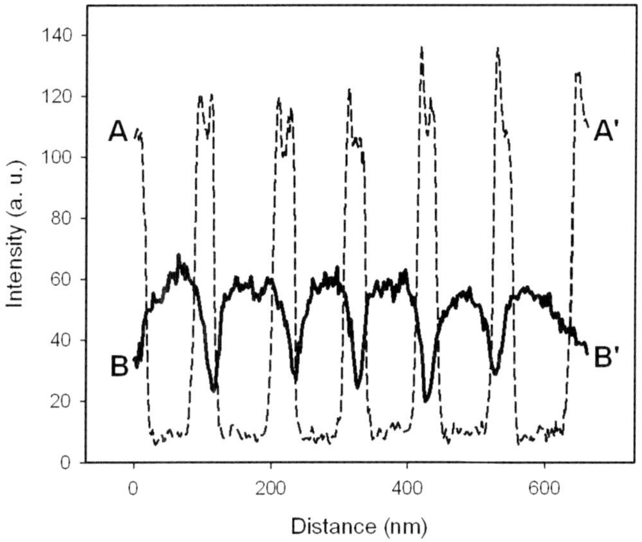

Fig. 4.20 Intensity profiles of the AAO template (*dashed line*) and the Ni nanodot arrays on the substrate (*solid line*), which are corresponding to the *dotted lines* as indicated in Fig. 4.19**b**, **c**, respectively. Reprinted with permission from [107] by The Korean Chemical Society

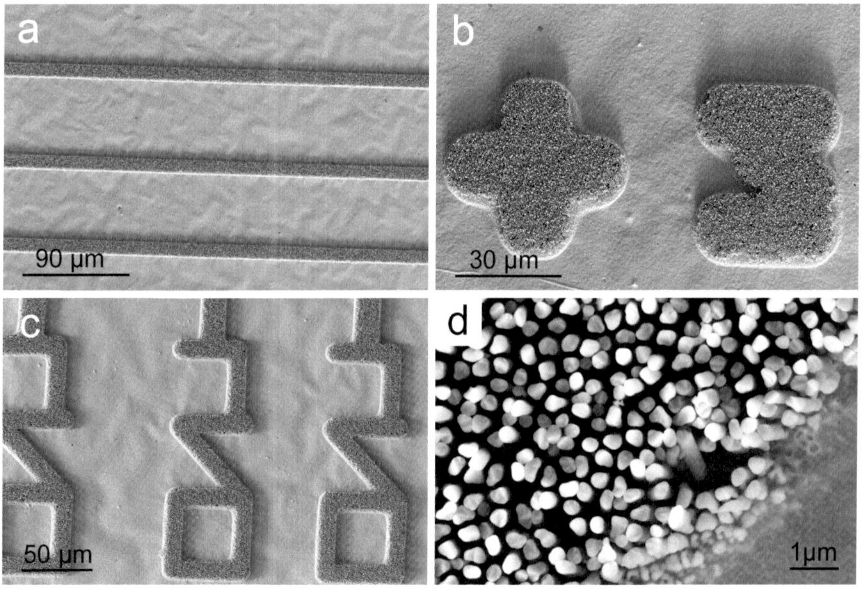

Fig. 4.21 FESEM images of (**a–c**) patterned Au nanowire with different features and different sizes and (**d**) higher magnification image indicating Au nanowire arrays with about 200 nm diameters. Reprinted with permission from [117] by American Chemical Society (ACS)

Electron beam lithography (e-beam lithography) is a sophisticated nanofabrication tool based on a highly focused electron beam for very high-resolution direct-write patterning on electron beam-sensitive resists [118]. e-Beam lithography systems can regularly achieve patterning of sub-micron features and can easily produce designed nanostructures on a substrate for applications such as field emitters [119], transistors [120], and sensor devices [121]. Patterning lines with small width is a key objective of nanofabrication. In several cases, however, both small widths and large aspect ratio were required [122]. In general, for fabrication of patterned arrays, most people used a lift-off process in which the metal is vapor deposited through a line mask. The width and length of metal are determined lithographically, and the thickness of the metal is controlled during deposition process [123]. To increase the length of metal, the use of metal lift-off techniques is limited owing to the restricted length of polymer resist. Xu et al. [123] prepared a highly anisotropic metal nanostructures through the combination of e-beam lithography and electrodeposition. The procedure of electrodeposition in patterned PMMA template is schematically summarized in Fig. 4.22. Mushroom-shaped nanostructures were formed when Ni nanowires were overgrown on top of the PMMA template. Duvail et al. [124] employed this method for fabrication of a single multilayered nanopillar with large aspect ratio. Burek and Greer [125] reported vertically oriented gold and copper nanopillars based on e-beam lithography and subsequent electroplating into the prescribed template for mechanical testing of specimens with nanometer dimensions. Simple and mushroom-shaped Cu nanostructures were observed by using focused ion beam (FIB) as shown in Fig. 4.23. Like in photolithography, e-beam lithography is able to prepare a pre-patterned template for creation of patterned nanowire structures or devices. When templates were utilized as a substrate during lithography, it is easy to make vertically aligned nanomaterials. The uniformity and diameter of nanomaterials highly depend on the pore channels [126].

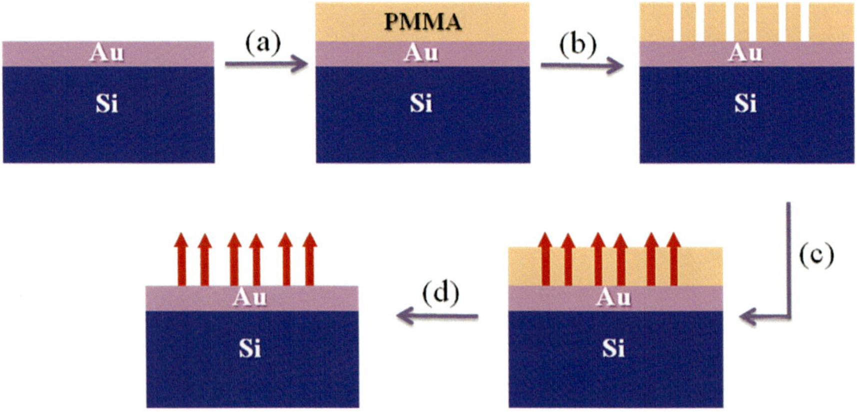

Fig. 4.22 Schematic diagram of nanowires in PMMA template on silicon wafer. (**a**) PMMA coating on the substrate; (**b**) patterning of PMMA by e-beam lithography; (**c**) electrodeposition of Ni nanowire; and (**d**) PMMA removal

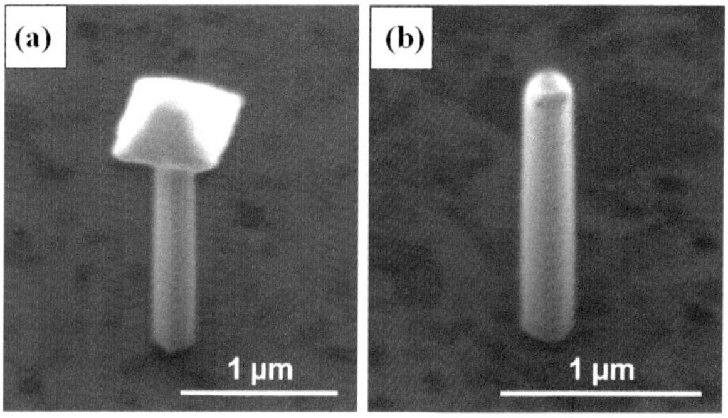

Fig. 4.23 FIB images of (**a**) overplated mushroom-shaped Cu nanopillar and (**b**) simple Cu nanopillar with 250 nm diameters. Reprinted with permission from [125] by American Chemical Society (ACS)

A third method known as nanosphere lithography, an alternative to e-beam lithography, was reported for the formation of periodic nanosized metal dots [127, 128]. A self-organized layer of colloidal spheres is used as a mask for a lithographic step-like illumination, deposition, or etching. The use of self-assembled nanospheres has advantages over more conventional lithography, in that its cost is low and it is easy to control the patterned morphology. However, the size of holes in the mask is determined by the size of the colloidal particles that form the mask. Highly ordered sphere arrays can be used to create interstitial triangular structures [129], metallic rings [130], nanopillars [131], and multilayers with modified topography [132]. Vossen et al. [133] modified the hole sizes of colloidal masks to prepare a thin layer of silica by a simple wet chemical method. Hexagonal-patterned Si nanowires [127], ZnO nanorods [128], or carbon nanotubes [134] have been produced by this approach.

Acknowledgments The support from the National Science Foundation (NSF-0403673) is gratefully acknowledged. And special thanks to Dr. G. S. Chaubey, J. Zhang, Y. Yao, D. Mohanty, and J. Tripathy for helpful suggestions.

References

1. H. Pan, B. Liu, J. Yi, C. Poh, S. Lim, J. Ding, Y. Feng, C.H.A. Huan, J. Lin, J. Phys. Chem. B **109**, 3094 (2005)
2. Y. Ren, Q.F. Liu, S.L. Li, J.B. Wang, X.H. Han, J. Magn. Magn. Mater. **321**, 226 (2009)
3. J. Qin, J. Nogués, M. Mikhaylova, A. Roig, J.S. Muñoz, M. Muhammed, Chem. Mater. **17**, 1829 (2005)
4. X.-F. Han, S. Shamaila, R. Sharif, J.-Y. Chen, H.-R. Liu, D.-P. Liu, Adv. Mater. **21**, 1 (2009)
5. J.M. Baik, M. Schierhorn, M. Moskovits, J. Phys. Chem. C **112**, 2252 (2008)
6. X.H. Huang, G.H. Li, X.C. Dou, L. Li, J. Appl. Phys. C **105**, 084306 (2009)
7. M. Zheng, L. Menon, H. Zeng, Y. Liu, S. Bandyopadhyay, R.D. Kirby, D.J. Sellmyer, Phys. Rev. B **62**, 12282 (2000)
8. D.J. Sellmyer, M. Zheng, R. Skomski, J. Phys.: Condens. Mater. **13**, R443 (2001)
9. F. Li, T. Wang, L. Fen, J. Sun, J. Phys.: Condens. Mater. **16**, 8053 (2004)
10. X. Huang, L. Li, X. Luo, X. Zhu, G. Li, J. Phys. Chem. C **112**, 1468 (2008)
11. J.S. Wright, W. Lim, D.P. Norton, S.J. Pearton, F. Ren, J.L. Johnson, A. Ural, Semicond. Sci. Technol. **25**, 024002 (2010)

12. M. Lai, J.-H. Lim, S. Mubeen, Y. Rheem, A. Mulchandani, M.A. Deshusses, N.V. Myung, Nanotechnology **20**, 185602 (2009)
13. S. Mathur, S. Barth, H. Shen, J.-C. Pyun, U. Werner, Small **1**, 173 (2005)
14. Y. Liu, M. Liu, Adv. Funct. Mater. **15**, 57 (2005)
15. Z.R. Dai, J.L. Gole, J.D. Stout, Z.L. Wang, J. Phys. Chem. B **106**, 1274 (2002)
16. M. Zheng, G. Li, X. Zhang, S. Huang, Y. Lei, L. Zhang, Chem. Mater. **13**, 3859 (2001)
17. X.Y. Kong, Z.L. Wang, Appl. Phys. Lett. **84**, 975 (2004)
18. L. Li, S. Pan, X. Dou, Y. Zhu, X. Huang, Y. Yang, G. Li, L. Zhang, J. Phys. Chem. C **111**, 7288 (2007)
19. C. Jagadish, S.J. Pearton, *ZnO Bulk, Thin Films and Nanostructures*. (Elsevier, Maryland Heights, MA, 2006)
20. D. Gong, C.A. Grimes, O.K. Varghese, W.C. Hu, R.S. Singh, Z. Chen, E.C. Dickey, J. Mater. Res. **16**, 3331 (2001)
21. M. Paulose, K. Shankar, S. Yoriya, H.E. Prakasam, O.K. Varghese, G.K. Mor, T.A. Latempa, A. Fitzgerald, C.A. Grimes, J. Phys. Chem. B **110**, 16179 (2006)
22. H.E. Prakasam, K. Shankar, M. Paulose, O.K. Varghese, C.A. Grimes, J. Phys. Chem. C **111**, 7235 (2007)
23. S. Banerjee, S.K. Mohapatra, P.P. Das, M. Misra, Chem. Mater. **20**, 6784 (2008)
24. S. Berger, H. Tsuchiya, A. Ghicov, P. Schmuki, Appl. Phys. Lett. **88**, 203119 (2006)
25. X. Qiu, J.Y. Howe, M.B. Cardoso, O. Polat, W.T. Heller, M.P. Paranthaman, Nanotechnology **20**, 455601 (2009)
26. A. Majumdar, Science **303**, 777 (2004)
27. N.B. Duarte, G.D. Mahan, S. Tadigadapa, Nano Lett. **9**, 617 (2009)
28. L.D. Hicks, M.S. Dresselhaus, Phys. Rev. B **47**, 16631 (1993)
29. A.L. Prieto, M. Martin-González, J. Keyani, R. Gronsky, T. Sands, A.M. Stacy, J. Am. Chem. Soc. **125**, 2388 (2003)
30. A.L. Prieto, M.S. Sander, M.S. Martin-González, R. Gronsky, T. Sands, A.M. Stacy, J. Am. Chem. Soc. **123**, 7160 (2001)
31. C.R. Martin, Science **266**, 1961 (1994)
32. L. Li, Y. Zhang, G. Li, L. Zhang, Chem. Phys. Lett. **378**, 244 (2003)
33. L. Li, Y.W. Yang, X.H. Huang, G.H. Li, R. Ang, L.D. Zhang, Appl. Phys. Lett. **88**, 103119 (2006)
34. X.C. Dou, Y.G. Zhu, X.H. Huang, L. Li, G.H. Li, J. Phys. Chem. B **110**, 21572 (2006)
35. Y. Zhang, L. Li, G.H. Li, Nanotechnology **16**, 2096 (2005)
36. F.H. Xue, G.T. Fei, B. Wu, P. Cui, L.D. Zhang, J. Am. Chem. Soc. **127**, 15348 (2005)
37. M.S. Sander, R. Gronsky, T. Sands, A.M. Stacy, Chem. Mater. **15**, 335 (2003)
38. Y. Berdichevsky, Y.-H. Lo, Adv. Mater. **18**, 122 (2006)
39. A. Kros, R.J.M. Nolte, N.A.J.M. Sommerdijk, Adv. Mater. **14**, 1779 (2002)
40. J.C.W. Chien, J.B. Schlenoff, Nature **311**, 362 (1984)
41. F. Roussel, R. Chan-Yu-King, J.-M. Buisine, Eur. Phys. J. E. **11**, 293 (2003)
42. K. Ramanathan, M.A. Bangar, M. Yun, W. Chen, A. Mulchandani, N.V. Myung, Nano Lett. **4**, 1237 (2004)
43. K. Ramanathan, M.A. Bangar, M. Yun, W. Chen, N.V. Myung, A. Mulchandani, J. Am. Chem. Soc. **127**, 496 (2004)
44. S.I. Cho, S.B. Lee, Acc. Chem. Res. **41**, 699 (2008)
45. P.N. Bartlett, P.R. Birkin, M.A. Ghanem, C.-S. Toh, J. Mater. Chem. **11**, 849 (2001)
46. L.B. Groenendaal, F. Jonas, D. Freitag, H. Pielartzik, J.R. Reynolds, Adv. Mater. **12**, 481 (2000)
47. F. Yue, T. Ngin, G. Hailin, Sens. Actuators B **32**, 33 (1996)
48. M. Yun, N.V. Myung, R.P. Vasquez, C. Lee, E. Menke, R.M. Penner, Nano Lett. **4**, 419 (2004)
49. M. Lahav, E.A. Weiss, Q. Xu, G.M. Whitesides, Nano Lett. **6**, 2144 (2006)
50. D.H. Park, B.H. Kim, M.K. Jang, K.Y. Bae, S.J. Lee, J. Joo, Synthetic Met. **153**, 341 (2005)

51. J. Joo, D. Hyuk Park, M.-Y. Jeong, Y.B. Lee, H.S. Kim, W.J. Choi, Q.-H. Park, H.-J. Kim, D.-C. Kim, J. Kim, Adv. Mater. **19**, 2824 (2007)
52. M.V. Kamalakar, A.K. Raychaudhuri, Adv. Mater. **20**, 149 (2008)
53. T.A. Crowley, B. Daly, M.A. Morris, D. Erts, O. Kazakova, J.J. Boland, B. Wu, J.D. Holmes, J. Mater. Chem. **15**, 2408 (2005)
54. J. Bachmann, J. Jing, M. Knez, S. Barth, H. Shen, S. Mathur, U. Gősele, K. Nielsch, J. Am. Chem. Soc. **129**, 9554 (2007)
55. D. Lee, R.E. Cohen, M.F. Rubner, Langmuir **23**, 123 (2007)
56. Z. Hua, S. Yang, H. Huang, L. Lv, M. Lu, B. Gu, Y. Du, Nanotechnology **17**, 5106 (2006)
57. C. Barrett, D. Iacopino, D. O'Carroll, G.D. Marzi, D.A. Tanner, A.J. Quinn, G. Redmond, Chem. Mater. **19**, 338 (2007)
58. M. Mertig, R. Kirsch, W. Pompe, Appl. Phys. A: Mater. Sci. Process. **66**, S723 (1998)
59. C. Mu, Y. Yu, R. Wang, K. Wu, D. Xu, G. Guo, Adv. Mater. **16**, 550 (2004)
60. K.K. Nanda, F.E. Kruis, H. Fissan, Nano Lett. **1**, 605 (2001)
61. W. Lee, R. Scholz, K. Nielsch, U. Gősele, Angew. Chem. Int. Ed. **44**, 6050 (2005)
62. Y. Fukunaka, M. Motoyama, Y. Konishi, R. Ishii, Electrochem. Solid-State Lett. **9**, C62 (2006)
63. H. Cao, L. Wang, Y. Qui, Q. Wu, G. Wang, L. Zhang, X. Liu, Chem.Phys.Chem. **7**, 1500 (2006)
64. J. Escrig, M. Daub, P. Landeros, K. Nielsch, D. Altbir, Nanotechnology **18**, 445706 (2007)
65. T.N. Narayanan, M.M. Shaijumon, P.M. Ajayan, M.R. Anantharaman, Nanoscale Res. Lett. **5**, 164 (2010)
66. J.-H. Lim, J.B. Wiley, Mater. Res. Soc. Symp. Proc. **1142**, 1142JJ0531 (2009)
67. C.C. Berry, A.S.G. Curtis, J. Phys. D **36**, R198 (2003)
68. D.H. Park, Y.B. Lee, M.Y. Cho, B.H. Kim, S.H. Lee, Y.K. Hong, J. Joo, H.C. Cheong, S.R. Lee, Appl. Phys. Lett. **90**, 093122 (2007)
69. S. Si, C. Li, X. Wang, Q. Peng, Y. Li, Sens. Actuators B **119**, 52 (2006)
70. H. Zeng, S. Sun, J. Li, Z.L. Wang, J.P. Liu, Appl. Phys. Lett. **85**, 792 (2004)
71. C. Jin, P.C. Searson, Appl. Phys. Lett. **81**, 4437 (2002)
72. Y.-Y. Yu, S.-S. Chang, C.-L. Lee, C.R.C. Wang, J. Phys. Chem. B **101**, 6661 (1997)
73. S.R. Nicewarner-Peña, R.G. Freeman, B.D. Reiss, L. He, D.J. Peña, I.D. Walton, R. Cromer, C.D. Keating, M.J. Natan, Science **294**, 137 (2001)
74. B. Nikoobakht, Z.L.Wang, M.A. El-Sayed, J. Phys. Chem. B **104**, 8635 (2000)
75. L. Liu, W. Lee, Z. Huang, R. Scholz, U. Gősele, Nanotechnology **19**, 335604 (2008)
76. S.H. Yoo, S. Park, Adv. Mater. **19**, 1612 (2007)
77. D. Van-Noort, C.F. Mandenius, Biosens. Bioelectron. **15**, 203 (2000)
78. H.-O. Lee, E.-M. Kim, H. Yu, J.-S. Jung, W.-S. Chae, Nanotechnology **20**, 325604 (2009)
79. H.-M. Bok, K.L. Shuford, S. Kim, S.K. Kim, S. Park, Nano Lett. **8**, 2265 (2008)
80. J. Weissmuller, R.N. Viswanath, D. Kramer, P. Zimmer, R. Wurschum, H. Gleiter, Science **300**, 312 (2003)
81. V. Zielasek, B. Jurgens, C. Schulz, J. Biener, M.M. Biener, A.V. Hamza, M. Baumer, Angew. Chem. Int. Ed. **45**, 8241 (2006)
82. Z. Liu, P.C. Searson, J. Phys. Chem. B **110**, 4318 (2006)
83. S. Chen, Z. Fan, D.L. Carroll, J. Phys. Chem. B **106**, 10777 (2002)
84. L. Gou, S.E. Hunyadi, T. Li, J. Phys. Chem. B **109**, 1385 (2005)
85. M.A. El-Sayed, Acc. Chem. Res. **34**, 257 (2001)
86. L. Manna, D.J. Milliron, A. Meisel, E.C. Scher, A.P. Alivisatos, Nat. Mater. **2**, 382 (2003)
87. T.S. Ahmadi, Z.L. Wang, T.C. Green, A. Henglein, M.A. El-Sayed, Science **272**, 1924 (1996)
88. R. Jin, Y.W. Cao, C.A. Mirkin, K.L. Kelly, G.C. Schatz, J.G. Zheng, Science **294**, 1901 (2001)
89. Y. Sun, Y. Xia, Science **298**, 2176 (2002)
90. M. Quinten, A. Leitner, J.R. Krenn, F.R. Aussenegg, Opt. Lett. **23**, 1331 (1998)

91. S.A. Maier, M.L. Brongersma, P.G. Kik, S. Meltzer, A.A.G. Requicha, H.A. Atwater, Adv. Mater. **13**, 1501 (2001)
92. L. Liu, W. Lee, R. Scholz, E. Pippel, U. Gösele, Angew. Chem. Int. Ed. **47**, 7004 (2008)
93. M.-C. Daniel, D. Astruc, Chem. Rev. **104**, 293 (2004)
94. F. Li, J. He, W.L. Zhou, J.B. Wiley, J. Am. Chem. Soc. **125**, 16166 (2003)
95. T. Gao, G. Meng, J. Zhang, S. Sun, L. Zhang, Appl. Phys. A **74**, 403 (2002)
96. H. Masuda, H. Asoh, M. Watanabe, K. Nishio, M. Nakao, Toshiaki, Adv. Mater. **12**, 189 (2001)
97. Y.B. Li, M.J. Zhang, L. Ma, Appl. Phys. Lett. **91**, 073109 (2007)
98. W. Lee, R. Ji, U. Gösele, K. Nielsch, Nat. Mater. **5**, 741 (2006)
99. N. Gao, H. Wang, E.-H. Yang, Nanotechnology **21**, 105107 (2010)
100. R.M. Hernández, L. Richter, S. Semancik, S. Stranick, T.E. Mallouk, Chem. Mater. **16**, 3431 (2004)
101. Z. Gu, H. Ye, D.H. Gracias, JOM **57**, 60 (2005)
102. J. Wang, Chem. Phys. Chem. **10**, 1748 (2009)
103. S.R. Nicewarner-Peña, R.G. Freeman, B.D. Reiss, L. He, D.J. Peña, I.D. Walton, R. Cromer, C.D. Keating, M.J. Natan, Science **294**, 5540 (2004)
104. J.B.-H. Tok, F.Y.S. Chuang, M.C. Kao, K.A. Rose, S.S. Pannu, M.Y. Sha, G. Chakarova, S.G. Penn, G.M. Dougherty, Angew. Chem. Int. Ed. **45**, 6900 (2006)
105. S. Wen, J.A. Szpunar, Micro Nano Lett. **1**, 89 (2006)
106. M. Tian, S. Xu, J. Wang, N. Kumar, E. Wertz, Q. Li, P.M. Campbell, M.H.W. Chan, T.E. Mallouk, Nano Lett. **5**, 697 (2005)
107. J.-S. Jung, E.-M. Kim, W.-S. Chae, L.M. Malkinski, J.-H. Lim, C. O'Connor, J.-H. Jun, Bull. Korean Chem. Soc. **29**, 2169 (2008)
108. N. Yasui, A. Imada, T. Den, Appl. Phys. Lett. **83**, 3347 (2003)
109. T.R.B. Foong, A. Sellinger, X. Hu, ACS Nano **2**, 2250 (2008)
110. A.I. Gapin, X.R. Ye, J.F. Aubuchon, L.H. Chen, Y.J. Tang, S. Jin, J. Appl. Phys. **99**, 08G902 (2006)
111. J. Zou, X. Qi, L. Ten, B.J.H. Stadler, Appl. Phys. Lett. **89**, 093106 (2006)
112. K. Kim, M. Kim, S.M. Cho, Mat. Chem. Phys. **96**, 278 (2006)
113. Y. Yang, H. Chen, Y. Mei, J. Chen, X. Wu, X. Bao, Solid State Commun. **123**, 279 (2002)
114. H.M. Saavedra, T.J. Mullen, P. Zhang, D.C. Dewey, S.A. Claridge, P.S. Weiss, Rep. Prog. Phys. **73**, 036501 (2010)
115. E.J. Bae, W.B. Choi, K.S. Jeong, J.U. Chu, G.-S. Park, S. Song, I.K. Yoo, Adv. Mater. **12**, 277 (2002)
116. S.E. Jee, P.S. Lee, B.-J. Yoon, S.-H. Jeong, K.-H. Lee, Chem. Mater. **17**, 4049 (2005)
117. F. Li, M. Zhu, C. Liu, W.L. Zhou, J.B. Wiley, J. Am. Chem. Soc. **128**, 13342 (2006)
118. H. Duan, J. Zhao, Y. Zhang, E. Xie, L. Han, Nanotechnology **20**, 135306 (2009)
119. S.E. Huq, L. Chen, P.D. Prewett, Microelectron. Eng. **27**, 95 (1995)
120. T. Hanrath, B.A. Korgel, J. Nanoeng. Nanosyst. **218**, 25 (2005)
121. K.J. Jeon, J.M. Lee, E. Lee, W. Lee, Nanotechnology **20**, 135502 (2009)
122. T.N. Lo, Y.T. Chen, C.W. Chiu, C.J. Liu, S.R. Wu, I.K. Lin, C.I. Su, W.D. Chang, Y. Hwu, B.Y. Shew, C.C. Chiang, J.H. Je, G. Margaritondo, J. Phys. D: Appl. Phys. **40**, 3172 (2007)
123. W. Xu, J. Wong, C.C. Cheng, R. Johnson, A. Scherer, J. Vac. Sci. Technol. B **13**, 2372 (1995)
124. J.L. Duvail, S. Dubois, L. Piraux, A. Vaurès, A. Fert, D. Adam, M. Champagne, F. Rousseaux, D. Decanini, J. Appl. Phys. **84**, 6359 (1998)
125. M.J. Burek, J.R Greer, Nano Lett. **10**, 69 (2010)
126. J.-H. Lim, W.L. Zhou, J.B. Wiley, unpublished work (2010)
127. B. Fuhrmann, H.S. Leipner, H. Höche, Nano Lett. **5**, 2524 (2005)
128. X. Wang, C.J. Summers, Z.L. Wang, Nano Lett. **4**, 423 (2004)
129. M. Winzer, M. Kleiber, N. Dix, R. Wiesendanger, Appl. Phys. A **63**, 617 (1996)
130. J. Boneberg, F. Burmeister, C. Schäfle, P. Leiderer, Langmuir **13**, 7080 (1997)
131. S.M. Weeks, F.Y. Oring, W.A. Murray, Langmuir **20**, 11208 (2004)

132. M. Albrecht, G. Hu, I.L. Guhr, T.C. Ulbrich, J. Boneberg, P. Leiderer, G. Schatz, Nat. Mater. **4**, 203 (2005)
133. D.L.J. Vossen, J.J. Penninkhof, A. Blaaderen, Langmuir **24**, 5967 (2008)
134. K.H. Park, S. Lee, K.H. Koh, R. Lacerda, K.B.K. Teo, W.I. Milne, J. Appl. Phys. **97**, 024311 (2005)

Chapter 5
Three-Dimensional Micro/Nanomaterials Generated by Fiber-Drawing Nanomanufacturing

Zeyu Ma, Yan Hong, Shujiang Ding, Minghui Zhang, Mainul Hossain, and Ming Su

5.1 Introduction

Fiber drawing nanomanufacturing (FDN) is a new high-yield method to make three-dimensional (3D) ordered micro- and nanostructured materials. The main equipment used in the FDN process is a fiber draw tower, which is similar to those used in industry to make optical fibers for telecommunications. In FDN, glassy materials such as silicate glasses and polymers either in rod form or in tube form with appropriate filling materials can be drawn into fibrous building blocks (FBBs), which can be assembled into ordered 3D structures or ordered bundles of fibers for the next drawing. By repeating the same drawing process several times, the diameters of FBBs can be reduced from centimeters stepwise down to sub-micrometers, while preserving the ordered structures of FBBs. Depending on the properties of glassy materials, the filling materials of choice can be semiconductor materials, glasses, metallic wires, or organic materials. This chapter briefly summarizes several aspects of the FDN technique including material selection, draw process, size control, and assembly and also discusses various examples of FDN.

5.2 Fiber Draw Tower

Compared with industrial draw tower used to make kilometer-long fibers, a major difference of draw tower used in the FDN process is that long fibers are cut into

M. Su (✉)
NanoScience Technology Center, University of Central Florida, Orlando, FL 32826, USA; Department of Mechanical, Materials and Aerospace Engineering, University of Central Florida, Orlando, FL 32826, USA; School of Electrical Engineering and Computer Science, University of Central Florida, Orlando, FL 32826, USA
e-mail: mingsu@mail.ucf.edu

W.L. Zhou, Z.L. Wang (eds.), *Three-Dimensional Nanoarchitectures*, DOI 10.1007/978-1-4419-9822-4_5,

Fig. 5.1 Glass draw tower

short and equal length for assembly. Figure 5.1 shows the major components of laboratory scale draw tower, which include structural frame, feeding motor, vacuum pump, furnace with designed heating zone, rubber tractor, laser micrometer, automatic carbide cutter, and computer. A preform is first formed using matching glassy tubing materials and filling materials and moved through the furnace at a controlled speed. The vacuum pump is used during the drawing process in order to avoid possible oxidation of filling materials and to make intimate contact between the glassy tube and the filling materials. The length and the diameter of preform are only limited by the spacing from ceiling to furnace entrance and the diameter of the furnace core. The feeding speed and the drawing speed can be controlled to adjust the diameter of fibers based on material conservation in fiber drawing. Equation (5.1) shows the relation between the geometrical parameters of preform and those of fibers. By changing the drawing conditions, fibers can be produced with diameters ranging from tens of micrometers to tens of millimeters. The length of FBBs can be varied by controlling the interval between consecutive cuts:

$$D = F \times \frac{\text{preform\_cross sectional\_area}}{\text{fiber\_cross sectional\_area}} \tag{5.1}$$

5.3 Materials Selections

Glassy materials including silicate glasses and some polymers do not have fixed melting temperatures and can be deformed at high temperatures (softening temperature). Successful fiber drawing could be achieved when the viscosity of softened materials is in a certain range. Figure 5.2 shows that silicate glasses can be drawn into fibers when the viscosity is between $10^{3.5}$ and 10^7 P, where the flow rate of molten glass can be controlled. Once a desired glassy material is selected, a major concern is the selection of filling materials. There are several criteria for choosing the right filling materials: (1) filling materials should have similar melting temperature as the softening temperature of glassy materials; (2) the thermal expansion coefficients of the filling materials and the glassy materials, at the drawing temperature, should be close to each other to avoid breaking of glassy materials due to stress concentration; (3) there should be no chemical reaction between the glassy and filling materials at the drawing temperature; (4) the filling materials should have certain wettability to the glassy materials to avoid bubble formation; and (5) the filling materials should not have high vapor pressure at the drawing temperature in order to ensure the continuity of filling materials. Currently available materials include glass and poly(methyl methacrylate) (PMMA) that have a large range of melting temperatures from 300 to 900°C (glass) and 100 to 300°C (PMMA). Borosilicate glass (Pyrex) contains silica and boron oxide, which have low coefficients of thermal expansion. By changing the ratio of boron oxide to silica, the defect level of glassy materials can be changed, so that the etching rate in hydrogen fluoride can be adjusted, which allows fabrication of different glass micro/nanostructures. The filling materials include copper phosphor alloy wires, lead telluride (PbTe) semiconductor powder, glass rods/tubes, salts, and paraffins.

5.4 Drawing Process

Figure 5.3 shows the procedure of drawing and assembling processes. The preform is drawn into fibers of small diameter when moving through a heating furnace. The

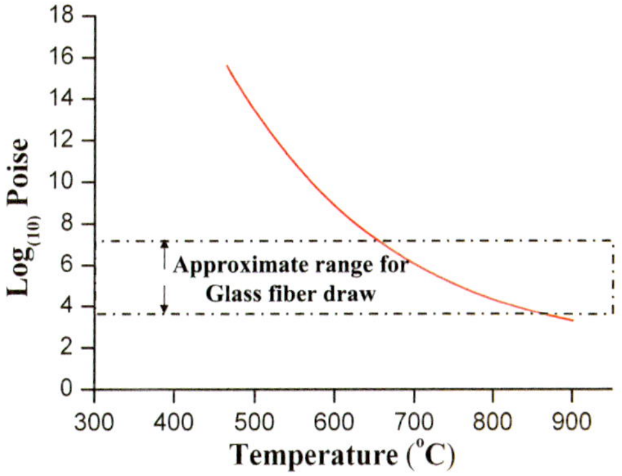

Fig. 5.2 Temperature-dependent glass viscosities

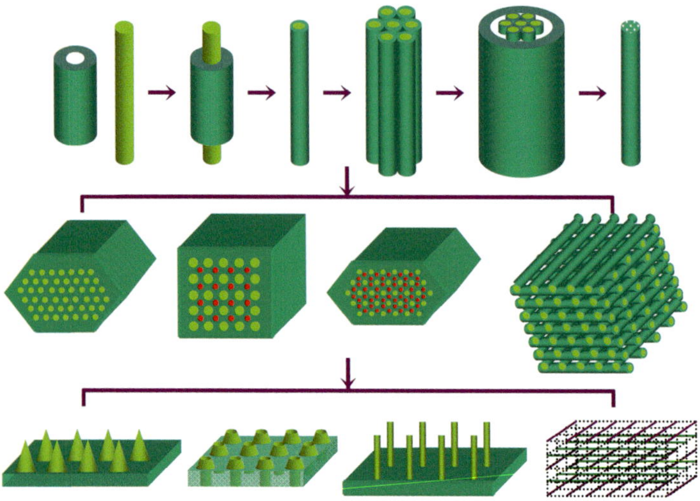

Fig. 5.3 Three-dimensional micro/nanostructure made by fiber drawing technique

fibers from the first drawing process are cut into short pieces with equal length as fibrous building blocks (FBBs). The FBBs can be assembled to form 3D structures with controlled morphologies or stacked to form hexagonal bundle for the next drawing. By repeating the same draw–cut–stack cycle, the outer and inner diameters, as well as the wall thickness of the glassy tube, can be decreased from centimeters to micrometers, while still preserving the ordered arrangement of fibers. When the required sizes are achieved after several drawing cycles, the fibers are stacked together and annealed around the lower limit of the softening temperature of glass to make a solid rod for further treatment. In a typical drawing process, using glass as vehicle tube material, the diameter of preform is 20 mm. The feed and draw speeds are in millimeters per minute and meters per minute range, respectively, depending on the desired diameter of the fibers. The diameter of fiber is controlled by combining three factors: drawing temperature, feed speed, and draw speed. In contrast to the Taylor wire process that has been practiced to make glass-encapsulated microwires for several decades, this method can achieve appropriate size reduction to draw fibers with nanometer diameter and assemble micro- or nanowires into ordered arrays.

5.5 Size Design

One-dimensional micro/nanowires with small diameters can be prepared by either bottom-up synthesis or top-down size reduction. In the bottom-up method, molecular or atomic precursors are assembled or deposited by physical or chemical vapor deposition, hydrothermal synthesis, and template-based deposition [1–6]. The top-down methods include electrospinning of polymer nanofibers and extruding wires through small diameter dies, thus achieving diameter control by adjusting operating parameters, such as voltage and viscosity (electrospinning), drawing temperature and speed (fiber drawing), and die diameter and pressure (extruding) [7–11]. In order to achieve desired functions, the positions, spacings, and orientations of micro/nanowires should be controlled with high precision. Such wires can be aligned in 2D planes by using electrical field, hydrodynamic flow, expanding balloons, or direct manipulation [12–16]. However, there is no controlled method currently available that can make ordered 3D micro/nanowire structures at high yield. Electrospinning makes randomly tangled nanowires, and extruding makes individual wires rather than an ordered network. Such a deficiency in fabrication techniques has consequently led to the lack of fundamental knowledge on the effect of 3D ordered structure. On the other hand, a variety of methods (vapor deposition or dielectrophoresis) can occasionally make random 3D networks of micro/nanowires, but these methods do not have sufficient control over the length and diameter of single wires, and the ordering (position, spacing, and orientation) of wires in 3D networks therefore cannot support a systematic study on the effects of structural ordering.

FDN can achieve controlled diameter and spacing of fibrous building blocks through simple geometric relation. Provided the thermophysical properties of the glass and the material are constant, the wire diameters after n times cycles can be derived by using the following equation:

$$d_n = \frac{(\Phi_{\mathrm{f}})^n r}{[2(R+T)]^{n-1}(r+t)} \tag{5.2}$$

where d_n is the diameter of wires after n times draws, Φ_{f} is the outer diameter of fiber after n times draws, r is the inner radius of the glass tube in the first draw, R is the inner radius of the glass tube used to encircle the fiber bundles obtained after the first draw, t is the thickness of the glass tube in the first draw, and T is the thickness of the glass tube used to encircle the fiber bundles obtained after the first draw. The inter-wire spacing during the fiber drawing process can be derived from

$$l_n = \frac{(\Phi_{\mathrm{f}})^n t}{[2(R+T)]^{n-1}(r+t)} \tag{5.3}$$

where l_n is the spacing between two wires after n times draws. Φ_{f}, R, T, r, and t have the same meaning as those in the previous equation. Assuming there is no material loss during each drawing with the scaling ratio of 20, 40, and 40 and starting with a core diameter of 1.6 mm, the diameters of core material become 80 μm after the first draw with a scaling ratio of 20. The fiber pieces are packed inside a glass tube that has an inner diameter of 15.6 mm and an outer diameter of 17.7 mm. The as-drawn fibers have diameters of 2 μm and 50 nm after the second and third draw, respectively, with an identical scaling ratio of 40. The same scaling ratio will be applied to wall thickness, i.e., the spacing between adjacent two microwires. The smallest feature that can be achieved in glass is about 20 nm [17]. Such limit is determined by cross-interface diffusion of glasses. Diffusion can be minimized by lowering fiber drawing temperature, but it requires large traction force to overcome the high viscosity.

5.6 Three-Dimensional Assembling

FDN can produce ultralong continuous micro/nanowires. The obtained fibers are used as FBBs to construct 1D, 2D, and 3D structures depending on the different functional applications (Fig. 5.3). In the first place, the FBBs are densely packed in a glassy tube. Then the bundle is annealed at a temperature at which the bundle has very high viscosity and cannot flow automatically; a vacuum pump is also connected at one end of the glassy tube to make intimate contact between the tube and the filling FBBs. The single-component FBBs are preferable in close-packed hexagonal structure after annealing. On the other hand, by introducing a spacing fiber, either same filling materials with different diameters or different filling materials, the fibers can be packed in tetragonal (R_{spacer}/R_{fiber}=0.414) and hexagonal structures (R_{spacer}/R_{fiber}=0.155) or other complex structures. Taking two-component FBB constructed structures as an example, the structure can be stacked, namely, as AB_1 (tetragonal) and AB_2 (hexagonal) structures in stoichiometry ratio. By combining with differential etching, more complex structures like spike array, well array, and volcano array can be formed. For 2D assembly, the FBBs can be constructed as heat blocking planes. For instance, some researches have been done to fabricate fibrous cloth that encapsulates phase change materials (paraffin wax), which can reduce the fluctuation of the temperature and keep a nearly constant temperature inside the cloth [18–20]. In the 3D assembly, the fibers can be stacked in a layer-by-layer manner, where the wire orientations in one layer are at a certain angle (θ) to those of adjacent layers, forming an orthogonal array of aligned fibers. The θ can be changed from 0° to 180°. In case of 0°, the nanowires are aligned in parallel. Such a method allows the manipulation of thousands of fibers at the same time.

5.7 Metallic Nanowires

FDN has been used to make metallic microwire arrays [21]. Copper phosphor alloy wire or powder (melting point 750°C) is inserted into a Pyrex glass tube to form a preform, which is drawn at 880°C. Figure 5.4a is an SEM image of a broken fiber containing an alloy microwire obtained after the first draw. The microwire forms good contact with the glass shell and is bent off the center, indicating its good ductility. The bundle of fiber pieces is annealed, and the glass–microwire composite is cut perpendicular to its axis into thin plates. Next, the plates are polished using grits of different grade. Figure 5.4b is the SEM image taken at the cross section of the plate, where the microwires form a hexagonal array. The size of the metal wires can be reduced to several micrometers by repeating the draw–cut–stack cycle for a second time. Figure 5.4c and its inset show the SEM images of the straight copper

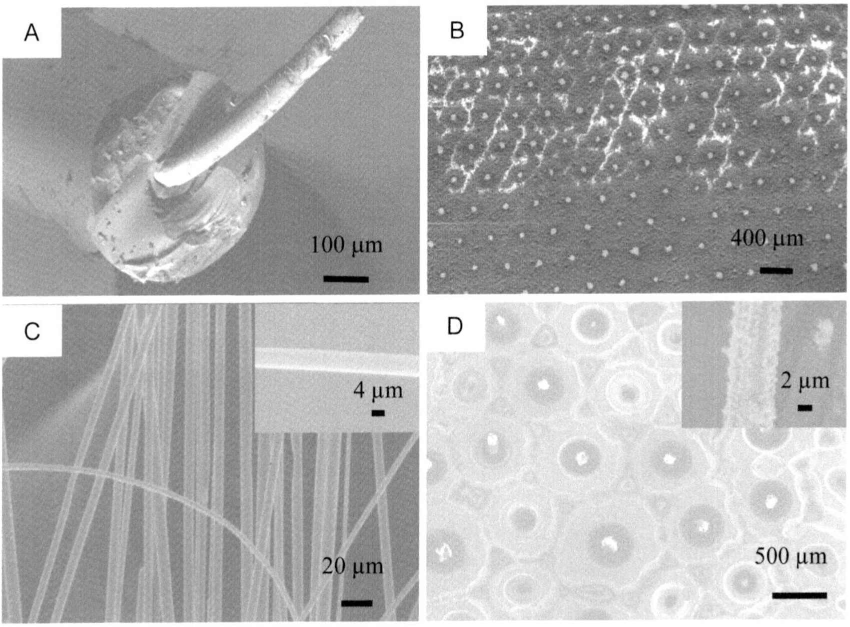

Fig. 5.4 Scanning electron microscope images of alloy micro/nanowires embedded in glass matrix: single wire after first drawing (**a**); copper phosphor alloy wire bundle after the first drawing (**b**); microwires after etching glass matrix (**c**); zinc nanowire bundle after first drawing (**d**); and single wire after removing glass matrix (**d** *inset*)

phosphor microwires, where the diameters are uniform over a length of several centimeters. Similar approach can be used to make micro/nanowires of other materials as long as the matching filling material and glass material can be identified. For instance, glass-encapsulated zinc and tin microwires have been made by filling their powders in a glass tube using FDN method. The FBBs have been packed in a glass tube to form a metal–glass composite rod as shown in Fig. 5.4d. The image shows that the continuity is not as good as copper phosphor alloy microwires. Bulk zinc and tin have melting points of 420 and 232°C, respectively. At the drawing temperature of 850°C, both materials are in liquid state, but have high vapor pressure, wherein the micropores can be seen and some air gaps form along microwires, thus making segments that are several centimeters long. Copper–tin alloy has a low melting temperature, but the ductility is so poor that it cannot be elongated. Conclusively, the filling materials must fulfill all of the criterions mentioned in Section 5.3 so as to be drawn into FBBs.

5.8 Semiconductor Nanowires

Theoretical and experimental studies have shown that large improvements in *ZT* (*figure of merit* for thermoelectric devices) could be achieved in micro/nanostructured systems. Significant enhancement in thermoelectric conversion efficiency was predicted in quantum-confined systems [22–25]. This enhancement was confirmed recently by the 100-fold improvement in *ZT* value of silicon nanowires over bulk Si at around room temperature [26, 27]. Lead telluride (PbTe) is particularly promising for high-*ZT* thermoelectric device, owing to its narrow band gap (0.31 eV at 300 K), face-centered cubic structure, and large average excitonic Bohr radius (46 nm) possessing strong quantum confinement within a large range of sizes [28, 29]. In order to keep the size factor thermoelectric property of PbTe, 3D nanostructures have been fabricated by encapsulating the PbTe micro/nanowires inside glass [30]. PbTe has a very high melting point of 905°C; therefore, the set draw temperature is 960°C, and the glass feeding and the drawing speeds are 0.008 and 1.8 m/s, respectively. Figure 5.5a shows a scanning electron microscopy (SEM) image of an array of hexagonally ordered microwires, which is generated by annealing a bundle of first drawn fibers at 800°C to make a rod, cutting the rod perpendicular to its axis to make plates, and polishing the plates using grits of different grade. Figure 5.5a inset shows a broken glass fiber that contains a 100 μm diameter wire obtained from the first draw cycle. The bundled glass–wire composite rod can be drawn several times to reduce the diameter. The conductivity has also been measured using conductive

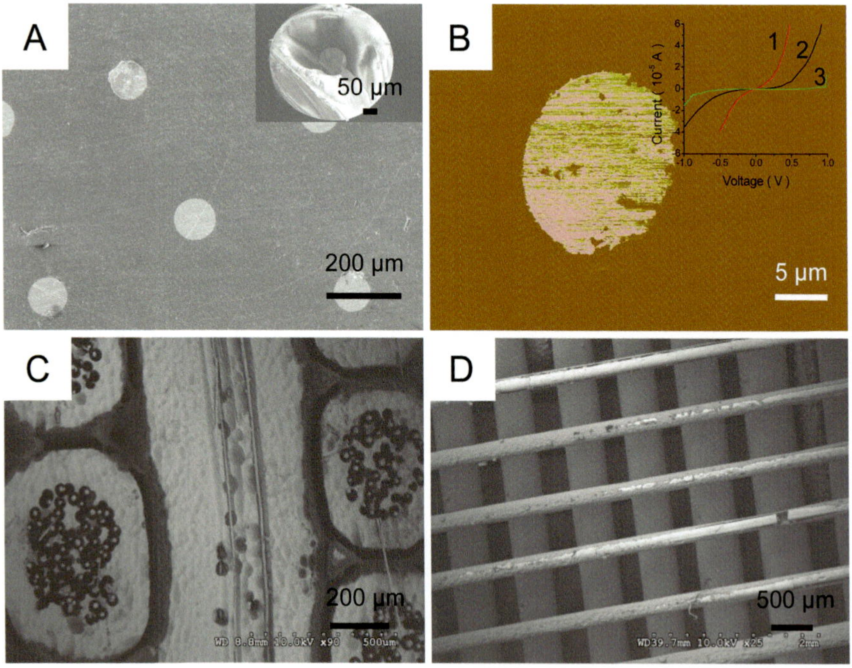

Fig. 5.5 Scanning electron microscope image of PbTe wires in glass matrix (**a**); conductive atomic force microscope image (**b**) of single wire and current–voltages collect after the first (1), second (2), and third (3) drawing (**b** *inset*); cross section of aligned nanowires (**c**); and microwires after removing glass shells (**d**)

atomic force microscopy (C-AFM), where the AFM tip, vertically aligned PbTe microwires, and the substrate are in a series circuit; when scanning on the cross section of the PbTe–glass array, the conductive areas show brighter contrast, as shown in Fig. 5.5b. The resistivity of the annealed slices after the first, second, and third draws shows typical semiconductive behavior (Fig. 5.5b inset). The short PbTe fibers can be assembled in a layer-by-layer manner. Figure 5.5c shows the SEM images of a 3D network of PbTe microwires from the second draw cycle. In this image, the wires perpendicular to the observation direction can be seen as circles. The FBBs containing nanowires are assembled into perpendicular configuration using the same rectangular mold. The glass shells around the PbTe wires from the first draw are etched by immersing the 3D structure into hydrofluoric acid (HF) for 24 h (Fig. 5.5d), where the PbTe wires and residual glass shell can be seen clearly. The SEM images show a higher brightness of the core than the residual shell, which further confirms the high electron conductivity of the PbTe wires.

5.9 Glass Microchannel Array

Salts such as sodium chloride (NaCl; melting point 801°C, boiling point 1465°C) and potassium bromide (KBr; melting point 734°C, boiling point 1435°C) are ionic bond chemicals that have high solubility in water and have high vapor pressure when molten. Salt can also be filled into glass tubes and drawn as preform. The draw usually takes place at 900°C, and several cycles can be repeated by the designed scaling ratios. A thin salt-containing array was obtained by cutting the annealed bundle perpendicular to its sidewall. Salt has a strong solubility in water, and after soaping the array in water for a couple of hours, the salt will be totally fluxed away by water. The SEM image of the glass nanochannels after soaping the slice of the second drawing cycle in water for 2 h is shown in Fig. 5.6. The average diameter can be as small as 400 nm, but cannot go down further. First, the inter-diffusion at drawing or annealing temperatures is ultrafast. The salt can diffuse hundreds of nanometers according to the theory of Fick's second law. Another point is that sodium chloride has a face-centered cubic (FCC) structure, and while cooling down naturally, it transforms from amorphous phase into FCC, but the glass will still remain amorphous.

5.10 Differential Etching of Glasses

Anisotropic wet chemical etching was used to etch single crystal silicon wafers by potassium hydroxide (KOH), with square mask openings, leading to pyramid-shaped pits on the wafer surface [31–33]. The anisotropic properties of different planes determine the structure of the obtained glass samples. But the structures of glasses after differential etching are determined by the preform compositions.

Fig. 5.6 SEM image of glass microhole array made by using salt as template

Different borosilicate glasses vary greatly in their resistivity to the etchant (usually HF) [34–36], and the variation of etching resistivity of the borosilicate glasses is determined by the content of B_2O_3. When B_2O_3 is doped in the SiO_2 lattice, B^{3+} will replace the Si^{4+}. Since the valence will not be balanced if one B^{3+} replaced one Si^{4+}, an extra B^{3+} is needed when three Si^{4+} have been replaced. The extra B^{3+} will result in more defects in the silica lattice and therefore the more B_2O_3, the lower etching resistivity the glass will have. Based on this theory, FDN can be used to make glass microspikes or microwells with well-defined size, location, and arrangement of each component. The keys to the success of this method are as follows: (1) the creation of bulk glass composites with ordered micro-features, (2) the identification of an etchant that shows large differences in the etching rates for two glasses, and (3) the diffusion-limited wet etching of glass composites. The structures of etched microstructures are determined by fiber drawing and wet etching. The spacing and the diameter of the structures are determined by the sizes of starting glasses and drawing and etching conditions. The HF concentration is proportional to the etch rate. Figure 5.7 illustrates the variation of the etch rate with the concentration of HF solution for Corning 7740 (same as Schott 8830) glass wafers [37].

5.11 Glass Microspike Array

A large scale of 3D ordered glass microspike/well array are generated by combining fiber drawing and differential chemical etching [38–40]. For spike array, the starting materials are two types of glasses: core glass is in rod form with high etchant resistance and clad glass is in tube form with low etchant resistance. The two glasses are chosen in such a way that they have similar softening temperatures

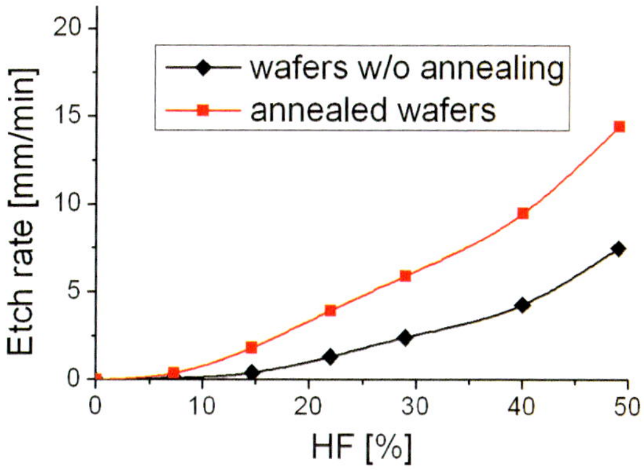

Fig. 5.7 Etch rate of glass (Corning 7740) in concentrated HF solution

and coefficients of thermal expansion, but dissimilar resistance to an etchant. Tubes of Schott 8487 glass that contain a rod of 8330 glass or salt (NaCl) powders are drawn into long microfibers, which are cut into short and even length FBBs of microspikes and microchannels, respectively. The 8487 glass (soft) contains more borate oxide (16.5 over 13.0 mass%) and reacts faster with hydrofluoric acid (HF) than the 8330 glass (hard). The FBBs of salt or hard glass or two types of FBBs can be packed to form a parallel bundle with ordered arrangements at cross section. The bundle is drawn again into long fibers and cut into short FBBs, where the diameter and spacing of hard glass or salt core reduce from millimeters to micrometers. The centimeter-long FBBs from the second draw are assembled to another glass bundle, which is annealed to a glass rod and cut into thin plates. After polishing, the plates are immersed in water to dissolve the salt and etched in HF to make microspikes. Figure 5.8a is a scanning electron microscopy (SEM) image of microspikes etched in 20% HF (volume) for 60 min, which makes microspikes with the cone angle of 40°. Figure 5.8b is an SEM image of microwells etched in 2% HF for 30 min. Figure 5.8c is an image of an array of microchannels after dissolving salt, where both the diameter and spacing are around 3 μm. The FBBs of microchannels and microspikes are closely packed to form a hexagonal pattern, following the rule of packing non-overlapping hard spheres with identical diameters.

The cone angles of glass structures can be controlled over a wide range from 5° to 140°, depending on conditions of the chemical etching. The micro-mirror array can deflect more than 97% of the incident light away from the incident direction, thus making novel substrates ideal to increase optical paths for a variety of optoelectronic applications. The wet etching is carried out in a quiescent solution in order to reduce convective flow, allowing the diffusion of molecules to be dominant. The buffered oxide etchant (BOE) maintains a nearly constant etching rate by keeping the concentration of fluorine ions constant. The chemical etching will start from the center, where the mass ratio of the etchable glass is largest. Based on the etching experiments, the curvatures of the microspike/well could be controlled by changing the etching time or the concentration of the etchant.

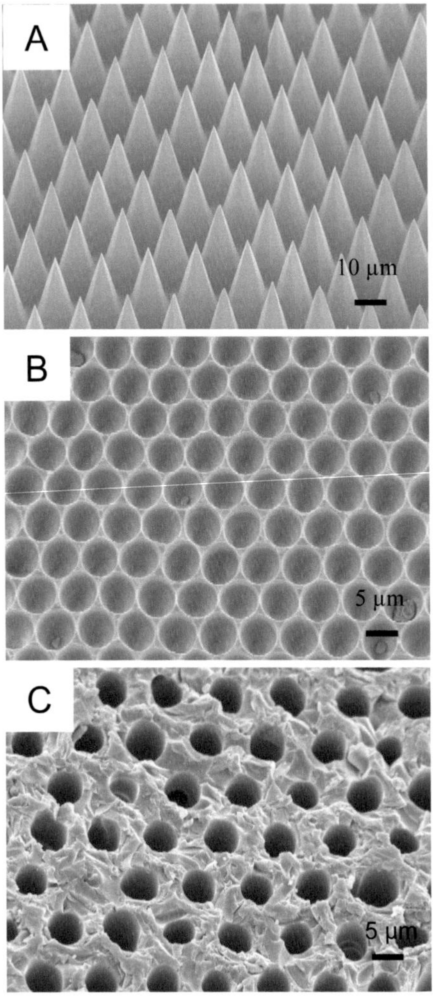

Fig. 5.8 SEM images of glass microspikes (**a**), microwells (**b**), and microchannels (**c**)

5.12 Hybrid Glass Membranes

Wetting is the ability of a liquid to maintain contact with a solid surface, resulting from intermolecular interactions when the two are brought together. The degree of wetting is determined by a force balance between adhesive and cohesive forces, where adhesive forces between a liquid and a solid cause a liquid drop to spread across the surface and cohesive forces within the liquid cause the drop to ball up and avoid contact with the surface. The adhesive force is determined by the surface energy of the solid and, in ideal case, depends on the thermodynamic stability between the solid, air, and liquid phases which is defined by the equilibrium state in Young's equation:

$$\gamma_{SG} = \gamma_{SL} + \gamma_{LG}\cos\theta \tag{5.4}$$

On non-ideal surfaces, the contact angle hysteresis is determined by the roughness of the surface textures. The rough texture of a surface is divided into two categories: homogeneous and heterogeneous, which is interpreted by Wenzel's [41] model and Cassie–Baxter's [42] model, respectively. By fine-tuning the surface roughness, it is possible to achieve a transition between superhydrophobic and superhydrophilic regions. Generally, the rougher the surface, the more hydrophobic it is [43]. On the other hand, by controlling the surface tension using surfactants, a partially or completely wetting surface can also be made to transit between superhydrophobic and superhydrophilic properties.

A superhydrophobic glass membrane has an integrated array of microspikes and micropores by packing FBBs, dissolving template material (i.e., salt) to form glass microchannels, differential etching of two glasses, and modifying etched glass using self-assembled monolayer [44]. The FBBs of microchannels and microspikes are closely packed to form a hexagonal pattern, following the rule of packing non-overlapping hard spheres with identical diameters. The FBBs of microspikes and microchannels can be packed to form multi-component ordered structures. Figure 5.9a is a scanning electron microscopy (SEM) image of microspikes etched in diluted HF, which makes microspikes with the cone angle of 30°. In order to generate superhydrophobic membranes, etched glass microspikes are modified by chloro-group silane. The modified membrane strongly repels water drop, which has to be held in place by a micropipette for contact angle measurements. Figure 5.9b, c shows a water drop sitting on a glass membrane with microspike cone angle of 30°. The advancing and receding angles are measured as 166° and 159°, respectively.

An array of microchannels with spiky edges and superhydrophobicity has been made using a preform with tube–tube–salt configuration, where salt powders are filled in a hard glass tube and the tube is inserted into an 8487 glass tube. This preform is processed by FDN, and the FBBs from each draw cycle are packed to ordered hexagonal pattern. After polishing plate and dissolving salt, an ordered array of microchannels with spiky edge is generated by chemical etching (Fig. 5.10a).

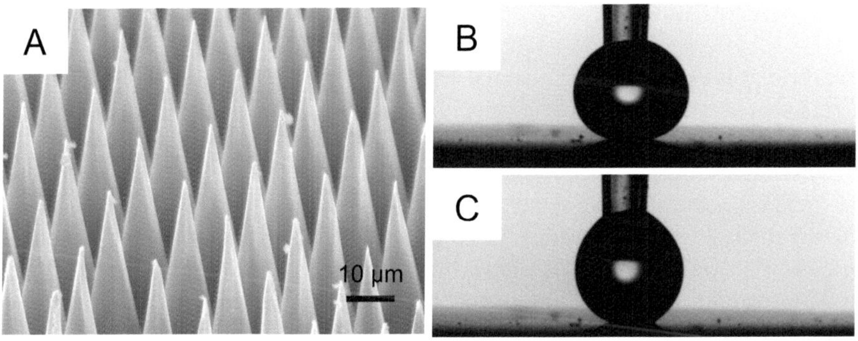

Fig. 5.9 High-resolution SEM image of glass microspikes (**a**) and water drops on superhydrophobic microspike array during compression (**b**) and retraction (**c**)

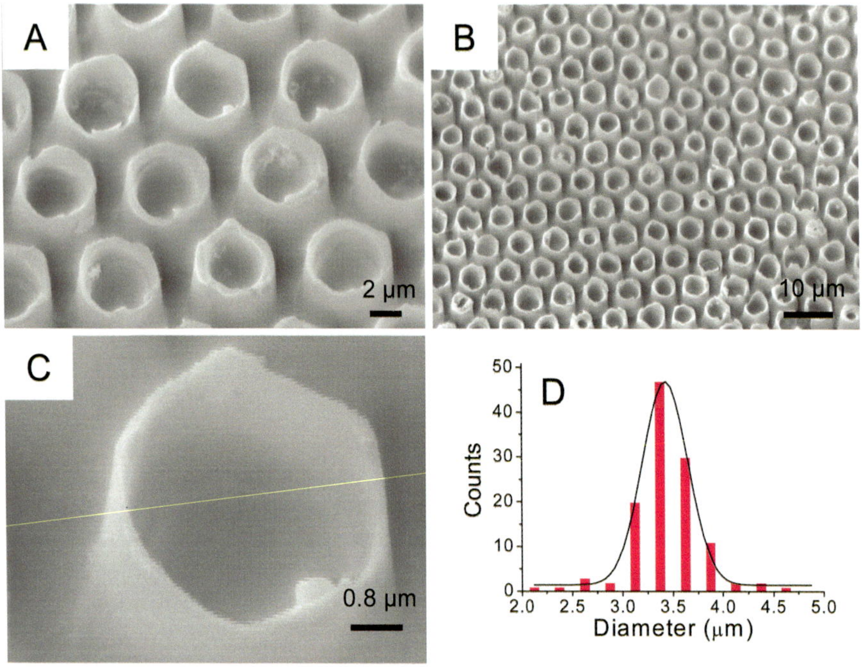

Fig. 5.10 Microchannels with sharp edges at different magnifications (**a–c**); diameter distribution of microchannels (**d**)

Figure 5.10b is an SEM image of the glass membrane. High-resolution image shows that the bottom and top width of the spiky edge are 2 μm and 100 nm, respectively (Fig. 5.10c). The pore diameter has shown a narrow distribution of about 3.4 μm with 95% of pore diameters falling in the range of 3–4 μm (Fig. 5.10d). The porosity is about 30%.

5.13 Textured Structure of Encapsulated Paraffin Wax Microfiber

Paraffin wax is a phase change material (PCM) that has high latent heat of fusion; the large latent heat can act as a heat reservoir absorbing heat when the outside is hotter and releasing heat when the outside is colder. Some researches have been done on fabricating cloth using polymer-encapsulated PCM for thermal regulating. But the yield using the wet coating method is not high enough to fulfill the requirement of real applications. Poly(methyl methacrylate) (PMMA) has a melting temperature of over 200°C and can be drawn into fibers of small diameters. Solid–liquid PCM of paraffin wax at melting temperature of 26°C is used as a filling material inside polymer tubes to form preform for fiber drawing. The drawing is performed at 190°C as described above. Figure 5.11a is an SEM image of a polymer fiber that

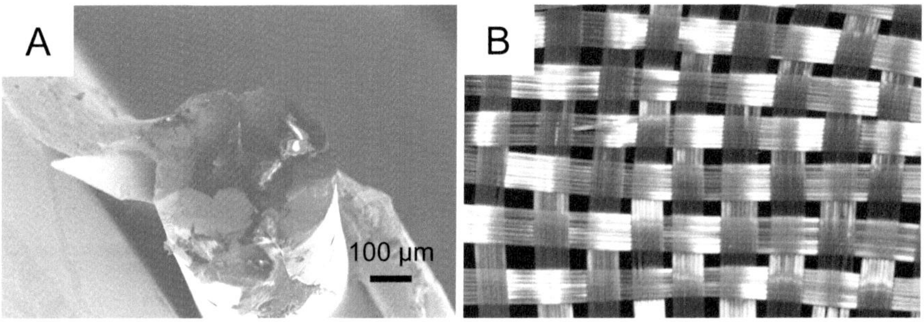

Fig. 5.11 SEM image of polymer-encapsulated paraffin microwires (**a**) and 3D structures of polymer-encapsulated paraffin microwires (**b**)

contains paraffin wax with diameter of 100 μm obtained after the first draw. The fibers are not transparent below the melting temperature of wax due to scattering of light. After heating above its melting temperature, the fibers become transparent. In addition, the flexible fibers can be weaved into a large sheet as shown in Fig. 5.11b. Owing to their large latent heat of fusion, PCM materials have been used widely for thermal storage and transfer applications. By making PCM fibers and their 3D textured structures, it is possible to make clothes that can absorb heat energy and decrease heat or cold fluctuations, thus providing a new superior level of comfort in clothing.

5.14 Conclusions

By using glassy tube with desired thermophysical, chemical, and surface properties, 3D micro/nanostructures can be made over a large area at high yield, controllability, and manufacturability. Several issues have to be considered in FDN: (1) softening temperature of glassy materials should be close to melting temperature of filling materials; (2) coefficients of thermal expansion of glassy materials and the filling material at the drawing temperature must be close to each other; (3) there should be no chemical reaction between the glassy and the filling materials at the drawing temperature; (4) the molten material should have a certain wettability with glass surface; (5) the filling materials should not have low vapor pressure at high temperature. Although these criteria seem rigorous, many choices are available with respect to glass (or polymer) composition and working conditions, making the method a general strategy.

References

1. M. Law, J. Goldberger, P. Yang, Annu. Rev. Mater. Res. **34**, 83 (2004)
2. J. Zhan, X. Yang, D. Wang, S. Li, Y. Xie, Y. Xia, Y. Qian, Adv. Mater. **12**, 1348 (2000)
3. G. Che, B.B. Lakshmi, E.R. Fisher, C.R. Martin, Nature **393**, 346 (1998)

4. T. Hanrath, B. Korgel, Adv. Mater. **15**, 437 (2003)
5. C.J. Novotny, E.T. Yu, P.K.L. Yu, Nano Lett. **8**, 775 (2008)
6. S.H. Lee, T. Minegishi, J.S. Park, S.H. Park, J.S. Ha, H.J. Lee, H.J. Lee, S. Ahn, J. Kim, H. Jeon, T. Yao, Nano Lett. **8**, 2419 (2008)
7. M.S. Gudiksen, J. Wang, C.M. Lieber, J. Phys. Chem. B **105**, 4062 (2001)
8. S. Zhao, H. Roberge, A. Yelon, T. Veres, J. Am. Chem. Soc. **128**, 12352 (2006)
9. J. He, Y. Wan, J. Yu, Polymer **46**, 2799 (2005)
10. I.W. Donald, J. Mater. Sci. **22**, 2661 (1987)
11. L. Tong, E. Mazur, J. Non-Cryst. Solids. **354**, 1240 (2008)
12. Y. Huang, X. Duan, Q. Wei, C.M. Lieber, Science **291**, 630 (2001)
13. K.M. Ryan, A. Mastroianni, K.A. Stancil, H. Liu, A.P. Alivisatos, Nano Lett. **6**, 1479 (2006)
14. G. Yu, A. Cao, C.M. Lieber, Nat. Nanotechnol. **2**, 372 (2007)
15. P.J. Pauzauskie, A. Radenovic, E. Trepagnier, H. Shroff, P. Yang, J. Liphardt, Nat. Mater. **5**, 97 (2006)
16. Y. Yang, S.C. Kung, D.K. Taggart, C. Xiang, F. Yang, M.A. Brown, A.G. Guell, T.J. Kruse, J.C. Hemminger, R.M. Penner, Nano Lett. **8**, 2447 (2008)
17. R.J. Tonucci, B.L. Justus, A.J. Campillo, C.E. Ford, Science **258**, 783 (1992)
18. S. Mondal, Appl. Thermal Eng. **28**, 1536 (2008)
19. S.X. Wang, Y. Li, J.Y. Hu, H. Tokura, Q.W. Song, Polym. Test. **25**, 580 (2006)
20. Y. Shin, D.I. Yoo, K. Son, J. Appl. Polym. Sci. **97**, 910 (2005)
21. X. Zhang, Z. Ma, Z. Yuan, M. Su, Adv. Mater. **20**, 1310 (2008)
22. Y. Li, J. Jiang, G. Xu, W. Li, L. Zhou, Y. Li, P. Cui, J. Alloys Compounds **480**, 954 (2009)
23. X. Zhao, S. Yang, Y. Cao, J. Mi, Q. Zhang, T. Zhu, J. Electron. Mater. **38**, 1017 (2009)
24. D. Teweldebrhan, V. Goyal, A.A. Balandin, Nano Lett. **10**, 1209 (2010)
25. Y. Lin, O. Rabin, S.B. Cronin, J.Y. Ying, M.S. Dresselhaus, Appl. Phys. Lett. **81**, 2403 (2002)
26. A.I. Boukai, Y. Bunimovich, J. Tahir-Kheli, J.K. Yu, W.A. Goddard Iii, J.R. Heath, Nature **451**, 168 (2008)
27. L. Weber, E. Gmelin, Appl. Phys. A Mater. Sci. Proc. **53**, 136 (1991)
28. X. Chen, T.J. Zhu, X.B. Zhao, J. Crystal Growth **311**, 3179 (2009)
29. J.R. Sootsman, J. He, V.P. Dravid, S. Ballikaya, D. Vermeulen, C. Uher, M.G. Kanatzidis, Chem. Mater. **22**, 869 (2010)
30. Y. Hong, Z. Ma, C. Wang, L. Ma, M. Su, ACS Appl. Mater. Interfaces **1**, 251 (2009)
31. E. van Veenendaal, K. Sato, M. Shikida, J. van Suchtelen, Sens. Actuators A **93**, 219 (2001)
32. S. Tan, M.L. Reed, H. Han, R. Boudreau, J. Microelectromech. Syst. **5**, 66 (1996)
33. A.J. Nijdam, E. van Veenendaal, J. van Suchtelen, M.L. Reed, J.G.E. Gardeniers, W.J.P. van Enckevort, E. Vlieg, M. Elwenspoek, J. Appl. Phys. **89**, 4113 (2001)
34. G.A.C.M. Spierings, J. Mater. Sci. **28**, 6261 (1993)
35. T. Diepold, E. Obermeier, J. Micromech. Microeng. **6**, 29 (1996)
36. T. Corman, P. Enoksson, G. Stemme, J. Micromech. Microeng. **8**, 84 (1998)
37. C. Iliescu, F.E.H. Tay, J. Miao, Sens. Actuators A **133**, 395 (2007)
38. B. D'Urso, J.T. Simpson, M. Kalyanaraman, J. Micromech. Microeng. **17**, 717 (2007)
39. Z. Ma, L. Ma, M. Su, Adv. Mater. **20**, 3734 (2008)
40. Z. Ma, Y. Hong, L. Ma, Y. Ni, S. Zou, M. Su, Langmuir **25**, 643 (2008)
41. R.N. Wenzel, Ind. Eng. Chem. **28**, 988 (1936)
42. A.B.D. Cassie, S. Baxter, Trans. Faraday Soc. **40**, 546 (1944)
43. B. D'Urso, J.T. Simpson, Appl. Phys. Lett. **90**, 044102 (2007)
44. Z. Ma, Y. Hong, L. Ma, M. Su, Langmuir **25**, 5446 (2009)

Chapter 6
One-Dimensional Metal Oxide Nanostructures for Photoelectrochemical Hydrogen Generation

Yat Li

6.1 Introduction

Hydrogen represents a promising solution that can simultaneously address energy crisis and environmental pollutions caused by carbon-containing energy carriers, which have been recognized as two major challenges human beings seek to overcome in the 21st century [1]. Hydrogen has very high gravimetric energy density of ~140 MJ/kg, which is three times higher than that of gasoline. More importantly, hydrogen is a portable fuel that can react with oxygen in a fuel cell device to generate electricity in an environmentally benign manner, with water as the only side product. Central to the success of hydrogen technology and economy, the efficient generation, transportation, and storage of hydrogen are the major challenges. Despite hydrogen being known as a clean energy carrier, ironically, most hydrogen (~95% in the United States) is presently produced from steam methane reforming and water-gas shift reaction. This approach still relies on fossil fuels or natural gas and produces undesired by-products including carbon monoxide and carbon dioxide [2]. The development of an efficient, low-cost, and scalable method for hydrogen generation from renewable and carbon-free energy sources is the first and may be the most important step. Enormous efforts have been made in developing new strategies for hydrogen production [3, 4], and photoelectrochemical (PEC) approach is one of the major research focuses [3, 5–11].

6.1.1 Photoelectrochemical Hydrogen Generation

Solar hydrogen generation from PEC water splitting has been a holy grail in chemistry and energy conversion research for decades, as water is the most abundant

Y. Li (✉)
Department of Chemistry and Biochemistry, University of California, Santa Cruz, CA 95064, USA
e-mail: yatli@ucsc.edu

W.L. Zhou, Z.L. Wang (eds.), *Three-Dimensional Nanoarchitectures*,
DOI 10.1007/978-1-4419-9822-4_6,

hydrogen source on the Earth. Water splitting is known to be a thermodynamically uphill process:

$$H_2O(l) \rightarrow H_2(g) + 1/2O_2(g)\,(\Delta G \sim 237.2\text{KJ/mol})$$

A minimal potential of 1.23 V is needed for the reaction to proceed. Due to the recombination loss of photogenerated electron–hole pairs and the imperfection of devices such as contact and electrode resistances, practically optimal energy required for water splitting is around 1.8–2 V. This energy can be obtained from renewable or non-renewable sources. It is well known that hydrogen can be generated from water electrolysis via an electrolysis cell given that enough potential is applied. However, this is a process that basically transforms electricity generated by other methods into chemical energy in the form of hydrogen. Furthermore, significant amount of energy loss is involved in the process since electrolysis cells have a typical cell efficiency limit of around 80% [12, 13]. Therefore, electrolytic hydrogen production is not a sustainable solution for energy problem. A better solution may be the combination of a photovoltaic (PV) cell with an electrolysis cell. The photovoltage created in a PV cell can be used to supply the requested energy for water electrolysis. Nevertheless, the relatively high cost of PV cell is a major drawback of this approach. In addition, the efficiency of this two-step process is expected to be limited by the energy loss in electrolysis and the efficiency of commercially available PV devices, which is typically around 10–15%. In this regard, a PEC cell consisting of semiconductor photoelectrodes that can efficiently harvest solar light and use this energy directly for water splitting can potentially be a more promising and cost-effective way for hydrogen generation.

A PEC cell is composed of a cathode and an anode, on which redox chemical reactions involving electron transfer take place. At least one of the electrodes should be a semiconductor that absorbs solar energy. Figure 6.1 shows a simple schematic of a conventional PEC cell for hydrogen generation that typically consists of a semiconductor photoanode and a platinum cathode. Upon light irradiation with the photon energy equal to or exceeding the band-gap energy of the semiconductor photoanode, electrons will be promoted from the valence band to the unoccupied conduction band and leave holes in valence band. The semiconductor will form a depletion layer at the semiconductor–electrolyte interface, which causes the energy bands to bend upward and downward in *p*-type and *n*-type semiconductors, respectively. In both cases, the energy band bending helps to separate the photogenerated electrons and holes. For *n*-type semiconductor photoanode, while the electrons will transport to the platinum cathode and reduce protons in the electrolyte solution to hydrogen, the holes will accumulate on the photoanode surface for water oxidation to produce oxygen, as illustrated in the following equations:

Platinum cathod: $2H^+(aq) + 2e^- \rightarrow H_2(g)$ $\quad E^0_{red} = 0V$

Semiconductor anode: $2H_2O(l) + 2h^+ \rightarrow 4H^+(aq) + 4e^- + O_2(g)$ $\quad E^0_{ox} = -1.23V$

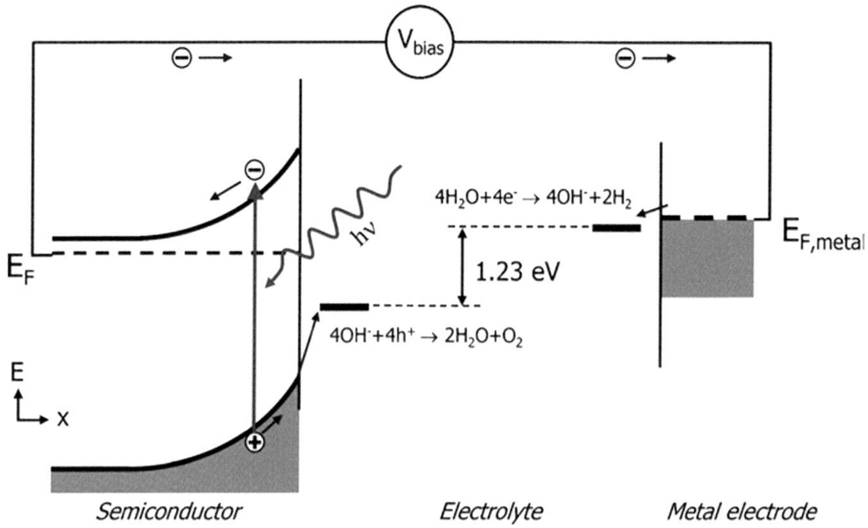

Fig. 6.1 Principle of operation of PEC cells for hydrogen generation via photo-cleavage of water. The PEC cell is composed of *n*-type semiconductor photoanode and a platinum cathode [11]

In some cases, sacrificial electrolyte solution, such as alcohols, will be used for protecting the oxidation of photoanode (will be discussed in more detail in Section 6.4); then the photogenerated holes would oxidize these hole scavengers without producing oxygen, if their oxidation potentials are above that of water.

To serve as a high-performance PEC photoanode, the semiconductor must satisfy several key criteria [5, 14, 15]: (i) photochemical stability with good corrosion resistance in aqueous solution; (ii) a conduction band edge that is more negative than the H_2 evolution potential and a valence band edge more positive than the O_2 evolution potential; (iii) a favorable band-gap energy that allows strong absorption of solar light; (iv) high-quality material with low density of defects that ensures efficient charge transfer and reduced electron–hole recombination; and (v) low cost. Unfortunately, to date, there is no such material that can meet all the requirements simultaneously. Among the various candidates for the photoanode, semiconductor metal oxides represented by TiO_2, ZnO, Fe_2O_3, and WO_3 have attracted much attention due to their superior chemical stability in oxidative environment and inexpensive material cost [5, 16–22].

6.1.2 Challenges in Metal Oxide-Based PEC Hydrogen Generation

Since the first report of using TiO_2 photoelectrode for PEC water splitting in 1972 [7], these metal oxide photoelectrode materials have been investigated extensively. There are several key challenges that have been identified. First, the conduction band of most metal oxide materials is less negative than the H_2 evolution potential, and therefore an external potential is required for photochemical hydrogen generation. Second, the inefficient charge carrier transportation in some metal oxide materials causes significant electron–hole recombination loss and therefore limits their overall conversion efficiency. A representative example is α-Fe_2O_3, which has a very short excited state lifetime ($\sim$10 ps) and a short hole diffusion length ($\sim$2–4 nm). Finally, the weak visible light absorption is the major limitation for all wide band-gap semiconductors such as TiO_2, ZnO and WO_3.

Considerable efforts have been made to address these issues in recent years [5, 6, 11, 23]. For instance, elemental doping and sensitization are two promising approaches that can increase the visible light absorption of wide band-gap metal oxides. For example, the enhanced photoresponse has been observed in metal oxides doped with elements such as nitrogen, sulfur, and carbon [14, 22, 24, 25]. Alternatively, metal oxides can be coupled to small band-gap semiconductors via sensitization [26–30]. The semiconductor needs to form a type II band alignment with metal oxide; the photoexcited electrons created in the sensitizer can therefore be injected into metal oxide and transferred to the counter-electrode for hydrogen generation. We will discuss these two approaches in more detail in Sections 6.3 and 6.4.

On the other hand, increasing the energy level of the conduction band and improving the carrier charge transportation in metal oxide are scientifically very challenging in conventional planar devices. For instance, a thick layer of metal oxide film is typically needed for complete absorption of solar light, especially for indirect band-gap semiconductor, as it reduces the charge-collection efficiency due to increased possibility of electron–hole recombination. The recent development of nanomaterials opens up new opportunities in addressing these fundamental scientific issues. In this chapter, we will review and discuss the recent developments in unique one-dimensional (1D) metal oxide nanowires and nanotubes for PEC hydrogen generation.

6.1.3 One-Dimensional Nanomaterials for Photoelectrodes

The concept of PEC solar hydrogen generation has been well demonstrated based on metal oxide nanocrystalline films and bulk materials [5–8]. One-dimensional metal oxide nanowires and nanotube arrays with unique structural morphology- and size-associated properties represent a new class of photoelectrode materials that could potentially address the abovementioned critical issues (Fig. 6.2). They have potential advantages over their bulk counterparts. First, nanostructured photoelectrodes provide an extremely large semiconductor–electrolyte interface whereon the redox reactions can take place, which could facilitate the charge separation. Second, in comparison to planar structures, 1D nanostructure with high aspect ratio provides not only a long path for efficient light absorption but also a much shorter diffusion length (nanowire radius) for minority carriers (Fig. 6.2, inset). When the nanowire diameter is comparable to the minority carrier diffusion length, the

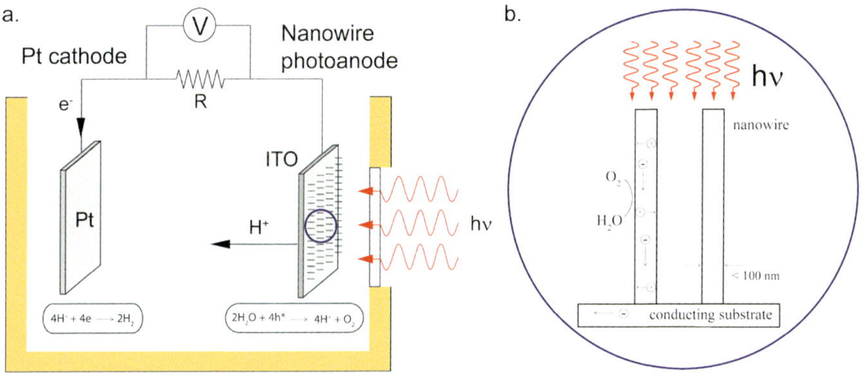

Fig. 6.2 (**a**) Schematic diagram of PEC cell with 1D nanowire-arrayed photoanode. (**b**) A schematic of nanowire photoanodes for water splitting. The small diameter of the nanowires ensures a short hole diffusion length

electron–hole recombination loss will be substantially reduced, which is especially important for semiconductors with low carrier mobilities such as α-Fe_2O_3. Third, the electron–hole overlap factor and electron–hole exchange interaction increase greatly due to quantum size confinement in nanomaterials, resulting in increased band-gap energy as well as increased oscillator strength or absorption coefficient, as compared to bulk materials. The increase of conduction band energy level and absorption coefficient should reduce the required external potential for hydrogen generation and enhance the light absorption efficiency of indirect band-gap metal oxide. Fourth, in comparison to nanoparticles (nanocrystalline films), single-crystal 1D nanostructure without grain boundaries is expected to have improved vectorial charge transportation and, thus, could potentially reduce the electron–hole recombination loss. Fifth, the unique bottom-up synthetic strategy allows the growth of single-crystal 1D nanomaterials on different substrates and prevents the formation of dislocations from the lattice mismatch between the growth substrate and the semiconductor. These high-quality nanostructured photoelectrodes with low density of defects could reduce trapping and improve the transport of charge carriers. Finally, the coating of nanomaterials on conducting substrates naturally forms an antireflection layer, the energy loss due to direct light reflection can be reduced compared to conventional planar shiny electrode. With these potential advantages, 1D nanostructured semiconductor photoelectrodes could fundamentally change the design of PEC cells and improve the solar-to-hydrogen conversion efficiency.

One-dimensional nanostructured photoelectrodes have been explored for a number of pristine metal oxide materials, with particular emphasis on WO_3 [31], ZnO [22, 32, 33], Fe_2O_3 [34–38], and more commonly TiO_2 [9, 19, 24, 39–49]. Among different approaches to improve the PEC performance of metal oxides, elemental doping [14, 22, 24, 25] and quantum dot (QD) sensitization [27, 50–52] have been extensively investigated. In this chapter, we will provide an overview of the recent research activities in the area of PEC solar hydrogen generation based on 1D nanostructured metal oxide photoelectrodes. The discussion will be presented in the order of structural complexity of nanostructured semiconductor photoelectrodes: (1) pristine metal oxides; (2) element-doped metal oxides; (3) QD-sensitized metal oxides; and (4) element-doped and QD-sensitized metal oxides. We will highlight the key scientific and technical challenges in each case and discuss the potential solutions.

6.2 Pristine Metal Oxide Nanowire/Nanotube-Arrayed Photoelectrodes

6.2.1 Nanowire-Arrayed Photoelectrodes

6.2.1.1 Hematite (α-Fe_2O_3)

The first demonstration of PEC water splitting using 1D nanostructures as photoelectrodes was reported by Beermann et al. [34]. In 2000, they reported the photoelectrochemistry of hematite nanorod arrays for the photoanode. Among many metal oxide semiconductors that have been proposed for PEC anode, α-Fe_2O_3 has received considerable attention due to its abundance, inexpensive cost, excellent chemical stability, and favorable band-gap energy of 2.1 eV. It has been predicted that a solar-to-hydrogen efficiency of 12.9% can be achieved by a material with this band gap, which makes PEC hydrogen generation potentially competitive with PV-assisted electrolysis cells in conversion efficiency. However, α-Fe_2O_3 has a very short excited state lifetime (~10 ps) [53, 54] and a small hole diffusion length (~2–4 nm) [55] that significantly limit its efficiency in charge separation and collection as PEC anode. Due to the substantial loss via electron–hole recombination, hematite photoanodes show a much lower solar-to-hydrogen efficiency compared to the theoretical value. In this regard, a major motivation of developing α-Fe_2O_3 nanorods is minimizing the distance for minority carrier (e.g., holes) diffusion to the semiconductor–electrolyte interface for oxidation reaction. Indeed, this pioneering work demonstrated that the thin films of aligned α-Fe_2O_3 nanorods allow more efficient transport and collection of photogenerated electrons through a designed path compared to films constituted of sintered spherical particles [34]. PEC measurements were carried out in a three-electrode system, with 0.1 M KI in water (pH 6.8) as electrolyte, illuminated either through the electrolyte–electrode interface or through the substrate–electrode interface, α-Fe_2O_3 nanowire photoanodes, showed a significant improvement of the incident photon-to-current conversion efficiency (IPCE) compared to the thin films of hematite spherical particles.

The α-Fe_2O_3 nanorod arrays prepared by the same method were also investigated for direct splitting of water using 0.1 M NaOH solution as electrolyte [35]. These α-Fe_2O_3 nanorod-arrayed electrodes were able to yield an improved IPCE of 8% at a wavelength of 350 nm and a light intensity of 0.1 mW/cm^2 without any applied voltage. The results support that the nanorod structure is one promising

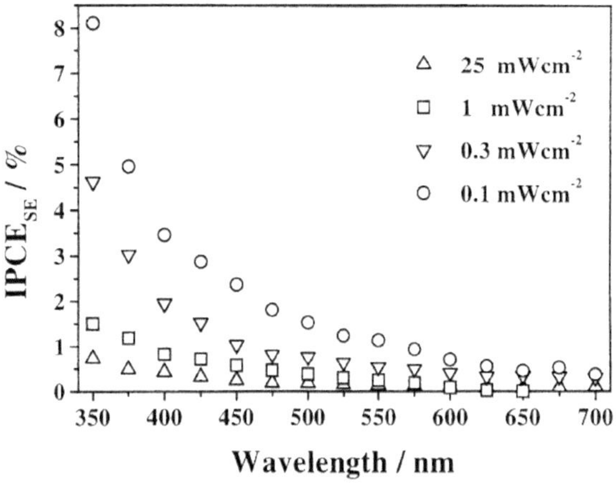

Fig. 6.3 Measured IPCE of α-Fe_2O_3 nanorod-arrayed photoanode at different light intensities [35]

way to reduce the electron–hole recombination rate of α-Fe_2O_3 material by reducing the hole diffusion length. However, the light intensity dependence studies of the IPCE revealed that the highest light intensity (25 mW/cm^2) yields the lowest IPCE (Fig. 6.3), indicating that the hole transfer from the valence band of α-Fe_2O_3 to the electrolyte was still slow and limited the conversion efficiency. The minor improvement of the semiconductor–electrolyte kinetics was ascribed to the relatively large diameter of hematite nanorods.

Since the report of these pioneering works, efforts have been made to reduce the diameter of α-Fe_2O_3 nanorods. Vayssieres et al. reported the synthesis and characterization of bundled ultrafine α-Fe_2O_3 nanorods of typically 4–5 nm in diameter [56, 57], which is comparable to the hole diffusion length of α-Fe_2O_3 (Fig. 6.4). This structure is expected to be able to substantially improve the collection of photoexcited holes and minimize the electron–hole recombination of α-Fe_2O_3. In addition, the lateral quantum confinement effect was observed in these ultrafine nanorods, which led to a significant band-gap increase of 0.3–0.6 eV [57]. The increase in band gap could potentially move the conduction band of α-Fe_2O_3 above the H^+/H_2 potential, which could allow the generation of hydrogen under visible light absorption without applied bias.

Since α-Fe_2O_3 is an indirect band-gap semiconductor that has relatively low absorptivity, a relatively thick film/long nanowires are necessary for complete light absorption. The low electron mobility is another major limitation for α-Fe_2O_3 photoanode. One of the current research focuses on the enhancement of electron transportation in hematite, including preparing element-doped α-Fe_2O_3 materials to improve the electronic and structural properties, which will be discussed in Section 6.3.3.

6.2.1.2 Titanium Oxide (TiO_2) and Zinc Oxide (ZnO)

The observed unique advantages offered by nanorod-arrayed photoanode have been observed in other semiconductor metal oxides as well, including TiO_2 [47] and

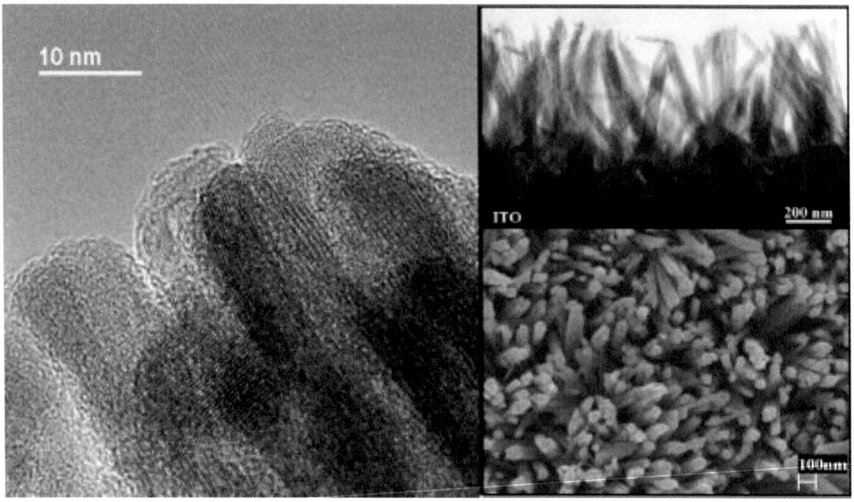

Fig. 6.4 Electron microscopy images of α-Fe_2O_3 arrays consisting of bundled ultrafine nanorods [57]

ZnO [32, 33]. TiO_2 is one of the most promising metal oxide materials for PEC hydrogen generation due to its excellent chemical stability and low material cost. In comparison to α-Fe_2O_3, TiO_2 has better hole diffusion length ($\sim 10^{-8}$ m) [58] and electron mobility (~ 1 cm^2/Vs) [59], and thereby, less electron–hole recombination loss is expected. Khan and coworkers compared the photoresponse of TiO_2 thin film and nanowire film as photoanodes toward water splitting. A more than twofold increase in maximum photoconversion efficiency was observed in the experiments when a single-layer thin film of TiO_2 was replaced by nanowires [47]. This result was consistent with the conclusion obtained in α-Fe_2O_3 studies and again proved the advantage of nanorod structures for PEC photoanode.

A number of new synthetic strategies have been develope to improve the density, morphology, and material quality of TiO_2 nanorod arrays. Wolcott et al. have reported the PEC water splitting using dense and vertically aligned anatase TiO_2 nanorod-arrayed photoanodes prepared by oblique angle deposition on indium tin oxide (ITO)-conducting substrates (Fig. 6.5a) [19]. The TiO_2 nanorods were measured to be 800–1100 nm in length and 45–400 nm in diameter, with a high density of 25×10^6 nanorods/mm^2. They showed pronounced photoresponse in the UV region. The light intensity-dependent studies confirmed that no saturation of photocurrent was observed at least up to 230 mW/cm^2 indicates very efficient charge transportation and redox reaction at the semiconductor–electrolyte interface (Fig. 6.5b). A recent study also reported the growth of single-crystal, rutile TiO_2 nanowire arrays on fluorine-doped tin oxide (FTO) using hydrothermal method [16, 60]. These aligned nanowires were uniform in diameters in the range from $\sim$100 to 150 nm and the typical nanowire lengths of 2–3 μm. These TiO_2 nanowires showed a pronounced photocurrent density of 0.5 mA/cm^2 at –0.2 V vs. Ag/AgCl

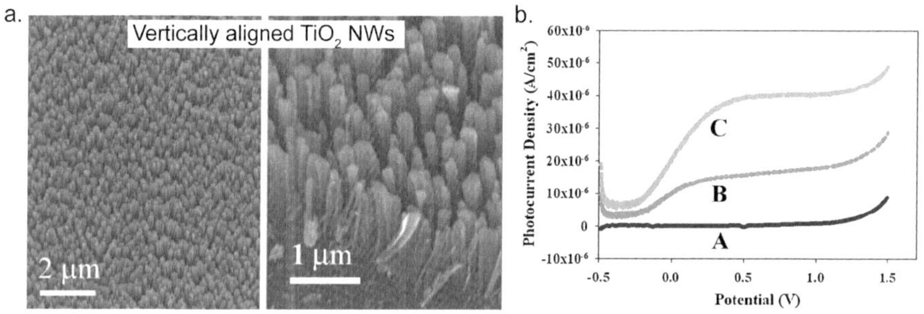

Fig. 6.5 (**a**) SEM images of vertically aligned nanorod arrays of n-TiO_2. (**b**) Linear sweep voltammograms taken at a 10 mV/s scan rate in a 0.5 M $NaClO_4$ electrolyte solution with an Ag/AgCl reference electrode, a Pt coiled counter-electrode, and a TiO_2 nanorod array working electrode (A) in the dark and under light intensity of (B) 100 mW/cm^2 and (C) 230 mW/cm^2 [19]

under light intensity of 100 mW/cm^2 (AM 1.5), in a mixture of 0.25 M Na_2S and 0.35 M Na_2SO_3 electrolyte solution [60]. The advancement in synthetic controllability opens up new opportunities in studying and understanding the interplay between the structure and the PEC performance of nanomaterials.

ZnO, a direct band-gap semiconductor with similar band-gap and band edge positions as TiO_2, represents another promising photoelectrode material. Despite ZnO being known to be not stable in acidic condition, the electron mobility of ZnO is typically 10–100-fold higher than that of TiO_2, which could lead to reduced electrical resistance and enhanced electron transportation. In 2008, Ahn et al. reported the enhancement of PEC response by aligned nanorods in ZnO thin films. ZnO thin films were deposited in pure Ar and mixed Ar and N_2 gas ambient at various substrate temperatures by rf sputtered ZnO targets. The authors find that the deposition in pure Ar ambient leads to polycrystalline ZnO thin films. The presence of N_2 in the deposition ambient promoted the formation of single-crystal aligned nanorods and impurity bands at temperatures above 300°C. ZnO films with aligned nanorods deposited at 500°C exhibit significantly enhanced photoelectrochemical response, compared to polycrystalline ZnO thin films grown at the same temperature (Fig. 6.6). The enhanced photocurrent was attributed to the improved crystallinity and/or the improved light absorption in long wavelength.

Recently, Li et al. also grew single-crystal, vertically aligned ZnO nanowire arrays on ITO substrate using previously reported hydrothermal method [61] and studied their PEC performance. These ZnO nanowire-arrayed photoanodes typically exhibited a photocurrent density of ~0.3 mA/cm^2 at 0.5 V vs. Ag/AgCl under light intensity of 100 mW/cm^2 in $NaClO_4$ electrolyte solution at pH 7 (Fig. 6.7), which is comparable to that observed in TiO_2 nanowires in the same electrolyte solution. Despite the nanostructured ZnO and TiO_2 electrodes show good PEC properties, they are wide band-gap semiconductors with very limited visible light absorption, and therefore, their overall conversion efficiencies are typically less than 1%. We will discuss the approaches to address this major challenge for ZnO and TiO_2 in Sections 6.3, 6.4, and 6.5.

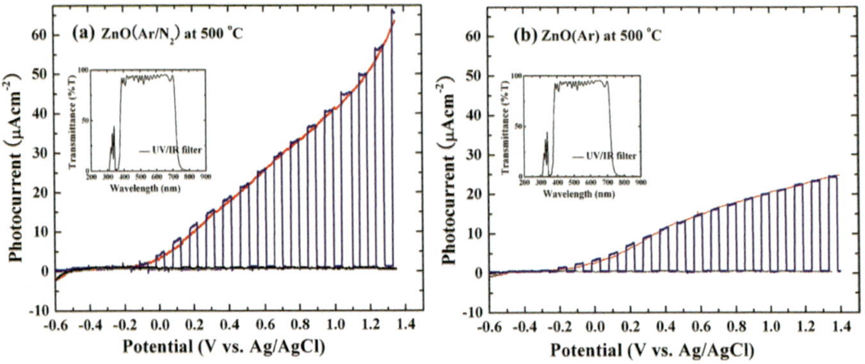

Fig. 6.6 Photocurrent–voltage curves of (**a**) $ZnO(Ar/N_2)$ nanorod and (**b**) ZnO(Ar) films, deposited at 500°C under (*red curve*) continuous illumination, (*black curve*) dark condition, and (*blue curve*) light on/off illumination with an UV/IR filter. Electrolyte and scan rate were 0.5 M Na_2SO_4 aqueous solution and 5 mV/s, respectively

6.2.1.3 Tungsten Trioxide (WO_3)

WO_3 with a band gap of 2.9 eV is another attractive metal oxide material and has been extensively studied. While most studies have been focused on planar structures [62–67], Sunkara et al. recently reported the hot filament-assisted synthesis of vertical arrays of tungsten oxide nanowires [68] and their PEC studies [31]. WO_3 nanowire-arrayed photoanode grown on FTO substrate showed a maximum solar-to-hydrogen conversion efficiency of 0.33% at the potential of 0.82 V vs. saturated calomel electrode (SCE) and IPCE of 85% at 370 nm at 1.2 V vs. SCE [31]. The high quantum efficiency was attributed to the nanowire architecture that enhanced charge transport. In addition, the photocurrent of WO_3 measured with increasing light intensity showed a linear increase with light intensity up to 500 mW/cm^2. This data indicates that the carrier generation was the rate-limiting step and the carrier

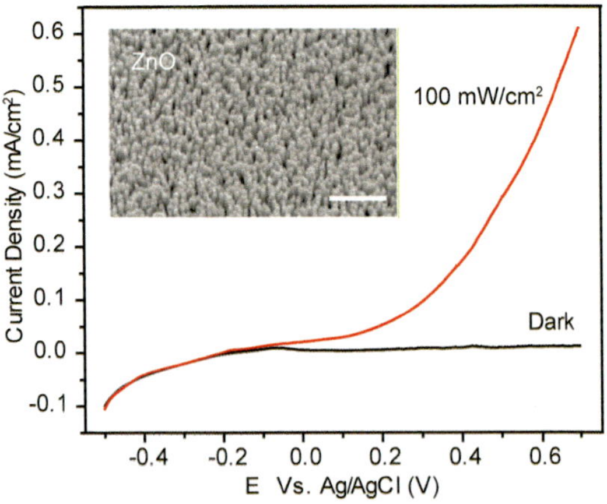

Fig. 6.7 Linear sweep voltammograms, collected at a scan rate of 10 mV/s at applied potentials from −0.5 to +0.7 V from ZnO nanowires in the dark (*black line*) and at 100 mW/cm^2

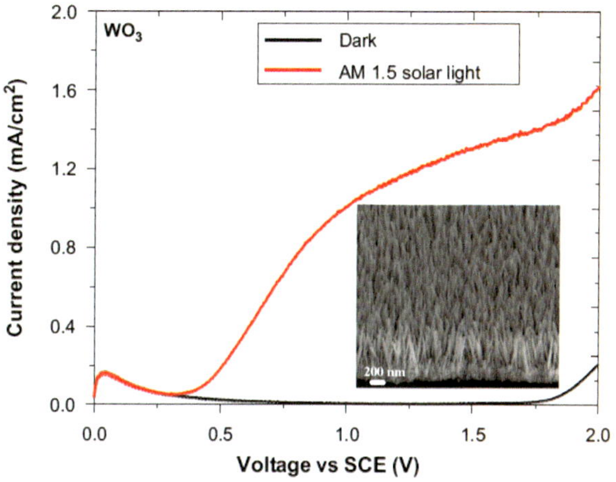

Fig. 6.8 Linear sweep voltammograms collected from PEC cells under light illumination of 100 mW/cm^2 with WO_3 nanowire-arrayed photoanodes at a scan rate of 25 mV/s. Insets: SEM images of WO_3 nanowire array on transparent conducting oxide substrates [31]

transportation was efficient in WO_3. The nanowire arrays were stable for up to 8 h of continuous gas evolution under light illumination (Fig. 6.8).

6.2.2 Nanotube-Arrayed Photoelectrodes

The general concept of using 1D nanostructures for PEC photoelectrodes has also been extended to metal oxide nanotubes [24, 40, 42, 43, 45, 69–76]. In comparison to nanorods or nanowires, nanotubes exhibit less semiconductor material for light absorption but larger surface area for interfacial redox reactions. The large effective surface area is in close proximity with the electrolyte and the thin tube wall, thus enabling diffusive transport of photogenerated holes to oxidizable species in the electrolyte. Water splitting using highly ordered TiO_2 nanotube arrays as photoelectrode was first reported in 2005 [69]. Vertically aligned titania nanotubes were prepared from a starting titanium sheet by potentiostatic anodization and followed by post-annealing at 500°C in oxygen atmosphere to improve the material crystallinity (Fig. 6.9a). The wall thickness was controlled by the anodization bath temperature. Mor et al. demonstrated that the nanotube wall thickness is a key parameter influencing the magnitude of the photoanodic response and the overall efficiency of the water splitting reaction. The titania nanotube array with 34 nm wall thickness showed much greater photocurrent density than a similar array with 9 nm walls. It was believed to be due to the enhanced band bending of the thicker walls decreasing surface recombination rates, and thus, increasing the photocurrent. Upon UV illumination (320–400 nm) at an intensity of 100 mW/cm^2, hydrogen gas was generated in a 1 M KOH electrolyte solution at the power-time normalized rate of 960 μmol/h/W (24 mL/h/W) at an overall conversion efficiency of 6.8% (Fig. 6.9b), which was the highest reported value in TiO_2-based PEC cells.

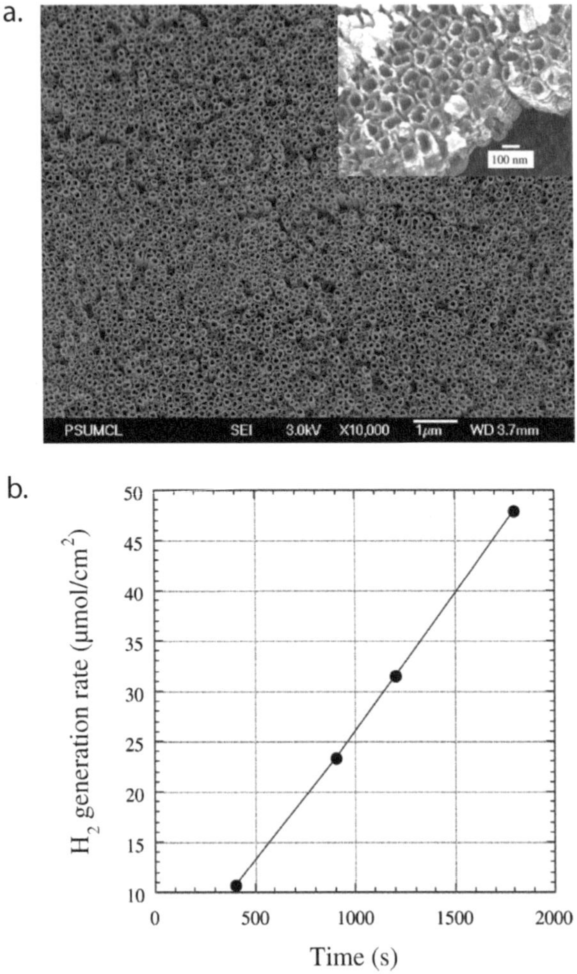

Fig. 6.9 (**a**) SEM image of TiO_2 nanotubes on titanium substrate prepared via anodization; inset: magnified SEM image of TiO_2 nanotubes. (**b**) Hydrogen photoproduction data collected from a PEC cell consisting of a titania nanotube-arrayed photoanode and Pt mesh cathode [69]

Using a similar anodization approach, α-Fe_2O_3 nanotubes have been prepared and studied for PEC hydrogen generation by Misra [77] and Grimes [78] groups. As mentioned above, one of the major limitations of α-Fe_2O_3 for photo-cleavage of water is the low mobility of minority carrier charge. The α-Fe_2O_3 nanotubes could address this issue by offering a very short diffusion length (thin tube wall) for holes and large semiconductor/electrolyte interfacial area for water oxidation. Latemp et al. reported the self-assembled fabrication and photoelectrochemical properties of α-Fe_2O_3 nanotube arrays, prepared by potentiostatic anodization (at a potential of 30–60 V) of iron foil in an ethylene glycol electrolyte containing 0.2–0.5 wt% NH_4F and 2–4% deionized water. The nanotube arrays have a pore diameter ranging between 30 and 80 nm, with a minimum wall thickness of ~10 nm. Crystallization of the as-synthesized, amorphous, iron oxide nanotube arrays to α-Fe_2O_3 was achieved through annealing in an oxygen-deficient ambient. IPCE measurements showed the visible light photoactivity of the α-Fe_2O_3 nanotubes, although the IPCE quickly dropped after ~400 nm. It was believed that an additional reduction in the wall thickness and improved crystallinity could further improve the photoresponse of α-Fe_2O_3 through the reduction of electron–hole recombination.

Mohapatra et al. reported a simple sonoelectrochemical anodization method to grow smooth and ultrathin (5–7 nm thick) α-Fe_2O_3 nanotube arrays (3–4 μm long) on Fe foil (Fig. 6.10a) [77]. A maximum photocurrent density of 1.41 mA/cm^2 (solar-to-hydrogen efficiency of 0.84%) was observed for PEC cell based on the hydrogen-annealed α-Fe_2O_3 nanotube-arrayed photoanode in 1 M KOH electrolyte solution at 0.5 V vs. Ag/AgCl under a light intensity of 100 mW/cm^2 (Fig. 6.10b). The development of novel nanotube structures provides a promising approach to address the fundamental limitation in bulk α-Fe_2O_3 materials.

The abovementioned examples have clearly demonstrated the unique advantages of 1D metal oxide nanostructures as photoanodes from a structural perspective. Meanwhile, the PEC performance of metal oxides depends strongly on the optical and electronic properties as well. For instance, one of the major challenges for large

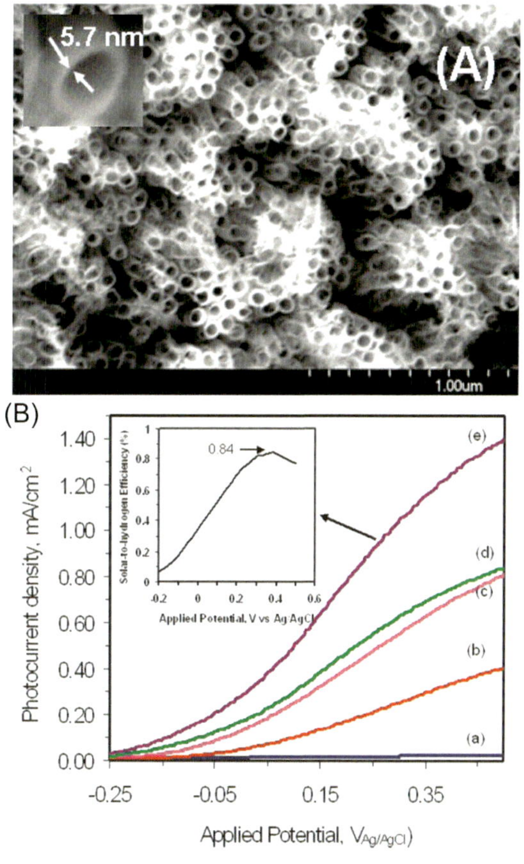

Fig. 6.10 (**a**) As-anodized Fe_2O_3 nanotubes fabricated on Fe foil at 50 V using 3 vol% water for 13 min; inset: magnified SEM image of an individual Fe_2O_3 showing the tube wall thickness. (**b**) Potentiodynamic (*I–V*) behavior of iron oxide nanotubes under AM 1.5 illumination (87 mW/cm^2): (**a**) as-anodized, (**b**) oxygen annealed, and (**c**) hydrogen annealed; and under visible light illumination: (**d**) oxygen annealed and (**e**) hydrogen annealed. Inset shows the efficiency of the nanotube arrays

band-gap metal oxides, such as TiO_2, ZnO, and WO_3, is the limited visible light absorption. In the following sections, we will review the recent research activities with the goal to address this issue by developing more sophisticated 1D nanostructures through elemental doping and quantum dot sensitization. These approaches offer new possibilities in manipulating the material structure optical and electronic properties.

6.3 Element-Doped Metal Oxide 1D Nanostructures

One promising approach to altering the properties of semiconductor nanomaterials is to dope them with other elements including metal ions such as iron [43, 73] and copper [79] and non-metal species such as silicon [36, 48, 80, 81], nitrogen [46, 82–84], carbon [74, 85], sulfur [86], and boron [87], which has been well demonstrated in PEC studies. Recently, it has been proved concept that doping can narrow the band-gap energy, suppress the electron–hole recombination, increase the conductivity of semiconductor, and modify material structure to increase effective surface area of nanomaterials, which all could lead to enhanced PEC performance compared to pristine semiconductors.

6.3.1 TiO_2 Nanostructures

In comparison to cationic metals, anionic non-metal species for doping has attracted more interest in recent years. One of the major purposes of doping is to enhance the photoactivity of wide band-gap metal oxide for hydrogen generation. Doping should create energy states in the band gap of semiconductors to enhance visible light absorption, while the conduction band edge of metal oxides, including subsequent impurity states, should be higher than the H_2 evolution potential to ensure its photoreduction activity. Unlike most metal ions, which often give quite localized d states deep in the band gap of the semiconductors and result in recombination centers of carriers, the *p* states of some anionic species could mix well with the corresponding states in the semiconductors. For instance, in the N-doped TiO_2 films prepared by sputtering, the *p* states of N were mixed well with the O *2p* states and contributed to the band-gap narrowing of TiO_2, resulting in noticeable absorption of visible light at wavelengths less than 500 nm [84]. Similarly, enhanced photoresponse was also observed in C-doped TiO_2 [14]. The C-doped TiO_2 film prepared by flame pyrolysis of a Ti metallic sheet in natural gas showed lower band-gap energy and absorbed appreciably at wavelengths less than 535 nm, whereas the reference *n*-TiO_2 samples showed absorption at 414 nm [14].

Carbon doping has been demonstrated in vertically aligned TiO_2 nanotube arrays [24]. Incorporation of carbon into TiO_2 nanotubes prepared by electrochemical anodization of Ti foil was achieved by annealing the tubes in CO atmosphere at 600°C. The band-gap energy of $TiO_{2-x}C_x$ nanotube arrays deduced from the variation of the absorption coefficient near the band edge was about 2.2 eV, and they

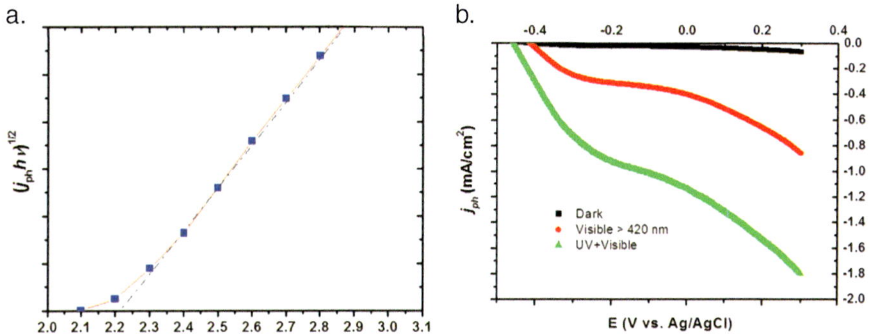

Fig. 6.11 (**a**) Band-gap determination of the $TiO_{2-x}C_x$ nanotube array prepared at 600°C from the $(j_{ph}h\nu)^{1/2}$ vs. $h\nu$ plot and (**b**) current–potential characteristics of the $TiO_{2-x}C_x$ nanotube array in the dark (*black*) and under visible light (*red*) and white light (*green*) illumination [24]

showed significant photoresponse under visible light illumination, (Fig. 6.11). These $TiO_{2-x}C_x$ nanotube arrays showed much higher photocurrent density and more efficient water splitting than pristine TiO_2 nanotube arrays, and the total photocurrent was 20 times that of a P-25 TiO_2 nanoparticulate film. The enhancement was attributed to the improved visible light absorption in C-doped TiO_2. This investigation demonstrated that a chemical approach to decrease the band-gap semiconductor and combined with the unique 1D nanostructure to improve PEC performance.

In addition, boron-doped and Si-doped TiO_2 nanotube arrays were also reported recently by Quan et al. [87, 88]. The dense and vertically aligned TiO_2 nanotube film was produced in a similar anodization process on a Ti sheet. Boron and silicon were doped into the nanotubes by chemical vapor deposition treatment using trimethyl borate and tetraethylorthosilicate as the boron and silicon sources, respectively. X-ray photoelectron spectroscopy studies confirmed the incorporation of dopants and the formation of Ti–B–O and Si–O–Ti bonds. In comparison to pristine TiO_2, the boron-doped TiO_2 showed a slight shift of the absorption edge to lower energy in the spectrum and a higher photoconversion efficiency under both UV and 400–620 nm visible light irradiation [87]. The enhanced photocurrent was due to the increased visible light absorption via B-doping. In contrast, the Si-doped TiO_2 showed a blue shift of the absorption edge [88], and surprisingly an enhanced photoresponse was observed in the UV region. The enhancement of photocurrent density was also believed to be due to the photo-induced hydrophilicity. The contact angle of Si-doped TiO_2 sample was decreased from 27.7° to ~0° under UV light irradiation, which facilitated the oxidation reaction at the semiconductor–electrolyte interface. Although both the B and Si doping improved the PEC performance of TiO_2 compared to pristine sample, these results demonstrated that the role of doping in metal oxide can be very different. The understanding of the chemical states and the function of doping is the key in optimizing the material properties for PEC hydrogen generation.

In addition, it is well known that doping is capable of changing the electronic properties of semiconductor. The pristine TiO_2 is of *n*-type semiconductor due to the presence of oxygen vacancies and typically serves as photoanode. In a previous study, Mor et al. demonstrated that TiO_2 can be doped with copper to form *p*-type TiO_2 materials for photocathode [45]. The vertically oriented *p*-type Cu–Ti–O nanotube array films were fabricated by anodization of copper-rich (60–74%) Ti metal films co-sputtered onto fluorine-doped tin oxide (FTO)-coated glass. Moreover, a *pn*-junction diode was fabricated using *p*-type Cu–Ti–O nanotube array films in combination with *n*-type TiO_2 nanotube array films for self-biased PEC water splitting (Fig. 6.12a). Nozik and coworkers have demonstrated this idea in 1994 by combining a single-crystal *n*-TiO_2 photoanode and a *p*-GaP photocathode through ohmic contacts [89]. To achieve hydrogen generation at zero bias, the conduction band of the *p*-type photocathode must be above the H_2 evolving potential and the valence band of the *n*-type photoanode must be below the water oxidation potential. The combination of two semiconductors relaxes the criteria of the energetic overlap for photoelectrodes, which increases the number of semiconductors that could be used as photoelectrodes water splitting [89–91]. Upon illumination, photons above the band gap will be simultaneously absorbed to generate photoexcited electrons and holes at both photoelectrodes. The majority carriers created at the anode (electrons) and cathode (holes) will recombine at the ohmic contact, while minority carriers generated at the anode (holes) and cathode (electrons) will be used for water splitting. Overall, this process absorbs two photons to produce one useful electron–hole pair, which limits the efficiency. With the glass substrates oriented back-to-back, light is incident upon the UV-absorbing *n*-TiO_2 side, with the visible light passing to the *p*-Cu–Ti–O side, as shown in Fig. 6.12. Under global AM 1.5 illumination, this *pn*-junction diodes generated photocurrents of ~0.25 mA/cm^2, with a photoconversion efficiency of 0.30% at zero bias. This work

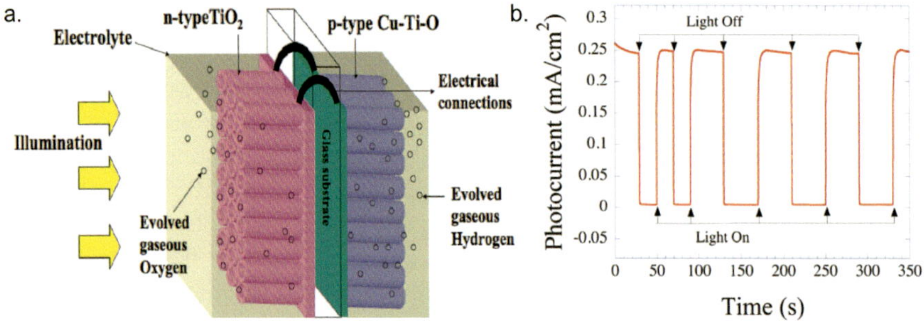

Fig. 6.12 (**a**) Illustration of PEC *pn*-junction diode for water splitting comprised of *n*-type TiO_2 and *p*-type Cu–Ti–O nanotube array films, with their substrates connected through an ohmic contact. (**b**) Photocurrent from the self-biased photochemical diode in (**a**) under global AM 1.5 illumination. The *n*-TiO_2 side of the diode is kept immersed in a 1 M KOH aqueous solution and the *p*-Cu–Ti–O side is kept in 0.1 M Na_2HPO_4 with a salt bridge linking the solution of the two sides [45]

provided important insight in developing self-biased PEC nanodevices for hydrogen generation.

6.3.2 ZnO Nanostructures

Non-metal doping was demonstrated in other semiconductor metal oxides such as ZnO and α-Fe_2O_3. Theoretical calculations of the densities of states by the full-potential linearized augmented plane wave formalism revealed that among the elements used for substitutional doping of O in metal oxide, N was one of the most effective dopants in narrowing its band gap because its *p* states have a good mix with the O *2p* states of metal oxide. Previous studies of *n*-type N-doped ZnO films have shown a significant red shift in the light absorption wavelength [32, 92, 93]. Recently, vertically aligned single-crystal ZnO nanowire arrays were synthesized by a hydrothermal method, followed by annealing in ammonia at 550°C to incorporate nitrogen as a dopant [22]. Yang et al. demonstrated that the N concentration of nanowires can be controlled by varying the annealing time. XPS studies confirmed the substitutional doping of N. PEC studies of N-doped ZnO nanowire arrays as photoanodes demonstrated an enhancement of photoresponse in the visible region and almost an order of magnitude increase in photocurrent density compared to undoped ZnO nanowires prepared under similar conditions (Fig. 6.13).

6.3.3 Hematite (α-Fe_2O_3) Nanostructures

α-Fe_2O_3 is an indirect band-gap semiconductor with a band gap of ~2.2 eV that can absorb a considerable part of the solar light up to 620 nm, which consists nearly 40% of the sunlight photons at global AM 1.5. However, the photoactivity of α-Fe_2O_3 is

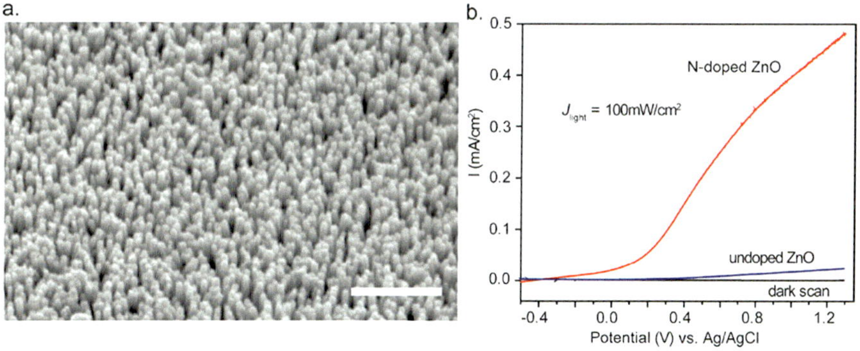

Fig. 6.13 (**a**) SEM image of ZnO nanowire array on ITO substrate. (**b**) Linear sweep voltammograms collected at a scan rate of 10 mV/s at applied potentials from –0.5 V to +1.3 V, from undoped ZnO nanowires in the dark (*black line*), undoped ZnO nanowires (*blue line*), and N-doped ZnO nanowires (*red line*) at 100 mW/cm^2 [22]

limited by its low charge carrier mobility and high resistivity, which cause substantial recombination between the photogenerated electron–hole pairs. In this regard, previous studies on the element doping of α-Fe_2O_3 were not intended to enhance the light absorption, but to improve the separation and collection of photoexcited charge carriers. Silicon has been known to be a good dopant for α-Fe_2O_3 nanomaterials for PEC water splitting [36, 80, 81, 94]. The nanocrystalline thin-film Si-doped α-Fe_2O_3 nanorod photoelectrodes prepared by atmospheric pressure chemical vapor deposition produce a photocurrent of up to 1.45 mA/cm^2 at 1.23 V vs. NHE, which is much higher than that with undoped α-Fe_2O_3 electrodes. Different mechanisms for the photoresponse enhancement by silicon doping have been proposed. First, silicon serves as an *n*-type dopant that improves the electrical conductivity in the α-Fe_2O_3 lattice. Second, the incorporation of Si dopant leads to the formation of dendritic α-Fe_2O_3 nanostructure with branches, which substantially increases the surface area of the photoanode (Fig. 6.14). Electron microscopy studies revealed that the highly developed nanorod-like dendritic nanostructure of 500 nm thickness has a feature size of only 10–20 nm at the surface. The reduction of the grain size to a level that is comparable to the hole diffusion length of α-Fe_2O_3 nanoparticles simultaneously minimizes the recombination of electrons and holes and increases the surface area of the photoanode. More detailed studies of the effect of Si doping

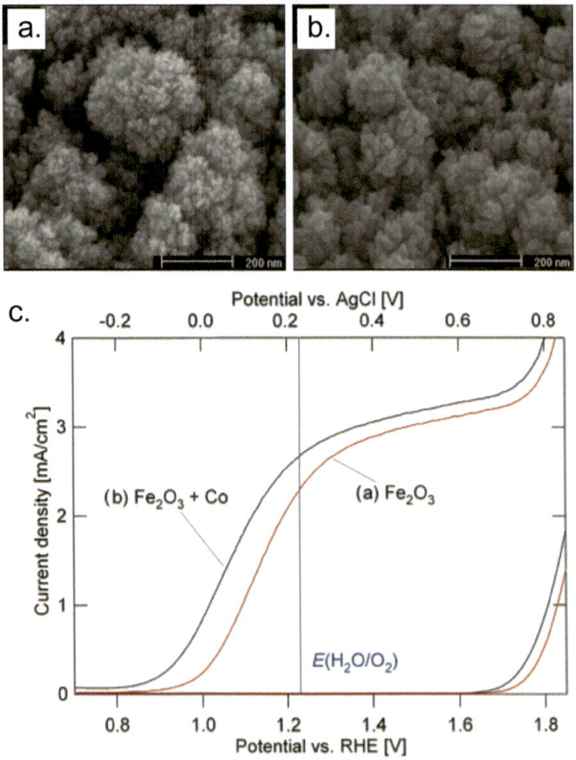

Fig. 6.14 Top view (45° tilted) SEM images of α-Fe_2O_3 films grown by atmospheric pressure chemical vapor deposition on FTO glass. (**a**) Si-doped α-Fe_2O_3 and (**b**) undoped α-Fe_2O_3 film. (**c**) Current–voltage characteristics in darkness and under simulated sunlight at a scan rate of 50 mV/s [81]

on the performance of nanostructured α-Fe_2O_3 photoanodes for water splitting have recently been reported [36, 81].

Frites et al. also demonstrated the α-Fe_2O_3 nanowires and carbon-modified α-Fe_2O_3 film (denoted as (CM)-*n*-Fe_2O_3) that showed pronounced photoresponse and conversion efficiencies [38]. The *n*-type α-Fe_2O_3 nanowire thin films were synthesized by thermal oxidation of Fe metal sheet in an electric oven at different temperatures from 550 to 850°C for 15 min. The nanowires showed certain orientation in growth direction, but the density was relatively low compared to TiO_2 or ZnO nanowire arrays. The synthesis of nanocrystalline (CM)-*n*-Fe_2O_3 photoelectrodes was carried out by flame oxidation of Fe metal sheet at different temperatures (700–850°C) and oxidation times (3–15 min). Natural gas provided the carbon source. The optimized α-Fe_2O_3 nanowire photoelectrode showed a photocurrent density of 1.32 mA/cm^2 at a measured potential of 0.0 V vs. SCE (Fig. 6.15) in a 5 M KOH electrolyte under a light intensity of 100 mW/cm^2 standard AM 1.5 illumination, which was the highest reported photocurrent density obtained from α-Fe_2O_3 nanowires. The enhancement can be attributed to the reduced transport distance of the photogenerated holes from the bulk to the semiconductor–electrolyte interface. Significantly, the photocurrent density was further improved to 3.14 mA/cm^2 in CM-*n*-Fe_2O_3 sample under the same potential, electrolyte, and light intensity. The carbon incorporation was believed to be responsible for increasing the conductivity as well as lowering the indirect band gap of CM-*n*-Fe_2O_3 film from 2.2 to 1.95 eV. The maximum photoconversion efficiency of 1.69 and 2.23% was observed in the nanowire and CM-*n*-Fe_2O_3 film, respectively. Doping was again proved to be effective in improving the PEC performance of α-Fe_2O_3. A long-term goal is combining the 1D materials with favorable doping to simultaneously enhance the transportation of photogenerated electrons and holes.

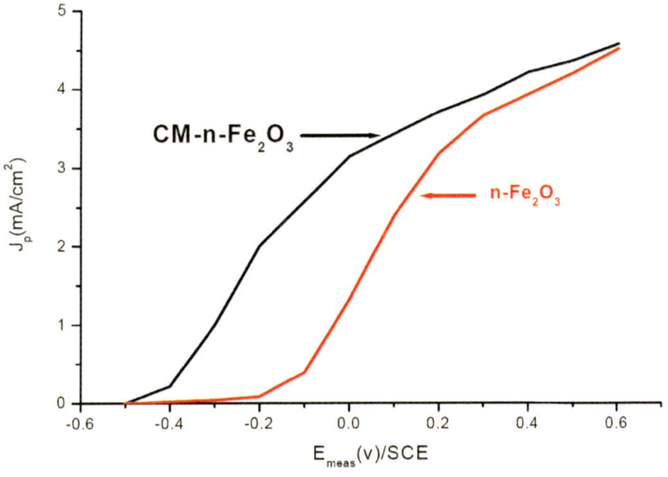

Fig. 6.15 Current–voltage characteristics of optimized CM-*n*-Fe_2O_3 nanocrystalline film and *n*-type α-Fe_2O_3 nanowires in a 5 M KOH electrolyte solution under white light intensity of 100 mW/cm^2 (AM 1.5)

6.4 Quantum Dot Sensitizations

6.4.1 Background

Quantum dot (QD) sensitization is a promising approach in increasing the visible light absorption of large band-gap semiconductors. This idea was motivated by the development of Grätzel cells. Grätzel and coworkers developed nanocrystalline, dye-sensitized solar cells that convert light to electrical energy with a highest reported efficiency of ~11%. The dye-sensitized solar cell consists of TiO_2 nanoparticles deposited on a conducting substrate. Dye molecules that are able to absorb visible light are adsorbed to the surface of TiO_2. Under irradiation, the photoexcited dye molecules inject electrons to the TiO_2 nanoparticles, which are transported through the porous TiO_2 layer and collected by a back contact. The oxidized dye is regenerated by a redox-active electrolyte. Since organic dye molecules are often unstable under heat or light, especially over an extended period of time, an alternative approach is replacement of the dye molecules with more stable inorganic crystalline semiconductor QDs in a variation of the Grätzel cell [95]. Small band-gap QDs can sensitize the metal oxide and absorb visible light; furthermore, the band gap can be controlled by varying QD size to potentially cover the entire spectrum of solar light [96, 97]. This device concept has been extended to PEC devices for hydrogen generation. Various studies have shown improved visible light photoactivity with coupled semiconductors with different band-gap energies, e.g., TiO_2/CdS [26, 42, 98], TiO_2/CdSe [99, 100], and ZnO/CdS [60]. These nanocomposite structures are typically composed of two semiconductors, one with a wide band gap and another with a smaller band gap. The small band-gap QD is primarily responsible for visible light absorption and sensitizing the large band-gap metal oxide through electron and/or hole injection. Efficient electron injection from QD to metal oxide requires that the bottom of the conduction band of the QD be above the bottom of the conduction band of the metal oxide. This electron transfer could enhance the charge separation and reduce the electron–hole recombination in QD by forming a potential gradient at the QD–metal oxide interface.

In order to effectively sensitize metal oxide with semiconductor QDs (e.g., CdSe or CdS), two common approaches have been employed to ensure good interfacial contact and, thereby, efficient charge transfer between the QD and the metal oxide. One is to grow/deposit QDs directly on the metal oxide surface using chemical bath deposition (CBD) methods [101, 102]. The other approach is to link QDs to metal oxide nanomaterials with bifunctional molecules *X*–R–*Y*, where *X* and *Y* are functional groups that bind to the QD and metal oxide, respectively [96, 103–105]. For example, a number of mercaptoalkanoic acids have been used as linker molecules, such as thioglycolic acid ($HSCH_2CO_2H$), that have both thiol and carboxylic groups binding to the QD and metal oxide, respectively. A linker molecule can affect both the electronic structure and the electron transfer rate at the interface.

6.4.2 Quantum Dot-Sensitized ZnO Nanowires

QD sensitization has been demonstrated in different 1D structures. Here we will review the recent developments in quantum dot-sensitized TiO_2 and ZnO nanowires. In a previously reported CdS QD/ZnO nanowire heterostructures [27], the small-sized CdS nanoparticles were deposited uniformly on the surface of ZnO nanowire arrays in a chemical bath containing $CdSO_4$, thiourea, and an aqueous ammonia solution (Fig. 6.16). The deposition of CdS nanoparticles increased the light absorption limit up to ~550 nm, as shown in Fig. 6.16b [27]. Since the bottom of the conduction band of CdS is located above the conduction band of ZnO, the photoexcited electrons in CdS can transfer easily to ZnO. To protect the CdS QDs from corrosion, 1 M Na_2S aqueous solution was used as electrolyte. The S^{2-} ions serving as sacrificial reagents will be oxidized at photoanode to form S, instead of evolving oxygen as expected for water splitting. The PEC cells based on these CdS/ZnO heterostructures as photoanodes showed enhanced photocurrent compared to those of a bare ZnO nanowire photoanode (Fig. 6.16c). The photocurrent density is proportional to loading amount of CdS QDs, which can be controlled via the CBD

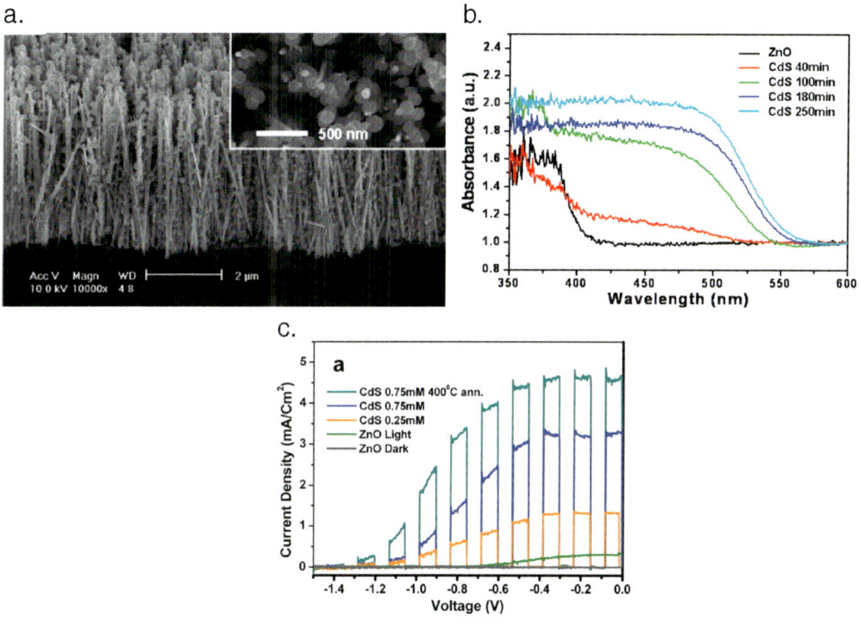

Fig. 6.16 (**a**) Tilted cross-sectional SEM image of the CdS NP/ZnO nanowire heterostructure array. Inset is the top view SEM image. (**b**) Light absorbance spectra of CdS nanoparticle/ZnO nanowire heterostructure arrays, depending on CdS deposition times. (**c**) Chopped illumination (100 mW/cm^2) current–voltage characteristics of bare ZnO nanowire and CdS nanoparticle-deposited ZnO nanowire arrays using two different concentrations of a CdS deposition solution in a 1 M NaS solution with a saturated calomel reference (SCE) electrode [27]

deposition time and reactant concentration. The results confirmed that the photogenerated electrons in QDs can be injected efficiently into the ZnO, and the substantially enhanced photocurrent was due to the improved visible light absorption.

QD-sensitized TiO_2 samples were also demonstrated by the Kamat group, although most of the work focused on their applications for solar cells [26, 106, 107]. Interestingly, it has been recently suggested that TiO_2 nanorod arrays can be sensitized by different sizes of CdSe nanoparticles to form rainbow solar structures [26]. The band-gap and band edge positions can be engineered by synthetically tailoring the particle size of QDs. This multiple-junction QD heterostructures could potentially be used in PEC cells to maximize the light absorption efficiency in the same way as has been demonstrated in tandem PV cells.

6.4.3 Quantum Dot-Cosensitized Nanowires

Motivated by the tandem cell structure, recently an efficient PV photoelectrode was prepared by sequentially assembled CdS and CdSe QDs onto a nanocrystalline TiO_2 film using CBD method [100]. The CdS/CdSe-cosensitized TiO_2 photoelectrode (TiO_2/CdS/CdSe) was found to have a complementary effect in light absorption and exhibited a significant enhancement in the current–voltage response, both in dark conditions and under light illumination. More importantly, the open circuit potentials measured in the dark for these electrodes indicated that a Fermi level alignment occurs between CdS and CdSe, causing downward and upward shifts of the band edges, respectively, for CdS and CdSe. A stepwise band edge structure was, therefore, constructed in the TiO_2/CdS/CdSe electrode, which is responsible for the performance enhancement of this photoelectrode.

Wang et al. extended this device concept to 1D metal oxide nanostructures [60]. Two QD-sensitized ZnO nanowire-arrayed photoanodes (denoted as CdS–ZnO and CdSe–ZnO) and two QD-cosensitized ZnO nanowire samples (denoted as CdSe–CdS–ZnO and CdS–CdSe–ZnO) were prepared by CBD method for PEC measurements. These cosensitized samples were illuminated from the backside of the substrate in all PEC measurements. The IPCE measurements of these samples displayed several important features (Fig. 6.17a). First, CdS–CdSe–ZnO and CdS–ZnO had similar IPCE characteristics, indicating that the intermediate CdSe layer in CdS–CdSe–ZnO sample had little or no contribution to the photocurrent. This can be understood by the band alignment between CdS and CdSe, where the photogenerated holes in CdSe QD cannot be transferred to electrolyte due to the large potential difference between the valence band edges of CdSe and CdS (Fig. 6.17b). Second, both CdSe–ZnO and CdS–ZnO samples exhibited IPCE of $\sim$25% in the wavelength above their band-gap energy, suggesting efficient electron transfer from QD to ZnO nanowire. CdSe–ZnO showed higher photocurrent density than CdS–ZnO and CdS–CdSe–ZnO samples as a result of strong absorption in a broader range of wavelength. Finally, the CdSe–CdS–ZnO sample presented the most pronounced photocurrent among all four samples due to the stepwise band alignment. As shown in Fig. 6.17b (right scheme), the band alignment

Fig. 6.17 (a) Measured IPCE spectra of single-sided nanowire samples in the wavelength region of 310–650 nm at a potential of 0 V vs. Ag/AgCl. (b) Energy band diagram of CdS–CdSe–ZnO and CdSe–CdS–ZnO nanowire photoanodes [60]

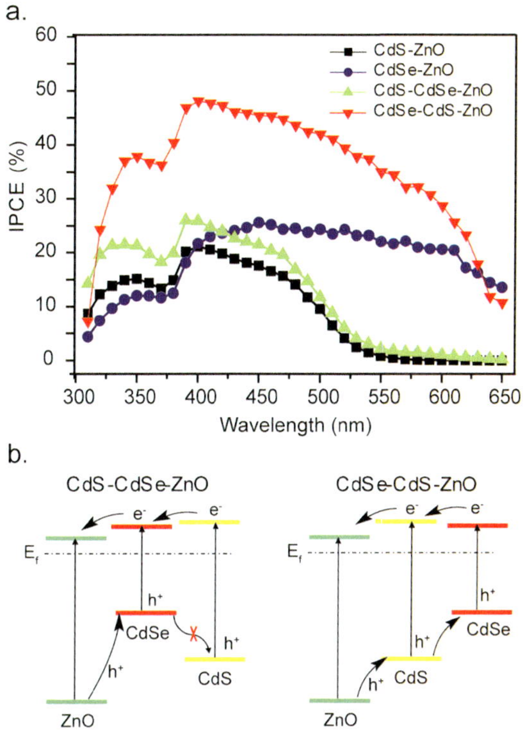

allows transportation of both electrons and holes in this composite structure. On the contrary, the performance of the reverse structure, CdS/CdSe/ZnO, was limited by potential barrier at the CdS/CdSe interface. The enhanced IPCE observed in the CdSe–CdS–ZnO sample compared to CdSe–ZnO can be attributed to the elevation of CBE of CdSe via band alignment between CdS and CdSe and the increase in light absorption below the wavelength of 550 nm due to the incorporation of CdS middle layer. However, the presence of CdS layer also increased the diffusion length for charge carriers, which may cause recombination loss and reduce the efficiency in charge collection. This effect was reflected by the gradual decrease of IPCE above 500 nm in CdSe–CdS–ZnO compared to CdSe–ZnO sample (Fig. 6.17a).

6.4.4 Double-Sided Quantum Dot Sensitization

PV studies have demonstrated that tandem structure made of a combination of semiconductors with optimal band-gap energies can maximize the absorption of solar light [26]. However, rational synthesis of the multi-junction nanostructures for PV or PEC devices has been technically challenging. Wang et al. recently reported the

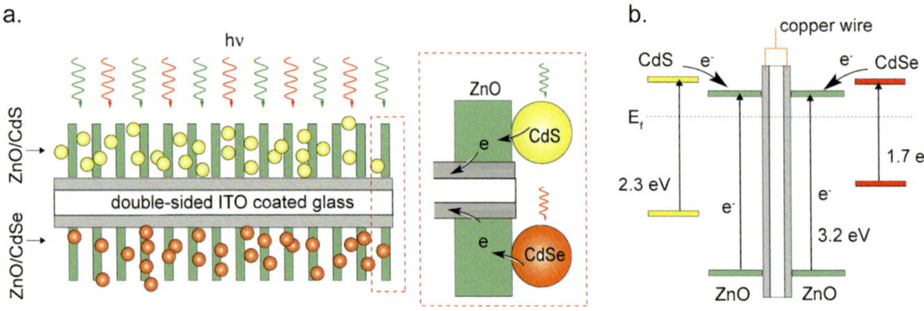

Fig. 6.18 Schematic diagrams illustrating (**a**) the architecture and (**b**) the corresponding energy diagram of double-sided CdS–ZnO–ZnO–CdSe nanowire-arrayed photoanode. The dashed box highlights the CdS–ZnO and CdSe–ZnO interfaces [60]

design and characterization of a novel double-sided CdS and CdSe quantum dot-cosensitized ZnO nanowire-arrayed photoanode for PEC hydrogen generation [60]. The double-sided design represents a simple analogue of tandem cell structure, in which the dense ZnO nanowire arrays were grown on an ITO substrate followed by sequential sensitization of CdS and CdSe quantum dots on each side (Fig. 6.18a). At equilibrium, the Fermi levels of the three semiconductors will be aligned in electrolyte solution as shown in the simplified energy diagram (Fig. 6.18b). Importantly, the conduction band edges of CdS and CdSe are located above the conduction band of ZnO, allowing efficient transfer of photoexcited electrons from CdS and CdSe QDs to ZnO nanowires. Double-sided illuminated TiO_2 nanotubes developed previously by Misra group require the two sides to be operated separately, and therefore the need for two energy sources. In contrast, this new double-sided QD-sensitized nanowire device can be operated using a single white light source by illuminating from the CdS side of the substrate. To allow enough light to penetrate through the first layer, the loading of CdS QDs was intentionally reduced, while the CdSe QDs were in maximum loading for complete light absorption.

In addition to the CdS–ZnO–ZnO–CdSe nanowire structure, three additional double-sided samples including ZnO nanowires (denoted as ZnO–ZnO), CdS QD-sensitized ZnO nanowires (denoted as CdS–ZnO–ZnO–CdS), and CdSe QD-sensitized ZnO nanowires (denoted as CdSe–ZnO–ZnO–CdSe) were prepared in similar growth and sensitization conditions and used as control samples for PEC measurements. The linear sweep voltammograms recorded from these samples showed that QD-sensitized nanowire samples exhibited substantially enhanced photocurrent compared to pristine ZnO nanowires, as expected. Significantly, on the basis of a single white light illumination of 100 mW/cm^2, the cosensitized CdS–ZnO–ZnO–CdSe nanowires showed a maximum photocurrent density of ~12 mA/cm^2 at 0.4 V vs. Ag/AgCl, which is about two times larger than those of single-sensitized samples (CdS–ZnO–ZnO–CdS and CdSe–ZnO–ZnO–CdSe). The data confirmed that the incident light transmitted through the CdS–ZnO layer to the CdSe–ZnO layer at the back of the substrate and the light was absorbed by both layers. More importantly, there was a synergistic effect in the cosensitized sample that led to the photocurrent enhancement.

IPCE measurements were performed to study their photoresponse as a function of incident light wavelength. The CdS–ZnO–ZnO–CdS sample had a pronounced IPCE of 35–45% at wavelengths up to 500 nm and then the IPCE dropped to a minimal level at wavelengths below the CdS band-gap energy of ~550 nm. On the other hand, the CdS–ZnO–ZnO–CdSe sample showed photoactivity at a broader range of wavelengths from 400 to 650 nm due to the smaller band gap of CdSe, with a relatively lower IPCE value of ~25%. At the same incident wavelength (400–500 nm), the higher IPCE revealed that the CdS–ZnO nanocomposite was more efficient than CdSe–ZnO in separating and collecting photoexcited electrons, which is in agreement with the larger potential difference between the conduction band edges of CdS and ZnO. It is noteworthy that the cosensitized CdS–ZnO–ZnO–CdSe nanowire photoanode showed a nearly constant IPCE of ~45% in the whole visible region from 400 to 650 nm, clearly demonstrating the advantage of this tandem structure. The high photoactivity at the incident wavelength from 400 to 500 nm is expected as both CdS and CdSe layers have substantial absorption in this region. More importantly, the IPCE of CdS–ZnO–ZnO–CdSe photoanode is about two times higher than that of CdSe–ZnO–ZnO–CdSe at the incident wavelength of 550–650 nm, where the CdS layer in the cosensitized sample has minimal light absorption. The enhanced IPCE was attributed to the Fermi level alignment between CdS and CdSe, which increased the potential difference between the conduction band edges of CdSe and ZnO in the cosensitized sample. The increased potential difference facilitated the electron separation and transfer from CdSe QD to ZnO nanowire in the cosensitized sample and thereby improved the IPCE.

Moreover, in comparison to single-sided cosensitized nanowire structure (CdS–CdSe–ZnO), this double-sided architecture that enables direct interaction between QD and nanowire showed improved charge-collection efficiency. As shown in Fig. 6.19a, they had similar current–voltage profile at a bias below –0.1 V. While the photocurrent of CdS–CdSe–ZnO saturated quickly with a maximum current density of ~6 mA/cm^2, there was no current saturation for the double-sided CdS–ZnO–ZnO–CdSe sample at large bias. The data suggested that electron transfer in the CdSe–CdS–ZnO is less efficient compared to CdS–ZnO–ZnO–CdSe due to the presence of this intermediate CdS layer that would increase the chance of electron–hole recombination and limit electron collection efficiency. The limited electron transfer efficiency from CdSe QD to ZnO nanowire was also reflected in the IPCE data. The IPCE of CdSe–CdS–ZnO sample at the wavelength above 500 nm was smaller compared to CdS–ZnO–ZnO–CdSe sample (Fig. 6.19b), indicative of relatively low collection efficiency of electrons created in CdSe. Importantly, the double-sided cosensitized sample addressed this limitation, with CdS and CdSe QDs that directly contact ZnO nanowires allowing electron transfer efficiently from QD to nanowire.

The results represented the first double-sided nanowire photoanode that integrates uniquely two semiconductor QDs of distinct band gaps for PEC hydrogen generation and can be possibly applied to other applications such as nanostructured tandem PV cells.

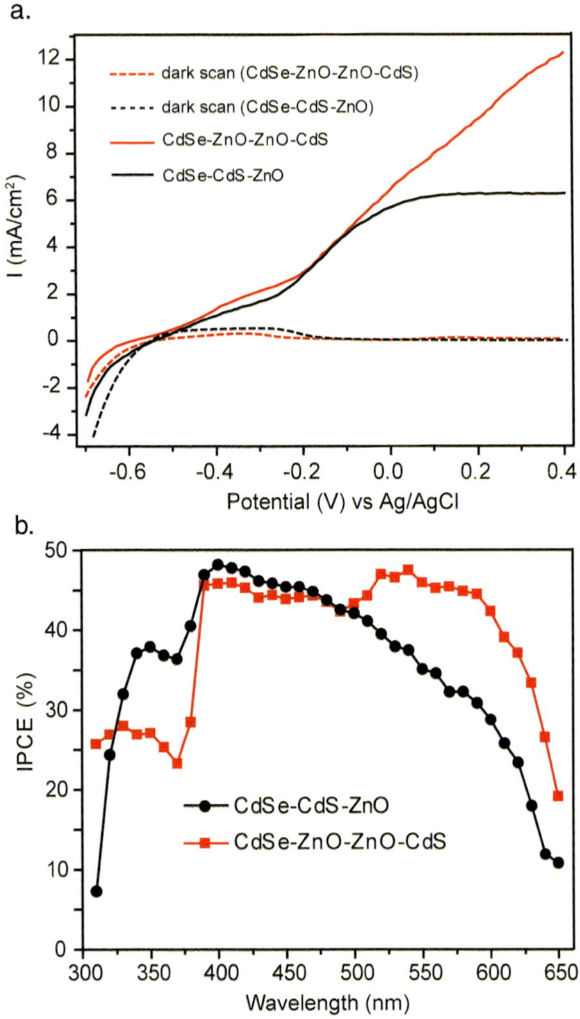

Fig. 6.19 Comparison of PEC performance of double-sided CdS–ZnO–ZnO–CdSe and single-sided CdSe–CdS–ZnO samples. (**a**) Linear sweep voltammograms, collected at a scan rate of 10 mV/s at applied potentials from –0.7 to +0.4 V vs. Ag/AgCl at 100 mW/cm^2; red and black dashed line are the dark scan of CdS–ZnO–ZnO–CdSe and CdSe–CdS–ZnO samples, respectively. (**b**) Measured IPCE spectra at the incident wavelength from 310 to 650 nm at a potential of 0 V vs. Ag/AgCl [60]

6.5 Synergistic Effect of Quantum Dot Sensitization and Elemental Doping

To date, both sensitization and doping of metal oxide nanostructures have been explored separately for PEC hydrogen generation, yet little work has been done on combining the two approaches. In 2008, Lopez-Luke et al. demonstrated a synergistic effect in combining CdSe QD sensitization with N-doping of TiO_2 for photovoltaic cells [103]. The device showed substantially enhanced PV response and was tentatively attributed to increased hole transport from the CdSe valence band to the N-dopant level. Recently, Hensel et al. reported the first example of CdSe QD-sensitized *and* N-doped TiO_2 nanomaterials for PEC hydrogen generation [99]. A significant synergistic effect between QD sensitization and N-doping was observed in both anatase TiO_2 nanoparticle films and vertically aligned rutile TiO_2 nanowire arrays. Here we will use TiO_2 nanowire as an example for discussing the PEC results.

The vertically aligned TiO_2 nanowire arrays were prepared via a non-polar solvent/hydrophilic substrate interfacial reaction under mild hydrothermal conditions, reported previously [16]. These nanowire arrays were N-doped by annealing in ammonia and sensitized with CdSe QDs using CBD method. The linear sweep voltammograms recorded from four samples (pristine TiO_2, N-doped TiO_2, QD-sensitized TiO_2, and QD-sensitized N-doped TiO_2) in dark and with light illumination of 100 mW/cm^2 are illustrated in Fig. 6.20. The results showed a general trend in photocurrent density of samples: CdSe–TiO_2:N > CdSe–TiO_2 > TiO_2 ~ TiO_2:N. Pristine TiO_2 and TiO_2:N samples exhibited a similar photocurrent density of ~0.5 mA/cm^2 at –0.2 V vs. Ag/AgCl, which indicated that N-doping has no obvious effect on photocurrent. Both TiO_2 and TiO_2:N nanowire samples showed a great enhancement in photocurrent after CdSe sensitization due to improved visible light absorption. Significantly, the CdSe–TiO_2:N nanowires showed the greatest photocurrent density of 2.75 mA/cm^2, which is almost two times enhancement compared to CdSe–TiO_2 nanowires. These results confirmed the synergistic effect in combining CdSe QD sensitization and N-doping since no enhancement was observed in N-doped TiO_2.

A theoretical model was developed to explain the synergistic effect. In comparison to the undoped TiO_2 sample, TiO_2:N has a higher density of partially occupied oxygen vacancy states, V_o, located at ~0.4 eV above the CdSe valence band edge (Fig. 6.21) that can facilitate hole transfer from CdSe to TiO_2 following photoexcitation of the CdSe QDs. This interfacial hole transfer was believed to improve the PEC photocurrent of CdSe–TiO_2:N nanoparticle films in two ways. First, it can reduce the electron–hole recombination in CdSe QDs. Second, the holes transferred to the V_o levels in TiO_2 can either oxidize the sacrificial reagent on site or be further transported through the TiO_2 network to other oxidation sites, the latter being especially important for thick nanocrystalline films. The recombination between the holes transferred to the V_o levels and the electrons in the conduction band of TiO_2 or CdSe is expected to be not significant, since the coupling between the localized V_o states and the delocalized conduction band should be weak. This model was also modified from the previous model proposed to explain the enhanced photoresponse in CdSe–TiO_2:N films for PV cells where N-doping was thought to directly facilitate hole transport [103]. These PEC results strongly suggest that the enhancement in photoresponse for the CdSe QD-sensitized *and* N-doped TiO_2 was due to improved hole transfer/transport enhanced by oxygen vacancy states mediated by N-doping. It provided useful insights for developing new nanostructures tailored for PEC hydrogen generation and other applications via controlled band engineering. A long-term goal would be the detailed kinetic studies on the interfacial carrier transfer between CdSe QD and TiO_2, which could lead to a better understanding of the carrier transfer.

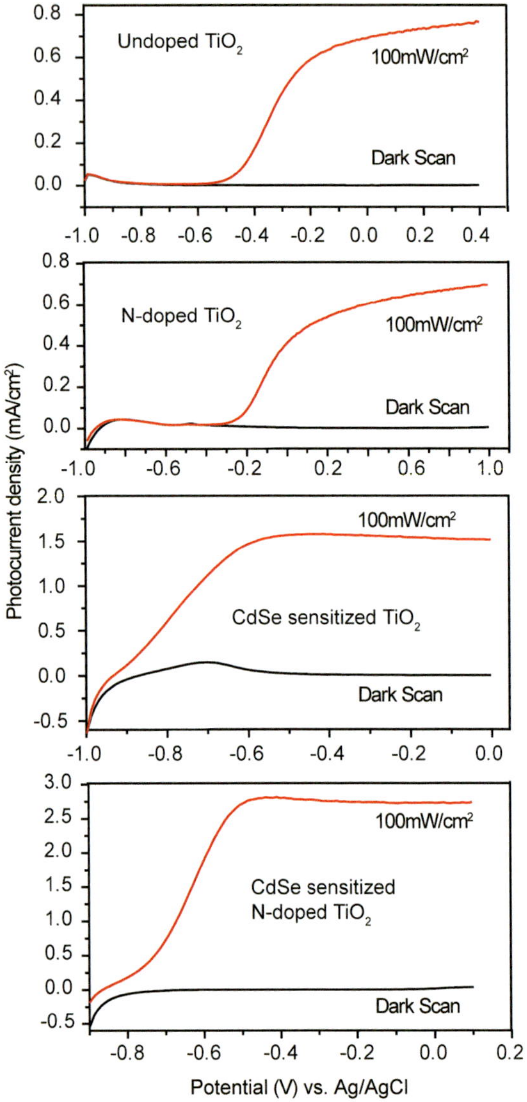

Fig. 16.20 Linear sweep voltammograms collected at a scan rate of 10 mV/s from TiO_2, TiO_2:N, CdSe–TiO_2, and CdSe–TiO_2:N nanowire arrays, in dark and with light illumination of 100 mW/cm^2 [99]

6.6 Concluding Remarks

This chapter has provided a brief overview of some of the recent research activities in the study of 1D semiconductor metal oxides for PEC hydrogen generation. With precise control over the size, morphology, and doping of nanomaterials as well as material and device structure, they open up new opportunities in addressing the fundamental issues in PEC hydrogen generation. Advantages of 1D nanomaterials

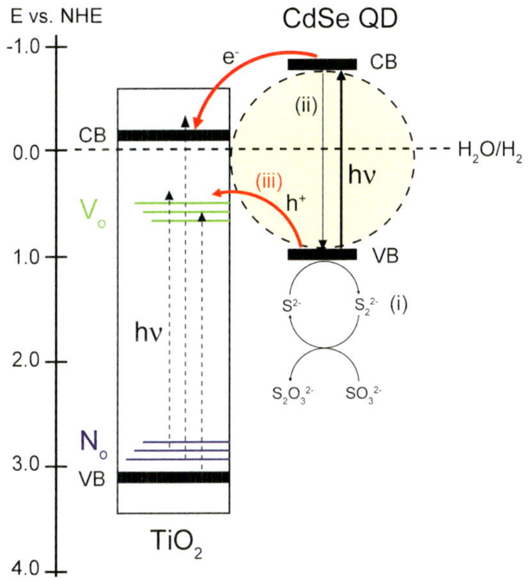

Fig. 6.21 Proposed model for the electron transfer at CdSe/TiO_2 interface in a CdSe–TiO_2:N sample. All the energy levels are referenced to NHE scale. CB and VB are conduction band and valence band, respectively. Green lines and blue lines represent the energy levels of V_o and N_o, respectively. The *horizontal dashed line* indicates H_2O/H_2 potential level. *Red arrows* highlight the hole and electron transfer from CdSe to TiO_2. *Black dashed arrows* highlight the possible electronic transitions between the different energy levels in TiO_2. The schematic diagram shows three possible competing pathways for the photogenerated holes in CdSe: (i) oxidization of S^{2-} to S_2^{2-}; (ii) recombination with electrons in the CB, and (iii) transfer to V_o levels in TiO_2 [99]

for PEC applications include large surface area, small size, significantly reduced distance for carrier diffusion, and low light reflectivity, as demonstrated in many studies. However, several challenging issues remain to be addressed in order for PEC hydrogen generation to become practical. One issue is efficiency that is currently still low, i.e., less than 1%. Another issue is long-term stability. Both issues are likely rooted in the high density of defect states, largely due to large surface-to-volume ratio, that act to trap charge carriers and thus lower useful photocurrent as well as result in undesired irreversible chemical or photochemical reactions that degrade the photoelectrodes. Possible solutions to these problems include improved material architectures and properties based on more advanced nanostructures and device architectures that can help to reduce defects and improve charge transport and transfer. It is highly hopeful that some of the fundamental problems of metal oxide semiconductors for PEC applications, such as light absorption, charge-collection efficiency, and photochemical stability, are expected to be solved by developing new nanomaterials and device architectures.

References

1. A. Yilanci, I. Dincer, H.K. Ozturk, A review on solar-hydrogen/fuel cell hybrid energy systems for stationary applications. Prog. Energy Combustion Sci. **35**(3), 231–244 (2009)
2. Y. Wang, A.E. Rodrigures, Hydrogen production from steam methane reforming coupled with in situ CO_2 capture: conceptual parametric study. Fuel **84**, 1778–1789 (2005)
3. C.A. Grimes, O.K. Varghese, S. Ranjan, *Light, Water, Hydrogen: The Solar Generation of Hydrogen by Water Photoelectrolysis* (Springer, New York, NY, 2008)
4. M. Ni, D.Y.C. Leung, M.K.H. Leung, K. Sumathy, An overview of hydrogen production from biomass. Fuel Process. Technol. **87**, 461–472 (2006)
5. T. Bak, J. Nowotny, M. Rekas, C.C. Sorrell, Photoelectrochemical hydrogen generation from water using solar energy. Materials-related aspects. Int. J. Hydrogen Energy **27**, 991–1022 (2002)
6. A.J. Bard, M.A. Fox, Artificial photosynthesis – solar splitting of water to hydrogen and oxygen. Acc. Chem. Res. **28**, 141–145 (1995)
7. K. Honda, A. Fujishima, Electrochemical photolysis of water at a semiconductor electrode. Nature **238**, 37–39 (1972)
8. N.S. Lewis, Light work with water. Nature **414**, 589 (2001)
9. M. Ni, M.K.H. Leung, D.Y.C. Leung, K. Sumathy, A review and recent developments in photocatalytic water-splitting using TiO_2 for hydrogen production. Renew Sust. Energy Rev. **11**(3), 401–425 (2007)
10. K. Rajeshwar, Hydrogen generation at irradiated oxide semiconductor-solution interfaces. J. Appl. Electrochem. **37**(7), 765–787 (2007)
11. R. van de Krol, Y.Q. Liang, J. Schoonman, Solar hydrogen production with nanostructured metal oxides. J. Mater. Chem. **18**(20), 2311–2320 (2008)
12. S. Dutta, Technology assessment of advanced electrolytic hydrogen production. Int. J. Hydrogen Energy **15**(6), 379–386 (1990)
13. D.L. Stojic, M.P. Marceta, S.P. Sovilj, S.S. Miljanic, Hydrogen generation from water electrolysis-Possibilities of energy saving. J. Power Sources **118**, 315–319 (2003)
14. S.U.M. Khan, M. Al-Shahry, W.B. Ingler, Efficient photochemical water splitting by a chemically modified n-TiO_2. Science **297**, 2243–2245 (2002)
15. O. Khaselev, J.A. Turner, A monolithic photovoltaic-photoelectrochemical device for hydrogen production via water splitting. Science **280**, (5362), 425–427 (1998)
16. X. Feng, K. Shankar, C.K. Varghese, M. Paulose, T.J. Latemp, C.A. Grimes, Vertically aligned single crystal TiO2 nanowire arrays grown directly on transparent conducting oxide coated glass: synthesis details and applications. Nano Lett. **8**(11), 3781–3786 (2008)
17. E.Y. Kim, J.H. Park, C.Y. Han, J. Power Sources **184**, 284 (2008)
18. B. Liu, E.S. Aydril, J. Am. Chem. Soc. **131**, 3985–3990 (2009)
19. A. Wolcott, W.A. Smith, T.R. Kuykendall, Y.P. Zhao, J.Z. Zhang, Photoelectrochemical water splitting using dense and aligned TiO2 nanorod arrays. Small **5**(1), 104–111 (2009)
20. A. Wolcott, W.A. Smith, Y.P. Zhao, J.Z. Zhang, Photoelectrochemical study of nanostructured ZnO thin films for hydrogen generation from water splitting. Adv. Funct. Mater. **19**(12), 1849–1856 (2009)
21. A. Wolcott, J.Z. Zhang, W.A. Smith, Y.P. Zhao, WO_3 and ZnO nanostructures for photoelectrochemical generation of hydrogen from water splitting. Appl. Phys. Lett. (2008), to be submitted
22. X. Yang, A. Wolcottt, G. Wang, A. Sobo, R.C. Fitzmorris, F. Qian, J.Z. Zhang, Y. Li, Nitrogen-doped ZnO nanowire arrays for photoelectrochemical water splitting. Nano Lett. **9**(6), 2331–2336 (2009)
23. J.R. Bolton, Solar photoproduction of hydrogen: a review. Sol. Energy **57**(1), 37–50 (1996)

24. J.H. Park, S. Kim, A.J. Bard, Novel carbon-doped TiO_2 nanotube arrays with high aspect ratios for efficient solar water splitting. Nano Lett. **6**(1), 24–28 (2006)
25. T. Umebayashi, T. Yamaki, H. Itoh, K. Asai, Band gap narrowing of titanium dioxide by sulfur doping. Appl. Phys. Lett. **81**(3), 454–456 (2002)
26. A. Kongkanand, K. Tvrdy, K. Takechi, M. Kuno, P.V. Kamat, Quantum dot solar cells. Tuning photoresponse through size and shape control of CdSe-TiO_2 architecture. J. Am. Chem. Soc. **130**(12), 4007–4015 (2008)
27. Y. Tak, S.J. Hong, J.S. Lee, K. Yong, Solution-based synthesis of a CdS nanoparticle/ZnO nanowire heterostructure array. Crystal Growth Des. **9**(6), 2627–2632 (2009)
28. Y.L. Lee, B.M. Huang, H.T. Chien, Chem. Mater. **20**(22), 6903 (2008)
29. Y.W. Tang, X.Y. Hu, M.J. Chen, L.J. Luo, B.H. Li, L.Z. Zhang, Electrochim. Acta **42**, 2742 (2009)
30. H. Zhang, X. Quan, S. Chen, H.T. Yu, N. Ma, Chem. Mater. **21**(14), 3090 (2009)
31. V. Chakrapani, J. Thangala, M.K. Sunkara, WO_3 and W_2N nanowire arrays for photoelectrochemical hydrogen production. Int. J. Hydrogen Energy **34**, 9050–9059 (2009)
32. K.S. Ahn, S. Shet, T. Deutsch, C.S. Jiang, Y.F. Yan, M. Al-Jassim, J. Turner, Enhancement of photoelectrochemical response by aligned nanorods in ZnO thin films. J. Power Sources **176**(1), 387–392 (2008)
33. K.S. Ahn, Y. Yan, S. Shet, K. Jones, T. Deutsch, J. Turner, M. Al-Jassim, ZnO nanocoral structures for photoelectrochemical cells. Appl. Phys. Lett. **93**, 163117 (2008)
34. N. Beermann, L. Vayssieres, S.E. Lindquist, A. Hagfeldt, Photoelectrochemical studies of oriented nanorod thin films of hematite. J. Electrochem. Soc. **147**(7), 2456–2461 (2000)
35. T. Lindgren, H.L. Wang, N. Beermann, L. Vayssieres, A. Hagfeldt, S.E. Lindquist, Sol. Energy Mater. Aqueous photoelectrochemistry of hematite nanorod array. Sol. Cells **71**(2), 231–243 (2002)
36. S. Saretni-Yarahmadi, K.G.U. Wijayantha, A.A. Tahir, B. Vaidhyanathan, Nanostructured alpha-Fe_2O_3 electrodes for solar driven water splitting: effect of doping agents on preparation and performance. J. Phys. Chem. C **113**(12), 4768–4778 (2009)
37. V.R. Satsangi, S. Kumari, A.P. Singh, R. Shrivastav, S. Dass, Int. J. Hydrogen Energy **33**, 312 (2008)
38. M. Frites, S.U.M. Khan, Photoelectrochemical splitting of water to H2 and O2 at n-Fe2O3 nanowire films and nanocrystalline carbon-modified (CM)-n-Fe2O3 thin films. ECS Trans. **19**(3), 137–145 (2009)
39. P.R. Mishra, P.K. Shukla, O.N. Srivastava, Study of modular PEC solar cells for photoelectrochemical splitting of water employing nanostructured TiO2 photoelectrodes. Int. J. Hydrogen Energy **32**(12), 1680–1685 (2007)
40. S. Chen, M. Paulose, C. Ruan, G.K. Mor, O.K. Varghese, D. Kouzoudis, C.A. Grimes, Electrochemically synthesized CdS nanoparticle-modified TiO_2 nanotube-array photoelectrodes: preparation, characterization, and application to photoelectrochemical cells. J. Photochem. Photobio. A: Chem. **177**, 177–184 (2006)
41. D. Chen, Y.F. Gao, G. Wang, H. Zhang, W. Lu, J.H. Li, Surface tailoring for controlled photoelectrochemical properties: effect of patterned TiO2 microarrays. J. Phys. Chem. C **111**(35), 13163–13169 (2007)
42. C.J. Lin, Y.T. Lu, C.H. Hsieh, S.H. Chien, Surface modification of highly ordered TiO2 nanotube arrays for efficient photoelectrocatalytic water splitting. Appl. Phys. Lett. **94**, 11 (2009)
43. D. Eder, M. Motta, A.H. Windle, Iron-doped Pt-TiO2 nanotubes for photo-catalytic water splitting. Nanotechnology **20**, 5 (2009)
44. Y.J. Hwang, A. Boukai, P.D. Yang, High density n-Si/n-TiO2 core/shell nanowire arrays with enhanced photoactivity. Nano Lett. **9**(1), 410–415 (2009)
45. G.K. Mor, C.K. Varghese, R.H.T. Wilke, S. Sharma, K. Shankar, T.J. Latemp, K.S. Choi, C.A. Grimes, p-Type Cu-Ti-O nanotube arrays and their use in self-biased heterojunction photoelectrochemical diodes for hydrogen generation. Nano Lett. **8**, 1906–1911 (2008)

46. X. Cui, M. Ma, W. Zhang, Y.C. Yang, Z.J. Zhang, Nitrogen-doped TiO_2 from TiN and its visible light photoelectrochemical properties. Electrochem. Commun. **10**(3), 367–371 (2008)
47. S.U.M. Khan, T. Sultana, Photoresponse of n-TiO2 thin film and nanowire electrodes. Sol. Energy Mater. Sol. Cells **76**(2), 211–221 (2003)
48. S. Takabayashi, R. Nakamura, Y. Nakato, A nano-modified Si/TiO2 composite electrode for efficient solar water splitting. J. Photochem. Photobio. A: Chem. **166**(1–3), 107–113 (2004)
49. J.H. Park, O.O. Park, S. Kim, Photoelectrochemical water splitting at titanium dioxide nanotubes coated with tungsten trioxide. Appl. Phys. Lett. **89**, 16 (2006)
50. J.L. Blackburn, D.C. Selmarten, A.J. Nozik, Electron transfer dynamics in quantum dot/titanium dioxide composites formed by in situ chemical bath deposition. J. Phys. Chem. B. **107**, 14154–14157 (2003)
51. M. Gratzel, Mesoscopic solar cells for electricity and hydrogen production from sunlight. Chem. Lett. **34**, 8–13 (2005)
52. W.T. Sun, Y. Yu, H.Y. Pan, X.F. Gao, Q. Chen, L.M. Peng, CdS quantum dots sensitized TiO2 nanotube-array photoelectrodes. J. Am. Chem. Soc. **130**(4), 1124–1125 (2008)
53. R.F.G. Gardner, F. Sweett, D.W. Tanner, J. Phys. Chem. Solids **24**, 1183 (1963)
54. A.G. Joly, J.R. Williams, S.A. Chambers, G. Xiong, W.P. Hess, D.M. Laman, J. Appl. Phys. **99**, 053521 (2006)
55. J.H. Kennedy, K.W. Frese, J. Electrochem. Soc. **125**(5), 709 (1978)
56. L. Vayssieres, N. Beermann, S.E. Lindquist, A. Hagfeldt, Controlled aqueous chemical growth of oriented three-dimensional crystalline nanorod arrays: application to iron(III) oxides. Chem. Mater. **13**(2), 233–235 (2001)
57. L. Vayssieres, C. Sathe, S.M. Butorin, D.K. Shuh, J. Nordgren, J.H. Guo, One-dimensional quantum-confinement effect in a-Fe2O3 ultrafine nanorod arrays. Adv. Mater. **17**, 2320–2323 (2005)
58. P. Salvador, Hole diffusion length in n-TiO2 single crystals and sintered electrodes: photoelectrochemical determination and comparative analysis. J. Appl. Phys. **55**(8), 2977–2985 (1983)
59. E. Hendry, F. Wang, J. Shan, T.F. Heinz, M. Bonn, Electron transport in TiO2 probed by THz time-domain spectroscopy. Phys. Rev. B **69**, 081101(R) (2004)
60. G. Wang, X.Y. Yang, F. Qian, J.Z. Zhang, Y. Li, Double-sided CdS and CdSe quantum dot co-sensitized ZnO nanowire arrays for photoelectrochemical hydrogen generation. Nano Lett. **10**, 1088–1092 (2010)
61. L. Greene, M. Law, J. Goldberger, F. Kim, J.C. Johnson, Y.F. Zhang, R.J. Saykally, P.D. Yang, Low-temperature wafer scale production of ZnO nanowire arrays. Angew. Chem. Int. Ed. **42**, 3031–3034 (2003)
62. I. Bedja, S. Hotchandani, R. Carpentier, K. Vinodgopal, P.V. Kamat, Electrochromic and photoelectrochemical behavior of thin WO3 films prepared from quantized colloidal particles. Thin Solid Films **247**, 195–200 (1994)
63. I. Saeki, N. Okushi, H. Konno, R. Furuichi, The photoelectrochemical response of TiO_2-WO_3 mixed oxide films prepared by thermal oxidation of titanium coated with tungsten. J. Electrochem. Soc. **143**(7), 2226–2230 (1996)
64. Y.Q. Wang, H.M. Cheng, L. Zhang, Y.Z. Hao, J.M. Ma, B. Xu, W.H. Li, The preparation, characterization, photoelectrochemical and photocatalytic properties of lanthanide metal-ion-doped TiO2 nanoparticles. J. Mol. Catal. A-Chem. **151**(1–2), 205–216 (2000)
65. U.O. Krasovec, M. Topic, A. Georg, A. Georg, G. Drazic, Preparation and characterisation of nano-structured WO3-TiO2 layers for photoelectrochromic devices. J. Sol-Gel. Sci. Technol. **36**(1), 45–52 (2005)
66. C.V. Ramana, S. Utsunomiya, R.C. Ewing, C.M. Julien, U. Becker, Structural stability and phase transitions in WO3 thin films. J. Phys. Chem. B **110**(21), 10430–10435 (2006)

67. A. Wolcott, T.R. Kuykendall, W. Chen, S.W. Chen, J.Z. Zhang, Synthesis and characterization of ultrathin WO3 nanodisks utilizing long-chain poly(ethylene glycol). J. Phys. Chem. B **110**(50), 25288–25296 (2006)
68. J. Thangala, S. Vaddiraju, R. Bogale, R. Thurman, T. Powers, B. Deb, M.K. Sunkara, Large-scale, hot-filament-assisted synthesis of tungsten oxide and related transition metal oxide nanowires. Small **3**(5), 890–896 (2007)
69. G.K. Mor, K. Shankar, M. Paulose, O.K. Varghese, C.A. Grimes, Enhanced photocleavage of water using titania nanotube arrays. Nano Lett. **5**(1), 191–195 (2005)
70. S.K. Mohapatra, M. Misra, V.K. Mahajan, K.S. Raja, Design of a highly efficient photoelectrolytic cell for hydrogen generation by water splitting: application of $TiO_{2-x}C_x$ nanotubes as a photoanode and Pt/TiO_2 nanotubes as a cathode. J. Phys. Chem. C **111**(24), 8677–8685 (2007)
71. S.K. Mohapatra, V.K. Mahajan, M. Misra, Double-side illuminated titania nanotubes for high volume hydrogen generation by water splitting. Nanotechnology **18**, 44 (2007)
72. S.K. Mohapatra, M. Misra, V.K. Mahajan, K.S. Raja, A novel method for the synthesis of titania nanotubes using sonoelectrochemical method and its application for photoelectrochemical splitting of water. J. Catal. **246**(2), 362–369 (2007)
73. G.K. Mor, H.E. Prakasam, O.K. Varghese, K. Shankar, C.A. Grimes, Vertically oriented Ti-Fe-O nanotube array films: toward a useful material architecture for solar spectrum water photoelectrolysis. Nano Lett. **7**(8), 2356–2364 (2007)
74. C.K. Xu, Y.A. Shaban, W.B. Ingler, S.U.M. Khan, Sol. Energy Mater. Nanotube enhanced photoresponse of carbon modified (CM)-n-TiO_2 for efficient water splitting. Sol. Cells **91**(10), 938–943 (2007)
75. Y.X. Yin, Z.G. Jin, F. Hou, Enhanced solar water-splitting efficiency using core/sheath heterostructure CdS/TiO_2 nanotube arrays. Nanotechnology **18**, 49 (2007)
76. O.K. Varghese, C.A. Grimes, Appropriate strategies for determining the photoconversion efficiency of water photo electrolysis cells: a review with examples using titania nanotube array photoanodes. Sol. Energy Mater. Sol. Cells **92**(4), 374–384 (2008)
77. S.K. Mohapatra, S.E. John, S. Banerjee, M. Misra, Water photoxidation by smooth and ultrathin a-Fe_2O_3 nanotube arrays. Chem. Mater. **21**, 3048–3055 (2009)
78. T.J. Latemp, X. Feng, M. Paulose, C.A. Grimes, Temperature-dependent growth of self-assembled hematite (a-Fe2O3) nanotube arrays: rapid electrochemical synthesis and photoelectrochemical properties. J. Phys. Chem. C **113**, 16293–16298 (2009)
79. W.B. Ingler, S.U.M. Khan, Int. J. Hydrogen Energy **30**, 821 (2005)
80. I. Cesar, A. Kay, J.A. Gonzalez Martinez, M. Gratzel, J. Am. Translucent thin film Fe2O3 photoanodes for efficient water splitting by sunlight: nanostructure-directing effect of si-doping. Chem. Soc. **128**, 4582–4583 (2006)
81. I. Cesar, K. Sivula, A. Kay, R. Zboril, M. Gratzel, J. Phys. Chem. C **113**, 772 (2009)
82. Y. Yan, K.S. Ahn, S. Shet, T. Deutsch, M. Huda, S.H. Wei, J. Turner, M.M. Al-Jassim, Band gap reduction of ZnO for photoelectrichemical splitting of water. Sol. Hydrogen Nanotechnol. II Proc. SPIE **6650**, 66500H (2007)
83. K.R. Reyes-Gil, E.A. Reyes-Garcia, D. Raftery, Nitrogen-doped In2O3 thin film electrodes for photocatalytic water splitting. J. Phys. Chem. C **111**(39), 14579–14588 (2007)
84. R. Asahi, T. Morikawa, T. Ohwaki, K. Aoki, Y. Taga, Visible-light photocatalysis in nitrogen-doped titanium oxides. Science **293**, 269 (2001)
85. S. Sakthivel, H. Kisch, Daylight photocatalysis by carbon-modified titanium dioxide. Angew. Chem. Int. Ed. **42**, 4908 (2003)
86. T. Ohno, T. Mitsui, M. Matsumura, Photocatalytic activity of S-doped TiO2 photocatalyst under visible light. Chem. Lett. **32**, 364 (2003)
87. N. Lu, X. Quan, J.Y. Li, S. Chen, H.T. Yu, G.H. Chen, Fabrication of boron-doped TiO2 nanotube array electrode and investigation of its photoelectrochemical capability. J. Phys. Chem. C **111**, 11836–11842 (2007)

88. Y. Su, S. Chen, X. Quan, H. Zhao, Y. Zhang, A silicon-doped TiO2 nanotube arrays electrode with enhanced photoelectrocatalytic activity. Appl. Surf. Sci. **255**, 2167–2172 (2008)
89. S.S. Kocha, J.A. Turner, A.J. Nozik, J. Electroanal. Chem. **367**, 27 (1994)
90. A.J. Nozik, Appl. Phys. Lett. **30**, 567 (1977)
91. H. Wang, T. Deutsch, J.A. Turner, Direct water splitting under visible light with nanostructured hematite and WO_3 photoanodes and a $GaInP_2$ photocathode. J. Electrochem. Soc. **155**(5), F91–F96 (2008)
92. K.S. Ahn, Y. Yan, S.H. Lee, T. Deutsch, J. Turner, C.E. Tracy, C.L. Perkins, M.M. Al-Jassim, Photoelectrochemical properties of N-incorporated ZnO films deposited by reactive RF magnetron sputtering. J. Electrochem. Soc. **154**(9), B956–B959 (2007)
93. K.S. Ahn, Y. Yan, S. Shet, T. Deutsch, J. Turner, M. Al-Jassim, Enhanced photoelectrochemical responses of ZnO films through Ga and N codoping. Appl. Phys. Lett. **91**, 231909 (2007)
94. A. Kay, I. Cesar, M. Gratzel, New benchmark for water photooxidation by nanostructured a-Fe2O3 films. J. Am. Chem. Soc. **128**, 15714–15721 (2006)
95. P.V. Kamat, Quantum dot solar cells. Semiconductor nanocrystals as light harvesters. J. Phys. Chem. C **112**(48), 18737–18753 (2008)
96. I. Robel, V. Subramanian, M. Kuno, P.V. Kamat, Quantum dot solar cells. Harvesting light energy with CdSe nanocrystals molecularly linked to mesoscopic TiO2 films. J. Am. Chem. Soc. **128**(7), 2385–2393 (2006)
97. R.S. Dibbell, D.F. Watson, Distance-dependent electron transfer in tethered assemblies of CdS quantum dots and TiO2 nanoparticles. J. Phys. Chem. C **113**(8), 3139–3149 (2009)
98. S. Yamada, A.Y. Nosaka, Y. Nosaka, Fabrication of US photoelectrodes coated with titania nanosheets for water splitting with visible light. J. Electroanal. Chem. **585**(1), 105–112 (2005)
99. J. Hensel, G. Wang, Y. Li, J.Z. Zhang, Synergistic effect of CdSe quantum dot sensitization and nitrogen doping of TiO2 nanostructures for photoelectrochemical applications. Nano Lett. **10**, 478 (2010)
100. Y.L. Lee, C.F. Chi, S.Y. Liau, CdS/CdSe co-sensitized TiO_2 photoelectrode for efficient hydrogen generation in a photoelectrochemical cell. Chem. Mater. **22**(3), 922–927 (2010)
101. L.J. Diguna, Q. Shen, J. Kobayashi, T. Toyoda, High efficiency of CdSe quantum-dot-sensitized TiO2 inverse opal solar cells. Appl. Phys. Lett. **91**(2), 023116 (2007)
102. O. Niitsoo, S.K. Sarkar, C. Pejoux, S. Ruhle, D. Cahen, G. Hodes, Chemical bath deposited CdS/CdSe-sensitized porous TiO2 solar cells. J. Photochem. Photobiol. A Chem. **181**(2–3), 306–313 (2006)
103. T. Lopez-Luke, A. Wolcott, L.P. Xu, S. Chen, Z. Wen, J.H. Li, E. De La Rosa, J.Z. Zhang, Nitrogen doped and CdSe quantum dot sensitized nanocrystalline TiO2 films for solar energy conversion applications. J. Phys. Chem. C **112**, 1282–1292 (2008)
104. K.S. Leschkies, R. Divakar, J. Basu, E. Enache-Pommer, J.E. Boercker, C.B. Carter, U.R. Kortshagen, D.J. Norris, E.S. Aydril, Photosensitization of ZnO nanowires with CdSe quantum dots for photovoltaic devices. Nano Lett. **7**, 1793–1798 (2007)
105. C. Ratanatawanate, C.R. Xiong, K.J. Balkus, Fabrication of PbS quantum dot doped TiO2 nanotubes. ACS Nano **2**(8), 1682–1688 (2008)
106. J.H. Bang, P.V. Kamat, Solar cell by design. Photoelectrochemistry of TiO2 nanorod arrays decorated with CdSe. Adv. Funct. Mater. **20**, 1970–1976 (2010)
107. D.R. Baker, P.V. Kamat, Photosensitization of TiO2 nanostructures with CdS quantum dots. Particulate versus tubular support architectures. Adv. Funct. Mater. **19**, 805–811 (2009)

Chapter 7
Helical Nanostructures: Synthesis and Potential Applications

Pu-Xian Gao and Gang Liu

7.1 Introduction

In nature, three-dimensional (3D) helical structure is the most fundamental structural configuration of DNAs, proteins, and bio-functional groups, such as cytoplasm and periplasm [1]. Synthetically, many 3D helical structures with micro- and nano-features have been fabricated from a number of inorganic materials. Typical examples include ZnO nanohelices/nanosprings [2–7], SiO_2 nanohelices [8, 9], carbon-based nano-/microcoils [10–12], and some other nanohelices based on III–V and II–VI semiconductors [13–16]. Because of their nanoscale 3D spiral symmetric geometry, as well as unique mechanical, electrical, and electromagnetic properties, helically nanostructured materials are attracting considerable attention and have potential applications in electronics, optics, nano- and micro-electromechanical system (NEMS and MEMS), energy and environment-related technologies, and biomedicine.

In the past decade, different methods, such as glancing angle deposition (GLAD) [17–18], thermal vapor deposition [2, 3], focused ion beam chemical vapor deposition [19], and template method [15], have been developed for the synthesis of 3D nanomaterials with helical structures. However, the rationale control of the chirality, length, pitch, shape, and orientation of the helical nanostructures is still a daunting task. In addition, compared with the synthesis of 1D structured nanotubes, nanowires, and nanorods, the yield of 3D nanohelices is usually low (mostly less than 30%).

In this chapter, we review the latest progress in the synthesis of different helical nanostructures such as semiconductor helices, carbon-related helical nanostructures, and other helical nanostructures. In the meantime, we review their unique and novel mechanical, electrical, and electromagnetic properties of individual nanohelices and their relevant applications.

P.-X. Gao (✉)
Department of Chemical, Materials and Biomolecular Engineering and Institute of Materials Science, University of Connecticut, Storrs, CT 06269-3136, USA
e-mail: puxian.gao@ims.uconn.edu

W.L. Zhou, Z.L. Wang (eds.), *Three-Dimensional Nanoarchitectures*,
DOI 10.1007/978-1-4419-9822-4_7,

7.2 Semiconductor Nanohelices

7.2.1 ZnO Nanohelices

Zinc oxide (ZnO) is a typical semiconducting oxide material, which can be synthetically fabricated in different helical forms including single-crystal nanohelices [3], nanorings [20], and nanobows [21] by bending/folding polar surface-dominated nanobelts. These growth processes are thermodynamically driven by a minimization of the electrostatic energy contributed by the cation- and anion-terminated surfaces in this ionic material, and the helical structured shapes are determined by a balance between the electrostatic energy and the elastic deformation energy.

7.2.1.1 Superlattice-Structured ZnO Nanohelices

In 2005, Gao et al. [2] discovered a distinctive helical structure of ZnO made of a superlattice-structured nanobelt that is formed spontaneously in a solid–vapor growth process. The superlattice nanobelt is a periodic, coherent, epitaxial, and parallel growth of two alternating nanostripes of ZnO crystals oriented with their *c*-axes perpendicular to each other.

The ZnO nanohelices were grown with high reproducibility via a vapor–solid process [2] by using temperature ramping strategy to control the growth kinetics. The experimental setup consists of a horizontal high-temperature tube furnace, an alumina tube, a rotary pump system, and a gas controlling system. First, 2 g of commercial ZnO powder (Alfa Aesar) was compacted and loaded into an alumina boat and positioned at the center of the alumina tube as the source material. The system was pre-pumped to $\sim 2 \times 10^{-2}$ mbar, and the ramp rate was controlled at 20–25°C/min when the temperature was raised from room temperature to 800°C. The furnace was then held at 800°C for 20 min and the temperature was ramped at 20°C/min from 800 to 1400°C. When the temperature reached 1000°C, argon was introduced as a carrier gas to raise the pressure from $\sim 2 \times 10^{-2}$ mbar to the desired synthesis pressure of 200–250 mbar within ~2.5 min. The solid–vapor deposition was carried out at 1400°C for ~2 h under a pressure of 200–250 mbar. The argon carrier gas was kept at a flow rate of 50 sccm (standard cubic centimeters per minute).

Figure 7.1a shows the scanning electron microscopy (SEM) images of the newly discovered ZnO helical nanostructure in the form of a crystal-orientation-modulated superlattice (Fig. 7.1b). The superlattice nanostructure is a periodic, coherent, epitaxial, and parallel assembly of two alternating stripes (I and II) of zinc oxide nanocrystals oriented with *c*-axes perpendicular to each other (bottom inset in Fig. 7.1b). Each nanostripe crystal has a width of ~1.8 nm, defining a period of ~3.5 nm. The nanostripes I and II have top and bottom surface pair of $\pm(0001)$ polar surfaces and $\pm\{01\bar{1}0\}$ non-polar surfaces, respectively. The nanohelix superlattice is about 300–800 nm in diameter, ~10–30 nm thick, and ~100–600 nm wide.

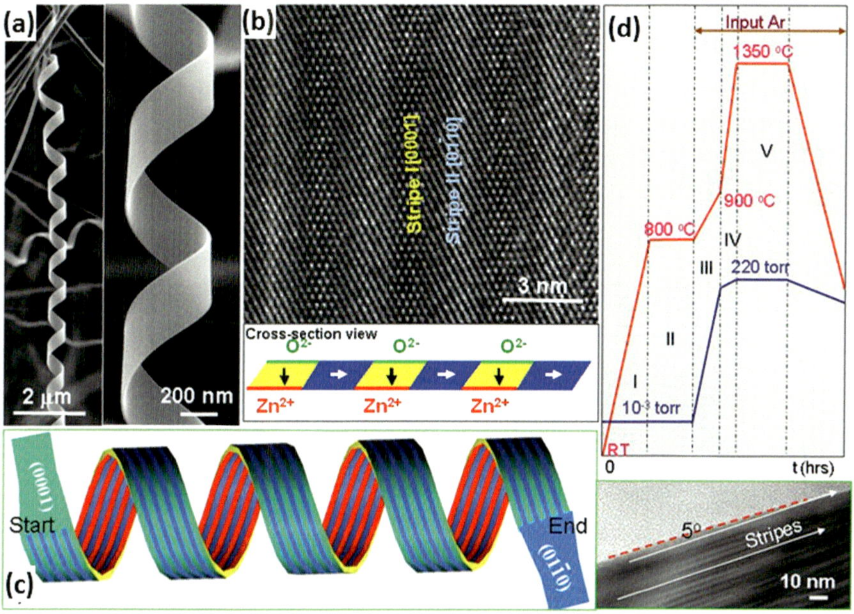

Fig. 7.1 (**a**) Low (*left*) and high magnification SEM images of a right-handed ZnO nanohelix superlattice; (**b**) nanohelix superlattice nanotripes I and II dominated by ±(0001) polar surfaces and $\pm\{01\bar{1}0\}$ non-polar surfaces, respectively; (**c**) superlattice nanohelix model with periodically running nanostripes; and (**d**) the synthesis parameter control profile of ZnO superlattice nanoehlices

The observed superlattice lengths are in the range of ~1–500 μm. The nanohelix growth starts with a structural transformation from a single-crystal ±(0001) dominated nanobelt into a $(01\bar{1}0)/(0001)$ superlattice nanohelix and then is terminated with a transformation into a $\pm\{01\bar{1}0\}$ dominated single-crystal nanobelt (Fig. 7.1c). It is suggested that reducing the polar surfaces could be the driving force for forming the superlattice structure, and the rigid structural rotation/twisting caused by the superlattice results in the initiation and formation of the nanohelix. It is worth noting that with a ~5° offset shown in the dark-field transmission electron microscopy (TEM) image in the right inset of Fig. 7.1c, the nanostripes are nearly parallel to the nanobelt growth direction and run along the length of the superlattice nanohelix.

During this growth, pure ZnO was used as the source materials, and the process was a vapor–solid (VS) process, rather than a vapor–liquid–solid (VLS) process. The five-step synthesis parameter control profile shown in Fig. 7.1d was used for this orientation-modulated superlattice growth, suggesting a necessary dynamic control over parameters for this type of growth. With the systematic control of the parameters, the yield of superlattice nanohelices can reach 10–20%, relatively lower than the high yield of single-crystalline nanohelices, which is above 50% [3].

7.2.1.2 Superelasticity, Nanobuckling, and Nonlinear Electronic Transport of Superlattice-Structured ZnO Nanohelices

To unravel the physical characteristics of superlattice nanohelices, Gao et al. conducted some experimental work on the nanomechanical manipulation and electrical characterization. During the mechanical manipulation process, some unusual nanoscale phenomena were observed including superelasticity (Fig. 7.2)

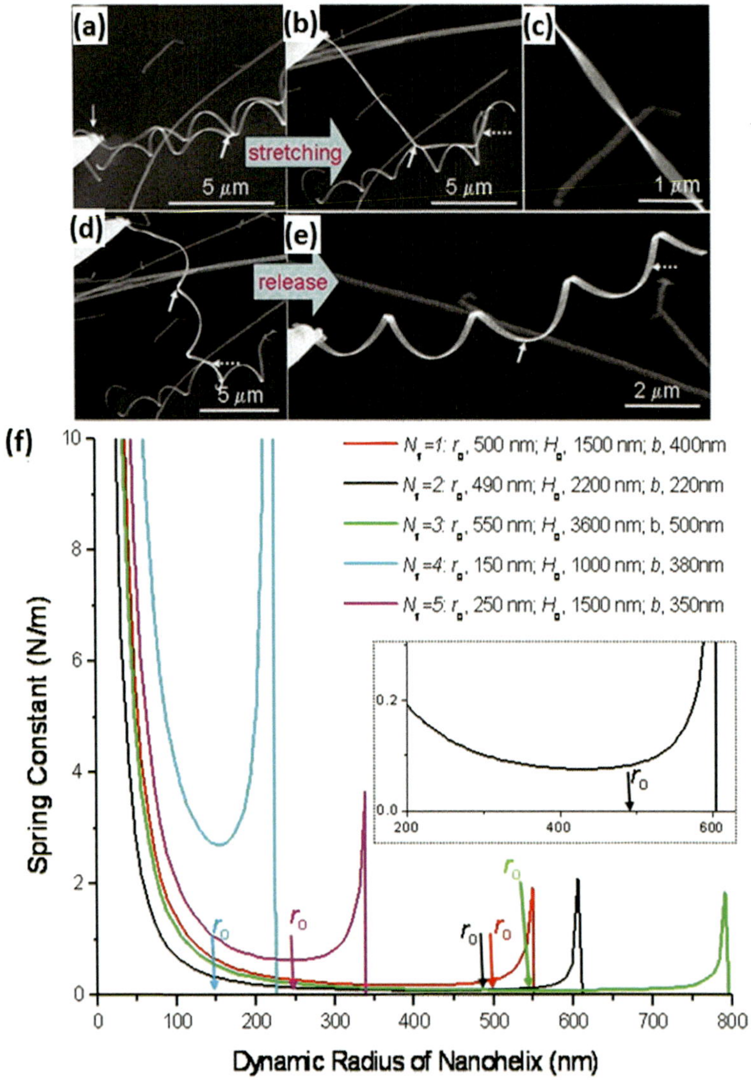

Fig. 7.2 An in situ mechanical nano-manipulation (**a–e**) of a superlattice nanohelix using a tungsten nanoprobe under a focused ion beam microscope; (**f**) a nonlinear function of dynamic spring constant versus radius of nanohelix during compression and tension

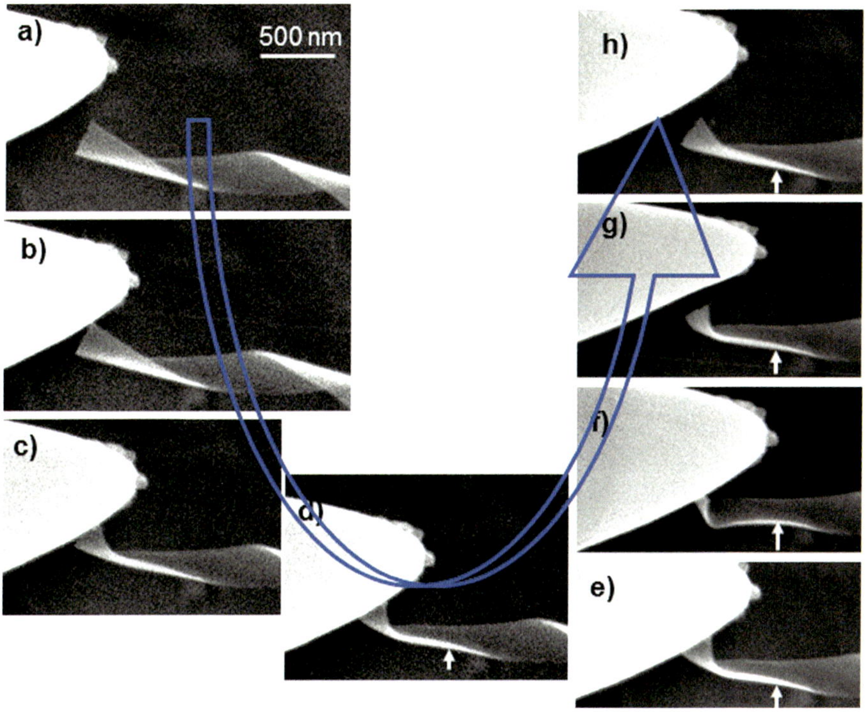

Fig. 7.3 Nanoscale buckling phenomenon observed in ZnO nanohelix by tip compression (**a–f**) and release (**g–h**)

and nanobuckling (Fig. 7.3). On the study of electronic transport of ZnO superlattice nanohelix, a non-linear transport characteristic was revealed (Fig. 7.5).

Superelasticity of Superlattice-Structured ZnO Nanohelix

Mechanically, the superlattice-structured ZnO nanohelix has recently been found to have unique superelastic characteristics [4]. During an in-situ nano-manipulation using a tungsten nanoprobe in a scanning electron (SE)/focused ion dual beam microscope, it was discovered that the nanohelix could elastically recover its shape (Fig. 7.2a–e) after an extremely large axial stretching to a degree close to the theoretical limit (Figs. 7.2b, c), while suffering little residual plastic deformation. As a result, the spring constant can be continuously increased up to 300–800%.

Based on the experimental data acquired from the nanohelix manipulation, five sets of nanohelix dimensions have been used for numerical analysis of the radius dependence of the spring constant upon extension or compression. Figure 7.2f shows plots of spring constant versus instant radius of the five manipulated nanohelices. Upon extension, the radius of the nanohelix decreases as the pitch is increased, while a compression would lead to an increase of the nanohelix radius as the pitch decreases. It is also noticed that upon an extremely large extension, the nanohelix spring constant increases nonlinearly with a dramatic increase.

Table 7.1 Increase in spring constant for the superlattice nanohelix at the maximum stretching: r_0, original radius without deformation; r_f, final radius at maximum axial stretching; H_0, pitch distance of nanohelix; a, thickness of superlattice nanobelt; b, width of superlattice nanobelt; K_0, spring constant at r_0; ε_{theory}, relative theoretical maximum elongation in length; ε_{exper}, relative experimentally observed maximum elongation; $\delta K/K_0$, increase in spring constant

N_1	$2r_0$ (nm)	$2r_f$ (nm)	H_0 (nm)	a (nm)	b(nm)	K_0(N/m)	$\delta K/K_0$ (%)	ε_{Theory} (%)	ε_{Exper} (%)
2	980	190	2200	20	220	0.082	810	72.0	69.8
3	1100	500	3600	20	500	0.072	311	38.6	34.1
4	300	80	1000	20	380	2.71	651	37.4	35.1

Table 7.1 summarizes the nanohelix elongation and spring constant increase as a result of maximum extensions for the manipulated nanohelices under the microscope. It is seen that upon extreme extensions, the nanohelix becomes stiff with the radius decrease. It is also noticed that the static spring constant of relatively tightly packed nanohelix (such as $N_1 = 4$) is relatively large, 2.7 N/m. Under a possible extreme extension condition, its spring constant would have a 650% increase up to 20.3 N/m.

Nanobuckling and Fracture of Superlattice-Structured ZnO Nanohelix

Similar to the superelastic extension and compression illustrated in previous section, a shape memory/recovery of the superlattice-structured ZnO nanohelix was also observed after being subjected to a buckling deformation (Fig. 7.3), which provides direct experimental evidence about the buckling mechanism for explaining the force–displacement measurement results by atomic force microscopy (AFM) tip [4]. Figure 7.3 records an elastic deformation and recovery process with a buckling deformation phenomenon for a nanohelix compressed by a tungsten nanoprobe. Starting from the edge of one end of a nanohelix lying on flat Si substrate (Fig. 7.3a), the nanoprobe tip was manipulated close to the edge (Fig. 7.3b). A compressing operation by manipulator enabled the probe tip to contact the nanohelix front edge and deform it significantly at the edge (Fig. 7.3c). As indicated by an arrowhead, the nanoscale local buckling and recovery of the nanohelix are shown in Fig. 7.3d–h. The hanged side of the coiled nanobelt was compressed to become flatter and flatter (Fig. 7.3d), until being bent over and buckled (Fig. 7.3f), where an inward arched deformation was formed from the side of the coiled nanobelt. The release of the load on the edge led to the recovery of the deformed nanohelix, especially the buckled position as indicated by an arrowhead. Compared to the original shape in Fig. 7.3a, the buckled position of the recovered nanohelix has an almost identical shape. Also there are unlikely to have dislocations being created by the deformation due to the small thickness (~20 nm) of the nanohelix; the data suggest a possible nanoscale elastic buckling and recovery process of the nanohelix when subject to a transverse tip compression.

Nanofracture experiments were carried out using an AFM, which was used to transversely compress a nanohelix until fracture occurred. Figure 7.4a, b shows an

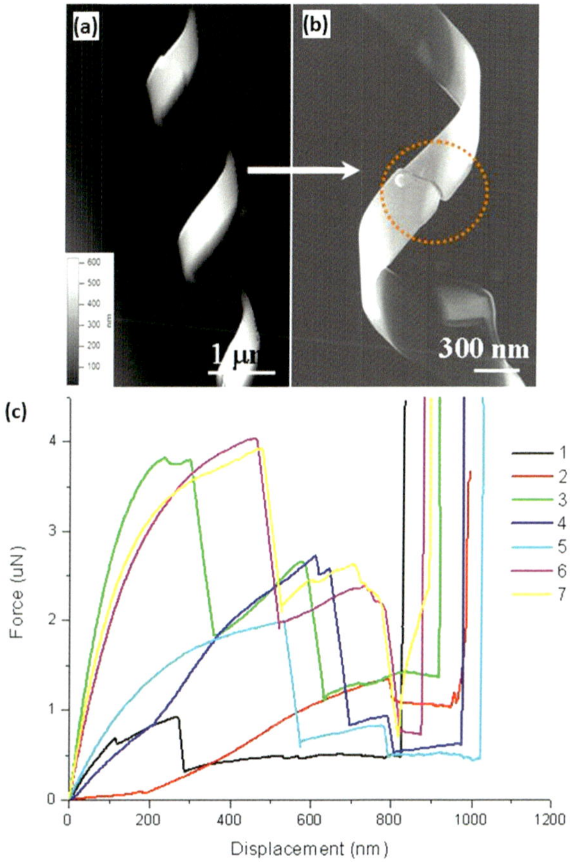

Fig. 7.4 Nanohelix fracture mechanics under AFM: (**a**) and (**b**) are respectively the AFM topography image of original ZnO nanohelix and corresponding fractured ZnO nanohelix SEM image after transversal AFM tip compressing; (**c**) the force-displacement profiles of 7 different nanohelices during AFM transversal compression and release manipulation

AFM topography image of a nanohelix after being compressed by the AFM tip and a corresponding SEM image, respectively. The measured force–displacement curves for a group of nanohelices are shown in Fig. 7.4c. A common factor for the fracture of the nanohelices is that the force–displacement presents two sharp drops at F_1 and F_2, the values of which depend on the size of the nanohelix. A complete fracture of the nanohelix follows the sharp drop at F_2.

In Fig. 7.4, we have observed that a nanoscale elastic buckling deformation is possible upon compression. It is reported that in the buckling process of a thin millimeter-scale elastic plate arch, a similar force drop to a smaller value was due to a transition from stretching deformation to pure bending deformation [23]. Therefore, for a transverse compression of the nanohelix, a two-step fracture process involving a nanoscale buckling phenomenon could be proposed similarly. The force drop might be resulting from a nanobuckling deformation associated with a transition from stretching-resistant deformation to bending-dominant one.

Nonlinear Electronic Transport of Superlattice-Structured ZnO Nanohelix

Nonlinear transport characteristics have been observed (Fig. 7.5) during the electrical characterization of the superlattice-structured ZnO nanohelices, as compared to the single-crystal nanobelt, which shows a linear ohmic transport property (right inset of Fig. 7.5), similar to that in pure bulk polycrystalline ZnO [5, 24]. This indicates that the nonlinear *I–V* curve for the nanohelix is likely due to the intrinsic superlattice structure of the nanohelices, i.e., by involving a periodically built-in electrostatic potential barrier across interfaces. The possible sources from the contact resistances and instrumentation have been excluded. The forward sweeping at high field led to a regular voltage oscillation, suggesting a possible resonant tunneling of electrons at high field, with the large thermionic electron emission being suppressed at low temperature.

Figure 7.6a shows a possible electron transfer path across two adjacent nanostripes from II to I as indicated by a white arrowhead. At the same time, across the nanostripe interfaces, the surface paths (Fig. 7.6b) will be important for the electron transport. In the nanostripe II, electrons will flow through each column of Zn–O stacks along *c*-axis. When electrons pass through to the nanostripe I, which is a polar nanostripe, they will follow the *c*-axis first to reach the surfaces and then the surface conductivity of <0001> will be another major source of the electron transport.

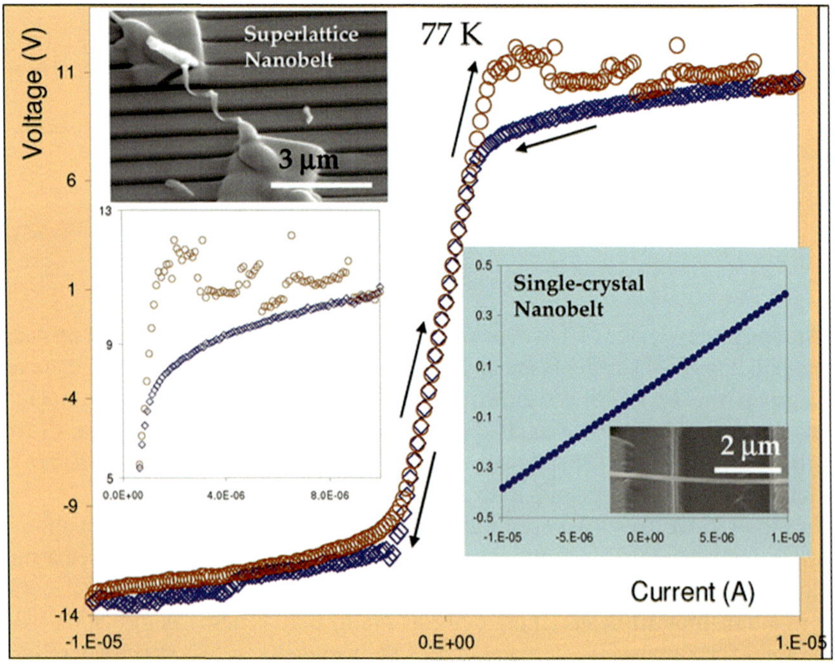

Fig. 7.5 Nonlinear *I–V* characteristics of a superlattice nanohelix and linear characteristics of a single-crystal nanobelt (*right inset*) at 77 K in vacuum. The superelasticity and nonlinear transport properties pave foundations for smart electronic and optoelectronic nanodevice enabling

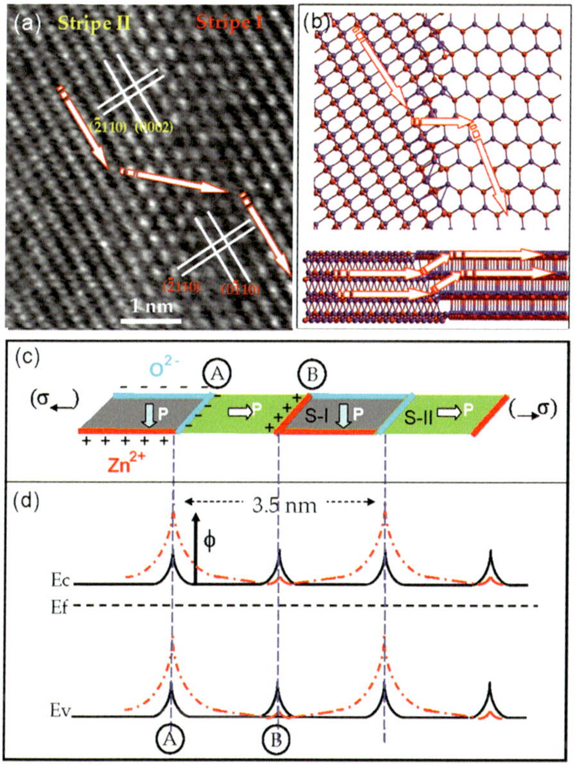

Fig. 7.6 (**a**) HRTEM recorded with the electron beam perpendicular to the nanobelt surface, showing the interface structure between two nanostripes. A possible electron path is indicated. (**b**) Schematic atomic models showing the planar and cross-sectional structures between the two nanostripes. Possible electron paths are indicated. The cross-nanostripe boundary electron flow as well as the surface conduction from $\pm\{01\bar{1}0\}$ in nanostripe II to $\pm\{0001\}$ in nanostripe I (**a**) and (**b**); (**c**) schematic diagram of ZnO superlattice nanostripe domains, corresponding polarization directions (P). (**d**) The interface back-to-back potential energy barrier modulation in band structure due to the polar charges, anisotropic surface electronic structures, as well as piezoelectric effect; *solid curve*: stress-free barrier profile, *dashed curve*: modified profile due to stress-induced polar charges

The nonlinear transport of the nanohelix may be due to the electronic potential introduced by polar charges from ZnO {0001} planes and other polar surfaces as well as the stress-induced piezoelectric field [25, 26] across oriented nanostripes. As shown in Fig. 7.6c, d, a possible potential energy barrier profile at zero bias can be introduced and modified by polar charges and the internal/external stresses induced by nanostripe lattice mismatch and external thermal or mechanical stimuli. In Fig. 7.6d, the dark solid line indicates the stress-free conduction band, and the red dashed line indicates the conduction band energy after considering the electronic potential introduced by interface polar charges. This periodic electrostatic potential modulation due to the anisotropic nanostripe boundaries and surfaces represents a new path for band-gap engineering with impact parallel to that of the traditional heterostructured superlattices, as mentioned in Fig. 7.1. When coupled with their piezoelectric and semiconducting characteristics, these superlattices can be tuned through external bias, mechanical stimuli, or optical injection. Therefore, these entirely new single-phase superlattices can potentially form a new class of useful electronic and optoelectronic nanomaterials.

7.2.1.3 Other ZnO Nanohelices

In Section 7.2.1.1, superlattice-structured ZnO nanohelices were synthesized by a vapor–solid process with pure ZnO as the source material. In the literature, some other ZnO nanohelices were synthesized by thermal evaporation of a mixture of source materials consisting of ZnO and some other impurities such as Li_2CO_3 and Ga_2O_3 [6], and Sb [27].

Yang et al. [6] reported a deformation-free, single-crystal ZnO nanohelix synthesized by thermal evaporation method using ZnO together with Li_2CO_3 and Ga_2O_3 as the source materials. The typical synthesis procedure is as follows. First, 0.6 g of ZnO, 0.3 g of Li_2CO_3, and 0.1 g of Ga_2O_3 powder were placed at the center of an alumina tube that was inserted in a horizontal tube furnace. The tube furnace was heated to 1000°C with a ramp rate of 30°C/min. The solid–vapor deposition was carried out at 1000°C for 2 h under a pressure of 200 torr with Ar flux at about 25 sccm. The nanohelices were deposited onto an alumina substrate placed at the downstream end of the alumina tube, where the deposition temperature was 250–350°C. The introduction of Li_2CO_3 and Ga_2O_3 has been found helpful for the synthesis of polar surface-dominated ZnO nanobelts [7].

The low-magnification SEM image (Fig. 7.7a) shows that the yield of the synthesized nanohelices is high. High-resolution SEM images clearly revealed the high degree of uniformity of the nanohelices, which can be left-handed (Fig. 7.7c) or right-handed (Fig. 7.7b). Their population ratio is nearly 1:1, which resulted from the statistics results. The synthesized nanohelices have a typical diameter of about 30 nm, which is much smaller than the ZnO nanosprings as reported in [6, 7]. Energy-dispersive X-ray (EDX) spectroscopy analyses indicated that the nanohelices are ZnO without detectable impurity of other elements in the sample.

The TEM images show that the synthesized ZnO nanohelices have intrinsic crystal structures with uniform shape and contrast (Fig. 7.8a). High-resolution TEM (HRTEM) images reveal that the nanohelix has an axial direction of [0001], although the growth direction of the nanowire changes periodically along the length. Detailed HRTEM images from the regions labeled c and d in Fig. 7.8a are displayed in Fig. 7.8c, d, respectively. The detailed HRTEM images show that the nanowire that constructs the nanohelix grows along $[01\bar{1}1]$. Because the incident electron beam is parallel to $[2\bar{1}\bar{1}0]$, the two side surfaces of the nanowire are $\pm(01\bar{1}\bar{2})$. No dislocations were found in the nanohelices. It is important to note that the image recorded from the "twist" point (c) of the nanohelix shows no change in crystal lattice (Fig. 7.8c), and the traces of the two sides are visible, indicating the nontwisted single-crystal structure of the entire nanohelix.

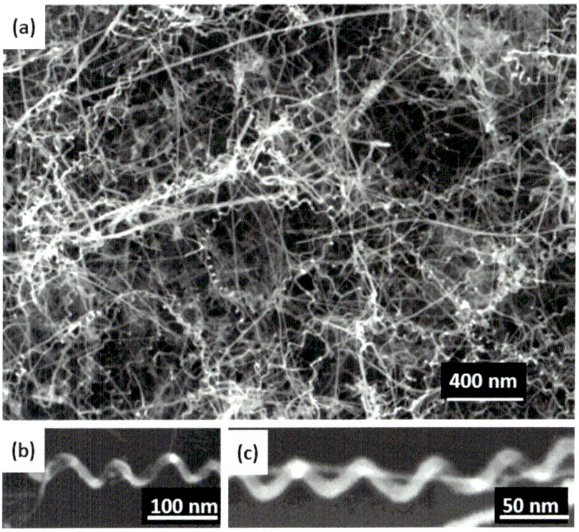

Fig. 7.7 (**a**) Low-magnification SEM image of the as-synthesized nanohelices of ZnO, showing their uniform sizes and high yield. (**b**) An enlarged right-handed nanohelix. (**c**) A small nanohelix with pitch distance of 60 nm and radius of 40 nm, which grows around a straight nanowire

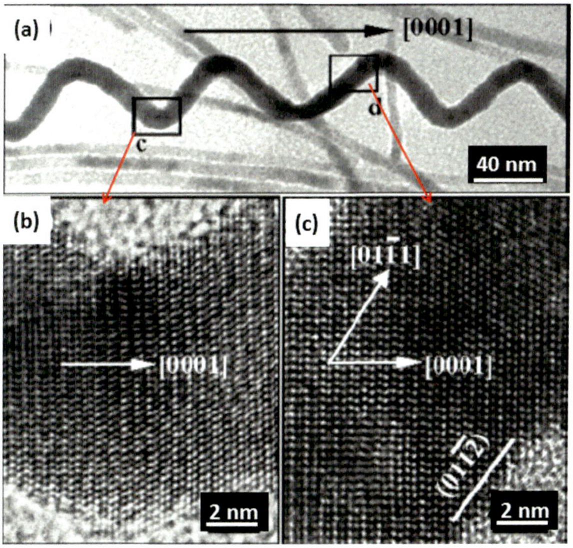

Fig. 7.8 (**a**) A bright-field TEM image of a nanohelix. No significant strain contrast is found (apart from the overlap effect between the nanohelix and the nanowires). (**c**, **d**) HRTEM images recorded from the *c* and *d* areas labeled in (**a**), respectively, showing the growth direction, side surfaces, and dislocation-free volume

According to the literature, elastic deformation usually occurred in forming the nanorings and nanosprings due to the different structural configurations. However, the as-discussed ZnO nanohelices are distinct from the previously reported nanorings and nanosprings [7, 20, 21], which are single crystals without elastic deformation. Since there is no elastic/plastic deformation introduced in forming the nanohelix, it is the electrostatic energy that dominates the entire formation process, making it possible to form the nanohelices much smaller than the diameters of the nanorings and nanosprings. In addition, as shown by the HRTEM images, the new polar surfaces of $\pm\{01\bar{1}\bar{1}\}$ are responsible for forming the nanohelices reported here.

Gao et al. [27] prepared a kind of super-uniform crystalline ZnO nanohelices fabricated with the addition of Sb by thermal evaporation. Typically, a mixture of ZnO powder and Sb in a weight ratio of 100:1 was used as the source material and placed in the center of a horizontal tube furnace. The tube furnace was pre-pumped to 6.0 Pa to remove the oxygen in the furnace. Then the temperature was raised up to 1300°C and kept for 5–10 min under a constant pressure between 75 and 3000 Pa. Nitrogen gas was used as carrier gas with a flow rate of 75–100 sccm. The nanohelices were deposited onto the Au-coated Si substrate placed at the downstream of the source materials.

Figure 7.9a is a typical SEM image showing that the as-synthesized samples were composed of nanohelices and nanowires with lengths of several tens of micrometers. High-resolution SEM image (Fig. 7.9b) shows that the nanohelix is super-uniform. The cross sections of the helix are perfect hexagonal, suggesting that axis of the helix may be along [0001] of ZnO. Six blocks of equal length build up one period of the helix. The pitch distance, the mean diameter, and the thickness of the nanohelices are typically 600, 400, and 100 nm, respectively. Nanohelices of both left-handed (Fig. 7.9b) and right-handed (Fig. 7.9d, f) chiralities were observed and one helix can switch chiralities (Fig. 7.9c, e).

Figure 7.10a is a low-magnification TEM image of a ZnO helix showing that there is an angle of about 47° between the growth direction and the axial direction. Figure 7.10b, c shows the HRTEM images recorded near the outer and inner surfaces of the connection, respectively, between two neighboring blocks of the helix (marked in Fig. 7.10a) which confirm that the helix is hexagonal single-crystal ZnO with no deformation and that the helix has an axial direction of [0001]. Like the function of Li_2CO_3 and Ga_2O_3 [6], the addition of Sb helps the synthesis of super-uniform ZnO nanohelices.

7.2.2 SiO_2 Nanohelices

Wang et al. [8] reported a chemical vapor deposition (CVD) method for the synthesis of silica nanosprings onto a variety of substrates using a gold catalyst. First, they sputtered a thin layer of gold onto the support substrate with a thickness of 15–90 nm. The silicon precursor is proprietary. The vapor–liquid–solid deposition was carried out at 350–1000°C for ~30 min under a constant O_2 flow. The yield of the nanosprings is as high as 90%. X-ray photoelectron spectroscopy (XPS) analysis shows that the as-grown nanosprings have components of silicon and oxygen with an atomic ratio of silicon to oxygen close to ½, which confirms the synthesis of SiO_2 nanosprings.

Figure 7.11 shows the typical SEM images of SiO_2 nanosprings grown at 350, 650, and 1000°C with a Au catalyst layer of 30 nm. As shown in Fig. 7.11, there are no visible changes in the geometries or sizes of the SiO_2 nanosprings synthesized at different temperatures. A magnified image of the SiO_2 nanosprings synthesized

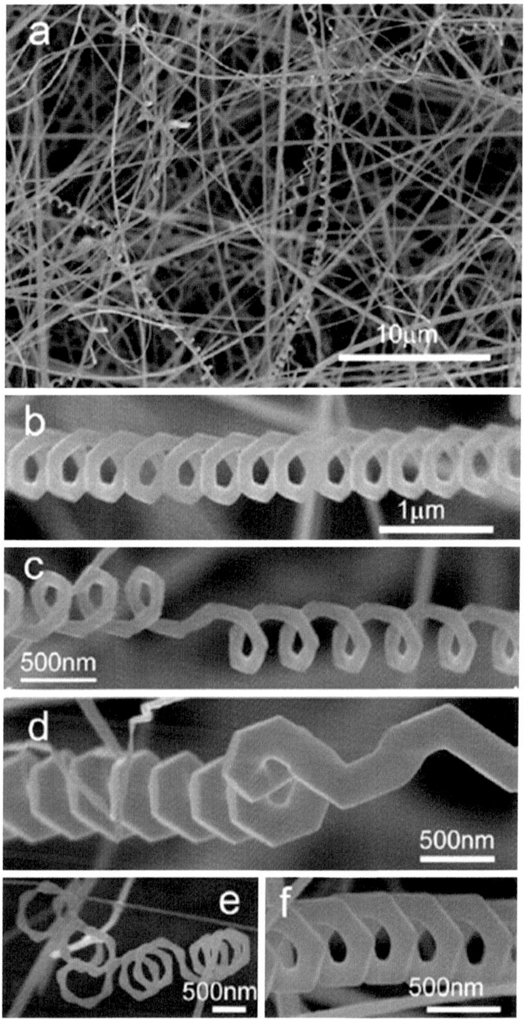

Fig. 7.9 (**a**) SEM image of ZnO nanostructures. (**b**) SEM image of a ZnO nanohelix with left-handed chirality. (**c**) A ZnO nanohelix switches its chirality. (**d**) A ZnO nanohelix loosens into a zigzag nanowire gradually. (**e**) SEM image of a ZnO nanohelix constructed by different shapes of blocks. (**f**) A uniform ZnO nanohelix with right-handed chirality

at 1000°C illustrates the extremely uniform helical structure that most of the nanosprings exhibit. TEM images (Fig. 7.12) revealed that two types of nanosprings are formed. The first type is formed from a single nanowire (Fig. 7.12a, b) and the second type is formed from multiple interwined nanowires (Fig. 7.12c, d). The multi-nanowire nanosprings are considerably larger in diameter and pitch than nanosprings formed from a single nanowire. However, the diameters of nanowires that form the multi-nanowire nanosprings are two to three times smaller.

Figure 7.13 shows a typical SEM image of silica nanospring mats grown with a 30 nm Au catalyst layer. The bright spots are the Au catalysts at the ends of the nanosprings. The image and others demonstrate that the silica nanosprings grow

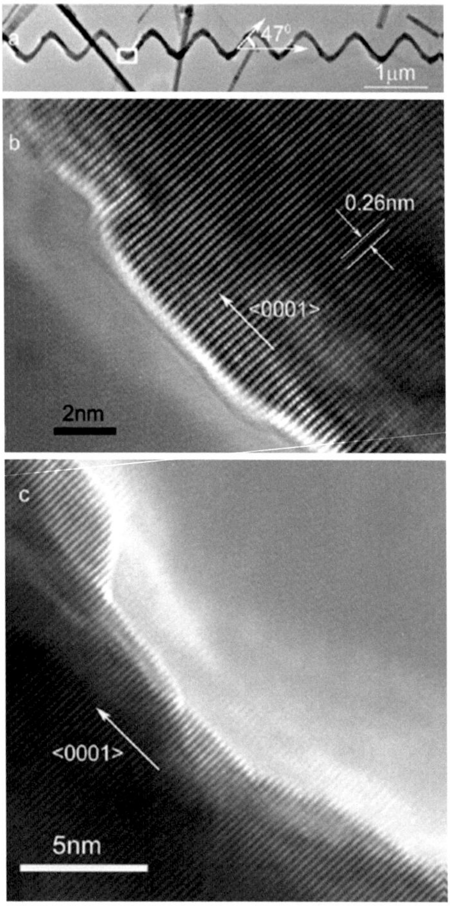

Fig. 7.10 (**a**) TEM image of a ZnO nanohelix. (**b**, **c**) HRTEM images recorded near the outer and inner surfaces of the connection place between two neighboring blocks marked in (**a**)

via the vapor–liquid–solid mechanism. In the case of nanosprings formed from a single nanowire, it is the existence of contact angle anisotropy (CAA) at the interface between the nanowire and the Au catalyst that introduces the asymmetry. For the synthesis of multi-nanowire nanosprings, CAA cannot be the mechanism driving asymmetric growth. An alternative model of multi-nanowire nanospring formation must take into account that the difference in the growth rate between nanowires of the multi-nanowire nanosprings produces torques on the catalyst, which in turn produce the helical structures.

Delclos et al. [9] used assemblies of amphiphilic molecules as templates for the growth of inorganic silica with helical structures. The amphiphiles are cationic bis-quaternary ammonium gemini surfactants having the formula C_2H_4-1,2-$((CH_3)_2N^+C_{16}H_{33})_2$, noted hereafter 16-2-16, with tartrate as a counterion. The assemblies of these amphiphiles exhibit a wide diversity of morphologies, including twisted and helical ribbons, as well as nanotubes (Fig. 7.14). Because of their

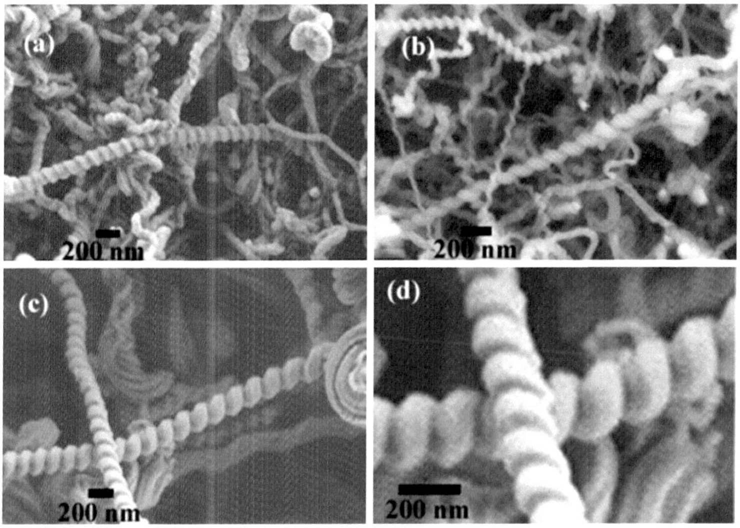

Fig. 7.11 SEM images of silica nanosprings using different deposition temperatures: (**a**) 350°C, (**b**) 650°C, (**c**) 1000°C, and (**d**) a zoom-in image from (**c**)

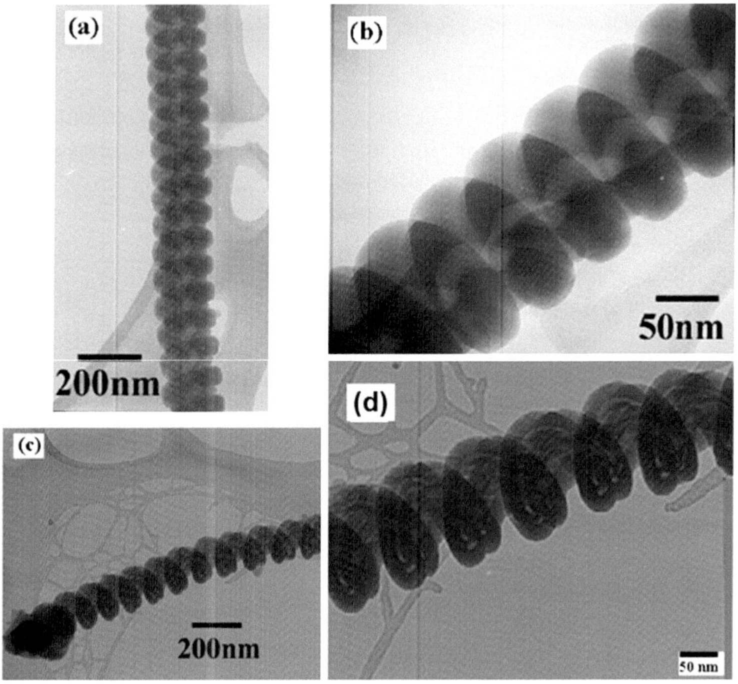

Fig. 7.12 Bright-field TEM images of two different types of silica nanosprings: (**a**, **b**) conventional types of nanosprings consisting of a single nanowire; (**c**, **d**) nanosprings formed from multiple nanowires

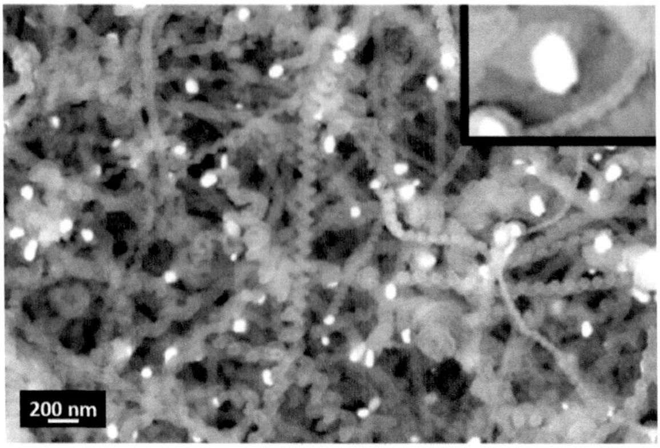

Fig. 7.13 SEM image of silica nanosprings grown with a 30-nm Au catalyst layer. The bright spots are the Au catalyst at the tips of the nanosprings. The *inset* is a magnification of the Au catalyst

structural diversity and tenability, these 16-2-16 tartrate assemblies represent an ideal template for inorganic silica transcription.

In a typical experiment, the 16-2-16 tartrate was first dispersed in water at a concentration of 5 mM, and the resulting mixture was heated up to 55°C and then cooled down to 22 or 40°C leading to a gel which was aged for 5–45 days before sol–gel replication. Tetraethoxysilane (TEOS) was used as silica precursor. TEOS was prehydrolyzed in water for 12 h and then mixed with the 16-2-16 tartrate gel for 36 h at 22°C for the sol–gel replication. After the sol–gel replication, the samples were thoroughly washed with ethanol to remove all the organic components including the templates.

Figure 7.15 shows the TEM images of the organic 16-2-16 tartrate gels after various aging times and the corresponding SiO_2 replicas. All the organic gels were aged for a longer time, which show similar morphologies with nanotubule structures, as previously described with the organic templates in Fig. 7.14. However, the structures of the replicated SiO_2 are different. The templates aged for 5, 21, and 45 days at room temperature were transcribed into twisted, helical, and nanotube structures, respectively.

Figure 7.16 shows the TEM images of the organic 16-2-16 tartrate gels aged at 22°C or 40°C for 21 days and the corresponding SiO_2 replicas. The organic gels aged at 22°C show a nanotubule structure and helical silica was obtained after sol–gel replication, while the organic gels aged at 40°C show a twisted structures and twisted silica was obtained after sol–gel replication.

As discussed above, the polymorphism (under different aging time and temperature) of the organic templates can be directly reflected on the morphology of inorganic silica structure. In addition, it was also found that the variation of

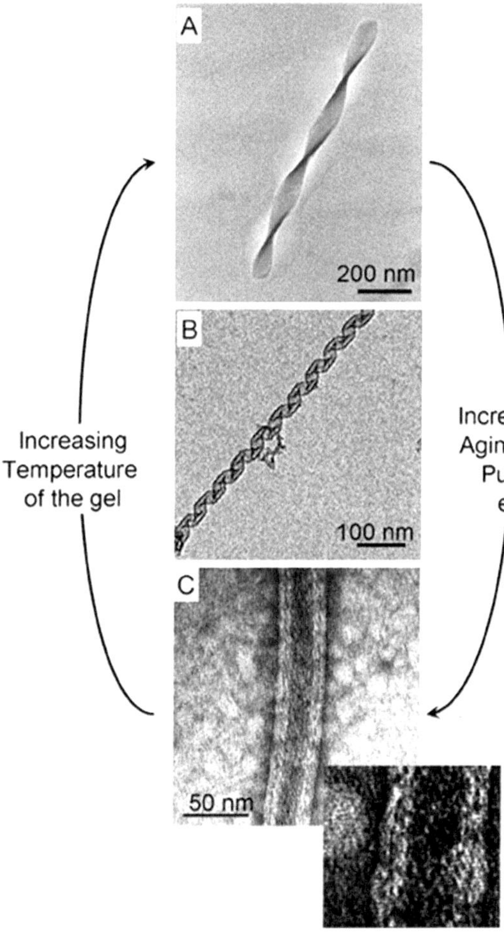

Fig. 7.14 TEM images showing the morphological diversity of the self-assembly of gemini 16-2-16 tartrate surfactants: (**a**) twisted ribbon; (**b**) helical ribbons; and (**c**) tubules. Upon variation of diverse independent parameters, the morphology can be finely tuned. The *inset* in (**c**) shows the double layers of the tubule wall

the transcription conditions (such as replication temperature) proves to control the morphologies of the replicated silica materials as illustrated in Fig. 7.17.

7.2.3 CdS Nanohelices

Wang et al. [14] reported the synthesis of helical CdS nanowire ropes by simple aqueous chemical growth. A typical procedure for the synthesis of helical CdS nanoropes is as follows. First, appropriate amount of $Cd(CH_3COO)_2{\cdot}2H_2O$ was dissolved into a deaerated 35 mol.% aqueous solution of ethylenediamine (ED) in a flask at room temperature. Then, stoichiometric $Na_2S{\cdot}9H_2O$ was quickly added to the above solution under vigorous stirring. A milk-white sol was formed soon. Next, the resultant milk-white sol was heated to 50°C and kept on stirring at this

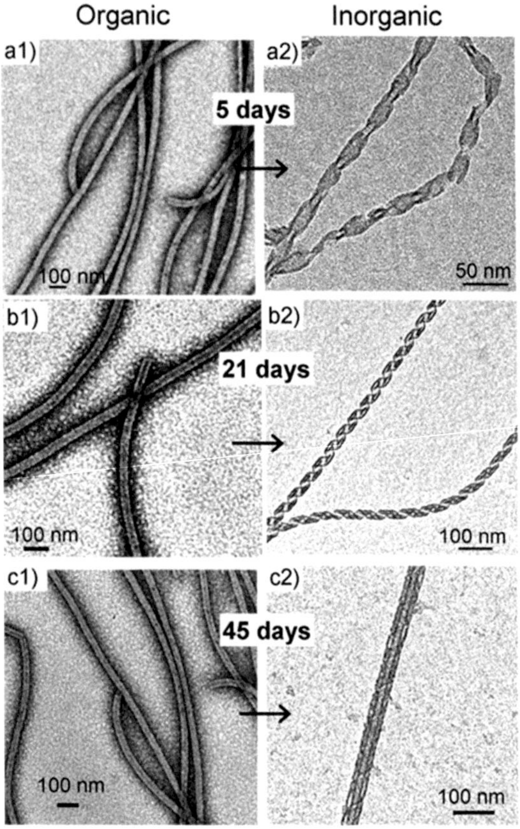

Fig. 7.15 TEM micrographs showing the morphology of inorganic ribbons obtained upon transcribing organic 16-2-16 tartrate gels after various aging times. Organic gels show the formation of entangled networks of nanotubules of similar morphology after 5, 21, or 45 days (**a1**, **b1**, and **c1**), respectively. Transcribing these structures at different aging times allowed controlling the morphology of inorganic chiral ribbons into twisted ribbons after 5 days (**a2**), helices after 21 days (**b2**), and nanotubes after 45 days (**c2**)

temperature for about 1 week until the milk-white sol turned to a yellow color. The final product was obtained by centrifugation and washed with distilled water and ethanol.

Figure 7.18 shows the TEM images of the as-synthesized CdS nanoropes. It can be seen that the nanoropes were composed of spirally twisted and interstranded nanowires. The individual nanowires have diameters of 6–10 nm and the nanoropes have width in the range of 50–200 nm. The length of the CdS nanoropes ranges from a few micrometers to several tens of micrometers, some of them even extending over 100 μm. As shown in Fig. 7.18a–c, the bundled monowire within each rope is spirally twisted rather than parallel to one another along their whole length. Thus, the self-assembled nanorope structure exhibits a helically wound morphological feature, though the helical pitch distances were relatively large and rather irregular. The inset of Fig. 7.18c shows the selected electron diffraction (SAED) pattern recorded on a single bundle of nanorope. The primary diffraction pattern can be indexed matching the [100] zone of hexagonal wurtzite structure. The HRTEM lattice image (Fig. 7.18d) further reveals that the nanowires grow along the [001] direction. The

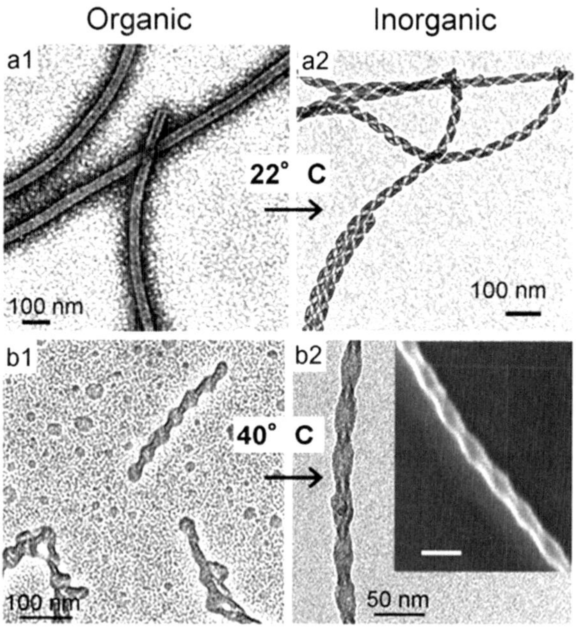

Fig. 7.16 TEM micrographs showing the effect of temperature on organic gel morphology and the transcribed inorganic fibers that result. Organic nanotubules of 16-2-16 tartrate formed at room temperature (**a1**) are transcribed into helical silica (**a2**), while organic twisted ribbons formed at 40°C (**b1**) resulted in transcription into inorganic ribbons with conserved twisted morphology (**b2**). The *inset* shows a high-resolution TEM image of a single twisted ribbon showing the 3D nature of the object and its pitch. Bar represents 50 nm

authors found that anisotropic growth of 1D wurtzite CdS nanocrystals along the [001] direction may involve preferential adsorption of the ED molecules onto the long-axis crystal faces of the growing crystallites, thus resulting in stabilization and growth inhibition of the sidewalls [28, 29]. As for the formation of the helical CdS nanowires, it is believed that the presence of high densities of stacking faults may play a role [28], and the stacking faults might be induced by strains while interweaving the growing nanowires through van der Waals interactions.

Sone et al. [15] synthesized single CdS nanohelices and double CdS nanohelices by a supramolecular template method. The template is the dendron rodcoil (DRC) molecules that self-assemble into a network of twisted ribbons 10 nm × 2 nm in cross section and up to 10 μm in length. In a typical synthesis, a gel of 3 wt% DRC in ethyl methacrylate (EMA) was prepared first. After that, appropriate amount of $Cd(NO_3)_2{\cdot}4H_2O$ in tetrahydrofuran (THF) was added to the above DRC gel. Finally, the above mixed suspension was exposed to H_2S gas for 5–15 min to make sure the nucleation and growth of CdS localized to the twisted ribbons. The schematic representation of the templating pathways is shown in Fig. 7.19.

Figure 7.20 shows the HRTEM images of the mineralized ribbons at different stages. At the early stage of the mineralization, CdS crystals have coalesced on

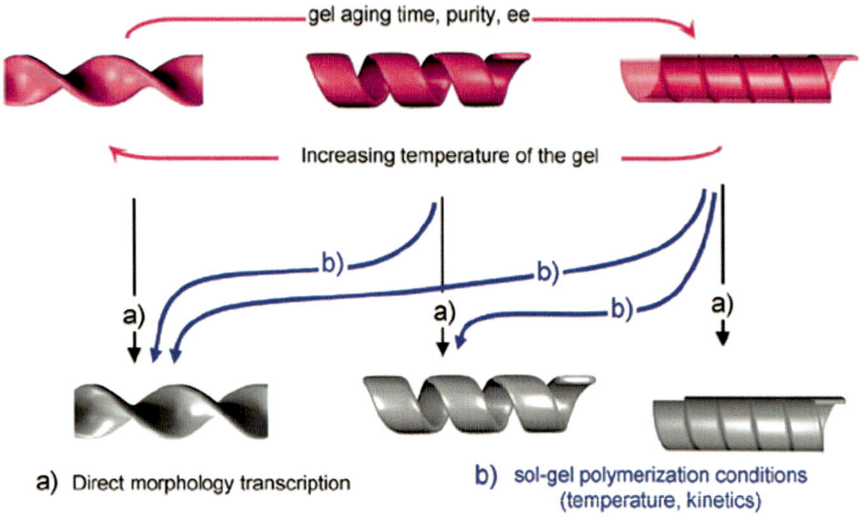

Fig. 7.17 Schematic representation of the diversity of inorganic silica chiral ribbons. Morphological diversity of the organic template (*pink arrows*), inorganic structures obtained by direct replication of organic morphologies leading to identical sol–gel transcription (*black arrows*, **a**). The fine-tuning of the transcription process: aging time variation as well as replication temperature widens the potentialities of morphology controls (*blue arrows*, **b**)

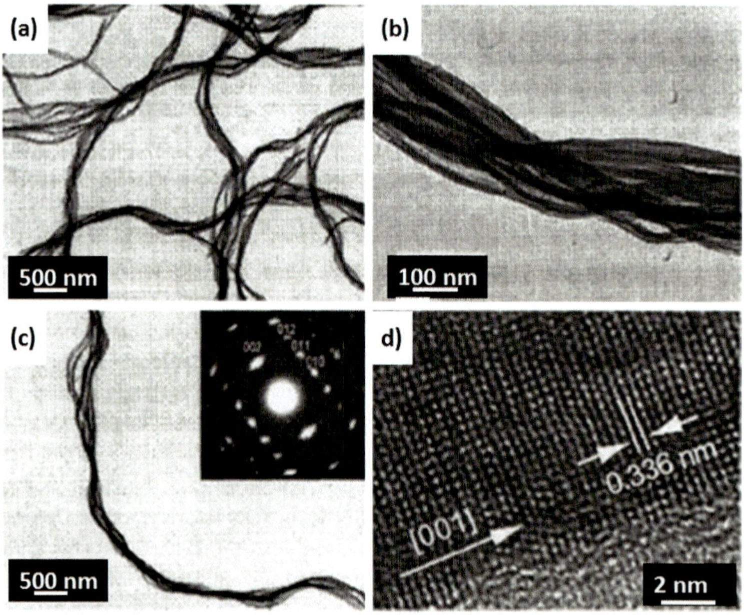

Fig. 7.18 (**a**)–(**c**) Typical TEM images of the newly synthesized CdS nanowire ropes. *Inset* of (**c**) SAED pattern recorded on a single bundle of CdS nanorope. (**d**) HRTEM image of a single-crystalline monowire comprising the nanoropes

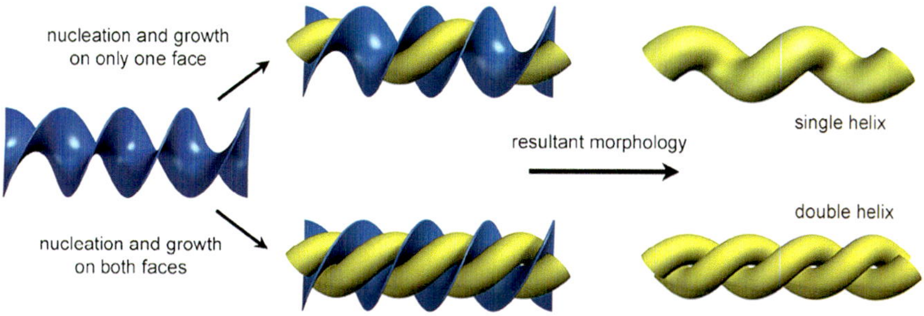

Fig. 7.19 Schematic representation of templating pathways. Nucleation and growth on one side of the twisted ribbons (*blue*) leads to single helices of CdS (*yellow*), while nucleation and growth on both sides of the ribbon leads to double helices

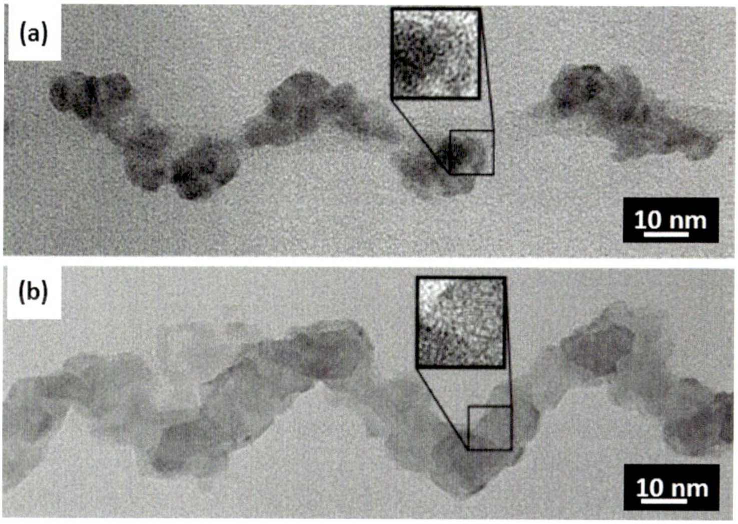

Fig. 7.20 HRTEM micrographs of mineralized ribbons: (**a**) a ribbon at an early stage of mineralization. The *inset* shows the absence of lattice fringes, indicating that the CdS is amorphous or only weakly crystalline at this stage; (**b**) a more fully mineralized ribbon. The *inset* shows the lattice fringes from the CdS

the DRC ribbon. However, at this stage, the helical structure is not complete. In addition, the HRTEM image (Fig. 7.20a) shows that the CdS is still amorphous or at least not strongly crystalline at this stage. As the CdS continues to grow, it eventually becomes a continuous helical structure (Fig. 7.20b). Meanwhile, at this point, the helical CdS is certainly crystalline, as evidenced by lattice fringes visible in Fig. 7.20b.

Figure 7.21 shows the TEM image of a double CdS nanohelix. The authors found that both the kinetics of CdS nucleation and growth, and the duration of mineralization had no effect on the relative portions of single and double helices.

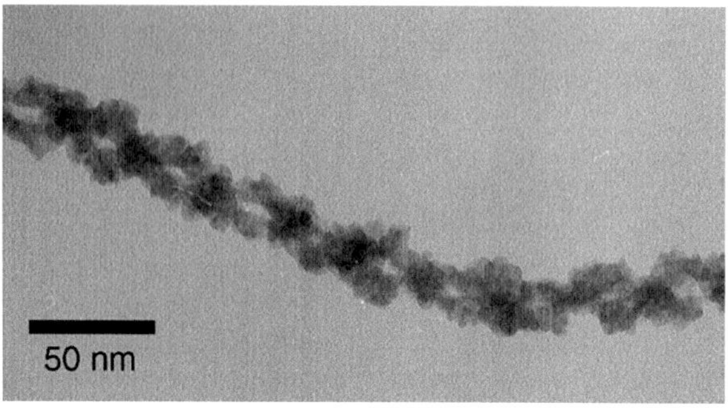

Fig. 7.21 TEM micrograph of CdS double helices. The *inset* shows an enlargement in which the expected braided appearance is clearly visible

It is probably due to a structural difference between different ribbons that leads to the growth of different CdS structures (single and double nanohelices). If a ribbon is with a perfect twisted structure, both faces of the ribbon would be equivalent and equally able to nucleate and grow CdS. Thus double CdS nanohelix is obtained. However, if a ribbon is not perfect and with a slightly coiled axis, this would make one face more exposed to the solvent and favorable to CdS nucleation and growth. Thus single CdS nanohelix is obtained. There is another possibility that the two faces of the ribbon are initially equivalent, but a nucleation event distorts the structure so as to render them nonequivalent and facilitate the synthesis of single CdS nanohelices.

7.2.4 InP Nanohelices

Shen et al. [16] reported the synthesis of single-crystalline, cubic structured InP nanosprings via a simple thermochemical process using InP and ZnS as the source materials. The InP nanosprings were synthesized in a vertical induction furnace consisting of a fused quartz tube and an induction-heated cylinder. Typically, a graphite crucible containing 1 g of InP and 0.2 g of ZnS was placed at the center cylinder zone. After evacuation of the quartz tube to ~20 Pa, a pure N_2 flow was introduced at a flow rate of 50 sccm. Then the furnace was rapidly heated to 1350°C and kept at this temperature for 1 h. After cooling to room temperature, a black powder-like product was collected at the bottom of a graphite crucible.

As shown by the SEM images of the synthesized product (Fig. 7.22), it is obvious that both left-handed and right-handed nanosprings are obtained through this thermochemical process. Each nanospring is formed by uniformly rolling up a single nanobelt. Most of the nanobelts have diameters of 30–200 nm with a thickness of ~80 nm. The energy-dispersive X-ray (EDX) spectroscopy reveals that the synthesized nanosprings are dominated by In and P elements at an In/P atomic ratio of 1.05:1, indicating the formation of InP. Figure 7.23a is a typical TEM image of the synthesized InP nanospring. The SAED patterns were taken in different spots along the constituent nanobelt. The spots on the pattern can be attributed to the [110] zone

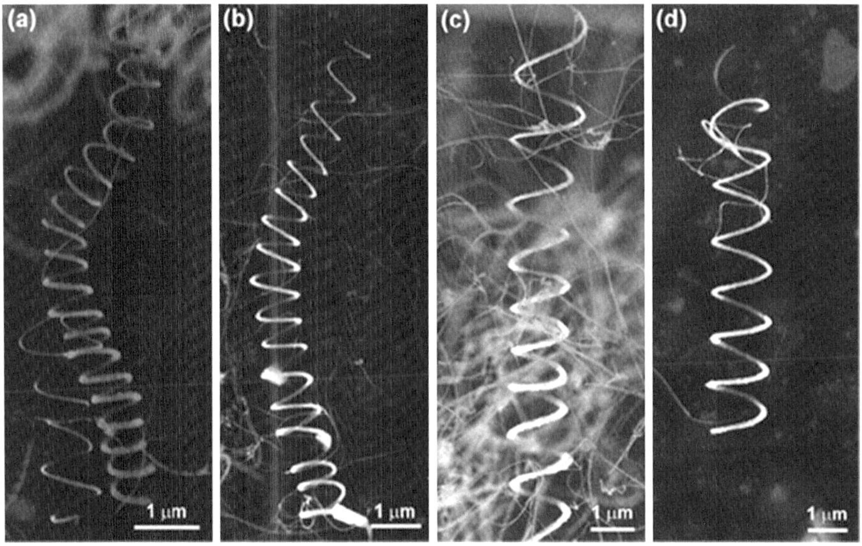

Fig. 7.22 SEM images of the as-synthesized InP nanosprings, which are left-handed (**a**, **b**) and right-handed (**c**, **d**)

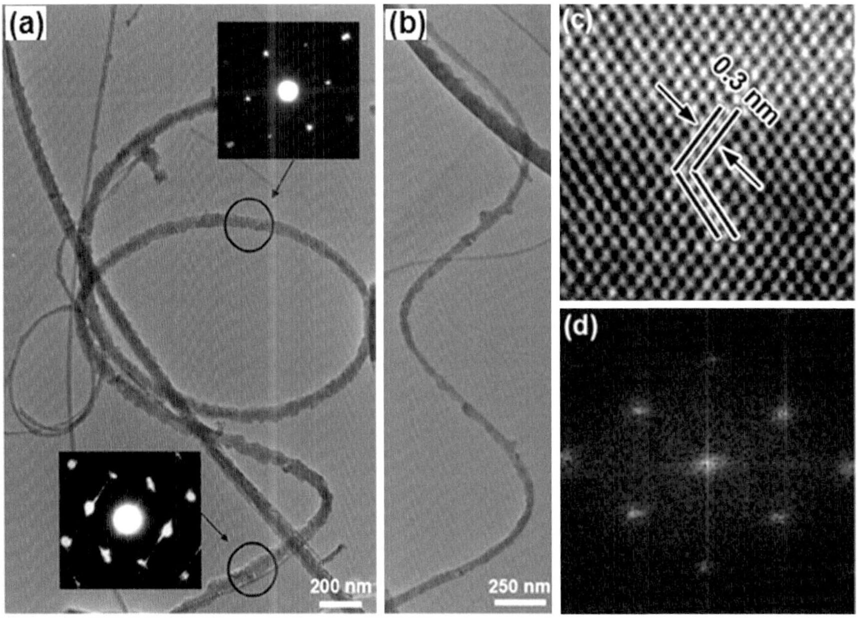

Fig. 7.23 (**a**) TEM image of a nanospring formed via curving of an InP nanobelt with a diameter of ~100 nm; (**b**) TEM image of two loops of an InP nanospring; (**c, d**) lattice-resolved HRTEM image and its corresponding fast Fourier transform image

axis of a cubic InP crystal. Figure 7.23b shows two loops of an InP nanospring. The clearly resolved lattice fringes with a 0.3 nm separation in Fig. 7.23c are typical for the [200] plane d spacing in cubic InP. The corresponding fast Fourier transform (FFT) pattern shown in Fig. 7.23d can be attributed to the [110] zone axis of a cubic InP crystal. Based on the FFT pattern and HRTEM images, it is confirmed that the nanosprings are single crystals with the preferential growth direction along the <111> orientation. The formation of the cubic InP nanosprings is mainly attributed to the minimization of the electrostatic energy due to the polar charges on the $\pm(002)$ side surfaces of cubic InP.

7.2.5 Ga_2O_3 Nanohelices

Zhan et al. [13] prepared single-crystalline, helical β-Ga_2O_3 nanostructures by thermal evaporation of gallium oxide in the presence of gallium nitride. The Ga_2O_3 nanostructures were synthesized in a vertical induction furnace consisting of a fused quartz tube and an induction-heated cylinder. Typically, a graphite crucible containing 0.37 g of Ga_2O_3 and 0.05 g of GaN was placed at the center cylinder zone. After evacuation of the quartz tube to ~0.1 torr, a pure N_2 flow was introduced at a flow rate of 200 sccm. Then the furnace was heated to 1200°C and kept at this temperature for 1 h. After cooling to room temperature, a white product was collected from the outlet of carbon induction-heated cylinder.

Figure 7.24a shows a typical zigzag Ga_2O_3 nanostructure with kinks of alternating sign. As shown in Fig. 7.24b, it is obvious that the diameter of the Ga_2O_3 nanostructure is about 40 nm. The periodicity along the nanostructure is ~150 nm long and ~80 nm wide. The ED pattern (Fig. 7.24d) of a single kink (Fig. 7.24c) reveals that the Ga_2O_3 is crystalline in nature. The spots on the pattern can be attributed to an overlay along the [010] and $[0\bar{1}0]$ zone axes of a β-Ga_2O_3 crystal. The spots marked with asterisks originate from the [010] zone axis and those marked with dots originate from the $[0\bar{1}0]$ zone axis. The same diffraction patterns were achieved in the whole zigzag nanostructure, which indicates the uniform crystalline nature of the nanostructures. In addition to the single-zigzag morphology discussed above, double-zigzag β-Ga_2O_3 nanostructures also occur in the products. The two sets of white lines in Fig. 7.24e suggest two zigzag lines. The corresponding ED pattern (Fig. 7.24f) is identical to that taken on a single-zigzag β-Ga_2O_3 nanostructure (Fig. 7.24d).

Apart from the zigzag nanowires growing along a [001] direction, 3D helical nanowires were also observed. Figure 7.25a is a typical TEM image of a helical Ga_2O_3 nanowire. The nanohelix has a periodicity of around 160 nm and a diameter of about 60 nm. A single periodicity of the nanohelix is shown in Fig. 7.25b. The inset of Fig. 7.25b is a Fourier-transformed pattern obtained from the HRTEM image of this single periodicity. The spots on the FT pattern can be indexed as those peculiar to the $[\bar{1}2\bar{1}]$ zone axis of β-Ga_2O_3, which suggests its single-crystalline nature. The HRTEM image (Fig. 7.25c) shows clearly resolved interplanar distances of 2.82 and 2.55 Å, which correspond to the $(\bar{2}02)$ and $(1\bar{1}1)$ lattice spacings,

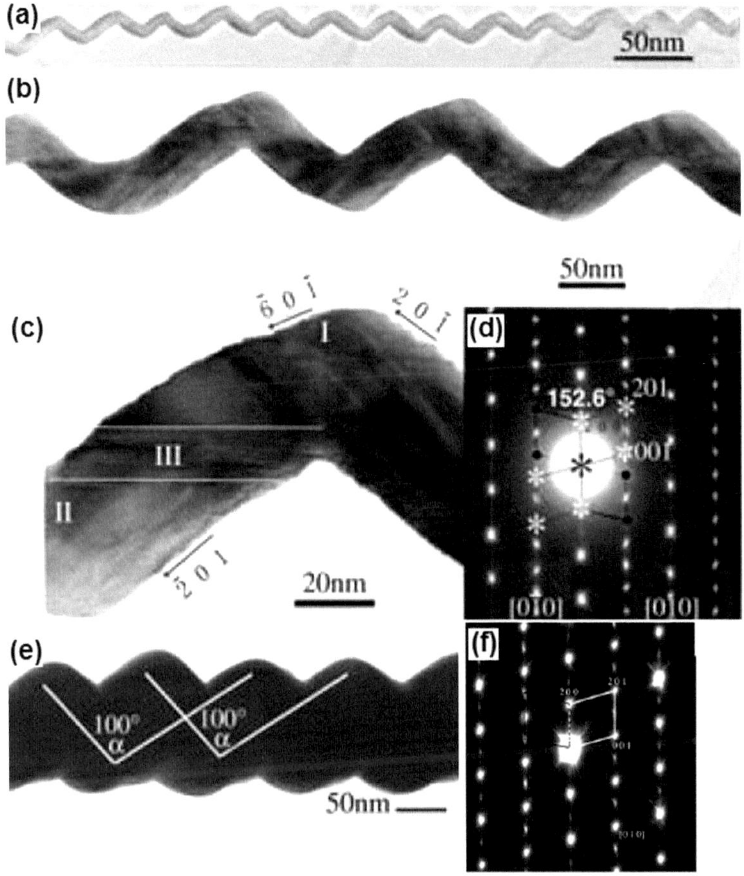

Fig. 7.24 TEM images (**a**) of a zigzag β-Ga_2O_3 nanostructure (**b**) showing three kinks; and (**c**) showing a single kink. (**d**) Selected area electron diffraction pattern; (**e**) TEM image of a double-zigzag β-Ga_2O_3 nanostructure; and (**f**) corresponding electron diffraction pattern of (**e**)

respectively. The image is consistent with the projection of a β-Ga_2O_3 crystal along the $[\bar{1}\bar{2}\bar{1}]$ orientation (inset of Fig. 7.25c). The HRTEM images and the FT pattern reveal that the nanohelices grow with their axial directions perpendicular to $(\bar{1}\bar{1}3)$. Since the incident electron beam is parallel to the $[\bar{1}\bar{2}\bar{1}]$ direction, the side surfaces of the nanowire are $\pm(0\bar{2}4)$ and $\pm(1\bar{1}1)$.

7.3 Carbon-Related Nanohelices

Carbon is such a mystical element that it can form a number of structures, ranging from zero-dimensional (0D) fullerenes to 3D diamond through 2D graphite and 1D nanotubes. The diversity of the carbon structure provides the flexibility to tailor

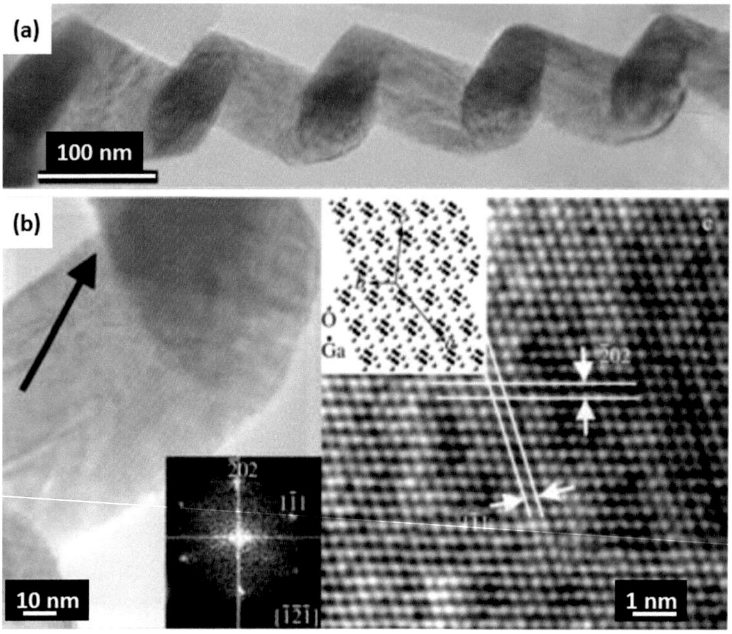

Fig. 7.25 (**a**) TEM image of a β-Ga_2O_3 nanohelix and (**b**) TEM image highlighting a single periodicity. The *arrow* denotes the axial direction of the nanostructure; the *inset* is its corresponding Fourier transform pattern; (**c**) HRTEM image of the nanohelix; the *inset* is a structural model of β-Ga_2O_3 projected along the $[\bar{1}2\bar{1}]$ orientation

their properties according to specific needs and potential applications. 3D helically structured carbon nanomaterials are attracting more and more attention due to their unique characteristics and wide potential applications. The helical structures and the micro-/nanometer-ordered dimensions directly affect these excellent characteristics. Therefore, the development of carbon materials with controlled helical patterns and dimensions is very important.

7.3.1 Helical Carbon Nanoribbon/Nanocoil

Chen et al. [10] reported that twisting carbon nanoribbons or carbon nanocoils with a helical nanostructure could be synthesized at a high purity by the catalytic pyrolysis of acetylene at 600 or 650°C using Fe-based alloy films as a catalyst. Typically, a thin layer of Fe-based alloy ($Fe_{74}Co_{18}Ni_8$ or $Fe_{71}Cr_{18}Ni_8Mo_3$) with a thickness of 20–50 nm was sputter coated on the glass substrate. The glass substrate was placed in the central part of a vertical reaction tube, which has an upper source gas inlet and a lower gas outlet. A source gas mixture of C_2H_2 and H_2, H_2S/H_2, and N_2 was introduced to the reaction tube at atmospheric pressure. The catalytic pyrolysis was carried out at 600 or 650°C for 30 min.

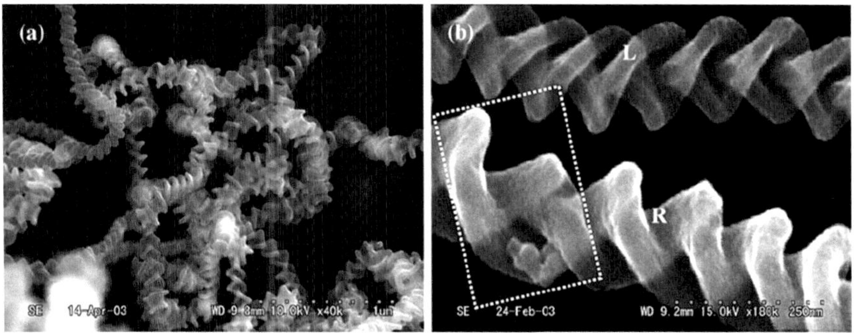

Fig. 7.26 (**a**) Representative twisting nanoribbon grown over an Fe alloy ($Fe_{74}Co_{18}Ni_8$) catalyst sputter coated on a glass plate and (**b**) their enlarged view; L, left-clockwise coiling; R, right-clockwise coiling. The part marked by a *rectangle* shows that the helix is composed of a single nanoribbon

Fig. 7.27 Representative twisting nanoribbons grown over an Fe alloy ($Fe_{71}Cr_{18}Ni_8Mo_3$) catalyst sputter coated on a glass substrate

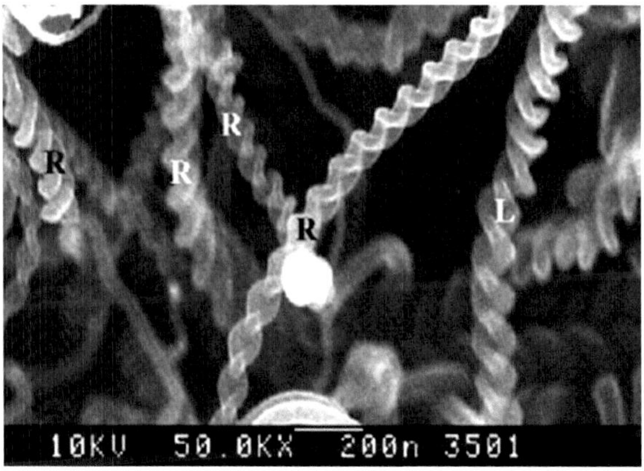

Figure 7.26 shows the representative SEM images of the twisting nanoribbon grown over $Fe_{74}Co_{18}Ni_8$ catalyst-coated glass substrate. It can be seen that the twisting carbon nanoribbon is formed by a single nanoribbon which continuously twists in a constant pitch. Both left-clockwise and right-clockwise coiling carbon helices are observed. The width and the thickness of the nanoribbon are about 100–200 and 50–10 nm, respectively. The length of the helical structured nanoribbon is up to several hundred micrometers. Figure 7.27 shows the SEM images of the twisting nanoribbon grown over $Fe_{71}Cr_{18}Ni_8Mo_3$ catalyst-coated glass substrate. It is obvious that the density of the twisted carbon nanoribbons in the deposits is low compared with that using $Fe_{74}Co_{18}Ni_8$ as catalyst. In addition, some straight-formed nanofibers co-grew under the pyrolysis process.

In this study, a metal catalyst particle was observed on the tip of the twisted nanoribbon. The growth of carbon microcoils/nanocoils was expected through a vapor–liquid–solid (VLS) mechanism and the catalytic anisotropy mechanism.

7.3.2 Helical Carbon Nanotube

Carbon nanotubes (CNTs) derived from arc-discharge and laser ablation are usually straight. A helically coiled CNT is constructed by periodically introducing heptagonal and pentagonal rings into the hexagonal network of the graphene layers along the tube axis [30]. Helically coiled nanotubes can be metallic, semiconducting, or semimetallic depending on the placement of the pentagonal and heptagonal rings, which is impossible for the straight tube [31].

Cheng et al. [11] reported a simple chemical vapor deposition synthesis of helical carbon nanotubes using naturally occurring marine manganese nodule as the catalyst. The major component of the mineral catalyst is manganese oxide as a porous phase, interlayered with various metallic cations and water molecules. The mineral catalyst also contains abundant Fe and a small quantity of Ni cations. Before the introduction of acetylene as a carbon resource, the mineral catalyst was heated at 750°C for 20 min. During this process, iron group cations, embedded in the mineral nanopores, migrate to the outer surface through the unstable pores and transform into fine metallic nanoparticles, which present to be the catalyst for the decomposition of C_2H_2 and nucleate the growth of helical CNTs.

Figure 7.28 shows a typical SEM image of the deposited CNT products. It can be seen that a large fraction of helical CNTs with different helix diameters and different helix pitch lengths was obtained. In addition, multi-helical carbon nanofibers were also synthesized as indicated by arrows in Fig. 7.28. Figure 7.29 shows the TEM images of the as-grown helical CNTs. More than 50% of the CNTs are regularly coiled with a variety of radii and helix pitches. There is no catalyst particle existing on the tip of the helical CNT. Therefore, the growth mechanism of the helical CNTs was proposed based on an asymmetric growth rate. In the present experiments, the collapse of the porous structure of the mineral catalyst happens simultaneously with the metallic catalyst formation at elevated temperature. The initially formed metallic nanoparticles exhibit irregular shapes. On prolonging the heat treatment time,

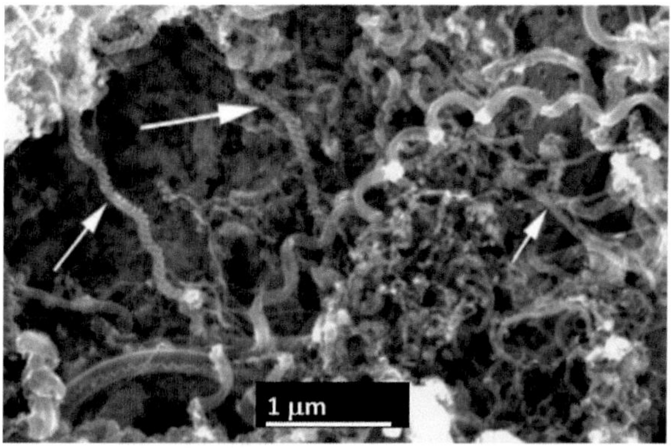

Fig. 7.28 SEM image of helically coiled CNTs in the product

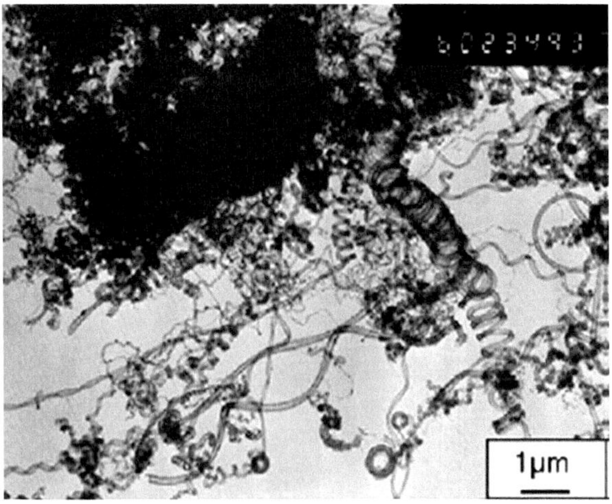

Fig. 7.29 TEM image of helical CNTs

the irregular nanoparticles will transfer to regular particles. It has been reported that regularly faceted catalyst particles are suited for the growth of regularly helical CNTs due to the asymmetric growth rate around the catalyst particles.

7.3.3 Tungsten-Containing Carbon (WC) Nanospring

Focused ion beam chemical vapor deposition (FIB-CVD) technique has two important capabilities: (I) several source gases can be used for the synthesis of 3D nanostructures of various materials and (II) the 3D nanostructures can be fabricated in arbitrary locations through in situ alignment. Therefore, FIB-CVD is a suitable method for the synthesis of nanosprings and other nanostructured materials. Nakamatsu et al. [19] applied this FIB-CVD method to fabricate WC nanosprings using phenanthrene (C_4H_{10}) and tungsten hexacarbonyl ($W(CO)_6$) as the source materials. The WC nanosprings were fabricated in a FIB apparatus (SMI9200: SII Nano Technology, Inc.) using a beam of 30-Ke Ga^+ ions. The beam was focused to a 7-nm spot size at a 1-pA beam current. The pressure inside the chamber was about 1×10^{-3} Pa after the precursor gas mixture was introduced. The WC nanosprings were grown for 15 min by a circular scan of the ion beam to control the spring diameter. Meanwhile, the waveform-generating equipment was used to control the scanning speed, which therefore controls the pitch distance of the nanosprings.

Figure 7.30 shows the SEM images of two WC nanosprings synthesized by FIB-CVD. The height, the coil diameter, and the section diameter were about 13.7 μm, 1.1 μm, and 200 nm, respectively, for the nanospring in Fig. 7.30a and 6.3 μm, 1.6 μm, and 200 nm, respectively, for the nanospring in Fig. 7.30b. As shown, the heights, pitches, spring diameters, and spring-section diameters of the nanosprings

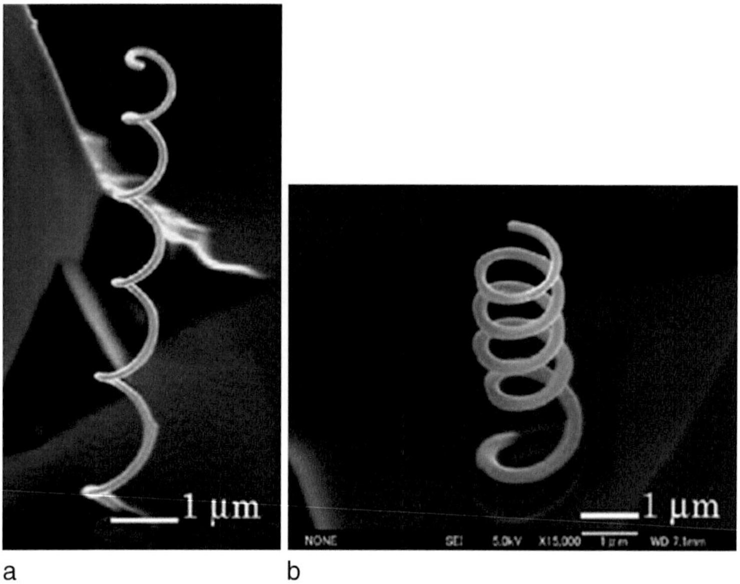

Fig. 7.30 Tungsten-containing carbon nanosprings grown by FIB-CVD. (**a**) WC spring with 13.7 μm height, 1.1 μm coil diameter, and 200 nm coil section diameter; (**b**) WC spring with 6.3 μm height, 1.6 μm coil diameter, and 200 nm coil section diameter

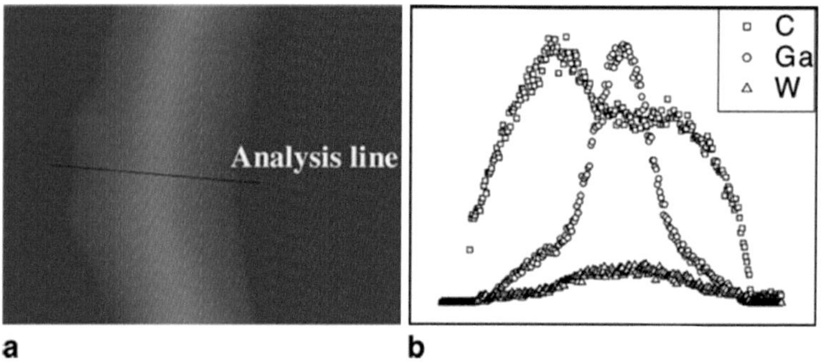

Fig. 7.31 (**a**) TEM image of a WC spring; (**b**) TEM-EDX line analysis result for a WC spring

can be controlled by using FIB-CVD method. The element analysis of the synthesized WC nanospring was conducted by TEM-EDX line analysis. As shown in Fig. 7.31, the C and W were uniformly distributed throughout the entire spring, while the Ga was concentrated in the middle of the spring. The C, W, O, and Ga contents of the nanospring, determined by TEM-EDX, were about 86, 3, 5, and 6 at.%, respectively. The impurity O atoms are attributed to the tungsten hexacarbonyl gas and the Ga to the ion beam irradiation.

7.4 Other Nanohelices

7.4.1 Helical SiC/SiO_2 Core–Shell Nanowires and Si_3N_4 Microcoils

Zhang et al. [28] synthesized helical crystalline SiC nanowires covered with a SiO_2 sheath by a chemical vapor deposition method. The helical composites were synthesized in a flow tube furnace through iron-catalyzed decomposition of methane at 1100°C. Iron powders were held in an alumina boat upstream in the flow tube and used as the catalyst to decompose methane. A pre-scratched silicon wafer located downstream was used as the substrate. First, the temperature was increased to 1100°C with a 200 sccm Ar flow and held at this temperature for 30 min. Then, a methane flow at 40 sccm together with an Ar flow at 160 sccm was introduced for 5–10 min. The helical composite growth was proceeded for another 1 h at 1100°C with only Ar flow.

Low-magnification TEM image (Fig. 7.32a) shows that there are straight and curled wire-like and helical core–shell structures in the product. Using HRTEM, electron diffraction, and elemental analyses, three types of nanowires were found. Those are pure amorphous SiO_2 nanowires, linear β-SiC/SiO_2 core–shell nanowires, and helical β-SiC/SiO_2 core–shell nanowires. A typical HETEM image of helical β-SiC/SiO_2 core–shell nanowire is shown in Fig. 7.32b. The SiC core typically has diameters of 10–40 nm with a helical periodicity of 40–80 nm and is covered by a uniform layer of 30–60-nm-thick amorphous SiO_2.

Cao et al. [32] reported a large-scale synthesis of Si_3N_4 microcoils by CVD method. The Si_3N_4 microcoils were prepared in a horizontal tube furnace. Pure silicon and silica powders were used as raw materials, which were held in an alumina boat and placed in the center of a ceramic tube. The tube was first purged with NH_3

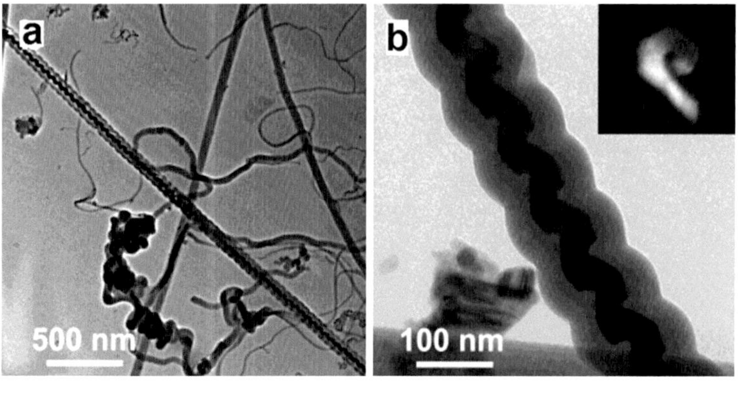

Fig. 7.32 (**a**) TEM image showing a long helical SiC/SiO_2 core–shell nanostructure; (**b**) HRTEM image of a helical SiC nanowire with an amorphous SiO_2 sheath. The *inset* shows a dark-field TEM image of a helical nanowire taken along the helical axis (wire diameter, ~ 22 nm)

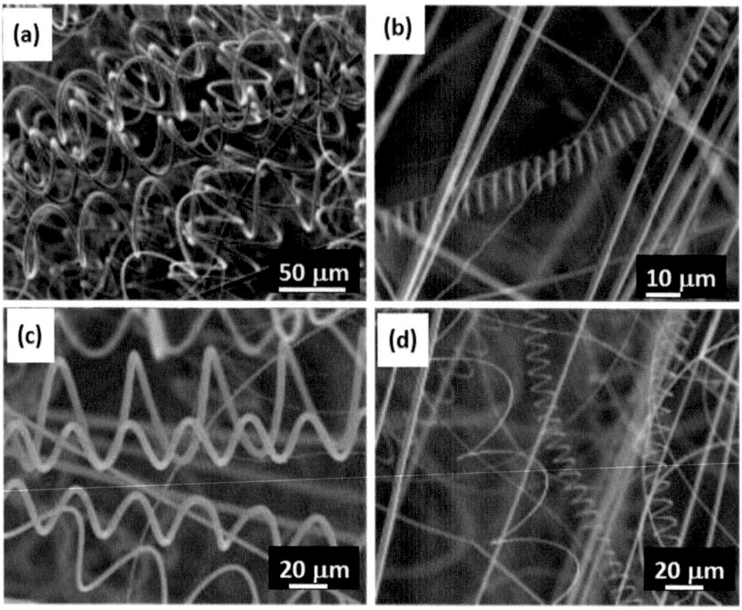

Fig. 7.33 SEM images of the as-synthesized silicon nitride microcoils: (**a**) equal pitch/diameter mcrocoils; (**b**) well-formed microspring; (**c**) coexistent left and right handed microcoils; (**d**) large dimension variation in microcoils

for 15 min. Then the NH_3 flow rate was set at 70 sccm during the whole experiment. The synthesis was conducted at 1350°C for 4 h. After the furnace was cooled to room temperature, a quantity of gray-white, wool-like product was deposited on the top of the source materials.

XRD analysis of the as-synthesized products revealed that they are mainly α-Si_3N_4. Figure 7.33 shows the SEM images of the as-synthesized Si_3N_4 microcoils. Detailed analyses indicated that the diameter of the Si_3N_4 fiber, the diameter of the microcoil, and the pitch of the microcoil are 0.5–4, 8–300, and 6–500 μm, respectively. The length of the microcoil is up to several millimeters. From Fig. 7.33a, it is obvious that most of the Si_3N_4 microcoils are regular coils with the coil pitch and coil diameter constant through the coils. Figure 7.33b shows a well-formed Si_3N_4 spring. Figure 7.33c, d shows that both left-handed and right-handed chiralities are obtained in the products. The surface of the coils is very smooth and no grains or impurities are observed.

7.4.2 MgB_2 Nanohelices

Nath et al. [33] reported the growth of MgB_2 nanohelices by a combination of physical and chemical vapor depositions on Si and other substrates by the reaction of Mg metal with diborane at 770–800°C under a flow of N_2 and H_2. Figure 7.34 shows

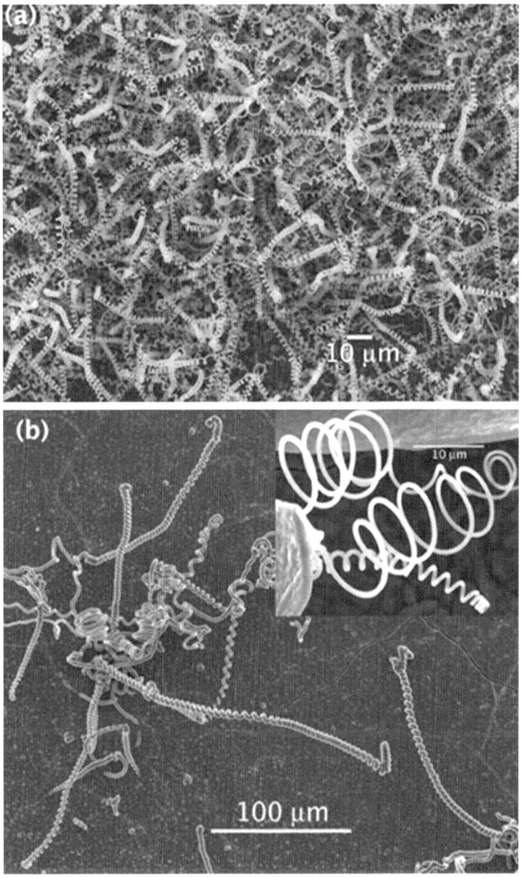

Fig. 7.34 SEM images of MgB_2 nanohelices grown on Si substrate: (**a**) large area on the Si substrate covered with nanohelices; (**b**) tightly wound nanospring, *inset* shows loosely wound coils

the SEM images of the as-synthesized products. As shown in Fig. 7.34a, it can be seen that a high density of nanohelices were obtained. The nanohelices typically exhibited round tips and circular cross section with diameters of 100–600 nm and lengths of 50–100 μm. The EDX analysis revealed that the nominal atomic ratio of Mg to B is 1:2. Together with the XRD analysis of the synthesized products, it confirms the growth of MgB_2 materials. As shown in Fig. 7.34b, both tightly wound springs and loosely wound coils (inset) are produced.

7.4.3 Si Spirals

Zhao et al. [34] reported the synthesis of Si spirals by the glancing angle deposition (GLAD) technique. Figure 7.35 presents the experimental setup for GLAD. In the GLAD system, the substrate is controlled by two stepper motors. One motor

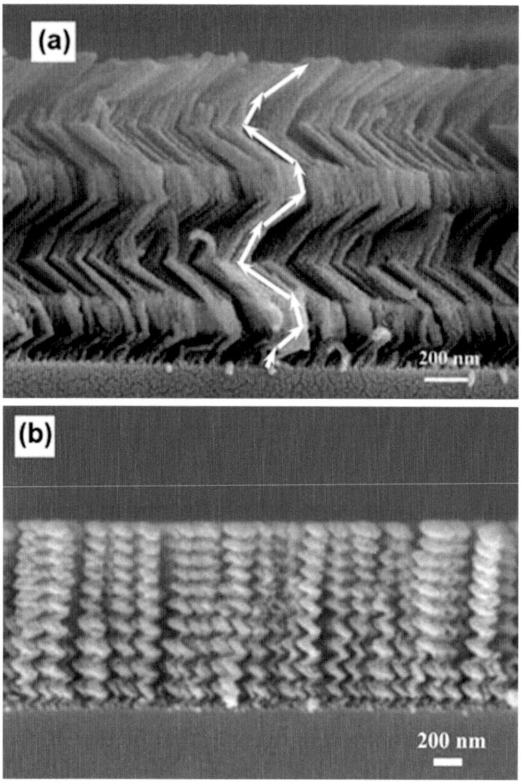

Fig. 7.35 SEM cross-sectional images of (**a**) 2-turn and (**b**) 10-turn square Si spirals

controls the incident angle, while another motor controls the azimuthal rotation of the substrate with respect to substrate surface normal. By changing the speed and phase of the azimuthal rotation, polar rotation, the combination of the two rotations, as well as the deposition rate, helically structured nanocolumns can be obtained. In this experiment, the Si spiral synthesis was conducted in a high-vacuum chamber with a background pressure of 2×10^{-4} Pa. The Si (99.9995%, Alfa Aeser) was evaporated by an electron beam bombardment method or the e-beam evaporation method. The pressure during the deposition was less than 1×10^{-3} Pa. The distance between the evaporation source and the substrate was about 30 cm. During the deposition, the deposition rate was fixed, while the rotation of the two motors was controlled by a computer.

Figure 7.35 shows the SEM images of the fabricated square Si spiral arrays on bare Si substrates. As shown in Fig. 7.35a (two-turn spiral), it can be seen that the spirals are uniformly distributed across the whole surface with almost the same length and each spiral is closely packed to the adjacent spiral. The diameter of the Si arm is about 50 nm, and the diameter of the spiral is about 200–500 nm. By adjusting the parameter of circumference for one pitch, smaller sized spiral nanostructure can be formed (Fig. 7.35b). It is a 10-turn spiral film.

7.5 Potential Applications

As discussed above, helically structured nanomaterials, especially semiconductors and carbon-related materials, have been well developed in the past decade. Because of their 3D symmetry morphology, as well as unique mechanical, electrical, and electromagnetic properties, helically nanostructured materials have very wide potential applications in micro-/nanosensors [18, 35] and actuators [36, 37], nano- and micro-electromechanical system (NEMS and MEMS) [38], energy and environment-related technology [39, 40], and biomedicine [38]. In the following, we will discuss the potential applications of 3D helical nanomaterials in detail, especially focusing on the most developed ZnO and carbon-based nanohelices.

Helical nanostructures of piezoelectric and semiconducting ZnO have drawn extensive research interest. The new discovery of the superlattice-structured ZnO nanohelices presents a brand-new helical nanostructure that is composed of two types of alternating and periodically distributed long crystal stripes, oriented with their *c*-axes normal to each other. As discussed in Section 7.2.1.2, a superelasticity (shape memory) effect has been discovered for the superlattice-structured ZnO nanohelices. By in-situ manipulation using a nanoprobe, the ZnO nanohelix could elastically recover its shape after an extremely large axial stretching, with an elastic elongation close to the theoretical limit, while suffering little residual plastic deformation. In addition, a shape memory/recovery of the ZnO nanohelix was observed after being subjected to a buckling deformation. The superelastic effect discovered in superlattice-structured ZnO nanohelices may be a new category of shape memory nanostructures made of ceramics, which could be of great interest for investigating the nanoscale fracture process and application in MEMS and NES. The elastic recovery of the nanohelix after extremely large deformation makes it a potential structure for nanoscale elastic energy storage. With an available high resonance frequency, the superelastic ZnO nanohelices can be made into smart microwave nanoantennas possibly with large bandwidths.

A nonlinear current–voltage characteristic has been observed for superlattice-structured ZnO nanohelices. The nonlinear electronic transport behavior of ZnO nanohelix might be due to a major contribution from nanostripe boundaries and surfaces, where a built-in periodic back-to-back energy barrier modulation might occur across the nanostripe interfaces as a result of polar charges and interface strain-induced piezoelectric effect. The superlattice nanohelices with nonlinear electronic behaviors could be used as nanoscale nonlinear electronic devices in varistors, lasers, sensors, and actuators. Other helically structured materials such as wurtzite ZnS, CdS, CdSe, AlN, InN, and GaN could also process this nonlinear transport property and have the potential applications as superlattice-structured ZnO nanohelices.

Helically structured, carbon-based materials are attracting more and more attentions because of their combination of 3D helical structures and other unique characteristics, such as good chiral conductivity, high surface area, high superelastic property, and high interaction ability with electromagnetic waves. Therefore, carbon-based helical nanomaterials show attractive potential applications in micro-/nanosensors and actuators, such as micro-magnetic sensors, electromagnetic wave absorbers, mechanical microsprings or actuators, and high elastic nano-electric conductors, and some other potential applications.

7.6 Summary

In this chapter, we mainly focused on reviewing the latest progresses on the growth of different nanohelices. Some unique properties of individual nanohelices were discussed and their potential applications were indicated. Although different nanohelices were produced from many kinds of inorganic materials by different synthesis methods, the synthesis of nanohelices with high uniformity and high yield is still an unresolved problem. Meanwhile, fundamental understanding about the properties, the testing and measuring techniques, and novel devices of the helically structured nanomaterials need to be further developed.

Acknowledgment The authors would like to thank the contributors for the materials used in this chapter, including Dr Z.L. Wang, Dr Y. Ding, Dr W.J. Mai, and Dr R.S. Yang. The authors also thank the financial support from the University of Connecticut New Faculty start-up funds, the University of Connecticut large faculty research grant, and the Department of Energy. Acknowledgment is also made to the Donors of the American Chemical Society Petroleum Research Fund for partial support of this work.

References

1. T. Murata, I. Yamato, Y. Kakinuma, A.G.W. Leslie, J.E. Walker, Structure of the rotor of the V-type Na^+-ATPase from *Enterococcus hirae*. Science **308**, 654–659 (2005)
2. P.X. Gao, Y. Ding, W.J. Mai, W. L. Hughes, C.S. Lao, Z.L. Wang, Conversion of zinc oxide nanobelts into superlattice structured nanohelices. Science **309**, 1700–1704 (2005)
3. P.X. Gao, Z.L. Wang, High-yield synthesis of single-crystal nanosprings of ZnO. Small **1**, 945–948 (2005)
4. P.X. Gao, W.J. Mai, Z.L. Wang, Superelasticity and nanofracture of ZnO nanohelix. Nano Lett. **6**, 2536–2543 (2006)
5. P.X. Gao, Y. Ding, Z.L. Wang, Electronic transport in superlattice-structured ZnO nanohelix. Nano Lett. **9**(1), 137–143 (2009)
6. R. Yang, Y. Ding, Z.L. Wang, Deformation-free single-crystal nanohelixes of polar nanowires. Nano Lett. **4**, 1309–1312 (2004)
7. X.Y. Kong, Z.L. Wang, Spontaneous polarization-induced nanohelixes, nanosprings, and nanorings of piezoelectric nanobelts. Nano Lett. **3**, 1625–1631 (2003)
8. L. Wang, D. Major, P. Paga, D. Zhang, M.G. Norton, D.N. McIlroy, High yield synthesis and lithography of silica-based nanospring mats. Nanotechnology **17**, S298–S303 (2006)
9. T. Delclos, C. Aime, E. Pouget, A. Brizard, I. Huc, M. Delville, R. Oda, Individualized silica nanohelices and nanotubes: Tuning inorganic nanostructures using lipidic self-assemblies. Nano Lett. **8**, 1929–1935 (2008)
10. X. Chen, S. Yang, S. Motojima, M. Ichihara, Morphology and microstructure of twisting nanoribbons prepared using sputter-coated Fe-based alloy catalysts on glass substrates. Mater. Lett. **59**, 854–858 (2005)
11. J. Cheng, X. Zhang, J. Tu, X. Tao, Y. Ye, F. Liu, Catalytic chemical vapor deposition synthesis of helical carbon nanotubes and triple helices carbon nanostructure. Mater. Chem. Phys. **95**, 12–15 (2006)

12. J.H. Xia, X. Jiang, C.L. Jia, C. Dong, Hexahedral nanocementites catalyzing the growth of carbon nanohelices. Appl. Phys. Lett. **92**, 063121 (2008)
13. J. Zhan, Y. Bando, J Hu, F. Xu, D. Golberg, Unconventional gallium oxide nanowires. Small **8–9**, 883–888 (2005)
14. W. Wang, F. Bai, Helical CdS nanowire ropes by simple aqueous chemical growth. Appl. Phys. Lett. **87**, 193109 (2005)
15. E.D. Sone, E.R. Zubarev, S.I. Stupp, Supramolecular templating of single and double nanohelices of cadmium sulfide. Small **7**, 694–697 (2005)
16. G.Z. Shen, Y. Bando, C.Y. Zhi, X.L. Yuan, T. Sekiguchi, D. Golberg, Single-crystalline cubic structured InP nanosprings. Appl. Phys. Lett. **88**, 243106 (2006)
17. K. Robbie, D.J. Broer, M.J. Brett, Chiral nematic order in liquid crystals imposed by an engineered inorganic nanostructure. Nature **399**, 764–766 (1999)
18. S.V. Kesapragada, P. Victor, O. Nalamasu, D. Gall, Nanospring pressure sensors grown by glancing angle deposition. Nano Lett. **6**, 854–857 (2006)
19. K. Nakamatsu, J. Igaki, M. Nagase, T. Ichihashi, S. Matsui, Mechanical characteristics of tungsten-containing carbon nanosprings grown by FIB-CVD. Microelectron. Eng. **83**, 808–810 (2006)
20. X.Y. Kong, Y. Ding, R. Yang, Z.L. Wang, Single-crystal nanorings formed by epitaxial self-coiling of polar nanobelts. Science **303**, 1348–1351 (2004)
21. W.L. Hughes, Z.L. Wang, Formation of piezoelectric single-crystal nanorings and nanobows. J. Am. Chem. Soc. **126**, 6703–6709 (2004)
22. Z.W. Pan, Z.R. Dai, Z.L. Wang, Nanobelts of semiconducting oxides. Science **291**, 1947–1949 (2001)
23. A. Boudaoud, P. Patricio, Y. Couder, M. Ben Amar, Dynamics of singularities in a constrained elastic plate. Nature **407**, 718–720 (2002)
24. Y. Sato, F. Oba, T. Yamamoto, Y. Ikuhara, T. Sakuma, Current-voltage characteristics across (0001) twist boundaries in zinc oxide bicrystals. J. Am. Ceram. Soc. **85**, 2142–2144 (2002)
25. D.L. Smith, C. Mailhiot, Theory of semiconductor superlattice electronic structure. Rev. Mod. Phys. **62**(1), 173–234 (1990)
26. D.R. Clarke, Varistor ceramics. J. Am. Ceram. Soc. **82**(3), 485–502 (1999)
27. H. Gao, X. Zhang, M. Zhou, E. Zhang, Z. Zhang, Super-uniform ZnO nanohelices synthesized via thermal evaporation. Solid State Commun. **140**, 455–458 (2006)
28. H.F. Zhang, C.M. Wang, L.S. Wang, Helical crystalline SiC/SiO_2 core–shell nanowires. Nano Lett. **2**, 941–944 (2002)
29. N. Herron, J.C. Calabrese, W.E. Farneth, Y. Wang, Crystal structure and optical properties of $Cd_{32}S_{14}(SC_6H_5)_{36}$-DMF4, a cluster with a 15 angstrom CdS core. Science **259**, 1426–1428 (1993)
30. A. Fonseca, K. Hernadi, J.B. Nagy, P. Lambin, A.A. Lucas, Model structure of perfectly graphitizable coiled carbon nanotubes. Carbon **33**, 1759–1775 (1995)
31. K. Akagi, R. Tamura, M. Tsukada, Electronic structure of helically coiled cage of graphitic carbon. Phys. Rev. Lett. **74**, 2307–2310 (1995)
32. C. Cao, H. Du, Y. Xu, H. Zhu, T. Zhang, R. Yang, Superelastic and spring properties of Si_3N_4 microcoils. Adv. Mater. **20**, 1738–1743 (2008)
33. M. Nath, B.A. Parkinson, Superconducting MgB_2 nanohelices grown on various substrates. J. Am. Chem. Soc. **129**, 11302–11303 (2007)
34. Y.P. Zhao, D.X. Ye, G.C. Wang, T.M. Lu, Designing nanostructures by glancing angle deposition. Proc. SPIE **5219**, 59–73 (2003)
35. D.J. Bell, Y. Sun, L. Zhang, L.X. Dong, B.J. Nelson, D. Grutzmacher, Three-dimensional nanosprings for electromechanical sensors. Sens. Actuators A Phys. **130–131**, 54–61 (2005)
36. J.P. Singh, D.L. Liu, D.X. Ye, R.C. Picu, T.M. Lu, G.C. Wang, Metal-coated Si springs: Nanoelectromechanical actuators. Appl. Phys. Lett. **84**, 3657–3659 (2004)
37. C. Daraio, V.F. Nesterenko, S. Jin, W. Wang, A.M. Rao, Impact response by a foamlike forest of coiled carbon nanotubes. J. Appl. Phys. **100**, 064309 (2006)

38. J. Jagadish, S.J. Pearton (eds.), *Zinc Oxide Bulk, Thin Film and Nanostructures* (Elsevier, Amsterdam, Netherlands, 2006)
39. U. Sahaym, M.G. Norton, Advances in the application of nanotechnology in enabling a 'hydrogen economy'. J. Mater. Sci. **43**(16), 5395–5429 (2008)
40. J.G. Gibbs, Y.P. Zhao, Measurement of driving force of catalytic nanomotors in dilute hydrogen peroxide by torsion balance. Rev. Sci. Instrum. **79**, 086108 (2008)

Chapter 8
Hierarchical 3D Nanostructure Organization for Next-Generation Devices

Eric N. Dattoli and Wei Lu

8.1 Introduction

The emerging field of nanotechnology research has made a great deal of progress in broadening the depth of knowledge related to the material properties and device application potentials of nanostructures such as carbon nanotubes (CNTs) and nanowires (NWs). These nanomaterials, as opposed to traditional thin-film or wafer-based planar materials, offer an array of desirable electrical, optical, and mechanical properties enabled by their well-controlled, nanoscale sizes. However, so far the main focus of nanostructure research has been on the fabrication and characterization of single or small-scale device structures [1–2]. Although these "proof-of-concept" structures are useful for probing the intrinsic physical properties of the devices, they are not applicable to commercial or real-world applications. Practical nanostructure-based electronics must be able to be fabricated in a scalable fashion and in sizable quantities while maintaining a good uniformity in performance among different devices. The purpose of this chapter is to examine the different assembly and integration methods which could help realize the manufacture of next-generation, nanostructure-based devices. In particular, it illustrates how these integration techniques offer the opportunity to achieve multi-functional, three-dimensional (3D) integrated systems based on nanomaterials.

The bright prospects related to nanostructure-based devices can be attributed to the unique physical properties of nanowires and CNTs. Both types of materials possess attractive electrical properties which are of considerable interest. For example, CNTs possess very long mean free carrier paths which are typically in the micrometer range at room temperature due to their unique band structures. The bottom-up growth of nanowires in addition allows for the growth of a wide range of materials with reproducible electronic properties as required for large-scale integrated systems. The crystalline structure, smooth surfaces, and the ability to

W. Lu (✉)
Department of Electrical Engineering and Computer Science, University of Michigan, Ann Arbor, MI 48109-2122, USA
e-mail: wluee@eecs.umich.edu

W.L. Zhou, Z.L. Wang (eds.), *Three-Dimensional Nanoarchitectures*,
DOI 10.1007/978-1-4419-9822-4_8,

obtain radial and axial heterostructures in nanowires in turn result in higher carrier mobility compared with nanofabricated samples with similar size. For example, nanowires synthesized in heterostructure configurations have been shown to facilitate doping-free carrier transport with measured performance that exceeds those of state-of-the-art planar devices [3]. Finally, since the body thickness (diameter) of nanowires and CNTs can be controlled down to well below 10 nm [4], the electrical integrity of nanowire-based electronics can be maintained even as the transistor size is aggressively scaled, a feat that has become increasingly difficult to achieve in conventional planar-based devices [5].

Furthermore, as a result of their growth mechanisms, nanowires and CNTs may be synthesized in an independent step and later be transferred (post-growth) to chosen device substrates in a layer-by-layer fashion, thus allowing effective 3D device integration. Presently, there has been considerable research into the usage of 3D and multilayer structures in the integrated circuit (IC) electronics field. Multilayered devices may be achieved through wafer bonding [6] or epitaxial growth [7]. In addition, conventional memory and logic devices may be replaced with 3D devices, such as vertical transistors [8] or crossbar memories [9], as device miniaturization continues. Owing to their naturally non-planar growth geometries and large aspect ratios, nanowires or nanotubes provide an elegant route for realizing such 3D device structures. In this regard, the nanoelectronics approach provides a basis for achieving multilayer chips consisting of stacks of diverse devices and materials by exploiting layerable nanostructure transfer and assembly processes or directed nanostructure growth. Such a multilayer chip provides a practical route for achieving the integration of heterogeneous and multi-functional devices. For example, the on-chip implementation of logic circuits and optical sensors has already been demonstrated using nanostructured materials [10]. Another benefit that multilayered chips may bring about is a reduction in IC power requirements and delay times as a result of the associated increases in device density and reductions in interconnect lengths inherent in a 3D, interconnected device configuration [11].

With the purpose of achieving such 3D device structures, an overview of the most significant nanostructure integration and assembly methods is presented. Specifically, assembly methods based on fluidic flow, nematic liquid crystalline phases, the Langmuir–Blodgett technique, dielectrophoresis, chemical affinities, and contact transfer are discussed in detail [12–13]. Additionally, efforts to achieve controlled growth of nanowires in predefined horizontal or vertical directions for direct device integration are discussed. Finally, a few demonstrated examples of prototype nanostructure-based devices are covered in order to illustrate their potential usage in real-world applications. The device configurations to be reviewed are nanowire- and nanotube-based thin-film transistors (TFTs), multilayer nanostructure-based devices, vertical nanowire-based field-effect transistors (FETs), and an integrated nanowire-based optical detector circuit.

8.2 Fluidic Flow-Assisted Assembly

In this section, the fluidic flow-assisted assembly of nanostructures is discussed. The shared characteristic of all the assembly techniques to be examined will feature the usage of a shear force to uniformly orient the nanostructures during their deposition from a liquid suspension to a receiver substrate. The shear force originates from the action of an induced, unidirectional fluidic flow. The methods to be discussed include drop-drying, channel-confined fluidic flow, and blown bubble films.

8.2.1 Drop-Drying

The conceptually simplest method used to perform the fluidic flow-assisted assembly of nanostructures is by merely allowing for a droplet of a homogeneous mixture of nanostructures (i.e., a suspension of nanoparticles, nanowires, or CNTs) to dry by solvent evaporation on a receiver substrate. After the entire drop dries, a single ring-shaped line of deposition will be noticeable in the region where the outer diameter of the droplet was previously located. This phenomenon is common in everyday life and is recognizable in coffee ring stains. The particular aspect of this phenomenon that requires closer examination is the uniformly distributed, preferential deposition of solute particles along the outer diameter of the droplet. It is curious to note that this ring-shaped deposition occurs rather than the uniform deposition of solute particles over the entire droplet/substrate interfacial area. This result can be explained by noting that for most solvent droplets which contain a homogeneous mixture of particles, the exterior contact line of the droplet at the substrate surface is pinned during the entire evaporation process, i.e., the surface area of the droplet at the substrate interface remains constant during evaporation. Such contact line pinning does not usually occur for pure solvent droplets and the pinning phenomenon can be explained by taking into account the effects of surface roughness and chemical heterogeneities that the first (and the outermost) deposited solute particles give rise to at the contact line of the droplet [14]. As the evaporation process proceeds, due to the contact line pinning, solvent must flow from the interior of the droplet to the contact perimeter in order to replenish evaporated solvent and maintain the constant droplet surface area. These solvent flows will cause a shear force that will carry and align nanostructures in this radial direction toward the perimeter of the droplet, schematically illustrated in Fig. 8.1b. Deposition of the nanostructures on the substrate can occur by two mechanisms (which will be discussed separately): by simple adherence of the nanostructures due to random physical contact and subsequent van der Waals interactions or by a dense concentration of nanostructures forming a nematic liquid crystal phase (to be discussed in a later section). It is critical to note that due to the radial liquid shear flow, the nanostructures will nearly be uniaxially aligned in this radial direction following deposition (Fig. 8.1a–c).

The first deposition mechanism (random collision with and subsequent adherence to the substrate) has been exploited in the assembly of both single-wall CNTs (SWNTs) [15–17] and nanowires (Ag, Si, and ZnO) [18]. In the latter study after noting that nanowire assembly occurred by simple drop-drying, Yang et al. proceeded to demonstrate a larger scale, manufacturing-compatible design through the use of a programmable dip coater (Fig. 8.1d–f). In this setup, a wafer was vertically dipped into a pool containing a nanowire liquid suspension. At the surface of the liquid, a meniscus formed where the liquid adhered to a portion of the deposition

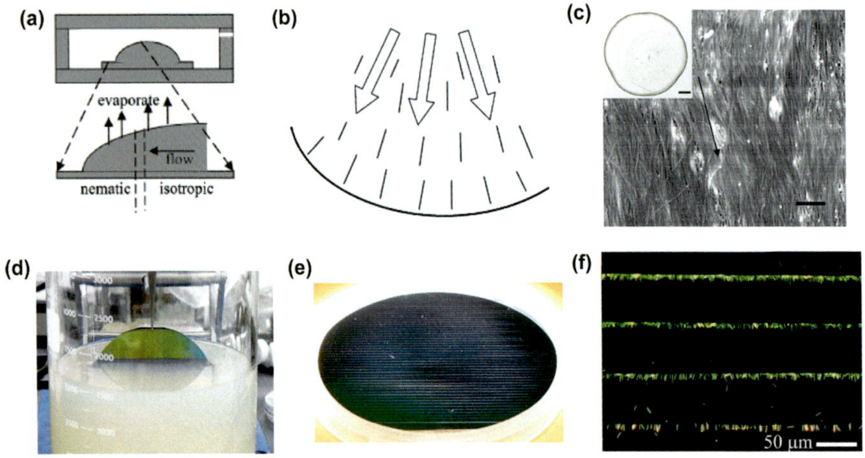

Fig. 8.1 Drop-drying-induced nanostructure assembly. (**a**) Cross-sectional schematic of drop-drying-induced nanostructure (i.e., nanowire and/or nanotube) assembly. Ordering of colloidal nanostructures is induced at the edge of the evaporating droplet. Adapted with permission from [17]. Copyright 2009 American Chemical Society. (**b**) Top-down schematic illustration of the deposition of aligned nanostructures at the edge of a drying droplet on a horizontal substrate. The pinning of the contact line (*black line*) induces an outward capillary flow (*block arrows*) to compensate for the loss of solvent at the perimeter by evaporation. This flow aligns the nanostructures (*gray lines*) and carries them toward the contact line, which leads to the final ring-shaped stain. Reproduced with permission from [18]. Copyright 2007 Wiley-VCH Verlag GmbH & Co. KGaA. (**c**) SEM images of the SWNT deposit formed at the edge of the dried droplet shown in (**b**). The *arrow* indicates the orientation direction of SWNT assembly. The *scale bar* corresponds to 100 nm. *Inset*, optical microscopy image of a ring formed by drying a drop of SWNT suspension on a glass surface at room temperature. The scale bar indicates 1 mm. Adapted with permission from [16]. Copyright 2006 American Chemical Society. (**d**) One-step patterning of aligned nanowire arrays by programmed dip coating of an oxygen plasma-cleaned silicon wafer that was immersed in a Ag nanowire dispersion in methylene chloride and pulled out by a programmable mechanical dipper to control the positioning of the solvent–substrate contact line. (**e**) Nanowire arrays with tunable density and arbitrary spacing over the entire 4 in. wafer were obtained. (**f**) Four equally spaced arrays with decreased nanowire density. Reproduced with permission from [18]. Copyright 2007 Wiley-VCH Verlag GmbH & Co. KGaA

wafer above the surface line and where additionally solvent evaporation occurred. The researchers noted that aligned nanowire deposition occurred at this meniscus line. They also observed that the deposition occurred at a roughly linear rate, leading one to conclude that random adherence of the NWs was the responsible mechanism. Through the use of a programmable dip coater, Yang et al. demonstrated the controlled deposition of arrays of uniaxially aligned nanowires at certain predetermined rows along the entire receiver wafer.

8.2.2 Channel-Confined Fluidic Flow

In a manner similar to drop-drying, laminar fluidic flow through channels gives rise to a shear force which can induce the uniaxial alignment of nanowires or nanotubes in the direction of the fluidic flow. The channels provide a dual function: to limit nanostructure deposition in only certain, desired areas and also to provide a restricted path for fluidic flow so that flow rates can be constantly controlled. By controlling the orientation of the fluidic channels, arbitrary alignment orientations relative to the receiver substrate can be achieved for the deposited nanostructures (Fig. 8.2). The technique has been demonstrated on InP and Si NWs while utilizing fluidic flow through PDMS molded channels [19–20]. A slightly modified fluidic flow technique utilizing controlled flocculation was shown to achieve the controlled deposition of single-wall CNTs (SWNTs) [21].

As with the drop-drying method, nanostructures are deposited onto the substrate as a result of their collisions with and subsequent adherence to the substrate due to

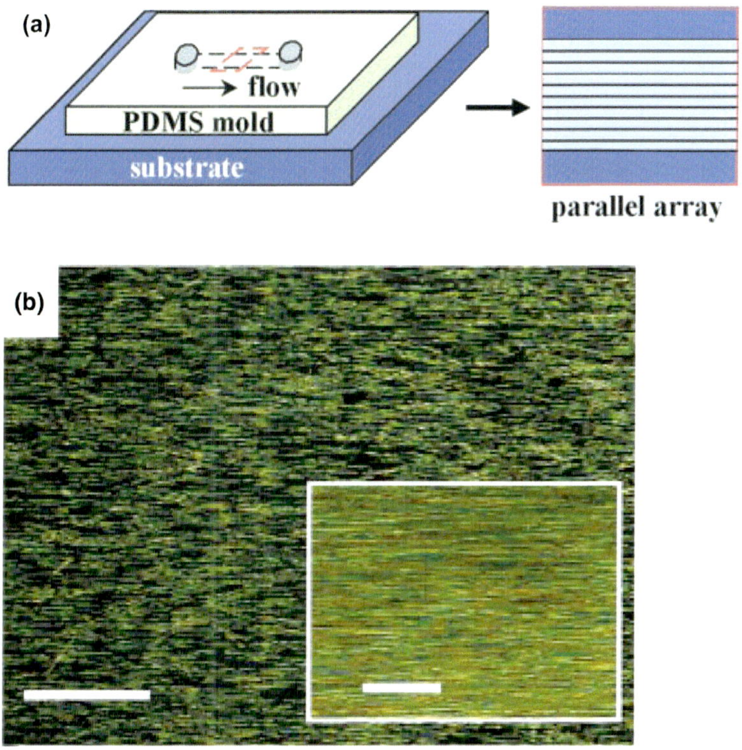

Fig. 8.2 Fluidic flow-assisted nanostructure assembly. (**a**) Schematic of fluidic channel structures for flow assembly. A channel is formed when the PDMS mold was brought in contact with a flat substrate. NW assembly was carried out by flowing an NW suspension inside the channel with a controlled flow rate for a set duration. Reprinted with permission from [19]. Copyright 2001 AAAS. (**b**) Optical micrograph of flow-aligned NW thin film. Scale bar, 80 μm. *Inset*, a picture at higher magnification. Scale bar, 20 μm. The micrographs show that the NW thin film is nearly a monolayer of NWs, but occasionally a few NWs cross over each other. The average space between parallel NW arrays is estimated to be ~540 nm. Reprinted with permission from [20]. Copyright 2003 Macmillan Publishers Ltd: Nature

van der Waals interactions. As a result of this, the deposition density in the channel-confined fluidic flow method can be controlled by flow duration, i.e., a longer flow time provides for additional opportunities for nanostructure adsorption. Huang et al. found that the deposited density of NWs increases systematically with the flow duration at a constant flow rate [19]. For example, a flow duration of 30 min produced a density of about 250 NWs per 100 mm or an average NW–NW separation of $\sim$400 nm, and NW spacing on the order of 100 nm or less can be achieved with an extended deposition time in excess of 40 min.

The channel-confined fluid flow method was shown to be capable of depositing multiple layers of nanowires on the same substrate with arbitrary alignments by carrying out sequential fluidic flow deposition steps. For example, crossbar structures with a high yield have been demonstrated by alternated flow in orthogonal directions in a two-step assembly process [19]. Moreover, this technique was found to be compatible with chemical surface patterning methods (to be discussed later) that can allow template-assisted deposition of both NWs and CNTs in a controlled alignment and pattern [22].

8.2.3 Blown Bubble Film Transfer

A third technique that utilizes the shear flow-induced alignment of nanostructures is the blown bubble film (BBF) transfer method [23–24], which has been demonstrated on CNTs, NWs, and nanoparticles. The BBF technique is carried out by utilizing gas blown spherical-shaped bubbles of nanostructure-embedded epoxy films which can be subsequently deposited conformally onto a wealth of substrates ranging from silicon wafers to even plastic sheets. The inspiration for this method comes from the widespread use of blown polymer film extrusion techniques in industry for the manufacture of plastic products, like bags or films. A key feature of this assembly method is the use of a pressurized gas flow which serves two purposes: (1) it induces a uniform shear flow of liquid (the epoxy) which in turn induces the uniaxial alignment of embedded nanostructures in the film and (2) it allows for the formation of bubbles which may be deposited onto large substrates (deposition onto 200 mm wafers has been demonstrated). For the sake of completeness, it needs to be noted that the gas flow-induced, liquid shear force alignment of nanostructures has also been demonstrated on CNTs and NWs from liquid suspensions independently of the BBF technique and without the use of epoxy films [25–26].

A brief description of the BBF technique will now be given as shown in Fig. 8.3. First, a stabilized suspension of nanowires or CNTs is prepared. A proper chemical treatment is required in the suspension in order to functionalize the surface of the nanostructures and to prevent their aggregation (i.e., flocculation or coagulation). The density of nanostructures in the prepared suspension was found to directly correlate with the density of the transferred nanostructures later deposited on receiver substrates. Next, the nanostructure suspension is mixed with an epoxy resin and then cured for the appropriate time duration to achieve the optimal viscosity for bubble formation. Bubbles are then formed by dispensing the epoxy–nanostructure

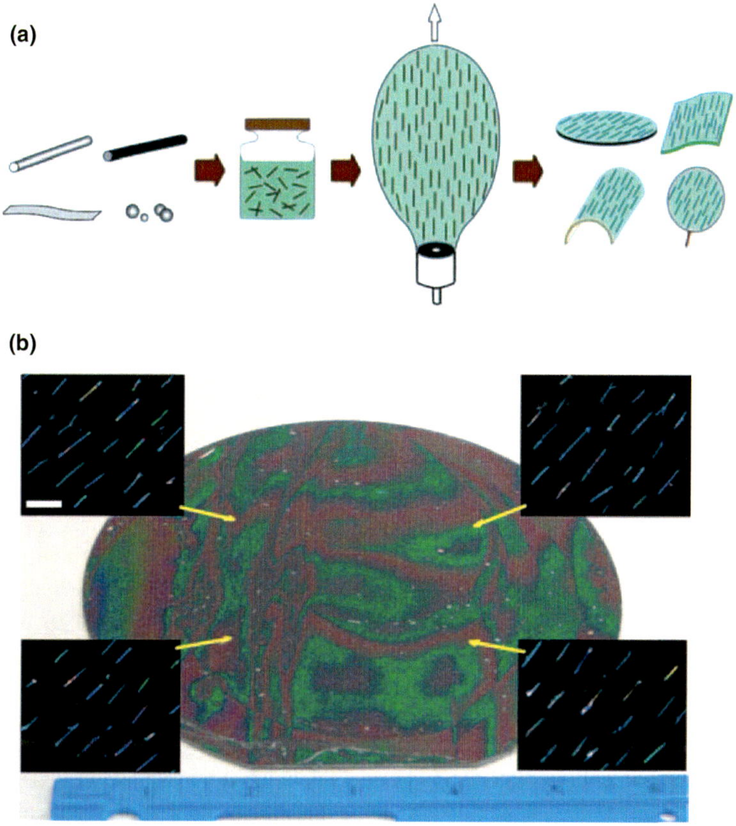

Fig. 8.3 Blown bubble film (BBF) approach. (**a**) Nanomaterials (e.g., nanotubes, nanowires, nanobelts, and nanoparticles) are dispersed in a polymer solution, a volume of solution is expanded as a bubble, and then BBFs are transferred to substrates, including crystalline wafers, plastic sheets, curved surfaces, and open frames. The *black straight lines* illustrated in the solution and bubble films represent 1D nanomaterials such as nanowires or nanotubes. Adapted with permission from [24]. Copyright 2008 The Royal Society of Chemistry. (**b**) Optical image of a 0.10 wt.% Si NW-BBF on 150 mm Si wafer. *Insets*, dark-field (DF) optical images showing aligned Si NWs at different locations. Scale bar, 10 μm. Reprinted with permission from [23]. Copyright 2007 Macmillan Publishers Ltd: Nature Nanotechnology

solution on a circular die with a gas outlet situated at its center. A controlled pressure of gas is allowed to flow through the outlet to cause bubble formation and elongation. Bubbles of diameters greater than 30 cm have so far been demonstrated. Once the bubble reaches a desired diameter, a portion of the bubble film may be conformally coated onto a receiver substrate and subsequently processed as a deposited thin film. Device fabrication has been demonstrated on the transferred nanostructures after utilizing post-deposition dry etching of the epoxy in order to expose the embedded nanostructures.

The BBF method has been able to realize the assembly of aligned nanowires with separations as low as 3 μm between each NW. Such a deposition density is lower than that in other transfer methods, like contact printing discussed later, but may be suitable for certain device applications like sensors. Attempts at producing higher deposition densities by utilizing higher density NW suspensions resulted in the aggregation of the nanowires. One unique advantage of the BBF method is that it has demonstrated large-scale deposition dimensions (i.e., transfer onto a 200 mm wafer); such size lengths have not yet been matched by other assembly techniques. Moreover, the use of a PMMA thin film in place of the epoxy resin has been shown to be a suitable material for blown bubble processing. PMMA is used as an e-beam and deep UV resist, and its usage in the BBF technique could help to streamline additional device fabrication steps for practical nanostructure-based devices.

8.3 Nematic Liquid Crystal-Induced Assembly

Nematic liquid crystal (LC)-induced transfer and alignment relies upon the use of dense suspensions of NWs or CNTs which assemble into a nematic LC film either in solution or upon application to a substrate. To explain the mechanism, first some background information on the phase of matter known as liquid crystals is required. The distinguishing feature of LCs is that they possess a degree of order in between that of liquids and solids. Specifically, liquid crystals can be classified under different mesophases depending on their degree of molecular alignment. The nematic LC phase possesses a large degree of orientational order along a single axis, e.g., uniaxial alignment. Nematic LCs do not possess any positional order, that is, there is no exact relative placement of each unit in the LC as compared to adjacent units and as a result the end-to-end registry of nanostructures in nematic LCs is poor.

Materials that are capable of adopting the LC phase are electrically polarizable and have a rigid, rod-shaped molecular structure. Liquid crystalline materials are called lyotropic if ordering can be induced resulting from certain concentrations of the material within a solvent. It has been shown that CNTs [27] and nanorods [28] are indeed lyotropic; in concentrated suspensions, they behave as nematic LCs.

LCs are readily aligned in a desired uniaxial direction using a variety of methods. Dense, liquid crystalline suspensions of CNTs have been aligned using methods such as surface roughness or electric field [29]. Alternatively, it has been shown that dilute suspensions of CdSe nanorods [30] or SWNTs [31] may be aligned by drop-drying. It is thought that the high evaporation rate at the air–liquid interface results in a highly concentrated solution, thus promoting liquid crystalline ordering (Fig. 8.4a, b) in the latter cases. This uniaxial ordering typically occurs in random directions and over a fairly short length (tens of micrometers). In order to achieve uniaxial ordering along a controlled direction and over larger distances, it has been shown that it is possible to combine the use of channel-confined fluid flow and drop-drying to achieve the long-range ordering of nematic-phase CNTs (Fig. 8.4c, d). Photoresist-defined channels were used to induce a unidirectional flow of a CNT suspension by simply tilting the receiver substrate during drop-drying.

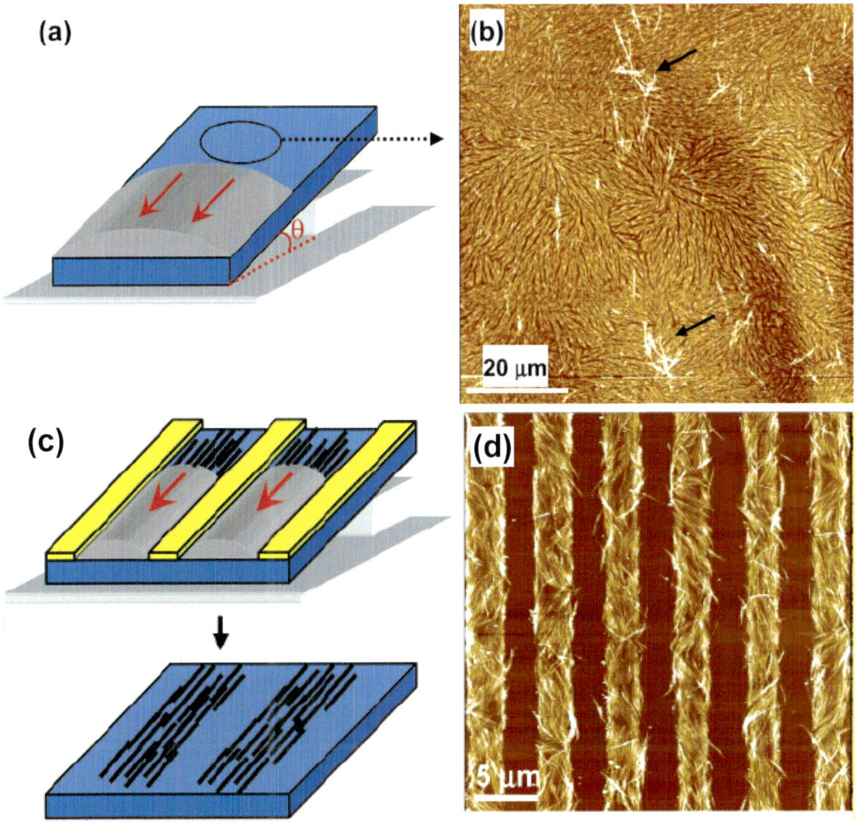

Fig. 8.4 Surface ordering of carbon nanotube films by induced nematic liquid crystal phase as a result of the slow evaporation of a carbon nanotube solution during tilted-drop casting. (**a**) Scheme for the tilted-drop fabrication routine without physical confinement. (**b**) AFM topographical images of a CNT surface showing liquid crystalline texture and ordering along with characteristic topological defects. (**c**) Schemes for the tilted-drop fabrication of a thin film on an amine-terminated SAM surface micropatterned with photoresist polymer stripes. (**d**) AFM topographical image of carbon nanotube films showing uniaxially oriented, densely packed CNT bundles. Adapted with permission from [31]. Copyright 2006 American Chemical Society

Unidirectional alignment over a single defined direction was achieved for lengths exceeding 40 μm.

8.4 Langmuir–Blodgett Assembly

The Langmuir–Blodgett (LB) technique refers to the formation of a monolayer film of molecules or particles on an aqueous surface and the subsequent transfer of the layer to a solid substrate which is vertically drawn through the air–water interface [32]. The LB technique is commonly used in biological or chemical applications, and recently the method has been used to achieve the transfer of aligned arrays of nanostructures (i.e., nanowires, nanotubes, and nanoparticles) [32–33]. The Langmuir–Blodgett method is performed in an apparatus termed the LB trough which is a water-filled trough fitted with moving barrier mechanisms that allow for the modification of the surface pressure of a water surface monolayer.

An overview of the methods used to apply the LB technique to nanostructure assembly will now be detailed. To begin with, a stabilized suspension of nanostructures in a nonpolar solvent/surfactant mixture is dispersed on top of an aqueous subphase in a LB trough. One purpose of the surfactant molecules is to prevent the agglomeration of the nanostructures. On the water surface, the nanostructures form a monolayer where the nanostructures are supported by the effects of water surface tension. Subsequently, a compressive force is applied to the monolayer by narrowing the barriers of the LB trough. Although some surfactant molecules may dissolve into the aqueous layer as time proceeds, increasing the compressive force tends to increase the surface pressures in most experimental situations.

As the nanostructures compress due to higher surface pressures, the monolayer takes on the properties of a nematic liquid crystal where the axial alignment direction is parallel to the barrier. The underlying physical reasons for the induced uniform alignment of the individual nanostructures can be attributed to thermodynamic considerations once the compressive forces bring about the formation of a liquid crystal phase. The nanostructure-to-nanostructure spacing can be controlled by adjusting the magnitude of the compressive force. As the surface pressure increases, the nanostructure-to-nanostructure spacing decreases; finally after a critical buckling pressure is reached, the formation of a multilayer film can be induced [34]. The aligned nanostructure layer is typically transferred to a desired receiver substrate by vertically dipping the receiver substrate in the trough along the direction of the aligned NWs or CNTs and the moving trough barriers (Fig. 8.5a). Van der Waals attractions cause the nanostructures to adhere onto the receiver substrate as it is lifted away from the surface, and the nanostructure film is transferred to the receiver substrate while maintaining the film's density and alignment direction. Deposition with an arbitrary alignment direction with respect to the receiver substrate may be obtained by rotating the receiver substrate at an appropriate angle relative to the water surface [35].

As a result of the distinct surface properties that are associated with different types of nanostructures, the optimal conditions for carrying out the Langmuir–Blodgett assembly of various nanostructures have been explored in many studies. A survey of the most significant reports is given below. Monolayer films of Si/SiO_2 core–shell NWs were obtained with NW diameters of 45 and 90 nm and for transfer areas up to 20 cm^2 [36]. The wire-to-wire spacings were found to be readily controlled by the amount of applied compressive force; spacings varying from 0.8 μm to completely close-packed monolayer films were demonstrated. Nearly completely close-packed monolayer films have also been obtained using Ge nanowires with diameters of $\sim$10–20 nm [37]. Similarly structured close-packed films of SWNTs that possess a diameter less than 2 nm and lengths between 200 nm and 1 μm have also been obtained using Langmuir–Blodgett assembly (Fig. 8.5d) [33]. This result

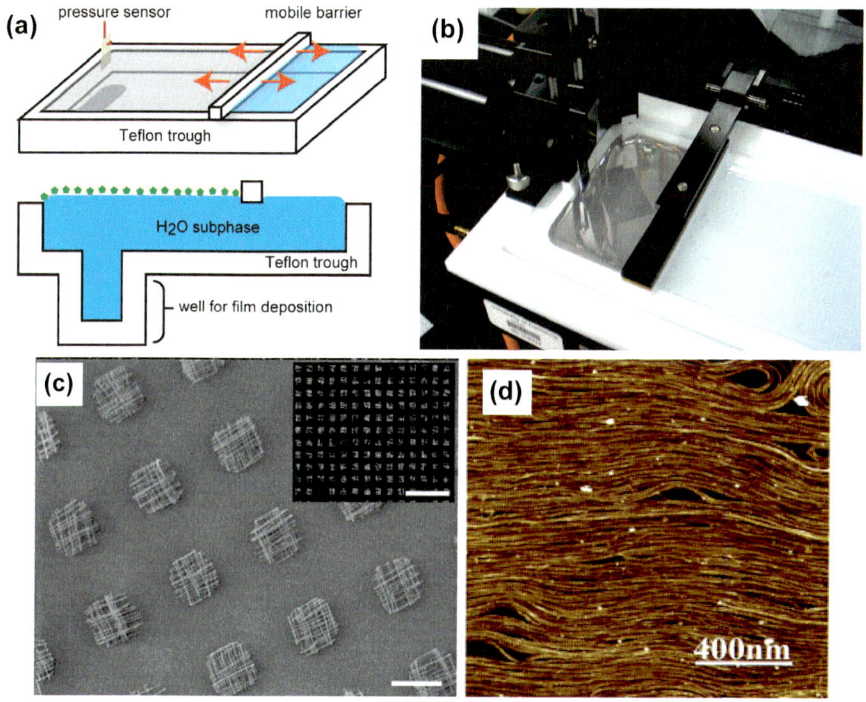

Fig. 8.5 Langmuir–Blodgett assembly of nanostructures. (**a**) Schematic of a water-filled, Langmuir–Blodgett trough from the top and side views. (**b**) Image of a substrate being pulled vertically through a Langmuir monolayer of silver nanowires. Adapted with permission from [32]. Copyright 2008 American Chemical Society. (**c**) Scanning electron microscopy image of patterned crossed NW arrays formed by multiple dipping of the receiver substrate; scale bar, 10 μm. *Inset*, large-area, dark-field optical micrograph of the patterned crossed NW arrays; scale bar is 100 μm. Adapted with permission from [36]. Copyright 2003 American Chemical Society. (**d**) AFM image of an LB film of SWNTs. Adapted with permission from [33]. Copyright 2007 American Chemical Society

indicates that even nanostructures with small diameters and large aspect ratios may be assembled with this method.

8.5 Dielectrophoresis Assembly

Dielectrophoresis (DEP) is a technique that has gained widespread adoption throughout the field of biochemistry as a method for manipulating the position of molecules inside a liquid medium. The same technique has also been successfully applied to the task of assembling either individual or massive numbers of nanostructures. It has been applied to a wide variety of nanostructures, including CNTs (single and bundles) and both metallic and semiconducting NWs. A notable feature of this assembly method is that it offers the possibility of achieving fine control over

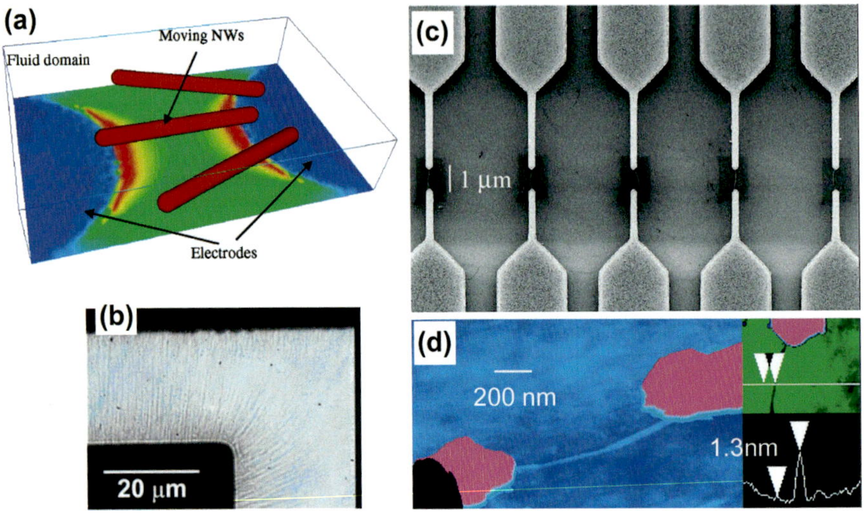

Fig. 8.6 Dielectrophoretic assembly of nanostructures. (**a**) Schematic of the dielectrophoretic alignment process for 1D nanostructures. *Redder colors* indicate regions of stronger electric fields and *bluer colors* represent regions of weaker fields. Adapted with permission from [41]. Copyright 2006 American Chemical Society. (**b**) Bright field image of dielectrophoretically aligned CdSe NWs using an AC electric field (10 V) with electrodes separated by a 20 μm gap. Reprinted with permission from [40]. Copyright 2007 American Institute of Physics. (**c**) SEM image of an electrode array showing five adjacent devices, with each electrode pair bridged by one carbon nanotube, visible as fine white lines within the dark central areas. (**d**) Atomic force microscopy image of one such device. The height profile confirms the bridging by an individual nanotube. Adapted with permission from [46]. Copyright 2007 American Chemical Society

both the location and the orientation of the assembled objects. The physical process that dielectrophoresis relies upon relates to the ability of an applied electric field to induce a dipole moment on a neutrally charged nanostructure and to simultaneously exert an electric force on that dipole (Fig. 8.6a). The polarized nanostructures respond to applied electric fields like any other dipole: their poles experience forces (termed DEP forces) which cause the dipole to both align along the E-field direction and migrate toward the regions where the gradient of the electric field is the strongest. The general approach and important experimental parameters for carrying out dielectrophoretic assembly are first detailed and then specific examples of nanostructure alignment are discussed.

The most common setup for carrying out the dielectrophoretic assembly of nanostructures consists of a pair of metallic electrodes situated on a receiver substrate. The procedure itself is performed by first exposing the receiver substrate to a liquid suspension of nanostructures (usually by simple pipette dropping) and then applying a certain voltage across the pair of electrodes to initiate the dielectrophoretic assembly process. Nanostructures that move within the vicinity of the electric field by random motion will become polarized and be subject to dielectrophoretic forces. The ability of an object to become polarized is given by its polarizability, a physical property which derives from certain characteristics including conductivity and dielectric permittivity, and is intrinsic to the object's material properties. Due to these electric forces, the object will migrate toward the area between the electrodes where the gradient of the electric field is strongest while concurrently orientating in a parallel direction along the field lines. This alignment process typically occurs within seconds or minutes and finally concludes by the affixation of the nanostructure on the receiver substrate in the region between the electrodes. The liquid suspension is then removed from the substrate, most commonly by allowing the droplet to evaporate. The electric field is held in place during the liquid removal so as to prevent capillary forces from disturbing the position of the aligned nanostructure.

The applied electric fields may be either DC or AC; DEP can be carried out in either condition. However, typically AC fields are preferred since at low frequencies, polar molecules within a liquid dielectric medium have the ability to screen out the charge separation on polarized nanostructures and reduce the DEP force; these molecules are not able to respond at higher frequencies due to their long relaxation times [38]. Additionally, DC electric fields emanating from the electrodes are capable of attracting impurities which possess surface charges by simple electrostatic interactions which occur concurrently with the DEP process [39]. Another advantage of AC DEP is that it does not produce a net current through the medium and as a result also does not produce faradic products, i.e., through electrolytic reactions [40].

The spacing between the electrodes is also an important parameter. The strength of the DEP force depends on the ratio of the electrode gap size to the length of the nanostructure. There exists an optimal ratio of gap size to nanostructure length so as to maximize the DEP force in a certain DEP structure. For instance, simulations on the DEP alignment of nanowires in a particular DEP configuration found that the optimal ratio was 0.85 [41]. Moreover, it should be noted that the DEP force is also directly proportional to the applied field strength between the electrodes and the polarizability of the nanostructure, among other factors.

During the DEP process, the electrical potential between the electrodes may be either applied directly to the electrodes by electrical routing or capacitively coupled to the electrodes via underlying bus bars which are used for voltage biasing. Although direct electrical connection is the simplest means to apply a voltage, there are a couple of associated downsides: the nanostructures post-assembly end up in direct contact with the DEP electrodes, thus preventing electrical isolation of different devices; additionally, there is a possibility of large current spikes arising between the electrode pair if a nanostructure bridges the pair during the DEP process [42]. In the alternate approach, DEP electrodes at the surface of the receiver substrate may be electrically isolated from the underlying bus bars by the presence of a sandwiched insulating layer. The purpose of the insulator layer is to both prevent significant levels of current draw and allow for the capacitive coupling of applied AC voltage signals to the DEP electrodes that are situated on the surface [38].

The geometrical design of the DEP electrodes varies depending on the application. Two types of electrode designs are common: wide bus bar electrode designs allow for the alignment of large numbers of nanostructures between each electrode pair (Fig. 8.6b), while alternatively, arrays of narrow finger-shaped pairs, which may be biased at the same voltage potentials, are used to align individual nanostructures between single electrode pairs simultaneously along the entire electrode array (Fig. 8.6c, d). The alignment of large numbers of nanostructures has been demonstrated with CdSe and InP nanowires [40, 43]. A disadvantage of this method is that it does not provide fine spatial control; although large numbers of nanostructures can be made to line up perpendicular to a particular electrode, there is no control of the exact location of each nanostructure along the width of the electrode. To obviate this problem, the width of the pair of electrodes must be narrowed to roughly the same width as the nanostructure, i.e., narrow electrodes have to be used.

Using such a technique, the one-to-one DEP alignment of nanostructures between pairs of electrodes has been demonstrated with CNTs, Si, and Rh nanowires [44–46]. The physical reason that explains the one-to-one assembly of nanostructures on narrow electrodes relies upon the phenomenon that as soon as a single nanostructure occupies the region between an electrode pair, the E field in the area immediately surrounding the object is diminished due to the high electrical conductivity of most nanostructures as compared to the liquid medium. Due to this diminishment of the E field strength in the gap region where the nanostructure is situated, the attraction of additional nanostructures is suppressed toward the electrode pair. This procedure has been demonstrated in the alignment of individual single-wall CNTs (SWNTs): 90% transfer yield was obtained for 400 electrode pairs, sized 1 μm^2, in a 100×100 μm^2 area, equating to a transfer density of over 1 million single CNT bridges in 1 cm^2 and roughly 1 CNT/μm along a single row of the array.

In addition, a photoresist (PR) patterning technique has been demonstrated to improve the yield of single nanostructure DEP assembly. The technique has been employed in order to suppress the deposition of nanostructures in unwanted areas on the deposition substrate. During DEP assembly, it is typical that some nanostructures adhere to sites on the receiver substrate simply due to random collision events and subsequent van der Waals interactions. This effect can be mitigated by patterning of a PR protection layer onto the top of the receiver substrate and opening up holes in the PR layer only in the areas near the DEP electrodes. After DEP assembly is conducted, the PR layer is washed off by solvent, simultaneously lifting off the unwanted nanostructure deposition.

The DEP technique possesses its own advantages and disadvantages as compared to other assembly techniques. The main benefit of the DEP method is that it offers the ability to finely control the spatial location of individual nanostructures to an extent not easily matched by other techniques. A disadvantage of this method, however, is that the ultra-high-density assembly of nanostructures at submicrometer size lengths has yet to be demonstrated. Such a demonstration remains a challenge due to the deleterious effects of fringing electric fields between adjacent electrodes at such small size lengths. Another disadvantage of this method is that a dedicated DEP structure is required on-chip. The necessary fabrication of such a structure will inevitably increase the cost and complexity of manufactured devices.

8.6 Chemical Affinity and Electrostatic Interaction-Directed Assembly

The use of chemical affinities to drive the assembly of nanostructures on a receiver substrate from a liquid suspension has been extensively studied. Broadly speaking, this approach relies on the action of chemical and electrostatic attractive forces to direct the deposition and assembly of nanostructures onto patternable areas on a wide variety of receiver substrates. Moreover, in a similar manner, repulsive forces may be utilized to leave the substrate deposition-free in desired areas (Fig. 8.7a). The type of chemical interaction force which the nanostructures experience, either attractive or repulsive, is determined by utilizing the appropriate type of nanostructure or substrate surface functionalization. Besides chemical interactions, electrostatic interactions between the nanostructures and the substrate may be modified by adjusting the electrical bias of the substrate.

The mechanism behind the selective deposition of nanostructures depends on whether the surface of the nanostructure is either charged or neutral. Certain nanowires, i.e., ZnO or V_2O_5, have been noticed to preferentially deposit on

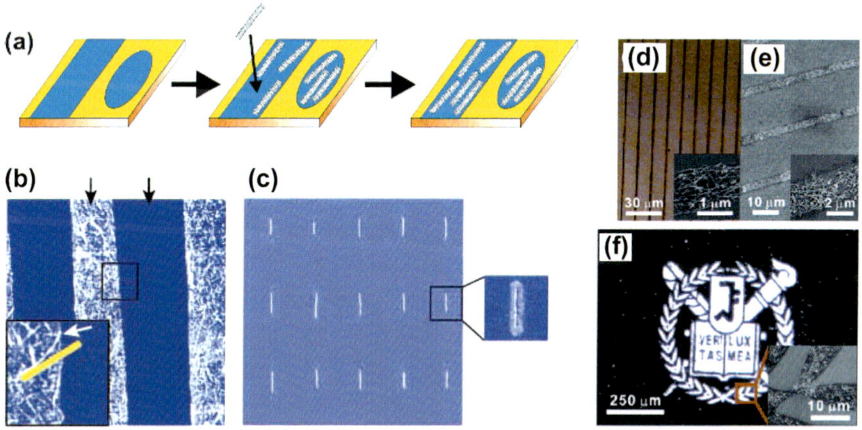

Fig. 8.7 Chemical affinity and electrostatic interaction-directed assembly of nanostructures. (**a**) Schematic showing the directed assembly of 1D nanostructures onto molecular patterns. Adapted with permission from [52]. Copyright 2006 American Chemical Society. (**b**) Atomic force micrograph (12×12 μm^2) showing large-scale self-assembly of SWNTs onto molecular patterns on a gold surface. (**c**) Topography (30×30 μm^2) of an array of individual SWNTs covering about 1 cm^2 of gold surface. The friction force image (*inset*) shows a single SWNT (*dark line*) and the regions containing molecular patterning of 2-mercaptoimidazole (*bright area*) and ODT (*dark area*). Reprinted with permission from [49]. Copyright 2003 Macmillan Publishers Ltd: Nature. (**d**) Optical micrograph image of Si NWs assembled on a Au substrate. The *inset* shows the scanning electron microscope (SEM) image of the adsorbed Si NWs. (**e**) SEM image of Si NWs assembled on a SiO_2 substrate. (**f**) Dark-field optical micrograph image of functionalized Si NWs assembled onto complex patterns with arbitrary orientations. The Si NWs were adsorbed onto bare Au surface regions, while hydrophobic ODT SAM prevented their adsorption. Adapted with permission from [53]. Copyright 2008 American Chemical Society

negatively biased substrates while having their deposition suppressed at positive biases. Such behavior is presumed to be the result of a positive surface charge that these unmodified nanowires possess; the surface charge may possibly be a result of oxygen vacancies in the materials. In contrast, it has been determined that unmodified SWNTs do not possess a significant surface charge; their deposition is only moderately affected by substrate bias [47]. On the other hand, the selective deposition of SWNTs may be carried out by exploiting the fact that SWNTs adhere to substrates with hydrophilic (i.e., polar) surfaces while being repelled away from hydrophobic (i.e., nonpolar) surfaces. Experimental and theoretical analyses have explained this effect as a consequence of van der Waals interactions [48].

The selective deposition of CNTs based on chemical affinity properties has been carried out on a wide variety of substrates including gold and silicon oxide. Surface functionalization patterning has been demonstrated through direct deposition of organic molecules by dip-pen nanolithography or by microcontact stamping. Alternatively, standard microelectronics-compatible photolithography may be used to selectively expose certain substrate regions to self-assembled monolayer (SAM) deposition. SAM patterns are typically deposited by simply immersing substrates in the appropriate SAM solution. The SAM molecules may be functionalized with either nonpolar groups (i.e., methyl $-CH_3$) or polar groups (i.e., amino $-NH_2$/$-NH_3^+$ or carboxyl $-COOH$/$-COO^-$) to produce hydrophobic or hydrophilic areas of the substrate, respectively. Using the distinct nonpolar and polar chemical functionalizations of patterned Au substrates, the selective depositions of large area patterns of SWNTs have been obtained (Fig. 8.7b) [48–49]. SWNTs were found to be repelled by adjacent nonpolar regions and, in response, were observed to actively bend their structure away from these regions. By carrying out the patterning of polar regions with dimensions approaching that of an SWNT, single SWNT depositions with controlled orientation and location were also achieved (Fig. 8.7c). Moreover, another contributing factor to the realization of controllable single SWNT depositions is the fact that CNT adherence to small polar patterns is self-limiting; presumably the hydrophobic surface of the SWNT passivates the polar pattern on the substrate surface and limits the adhesion of additional CNTs. In contrast, the use of large area patterned polar regions results in the deposition of CNTs which are randomly orientated and which possess irregular spacings.

A simplification to the previously described chemical affinity deposition method has also been demonstrated. In this case, suspensions of SWNTs were found to be able to natively deposit on a wide variety of bare, unfunctionalized surfaces (including Au, Si, Al, SiO_2, and glass) without the aid of any additional polar surface functionalization. It was argued that this deposition was probably made possible due to the natural polarization state of pristine surfaces [47]. For instance, Au and SiO_2 have been noted to form a negative surface charge in deionized water. The method has also been extended to show that deposition is possible on high-k dielectrics like Al_2O_3 and HfO_2 [50]. The patterned deposition onto bare surfaces can be carried out by utilizing a nonpolar SAM layer as a mask where deposition is prevented.

By employing similar patterning techniques, the deposition of single or large amounts of nanowires can also be carried out on negatively charged bare or

SAM-covered surfaces (Fig. 8.7d–f) [51–52]. The preferred deposition of nanowires onto negatively charged surfaces was observed for nanowires with an innate positive surface charge (e.g., ZnO and V_2O_5) in aqueous solution and for positively functionalized Si nanowires [53] that were exposed to SAM treatment. The surface functionalization of Si nanowires was found to be necessary in order to prevent the quick aggregation of the nanowires while being suspended in deionized water. The total amount of deposited nanowires was found to be readily adjusted by varying the substrate bias: larger negative biases correlated with larger deposition numbers.

There are a number of benefits to this assembly approach as compared to other methods. The reliance on chemical processing (i.e., nanostructure solutions, photoresist patterning, and SAM treatments) makes the approach relatively simple and appropriate for large-scale and high-throughput settings. It is important to note that the deposited nanostructures resulting from this assembly method were found to completely adhere to the substrates when going through post-assembly microelectronic fabrication steps. A disadvantage of the method, which is analogous to the main disadvantage of dielectrophoresis, is that it is difficult to obtain the simultaneous deposition of a high density of spatially well-controlled and accurately orientated nanostructures. As opposed to other methods, such as contact printing, the deposition of aligned nanostructures with a density greater than one nanostructure per 5 μm over large areas (i.e., 1 cm^2) has not yet been demonstrated in the literature. This disadvantage can be attributed to the difficulty in patterning charged surface regions with the simultaneous requirements of small dimensional lengths, high densities, and large patterning areas.

8.7 Contact Transfer

NW and CNT assembly by contact transfer relies upon the placement of NWs or CNTs in direct physical contact with a receiver substrate, thereby bypassing the usage of an intermediary liquid suspension which the other transfer methods addressed in this chapter utilize. Two contact transfer methods have been demonstrated: contact printing which carries out the simultaneous separation and placement of nanostructures from a growth (donor) substrate to a receiver substrate by shear-assisted fracture and alignment and stamp transfer which divides the separation and placement of the nanostructures into two discrete steps: a stamp is used to pick up and separate the nanostructures from the growth substrate and subsequently the stamp is placed down on a receiver substrate in order to achieve final nanostructure placement.

8.7.1 Shear-Assisted Contact Printing

An overview of the shear-assisted contact printing method for 1D nanostructures is presented. The approach relies upon the simultaneous application of two forces (van der Waals and shear force) to carry out the uniaxial alignment and transfer of nanostructures like NWs and CNTs. The method is suited for the transfer of nanostructures at controllable densities (from high to low) while maintaining a common orientational alignment. A disadvantage of this method is that it does not allow for a high degree of registry between the end-to-end placements of the transferred nanostructures. The transfer of both nanowires and CNTs has been demonstrated with this technique [54–55].

A step-by-step summary of the contact printing method will now be presented as pictured schematically in Fig. 8.8a. The transfer sequence can be broken down into three steps: physical contact, application of the shear force, and nanostructure breakage.

Before the transfer process, the as-synthesized nanostructures are held onto their growth substrate via a mechanical anchor point at their initial growth site. During the transfer process, the following procedures can be carried out:

(1) If selective deposition of the nanostructures is desired, a patterned photoresist (PR) film may be used to selectively protect areas of the substrate where nanostructure deposition is to be avoided. To begin the transfer process, first the nanostructure growth substrate is turned over and brought into physical contact with the receiver substrate. A downward pressure is applied to the back of the growth substrate in order to exert the proper force so as to cause a significant amount of nanostructures to bend and come into direct contact with the receiver substrate surface [54]. At this point, due to surface interactions and resultant van

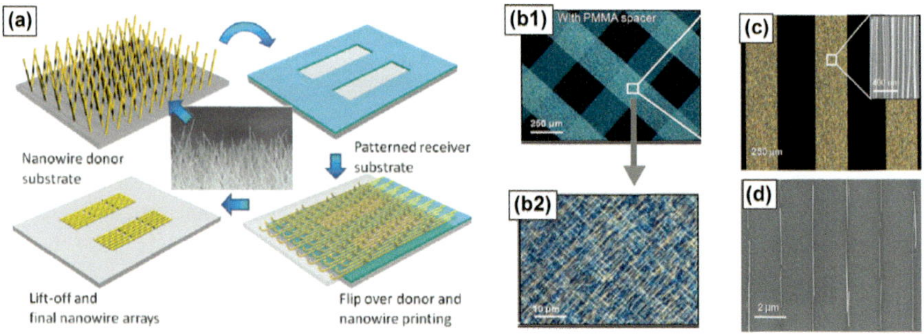

Fig. 8.8 Shear-assisted contact printing. (**a**) Schematic of the transfer and assembly method. The SEM image shows that grown NWs that are randomly oriented on a growth substrate serve as the donor material. During the transfer process, the NWs are aligned and then deposited onto the receiver substrate by application of a directional shear force, resulting in the printing of sub-monolayer NW arrays on the receiver substrate. Reproduced with permission from [54]. Copyright 2009 Wiley-VCH Verlag GmbH & Co. KGaA. (**b1, b2**) Dark-field optical images of Si NW crossbars formed by a two-step printing process with a PMMA buffer layer. (**c**) Patterned printing of NWs using polymer resists (PRs) on Si substrates. The dark-field optical image shows the assembled NWs after the printing and PR lift-off processes. *Inset* shows the SEM image of the well-aligned Ge NW arrays. (**d**) SEM image of the assembled NWs after NW printing and PMMA liftoff, showing single NW positioning on the substrate. Adapted with permission from [57]. Copyright 2007 American Chemical Society

der Waals forces, the nanostructures will adhere to the receiver substrate surface while simultaneously still being anchored to the growth substrate. Removal of the applied pressure in this step will generally result in very few nanostructures being transferred onto the receiver substrate. Practically speaking, the strength of the nanostructure anchor site on the growth substrate is stronger than the nanostructure adhesion to the receiver substrate.

(2) Next, a shear force is applied onto the growth substrate by carrying out the movement of the growth substrate in a lateral direction while the applied downward pressure is maintained. A shear force is simply defined as the force resulting from the movement of one object in physical contact against another object. In this situation, the applied shear forces are between the nanowires (or CNTs) situated on the growth substrate and the receiver substrate itself; the direction of the forces is parallel to the motion of the moving growth substrate. The applied shear force on the nanostructures will cause them to uniaxially orient themselves in the direction of the shear force in a similar way as discussed in Section 8.2. Note that in this step, the van der Waals and shear forces are acting on the nanostructures simultaneously and in different directions, downward and laterally, respectively. It is also important to note that the orientation of the synthesized nanostructures on the growth substrate is irrelevant, since the shear forces in this step will always force the alignment of the nanostructures into the direction of growth substrate movement.

(3) Eventually, the competing shear and van der Waals forces will cause a critical amount of stress on the nanostructure, resulting in the physical breakage (i.e., fracture) of the nanostructure and its deposition onto the receiver substrate surface. Typically the fracture point occurs somewhere in the midsection of the nanostructure, thereby resulting in transferred nanostructures that are shorter than the original growth length.

After nanostructure transfer, the growth substrate may be lifted off, resulting in a transferred nanowire or CNT "film" on the surface of the receiver substrate (Fig. 8.8c). If the transfer conditions were optimal, the nanostructure film coverage will nearly match the original contact area between the two substrates. Additionally, another consequence of the one-to-one nanostructure surface coverage is that sharp transitions between areas of high-density nanostructure coverage and bare areas may be obtained on the receiver substrate (Fig. 8.8c). An alternative method for achieving film patterning may be performed subsequent to nanostructure transfer by carrying out selective nanostructure removal in unwanted areas. This removal may be achieved by photoresist patterning followed by either sonication or etching of the exposed nanostructures. Conventional semiconductor processing may then be performed on the isolated device areas.

In order to adapt the contact printing method to large-scale manufacturing processes, a roll printing design has been developed which facilitates high-throughput transfers [56]. Instead of relying upon the lateral movement of a single growth substrate to facilitate the shear-assisted transfer, the nanostructures may instead be synthesized on a cylindrical roller and subsequently be transferred to a stationary receiver substrate by simply rotating the roller.

The length and density of the transferred nanostructures may be adjusted by varying the growth conditions of the nanostructures or by adjusting the downward pressure during the transfer process. For instance, sparser films may be obtained by synthesizing sparsely distributed nanostructures on the growth substrate. The contact printing process may also be repeated on a single substrate to produce multiple layers of overlapping nanostructures (Fig. 8.8b). Thin (~40 nm) polymer layers were found to enable multiple layer printing; they served as buffer layers for additional contact printing steps and could be removed without disturbing the transfer by dry etching.

Excessive breakage (and a resultant higher tendency for misalignment) has been observed to be a particular problem for the transfer of nanowires of diameters less than 50 nm and is presumed to be due to their weaker mechanical strength as compared to thicker nanowires. Moreover, at even higher pressures, the growth substrate may experience scratching due to abrasion from microscopic dust particles. It is thought that this excessive breakage and misalignment of transferred nanowires is a result of an excess of friction arising from NW–NW interactions. A strategy to mitigate these effects is through the use of wet lubricants during transfer [57]. In general, the purpose of a lubricant is to act as a buffer medium in order to reduce the friction between two sliding surfaces. Through the proper use of a lubricant in NW contact printing, the aforementioned sources of excess friction may be reduced, resulting in longer and straighter transferred nanowires. An image of an optimized transfer of ~20 nm diameter germanium nanowires is shown in Fig. 8.8c. The density of transferred ~20 nm diameter NWs along the direction perpendicular to the NW axial alignment may reach up to ~8 NWs/μm using this approach.

8.7.2 Stamp Transfer

An alternative method to achieve the assembly of nanostructures without the use of an intermediary liquid suspension involves the use of polymer stamps and is commonly termed stamp transfer. The transfer of CNTs and ZnO nanowires has been demonstrated with this method [58–61]. It relies upon the use of a polymer stamp (typically made up of PDMS or polyimide) and the action of van der Waals forces to promote the adhesion of nanostructures to first the stamp and subsequently the receiver substrate. The process, as shown in Fig. 8.9, is explained according to the following steps.

First, the stamp is placed in contact with a growth substrate so that the nanostructures adhere to the stamp. The surface energy of the stamp is modified in advance to promote the adhesion so that even after separating the stamp from the growth substrate, the nanostructures remain adhered to the stamp surface. The typical method used to modify the stamp's surface chemical state is by oxygen plasma treatment. Next, the stamp is placed in contact with a receiver substrate. By either employing

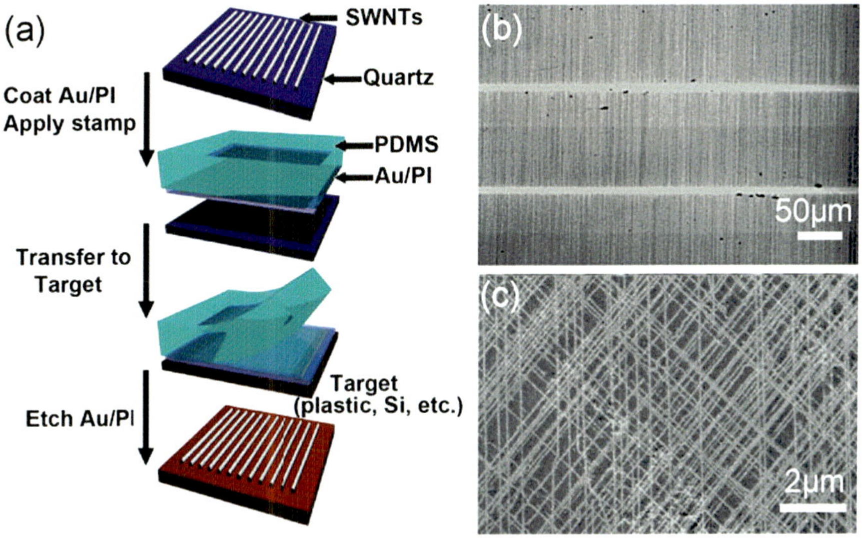

Fig. 8.9 Stamp transfer of SWNTs. (**a**) Schematic illustration of a process that uses polyimide (PI) and a gold (Au) film to transfer CVD-grown, aligned arrays of SWNTs from the quartz growth substrate to other receiver substrates. SEM images of (**b**) aligned SWNT arrays transferred from a single-crystal quartz growth substrate to a plastic substrate and (**c**) triple crossbar arrays of SWNTs formed by three consecutive transfer processes. Reproduced with permission from [13]. Copyright 2009 Wiley-VCH Verlag GmbH & Co. KGaA. Based on figures reprinted with permission from [64]. Copyright 2007 American Chemical Society

adhesive epoxy layers on the receiver substrates or by choosing a correct stamp peeling speed, the nanostructures can be made to preferentially adhere to the receiver substrate during stamp removal (i.e., stamp peel off). Polymers, such as PDMS, have certain viscoelastic properties which cause their adhesion strength to vary with peeling speeds [62].

Since the nanostructures are immobilized on the stamp during transfer, the final configuration (i.e., relative placement and orientation) of the nanostructures on the receiver substrate will match their original configuration when they were picked up by the stamp. For instance, CNTs synthesized in random directions on a growth substrate have been stamp transferred to produce randomly orientated CNTs on a receiver substrate [63]. More interestingly, CNTs synthesized with near-perfect uniaxial orientations have also been shown to preserve their alignment after stamp transfer onto a plastic (PET) substrate [60].

To achieve stamp transfer of CNTs with high yields, an additional step has been utilized in order to promote the release of the CNTs from their growth substrate. Such an extra step is not necessary for the transfer of larger nanostructures like nanowires. This step involves either the use of an underetch to release the CNTs subsequent to the initial stamp contact or the deposition of a thin carrier film (i.e., metal) on top of the synthesized CNTs prior to stamp contact [13, 64]. The deposited metal film is used to embed the CNTs in a material layer that is more readily picked up by the stamping surface as compared to the bare CNTs themselves. After the film and embedded CNTs are deposited on the receiver substrate, the metal film can be readily removed by immersion in a metal etchant solution, thereby leaving the CNTs alone on the receiver substrate surface.

The preceding two contact transfer methods have also been combined in order to carry out the transfer of vertically aligned ZnO nanowires from their growth substrate onto a PDMS stamp and then the receiver substrate in a uniform aligned, horizontal orientation. A PDMS stamp was placed in contact with the ZnO NW growth substrate and a uniform shear force was applied to the stamp. The ZnO NWs were observed to stick onto the stamp with the same orientation as the applied shear force. Finally, the PDMS stamp was pressed onto the receiver substrate (with only a downward force) and the ZnO NWs were deposited while maintaining the same uniform orientation.

A related stamp transfer technique using spun-on PMMA films as the stamp material has demonstrated the transfer of randomly orientated and aligned arrays of CNTs and also ZnO nanowires from a growth substrate [65]. The technique relies upon the spin casting of a thin PMMA layer on top of the nanowire or the CNT growth substrate in order to structurally embed the nanostructures in the PMMA film formed after solvent evaporation of the spin-casting solution. After the PMMA solidification, the entire layer (including the embedded nanostructures) may be peeled off the growth substrate, thus achieving nanostructure removal. Finally, the PMMA film mediator may be placed on a receiver substrate and the PMMA can be removed by thermal decomposition of the organic PMMA film which does not disturb the position or the alignment of the nanostructures. An advantage of this method is in multilayer stamp depositions. Successive layers of PMMA "stamps" can be deposited onto a receiver substrate without causing any possible disturbance (i.e., pickup) of the previously deposited nanostructure layers which can possibly occur in the traditional stamp transfer method.

8.8 Directed Growth

8.8.1 Horizontal Growth

One possible method that has been proposed for producing large-scale, addressable, nanostructure-based devices is to utilize the horizontal growth of nanostructures directly between pairs of electrodes. One-dimensional nanostructures (nanowires or CNTs) would effectively act as a mechanical and electrical "bridge" between the electrode pairs where the nanostructure is physically connected at each of its ends to a separate electrode. Due to this double (two-ended) connection, such a nano-sized "bridge" would be both rigid and suspended and could serve as a conduction path between the electrodes. Such a structure could serve as a backbone for the fabrication of nanostructure-based electronic devices. The demonstration of

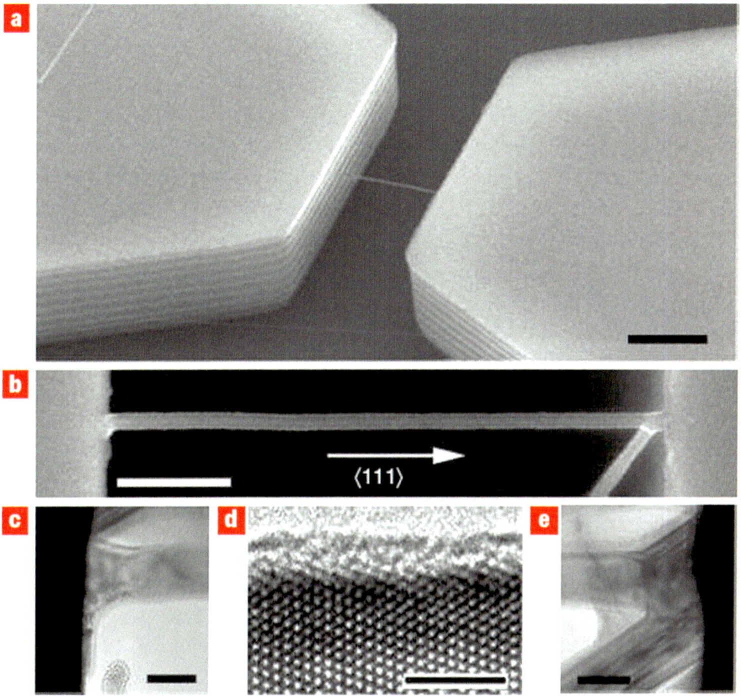

Fig. 8.10 Direct growth of Si nanowire bridges on SOI substrates. (**a**) Single nanowire with the <111> growth direction bridges a trench confined by vertical {111} faces created on a <110> oriented SOI substrate. The parallel lines on the sidewalls with alternating contrast are scallops formed during deep reactive ion etching. (**b**) General morphology of a bridged nanowire, which grew from the left sidewall along the <111> direction and impinged upon the opposite sidewall. The nanowire grew backward after self-welding into the sidewall. (**c, e**) Cross-sectional transmission electron microscope images for the two joints between the nanowire and trench sidewalls. (**d**) A high-resolution electron microscopy image confirms the <111> growth direction and reveals a thin oxide layer on the surface of the nanowire. The scale bars in (**a**–**e**) are 2 μm, 500 nm, 100 nm, 3 nm, and 100 nm, respectively. Reprinted with permission from [68]. Copyright 2006 Macmillan Publishers Ltd: Nature Nanotechnology

horizontally grown and doubly connected nanostructures in between electrodes has been achieved with silicon NWs (Fig. 8.10) [66–68], ZnO NWs [69], and single-wall carbon nanotubes (SWNTs) [70].

The use of epitaxially grown, horizontally orientated nanowires has garnered much research attention. Specifically, due to its prevalence in microelectronics fabrication, silicon-based horizontal nanowire growth has already been extensively researched and will serve as the basis for the following discussion. In epitaxial nanowire growth, the growth direction of the nanowire can be controlled in a specified orientation when synthesis occurs on a lattice-matched growth substrate. Thus, by properly choosing the composition and crystal orientation of the underlying growth substrate, directed nanowire growth in a desired direction can be achieved. Due to the fact that vapor–liquid–solid (VLS)-grown Si nanowires tend to grow in the <111> direction, the epitaxial growth of Si nanowires on a (111) Si surface will result in grown nanowires which are orientated in a vertical and perpendicular direction relative to the growth substrate surface. Horizontal Si nanowire growth can be readily obtained by creating (111) horizontally facing sidewalls by performing an anisotropic etching of (110) Si wafers. Moreover, a doubly connected bridging nanowire can be obtained if the Si nanowire grows into the opposing (111) Si sidewall during the VLS growth process (Fig. 8.10). As a result of the nearly exact epitaxial relationship between the nanowire and the facing sidewall, the impinging nanowire is presumed to "fuse" with the sidewall and create a second mechanical connection point at its surface.

The nature of the two end connections of horizontally bridging Si nanowires is of utmost concern in device applications. These connections serve two important functions: (a) as mechanical support points for the suspended structure and (b) as electrical connection points to the nanowire. Furthermore, these two connections are not identically formed: the first connection is formed at the nucleation site of the nanowire at the beginning of its growth (i.e., base end), and the second connection is created when the growing nanowire impinges on an oppositely facing substrate surface (i.e., impinging end). Due to the epitaxial manner of the nanowire growth, the properties of the base end of the nanowire are already well understood; however, it is not obvious whether the connection at the impinging end of the nanowire would be of comparable quality. TEM, mechanical, and electrical studies have been conducted on this topic and it has convincingly been shown that the nanowire sidewall connection at the impinging end of the Si nanowire possesses a high interfacial material quality which is of comparable quality to that of the base end [71–73].

To elaborate, the material and electrical properties of both of the nanowire connection points have been shown to possess superior qualities. TEM studies, including high-resolution TEM and EDS measurements, have shown that the impinging end interfacial connection region is both single-crystalline and free of catalyst (Au) atoms and can be considered of an epitaxial quality [71]. Moreover, atomic force measurements on Si nanowire bridges have revealed that the mechanical connections at both ends of the structure act as rigid anchor points which results in a strongly suspended nanowire that possesses beam-like mechanical behavior [72]. The electrical properties of the connection points have also been characterized: the specific contact resistance of the bridged Si nanowires was found to be in the $10^{-6}\ \Omega\ \text{cm}^2$ range, a value that compares favorably to the contacts in planar Si devices [73].

8.8.2 Vertical Growth

Vertically grown arrays of NWs and CNTs have been utilized to produce field emitters, FETs, and solar cells. The vertical geometry of these 1D nanostructures is well suited for the aforementioned applications. Single and large numbers of vertical MW-CNTs have been obtained by using optimized CVD growth conditions [74–75]. A variety of vertically grown nanowires have also been realized; the most widely used method for obtaining the vertical growth of nanowires is through lattice-matched epitaxial growth and it will now be examined.

Vertical epitaxial growth has been obtained for silicon [76–77], ZnO [78], and In_2O_3 nanowires [79–80] on lattice-matched substrates. Moreover, as a result of the ability of nanoscale material systems to handle larger magnitudes of mismatch strain as compared to planar systems, the heteroepitaxial growth of vertical Ge nanowires on Si substrates has also been demonstrated [81].

To illustrate a typical growth procedure, the steps taken to synthesize vertical Si nanowires are examined as shown in Fig. 8.11a [77]. The epitaxial growth procedure for <111> orientated Si nanowires is performed in a manner similar to the method in the previous section which details epitaxial, horizontal Si nanowire growth. To obtain patterned growth, Au colloidal catalyst particles were first selectively deposited onto molecularly patterned regions on the (111) Si substrate. Subsequently, the CVD-based VLS growth of the Si nanowire was carried out to produce vertical Si NWs (Fig. 8.11b). As evidenced in Fig. 8.11c, after the growth procedure a high density of nanowires is readily obtained.

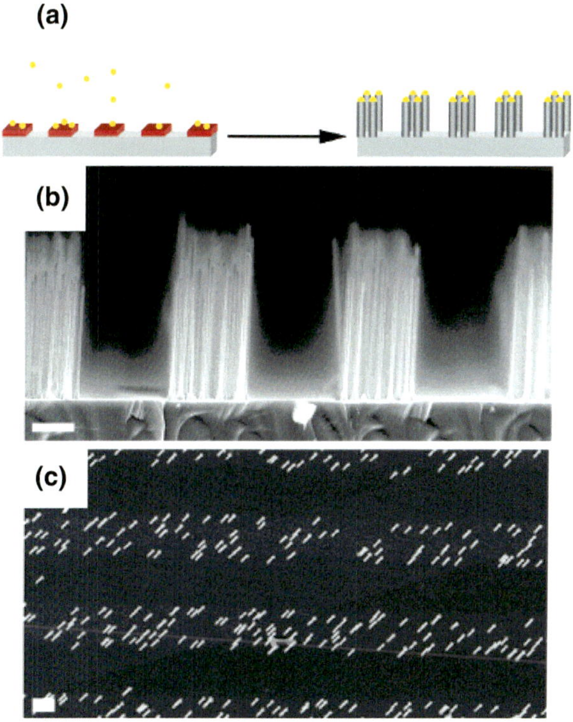

Fig. 8.11 Vertical nanowire growth. (**a**) Schematic of patterned Si NW growth. Au colloid solution adsorbs preferentially to molecularly patterned regions (*red*) on the substrate. The colloid-patterned substrate is grown using the conventional CVD synthesis, resulting in a corresponding pattern of Si NW arrays. (**b**) Cross-sectional SEM image of patterned Si NW growth and (**c**) plane-view SEM image of the same. Scale bars are 1 μm. Adapted with permission from [77]. Copyright 2005 American Chemical Society

8.9 Device Applications

8.9.1 Thin-Film Transistors

A possible use for the assembled NW/NT arrays so far mentioned in this chapter is to serve as the semiconducting channel region in thin-film transistors (TFTs). The main novelty of a nanowire- or a CNT-based TFT as compared to traditional TFTs using planar films is that the active element, i.e., semiconductor channel material, is made up of an array of aligned single-crystalline nanostructures. Such nanostructure arrays may in addition be produced on a wide variety of receiver (device) substrates, like glass or plastics, in a controlled fashion by the previously detailed assembly methods. This approach addresses the limited thermal budget dilemma facing conventional TFT devices in which the films are directly grown or deposited on the device substrates, namely the growth of high-performance films typically requires high growth temperatures that unfortunately are not compatible with plastic or glass substrates. In the NW-TFT or CNT-TFT approach, the high-temperature material growth process is completely de-coupled from the low-temperature device fabrication process by using the transfer and assembly processes discussed earlier. Following the nanostructure transfer to the receiver substrate, device fabrication can be carried out using established semiconductor processing methods (at low temperatures), while the use of aligned crystalline nanowire or CNTs as the channel material ensures the device will have superior performance. The fabrication of high-performance nanostructure-based TFTs on transparent substrates like glass or flexible substrates like plastic in turn opens up the possibility of achieving transparent and/or flexible electronic devices.

It is important to first establish the practicality of using NWs or CNTs in the fabrication of TFTs. Specifically, it is necessary to consider the effect of the non-ideal aspects, originating from the transfer and assembly process, of these arrays on TFT performance. Once a theoretical basis for evaluating their performance is established, the fabrication and operation of TFTs based on assembled arrays of nanostructures, including transparent nanowire-based TFTs and flexible CNT-based TFTs, will be examined in this section.

8.9.1.1 Performance Considerations for NW- or NT-Based TFTs

The main architectural difference which exists between conventional TFTs based on continuous films and NW- or NT-TFTs based on arrays of discrete conduction channels raises the question of how a discontinuous channel region will affect the TFT performance. Intuitively, one could surmise that as the nanostructure density in the TFT channel region increases, so would the current-carrying ability of the TFT and in turn its performance such as transconductance. Although this relationship is correct, it is not a linear one, and it depends crucially on understanding how the gate capacitance of the nanostructure-based TFT varies with nanostructure transfer density.

In order to simplify the theoretical analysis, a NW-based TFT is considered, but the same analysis may also be applied to CNT-based TFTs [60]. Here the term nanowire coverage denotes the ratio of the area covered by the transferred nanowires which bridge the source and drain electrodes as compared to that of the physical channel area of the TFT ($W \times L$). In order to predict the effect of varying nanowire coverages on TFT performance, electrostatic simulations (Fig. 8.12) of a specific NW-TFT device structure were performed. The simulations show that for NW-TFTs having a 130 nm thick SiO_2 gate dielectric, the simulated value for the total gate channel capacitance C_{gs} of the NW-TFT with a NW coverage of greater than 25% is in fact within 95% of the value estimated by using the parallel plate model assuming a complete coverage. This result reveals that when nanowire coverage exceeds 25% for this device geometry, the nanowire arrays are able to achieve a capacitive coupling between the gate electrode and the NW channel that is at least 95% as much as it is for a TFT device based on a continuous film. It also suggests that for nanowire coverage above this threshold, nanowire density fluctuations (which are likely unavoidable using the transfer and assembly techniques discussed earlier) will have little effect on the gate capacitance and hence the device performance such as drive current and transconductance. This result can be appreciated by examining the E-field profile of the simulated NW-TFT in Fig. 8.12a. Note that the E-field strength near the gate electrode is close to 7.7 MV/m along the entire width of the channel, i.e., the same E-field magnitude as would be predicted for a conventional TFT with the same dielectric thickness.

The nonlinear relationship between NW-TFT performance and the NW coverage can be explained by looking at this problem from a different angle. It is instructive to

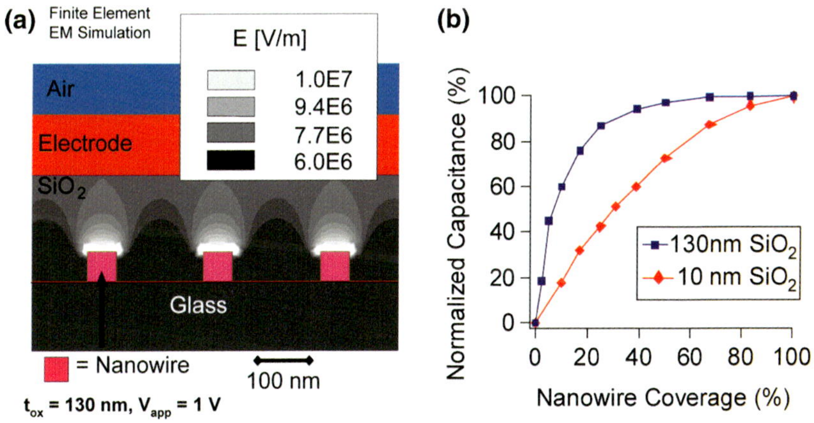

Fig. 8.12 Gate capacitance considerations for a NW-TFT. (**a**) Simulated electrostatic potential distribution in the cross section of a NW-TFT with $t_{ox} = 130$ nm and $V_{app} = 1$ V. (**b**) Plot of the gate capacitance of the NW-TFT vs. the percentage of nanowire coverage in the channel region. The gate capacitance is normalized to the maximum possible capacitance when the nanowire coverage is 100% (i.e., a conventional TFT). Adapted with permission from [84]. Copyright 2007 American Chemical Society

let the transconductance in the linear regime for a NW-TFT composed of N parallel NWs to be defined as the sum of N individual NW devices:

$$g_{\mathrm{m}} = N\frac{\mu_{\mathrm{fe}}C_{\mathrm{i}}}{L^2}V_{\mathrm{ds}} \tag{8.1}$$

assuming the nanowires are identical. Here C_{i} corresponds to the effective gate capacitance for a single nanowire.

Compared to the classical definition of a TFT's transconductance (in the linear regime), where C_{ox} is the capacitance as predicted by the parallel plate model,

$$g_{\mathrm{m}} = N\frac{W}{L}\mu_{\mathrm{fe}}C_{\mathrm{ox}}V_{\mathrm{ds}} \tag{8.2}$$

one is able to write a proportional relation such that

$$NC_{\mathrm{i}} \propto W{\cdot}L{\cdot}C_{\mathrm{ox}} \tag{8.3}$$

Thus one can notice that the fixed value of C_{ox} for a given TFT geometry places an upper bound on $N \times C_{\mathrm{i}}$ as N increases. Qualitatively, in the extreme case of very low NW coverage, C_{i} is determined by examining the capacitive contribution of each nanowire separately via the approximately correct cylinder-on-a-plane EM model, so the individual C_{i} value is large but N is small. As NW coverage increases, the field lines from the gate are shared by more nanowires, so the effective coupling to each nanowire (hence C_{i}) decreases. Above a certain nanowire coverage, the effects of increasing N and decreasing C_{i} almost completely cancel each other, and the transconductance of the NW-TFT device no longer improves with increased nanowire coverage.

This observation has important consequences in nanowire-based electronics, as it shows that nanowire arrays produced using less-than-optimal assembly techniques with relatively low coverage can still act effectively as thin-film devices. Figure 8.12b shows the calculated gate capacitance C_{gs} as a function of the nanowire coverage, at different oxide thickness conditions. For relatively thick gate dielectric thicknesses (blue squares) that can be readily produced with inexpensive, scalable techniques such as sputtering, a nanowire film with low surface coverage will electronically behave similarly to a complete planar single-crystalline thin film of the same material and thickness as the channel in TFT devices. To take advantage of a higher coverage (denser) nanowire channel, the use of thinner gate dielectrics or high-k materials are required in order to achieve more effective capacitive coupling to the nanowire array (red diamonds, Fig. 8.12b).

One reported strategy to achieve higher gate coupling in a NW-TFT relies upon a nanowire core–shell–shell structure [82]. In the study by Duan et al., silicon nanowires were chosen as the semiconducting material, and a high-quality SiO_2 shell was grown around the nanowires using thermal oxidation following growth. A conformal, metal gate electrode was next deposited onto the nanowire by using the atomic layer deposition (ALD) of WN. An improvement in the gate coupling was noted through a reduction of the subthreshold slope in the fabricated core–shell–shell Si NW-TFTs down to a value as low as 80–100 mV/dec which surpasses even the values obtained for typical poly-Si TFTs (~200 mV/dec).

8.9.1.2 Transparent Nanowire-Based TFTs

In this section, a nanowire-based transparent thin-film transistor that is fabricated on a glass substrate using the contact printing transfer method is presented [83–84]. Instead of directly growing semiconductor materials on glass or plastic as with typical TFT designs [85], the low-temperature requirements of glass and plastic substrates can be satisfied by transferring the semiconductor material from the nanowire growth substrate to a separate device substrate, thus decoupling the high-temperature processes required by high-quality material growth from the low-temperature processes required by device fabrication. Using this approach, Dattoli et al. and Ju et al. have recently demonstrated that it is possible to fabricate transparent, high-performance nanowire devices on glass and plastic substrates [83–84]. The transparent NW-TFTs to be detailed in this section possess DC performance levels comparable to single-crystalline metal oxide TFTs and importantly operate at frequencies above 100 MHz with tightly distributed performance metrics among different devices [86].

In the UM study by Dattoli et al., transparent NW-TFTs were fabricated on Pyrex glass substrates utilizing contact printing transferred, VLS grown SnO_2 nanowires and patterned ITO source, gate, and drain electrodes. A 75 nm thick SiO_2 gate dielectric was deposited by PECVD. A two-finger interdigitated gate design was chosen to facilitate RF measurements using ground–signal–ground (GSG) probes. The spacing between the source and drain electrodes for each finger was 2.5 μm and the channel region was 2 × 50 μm wide.

The NW-TFTs fabricated on Pyrex substrates are highly transparent with a transmittance of 80% measured at 550 nm. Figure 8.13d shows an optical photograph of a chip containing an array of 300 TFT devices, the high transparency of the chip is evident.

Figure 8.14a shows the I_{ds}–V_{ds} family curves and transfer (I_{ds}–V_{gs}) characteristics of a typical device. The gate sweep of the device in the saturation region ($V_{ds} = 6$ V) showed a large on-current of 2.7 mA (27 mA/mm) and a peak DC transconductance of 608 μS (6 mS/mm) with a peak saturation field-effect mobility of 210 cm^2/V s (assuming a parallel plate capacitance model). As discussed in the previous section, due to the high NW coverage in the channel region (estimated to be 25–50% in the channel region for this SnO_2 NW-TFT) and the relatively thick insulator thickness, the parallel plate capacitance model is appropriate for this device structure.

Moreover, RF measurements were carried out with the aim of circumventing problems associated with DC characterization such as the effects of mobile charges or surface states in deriving mobility values for nanowire- and nanotube-based electronics [87]. The high-frequency measurements, on the other hand, will be able to unambiguously attest to the NW-TFT's viability as high-speed electronic devices.

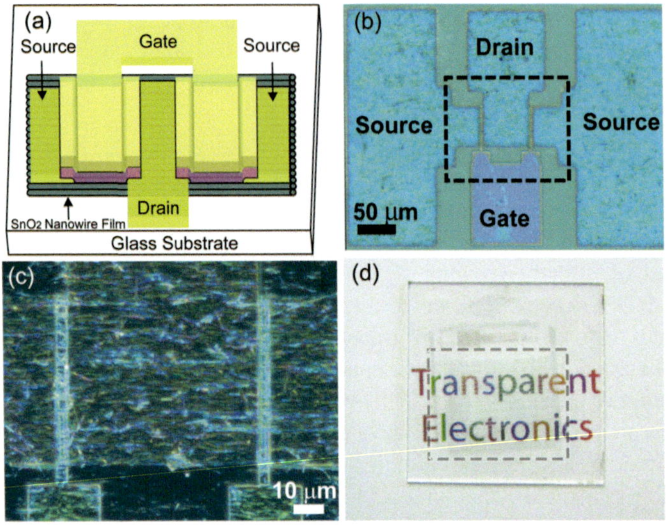

Fig. 8.13 Transparent NW-TFT. (**a**) Schematic of the transparent nanowire-based TFT using a two-finger gate design. (**b**) Optical micrograph image of an entire TFT device showing the device layout and the GSG testing pads. (**c**) Optical micrograph image of the active area of the device showing the aligned nanowires. (**d**) Digital photograph of the transparent TFT array on a glass substrate. The device area contains 300 test structures and is marked by a *dashed border*. Adapted with permission from [83]. Copyright 2009 IEEE

Figure 8.14c shows the current gain and maximum stable gain (MSG) as a function of frequency measured for the same device. The TFT was biased at $V_{ds} = 6$ V and $V_{gs} = 2$ V. The current gain and MSG curves can be well fitted using the ideal 20 dB/dec roll-off slope (dotted lines). The unity current gain cut-off frequency f_T was estimated to be 35 MHz, and the power gain cut-off frequency f_{max} was estimated to be 110 MHz. These values represent the highest operation speed reported for transparent electronics to date and are significantly higher compared with previous studies on organic semiconductor and conventional TOS thin film-based TFTs [88–89].

A key advantage of the NW-TFT approach is that the large number of nanowires that make up each TFT helps to suppress the performance variations among separate devices as compared to transistors based upon a single nanowire. The main device uniformity concern in the NW-TFT approach is the unavoidable fluctuation of the nanowire density at different locations and the resulting fluctuation of the number of nanowires that bridge the channel region in separate NW-TFT devices. However, as previously discussed, simulations suggest that the TFT performance can be insensitive to nanowire density fluctuations above a certain NW surface coverage for a given device geometry. The RF characteristics of an extensive number of TFTs were measured in order to examine whether the predicted high degree of uniformity would hold true experimentally.

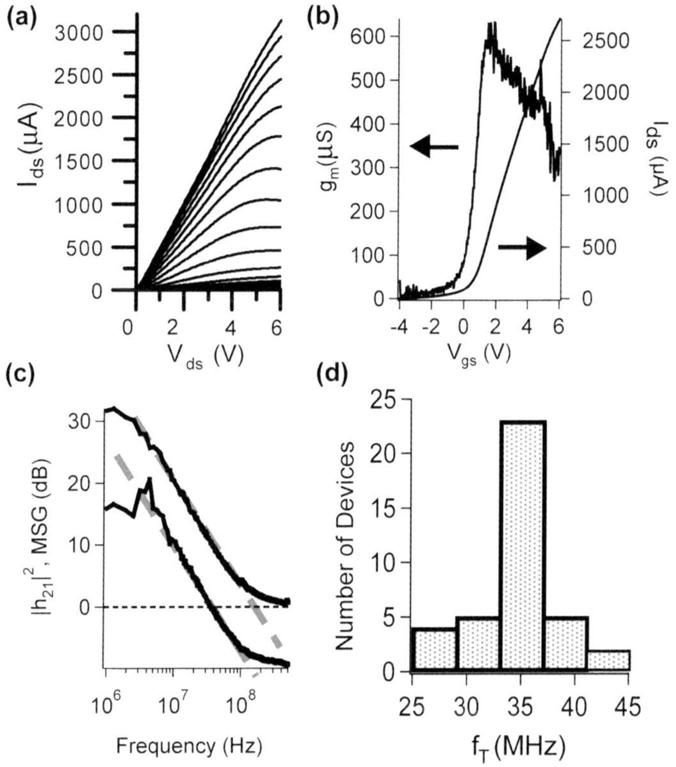

Fig. 8.14 Electrical characteristics of SnO_2 NW-TFTs. (**a**) Family of I_{ds}–V_{ds} curves of a typical device. (**b**) Transfer characteristics of the device measured at $V_{ds} = 6$ V. (**c**) Frequency dependence of the current gain ($|h_{21}|^2$) and MSG ($|S_{21}/S_{12}|$) of the same device measured at $V_{ds} = 6$ V and $V_{gs} = 2$ V. (**d**) Histogram of the extracted f_T values for 39 devices measured at the same bias conditions. Adapted with permission from [83]. Copyright 2009 IEEE

Figure 8.14d shows a histogram of the extracted f_T values for 39 NW-TFT devices fabricated using identical processes. All devices were biased at the same voltages ($V_{ds} = 6$ V and $V_{gs} = 2$ V), and no manual adjusting of the working points was performed. All devices measured show similar characteristics and the f_T data exhibit a very narrow distribution with an average of 34.3 MHz and a standard deviation of only 3.7 MHz. The narrow distribution clearly demonstrates that high device uniformity can be obtained even in the presence of nanowire density fluctuations and paves the way for the design and application of high-performance integrated circuits based on the NW-TFT approach.

8.9.1.3 CNT-Based TFTs

In a manner similar to nanowire-based TFTs, carbon nanotube-based TFTs may be realized through the use of either random or aligned arrays of CNTs as the channel material in a TFT. Two separate assembly methods, specifically stamp transfer and dielectrophoresis assembly, have been utilized to carry out the fabrication of these CNT arrays on a diverse range of substrates, such as polyimide or PET films. The fabricated TFTs display high performance levels; for instance, the unity current gain frequency (f_T) of the devices has been measured in the gigahertz range. Moreover, functioning prototype digital and analog circuits have also been demonstrated.

In a study carried out by Happy et al., high-density arrays of SWNTs from suspension have been assembled and aligned between pairs of guiding electrodes on flexible PET substrates (250 μm thick) using AC dielectrophoresis [90]. The SWNTs were assembled on top of a prefabricated bottom insulator/metal gate stack. Complete TFT structures were achieved by depositing over-lying source and drain contact metals on the SWNTs after assembly. Measured values of f_T for the fabricated devices were found to be ~1 GHz, while devices fabricated using a similar process on rigid substrates were found to possess values of $f_T$in excess of 4 GHz [91]. The authors attribute this discrepancy in performance capability to a difference in channel lengths between the two sets of devices. The authors also tested device performance, while the substrate was statically stressed into a certain bending curvature. The device characteristics, i.e., transconductance, were noted to remain stable down to bending curvatures as low as 3.3 mm. A reduction in transconductance was noted for smaller bending radii and can be attributed to a deterioration in the structure of the source and drain metal contact films. However, one key drawback in this study was the low on/off ratio by the CNT-TFTs due to the mixture of metallic and semiconducting CNTs in the channel.

The other technique that has been shown to achieve transferred arrays of highly aligned and dense CNTs suitable for high-performance TFT operation is the stamp transfer method. In a study carried out by Rogers' group at UIUC, aligned arrays of SWNTs, with a density of ~3 NTs/μm, were transferred post-synthesis from quartz growth substrates onto PET film device substrates (180 μm thick) by stamp transfer. Subsequently, microelectronic fabrication procedures were carried out to produce bottom-gated NT-TFTs [60]. Field-effect mobilities of ~480 cm^2/V s were observed for bending radii as small as 4 mm. Device performance reduction due to electrode deterioration was again observed for smaller radii of curvature. Using a similar stamp transfer technique, fully transparent NT-TFTs utilizing aligned arrays of SWNTs were realized on glass and PET substrates at comparable DC performance levels [92]. Furthermore, complementary logic inverters were demonstrated using transferred SWNT arrays on Si/SiO_2 substrates [93]. A DC gain of ~5 was measured for the complementary inverter consisting of p- and n-type bottom-gated CNT-TFT elements.

To probe the high-speed electronic capability of the SWNT arrays, RF measurements have been carried out on TFTs fabricated on the top surface of quartz CNT growth substrates [94]. The TFTs possessed submicrometer channel lengths and high-k gate dielectrics. A maximum value of f_T as high as ~5 GHz was measured for SWNT arrays consisting of tube densities of ~5 NTs/μm. Using a similar fabrication scheme, analog electronic circuits consisting of SWNT-TFTs were configured into a fully functioning AM radio receiver [95]. Once again limited on/off ratio was obtained in these studies due to the existence of metallic CNTs directly bridging the S/D contacts.

For device applications that necessitate low cost levels and that possess less demanding performance requirements, the stamp transfer of random networks of SWNTs onto thin plastic (polyimide, 50 μm thick) sheets has been demonstrated [63]. Simpler fabrication parameters were chosen to meet the aforementioned device requirements; specifically, long channel lengths (100 μm) and unaligned, randomly orientated SWNT networks were utilized (Fig. 8.15). In addition, etches along the channel direction were made in the transferred CNT films to break the metallic tubes and an on/off ratio as high as 10^5 was obtained. Peak field-effect motilities for the fabricated TFTs were limited to ~80 cm^2/V s due to the lengthier conduction paths which are a result of carrier percolation through long, unaligned CNT-TFT channels. Logic gates, such as NOT, NAND, and NOR, were fabricated using these p-type SWNT-TFTs as the building blocks. In order to demonstrate a more practical circuit application and the ability of the CNT-TFTs to be integrated together with a complicated set of interconnects, a four-bit row decoder was also successfully implemented. The decoder consisted of 88 CNT-TFTs, all operating with p-type logic. Accurate decoder operation was observed at clock frequencies in excess of 1 kHz.

8.9.2 3D Multilayer Device Structures

The demonstration of nanowire- or CNT-based devices in multilayer and interconnected (3D) electronic devices is now detailed. As discussed earlier in this chapter,

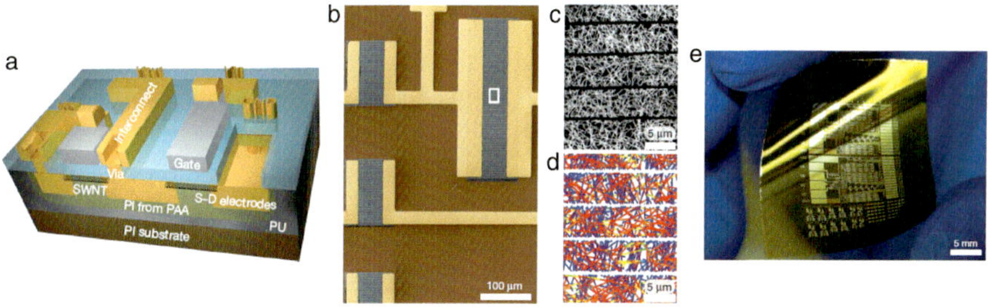

Fig. 8.15 Flexible SWNT integrated circuits on plastic fabricated using stamp transfer. (**a**) Cross-sectional diagram of an SWNT PMOS inverter on a PI substrate. PI, polyimide; PU, polyurethane; PAA, polyamic acid. (**b**) Scanning electron microscope image of part of the SWNT circuit, made before deposition of the gate dielectric, gate, or gate-level interconnects. The S–D electrodes (*gold*) and substrates (*brown*) had been colorized to highlight the SWNT network strips (*black and gray*) that form the semiconductor. (**c**) Magnified view of the network strips corresponding to a region of the device channel highlighted with the white box in (**b**). (**d**) Theoretical modeling results for the normalized current distribution in the on-state of the device (view as in (**c**)), where *color* indicates current density (*yellow*, high; *red*, medium; *blue*, low). (**e**) Photograph of a collection of SWNT transistors and circuits on a thin sheet of plastic (PI). Reprinted with permission from [63]. Copyright 2008 Macmillan Publishers Ltd: Nature

nanostructures possess the ability to be readily assembled post-growth on a variety of desired receiver substrates while still retaining their unique and desired electronic and material properties independent of the receiver substrate. These assembly processes may be carried out at low temperature, thus circumventing the previously encountered problems associated with traditional methods for carrying out the direct growth of high-quality, thin-film materials which require high growth temperatures that are incompatible with substrates that have low glass transition temperatures (i.e., glass or plastic) or underlying material layers that possess a large difference in thermal expansion coefficients or that possess prior metallization layers that have low melting points (i.e., Al) [96].

So far, due to the aforementioned problems, there has been only a limited success in achieving multilayer, 3D electronics using traditional microelectronics deposition techniques [11]. On the other hand, there have been some notable demonstrations of multilayer electronic devices using assembled layers of either nanowires or CNTs.

For example, using the shear-assisted contact printing method, the Lieber group at Harvard has demonstrated 10 layers of multiple-nanowire Ge/Si core–shell NW-FETs [97] on a single device substrate (Fig. 8.16a, b). Notably, the electrical characteristics of the top-gated NW-FETs were found to be nearly identical regardless of the device's location in the multilayer stack (Fig. 8.16c). The relative insensitivity of the nanowire devices to a multilayer fabrication process shows that the nanostructure-based approach to 3D electronics is well suited for producing well-controlled, manufacturable devices. Notably all processing was carried out at low-temperature conditions. Even though the authors chose to refer to their device structure with a different name in this report, the device structure of these multiple NW-based FETs in fact identically resembles the previously detailed NW-TFT devices.

In addition, identical fabrication processes were used to fabricate a two-layer transistor load inverter and floating gate memory structure on a plastic Kapton film. A PECVD-deposited SiO_2 film was used as the separation layer; etching techniques were utilized to form via interconnects between the separate logic and memory layers.

The Ge/Si core–shell NW heterostructures were chosen due to their established high-performance capabilities in field-effect devices [3]. Ge/Si NWs are p-type semiconductors; due to a band alignment mismatch, holes will be confined in the Ge core as the majority carriers. They are an attractive option for electronic devices since there are no dopants present in the nanowires, thereby avoiding the problem of impurity scattering and its associated lowering of the mobility [3].

As an extension to their first report on multilayer, NW-based electronics, the Harvard group recently demonstrated a two-layer NW-based complementary logic inverter [98]. The inverter was composed of n-type InAs NW-FETs on the bottom layer and p-type Ge/Si NW-FETs on the top layer. Using contact printing transfer, multiple aligned nanowires were utilized in each FET to provide for sufficient current drive in the inverter. A large DC gain of $\sim$45 was obtained for a single inverter with a nearly full output swing when biased at $V_{DD} = 4$ V. Using this two-layer device structure, a three-stage ring oscillator was also demonstrated. The gain of the

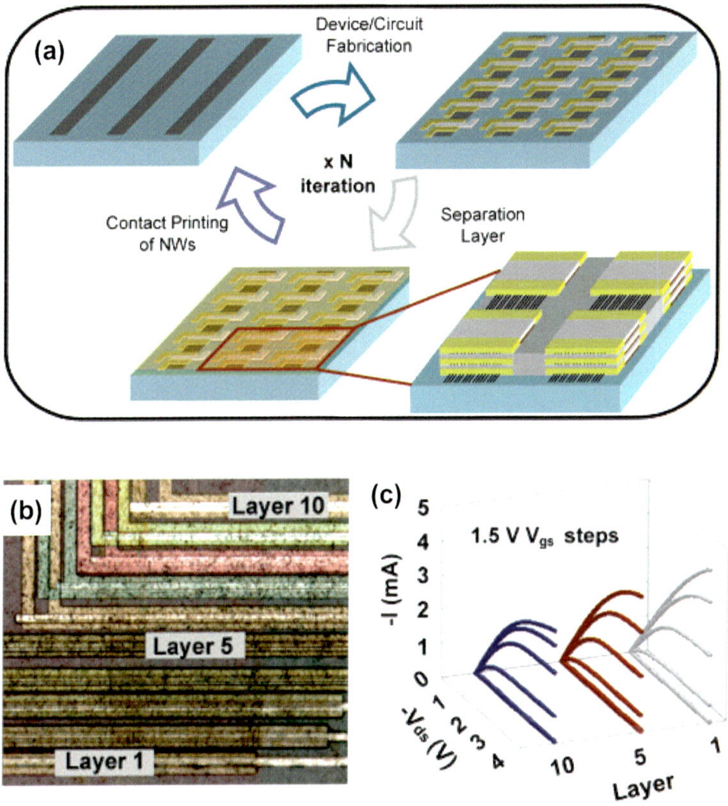

Fig. 8.16 Multilayer Ge/Si NW-FET device by contact printing. (**a**) Three-dimensional NW circuit fabricated by the iteration of the contact printing, device fabrication, and separation layer deposition steps *N* times. (**b**) Optical microscope image of 10 layers of Ge/Si NW-FETs. Each device is offset in *x* and *y* to facilitate imaging. (**c**) Current vs. drain–source voltage characteristics (with 1.5 V gate step) for NW-FETs from layers 1, 5, and 10. Adapted with permission from [97]. Copyright 2007 American Chemical Society

ring oscillator was large enough to readily sustain oscillations, and a max oscillation speed was obtained at 108 MHz at a supply voltage of $V_{DD} = 8$ V. Although the inverter structure was fabricated on a standard Si wafer, the multilayer fabrication process is in theory compatible with a diverse range of substrates such as glass or plastic. The multilayer structure should also be capable of being fabricated on top of conventional wafer-based devices, possibly allowing for heterogeneous device integration.

Lastly, the UIUC group has demonstrated a three-layer heterogeneous, flexible device structure using the stamp transfer of random networks of SWNTs and nanowires (Fig. 8.17a–c) [10]. The device structure was fabricated on a 25 μm thick polyimide (PI) substrate. Thin, 1.5 μm thick, spun-on PI layers served as the spacer layers. The three-layer structure consisted of a separate heterogeneous device that

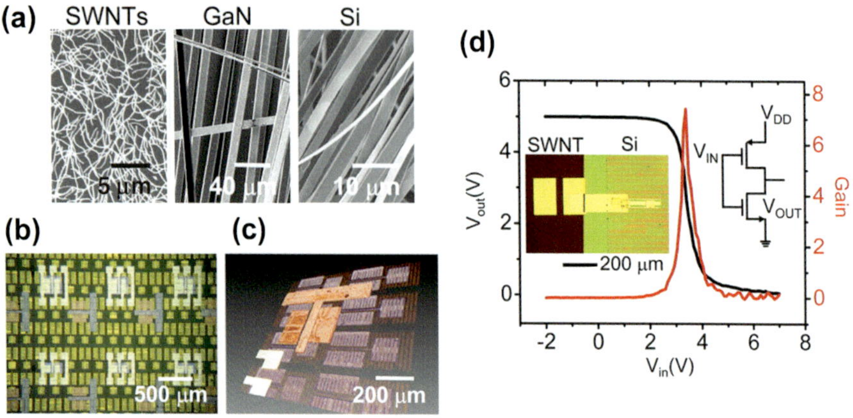

Fig. 8.17 Multilayer device by stamp transfer. (**a**) SEM images of the nanostructured materials assembled by stamp transfer onto a polyimide (PI) substrate. (**b**) Optical micrograph of 3D heterogeneously integrated electronic devices, including GaN nanoribbon HEMTs, Si nanoribbon MOSFETs, and SWNT network TFTs, in a three-layer stack. (**c**) 3D image collected by confocal microscopy. The layers are colorized (*gold*: top layer, Si MOSFETs; *red*: middle layer, SWNT-TFTs; *pink*: bottom layer) for ease of viewing. (**d**) Transfer characteristics of a printed complementary inverter that uses a p-channel SWNT-TFT (channel length and width of 30 and 200 μm, respectively) and an n-channel Si MOSFET (channel length and width of 75 and 50 μm, respectively). The *insets* provide an optical micrograph of an inverter (*left*) and a circuit schematic (*right*). Reprinted with permission from [10]. Copyright 2006 AAAS

was fabricated on each layer: a randomly orientated SWNT-TFT, Si nanoribbon MOSFET, and a GaN nanoribbon HEMT. The nanoribbons were formed by carrying out the etching and subsequent release of the materials from single-crystalline source wafers by an underetch and stamp transfer process. The field-effect mobility of the p-type SWNT-TFT was determined to be 5.9 cm^2/V s; this low value was attributed to the longer conduction paths present in randomly orientated CNT-TFTs as compared to those with channels comprised of aligned CNT arrays. No differences in electrical performance could be found due to the position of the devices in the multilayer structure. Bending tests on the devices show that their performance was relatively stable for bending radii of curvature down to 3.7 mm. Moreover, a two-layer complementary inverter was also demonstrated using an n-channel Si nanoribbon MOSFET and a p-type SWNT-TFT. The extracted gain of the complementary inverter was found to be ~7 at a supply voltage of 5 V (Fig. 8.17d).

8.9.3 Sensors

The 3D integration of nanostructures can potentially lead to heterogeneous and multi-functional systems. In this section we discuss recent progress on nanowire-based sensors. For example, Fan et al. have recently fabricated a nanowire-based photosensor with integrated nanowire amplifier circuitry [99]. The shear-assisted contact printing method was utilized to fabricate both CdSe multiple NW-based photosensors and Ge/Si core–shell multiple NW-FETs alongside each other on the surface of a single device substrate (Fig. 8.18). The CdSe NWs were found to possess a direct electrical band gap of ~1.76 eV and were employed in a two-terminal device configuration utilizing Schottky contacts with 5–10 CdSe nanowires in each detector. The use of multiple CdSe NWs was found to reduce the variability of the sensor's photoresponse among different devices due to the averaging effect. The resistance of the CdSe NW Schottky devices was found to drop by a factor of ~100 with exposure to light. Ge/Si NW-FETs, being found insensitive to white light, were utilized in an integrated two-transistor circuit as an analog amplifier. The resistance of the Ge/Si multiple NW-based FETs (i.e., NW-TFTs) was optimized for the circuit application by appropriately adjusting the width of each NW-FET in the two-transistor circuit, i.e., from ~1 to 300 μm.

A large array (13×20) of photosensors with integrated amplifiers were fabricated on a single substrate over a sizable area (~1 cm^2). A high device yield (~80%) was attained, and a circular light source was able to be imaged over a 2D area. These preliminary results help to show that the heterogeneous incorporation of nanostructure materials on a single substrate can indeed realize multi-functional devices with practical applications.

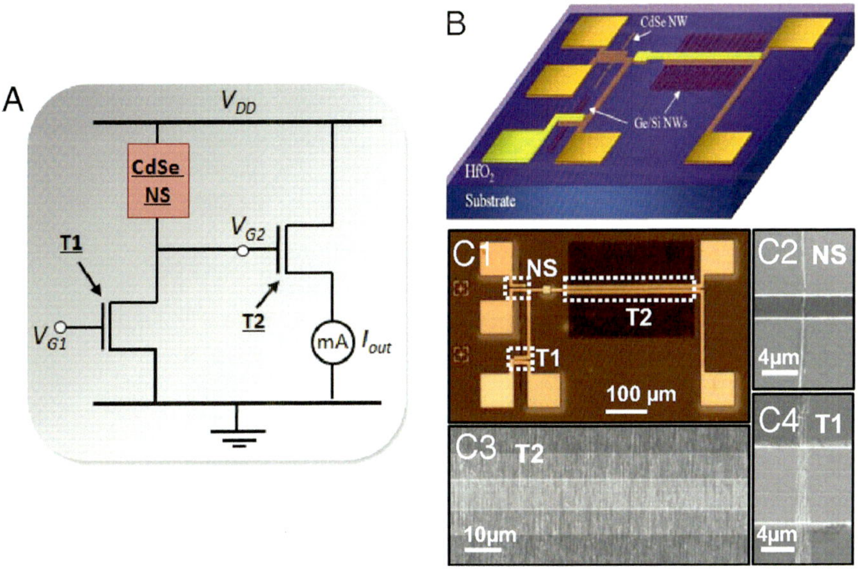

Fig. 8.18 Heterogeneous NW-based integrated sensor circuitry. (**a**) Circuit diagram for the all-nanowire photodetector, with high-mobility Ge/Si NW-FETs (T1 and T2) amplifying the photoresponse of a CdSe nanosensor. (**b**) Schematic of the all-nanowire optical sensor circuit based on ordered arrays of Ge/Si and CdSe NWs. **(C1)** An optical image of the fabricated NW circuitry, consisting of a CdSe nanosensor [NS **(C2)**] and two Ge/Si core/shell NW-FETs [T2 and T1, **(C3)** and **(C4)**] with channel widths ~300 and 1 μm, respectively. Adapted with permission from [99]. Copyright 2008 National Academy of Sciences, USA

8.9.4 Vertical Nanowire Field-Effect Transistors (FETs)

Finally, we want to point out a different 3D integration approach – field-effect transistors integrated vertically from a substrate using a vertical nanowire or multiwall CNT as the semiconducting channel region and a surround gate structure [76, 78, 100]. This 3D device structure has attracted much attention since it affords two major benefits over planar devices:

- *A fully wraparound gate* – the freestanding nanowire or CNT structure naturally supports the surround gate geometry (also termed gate-all-around, or GAA structure), resulting in increased switching efficiency and reduced power consumption. Achieving similar GAA transistor structures in planar devices typically involves non-conventional substrates (such as silicon-on-insulator wafers) and challenging processing steps.
- *Large integration density due to vertical geometry* – vertical FETs built on nanowire arrays provide increased drive current in a smaller area, resulting in a reduced switching time.

In particular, a vertical silicon nanowire FET structure was first demonstrated by the Yang group at UC-Berkeley (Fig. 8.19) [76]. The fabrication process was

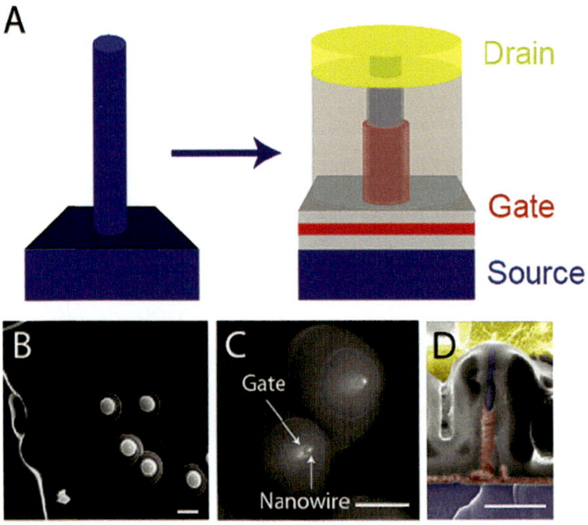

Fig. 8.19 Vertical NW-FET device configuration. (**a**) Schematic of vertical NW-FET device (*right*) fabricated from vertical silicon nanowires (*left*). (**b**) Top view SEM image of a completed device, scale bar is 2 μm. (**c**) Top view SEM image of the midsection of vertical NW-FET device, highlighting the conformal gate surrounding the nanowire channel, scale bar is 1 μm. SEM images (**b**) and (**c**) are obtained with a 30° tilt. (**d**) Cross-sectional SEM image of a vertical NW-FET device. Scale bar is 500 nm. False color is added to image (**d**) for clarity. In (**a**) and (**d**), *blue* corresponds to the Si source and nanowire, *gray* corresponds to the SiO_2 dielectric, *red* corresponds to the gate material, and *yellow* corresponds to the drain metal. Adapted with permission from [76]. Copyright 2006 American Chemical Society

carried out on epitaxial-grown silicon nanowires synthesized on a planar silicon substrate. It should be noted that the actual fabricated devices consisted of multiple-nanowire FETs in parallel. This is attributed to the difficulty in making individual contact to silicon nanowires (with an ~60 nm diameter) grown in a high density (~10–100/μm^2 in this case).

Electrical measurements of this particular device structure reveal that the Si vertical NW-FET behaves as a p-type device. The reported performance, i.e., transconductance and saturation current, of the device is however lower as compared to FETs fabricated with horizontally lying nanowires. Poor performance has normally been a problem for previously demonstrated vertical nanowire and multi-wall nanotube transistors that have been reported [76, 78–79, 101–102]. A possible cause of this poor performance may be due to a large contact resistance at the source (top) contact. For example, in the Berkeley study, Schottky-like I_{ds}–V_{ds} behavior was observed, giving credence to this possibility. Compared to the relative success of vertical FETs based on nanowires, less progress has been made so far toward achieving a capably functioning vertical CNT-FET due to the lack of rigidity and the small diameters of SWNTs as compared to nanowires or MW-CNTs [103].

8.10 Conclusion

A broad overview of the methods and device configurations which one day may be used in 3D nanostructure-based device structures has been presented. It is exciting to note that practical devices, including circuits and sensors, have already been demonstrated which clearly illustrate the future potential of nanostructures in next-generation devices. For instance, high-performance and optically transparent nanostructure-based TFTs have been successfully fabricated on flexible substrates; such an accomplishment already represents a significant achievement over what is possible with conventional devices and materials. Moreover, the development and integration of multi-functional and multilayered 3D circuits based on such nanostructures has begun to be reported. Through the continued development of nanostructure growth, assembly, and fabrication techniques, the achievement of practical 3D devices may one day be a reality.

References

1. P. Avouris, J. Chen, Nanotube electronics and optoelectronics. Mater. Today **9**, 46 (2006)
2. W. Lu, C.M. Lieber, Semiconductor nanowires. J. Phys. D. Appl. Phys. **39**, R387 (2006)
3. J. Xiang, W. Lu, Y. Hu, Y. Wu, H. Yan, C.M. Lieber, Ge/Si nanowire heterostructures as high-performance field-effect transistors. Nature **441**, 489 (2006)
4. Y. Wu, Y. Cui, L. Huynh, C.J. Barrelet, D.C. Bell, C.M. Lieber, Controlled growth and structures of molecular-scale silicon nanowires. Nano Lett. **4**, 433 (2004)
5. L. Wei, X. Ping, C.M. Lieber, Nanowire transistor performance limits and applications. IEEE Trans. Electron Devices **55**, 2859 (2008)

6. J.A. Burns, B.F. Aull, C.K. Chen, C. Chang-Lee, C.L. Keast, J.M. Knecht, V. Suntharalingam, K. Warner, P.W. Wyatt, D.R.W. Yost, A wafer-scale 3-D circuit integration technology. IEEE Trans. Electron Devices **53**, 2507 (2006)
7. Y. Hyun-Yong, M. Ishibashi, P. Jin-Hong, M. Kobayashi, K.C. Saraswat, p-Channel Ge MOSFET by selectively heteroepitaxially grown Ge on Si. IEEE Electron Devices L. **30**, 675 (2009)
8. J.A. Hutchby, G.I. Bourianoff, V.V. Zhirnov, J.E. Brewer, Extending the road beyond CMOS. IEEE Circuit Devices **18**, 28 (2002)
9. W. Lu, C.M. Lieber, Nanoelectronics from the bottom up. Nat. Mater. **6**, 841 (2007)
10. J.-H. Ahn, H.-S. Kim, K.J. Lee, S. Jeon, S.J. Kang, Y. Sun, R.G. Nuzzo, J.A. Rogers, Heterogeneous three-dimensional electronics by use of printed semiconductor nanomaterials. Science **314**, 1754 (2006)
11. K. Banerjee, S.J. Souri, P. Kapur, K.C. Saraswat, 3-D ICs: a novel chip design for improving deep-submicrometer interconnect performance and systems-on-chip integration. Proc. IEEE **89**, 602 (2001)
12. M.C.P. Wang, B.D. Gates, Directed assembly of nanowires. Mater. Today **12**, 34 (2009)
13. Q. Cao, J.A. Rogers, Ultrathin films of single-walled carbon nanotubes for electronics and sensors: a review of fundamental and applied aspects. Adv. Mater. **21**, 29 (2009)
14. R.D. Deegan, O. Bakajin, T.F. Dupont, G. Huber, S.R. Nagel, T.A. Witten, Contact line deposits in an evaporating drop. Phys. Rev. E **62**, 756 (2000)
15. L. Huang, X. Cui, G. Dukovic, S.P. O'Brien, Self-organizing high-density single-walled carbon nanotube arrays from surfactant suspensions. Nanotechnology **15**, 1450 (2004)
16. Q. Li, Y.T. Zhu, I.A. Kinloch, A.H. Windle, Self-organization of carbon nanotubes in evaporating droplets. J. Phys. Chem. B **110**, 13926 (2006)
17. S. Zhang, Q. Li, I.A. Kinloch, A.H. Windle, Ordering in a droplet of an aqueous suspension of single-wall carbon nanotubes on a solid substrate. Langmuir **26**, 2107 (2009)
18. J. Huang, R. Fan, S. Connor, P. Yang, One-step patterning of aligned nanowire arrays by programmed dip coating. Angew. Chem. Int. Ed. **46**, 2414 (2007)
19. Y. Huang, X. Duan, Q. Wei, C.M. Lieber, Directed assembly of one-dimensional nanostructures into functional networks. Science **291**, 630 (2001)
20. X. Duan, C. Niu, V. Sahi, J. Chen, J.W. Parce, S. Empedocles, J.L. Goldman, High-performance thin-film transistors using semiconductor nanowires and nanoribbons. Nature **425**, 274 (2003)
21. J.-U. Park, M.A. Meitl, S.-H. Hur, M.L. Usrey, M.S. Strano, P.J.A. Kenis, J.A. Rogers, In situ deposition and patterning of single-walled carbon nanotubes by laminar flow and controlled flocculation in microfluidic channels. Angew. Chem. **118**, 595 (2006)
22. T.D. Yuzvinsky, A.M. Fennimore, A. Kis, A. Zettl, Controlled placement of highly aligned carbon nanotubes for the manufacture of arrays of nanoscale torsional actuators. Nanotechnology **17**, 434 (2006)
23. G. Yu, A. Cao, C.M. Lieber, Large-area blown bubble films of aligned nanowires and carbon nanotubes. Nat. Nanotechnol. **2**, 372 (2007)
24. G. Yu, X. Li, C.M. Lieber, A. Cao, Nanomaterial-incorporated blown bubble films for large-area, aligned nanostructures. J. Mater. Chem. **18**, 728 (2008)
25. D. Wang, R. Tu, L. Zhang, H. Dai, Deterministic one-to-one synthesis of germanium nanowires and individual gold nanoseed patterning for aligned nanowire arrays. Angew. Chem Int. Ed. **44**, 2925 (2005)
26. H. Xin, A.T. Woolley, Directional orientation of carbon nanotubes on surfaces using a gas flow cell. Nano Lett. **4**, 1481 (2004)
27. W. Song, I.A. Kinloch, A.H. Windle, Nematic liquid crystallinity of multiwall carbon nanotubes. Science **302**, 1363 (2003)
28. N.R. Jana, L.A. Gearheart, S.O. Obare, C.J. Johnson, K.J. Edler, S. Mann, C.J. Murphy, Liquid crystalline assemblies of ordered gold nanorods. J. Mater. Chem. **12**, 2909 (2002)
29. M.D. Lynch, D.L. Patrick, Organizing carbon nanotubes with liquid crystals. Nano Lett. **2**, 1197 (2002)

30. L.-S. Li, A.P. Alivisatos, Semiconductor nanorod liquid crystals and their assembly on a substrate. Adv. Mater. **15**, 408 (2003)
31. H. Ko, V.V. Tsukruk, Liquid-crystalline processing of highly oriented carbon nanotube arrays for thin-film transistors. Nano Lett. **6**, 1443 (2006)
32. A.R. Tao, J. Huang, P. Yang, Langmuir–Blodgettry of nanocrystals and nanowires. Acc. Chem. Res. **41**, 1662 (2008)
33. X. Li, L. Zhang, X. Wang, I. Shimoyama, X. Sun, W.-S. Seo, H. Dai, Langmuir–Blodgett assembly of densely aligned single-walled carbon nanotubes from bulk materials. J. Am. Chem. Soc. **129**, 4890 (2007)
34. F. Kim, S. Kwan, J. Akana, P. Yang, Langmuir–Blodgett nanorod assembly. J. Am. Chem. Soc. **123**, 4360 (2001)
35. J. Park, G. Shin, J.S. Ha, Controlling orientation of V2O5 nanowires within micropatterns via microcontact printing combined with the gluing Langmuir-Blodgett technique. Nanotechnology **19**, 395303 (2008)
36. D. Whang, S. Jin, Y. Wu, C.M. Lieber, Large-scale hierarchical organization of nanowire arrays for integrated nanosystems. Nano Lett. **3**, 1255 (2003)
37. D. Wang, Y.-L. Chang, Z. Liu, H. Dai, Oxidation resistant germanium nanowires: bulk synthesis, long chain alkanethiol functionalization, and Langmuir–Blodgett assembly. J. Am. Chem. Soc. **127**, 11871 (2005)
38. P.A. Smith, C.D. Nordquist, T.N. Jackson, T.S. Mayer, B.R. Martin, J. Mbindyo, T.E. Mallouk, Electric-field assisted assembly and alignment of metallic nanowires. Appl. Phys. Lett. **77**, 1399 (2000)
39. T.H. Kim, S.Y. Lee, N.K. Cho, H.K. Seong, H.J. Choi, S.W. Jung, S.K. Lee, Dielectrophoretic alignment of gallium nitride nanowires (GaN NWs) for use in device applications. Nanotechnology **17**, 3394 (2006)
40. R. Zhou, H.-C. Chang, V. Protasenko, M. Kuno, A.K. Singh, D. Jena, H. Xing, CdSe nanowires with illumination-enhanced conductivity: induced dipoles, dielectrophoretic assembly, and field-sensitive emission. J. Appl. Phys. **101**, 073704 (2007)
41. Y. Liu, J.-H. Chung, W.K. Liu, R.S. Ruoff, Dielectrophoretic assembly of nanowires. J. Phys. Chem. B **110**, 14098 (2006)
42. C.S. Lao, J. Liu, P. Gao, L. Zhang, D. Davidovic, R. Tummala, Z.L. Wang, ZnO nanobelt/nanowire Schottky diodes formed by dielectrophoresis alignment across Au electrodes. Nano Lett. **6**, 263 (2006)
43. X. Duan, Y. Huang, Y. Cui, J. Wang, C.M. Lieber, Indium phosphide nanowires as building blocks for nanoscale electronic and optoelectronic devices. Nature **409**, 66 (2001)
44. M. Li, R.B. Bhiladvala, T.J. Morrow, J.A. Sioss, K.-K. Lew, J.M. Redwing, C.D. Keating, T.S. Mayer, Bottom-up assembly of large-area nanowire resonator arrays. Nat. Nanotechnol. **3**, 88 (2008)
45. T.J. Morrow, M. Li, J. Kim, T.S. Mayer, C.D. Keating, Programmed assembly of DNA-coated nanowire devices. Science **323**, 352 (2009)
46. A. Vijayaraghavan, S. Blatt, D. Weissenberger, M. Oron-Carl, F. Hennrich, D. Gerthsen, H. Hahn, R. Krupke, Ultra-large-scale directed assembly of single-walled carbon nanotube devices. Nano Lett. **7**, 1556 (2007)
47. M. Lee, J. Im, B.Y. Lee, S. Myung, J. Kang, L. Huang, Y.K. Kwon, S. Hong, Linker-free directed assembly of high-performance integrated devices based on nanotubes and nanowires. Nat. Nanotechnol. **1**, 66 (2006)
48. Y. Wang, D. Maspoch, S. Zou, G.C. Schatz, R.E. Smalley, C.A. Mirkin, Controlling the shape, orientation, and linkage of carbon nanotube features with nano affinity templates. Proc. Natl. Acad. Sci. USA **103**, 2026 (2006)
49. S.G. Rao, L. Huang, W. Setyawan, S. Hong, Nanotube electronics: large-scale assembly of carbon nanotubes. Nature **425**, 36 (2003)
50. J.B. Hannon, A. Afzali, C. Klinke, P. Avouris, Selective placement of carbon nanotubes on metal-oxide surfaces. Langmuir **21**, 8569 (2005)

51. S. Myung, M. Lee, G.T. Kim, J.S. Ha, S. Hong, Large-scale "surface-programmed assembly" of pristine vanadium oxide nanowire-based devices. Adv. Mater. **17**, 2361 (2005)
52. S. Myung, J. Im, L. Huang, S.G. Rao, T. Kim, D.J. Lee, S. Hong, "Lens" effect in directed assembly of nanowires on gradient molecular patterns. J. Phys. Chem. B **110**, 10217 (2006)
53. K. Heo, E. Cho, J.-E. Yang, M.-H. Kim, M. Lee, B.Y. Lee, S.G. Kwon, M.-S. Lee, M.-H. Jo, H.-J. Choi, T. Hyeon, S. Hong, Large-scale assembly of silicon nanowire network-based devices using conventional microfabrication facilities. Nano Lett. **8**, 4523 (2008)
54. Z. Fan, J.C. Ho, T. Takahashi, R. Yerushalmi, K. Takei, A.C. Ford, Y.-L. Chueh, A. Javey, Toward the development of printable nanowire electronics and sensors. Adv. Mater. **21**, 3730 (2009)
55. J. Im, I.-H. Lee, B.Y. Lee, B. Kim, J. Park, W. Yu, U.J. Kim, Y.H. Lee, M.-J. Seong, E.H. Lee, Y.-S. Min, S. Hong, Direct printing of aligned carbon nanotube patterns for high-performance thin film devices. Appl. Phys. Lett. **94**, 053109 (2009)
56. R. Yerushalmi, Z.A. Jacobson, J.C. Ho, Z. Fan, A. Javey, Large scale, highly ordered assembly of nanowire parallel arrays by differential roll printing. Appl. Phys. Lett. **91**, 203104 (2007)
57. Z. Fan, J.C. Ho, Z.A. Jacobson, R. Yerushalmi, R.L. Alley, H. Razavi, A. Javey, Wafer-scale assembly of highly ordered semiconductor nanowire arrays by contact printing. Nano Lett. **8**, 20 (2007)
58. Y.-K. Chang, F.C.-N. Hong, The fabrication of ZnO nanowire field-effect transistors by roll-transfer printing. Nanotechnology **20**, 195302 (2009)
59. S.-H. Hur, O.O. Park, J.A. Rogers, Extreme bendability of single-walled carbon nanotube networks transferred from high-temperature growth substrates to plastic and their use in thin-film transistors. Appl. Phys. Lett. **86**, 243502 (2005)
60. S.J. Kang, C. Kocabas, T. Ozel, M. Shim, N. Pimparkar, M.A. Alam, S.V. Rotkin, J.A. Rogers, High-performance electronics using dense, perfectly aligned arrays of single-walled carbon nanotubes. Nat. Nanotechnol. **2**, 230 (2007)
61. X. Liu, S. Han, C. Zhou, Novel nanotube-on-insulator (NOI) approach toward single-walled carbon nanotube devices. Nano Lett. **6**, 34 (2005)
62. A.J. Baca, J.-H. Ahn, Y. Sun, M.A. Meitl, E. Menard, H.-S. Kim, Won M. Choi, D.-H. Kim, Y. Huang, J.A. Rogers, Semiconductor wires and ribbons for high- performance flexible electronics. Angew. Chem Int. Ed. **47**, 5524 (2008)
63. Q. Cao, H.-S. Kim, N. Pimparkar, J.P. Kulkarni, C. Wang, M. Shim, K. Roy, M.A. Alam, J.A. Rogers, Medium-scale carbon nanotube thin-film integrated circuits on flexible plastic substrates. Nature **454**, 495 (2008)
64. S.J. Kang, C. Kocabas, H.-S. Kim, Q. Cao, M.A. Meitl, D.-Y. Khang, J.A. Rogers, Printed multilayer superstructures of aligned single-walled carbon nanotubes for electronic applications. Nano Lett. **7**, 3343 (2007)
65. L. Jiao, B. Fan, X. Xian, Z. Wu, J. Zhang, Z. Liu, Creation of nanostructures with poly(methyl methacrylate)-mediated nanotransfer printing. J. Am. Chem. Soc. **130**, 12612 (2008)
66. R. He, D. Gao, R. Fan, A.I. Hochbaum, C. Carraro, R. Maboudian, P. Yang, Si nanowire bridges in microtrenches: integration of growth into device fabrication. Adv. Mater. **17**, 2098 (2005)
67. M.S. Islam, S. Sharma, T.I. Kamins, R.S. Williams, Ultrahigh-density silicon nanobridges formed between two vertical silicon surfaces. Nanotechnology **15**, L5 (2004)
68. R. He, P. Yang, Giant piezoresistance effect in silicon nanowires. Nat. Nanotechnol. **1**, 42 (2006)
69. J.S. Lee, M.S. Islam, S. Kim, Direct formation of catalyst-free ZnO nanobridge devices on an etched Si substrate using a thermal evaporation method. Nano Lett. **6**, 1487 (2006)
70. W.Y. Lee, C.H. Weng, Z.Y. Juang, J.F. Lai, K.C. Leou, C.H. Tsai, Lateral growth of single-walled carbon nanotubes across electrodes and the electrical property characterization. Diam. Relat. Mater. **14**, 1852 (2005)

71. S. Sharma, T.I. Kamins, M.S. Islam, R.S. Williams, A.F. Marshall, Structural characteristics and connection mechanism of gold-catalyzed bridging silicon nanowires. J. Crystal Growth **280**, 562 (2005)
72. A.S. Paulo, J. Bokor, R.T. Howe, R. He, P. Yang, D. Gao, C. Carraro, R. Maboudian, Mechanical elasticity of single and double clamped silicon nanobeams fabricated by the vapor-liquid-solid method. Appl. Phys. Lett. **87**, 053111 (2005)
73. A. Chaudhry, V. Ramamurthi, E. Fong, M.S. Islam, Ultra-low contact resistance of epitaxially interfaced bridged silicon nanowires. Nano Lett. **7**, 1536 (2007)
74. S. Fan, M.G. Chapline, N.R. Franklin, T.W. Tombler, A.M. Cassell, H. Dai, Self-oriented regular arrays of carbon nanotubes and their field emission properties. Science **283**, 512 (1999)
75. L. Gangloff, E. Minoux, K.B.K. Teo, P. Vincent, V.T. Semet, V.T. Binh, M.H. Yang, I.Y.Y. Bu, R.G. Lacerda, G. Pirio, J.P. Schnell, D. Pribat, D.G. Hasko, G.A.J. Amaratunga, W.I. Milne, P. Legagneux, Self-aligned, gated arrays of individual nanotube and nanowire emitters. Nano Lett. **4**, 1575 (2004)
76. J. Goldberger, A.I. Hochbaum, R. Fan, P. Yang, Silicon vertically integrated nanowire field effect transistors. Nano Lett. **6,** 973 (2006)
77. A.I. Hochbaum, R. Fan, R. He, P. Yang, Controlled growth of Si nanowire arrays for device integration. Nano Lett. **5**, 457 (2005)
78. H.T. Ng, J. Han, T. Yamada, P. Nguyen, Y.P. Chen, M. Meyyappan, Single crystal nanowire vertical surround-gate field-effect transistor. Nano Lett. **4**, 1247 (2004)
79. P. Nguyen, H.T. Ng, T. Yamada, M.K. Smith, J. Li, J. Han, M. Meyyappan, Direct integration of metal oxide nanowire in vertical field-effect transistor. Nano Lett. **4**, 651 (2004)
80. Q. Wan, E.N. Dattoli, W.Y. Fung, W. Guo, Y. Chen, X. Pan, W. Lu, High-performance transparent conducting oxide nanowires. Nano Lett. **6**, 2909 (2006)
81. J.H. Woodruff, J.B. Ratchford, I.A. Goldthorpe, P.C. McIntyre, C.E.D. Chidsey, Vertically oriented germanium nanowires grown from gold colloids on silicon substrates and subsequent gold removal. Nano Lett. **7**, 1637 (2007)
82. X. Duan, Nanowire thin-film transistors: a new avenue to high-performance macroelectronics. IEEE Trans. Electron Devices **55**, 3056 (2008)
83. E.N. Dattoli, K.H. Kim, W.Y. Fung, S.Y. Choi, W. Lu, Radio-frequency operation of transparent nanowire thin-film transistors. IEEE Electron Devices L. **30**, 730 (2009)
84. E.N. Dattoli, Q. Wan, W. Guo, Y. Chen, X. Pan, W. Lu, Fully transparent thin-film transistor devices based on SnO2 nanowires. Nano Lett. **7**, 2463 (2007)
85. K. Nomura, H. Ohta, K. Ueda, T. Kamiya, M. Hirano, H. Hosono, Thin-film transistor fabricated in single-crystalline transparent oxide semiconductor. Science **300**, 1269 (2003)
86. S. Ju, A. Facchetti, Y. Xuan, J. Liu, F. Ishikawa, P. Ye, C. Zhou, T.J. Marks, D.B. Janes, Fabrication of fully transparent nanowire transistors for transparent and flexible electronics. Nat. Nanotechnol. **2**, 378 (2007)
87. S.A. Dayeh, C. Soci, P.K.L. Yu, E.T. Yu, D. Wang, Influence of surface states on the extraction of transport parameters from InAs nanowire field effect transistors. Appl. Phys. Lett. **90**, 162112 (2007)
88. T.W. Kelley, P.F. Baude, C. Gerlach, D.E. Ender, D. Muyres, M.A. Haase, D.E. Vogel, S.D. Theiss, Recent progress in organic electronics: materials, devices, and processes. Chem. Mater. **16**, 4413 (2004)
89. M. Ofuji, K. Abe, H. Shimizu, N. Kaji, R. Hayashi, M. Sano, H. Kumomi, K. Nomura, T. Kamiya, H. Hosono, Fast thin-film transistor circuits based on amorphous oxide semiconductor. IEEE Electron Devices L. **28**, 273 (2007)
90. N. Chimot, V. Derycke, M.F. Goffman, J.P. Bourgoin, H. Happy, G. Dambrine, Gigahertz frequency flexible carbon nanotube transistors. Appl. Phys. Lett. **91**, 153111 (2007)
91. A. Le Louarn, F. Kapche, J.M. Bethoux, H. Happy, G. Dambrine, V. Derycke, P. Chenevier, N. Izard, M.F. Goffman, J.P. Bourgoin, Intrinsic current gain cutoff frequency of 30 GHz with carbon nanotube transistors. Appl. Phys. Lett. **90**, 233108 (2007)

92. F.N. Ishikawa, H.-K. Chang, K. Ryu, P.-C. Chen, A. Badmaev, L. Gomez De Arco, G. Shen, C. Zhou, Transparent electronics based on transfer printed aligned carbon nanotubes on rigid and flexible substrates. ACS Nano **3**, 73 (2008)
93. C. Wang, K. Ryu, A. Badmaev, N. Patil, A. Lin, S. Mitra, H.S.P. Wong, C. Zhou, Device study, chemical doping, and logic circuits based on transferred aligned single-walled carbon nanotubes. Appl. Phys. Lett. **93**, 033101 (2008)
94. C. Kocabas, S. Dunham, Q. Cao, K. Cimino, X. Ho, H.-S. Kim, D. Dawson, J. Payne, M. Stuenkel, H. Zhang, T. Banks, M. Feng, S.V. Rotkin, J.A. Rogers, High-frequency performance of submicrometer transistors that use aligned arrays of single-walled carbon nanotubes. Nano Lett. **9**, 1937 (2009)
95. C. Kocabas, H.-S. Kim, T. Banks, J.A. Rogers, A.A. Pesetski, J.E. Baumgardner, S.V. Krishnaswamy, H. Zhang, Radio frequency analog electronics based on carbon nanotube transistors. Proc. Natl. Acad. Sci. USA **105**, 1405 (2008)
96. A.W. Topol, D.C.J. La Tulipe, L. Shi, D.J. Frank, Three-dimensional integrated circuits. IBM J. Res. Dev. **50**, 491 (2006)
97. A. Javey, S. Nam, R.S. Friedman, H. Yan, C.M. Lieber, Layer-by-layer assembly of nanowires for three-dimensional, multifunctional electronics. Nano Lett. **7**, 773 (2007)
98. S. Nam, X. Jiang, Q. Xiong, D. Ham, C.M. Lieber, Vertically integrated, three-dimensional nanowire complementary metal-oxide-semiconductor circuits. Proc. Natl. Acad. Sci. USA **106**, 21035 (2009)
99. Z. Fan, J.C. Ho, Z.A. Jacobson, H. Razavi, A. Javey, Large-scale, heterogeneous integration of nanowire arrays for image sensor circuitry. Proc. Natl. Acad. Sci. USA **105**, 11066 (2008)
100. C. Thelander, L.E. Froberg, C. Rehnstedt, L. Samuelson, L.E. Wernersson, Vertical enhancement-mode InAs nanowire field-effect transistor with 50-nm wrap gate. IEEE Electron Devices L. **29**, 206 (2008)
101. W.B. Choi, B.-H. Cheong, J.J. Kim, J. Chu, E. Bae, Selective growth of carbon nanotubes for nanoscale transistors. Adv. Funct. Mater. **13**, 80 (2003)
102. V. Schmidt, H. Riel, S. Senz, S. Karg, W. Riess, U. Gösele, Realization of a silicon nanowire vertical surround-gate field-effect transistor. Small **2**, 85 (2006)
103. A.D. Franklin, R.A. Sayer, T.D. Sands, T.S. Fisher, D.B. Janes, Toward surround gates on vertical single-walled carbon nanotube devices. J. Vac. Sci. Technol. B **27**, 821 (2009)

Chapter 9
Strain-Induced, Self Rolled-Up Semiconductor Microtube Resonators: A New Architecture for Photonic Device Applications

Xin Miao, Ik Su Chun, and Xiuling Li

9.1 Introduction

A semiconductor heterojunction is a junction between two chemically different semiconductors, such as GaAs and $In_xGa_{1-x}As$, or silicon (Si) and germanium (Ge). Semiconductor heterostructure-based electronics and photonics have been widely used in high-power lasers, light-emitting diodes, heterojunction bipolar transistors, and high-efficiency solar cells. The versatile heterojunctions, especially in compound semiconductors, allow the control of fundamental semiconductor parameters such as electronic band structure, strain, and mobility.

Strain–induced, self rolled-up semiconductor micro- or nanotube is a new type of architecture involving semiconductor heterojunctions for manipulating photons and electrons. These tubes are formed spontaneously as a result of energy minimization when a strained planar membrane deforms into curved surfaces by strain relaxation, first discovered by Prinz et al. in 2000 [1]. Figure 9.1 illustrates an example of such self-rolling phenomenon using a GaAs–$In_xGa_{1-x}As$ bilayer system as an example. $In_xGa_{1-x}As$ layer is compressively strained when pseudomorphically deposited on

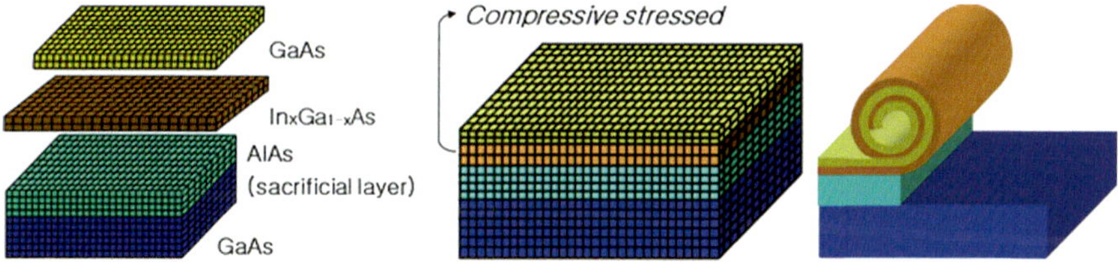

Fig. 9.1 Schematic illustration of the formation mechanism of self rolled-up GaAs–$In_xGa_{1-x}As$ tubes from pseudomorphically grown epitaxial layers to the epitaxial liftoff and self-rolling process by selectively removing the AlAs sacrificial layer. Adapted from [14] with permission

X. Li (✉)
Micro and Nanotechnology Laboratory, Department of Electrical and Computer Engineering, University of Illinois, Urbana, IL 61801, USA
e-mail: xiuling@illinois.edu

W.L. Zhou, Z.L. Wang (eds.), *Three-Dimensional Nanoarchitectures*,
DOI 10.1007/978-1-4419-9822-4_9,

Fig. 9.2 An array of highly ordered $In_{0.3}Ga_{0.7}As$–GaAs bilayer rolled-up tubes that are 0.6 μm in diameter, 50 μm in length, and 12 nm in wall thickness. Adapted from [10] with permission

GaAs substrate. Upon releasing the GaAs–$In_xGa_{1-x}As$ bilayer from the substrate by selectively removing the AlAs sacrificial layer, the $In_xGa_{1-x}As$ layer has the tendency to expand to its unstrained state, while the GaAs layer resists the expansion. The opposite force from each of the bilayers generates a net momentum driving the planar membrane to scroll up and continue to roll into a tubular spiral structure, as the sacrificial layer is etched laterally.

The epitaxial growth of the strained layers with defined thickness and mismatch strain determines the tube diameter. The epitaxial lift-off process through modern lithographical patterning and etching enables positioning and alignment control of these tubes. Shown in Fig. 9.2 is an array of large area of highly ordered, $In_{0.3}Ga_{0.7}As$–GaAs bilayer rolled-up tubes that are 0.6 μm in diameter, 50 μm in length, and 12 nm in wall thickness. Note that in addition to strained semiconductor heterojunctions, strained single-material membranes such as strained Si can also be used to form self rolled-up tubes [2]. In this chapter, the formation process of strain-induced, self rolled-up semiconductor micro- and nanotubes, and their optical properties and applications in photonic devices will be summarized. The goal is to highlight what the rolled-up tube structure has to offer in terms of the unique optical properties and devices associated with its architecture, not meant to be a comprehensive review of the literature.

9.2 Formation Process

Metalorganic chemical vapor deposition (MOCVD) and molecular beam epitaxy (MBE) have been established as the methods of choice to grow pseudomorphically strained epitaxial films with precisely controlled thickness and composition [3, 4].

The diameter of the self rolled-up semiconductor tube is proportional to total thickness and inversely proportional to mismatch strain of the strained membrane. For a tube with a strained bilayer with a total thickness of $d = d_1 + d_2$ and lattice mismatch of $\varepsilon = (a_2 - a_1)/a_1$, the diameter can be estimated by a classical continuum theory [5, 6], as described in Eq. (9.1):

$$D = \frac{d\left[3(1+m)^2 + (1+m\times n)\left[m^2 + (m\times n)^{-1}\right]\right]}{3\varepsilon(1+m)^2} \tag{9.1}$$

where m is the thickness ratio (d_1/d_2) and n is the Young's modulus (Y_1/Y_2) ratio. As reported before [7], the experimentally measured tube diameter is 15–20% smaller than that calculated using the continuum model. The smallest diameter nanotube that has been demonstrated is ~3 nm using one monolayer of InAs and one monolayer of GaAs which has a mismatch strain of 7.16% [1, 8]. In III–V material system, the typical mismatch strain is 1–7%, thus for total film thickness of 1–100 nm, the tube diameter would be in the range of 10 nm–10 μm.

The preferred rolling direction of strained membrane is governed by the Young's modulus. For cubic crystals such as GaAs and Si, the Young's modulus along <100> is the smallest compared to <110> and <111> directions. For a rectangular-shaped membrane, when the sides are aligned to <100>, the strain-induced self-rolling will lead to the formation of tubes; when the corners are aligned to <100>, helical structures can be formed [9, 10]. Furthermore, there is a geometry effect on the strain-driven deformation behavior. Depending on the actual dimension of the rectangular mesa (sides a and b, and ratio a/b), tube diameter, lateral etching anisotropy, and certain kinetic variations, the final rolling direction could end up occurring from the short side or the long side, both of which are crystallographically equivalent [11]. This is mostly driven by the energetics of the final configuration but can be changed by kinetic factors during the formation process [11]. Shown in Fig. 9.3 are examples of three rolling configurations from strained membranes (InGaAs–GaAs) with different geometries, along with finite element method (FEM)-simulated intermediate energy state of one-fourth of the membrane (due to center symmetry, the rest of the membrane is not shown) [11]. In addition to rolling up strained solid

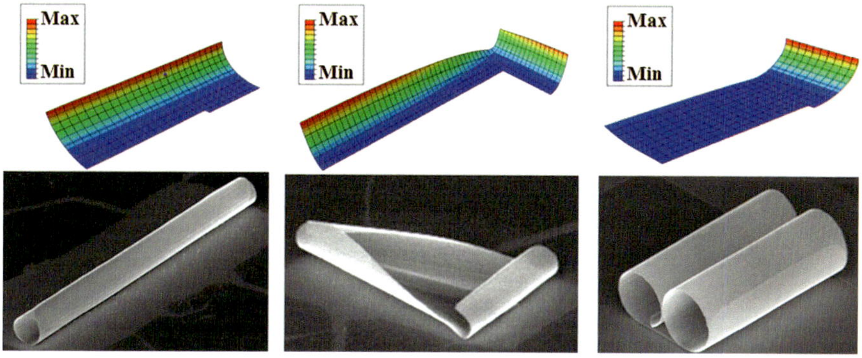

Fig. 9.3 SEM images of self-rolling configurations of strained rectangular membranes (InGaAs–GaAs) with different geometries, along with the corresponding FEM-simulated intermediate energy state of one-fourth of the membrane. From left to right, the long-side rolling, mixed rolling from both sides, and short-side rolling are illustrated. Reproduced from [11] with permission

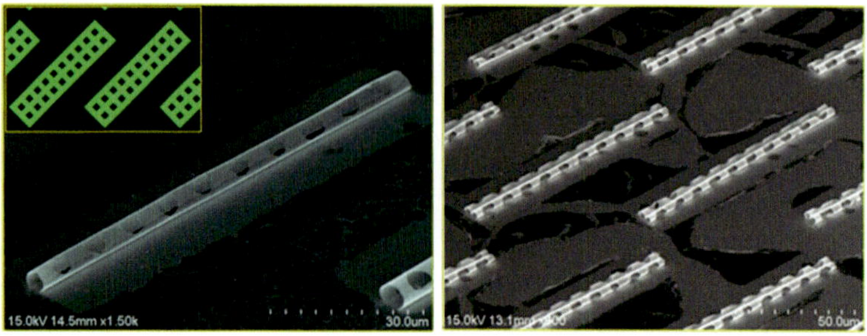

Fig. 9.4 SEM images of holey tubes formed using 20 μm × 105 μm strained membranes patterned with arrays of 5 μm × 5 μm holes. The tube diameter shows negligible change compared to solid membrane of the same materials. Reproduced from [15] with permission

membranes, membranes with patterns at the edges and throughout the entire film can be deformed by the same mechanism. Shown in Fig. 9.4 are tubes formed from patterned membrane with periodic arrays of holes.

9.3 Photonic Applications of Rolled-Up Semiconductor Tubes

Tubular structures can be formed from not only a simple strained bilayer such as InGaAs–GaAs but also structures that include active optical gain media and plasmonic structures [12], provided there is enough driving force from the built-in strain in the epitaxial film [13]. Illustrated in Fig. 9.5 are two kinds of active structures that can be rolled up by using the strained $In_xGa_{1-x}As$ layer as a wrapper while keeping the rest of the layers either lattice matched to the substrate (GaAs quantum well (QW)) or involving discrete structures (quantum dots (QDs)). The film growth for the QW and QD structures is straightforward as in the growth of planar laser structures, except that the design involves much thinner layers.

Compared to planar QW/QD structures, there are several distinct optical properties when the active medium is situated on a curved surface (tube wall), in a high index contrast environment (semiconductor–air interface), and forms new cavities (circular along tube periphery and longitudinal along tube axis).

9.3.1 Spontaneous Emission from Quantum Well Microtubes: Intensity Enhancement and Energy Shift

To illustrate the effect of curvature on optical properties, an epitaxial structure consisting of a 5-nm GaAs QW cladded by 10-nm $Al_{0.33}Ga_{0.67}As$ layers on both sides grown on a strained $In_xGa_{1-x}As$ layer is examined. The strained $In_xGa_{1-x}As$ layer composition and thickness are adjusted to change the tube curvature, while the active region (10-nm $Al_{0.33}Ga_{0.67}As$/5-nm GaAs/10-nm $Al_{0.33}Ga_{0.67}As$) is kept the

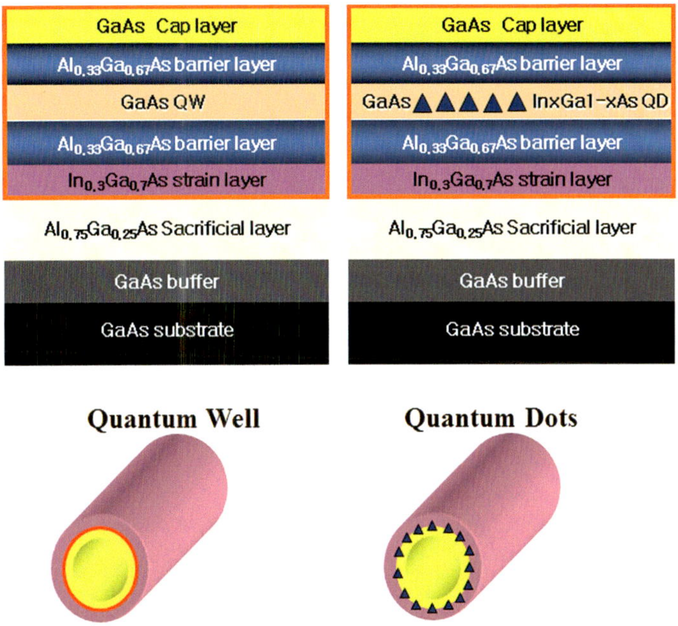

Fig. 9.5 Illustration of two types of structures with epitaxially embedded active gain media that can be rolled up by using the strained $In_xGa_{1-x}As$ layer as a wrapper while keeping the rest of the layers lattice matched to the substrate (GaAs QW, *left*) or involving discrete structures ($In_xGa_{1-x}As$ QDs, *right*). Adapted from [14] with permission

same. By removing the strained $In_xGa_{1-x}As$ layer in the structure, planar membranes with the same active region can be produced.

Shown in Fig. 9.6a is the room temperature photoluminescence (PL) spectra taken from planar membranes, rolled-up solid-wall tubes, and rolled-up patterned tubes of this structure. A strong and distinct peak around 821 nm is observed from the ultrathin-walled (38 nm) GaAs QW microtubes. Remarkably, the PL intensity from the rolled-up QW tube is enhanced dramatically ($\sim 10\times$) compared to the planar counterpart. Further intensity increase can be seen from the holey-patterned, rolled-up tubes, attributed to higher coupling efficiency [15]. Another unique optical effect of the rolled-up QW tubes is the strain-induced peak shift as a function of tube curvature. This can be seen from Fig. 9.6b where a consistent red shift is observed when the tube diameter becomes smaller. When the epitaxial stack is rolled up, GaAs QW layer experiences tensile strain in order to compensate for the partial relaxation of compressive strain in the InGaAs layer. By analyzing the strain distribution, the amount of bandgap energy shift induced by strain can be calculated. Reasonable agreement with experimental data has been found, confirming that the PL peak shift is a result of strain-induced bandgap change [15, 16]. It should be emphasized that the bandgap shift is not due to chemical composition change or thickness-related quantum confinement effect. *The rolled-up architecture introduces*

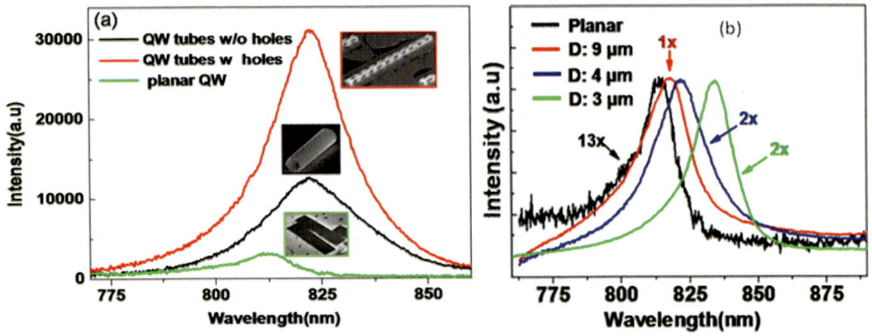

Fig. 9.6 Room temperature photoluminescence spectra taken from (**a**) planar GaAs QW membranes, rolled-up, solid-wall QW tubes (without holes), and rolled-up patterned QW tubes (with holes) under the same condition; (**b**) rolled-up GaAs QW tubes of various diameters. Reproduced from [15] with permission

another degree of freedom, the curvature, to manipulate fundamental semiconductor parameters, including bandgap, effective mass, and mobility.

9.3.2 Optical Resonance Modes in Rolled-Up Microtube Ring Cavity

Optical ring resonance whispering gallery modes were first reported in the pioneering work carried out by Kipp et al. [17] at 5 K, where a U-shaped mesa was defined to suspend the middle of the rolled-up InAs QD microtube. As shown in Fig. 9.7, sharp polarized and regularly spaced optical modes are clearly seen in the

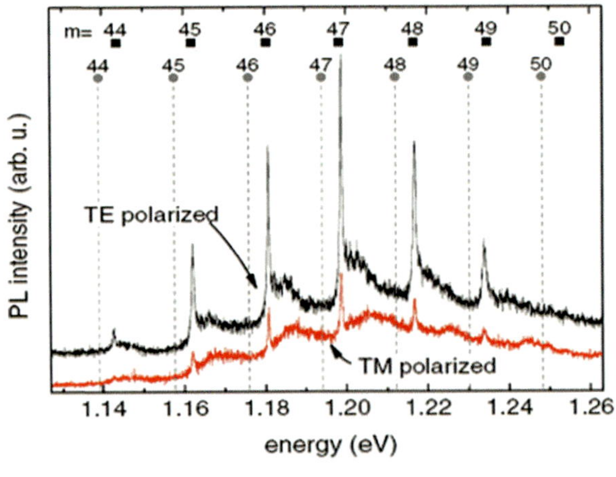

Fig. 9.7 Micro-PL spectra at 5 K taken from a suspended microtube with InAs quantum dots incorporated as the gain media. Azimuthal mode number m calculated using two different models are labeled, showing reasonable agreement with experimental data. Reproduced from [17] with permission

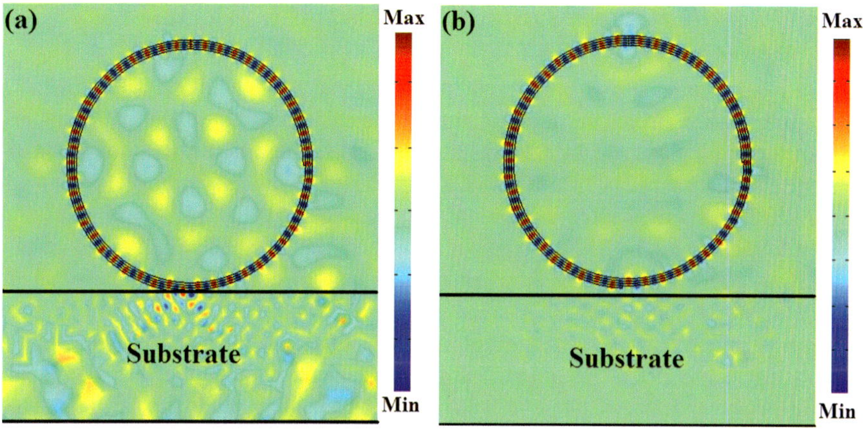

Fig. 9.8 Simulated electric field distribution of TE fundamental mode in a microtube (4.8 μm in diameter with a total wall thickness of 150 nm) attached to (**a**) and suspended from (**b**) the substrate by 150 nm; the color from *dark red* to *dark blue* shows the amplitude of electric field from crest to trough. Unpublished data by Xin Miao et al.

PL spectra. The excellent agreement found between experimental data and modeled mode peak positions confirms the optical ring resonator nature of the microtube. Resonance modes have soon after been reported at room temperature from Si–SiO_x rolled-up microtubes by the Schmidt group [18]. The active gain medium in this interesting hybrid system is Si nanoclusters embedded in the SiO_x matrix. More recently, optical modes excited by the evanescent field from PbS nanocrystals inside a rolled-up passive microtube consisting of $Al_{0.31}In_{0.15}Ga_{0.54}As$ and $Al_{0.20}Ga_{0.80}As$ have been observed. The PbS nanocrystals are brought into the passive microtube core by fluid filling and serve as active emitters [19].

In general, several requirements need to be met in order to produce mode-like peaks from these rolled-up tubes. First, the microtube should be suspended to eliminate light leakage to substrate. Shown in Fig. 9.8a, b are the simulated electric field distribution at resonance, using a 2D finite element method (FEM), in a microtube attached to and suspended from the substrate, respectively. It is clear that microtube attached to substrate would suffer from light leakage problem causing serious feedback loss in the ring cavity. In fact, for 1.05 μm wavelength light propagating in a microtube ring cavity that is 150 nm thick and 4.8 μm in diameter, the calculated Q factor increases from 1390 to 5260 when suspended above the substrate by 150 nm. Second, the tube wall needs to be thicker than the cutoff thickness of the modes to reduce radiation loss outside of the tube wall and enhance the Q factor [20]. This should be intuitive since the thicker walls correspond to higher effective refractive index for better optical confinement. Similarly, higher order resonance modes can be supported only by increasing the tube wall thickness [17]. Further optical confinement, thus higher Q, in the microtube ring cavity can be achieved by suppressing wave propagation along the tube axis.

9.3.3 Optically Pumped Lasing from Rolled-Up Microtube Ring Cavity

To suppress wave propagation along the tube axis, the strained membrane to be rolled up can be patterned by adding intentionally created bottle-like notches and periodic undulations [21–23]. Following this strategy, optically pumped rolled-up InGaAs QD microtube laser has been demonstrated [24]. Figure 9.9 shows mesa used to fabricate and the SEM images of a suspended microtube with defined undulating outer edge. Figure 9.10a shows the emission spectrum above and below lasing threshold, while Fig. 9.10b shows light intensity versus excitation power curve of this microtube laser [24]. An ultralow lasing threshold of ~4 μW and Q factor of ~3500 have been obtained. Furthermore, a microtube laser consisting of InAlGaAs/GaAs/AlGaAs quantum well in the wall using a predefined ridge as the axial confinement has also been realized at 4 K with lasing threshold between 260 μW and 595 μW [25].

One unique feature of rolled-up microtube resonator is the preferential and directional emission. As clearly demonstrated experimentally [21, 26], the inside edge of the rolled-up tube predominantly emits optical resonance modes, while the outside edge emits only leaky modes. Such directional emission can be visualized by the simulated electric field distribution of the resonance mode in a microtube where the inner edge and the outer edge do not overlap. Shown in Fig. 9.11a is the simulated electrical field distribution of the TE fundamental mode for a 1.8-turn microtube (50 nm thick for each turn) with diameter of 4.8 μm at resonant frequency (2.9×10^{14} Hz) at azimuthal number m of 35. The accumulated emission power along a circle of diameter 8 μm around the tube center is shown in the polar plot

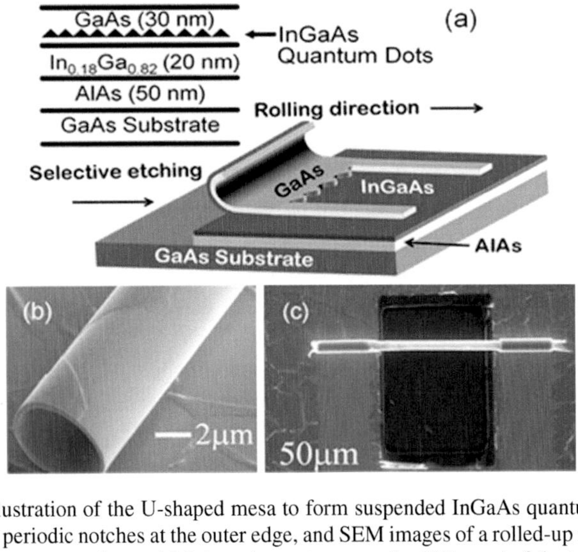

Fig. 9.9 (**a**) Illustration of the U-shaped mesa to form suspended InGaAs quantum dot rolled-up microtube with periodic notches at the outer edge, and SEM images of a rolled-up microtube showing (**b**) the uneven outer edge and (**c**) the scheme to suspend middle part of the tube. Reproduced from [24] with permission

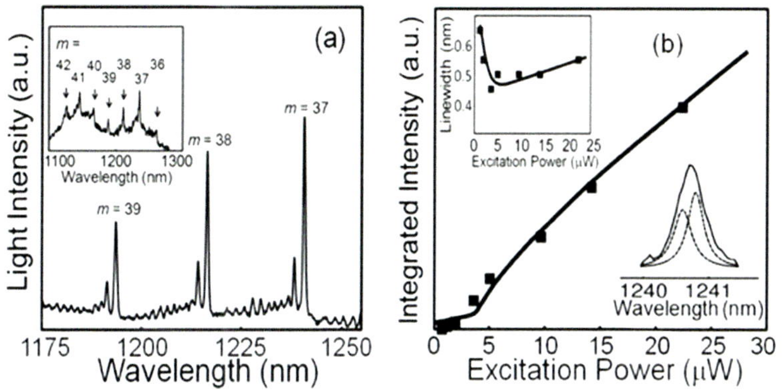

Fig. 9.10 (**a**) Emission spectrum of InGaAs/GaAs quantum dot microtube lasers measured above threshold. The emission spectrum measured at a pump power below threshold is shown in the *inset*. (**b**) The integrated light intensity for lasing mode at 1240.7 nm versus pump power at room temperature. Variation of the line width of the mode versus pump power is shown in the *upper inset*. Reproduced from [24] with permission

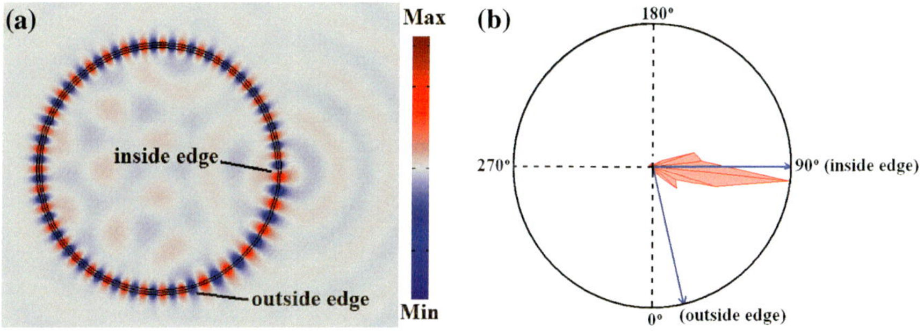

Fig. 9.11 (**a**) Simulated electric field distribution of a TE fundamental resonant mode at $m = 35$ for a 1.8-turn microtube of diameter 4.8 μm and single wall thickness 50 nm; the color from *dark red* to *dark blue* shows the amplitude of electric field from crest to trough and (**b**) polar plot of emission power distribution around the tube center. Unpublished data by Xin Miao et al.

(Fig. 9.11b). The maximum peak position is at 82° which accounts for the emission from the inside tube edge. This property makes microtube ring laser a good candidate for directional light source on chip, since the direction of laser light can be adjusted by controlling the relative position of inside and outside edges.

In summary, this chapter presented an overview of a new tubular architecture that is formed by self-rolling of strained semiconductor heterojunction membranes and its applications in photonics. The optical gain medium is usually embedded in the tube wall. The wall thickness is much thinner than the emission wavelength, while the tube diameter is on the micrometer scale. Rolled-up tubular structures show enhanced luminescence intensity and peak positions can be tuned continuously as a function of tube curvature due to strain-induced bandgap change. By engineering the tube geometry for better optical confinement, optical resonant modes in the microtube ring cavity can be produced, and optically pumped lasing has been achieved. Considering the relative ease to form n- and p- contact layers through epitaxial growth before rolling up, electrical injection lasing from this new architecture should not be beyond reach, although challenging. Potential application of this type of microtube light-emitting devices in optical interconnects, MEMS, metamaterials, and chemical and biological sensing can be envisioned.

Acknowledgment Xiuling Li acknowledges the support from NSF CAREER ECCS under Grant No. 0747178, NSF award under Grant No. 0749028 and DOE award under Grant No. DE-FG02-07ER46471. Technical assistance from Kevin Bassett and Archana Challa from the Li research group and Dr Jianguo Wen and Dr Julio Soares at the Materials Research Laboratory is highly appreciated.

References

1. V.Y. Prinz, V.A. Seleznev, A.K. Gutakovsky, A.V. Chehovskiy, V.V. Preobrazhenskii, M.A. Putyato, T.A. Gavrilova, Free-standing and overgrown InGaAs/GaAs nanotubes, nanohelices and their arrays. Phys. E: Low-Dimensional Syst. Nanostruct. **6**, 828–831 (2000)
2. R. Songmuang, D. Ch, O.G. Schmidt, Rolled-up micro- and nanotubes from single-material thin films. Appl. Phys. Lett. **89**, 223109 (2006)
3. G.B. Stringfellow, *Organometallic Vapor-Phase Epitaxy: Theory and Practice*, 2nd edn. (Academic Press, San Diego, 1999)
4. K. Ploog, *Molecular Beam Epitaxy of III–V Compounds* (Springer, Berlin, 1984)
5. P.O. Vaccaro, K. Kubota, T. Aida, Strain-driven self-positioning of micromachined structures. Appl. Phys. Lett. **78**, 2852 (2001)
6. Y.C. Tsui, T.W. Clyne, Analytical model for predicting residual stresses in progressively deposited coatings. Part 3: Further development and applications. Thin Solid Films **306**, 52 (1997)
7. I.S. Chun, V.B. Verma, V.C. Elarde, S.K. Kim, J.M. Zuo, J.J. Coleman, X. Li, InGaAs/GaAs 3D architecture formation by strain-induced self-rolling with lithographically defined rectangular stripe arrays. J. Crystal Growth **310**, 2353 (2008)
8. O.G. Schmidt, C. Deneke, S. Kiravittaya, R. Songmuang, H. Heidemeyer, Y. Nakamura, R. Zapf-Gottwick, C. Muller, N.Y. Jin-Phillipp, Self-assembled nanoholes, lateral quantum-dot molecules, and rolled-up nanotubes. IEEE J. Sel. Top. Quantum Electron. **8**, 1025–1034 (2002)
9. V.Y. Prinz, V.A. Seleznev, A.V. Prinz, A.V. Kopylov, 3D heterostructures and systems for novel MEMS/NEMS. Sci. Technol. Adv. Mater. **10**, 034502 (2009)
10. I.S. Chun, X. Li, Controlled assembly and dispersion of strain-induced InGaAs/GaAs nanotubes. IEEE Trans. Nanotechnol. **7**, 493–495 (2008)
11. I.S. Chun, A. Challa, B. Derickson, J.K. Hsia, X. Li, Geometry effect on the strain-induced self-rolling of semiconductor membranes. Nano Lett. **10**, 3927–3932 (2010)
12. S. Schwaiger, M. Broll, A. Krohn, A. Stemmann, C. Heyn, Y. Stark, D. Stickler, D. Heitmann, S. Mendach, Rolled-up three-dimensional metamaterials with a tunable plasma frequency in the visible regime. Phys. Rev. Lett. **102**, 163903 (2009)
13. X. Li, Strain induced semiconductor nanotubes: From formation process to device applications. J. Phys. D: Appl. Phys. **41**, 193001 (2008)
14. I.S. Chun, K. Bassett, A. Challa, X. Miao, M. Saarinen, X. Li, Strain-induced self-rolling III–V tubular nanostructures: Formation process and photonic applications. Proc. SPIE **7608**, 760810–760818 (2009)
15. I.S. Chun, K. Bassett, A. Challa, X.L. Li, Tuning the photoluminescence characteristics with curvature for rolled-up GaAs quantum well microtubes. Appl. Phys. Lett. **96**, 251106 (2010)

16. N. Ohtani, K. Kishimoto, K. Kubota, S. Saravanan, Y. Sato, S. Nashima, P. Vaccaro, T. Aida, M. Hosoda, *Uniaxial-Strain-Induced Transition from Type-II to Type-I Band Configuration of Quantum Well Microtubes* (Physica E, Netherlands, 2004), p. 732
17. T. Kipp, H. Welsch, C. Strelow, C. Heyn, D. Heitmann, Optical modes in semiconductor microtube ring resonators. Phys. Rev. Lett. **96**, 077403-1 (2006)
18. R. Songmuang, A. Rastelli, S. Mendach, O.G. Schmidt, SiO_x/Si radial superlattices and microtube optical ring resonators. Appl. Phys. Lett. **90**, 91905-1 (2007)
19. K. Dietrich, C. Strelow, C. Schliehe, C. Heyn, A. Stemmann, S. Schwaiger, S. Mendach, A. Mews, H. Weller, D. Heitmann, T. Kipp, Optical modes excited by evanescent-wave-coupled PbS nanocrystals in semiconductor microtube bottle resonators. Nano Lett. **10**, 627–631 (2010)
20. V.A.B. Quinones, G.S. Huang, J.D. Plumhof, S. Kiravittaya, A. Rastelli, Y.F. Mei, O.G. Schmidt, Optical resonance tuning and polarization of thin-walled tubular microcavities. Opt. Lett. **34**, 2345–2347 (2009)
21. C. Strelow, C.M. Schultz, H. Rehberg, H. Welsch, C. Heyn, D. Heitmann, T. Kipp, Three dimensionally confined optical modes in quantum-well microtube ring resonators. Phys. Rev. B (Condens. Matter. Mater. Phys.) **76**, 1–5 (2007)
22. C. Strelow, H. Rehberg, C.M. Schultz, H. Welsch, C. Heyn, D. Heitmann, T. Kipp, Optical microcavities formed by semiconductor microtubes using a bottlelike geometry. Phys. Rev. Lett. **101**, 127403 (2008)
23. F. Li, Z. Mi, S. Vicknesh, Coherent emission from ultrathin-walled spiral InGaAs/GaAs quantum dot microtubes. Opt. Lett. **34**, 2915–2917 (2009)
24. F. Li, Z.T. Mi, Optically pumped rolled-up InGaAs/GaAs quantum dot microtube lasers. Opt. Expr. **17**, 19933–19939 (2009)
25. C. Strelow, M. Sauer, S. Fehringer, T. Korn, C. Schuller, A. Stemmann, C. Heyn, D. Heitmann, T. Kipp, Time-resolved studies of a rolled-up semiconductor microtube laser. Appl. Phys. Lett. **95**, 22115 (2009)
26. C. Strelow, H. Rehberg, C.M. Schultz, H. Welsch, C. Heyn, D. Heitmann, T. Kipp, Spatial emission characteristics of a semiconductor microtube ring resonator. Phys. E **40**, 1836–1839 (2008)

Chapter 10
Carbon Nanotube Arrays: Synthesis, Properties, and Applications

Suman Neupane and Wenzhi Li

10.1 Introduction

Carbon nanotubes (CNTs) have become one of the most interesting allotropes of carbon since the discovery of multi-walled carbon nanotubes (MWNTs) by Iijima [1] in 1991. It took almost 2 more years until Iijima and Ichihashi [2] and Bethune et al. [3] synthesized simultaneously single-walled carbon nanotubes (SWNTs). Ever since, steady progress has been made to successfully synthesize vertically and horizontally aligned arrays of CNTs over a wide range of substrates by employing different techniques. CNTs have shown promising mechanical, electrical, optical, and thermal properties, rendering their applications in new structural and functional materials, electrical circuitry, energy storage, drug delivery, and many other devices of the future generation.

Several methods have been developed to synthesize CNTs with high purity and controllable diameter and length at desirable location over a wide variety of substrates. Among the available synthesis methods, arc discharge [1], laser ablation [4], chemical vapor deposition (CVD) [5], diffusion flame deposition [6], and electrochemical synthesis [7] are commonly used techniques for the synthesis of CNTs. A scalable device application requires the ability of control over the alignment of CNTs. In this chapter, we review the current state-of-the-art synthesis, properties, characterization, and applications of CNTs. In Section 10.2, the CNT synthesis techniques are discussed. Then Section 10.3 focuses on the fabrication mechanism of CNT arrays (CNTAs), Section 10.4 on mechanical properties, Section 10.5 on thermal properties, and Section 10.6 on electrical properties. In Section 10.7, we discuss some applications of CNTs, and finally, this review concludes with a summary.

W. Li (✉)
Department of Physics, Florida International University, Miami, FL 33199, USA
e-mail: liwenzhi@fiu.edu

W.L. Zhou, Z.L. Wang (eds.), *Three-Dimensional Nanoarchitectures*,
DOI 10.1007/978-1-4419-9822-4_10,

10.2 Carbon Nanotube Synthesis

10.2.1 Arc Discharge

MWNTs were synthesized originally by Iijima [1] using the arc discharge method (Fig. 10.1a). A high direct current (d.c.), typically of the order of 200 A, between two graphite electrodes at a potential difference of 20 V, was maintained inside a chamber filled with Ar gas at 100 torr resulting in MWNTs. These MWNTs had diameters between 4 and 30 nm and up to 1 μm in length with the separation of 0.34 nm between the graphite planes (Fig. 10.1b). Introduction of 10 torr methane and 40 torr Ar with Fe as catalyst was conducive for the growth of SWNTs [2] with diameters ranging between 0.7 and 1.6 nm and length as long as 700 nm (Fig. 10.1c). Large-scale synthesis of MWNTs with 75% conversion of graphite was achieved by Ebbesen and Ajayan [8] using helium gas at a pressure of 500 torr and electrical potential difference of 18 V between the electrodes for optimum results. Journet et al. [9] synthesized SWNTs on a large scale using a mixture of metallic catalysts in a He environment at 500 torr using arc discharge.

10.2.2 Laser Ablation

Guo et al. [4] pioneered the production of MWNTs by using Nd:YAG laser pulses over a graphite target heated to 1200°C inside a 50-cm-long, 2.5-cm-diameter quartz

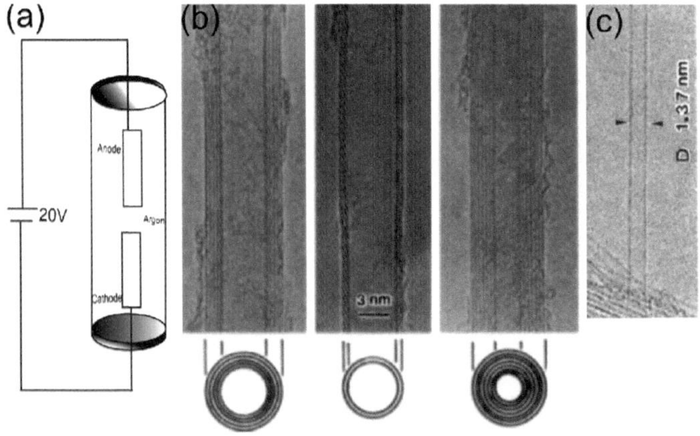

Fig. 10.1 (**a**) Schematic of arc discharge method. Two graphite electrodes are typically 1 mm apart inside a quartz tube maintained at an argon pressure of 100 torr. An electric discharge is produced by passing high current of the order of 100 A at 20 V. CNTs are collected at cathode. (**b**) Electron micrographs of CNTs having five, two, and seven walls with diameter 6.7, 5.5, and 6.5 nm, respectively. The CNT having seven layers has the smallest diameter [1]. (**c**) Electron micrograph of an SWNT showing the diameter of 1.37 nm produced by arc discharge [2] (reprinted with permission from Nature Publishing Group, Copyright 1991)

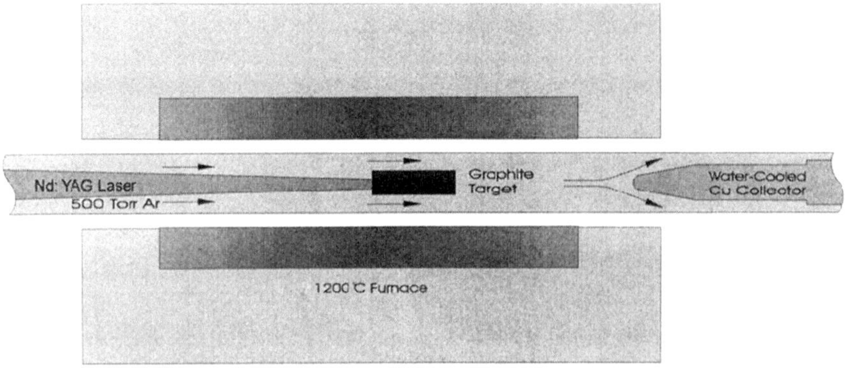

Fig. 10.2 Schematic of the oven laser vaporization chamber for the growth of CNTs. MWNTs are synthesized using Nd:YAG laser pulses over a graphite target heated to 1200°C inside a 50-cm-long, 2.5-cm-diameter quartz tube. The tube is maintained at the pressure of 500 torr by flowing Ar at a linear rate of 0.2–2 cm/s. Nanotubes are collected on the copper rod cooled by circulating water [4] (reprinted with permission from American Chemical Society, Copyright 1995)

tube (Fig. 10.2). The region inside the tube was maintained at 500 torr by flowing the Ar gas at a linear flow rate of 0.2–2 cm/s. Nanotubes are collected on the copper rod cooled by circulating water. The as-synthesized MWNTs consisted of 4–24 layers of graphite and were 300 nm long. The quality of the MWNTs declined as the oven temperature was reduced from 1200 to 900°C until no nanotubes were formed at 200°C. Thess et al. [10] optimized the process by adding transition metal catalysts to the graphite target to produce metallic SWNTs with yield greater than 70%. These uniform SWNTs self-organized into rope-like bundles of 5–20 nm in diameter and several micrometers in length. These bundles exhibit metallic transport property with resistivity less than 10^{-4} Ω cm at 300 K.

10.2.3 Electrochemical Synthesis

Matveev et al. [7] synthesized MWNTs from a C_2H_2 solution in liquid NH_3 below room temperature at 233 K without a metal catalyst. Laboratory-prepared pure dry C_2H_2 was mixed with liquid NH_3 formed by cooling gaseous ammonia to get 15–20 mol% C_2H_2 solution and poured in a glass vessel for electrolysis. A d.c. voltage of 150 V was applied for 5–10 h between n-type silicon (100) electrodes of dimensions 5 mm × 5 mm × 0.3 mm. After electrolysis, the immersed part of the cathode was covered by a light-gray porous layer with average thickness of 1–2 μm. The MWNTs consisted of 10–20 graphite layers and had an average diameter of 15 nm with a high aspect ratio greater than 1000. The atomic hydrogen generated on the cathode initiated chain radical reactions and liquid NH_3 promoted these reactions by stabilization of radicals; this process facilitates the growth of CNTs.

10.2.4 Diffusion Flame Synthesis

Vander Wal et al. [6] and Yuan et al. [11, 12] have demonstrated the synthesis of SWNTs via a less known simple laboratory-scale diffusion flame at temperatures between 1200° and 1500°C. A combined flow of CH_4 and air was used to ensure the production of stable, visible, laminar flame of height 65 mm at normal atmospheric pressure. CNTs were deposited on a stainless steel grid held in the flame and supported by a 0.4 mm wire of an alloy of Ni–Cr–Fe for 10–30 min. The spaghetti- and bamboo-shaped CNTs produced were between 20 and 60 nm in diameter.

10.2.5 Chemical Vapor Deposition

José-Yacamán et al. [5] used catalytic decomposition of carbon-containing gas over a metal surface to grow carbon filaments and CNTs at relatively lower temperatures than that in arc discharge and laser ablation methods. Fe catalyst uploaded on graphite substrate was obtained by impregnating the substrate in a 40 vol.% ethanol/60 vol.% water solution of iron(III) oxalate. The iron oxalate-impregnated graphite substrate was then reduced in the mixture of N_2 and H_2 at 350°C to convert the iron oxalate into metallic Fe catalyst particles and then CNTs were grown by the introduction of a mixture of N_2 and carbon source gas C_2H_2 at 700°C for several hours at standard atmospheric pressure. The as-grown CNTs measured 5–20 nm in diameter and 50 μm in length. The diameter distribution and length could be controlled by the variation of concentration of catalysts and the time of reaction for the synthesis of CNTs. Transition metals Ni [13], Co [14], and Fe [15] catalysts were used successfully to synthesize CNTs using carbon precursors like CH_4, C_2H_2, and C_2H_4. Li et al. [16] used a CVD technique to grow aligned CNTs perpendicular to a silica substrate in a large scale. Mesoporous silica-containing iron nanoparticles were prepared by a sol–gel process from tetraethoxysilane hydrolysis in an iron nitrate aqueous solution. C_2H_2, diluted by N_2, was used as the carbon precursor gas to grow CNTs on Fe catalyst nanoparticles formed after the reduction of the iron oxide nanoparticles. Scanning electron microscopy (SEM) images revealed the vertically aligned CNTs having diameters ~30 nm with spacing of ~100 nm (Fig. 10.3). The length of individual CNTs in the films is approximately 50 μm. The high-resolution TEM images show the presence of around 40 concentric shells of graphite in an individual CNT with a spacing of ~0.34 nm between the layers. The array consists of pure CNTs without catalyst particles or amorphous carbon.

Kong et al. [17] synthesized horizontally oriented SWNTs over a patterned Si substrate using CVD. A thin film of 0.25 μm polymethylmethacrylate (PMMA) was deposited on the Si substrate using spin coating at 4000 rpm. Square holes were fabricated on the PMMA film by an electron beam lithography. The exposed PMMA was removed by using organic solvents. Different solutions containing catalysts of Fe, Mo, or Al were deposited on the patterned PMMA substrate. Finally, the PMMA film was removed by heating and subsequently treating with 1,2-dichloroethane. The catalyst islands formed squares of 3 or 5 μm spaced at 10 μm on the Si substrate

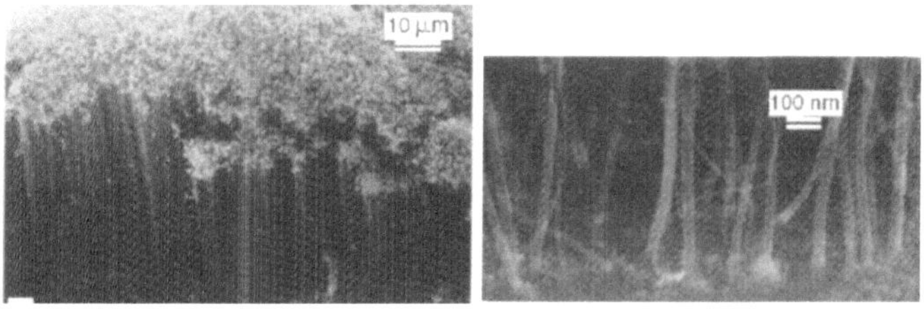

Fig. 10.3 SEM images of vertically aligned CNTs. *Left panel*: Low magnification showing CNT film of thickness 50 μm. *Right panel*: High magnification with CNTs of diameter ~30 nm and spacing ~100 nm [16] (reprinted with permission from AAAS, Copyright 1996)

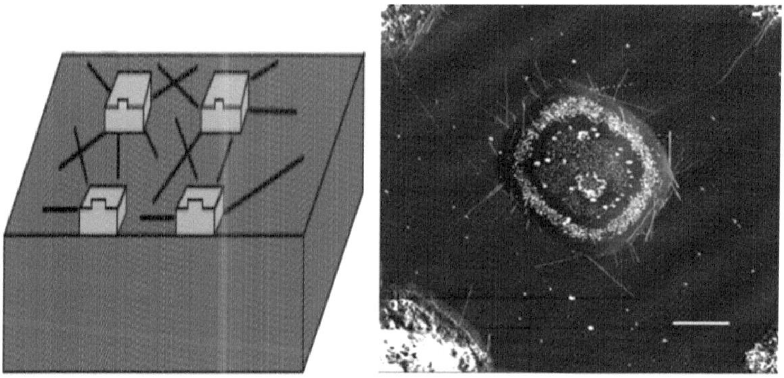

Fig. 10.4 *Left panel*: Schematic of fabrication of catalytic island and CVD growth of aligned CNTs. *Right panel*: Large-scale phase image recorded by tapping mode AFM showing CNTs grown from the patterned islands and bridging between islands. The scale bar is 2 μm [17] (reprinted with permission from Macmillan Publishers Ltd. Nature, Copyright 1998)

(Fig. 10.4). The SWNTs were synthesized using CH_4 as the carbon precursor gas at 1000°C. The as-synthesized SWNTs were 1–3 nm in diameter and ran several micrometers in length (Fig. 10.4).

10.3 Carbon Nanotube Arrays

The unique stability and structural, electrical, and mechanical properties render the possibility of using CNTs in a number of applications such as advanced scanning probes [18], nanoelectronic devices [8, 9], and electron field emission sources [19, 20]. However, for electronic applications, it is desirable to have high-quality CNTs in a controlled pattern in order to avoid post-growth treatments which generally give rise to defects and impurities. With this view point, CNT arrays (CNTAs) have been fabricated directly over silicon [21], quartz [22], steel [23], nickel [24], titanium [25], copper [26], platinum [27], sapphire [28], silicon carbide [29], and others. The process of production of CNTAs starts with pre-positioning the catalyst on the substrates. The control of CNT production has been achieved by the deposition of catalyst in a predetermined pattern using pulsed laser deposition (PLD) [30], anodic aluminum oxide (AAO) templates [21, 31], reverse micelle method [32], photolithography [33], electrochemical etching [34], sputtering [35, 36], nanosphere lithography [37], sol–gel method [38], and other methods. CNTA synthesis has been carried out by CVD and modified forms of CVD like d.c. plasma-enhanced chemical vapor deposition (PECVD) [22] and microwave PECVD [39], d.c. bias sputtering [40], electrophoretic deposition [41], screen printing [42], etc., have also been used to form well-aligned CNTAs. Horizontal and vertical alignments of CNTs have been successfully achieved using the aforementioned methods.

10.3.1 CNTA Synthesis Using Patterned Catalyst Arrays

The most efficient method of forming well-aligned CNTA is to deposit catalysts in a predetermined pattern to grow CNTs selectively. Commonly used methods for catalyst patterning are described below.

10.3.1.1 Pulsed Laser Deposition

Pulsed laser deposition (PLD) utilizes laser signals of predetermined pulse width of several tens of nanoseconds to strike on a rotating target material. Saurakhiya et al. [30] used a 248-nm KrF laser with a pulse width of 23 ns and a repetition rate of 10 Hz on a rotating iron target inside a vacuum chamber with base pressure of $\sim 10^{-6}$ torr. Square or hexagonal arrays of aligned CNTs were synthesized on the appropriately shaped catalyst patterns. The square or hexagonal pattern of catalyst on silicon and quartz substrates was obtained by utilizing TEM copper grids as masks during the PLD catalyst deposition. An aluminum sheet was used to hold and press the Cu grids to keep them as close as possible to the substrates. The thickness of the Fe film was controlled by varying the deposition time, laser power, and the distance between the target and the substrate. Vertically aligned CNTs were produced by PECVD using C_2H_2 and H_2 at 700°C.

10.3.1.2 Anodic Aluminum Oxide (AAO) Templates

Anodic aluminum oxide (AAO) templates with ordered nanohole arrays have been made by a two-step anodization of aluminum (Al) [21, 31]. During a typical double anodization process, a clean Al sheet is first anodized at 40 V in a 0.3 M oxalic acid for 5–6 h at room temperature. Then the disorderly anodic oxide layer formed during the process is removed in a mixture of phosphoric acid and chromic acid. The anodization process is repeated under the same conditions for 3–4 h which results in the formation of highly ordered porous AAO templates. The pores have diameters of ~50 nm and are ~100 nm apart.

10.3.1.3 Reverse Micelle Method

In a reverse micelle method, metal salts are reduced to metal nanoparticles in a nanoscale water pool inside a glove box filled with nitrogen gas to prevent oxidation.

Ago et al. [32] used didodecyldimethylammonium bromide (DDAB) as the cationic surfactant and sodium borohydride ($NaBH_4$) as a reducing agent to obtain catalyst cobalt particles from a cobalt chloride ($CoCl_2{\cdot}6H_2O$) solution. The surfactants help to stabilize the nanoparticles. DDAB was dissolved in toluene with a 10 wt% concentration, followed by dissolving $CoCl_2{\cdot}6H_2O$ to a concentration of 0.005 M. With the addition of 5 M $NaBH_4$ aqueous solution and continuous stirring, the solution turned from light blue to black due to the formation of colloidal dispersion of Co nanoparticles. The as-prepared Co nanoparticles were purified further by repeating centrifugation and redispersion in toluene and acetone. The colloidal dispersion of the Co nanoparticles was finally cast on Si substrate and dried at room temperature. The TEM image showed that the Co nanoparticles with average diameter of 4 nm were well separated due to the presence of surfactants that covered the surface of the nanoparticles. The size of the nanoparticles could be altered by tuning the concentration of $NaBH_4$ solution. Vertically aligned CNTs were synthesized using CVD.

10.3.1.4 Photolithography

Photolithography is the mechanism of transferring a geometrical pattern onto a substrate using light. The photosensitive material called resist will form desired patterns upon light exposure through photomasks. Catalysts are preferentially deposited by a suitable approach and the resist is generally removed by chemical etching (Fig. 10.5). The process of photolithography starts with the application of a thin and uniform layer of photoresist to the surface of the wafer by spin coating at the speed of 1000–5000 rpm for a period of 30–60 s. The photoresist-coated wafer is then prebaked at a temperature of 100°C to remove excess photoresist solvent for 30–60 s. The prebaking is followed by exposure of the photoresist to ultraviolet (UV) light through a suitably designed mask. A positive resist becomes soluble to the developer solution upon exposure to the UV while a negative resist becomes insoluble upon

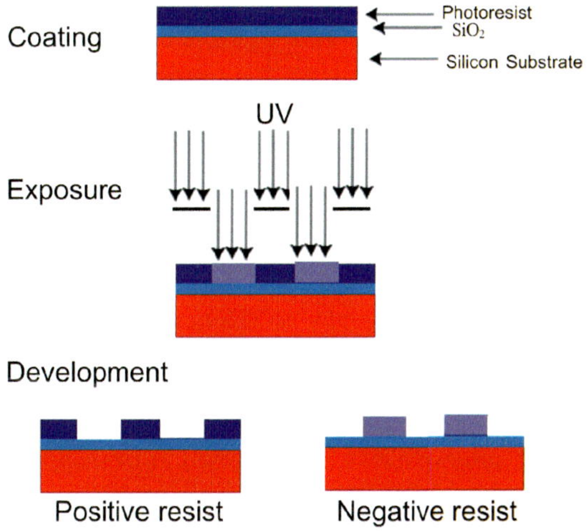

Fig. 10.5 Schematic of the process of photolithography. A light-sensitive material called photoresist is first coated on a silicon substrate. The spin coating is followed by exposure to UV light through a mask. The exposed area of a positive resist material dissolves in the developer solution, while the unexposed region dissolves in the case of negative resist, leaving behind the predetermined patterns of the mask

similar UV treatment. The UV-exposed substrate is immersed in a developer solution to remove the photoresist and obtain the desired pattern. The resulting wafer is then hard baked at around 200°C to solidify the photoresist. A suitable layer of the catalyst is then deposited by e-beam evaporation, sputtering, or other methods. Finally, the photoresist is removed by wet etching in chemical solutions, oxygen plasma etching, or other methods. Wei et al. [33] synthesized well-aligned CNTs by CVD on patterned catalyst nanoparticles formed by using photolithography. The density of the CNTAs can be controlled by the pattern geometry on the masks used in the photolithography process.

10.3.1.5 Electrochemical Etching

Xu et al. [34] have synthesized CNTA using electrochemical etching and by selectively depositing catalysts in a predefined pattern. The micro-, meso-, and macro-porous Si substrates were produced in an electrochemical etching cell containing an aqueous HF solution. Platinum wire was used as a cathode and the crystalline Si wafer acted as an anode. Thin Al films were evaporated on the back of the wafers to ensure good ohmic contact before anodization. The electrolyte for the anodization was made using 48% HF and ethanol in a ratio of 1:1. A current density of 1–80 mA/cm^2 was maintained in darkness for 1–10 min depending on the desired thickness and porosity. Nickel catalyst was deposited on the pores by immersing the substrates in a nickel acetate solution for 24 h. Then the CVD method was adopted to synthesize CNTA by using H_2 and Ar as reducing gases and C_2H_4 as the carbon precursor at 880°C on the patterned Ni nanodots on the Si substrate.

10.3.1.6 Sputtering

CNTAs have been synthesized using thin films of catalysts deposited by ion beam sputtering [36], d.c. magnetron sputtering [40], r.f. magnetron sputtering [43], and reactive sputtering [44]. The target material, which is usually a circular disc, is placed inside a high vacuum chamber at a certain distance from a substrate. Sputtering is usually carried out in an inert Ar atmosphere in the presence of a d.c. power supply or an r.f. generator. Wang et al. [35] have used d.c. magnetron sputtering to form 20-nm thin films of Ni on Si and then have synthesized well-aligned CNTA using PECVD using the Ni thin film as catalyst. C_2H_2 was used as the carbon precursor gas to produce vertically aligned CNTs around 1 μm long in less than 5 min at ~550–600°C.

10.3.1.7 Nanosphere Lithography

Nanosphere lithography (NSL) is another powerful tool to form patterns of catalysts for well-controlled growth of CNTAs. The process starts with making Si wafers hydrophilic by treating them with an RCA solution which is a mixture of NH_4OH, H_2O_2, and water in the volume ratio of 1:1:5. Then sub-micrometer-sized polystyrene spheres (PS) dispersed in methanol is spin coated onto Si substrate to obtain a monolayer of PS. The catalyst layer is deposited by e-beam evaporation or other methods followed by the removal of the PS layers through wet etching. The hexagonal patterns of catalyst obtained can be used to grow CNTAs by CVD and PECVD methods [37, 45, 46].

10.3.1.8 Sol–Gel Method

The sol–gel method has been employed to form mesoporous silica-containing Fe catalysts to grow CNTAs by CVD [16, 38]. Li et al. [16] formed a mixture of tetraethoxysilane (TEOS) in ethyl alcohol and aqueous solution of iron nitrate by magnetic stirring for ~30 min. A few drops of concentrated hydrogen fluoride were then added to form a gel. The gel was dried for 1 week at 60°C to remove the excess water and other solvents, followed by calcination for 10 h at 450°C at 10^{-2} torr. A uniform porous silica network was obtained with iron oxide nanoparticles embedded in the pores. The iron oxide nanoparticles were then reduced to Fe nanoparticles at 550°C in the continuous flow of N_2 and H_2 for 5 h. Acetylene (C_2H_2) was used as the carbon source for the CVD process of CNT synthesis to obtain vertically aligned arrays of CNTs. The well-graphitized tubes as long as 50 μm were synthesized with the spacing of 100 nm between the tubes (Fig. 10.3).

10.3.2 CNTA Synthesis by Other Methods

Joselevich and Lieber [47] have investigated the application of an electric field to produce aligned SWNTs. The SWNTs align in the direction of an electric field and perpendicular to the direction of gas flow by minimizing the van der Waals interactions with the nearby surfaces. The electric field between the electrodes will induce a dipole in each growing SWNT. The electric field then exerts a torque on those induced dipoles forcing CNTs to grow parallel to the direction of the electric field. Zhang et al. [48] have used d.c. (0–200 V) or a.c. (30 MHz, 10 V peak to peak) voltage between catalyst islands to produce aligned SWNTs on quartz by CVD. The optimum electric fields for the directed growth of suspended SWNTs were in the range of 0.5 V/μm. The absence of an electric field resulted in random CNTs.

Lee et al. [49] demonstrated the lateral alignment of CNTs as a result of applying magnetic field during the process of catalyst dispersion on silicon substrates. The growth direction of CNTs was found to be perpendicular to the direction of applied magnetic field. Kumar et al. [50] used a bacterium *Magnetospirillum magnetotacticum*, which synthesizes intracellular, linear, single-domain magnetic nanoparticles through highly regulated biomineralization, to produce highly orientated MWNTs on silicon oxide substrates by CVD. A magnetic bar of strength 17 mT was used to investigate the effects of magnetic field on CNT growth. The average diameter of MWNTs was 13 ± 3.6 nm and the samples grown on the magnetotactic bacteria show the preferential direction of growth along the magnetic field. This suggests the possibility of synthesis of dimensionally controlled and spatially oriented CNTs by exploiting various magnetotactic bacteria as catalyst carriers with an applied magnetic field.

10.3.3 Horizontal Arrays of CNTs

Horizontal arrays of CNTs have also shown a lot of potential for device fabrication and have been successfully grown by different groups. Huang et al. [51–53] have grown millimeter-long and horizontally aligned SWNTs on silicon substrates by the fast heating CVD process (Fig. 10.6a). A slow heating process resulted in random alignment of CNTs as compared to good horizontal alignment obtained by rapid heating. A two-dimensional network of SWNTs has been directly grown on the substrates by a two-step growth process (Fig. 10.6b) by altering the direction of flow of gases in subsequent stages. The well-defined crossed network structure of SWNTs on a large scale enables the fabrication of multiterminal devices and complex circuits necessary for various applications.

10.4 Mechanical Properties

CNTs exhibit tremendous strength as a consequence of carbon–carbon bonding which is considered to be the strongest bonding in nature. This property along with the low density and fibril shape leads to the exploration of the viability of CNTs as a reinforcement material in composites. Theoretical calculations by Sanchez-Portal et al. [54] predicted the exceptionally high Young's modulus of SWNTs and MWNTs. Young's modulus depends upon the radius of the CNTs considered. Treacy et al. [55] measured the Young's modulus of individual CNTs by measuring the amplitude of their intrinsic thermal vibrations inside a transmission electron microscope. The measurement of 11 CNTs ranging from 0.66 to 5.81 μm in length yielded an average Young's modulus of 1.8 TPa with higher modulus for thinner CNTs.

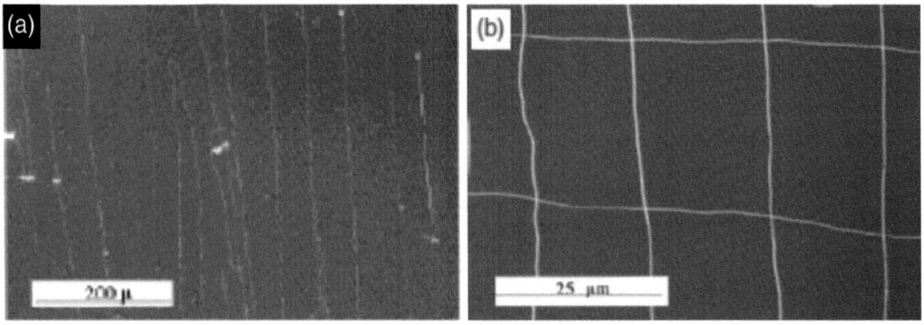

Fig. 10.6 SEM images of (**a**) horizontally aligned SWNTs prepared by fast heating process and (**b**) networks formed by horizontal CNTs as a result of two-step growth procedure [51, 52] (reprinted with permission from American Chemical Society, Copyright 2003 and Wiley-VCH Verlag GmbH & Co., Copyright 2003)

Wong et al. [56] employed an atomic force microscope (AFM) to measure the mechanical properties of the CNTs and found the Young's modulus of MWNTs was about 1.3 TPa. Gao et al. [57] and Hernandez et al. [58] estimated theoretically that the Young's modulus of individual SWNTs was in the range of ~0.4–0.6 TPa. Experimentally, Yu et al. [59] obtained an average value of 1 TPa and Zhu et al. [60] obtained an average value of ~100 GPa for strands of SWNTs of diameter ~5–20 μm. These values are dependent on the crystallinity of the materials and the number of defects introduced during the process of synthesis and measurement. Arrays of CNTs exhibit a high adhesive force of the magnitude of 100 N/cm^2 and much stronger shear adhesion force than normal adhesion force [61]. Based on the theoretical and experimental results, CNTs are found to exhibit large Young's modulus of elasticity, making them a viable supplement for rigid materials.

10.5 Thermal Properties

The specific heat capacities and thermal conductivities of carbon nanotube are due to contributions of phonons. The behavior of phonons at different temperatures completely describes the thermal transport properties of CNTs. The specific heat capacity of SWNTs exhibits the linear dependence on temperature from 300 to 1 K [62, 63] while maintaining $\sim T^{0.62}$ dependence below 1 K [64]. The linear temperature dependence is due to the linear k-vector dependence of the frequency of the longitudinal and twist acoustic phonons. The specific behavior of the specific heat below 1 K can be attributed to the transverse acoustic phonons with quadratic k dependence [65]. The results indicate that inter-wall coupling in MWNTs is rather weak compared with its parent form, graphite, so that one can treat an MWNT as a few decoupled, two-dimensional, single-walled tubules [63].

The thermal conductivity of MWNTs is roughly linear above ~120 K and becomes quadratic at lower temperatures, Fig. 10.7 [63]. Kim et al. [66] measured thermal conductivity using a suspended microdevice (Fig. 10.8). The observed thermal conductivity is more than 3000 W/K m at room temperature, which is two orders of magnitude higher than the estimation from macroscopic mat samples [63]. The temperature dependence of the thermal conductivity of nanotubes exhibits a peak at

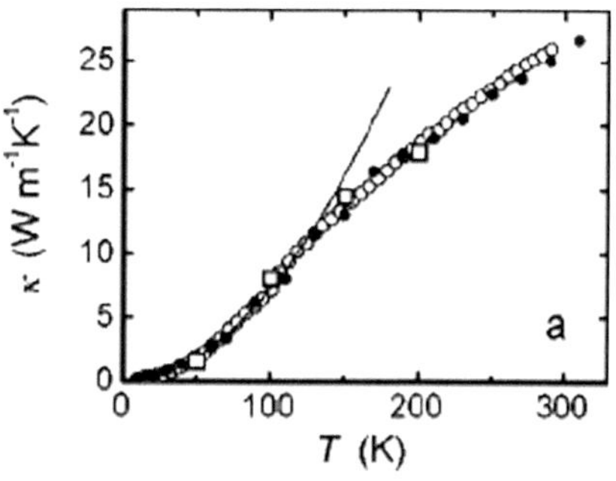

Fig. 10.7 The temperate dependence of the thermal conductivity of MWNT samples. It is almost quadratic ($k \propto T^{1.98 \pm 0.03}$) at temperatures below 120 K and roughly linear above 120 K [63] (reprinted with permission from Lu et al., Copyright 1999: American Chemical Society)

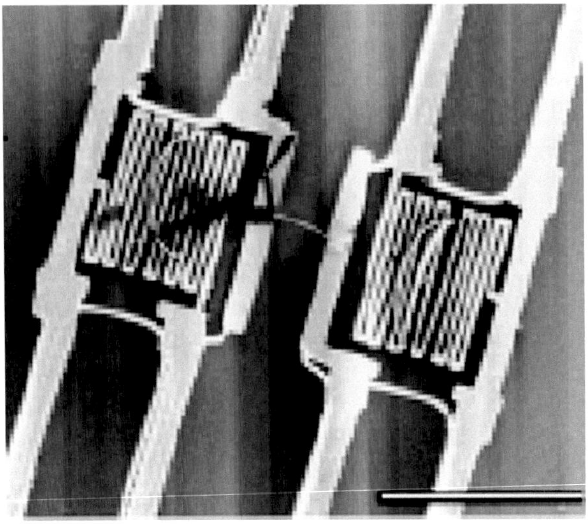

Fig. 10.8 SEM image of a microfabricated device. The islands of two silicon nitride membranes are suspended on silicon nitride beams. A platinum thin film resistor serves as a heater on each of the islands. A small bundle of CNTs form a bridge between the islands to form a thermal contact. The scale bar is 10 μm [66] (reprinted with permission from McEuen et al., Copyright 1999: The American Physical Society)

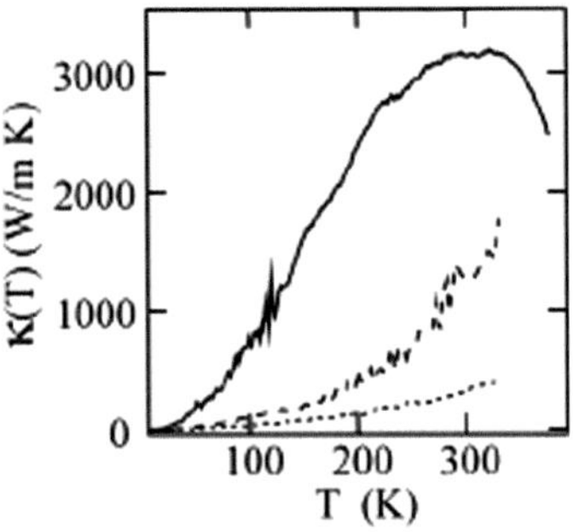

Fig. 10.9 The temperate dependence of the thermal conductivity of MWNT samples [66]. It is almost quadratic ($k \propto T^{1.98 \pm 0.03}$) at temperatures below 120 K and roughly linear above 120 K [63]. *Solid lines* represent κ (T) of an individual MWNT of diameter 14 nm. *Broken* and *dotted lines* are for bundles of diameters 80 and 200 nm, respectively (reprinted with permission from McEuen et al., Copyright 1999: The American Physical Society)

320 K due to the onset of Umklapp phonon scattering (Fig. 10.9). Berber et al. [67] theoretically determined an unusually high value of 6600 W/K m for an isolated (10, 10) nanotube at room temperature, comparable to the thermal conductivity of a hypothetical isolated graphene monolayer or diamond by combining equilibrium and non-equilibrium molecular dynamics simulations. These high values are associated with the large phonon mean free paths in the CNTs, graphene monolayer, and diamond systems, while substantially lower values are predicted and observed for the basal plane of bulk graphite. The numerical data indicate that in the presence of interlayer coupling, the thermal conductivity of the CNTs is reduced significantly to fall into the experimentally observed value range.

Shaikh et al. [68] have demonstrated the high thermal conductivity (8.3082 W/m K) of CNT film consisting of vertically aligned CNT arrays prepared using CVD on a glass substrate in comparison to using CNT composite film (1.2 W/m K) by Huang et al. [69]. This proves the superiority of CNTAs over films for thermal interface material. Owing to the superior thermal conductivity, Xu et al. [70] have demonstrated the possibility of using CNTAs for integrated circuit cooling.

10.6 Electrical Properties

CNTs are perfect one-dimensional conductors and exhibit interesting phenomena such as single-electron charging, resonant tunneling through discrete energy levels, and proximity-induced superconductivity. Langer et al. [71] reported on the electrical resistance of an MWNT bundle from room temperature down to 0.3 K under magnetic fields of up to 14 T. The nanotubes exhibited semi-metallic behavior analogous to rolled graphene sheets with a similar band structure. A magnetic field applied perpendicular to the sample axis decreases the resistance. Langer et al. [72] later reported on the electrical resistance measurements of an individual CNT down to a temperature of 20 mK. The conductance exhibits logarithmic temperature dependence and saturates at low temperatures. A magnetic field applied perpendicular to the tube axis increases the conductance and produces aperiodic fluctuations.

Bockrath et al. [73] measured the electrical properties of bundles of SWNTs. A gap due to suppressed conductance at low bias is observed in the current–voltage curves at low temperatures. Further, several prominent peaks are observed in the conductance as a function of a gate voltage which can be explained considering Coulomb blockade transport in quantum wires and dots considering CNTs as extended quantum dots. Theoretical calculations predict the metallic conductivity of individual SWNTs [74].

Tans et al. [75] measured the electrical characteristics of individual SWNTs at different gate voltages. Figure 10.10 shows the *I–V* curves at gate voltage of 88.2 mV for trace A, 104.1 mV for trace B, and 120.0 mV for trace C. The inset shows similar *I–V* curves with gate voltage ranging from 50 (bottom) to 136 mV (top). The *I–V* curves showed a clear gap around zero bias voltage. For higher voltages, the current increases in steps. The gaps were suppressed for certain gate voltages and have the maximum value corresponding to zero bias voltage. This variation of the gap with gate voltage around zero bias voltage implies Coulomb charging of the tube. Zhu et al. [76] used a lift-off process to pattern catalysts and synthesize vertical CNT arrays using CVD. Two-probe electrical measurements of the CNT arrays indicate a resistivity of 0.01 Ω cm compared to 8×10^{-4} and 12×10^{-4} Ω cm of individual SWNTs, and the capacitance of the nanotube bundle was ~2.55 pF as the voltage was scanned from –1 to 1 V, suggesting possible use of these CNTs as interconnect materials [76].

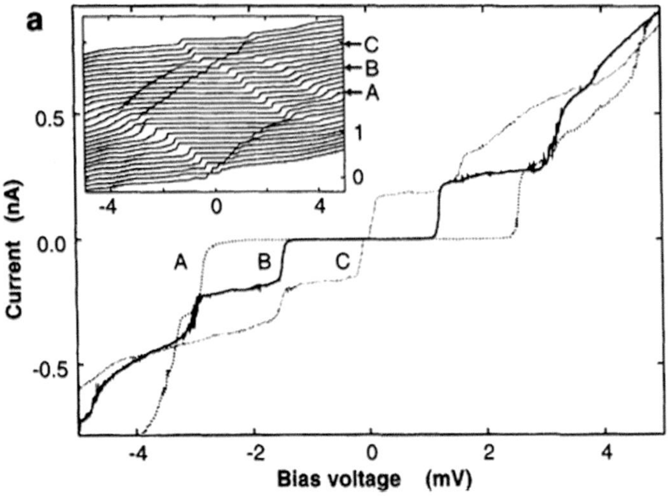

Fig. 10.10 *I–V* characteristic of the nanotube at a gate voltage of 88.2 mV (trace A), 104.1 mV (trace B), and 120.0 mV (trace C). *Inset* shows more *I–V* curves with gate voltage ranging from 50 (*bottom*) to 136 mV (*top*) [75] (reprinted with permission from Cees Dekker, Copyright 1997: Nature)

CNTs are a good source of electrons through the process of field emission. The experimental setup to measure the field emission property is shown in Fig. 10.11. CNTs are used as cathodes and a high electric field of the order of ~2 V/μm is created between a metal anode and the CNT cathode to measure electron emission in a low vacuum of about $\sim 10^{-7}$ torr. de Heer et al. [77] determined the emission characteristics of films of oriented nanotubes [78] and as-grown CNTAs [20, 34, 79]. Zhang et al. [80] demonstrated well agreement between the experimental

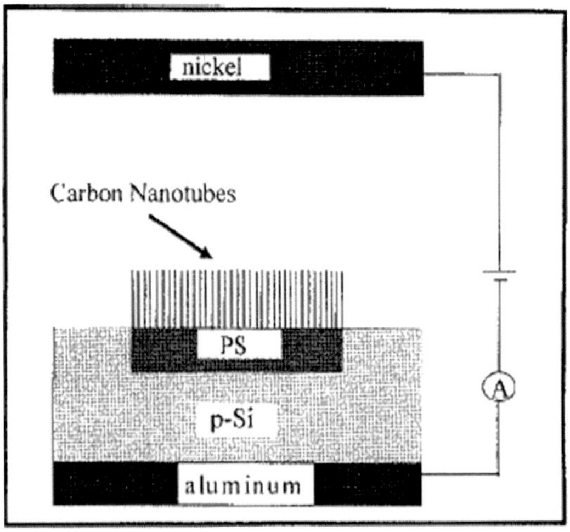

Fig. 10.11 Experimental setup of measurement of field emission property. CNTs are used as cathodes [34]. A high electric field of the order of ~2 V/μm is created between a metal anode and a CNT cathode to measure electron emission in low vacuum of $\sim 10^{-7}$ torr (reprinted with permission from American Institute of Physics, Copyright 1999)

and theoretical values of emission behavior of CNTAs. The total emission current depends upon the radius (*r*), the height (*h*), and the spacing (*d*) of the CNTs. The optimum space between two neighboring CNTs is about 75*r* and the height should be larger than 2.6*d* to obtain a large average current density. Hazra et al. [81] have observed the dramatic enhancement in the emission current density by a factor >106 with the onset field as low as 0.16 V/μm by using the plasma-sharpened tips of nanotubes containing only a few tubes at the apex of the structure (Fig. 10.12). Saturation in the emission current density is proposed due to the significant change in the tunneling barrier for a nanosized tip in a very high local electric field. Saurakhiya et al. [30] observed similar decrease in the threshold voltage of the as-grown CNTAs by laser pruning. In laser pruning process, the as-grown CNTs were irradiated by a He–Ne laser light with a wavelength of 632.8 nm and a power of 30 mW. Laser pruning resulted in the decrease of length of the CNTs by around 2 μm and better alignment of the CNTs. Zhao et al. [82] demonstrated that arrays of CNTs with large wall numbers exhibited lower threshold voltages. To achieve a lower threshold voltage, an array of small diameter nanotubes with large intertube spacing (two times the height) would be ideal. However, this situation was not easily achievable, as nanotube diameter and intertube spacing were in competition. In this case, the intertube spacing appeared dominant because the threshold voltage decreased despite increasing diameter. This means that intertube screening effects, which reduce the local electric field, are more dominant than the diameter on the resulting threshold voltage. Suh et al. [31] studied the field-screening effect of highly ordered CNTAs and concluded that the field emission was optimal when the tube height was similar to the intertube distance in agreement with the predictions by Nilsson et al. [83]. Charlier et al. [84] demonstrated that boron-doped CNTs exhibited better field emission with lower threshold voltage than did pristine CNTs. In general, the straight line observed in the Fowler–Nordheim plot is the evidence of field emission (Fig. 10.13).

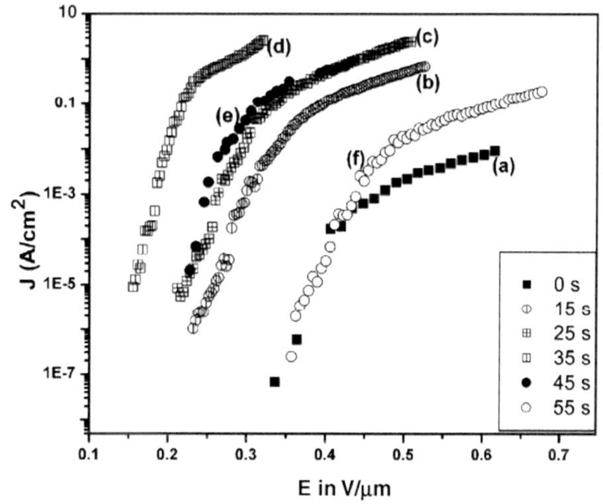

Fig. 10.12 Typical field emission curves showing current density versus the electric field [81]. Emission current for CNTs with tips pruned for different times shows different values at different electric fields. There is no electron emission below a threshold voltage (reprinted with permission from Misra et al., Copyright 2009: American Chemical Society)

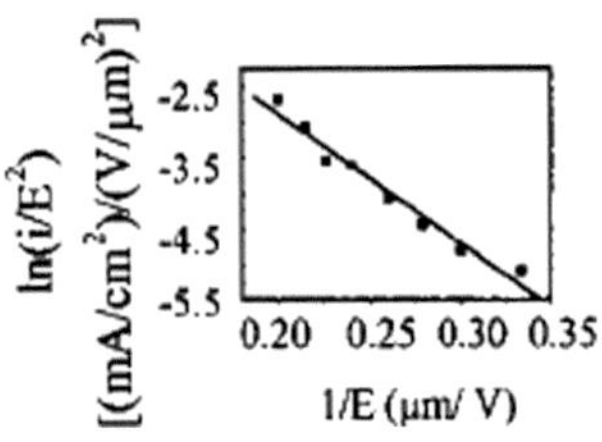

Fig. 10.13 The straight line graph of logarithm of (i/E^2) versus $1/E$, popularly called as Fowler–Nordheim (F–N) plot, is the evidence of field emission phenomenon [34] (reprinted with permission from American Institute of Physics, Copyright 1999)

10.7 Applications of CNTs and CNTAs

10.7.1 Hydrogen Storage

CNTs have high potential for hydrogen storage as the gas can effectively condense to a high density inside narrow SWNTs. Dillon et al. [85] compared the hydrogen storage capacity of carbon soot containing only about a 0.1–0.2 wt% of SWNTs to that of the activated carbon (AC) at 133 K. The SWNTs with diameter of 1.2 nm were synthesized in an electric arc discharge process. The adsorption of hydrogen on the SWNTs was probed with temperature-programmed desorption (TPD) spectroscopy in an ultrahigh vacuum chamber inside a liquid nitrogen cryostat. H_2 desorbs from SWNTs and activated carbons within the same temperature range but with different intensities. The signal from SWNTs is ~10 times greater than the signal from AC with the gravimetric storage density in SWNT ranging from ~5 to 10 wt% (Fig. 10.14).

Liu et al. [86] studied the hydrogen storage of SWNTs with mean diameter 1.85 nm and purity in the range of 50–60 wt% at room temperature for three types of samples. Sample 1 was used as-synthesized; sample 2 was soaked in 37% HCl acid for 48 h to partly eliminate the residual catalysts, rinsed with deionized water, and dried at 423 K; sample 3 was heated in vacuum at 773 K for 2 h after receiving the same treatment as sample 2. The heat treatment is to evaporate the organic compounds and functional groups formed in SWNTs during the synthesis procedure. A hydrogen storage capacity of 4.2 wt% was achieved at the pressure of 10 MPa. Furthermore, about 78.3% of the adsorbed hydrogen (3.3 wt%) could be released under ambient pressure at room temperature. Ye et al. [87] demonstrated the adsorption of H_2 exceeding 8 wt% on highly purified crystalline ropes of SWNTs at temperature of 80 K and pressure of ~100 Pa.

Zhu et al. [88] measured the H_2 absorption of well-aligned and randomly ordered MWNTs produced by catalytic pyrolysis on quartz substrate at 290 K and pressure between 3 and 10 MPa. They observed that the bundles of aligned MWNTs are better suitable for hydrogen adsorption as compared to randomly ordered MWNTs under similar conditions. This higher H_2 absorption capacity of 3.4 wt% of CNTA

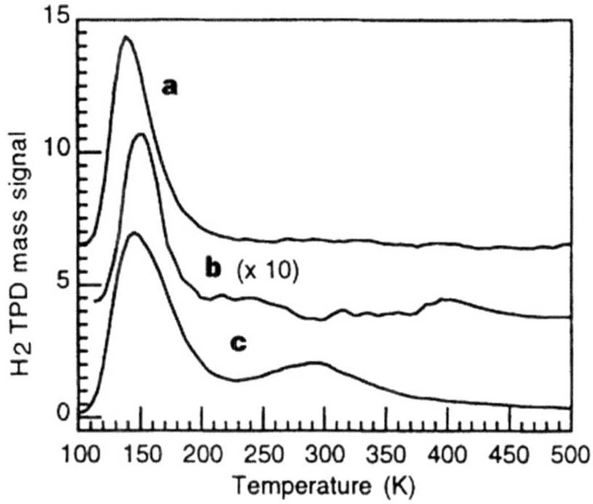

Fig. 10.14 Temperature-programmed desorption (TPD) spectra of hydrogen desorption [85]. (**a**) TPD spectrum of as-produced SWNT sample after standard hydrogen exposure. (**b**) TPD spectrum of activated carbon sample, magnified 10 times, after standard hydrogen exposure. (**c**) TPD spectrum of SWNT sample after heating in vacuum to 970 K and standard hydrogen exposure (reprinted with permission from Nature Publishing Group, Copyright 1997)

as compared to 0.5 wt% of the random MWNTs can be attributed to the strong interaction between the hydrogen molecules in the interstitial channels between the CNTs and the inter-layers of some cap-opened CNTs. Wang and Johnson [89] have performed classical grand canonical Monte Carlo simulations to calculate the absorption of hydrogen in tube arrays at 77 and 298 K. The tube lattice spacing has been varied to study the optimum hydrogen uptake using triangular and square lattices. The strength of the solid–fluid interaction potential has been increased in order to identify a combination of potential and geometry that will meet the Department of Energy (DOE) targets of 6.5 wt% for hydrogen storage for fuel cell vehicles. The DOE target values could not be reached even by tripling the fluid wall potential at ambient temperatures. However, it was possible to achieve the DOE targets at a temperature of 77 K if the strength of the interaction potential was increased by about a factor of 2 and the lattice spacing of the tubes was optimized.

Misra et al. [90] used electrically conducting surfaces of CNTAs as cathodes for H_2 generation and absorption by electrolyzing water. Figure 10.15 shows the experimental setup used for the electrolytic measurements on the CNTAs. An electrochemical cell was assembled by inserting a metal tip connected to a power supply with the deionized water bubble on top of CNTA acting as the cathode. d.c. measurements were performed using a Cascade M150 probe station, attached to a Keithley-2635 source inside a vacuum chamber. An application of external voltage (−10 V) between the electrodes resulted in collection of hydrogen gas near the surface of the CNTA due to the electrochemical deposition of water. The amount of H_2 measured ($2.2 \pm 0.35 \times 10^{-5}$ g) is less than the theoretical amount of hydrogen

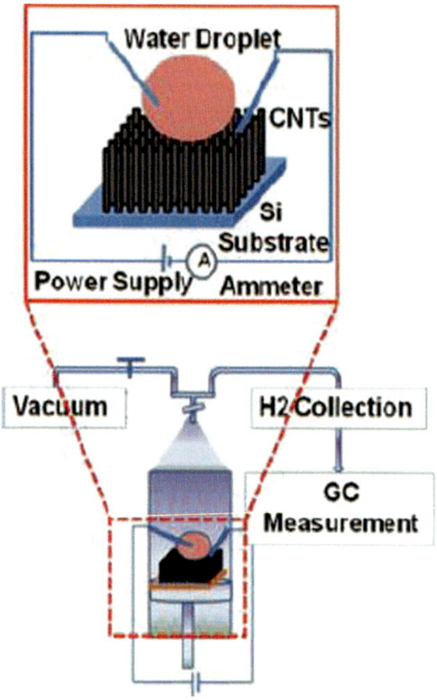

Fig. 10.15 Schematic diagram of the experimental setup used for measuring the electrolytic reaction measurements on the CNT forests. The top section is the enlarged view for clarity of the sample area with the details on the probe's positioning [90] (reprinted with permission from American Chemical Society, Copyright 2009)

generated (2.6×10^{-4} g) from flowing current over a period of 1 h, supporting the possibility of use of aligned CNTs as the H_2 storage materials.

10.7.2 CNTs as Sensors

CNTs can be used as chemical or biological sensors by exploiting their variation in optical, electrical, and electrochemical properties. For example, upon exposure to the gaseous molecules like NH_3, NO_2, and H_2O_2 or biological species such as enzymes, CNTs exhibit a dramatic increase or decrease in resistance. Sensitivity and recovery time are two key components for the sustainable use of CNTs in the detection of foreign elements.

Kong et al. [91] studied the electric response of semiconducting SWNTs before and after the introduction of NH_3 and NO_2. The I–V_g curve shifted by −4 or +4 V when NH_3 or NO_2 was introduced into the chamber, respectively. These shifts can be explained by the depletion or the enhancement of hole carriers brought about by the introduction of the respective gases [91]. Qi et al. [92] used arrays of electrical devices each comprised of multiple SWNT sensors with 100% yield for detecting gas molecules. Polymer functionalization was used to impart high sensitivity and selectivity to the sensors to fabricate n-type nanotube devices capable of detecting NO_2 at less than 1 ppb (parts per billion) concentrations while being insensitive to NH_3. CNTAs have been effectively used to detect glucose [27, 93], H_2O_2 [93], DNA

[94], protein [95], and others by tracking their electrical or optical response before and after the introduction of particular species.

10.7.3 CNTs for Battery and Supercapacitor Applications

CNTs can find their applications as electrode materials for highly efficient batteries due to their high electrochemical stability, large surface area, and unique electrical and electronic properties. SWNTs exhibit higher capacity of lithium (Li) intercalation than graphite and disordered carbon. Theoretical calculations [96] predicted the possibility of almost complete charge transfer between Li and SWNTs with relatively small deformation in the structure. Both the interstitial sites and the inner side of the tubes are energetically favorable sites for Li intercalation. Theoretical calculations predict the possibility of one Li atom intercalation for every two carbon atoms. Cyclic voltammograms [97] confirm that the reversible intercalation of Li^+ and presence of Fe, Pt, or Ru nanoparticles within the tube will double the intercalation capacity.

The cyclic efficiency of graphite as a function of added weight percent of CNTs was studied by Endo et al. [98]. The efficiency of graphite anodes increased continuously until the composition of 10 wt% CNTs which resulted in an efficiency of almost 100% up to 50 cycles. Wu et al. [99] demonstrated high Li ion storage capacity at 700 mAh/g by CNTs. Gao et al. [100] improved the storage capacity from 400 mAh/g for as-prepared SWNTs to 700 mAh/g after removing impurities and 1000 mAh/g by ball milling the SWNTs.

The super-capacitance property of CNTs has also been extensively studied because they are able to store and deliver energy rapidly and efficiently for a long life cycle via a simple charge separation process. Ma et al. [101] were able to construct electrochemical capacitors based on CNT electrodes with specific capacitances of about 25 F/cm^3 with 38 wt% sulfuric acid as the electrolyte. An et al. [102] reported a maximum specific capacitance of 180 F/g and a measured power density of 20 kW/kg at energy densities in the range of 7–6.5 Wh/kg by using SWNTs as electrode material in supercapacitors. Similar measurements also reported the specific capacity of 102 F/g for the electrodes using MWNTs [103].

10.7.4 CNTs for Photovoltaic Device

Ago et al. [104] fabricated a photovoltaic device (Fig. 10.16) using MWNTs as an electrode to collect holes and obtain an efficiency double that of the standard device with an indium–tin oxide (ITO) electrode. The visible light is shown through a semi-transparent Al electrode and made to pass through polyphenylene vinylene (PPV) of thickness 210 nm and MWNT film of thickness 140 nm in the photovoltaic device. The *I*–*V* characteristics (Fig. 10.17) expressed a clear diode rectification. Upon illumination of the device by light of wavelength 485 nm and

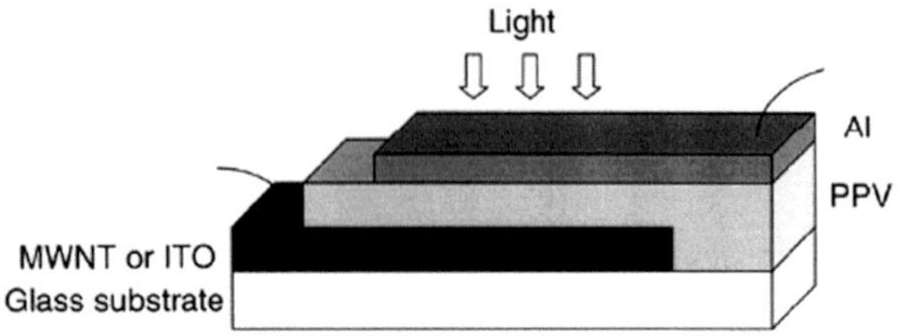

Fig. 10.16 Schematic of a photovoltaic device. Visible light is shown through semi-transparent Al electrode and made to pass through polyphenylene vinylene (PPV) of thickness 210 nm and MWNT film of thickness 140 nm [104] (reprinted with permission from Wiley-VCH Verlag GmbH & Co., Copyright 1999)

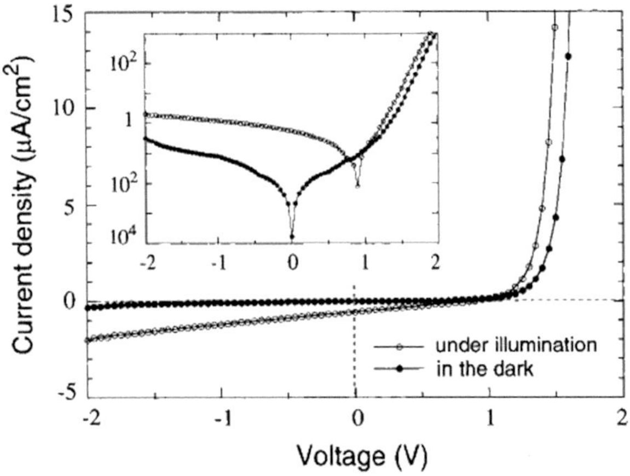

Fig. 10.17 *I–V* characteristics of an MWNT/PPV/Al device in the dark (*closed circles*) and under illumination at a wavelength of 485 nm and intensity of 37 μW/cm^2 (*open circles*). *Inset* is the representation of the same data with a logarithmic current axis [104] (reprinted with permission from Wiley-VCH Verlag GmbH & Co., Copyright 1999)

intensity 37 μW/cm^2, a photocurrent was observed with open-circuit voltage of 0.90 V and short-circuit current of 0.56 μA/cm^2. The external quantum efficiency of MWNT/PPA/Al was typically 1.5–2 times greater than the standard ITO device. The higher efficiency could be attributed to the complex interpenetrating network of PPV chains with the MWNT film and the relatively high work function of the MWNT film. Lagemaat et al. [105] have also reported similar successful replacement of In_2O_3:Sn by CNTs in an organic solar cell.

10.8 Conclusions

In this chapter, the synthesis techniques and the mechanical, electrical, and thermal properties of the CNTs, especially the CNTAs, have been reviewed. We have also discussed some promising applications of CNTs and CNTAs in hydrogen storage, biological and chemical sensors, lithium batteries, and photovoltaic devices. To explore the applications of CNTAs, CNT arrays with controllable nanotube diameter, density, spacing, and degree of defects are extremely important. In addition, for wide applications of the CNTAs, methods for synthesizing aligned CNTs at the conditions compatible with current fabrication techniques of nanodevices need to be developed.

References

1. S. Iijima, Helical microtubules of graphitic carbon. Nature **354**(6348), 56–58 7 Nov 1991
2. S. Iijima, T. Ichihashi, Single-shell carbon nanotubes of 1-nm diameter. Nature **363**(6430), 603–605 17 Jun 1993
3. D.S. Bethune, C.H. Kiang, M.S. Devries et al., Cobalt-catalyzed growth of carbon nanotubes with single-atomic-layer walls. Nature **363**(6430), 605–607 17 Jun 1993
4. T. Guo, P. Nikolaev, A.G. Rinzler et al., Self-assembly of tubular fullerenes. J. Phys. Chem. **99**(27), 10694–10697 6 Jul 1995
5. M. José-Yacamán, M. Mikiyoshida, L. Rendon et al., Catalytic growth of carbon microtubules with fullerene structure. Appl. Phys. Lett. **62**(2), 202–204 11 Jan 1993
6. R.L. Vander Wal, T.M. Ticich, V.E. Curtis, Diffusion flame synthesis of single-walled carbon nanotubes. Chem. Phys. Lett. **323**(3–4), 217–223 16 Jun 2000
7. A.T. Matveev, D. Golberg, V.P. Novikov et al., Synthesis of carbon nanotubes below room temperature. Carbon **39**(1), 155–158 Jan 2001
8. T.W. Ebbesen, P.M. Ajayan, Large-scale synthesis of carbon nanotubes. Nature **358**(6383), 220–222 16 Jul 1992
9. C. Journet, W.K. Maser, P. Bernier et al., Large-scale production of single-walled carbon nanotubes by the electric-arc technique. Nature **388**(6644), 756–758 21 Aug 1997
10. A. Thess, R. Lee, P. Nikolaev et al., Crystalline ropes of metallic carbon nanotubes. Science **273**(5274), 483–487 26 Jul 1996
11. L.M. Yuan, K. Saito, C.X. Pan et al., Nanotubes from methane flames. Chem. Phys. Lett. **340**(3–4), 237–241 1 Jun 2001
12. L.M. Yuan, K. Saito, W.C. Hu et al., Ethylene flame synthesis of well-aligned multi-walled carbon nanotubes. Chem. Phys. Lett. **346**(1–2), 23–28 28 Sep 2001
13. E.F. Kukovitsky, S.G. L'vov, N.A. Sainov et al., Correlation between metal catalyst particle size and carbon nanotube growth. Chem. Phys. Lett. **355**(5–6), 497–503 8 Apr 2002
14. J. Li, C. Papadopoulos, J.M. Xu et al., Highly-ordered carbon nanotube arrays for electronics applications. Appl. Phys. Lett. **75**(3), 367–369 19 Jul 1999
15. A.M. Cassell, J.A. Raymakers, J. Kong et al., Large scale CVD synthesis of single-walled carbon nanotubes. J. Phys. Chem. B. **103**(31), 6484–6492 5 Aug 1999
16. W.Z. Li, S.S. Xie, L.X. Qian et al., Large-scale synthesis of aligned carbon nanotubes. Science **274**(5293), 1701–1703 6 Dec 1996
17. J. Kong, H.T. Soh, A.M. Cassell et al., Synthesis of individual single-walled carbon nanotubes on patterned silicon wafers. Nature **395**(6705), 878–881 29 Oct 1998
18. H. Wang, J. Lin, C.H.A. Huan et al., Controlled synthesis of aligned carbon nanotube arrays on catalyst patterned silicon substrates by plasma-enhanced chemical vapor deposition. Appl. Surf. Sci. **181**(3–4), 248–254 21 Sep 2001
19. S.S. Fan, W.J. Liang, H.Y. Dang et al., Carbon nanotube arrays on silicon substrates and their possible application. Physica E **8**(2), 179–183 Aug 2000
20. Z.H. Yuan, H. Huang, H.Y. Dang et al., Field emission property of highly ordered monodispersed carbon nanotube arrays. Appl. Phys. Lett. **78**(20), 3127–3129 14 May 2001

21. W.C. Hu, D.W. Gong, Z. Chen et al., Growth of well-aligned carbon nanotube arrays on silicon substrates using porous alumina film as a nanotemplate. Appl. Phys. Lett. **79**(19), 3083–3085 5 Nov 2001
22. Z.F. Ren, Z.P. Huang, J.W. Xu et al., Synthesis of large arrays of well-aligned carbon nanotubes on glass. Science **282**(5391), 1105–1107 6 Nov 1998
23. C. Masarapu, B.Q. Wei, Direct growth of aligned multiwalled carbon nanotubes on treated stainless steel substrates. Langmuir **23**(17), 9046–9049 14 Aug 2007
24. Z.P. Huang, J.W. Wu, Z.F. Ren et al., Growth of highly oriented carbon nanotubes by plasma-enhanced hot filament chemical vapor deposition. Appl. Phys. Lett. **73**(26), 3845–3847 28 Dec 1998
25. Y. Gao, J. Liu, M. Shi et al., Dense arrays of well-aligned carbon nanotubes completely filled with single crystalline titanium carbide wires on titanium substrates. Appl. Phys. Lett. **74**(24), 3642–3644 14 Jun 1999
26. G. Li, S. Chakrabarti, M. Schulz et al., Growth of aligned multiwalled carbon nanotubes on bulk copper substrates by chemical vapor deposition. J. Mater. Res. **24**(9), 2813–2820 Sept 2009
27. S. Sotiropoulou, N.A. Chaniotakis, Carbon nanotube array-based biosensor Anal. Bioanal. Chem. **375**(1), 103–105 Jan 2003
28. H. Ago, N. Uehara, K. Ikeda et al., Synthesis of horizontally-aligned single-walled carbon nanotubes with controllable density on sapphire surface and polarized Raman spectroscopy. Chem. Phys. Lett. **421**(4–6), 399–403 15 Apr 2006
29. V. Derycke, R. Martel, M. Radosvljevic et al., Catalyst-free growth of ordered single-walled carbon nanotube networks. Nano Lett. **2**(10), 1043–1046 Oct 2002
30. N. Saurakhiya, Y.W. Zhu, F.C. Cheong et al., Pulsed laser deposition-assisted patterning of aligned carbon nanotubes modified by focused laser beam for efficient field emission. Carbon **43**(10), 2128–2133 Aug 2005
31. J.S. Suh, K.S. Jeong, J.S. Lee et al., Study of the field-screening effect of highly ordered carbon nanotube arrays. Appl. Phys. Lett. **80**(13), 2392–2394 1 Apr 2002
32. H. Ago, T. Komatsu, S. Ohshima et al., Dispersion of metal nanoparticles for aligned carbon nanotube arrays. Appl. Phys. Lett. **77**(1), 79–81 3 Jul 2000
33. B.Q. Wei, R. Vajtai, Y. Jung et al., Organized assembly of carbon nanotubes – cunning refinements help to customize the architecture of nanotube structures. Nature **416**(6880), 495–496 4 Apr 2002
34. D.S. Xu, G.L. Guo, L.L. Gui et al., Controlling growth and field emission property of aligned carbon nanotubes on porous silicon substrates. Appl. Phys. Lett. **75**(4), 481–483 26 Jul 1999
35. Y. Wang, K. Kempa, B. Kimball et al., Receiving and transmitting light-like radio waves: Antenna effect in arrays of aligned carbon nanotubes. Appl. Phys. Lett. **85**(13), 2607–2609 27 Sept 2004
36. C.V. Nguyen, L. Delzeit, A.M. Cassell et al., Preparation of nucleic acid functionalized carbon nanotube arrays. Nano Lett. **2**(10), 1079–1081 Oct 2002
37. K.M. Ryu, A. Badmaev, L. Gomez et al., Synthesis of aligned single-walled nanotubes using catalysts defined by nanosphere lithography. J. Am. Chem. Soc. **129**(33), 10104–10105 22 Aug 2007
38. Z.W. Pan, S.S. Xie, B.H. Chang et al., Direct growth of aligned open carbon nanotubes by chemical vapor deposition. Chem. Phys. Lett. **299**(1), 97–102 1 Jan 1999
39. J.F. AuBuchon, C. Daraio, L.H. Chen et al., Iron silicide root formation in carbon nanotubes grown by microwave PECVD. J. Phys. Chem. B. **109**(51), 24215–24219 29 Dec 2005
40. N. Hayashi, S. Honda, K. Tsuji et al., Highly aligned carbon nanotube arrays fabricated by bias sputtering. Appl. Surf. Sci. **212**, 393–396 15 May 2003
41. H.F. Zhao, H. Song, Z.M. Li et al., Electrophoretic deposition and field emission properties of patterned carbon nanotubes. Appl. Surf. Sci. **251**(1–4), 242–244 15 Sept 2005

42. N.S. Lee, D.S. Chung, I.T. Han et al., Application of carbon nanotubes to field emission displays. Diamond Relat. Mater. **10**(2), 265–270 Feb 2001
43. Y.M. Shin, S.Y. Jeong, H.J. Jeong et al., Influence of morphology of catalyst thin film on vertically aligned carbon nanotube growth. J. Crystal Growth **271**(1–2), 81–89 15 Oct 2004
44. R.M. Liu, J.M. Ting, J.C.A. Huang et al., Growth of carbon nanotubes and nanowires using selected catalysts. Thin Solid Films **420**, 145–150 2 Dec 2002
45. K. Kempa, B. Kimball, J. Rybczynski et al., Photonic crystals based on periodic arrays of aligned carbon nanotubes. Nano Lett. **3**(1), 13–18 Jan 2003
46. Z.P. Huang, D.L. Carnahan, J. Rybczynski et al., Growth of large periodic arrays of carbon nanotubes. Appl. Phys. Lett. **82**(3), 460–462 20 Jan 2003
47. E. Joselevich, C.M. Lieber, Vectorial growth of metallic and semiconducting single-wall carbon nanotubes. Nano Lett. **2**(10), 1137–1141 Oct 2002
48. Y.G. Zhang, A.L. Chang, J. Cao et al., Electric-field-directed growth of aligned single-walled carbon nanotubes. Appl. Phys. Lett. **79**(19), 3155–3157 5 Nov 2001
49. K.H. Lee, J.M. Cho, W. Sigmund, Control of growth orientation for carbon nanotubes. Appl. Phys. Lett. **82**(3), 448–450 20 Jan 2003
50. N. Kumar, W. Curtis, J.I. Hahm, Laterally aligned, multiwalled carbon nanotube growth using *Magnetospirillum magnetotacticum*. Appl. Phys. Lett. **86**(17), 173101 1–3 25 Apr 2005
51. S.M. Huang, X.Y. Cai, J. Liu, Growth of millimeter-long and horizontally aligned single-walled carbon nanotubes on flat substrates. J. Am. Chem. Soc. **125**(19), 5636–5637 14 May 2003
52. S.M. Huang, B. Maynor, X.Y. Cai et al., Ultralong, well-aligned single-walled carbon nanotube architectures on surfaces. Adv. Mater. **15**(19), 1651–1655 2 Oct 2003
53. S.M. Huang, Q. Fu, L. An et al., Growth of aligned SWNT arrays from water-soluble molecular clusters for nanotube device fabrication. Phys. Chem. Chem. Phys. **6**(6), 1077–1079 21 Mar 2004
54. D. Sanchez-Portal, E. Artacho, J.M. Soler et al., Ab initio structural, elastic, and vibrational properties of carbon nanotubes. Phys. Rev. B. **59**(19), 12678–12688 15 May 1999
55. M.M.J. Treacy, T.W. Ebbesen, J.M. Gibson, Exceptionally high young's modulus observed for individual carbon nanotubes. Nature **381**(6584), 678–680 20 Jun 1996
56. E.W. Wong, P.E. Sheehan, C.M. Lieber, Nanobeam mechanics: Elasticity, strength, and toughness of nanorods and nanotubes. Science **277**(5334), 1971–1975 26 Sep 1997
57. G.H. Gao, T. Cagin, W.A. Goddard, Energetics, structure, mechanical and vibrational properties of single-walled carbon nanotubes. Nanotechnology **9**(3), 184–191 Sept 1998
58. E. Hernandez, C. Goze, P. Bernier et al., Elastic properties of C and BxCyNz composite nanotubes. Phys. Rev. Lett. **80**(20), 4502–4505 18 May 1998
59. M.F. Yu, B.S. Files, S. Arepalli et al., Tensile loading of ropes of single wall carbon nanotubes and their mechanical properties. Phys. Rev. Lett. **84**(24), 5552–5555 12 Jun 2000
60. H.W. Zhu, C.L. Xu, D.H. Wu et al., Direct synthesis of long single-walled carbon nanotube strands. Science **296**(5569), 884–886 3 May 2002
61. L.T. Qu, L.M. Dai, M. Stone et al., Carbon nanotube arrays with strong shear binding-on and easy normal lifting-off. Science **322**(5899), 238–242 10 Oct 2008
62. J. Hone, B. Batlogg, Z. Benes et al., Quantized phonon spectrum of single-wall carbon nanotubes. Science **289**(5485), 1730–1733 8 Sept 2000
63. W. Yi, L. Lu, D.L. Zhang et al., Linear specific heat of carbon nanotubes. Phys. Rev. B. **59**(14), R9015–R9018 1 Apr 1999
64. J.C. Lasjaunias, K. Biljakovic, Z. Benes et al., Low-temperature specific heat of single-wall carbon nanotubes. Phys. Rev. B **65**(11), 113409 1–4 15 Mar 2002
65. V.N. Popov, Low-temperature specific heat of nanotube systems. Phys. Rev. B. **66**(15), 153408 15 Oct 2002
66. P. Kim, L. Shi, A. Majumdar et al., Thermal transport measurements of individual multiwalled nanotubes. Phys. Rev. Lett. **8721**(21), 225502 19 Nov 2001

67. S. Berber, Y.K. Kwon, D. Tomanek, Unusually high thermal conductivity of carbon nanotubes. Phys. Rev. Lett. **84**(20), 4613–4616 15 May 2000
68. S. Shaikh, L. Li, K. Lafdi et al., Thermal conductivity of an aligned carbon nanotube array. Carbon **45**(13), 2608–2613 Nov 2007
69. H. Huang, C.H. Liu, Y. Wu et al., Aligned carbon nanotube composite films for thermal management. Adv. Mater. **17**(13), 1652–1656 4 Jul 2005
70. Y. Xu, Y. Zhang, E. Suhir et al., Thermal properties of carbon nanotube array used for integrated circuit cooling. J. Appl. Phys. **100**(12), 074302 15 Dec 2006
71. L. Langer, L. Stockman, J.P. Heremans et al., Electrical-resistance of a carbon nanotube bundle. J. Mater. Res. **9**(4), 927–932 Apr 1994
72. L. Langer, V. Bayot, E. Grivei et al., Quantum transport in a multiwalled carbon nanotube. Phys. Rev. Lett. **76**(3), 479–482 15 Jan 1996
73. M. Bockrath, D.H. Cobden, P.L. McEuen et al., Single-electron transport in ropes of carbon nanotubes. Science **275**(5308), 1922–1925 28 Mar 1997
74. J.W. Mintmire, B.I. Dunlap, C.T. White, Are fullerene tubules metallic. Phys. Rev. Lett. **68**(5), 631–634 3 Feb 1992
75. S.J. Tans, M.H. Devoret, H.J. Dai et al., Individual single-wall carbon nanotubes as quantum wires. Nature **386**(6624), 474–477 3 Apr 1997
76. L.B. Zhu, J.W. Xu, Y.H. Xiu et al., Growth and electrical characterization of high-aspect-ratio carbon nanotube arrays. Carbon **44**(2), 253–258 Feb 2006
77. W.A. de Heer, A. Chatelain, D. Ugarte, A carbon nanotube field-emission electron source. Science **270**(5239), 1179–1180 17 Nov 1995
78. W.A. de Heer, W.S. Bacsa, A. Chatelain et al., Aligned carbon nanotube films – Production and optical and electronic-properties. Science **268**(5212), 845–847 12 May 1995
79. S.S. Fan, M.G. Chapline, N.R. Franklin et al., Self-oriented regular arrays of carbon nanotubes and their field emission properties. Science **283**(5401), 512–514 22 Jan 1999
80. Y.N. Zhang, W. Lei, X.B. Zhang et al., Calculation of the emission performance of the carbon nanotube array. Appl. Surf. Sci. **245**(1–4), 400–406 30 May 2005
81. K.S. Hazra, P. Rai, D.R. Mohapatra et al., Dramatic enhancement of the emission current density from carbon nanotube based nanosize tips with extremely low onset fields. ACS Nano **3**(9), 2617–2622 Sept 2009
82. B. Zhao, D.N. Futaba, S. Yasuda et al., Exploring advantages of diverse carbon nanotube forests with tailored structures synthesized by supergrowth from engineered catalysts. ACS Nano **3**(1), 108–114 Jan 2009
83. L. Nilsson, O. Groening, C. Emmenegger et al., Scanning field emission from patterned carbon nanotube films. Appl. Phys. Lett. **76**(15), 2071–2073 10 Apr 2000
84. J.C. Charlier, M. Terrones, M. Baxendale et al., Enhanced electron field emission in B-doped carbon nanotubes. Nano Lett. **2**(11), 1191–1195 Nov 2002
85. A.C. Dillon, K.M. Jones, T.A. Bekkedahl et al., Storage of hydrogen in single-walled carbon nanotubes. Nature **386**(6623), 377–379 27 Mar 1997
86. C. Liu, Y.Y. Fan, M. Liu et al., Hydrogen storage in single-walled carbon nanotubes at room temperature. Science **286**(5442), 1127–1129 5 Nov 1999
87. Y. Ye, C.C. Ahn, C. Witham et al., Hydrogen adsorption and cohesive energy of single-walled carbon nanotubes. Appl. Phys. Lett. **74**(16), 2307–2309 19 Apr 1999
88. H.W. Zhu, A.Y. Cao, X.S. Li et al., Hydrogen adsorption in bundles of well-aligned carbon nanotubes at room temperature. Appl. Surf. Sci. **178**(1–4), 50–55 2 Jul 2001
89. Q.Y. Wang, J.K. Johnson, Optimization of carbon nanotube arrays for hydrogen adsorption. J. Phys. Chem. B **103**(23), 4809–4813 Jun 10 1999
90. A. Misra, J. Giri, C. Daraio, Hydrogen evolution on hydrophobic aligned carbon nanotube arrays. ACS Nano **3**(12), 3903–3908 Dec 2009
91. J. Kong, N.R. Franklin, C.W. Zhou et al., Nanotube molecular wires as chemical sensors. Science **287**(5453), 622–625 28 Jan 2000

92. P.F. Qi, O. Vermesh, M. Grecu et al., Toward large arrays of multiplex functionalized carbon nanotube sensors for highly sensitive and selective molecular detection. Nano Lett. **3**(3), 347–351 Mar 2003
93. J.C. Claussen, A.D. Franklin, A. ul Haque et al., Electrochemical biosensor of nanocube-augmented carbon nanotube networks. ACS Nano **3**(1), 37–44 Jan 2009
94. H. Cai, X.N. Cao, Y. Jiang et al., Carbon nanotube-enhanced electrochemical DNA biosensor for DNA hybridization detection. Anal. Bioanal. Chem. **375**(2), 287–293 Jan 2003
95. R.J. Chen, S. Bangsaruntip, K.A. Drouvalakis et al., Noncovalent functionalization of carbon nanotubes for highly specific electronic biosensors. Proc. Natl. Acad. Sci. USA **100**(9), 4984–4989 29 Apr 2003
96. J. Zhao, A. Buldum, J. Han et al., First-principles study of Li-intercalated carbon nanotube ropes. Phys. Rev. Lett. **85**(8), 1706–1709 21 Aug 2000
97. G.L. Che, B.B. Lakshmi, E.R. Fisher et al., Carbon nanotubule membranes for electrochemical energy storage and production. Nature **393**(6683), 346–349 28 May 1998
98. M. Endo, Y.A. Kim, T. Hayashi et al., Vapor-grown carbon fibers (VGCFs) – basic properties and their battery applications. Carbon **39**(9), 1287–1297 Aug 2001
99. G.T. Wu, C.S. Wang, X.B. Zhang et al., Lithium insertion into CuO/carbon nanotubes. J. Power Sources **75**(1), 175–179 1 Sept 1998
100. B. Gao, A. Kleinhammes, X.P. Tang et al., Electrochemical intercalation of single-walled carbon nanotubes with lithium. Chem. Phys. Lett. **307**(3–4), 153–157 2 Jul 1999
101. R.Z. Ma, J. Liang, B.Q. Wei et al., Study of electrochemical capacitors utilizing carbon nanotube electrodes. J. Power Sources **84**(1), 126–129 Nov 1999
102. K.H. An, W.S. Kim, Y.S. Park et al., Electrochemical properties of high-power supercapacitors using single-walled carbon nanotube electrodes. Adv. Funct. Mater. **11**(5), 387–392 Oct 2001
103. C.M. Niu, E.K. Sichel, R. Hoch et al., High power electrochemical capacitors based on carbon nanotube electrodes. Appl. Phys. Lett. **70**(11), 1480–1482 17 Mar 1997
104. H. Ago, K. Petritsch, M.S.P. Shaffer et al., Composites of carbon nanotubes and conjugated polymers for photovoltaic devices. Adv. Mater. **11**(15), 1281 20 Oct 1999
105. J. van de Lagemaat, T.M. Barnes, G. Rumbles et al., Organic solar cells with carbon nanotubes replacing In2O3:Sn as the transparent electrode. Appl. Phys. Lett. **88**(23), 233503 1–3 5 Jun 2006

Chapter 11
Molecular Rotors Observed by Scanning Tunneling Microscopy

Ye-Liang Wang, Qi Liu, Hai-Gang Zhang, Hai-Ming Guo, and Hong-Jun Gao

11.1 Introduction

Molecules are important building blocks for bottom-up fabrication of functional nanostructures in the exciting field of nanotechnology [1, 2]. The most significant advantage of molecules lies in the fact that the well-developed molecular synthesis techniques can produce various molecular structures, which offers a high controllability and flexibility over single molecular properties and molecular nanostructures [3–5]. Moreover, most molecules can self-assemble into ordered nanostructures at surfaces, which can be precisely controlled through modifying the properties of molecules or surfaces [6–16]. Based on their significant advantages, functional molecules are undoubtedly becoming one of the most attractive candidates for the fabrication of nanodevices with integrated functions. Integrating device functions into single molecules is a crucial issue in molecule-based nanoengineering and is being considered as an ideal solution for the device miniaturization pushed by Moore's law.

Molecular machines are kinds of nanodevices which are expected to convert external chemical, electric, or optical energies into controlled mechanical movements at molecular levels. In fact, molecular machines exist widely in nature and play an important role in many biological processes [17, 18]. As we know, molecular machines can transport various cellular components in a large number of biological systems [19]. For instance, kinesin protein can burn adenosine triphosphate (ATP) and generate linear motion along cytoskeleton [20]. Other proteins, driven by proton gradients across membranes, mechanically rotate flagella to propel bacteria [21]. Nowadays, artificial biological machines (or so-called biological nanorobots) show many potential applications, such as acting as sensors for diagnosis, medical target identification, and even assistants in invasive surgery [22, 23].

H.-J. Gao (✉)
Institute of Physics, Chinese Academy of Sciences, Beijing 100190, China
e-mail: hjgao@iphy.ac.cn

W.L. Zhou, Z.L. Wang (eds.), *Three-Dimensional Nanoarchitectures*,
DOI 10.1007/978-1-4419-9822-4_11,

Molecular rotors are a type of molecular machines, and they can rotate with respect to their surrounding environments or involve one part that rotates relative to another part [24, 25]. In the past, molecular motor movements have been proved able to be powered by thermal [16, 26], optical [27–30], electrical [31, 32], and chemical means [33] and even by more complicated ratchetlike principles [34–36] and physical control techniques (for example, special substrate lattice) [16, 26, 37]. At the nanoscale, electric current driving could in principle be realized by electron tunneling [38, 39]. It was previously shown that tunneling can induce periodic vibrational [40] and translational motions in molecules [41]. Recently, Wang and Kral et al. have used molecular dynamics (MD) simulations to study possible activities of molecules powered by electron tunneling, and they demonstrate that electron tunneling could drive a rotation of single molecular rotor [42].

Molecular rotors at solid surfaces could be either controlled by molecule–substrate interactions or confined in a supramolecular assembly by intermolecular interaction [30, 37, 43–51]. Compared with molecular rotors in solutions or in bulk, molecular rotors at solid surfaces have advantages of being easily accessible by external fields [51, 52] and addressable by surface analysis methods [37, 45, 53] as well as being easily organized due to the reduced dimensions [16, 50].

The recent development of probe microscopy techniques, in particular, scanning tunneling microscopy (STM) [54, 55], has allowed the study of individual molecules with atomic-scale precision. Up to now, considerable efforts have been made to STM studies of single molecules at surfaces. With its capability of high spatial resolution and high energy resolution measurements, STM has helped reveal many interesting physics within single molecules, including electron transport [10, 56, 57], spin-flip excitations [58, 59], vibrational excitations [60], and mechanical motions [16, 36, 61]. Moreover, STM can offer scientists the opportunity to manipulate atomic and molecular events that operate not only on ensemble-averaged populations of species but also on single functional group of the chemical entities [62–67]. Without any doubt, STM has significantly provided a unique tool to investigate the molecular motor motions. With its further development as a powerful technique, it can be used to study the dynamic behavior in real time and motion mechanism of molecular motors/machines at single atom/molecule level. The feature of molecular movements can be recorded by the feedback loop signal in real time during STM manipulation [68].

In this review, we recall the recent advances of molecular rotors in the past years and then give a simple outlook about their perspective in the future. Our discussions are based on the most recent results obtained by the technique of STM. We mainly focus on the motions of molecular motors at solid surfaces. According to their motion dimension and movement styles at solid surfaces, the reported molecular motor motions are simply classified into two groups, that is, of single molecule and of array comprised of a plenty of molecules, as illustrated by a schematic in Fig. 11.1. In the literature, the motions of a single molecule are commonly used as models to introduce new design perception or control approaches of molecular movements. They are further divided into two groups, that is, in-plane motion and off-plane motion. The self-rotation of molecules, with rotation axis perpendicular

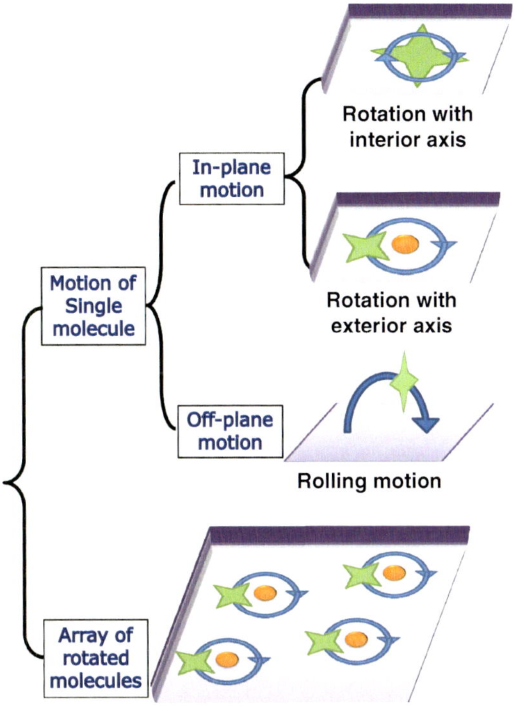

Fig. 11.1 Schematic of typical molecular rotor motions at solid surfaces

to the surface, is a classical example of the in-plane motion. In such cases the molecules ordinarily lie flat on the surface, and they show a rotation with either an interior rotation axis or an exterior rotation axis. As for the off-plane motion, a typical example is the rolling motion illustrated by some specially designed molecules, which usually hold some ball-like functional groups. These special groups normally serve as nanoscale "wheels" to reduce the motion barrier of rolling and facilitate molecular movements on surfaces. The lower barrier of the motions can be easily overcome by the electrical stimulation of an STM tip or by thermal driving of increasing substrate temperature.

11.2 Solution-Based and Surface-Mounted Molecular Machines

In the past 20 years, synthetic molecular machines in solution have been reported by several research groups, such as Feringa, Leigh, Kelly, Tour, and others [33, 69–76]. They demonstrated repeatedly some very elegant examples of synthetic molecular machines by specially designed molecular pieces. The general importance and broad appeal of synthetic molecular machines as well as the design principles in synthetic chemistry have been described in previous review articles [77, 78]. Besides a detailed discussion of design principles used to control linear and rotary motion in solution, these reviews also valuated the approaches to construct synthetic molecular machines. Several conceptual models of solution-based molecular nanodevices were illustrated, for example, molecular rotors, elevators, valves, transporters, muscles, and other motor functions used to develop smart materials. As a concerted and collection behavior of a large number of molecules, the molecular motions are even demonstrated on systems capable of affecting macroscopic movement in millimeter scale [30].

Clearly, the molecular motors proposed and studied in solution have many degrees of freedom. The exhibited behavior is the average effect of a large ensemble of molecules but not the behavior of a single molecule. By contrast, the molecular motors reflected here will focus on the motion of a single molecule mounted at solid surfaces. The molecules have decreased degrees of freedom on surfaces. It provides a platform to study and manipulate molecular motions individually, like by an STM tip.

11.3 Single Molecular Rotors at Surfaces

11.3.1 A Monomolecular Rotor in Supramolecular Network

In 1998, a research group headed by Prof. Christian Joachim in France reported for the first time a real molecular rotor based on their STM observations [36]. Typical results are selected and represented in Fig. 11.2. The used molecule, named as hexa-*tert*-butyl-decacyclene (HTB), is constituted of a polyaromatic ring of 1.5 nm in diameter and six *tert*-butyl legs. The legs are saturated hydrocarbon groups and allow the isolation of the molecular plane from the metallic surface. STM is used not only to image the molecule but also to manipulate the molecule by locating precisely at the leg position of the molecule.

At monolayer coverage deposited on a pre-cleaned Cu(100) surface, HTB molecules form a two-dimensional (2D) well-ordered lattice-like structure, which stabilized mainly by intermolecular interaction, for example, van der Waals force. At coverages significantly below one monolayer, random thermal motions are extremely fast and the molecules are not observable by STM. When coverage is a little less than a full monolayer, however, in the 2D molecular lattice some small nanoscale “cavities” have been observed (corresponding to a black zone in the monolayer, see Fig. 11.2c, d). The molecules at the borders of these free spaces can sit in one of two positions: a highly symmetrical one, following the order of the adjacent lattice, and a less symmetrical one. In the less symmetrical site, the molecule is still constrained by its neighbors, but is disengaged from some of the intermolecular interactions, so that it freely rotates in a random fashion at high speeds (Fig. 11.2d) [79]. Movement between these two sites can be affected by an STM tip. In the experiment, the molecule located in the middle part of the STM images is the manipulation target of the STM tip. Moving this molecule slightly toward the hole position (nanocavity) induces a change in the image. At this nanocavity position, the molecule has lower barrier to get into movement. After its translational motion, the molecule, initially fixed on the surface by incarceration among its neighbors, started

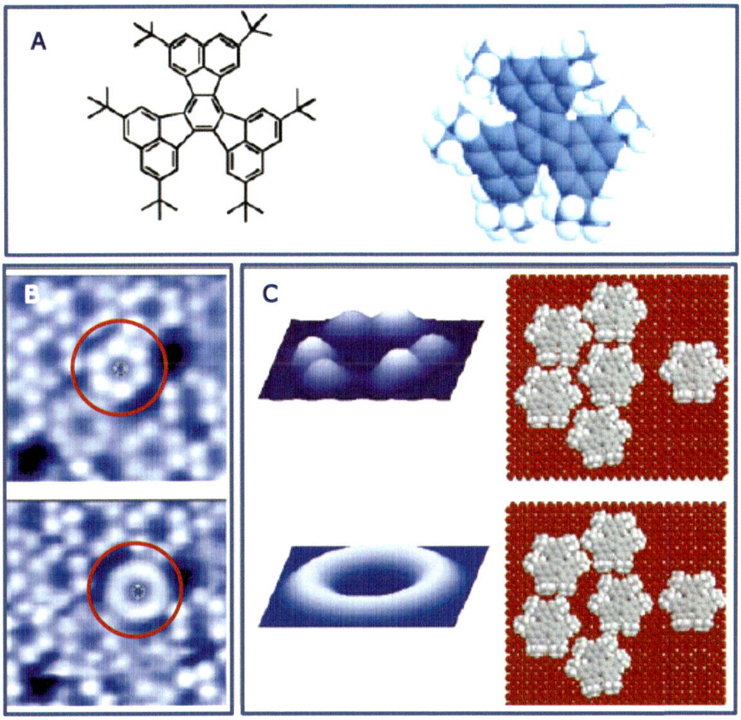

Fig. 11.2 (**a**) Chemical structure of rotary hexa-*tert*-butyl-decacyclene (HTB) molecule. (**b**) STM images of a HTB monolayer on a Cu(100) surface. One HTB molecule, marked by *red cycle*, is stationary at the *upper panel* and it appears as a *circular* arrangement of six lobes. The marked molecule has been moved by the STM tip and is no longer bound tightly by its neighbors and it therefore rotates, which is observed as a blurring of the lobed groups, and it becomes rotary (*down panel*). (**c**) Calculated models showing the rotated molecule located in the middle part of the monolayer. This molecule is fixed (*upper*) and in rotation (*down*). Two panels on the right show two calculated representations of the atomic arrangement of the molecules. Reprinted with permission from [36]. Copyright 1998 by the AAAS and 2005 by the Royal Society of Chemistry

to rotate. The white spots, corresponding to the legs of motionless molecules, are not localized any more. The molecule in the center is like a torus, indicating a rotation of the whole molecule.

This molecule really acts as a single molecular rotor and its rotation can be controlled by its surrounding environment. Without the supramolecular network, the molecule loses rotary property. The data demonstrated that STM could induce single-molecule positional change and control thermally driven motions of the molecules at surfaces. This kind of manipulation of organic molecules requires careful balancing of adsorbate–adsorbate interactions. In addition, the molecule–substrate interaction must be strong enough to prevent thermal-driven motion at the temperature used [65, 80].

11.3.2 Gear-Like Rotation of Molecular Rotor Along the Edge of the Molecular Island

At the end of the twentieth century and the beginning of the twenty-first century, the motions or rotations of molecules at surfaces reported are only random and uncontrolled rotations [36, 81, 82] or indirect signatures [49, 83] of a rotation. In order to control the rotation at single-molecule level, Prof. Joachim and his colleagues explored a new kind of molecule in 2006 [84]. The new molecule (HB–NPB, $C_{64}N_2H_{76}$) is specially designed, and it is comprised of five phenyl groups and a pyrimidine group (Fig. 11.3) with a size of 1.8 nm in diameter. Each aryl group is substituted by a *t*-butyl group in *para* position to the central benzene. The *t*-butyl groups acting as propellers have the function of lifting the molecule to reduce the interaction of the aromatic parts with the substrate. In addition, the replacement of one phenyl substituent by a pyrimidyl group, serving as a chemical tag that can be detected by an STM tip, provides an orientation axis to the molecule.

In the experiment, the HB–NPB molecules are deposited onto a pre-cleaned Cu(111) substrate. The molecules form large well-ordered islands on Cu(111) surface. The step edge of the molecular island presents a jagged shape. By applying high bias voltages above 2.2 V in STM measurements, the pyrimidine group of HB–NPB molecules is visible as brighter spots inside the molecule. It provides an excellent marker for the determination of the orientation of each molecule on the substrate (Fig. 11.3d).

They conducted decent manipulations to demonstrate how to control a single-molecule rotation by an STM tip with atomic-scale precision. They called the rotated molecule at the border a molecular rack-and-pinion device according to the observed structure (Fig. 11.3b). One rotary HB–NPB molecule functioning as a six-toothed wheel was named as a pinion, and a self-assembled molecular island was coined a rack. The pinion molecule interlocks at the edge of the stationary rack. In their low-temperature experiment, an STM tip is used to drive rotation of a single pinion molecule. The rotation of the pinion molecule teeth by teeth along the border of molecular island (the rack) is successfully observed.

At each step of the manipulation series as denoted in Fig. 11.3d, the orientation of the molecule in the Cu(111) surface plane changes by 60° or 120°. The STM tip approaches the molecule at its center, and the rotational movement is determined by the serrated shape of the border of the molecular island. The interlocking of the rack with the pinion opposes the sliding motion of the pinion molecule, resulting in a teeth-by-teeth rotary motion. The researchers claimed that the success rate of the pinion motion along the rack is 80% in their 150 manipulation experiments.

11.3.3 Thermal-Driven Rotation on Reconstructed Surface Template

It is well known that a clean Au(111) surface can reconstruct into a herringbone structure after an annealing treatment at a temperature of about 800 K. Figure 11.4a

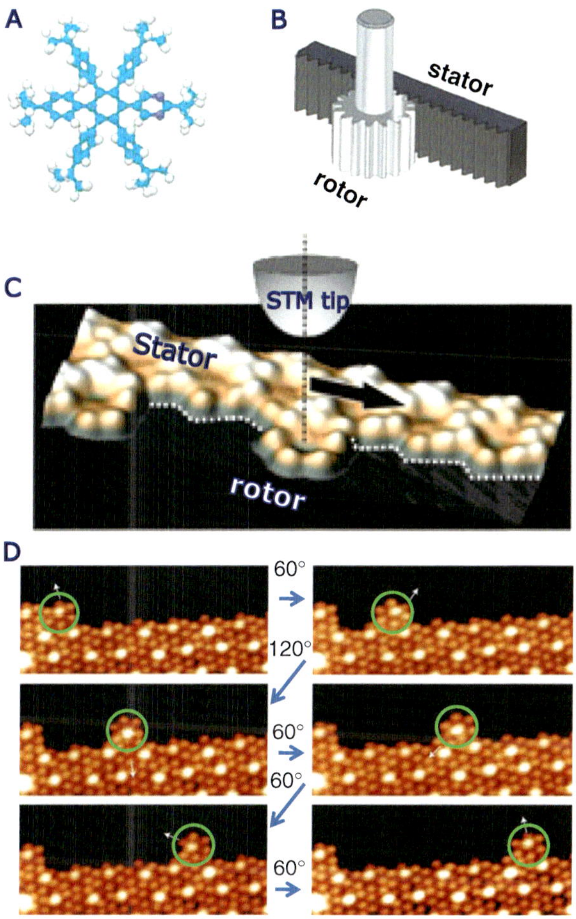

Fig. 11.3 (**a**) Chemical structure of HB–NPB molecule; two nitrogen atoms are in *dark blue*, carbon atoms in *light blue*, and hydrogen atoms in *white*. (**b**) Macroscopic "rack-and-pinion" model, showing a teeth-by-teeth rotation of the gear-like rotor along the linear edge of molecular island. (**c**) STM image of the border of a HB–NPB island with a single molecule adsorbed on its side, which acts as the rack and the pinion, respectively. STM tip illustrates the manipulation position (molecular center) for driving a rotation of the single molecule. The *black arrow* indicates the direction (parallel to the island border) of molecular movement. (**d**) Successive experimental images (13 nm × 6 nm, 0.1 nA, 2.2 V) showing a 360° rotation of the molecular rotor (marked by *green cycle*) along the island border. The *white arrows* mark the same lobe and denote the direction after reorientation of the molecule. Reprinted with permission from [84]. Copyright 2007 by Nature Publishing Group

shows typical topographic images of this reconstruction observed by STM. The herringbone structure can be divided into four types of regions with different arrangements of surface atoms [85, 86], that is, face-centered cubic (fcc), hexagonal close packed (hcp), corrugation ridges, and elbow sites. The elbow site is the connection position of ridges. An image with atomic resolution of three former

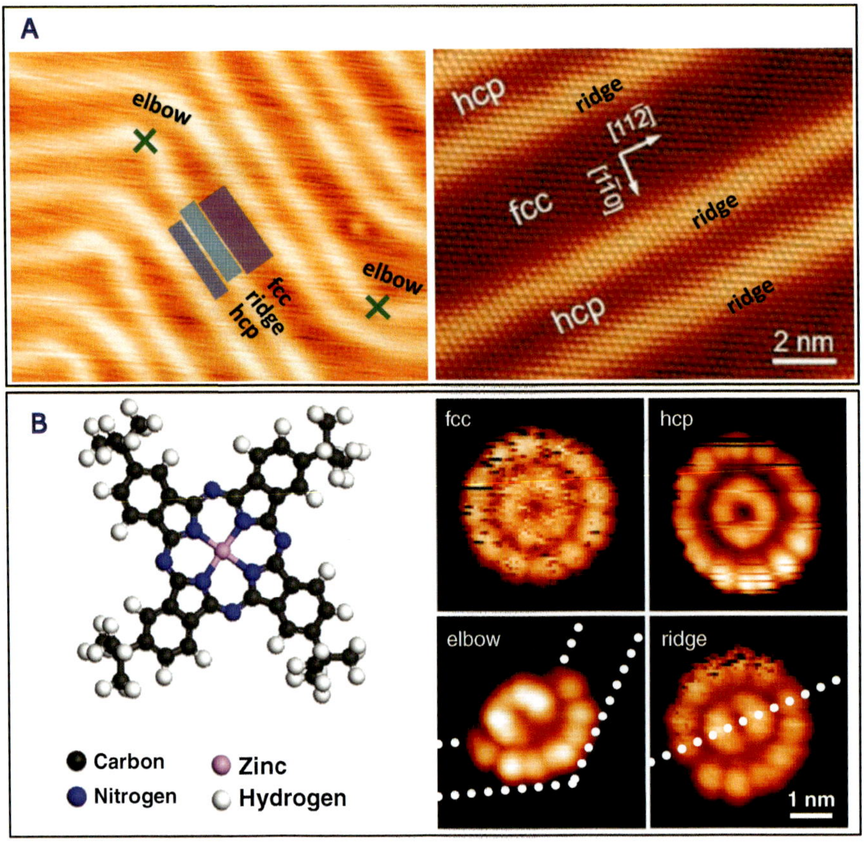

Fig. 11.4 (**a**) A typical STM image of herringbone-like topography of Au(111) reconstruction surface, in which four special regions with different arrangements of surface atoms are marked. *Right panel*: atomic resolution image of Au(111) reconstruction. (**b**) Chemical structure of tetra-*tert*-butyl zinc phthalocyanine (TBZP) molecule. High-resolution STM images of single molecular rotors, formed by TBZP molecules at different substrate regions, show different features due to the modulation by corrugation ridges (0.07 nA, –1.3 V). Images were taken at 78 K

arrangements can be found in the right panel of Fig. 11.4a. Although all of them show a parallel alignment, there is a clear contrast between the ridges and fcc/hcp regions. Obviously, the ridges are brighter than the others. In addition, the width of fcc zones is normally larger than that of hcp zones. Based on these features, we can easily recognize Au(111) surface as well as its special regions from STM topography. Besides these four features (can even be called as fingerprint) of Au(111) surface, another interesting phenomenon is the existence of Au adatoms at terraces. Metal adatoms are normally considered as evaporates from kinks and steps onto the terraces of the homogeneous metal surfaces [87, 88]. Analysis of similar metal systems by X-ray photoelectron spectroscopy (XPS) has already revealed the existence of copper adatoms as "background gas" on Cu substrate [89, 90]. Adatoms are hard to observe alone by STM due to their very high mobility on the metal surface, but they can be trapped by some molecules, like the gold adatoms on Au(111) surface captured by the deposited thiol-group-terminated molecule [91].

In a recent study [16] using STM, researchers have shown that single tetra-*tert*-butyl zinc phthalocyanine (TBZP) molecule on reconstructed Au(111) surface possesses a well-defined rotation axis fixed on the surface. Gold adatoms [91, 92] serve as the stable contact of the molecules to the surface. A rotation axis is formed by chemical bonding between a nitrogen atom of the molecule and a gold adatom on the surface, which gives them a well-defined contact while the molecules can have rotation-favorable configurations.

The molecular formula of the used TBZP is $C_{48}H_{48}N_8Zn$, and its chemical structure is shown in Fig. 11.4b; it looks like a windmill with four protruding vanes surrounding a central zinc atom. After depositing few of such molecules onto a clean Au(111) surface, the authors find that the appearance of the adsorbates depends to a large extent on the adsorbed surface sites with different atomic arrangements. The molecules located at the elbow sites show a folding-fan feature. In contrast, the STM images of the molecule located in the fcc region, in the hcp region, and on the corrugation ridges show "flower" features. More specifically, STM images of molecules in the fcc and hcp regions are composed of 2 concentric cirques: 12 bright lobes form the outer torus, just like 12 "petals," while the inner torus has no obvious divisions. Similarly but not completely the same, the STM image of the molecular rotors at the corrugation ridges is composed of an outer torus of 12 bright lobes, but with 2 inner elliptic protrusions. Both the folding-fan and flower features can only be seen at a temperature of 78 K and cannot be observed at 5 K.

In order to verify that the "folding fan" is caused by molecular instability with respect to the substrate surface, the authors monitored the tunneling current versus time by locating the STM tip at a fixed point on the "folding fan" (Fig. 11.5). They applied a constant bias voltage of –1.8 V to the sample and recorded the tunneling current as a function of time. Figure 11.5b shows the recorded tunneling current within a time interval of 80 ms. The amplitude of the tunneling current oscillates

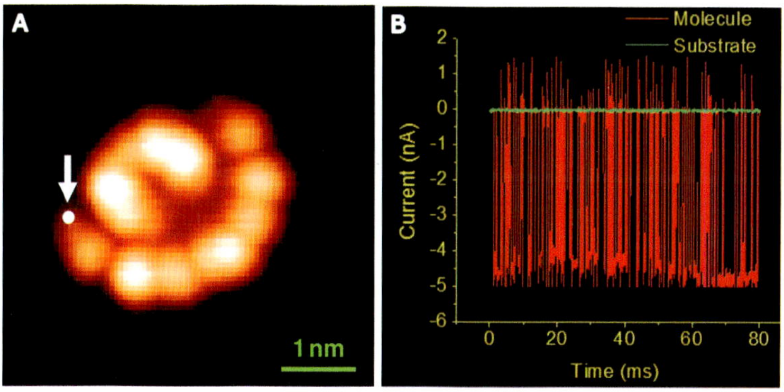

Fig. 11.5 (**a**) A high-resolution STM image of single TBZP molecular rotor showing a "folding-fan" feature (0.05 nA, –2 V). (**b**) Curves of tunneling current versus time, measured on the molecular rotor (*red curve*) and the substrate (*green curve*), and the *I*-*t* spectroscopy of the molecule was measured at the position indicated by the *arrow* in (**a**). Reprinted with permission from [16]. Copyright 2008 by the American Physical Society

frequently between 0 and 5 nA. The oscillation provides a direct evidence that the "folding-fan" feature is really due to rapid molecular motions.

Another further question raised is, how many molecules are there in the "folding fan?" The authors proved that this feature involved only one TBZP molecule. They observed different aggregates of stationary dimers, trimers, tetramers, and larger clusters of TBZP molecules at 78 K. In contrast, a stationary single TBZP molecule, whose STM image should be composed of four lobes, cannot be observed at 78 K. This indicates that single molecules are not stationary but unstable on the surface at this temperature. Besides, they isolated one molecule (marked by a white arrow in Fig. 11.6) which was adsorbed at the elbow position and attached by two large adjacent molecular clusters. By shifting its neighboring molecules in one cluster with an STM tip, this isolated molecule started to rotate, as shown in Fig. 11.6b. It is clear that the molecule remained stationary when it was attached by the clusters, but it became unstable showing the folding-fan feature as soon as it was released from the adjacent clusters.

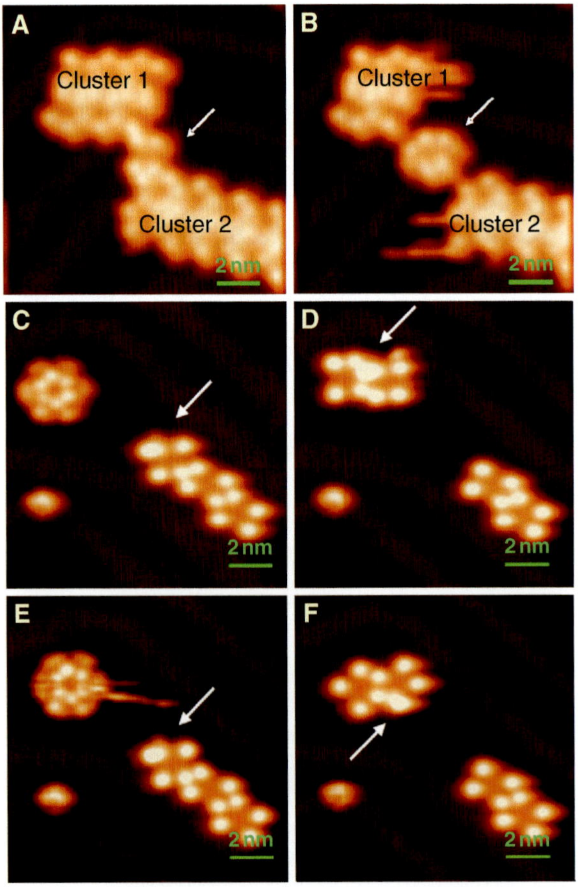

Fig. 11.6 Sequences of STM images show the manipulation of TBZP molecular rotor on an Au(111) surface by an STM tip. The molecule shown with an *arrow* in (**a**) remains stationary, due to the attachment by two adjacent molecular clusters formed by same molecules. It possesses rotary status while it is released when the blocking molecules were scanned off in (**b**). (**c**)–(**f**) A repeatable process in which the molecule shown with an *arrow* bounced between a rotating molecule (*top left*) and a dimer (*bottom right*), switching the rotor (*top left*) between rotation and stop

In another experiment, a sequence of images showed that such a kind of start–stop process is reversible. The authors observed that a single molecule (marked with white arrow in Fig. 11.6c–f) bounced between the molecular dimer at the bottom right and a single molecule at top left. The molecule shown with an arrow can block the rotating molecule (top left) by forming a new molecular dimer (Fig. 11.6d), and then this new dimer was dismantled if detaching the molecule shown with an arrow by the STM tip, the original rotary molecule was released and restarted to rotate (Fig. 11.6e). Reversibly, it remained stationary if it was attached again by the molecules shown with an arrow (Fig. 11.6f). These manipulations also demonstrate clearly that the folding-fan and flower features are due to the instability of a single TBZP molecule on the reconstructed Au(111) surface.

The existence of a rotation center is the prerequisite for rotation; otherwise a lateral diffusion of TBZP molecules along the surface would be hard to block at elevated temperatures. The rotation center cannot be at the position of the *tert*-butyl groups which appear as bright protrusions in STM measurements. Based on further STM observations combined with the first-principle calculations, the researchers revealed that the rotation center is a trapped gold adatom at the surface. By manipulation with an STM tip, one molecule of the "folding-fan" rotor, located at the elbow position of the Au(111) substrate (Fig. 11.7a), was removed. A small bright spot at the center position of the molecular rotor was observed [16], as shown in Fig. 11.7b. This bright spot, observed after the removal of a single molecule, is proposed as a gold adatom. As presented above, gold adatoms on the reconstructed gold surface are stable and prefer to adsorb at the elbow sites at 78 K [91] and capable of enhancing the interaction between the adsorbed molecule and the surface, forming a potential energy well that prevents lateral diffusion of the molecule along the surface.

The existence of a gold adatom is further proved based on the calculated results. Figure 11.7c, d is the top and side view, respectively, of the optimized configuration of a single TBZP molecule adsorbed on a gold adatom. The calculations show that the distance between the zinc atom and its nearest-neighbor gold atom is 4.60 Å, the distance between the bottom nitrogen (colored in yellow) and the gold atom is 2.25 Å; the adsorption energy of this configuration in this case is 804 meV. In contrast, in the case of lack of Au adatom, the calculation results [16] for a single TBZP molecule adsorbed directly on Au(111) show that the distance between the zinc atom and its nearest-neighbor gold atom is 4.35 Å, the distance between the bottom nitrogen atom and its nearest-neighbor gold atom is 4.40 Å; the adsorption energy of this configuration is 219 meV.

Obviously, the gold adatom significantly enhances the molecular bonding, which is most likely due to the surface dipole originating from smeared out electron charge at the position of the adatom [92]. Thus the strong chemical bond between nitrogen and the gold adatom prevents lateral molecular diffusion along the surface and in particular offers a fixed off-center axis for the rotation of single TBZP molecules at 78 K. The model is in good agreement with experimental measurements. The experimentally measured distance between the rotor center and the bright lobes on the outer torus is 1.3–1.4 nm, in reasonable agreement with the distance between

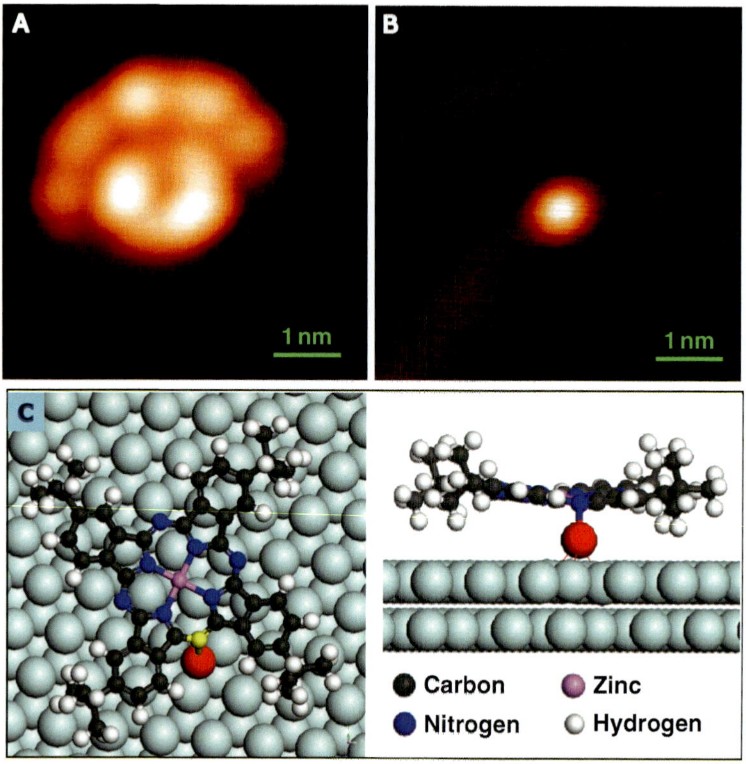

Fig. 11.7 (**a**) STM image of TBZP molecular rotors, locating at the elbow position of the substrate. (**b**) STM image after STM manipulation, *bright spot* indicates a gold adatom after removing the attached molecule. (**c**) *Top view* and *side view* of the optimized configuration of a TBZP molecule adsorbed on an Au(111) surface via a gold adatom (colored by *red*). The adatom acts as an off-center rotation axis for molecular rotation

the nitrogen atom and the *tert*-butyl groups (1.10 ± 0.05 nm), considering that the rotation center is the gold adatom which is not exactly under the nitrogen atom.

For a single TBZP molecular rotor on a flat Au(111) surface, calculations show that there are 12 stable adsorption configurations, which are 30° apart from each other and can be interpreted as intermediate states. The differences in adsorption energies between these stable configurations are only tens of millielectronvolts. The molecule switches between them with high frequency under thermal excitation. Since four *tert*-butyl groups are imaged as the bright lobes in STM measurements (see Fig. 11.8a), the ensuing STM image is the "flower." The proposed STM image for 360° rotation is in good agreement with the "flower" features observed in the experiments (Fig. 11.8c). The hcp and fcc regions of the surface have a similar symmetry with respect to the rotation axis; thus the molecular rotors in the two regions have almost identical STM images.

The rotation of a single TBZP molecule at the elbow sites is interpreted based on the model for the rotation in the fcc region. The corrugation of the ridges forms

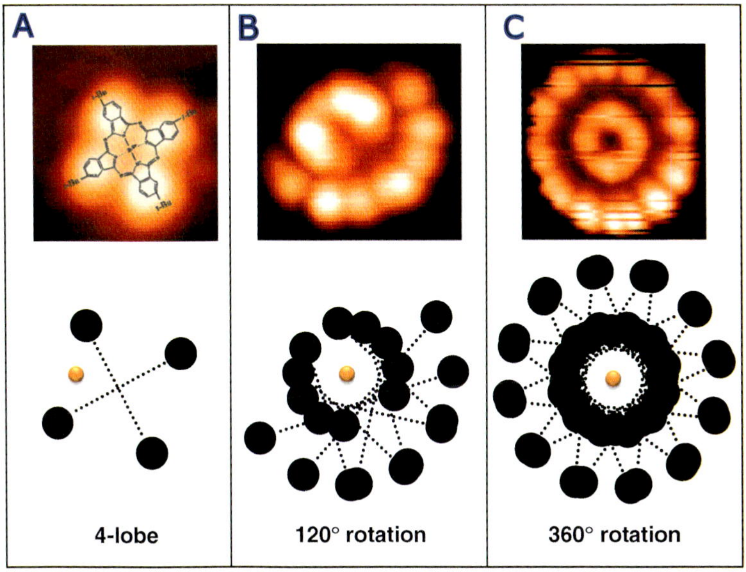

Fig. 11.8 STM images and schematic drawing of a single TBZP molecule on an Au(111) surface. (**a**) Stationary molecule looks like a four-lobed flower, the image is taken at the sample temperature of 5 K. (**b, c**) Rotors with rotation angles of 120° and 360° with the appearance of folding fan, and flower, respectively. These two images are obtained at 78 K. The *black solid circles* depict the bright lobes for stationary single molecules, and the *yellow circle* represents the rotation center (Au adatom)

barriers for the molecular rotation at the elbow sites. The molecular rotation is limited within an angle of 120° due to the bending of the corrugation of the ridges, which leads to the "folding-fan" feature. The proposed STM image for 120° rotation is in great agreement with the experimental STM image of a single molecule at the elbow sites (Fig. 11.8b). Here, the variation in the position of surface atoms leads to a redistribution of potential barriers for molecular rotation.

The researchers further revealed that such kind of molecular rotor exists widely for the molecules of phthalocyanine (Pc) families [93]. And by modifying either the central metal atoms or functional groups linked to the planar molecular backbones, they get different kinds of molecular rotors which have different diameters and rotation-favorable configurations, as indicated in Fig. 11.9. Three different folding-fan structures constituted of TBZP, ZnPc, and FePc molecules are demonstrated. It is clear that all of them have similar rotating mechanisms; however, due to holding tertiary butyl groups, TBZP rotor has a wider rotating angle than ZnPc (Fig. 11.9, top). For the FePc molecule, the central iron atom shows a bright spot in the STM images (Fig. 11.9, bottom); thus the center of its folding-fan structure, which is contributed by both the central atom and benzene lobes, shows a low contrast compared with ZnPc folding-fan structure, for which the zinc atom is shown as a dark dip in the molecular center (Fig. 11.9, middle).

In addition to direct STM tip manipulation on the molecular position, the STM tunneling currents can also affect the rotation of molecular rotors. Figure 11.10

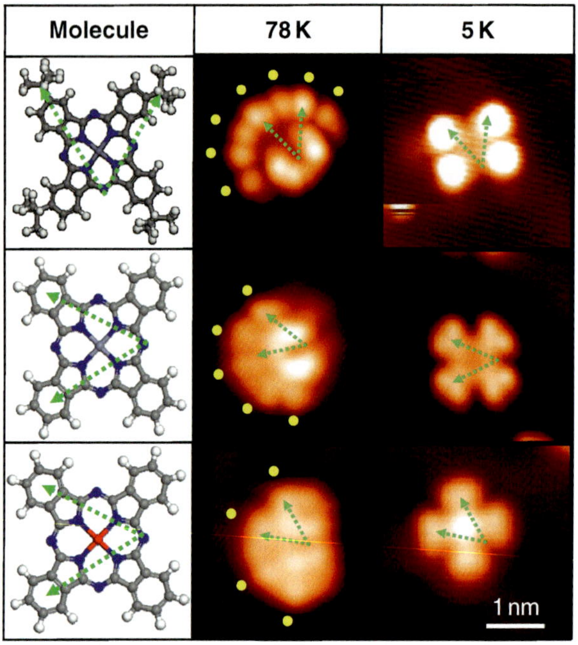

Fig. 11.9 STM images showing structural effect on molecular rotors formed by different molecules. TBZP (*top panel*), ZnPc (*middle panel*), and FePc (*bottom panel*) are located at the elbow positions of the Au(111) surface. In each panel, *left*, molecular structure; middle, rotation at 78 K; *right*, frozen at 5 K

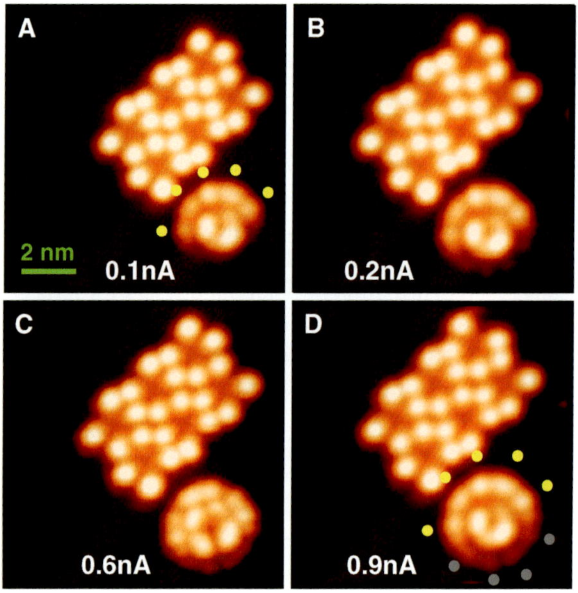

Fig. 11.10 Sequences of STM images show tunneling current effect on the TBZP molecular rotor when the sample bias is kept at –1.3 V. Tunneling current for each image is (**a**) 0.1, (**b**) 0.2, (**c**) 0.6, and (**d**) 0.9 nA. The molecular rotor can sweep faster and more degree ranges as the tunneling current increases

shows the current dependence of the TBZP molecular rotors. When increasing the tunneling current from a small current of 0.1 nA (Fig. 11.10a) to a larger one of 0.2 nA (Fig. 11.10b), the molecule rotates more actively so that the degree of circular arc of the rotation pattern becomes larger, which became more obvious when the current is kept static for a period of time (Fig. 11.10c). When the current is increased to 0.9 nA, a full circle of rotation pattern can be clearly observed (Fig. 11.10d). This gradual increase in the degree of the circular arc means that the molecular rotation movement becomes more active and can gradually override the barrier formed by the Au(111) reconstruction and covers the whole circle eventually. The above result shows a gradual activation process while increasing the current. Since the molecular rotation can be driven by thermal energy, the STM current provides a certain amount of energy by locally heating the rotating molecules, thus remarkably enhancing the rotating behavior of the molecular rotor.

11.3.4 STM-Driven Rotation on Reconstructed Surface Template

Besides the TPZP molecule and its analogues, another self-designed molecule (HB–NPB, $C_{64}N_2H_{76}$) shows a step-by-step rotation like a gear at herringbone Au(111) substrate [37], as demonstrated by Prof. Joachim and his colleagues. After depositing HB–NPB molecules onto a clean Au(111) surface, the molecules are anchored on top of an atomic defect, which holds a large potential energy barrier and acts as an atomic-scale axis for molecular rotations. HB–NPB was rotated clockwise step by step by gently pushing with an STM tip on one leg of the six legs of the HB–NPB molecule (see Fig. 11.11b). Impressively, the step-by-step rotation of the molecule is controllable. The molecular gear has nine stable stop stations in both directions as reported.

Adsorbed on the Au(111) surface, one HB–NPB molecule is imaged as a dented wheel where the six bright lobes correspond to the *t*-butyl groups of the molecule (see Fig. 11.11a). The appearance of HB–NBP molecules on Au(111) is different from that on Cu(111) surface in the STM images. Here the pyrimidine ring of the molecule appears in between two molecular legs, whereas on Cu(111) it appears collinear to one leg [84]. Thus the bright protrusion inside the wheel corresponds to the pyrimidine group of the HB–NBP molecule. On the Au(111) surface, due to the weak molecule–metal interactions, the molecular gear is very mobile and can be easily manipulated by an STM tip.

Researchers have mounted HB–NPB on three types of atomic-scale locations of Au(111) surface: a herringbone elbow, a single Au adatom, and a single atomic-sized defect natively bound to a herringbone elbow. At the herringbone elbow site, it is possible to rotate the molecule by pushing one of its legs using STM tip, but small lateral displacements are also observed during its rotation. This lateral motion is ascribed to the weak interaction between the aromatic core and the Au(111) surface. When HB–NPB is mounted on top of a single Au atom, it does not rotate concentrically with the Au adatom serving as a shaft because this adatom is always trapped in the middle of two legs of the molecule [94]. Thus neither the elbow site nor the Au atom site is suitable for a centered step-by-step rotation.

Fortunately, while the molecule is mounted on top of an atomic defect bound to one elbow, it is rotated clockwise step by step by gently pushing with the STM tip on one leg, generally the one associated with a chemical tag. After each manipulation,

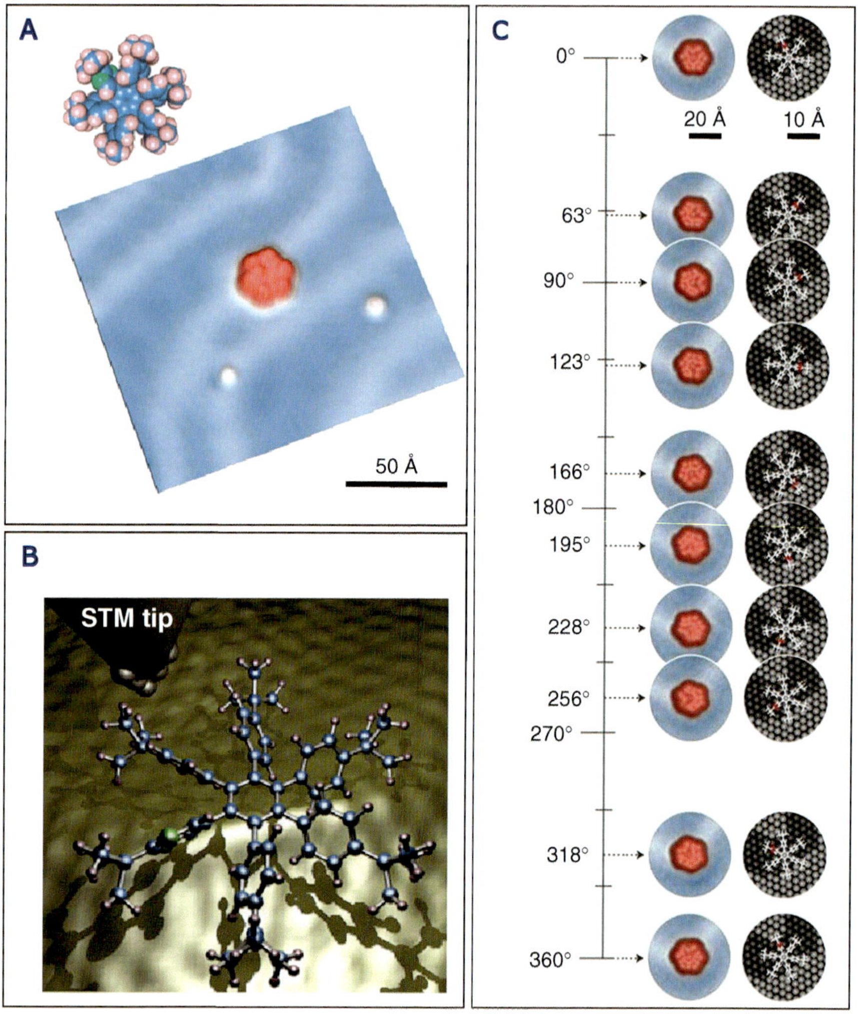

Fig. 11.11 (**a**) Chemical structure of a HB–NPB molecule and an STM image (31 pA, 0.1 V) of a single HB–NPB molecule adsorbed on Au(111). The herringbone reconstruction of the substrate is clearly seen. The molecule is located in the vicinity of two atomic-scale impurities. (**b**) Schematic drawing of HB–NPB on Au(111) surface, which shows that the molecule is located at an elbow site. An STM tip was used to gently push the chemically tagged leg and drive the rotation of the HB–NPB molecule. (**c**) Full step-by-step molecular gear rotation. STM images (5 pA, 0.1 V) showing orientation and position changes of a HB–NPB. The *right image* shows the initial configuration imaged before manipulation. The following images, taken after each tip manipulation, show the molecular gear stabilized at different stations. The tagged leg was used as a reference to determine the angle rotated by the gear with respect to the initial conformation. Reprinted with permission from [37]. Copyright 2009 by Nature Publishing Group

an image capturing the new in-plane configuration of the molecular gear was taken. Figure 11.11c shows a sequence of images illustrating a fully reproducible step-by-step clockwise HB–NBP rotation. The images show the in-plane conformations adopted by the gear after a full manipulation series. Similar to the Cu(111) case, the chemically tagged leg (interior bright spot) was used as a reference to determine the angle rotated by the molecule after each manipulation. Each push of the STM tip rotated the gear in small increments of 30–60°. Eventually, full rotation was possible in both clockwise and counterclockwise directions.

Based on theoretical calculations, the researchers further expounded a step-by-step rotation of HB–NPB molecules on Au(111) surface, as presented in Fig. 11.11c. The rotation was deliberately controlled by molecule trapping and manipulation. The molecule was trapped to a defect of elbow site. As we know, this site is connected by herringbone ridges, leaving different potential energy barriers for molecular rotation; thus these ridges can hinder the rotation of the molecule by interacting with the *t*-butyl end groups (chemically tagged leg). The gentle manipulation by STM tip drives the molecule to overcome the energy barrier and start to rotate, but this rotary motion will be blocked while the molecule encounters the next energy barrier at herringbone ridge. Thus the atomic structure of the surface reconstruction around the atomic defects determines the number of stable molecule stop stations.

11.3.5 Molecular Rotors with Variable Rotation Radii

Recently, Zhong [26] et al. from Münster University, Germany, have demonstrated a very exciting method for the design of molecular motors with variable rotation radii. They proposed a concept for designing molecular rotors containing two end groups connected by a linear linker. One of the end groups can interact strongly with substrate surfaces, serving as anchor or stator, while the other group has weak interaction with the substrate, which allows the rotor to undergo a rotation movement around the anchor (Fig. 11.12a). The rotation radius can be adjustable by simply changing the length of the linker. Thus three independent parts of the molecule are responsible for different functions to support the rotating movements of the molecule as a whole with changeable rotation dimensions. The researchers have successfully obtained molecular rotors with radii from 1.88 nm upto nearly 4 nm. Figure 11.12b shows the molecular rotors by FeCp-$(CH_2)_n$-Fc molecules at Au(111) surface with $n = 12$, 14, 18, and 28. It demonstrated that smaller and larger molecular rotors are accessible by simply changing the length of the connectors (number of n). The rotation of molecules with adjustable sizes shows different rotation radii.

For the molecules used in the experiment, they are diferrocene derivatives Fc-$(CH_2)_n$-Fc, which contain two ferrocene (Fc) groups connected by an oligoethylene chain [95]. By thermal activation during sample heating process at the temperature range of 350–400 K, one cyclopentadienyl ring was removed from the diferrocene molecule and transited to monocyclopentadienyl iron (FeCp) group. On metal surfaces, ferrocene (Fc) group has low binding energy (<0.6 eV) and negligible

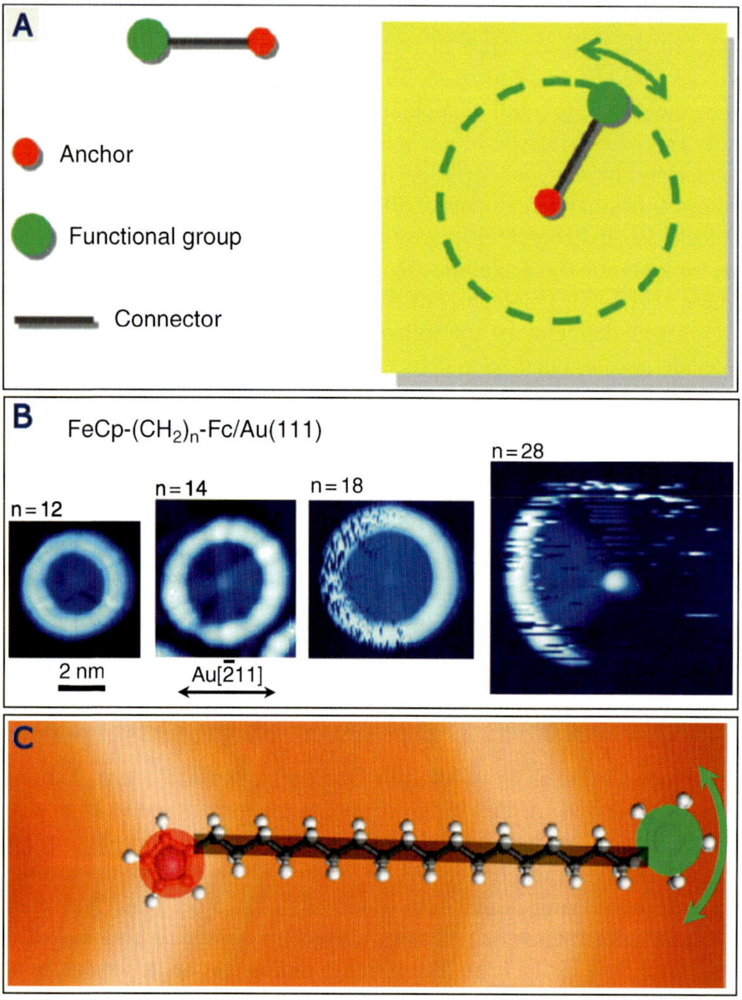

Fig. 11.12 (**a**) Molecular rotor (FeCp-$(CH_2)_n$-Fc) containing two end groups bridged by a linear linker. One end group strongly interacts with the surface serving as anchor while the other end group and the linker weakly interact with the surface, i.e., the FeCp terminal acts as anchor and the Fc group rotates around the anchor by thermal activation. (**b**) STM images demonstrate molecular rotors with variable rotation radii from 1.88 nm up to nearly 4 nm by FeCp-$(CH_2)_n$-Fc at the Au(111) surface. (**c**) Schematics showing the wheel-like molecule FeCp-$(CH_2)_{18}$-Fc anchored at elbow sites of the Au(111) surface. Reprinted with permission from [26]. Copyright 2009 by the American Chemical Society

surface diffusion barrier (~0.01 eV) and is mobile on the Au(111) surface even at 78 K. However, FeCp group is strongly bound to the surface with a binding energy of 3.66 eV. Compared with the Fc molecule, the FeCp group encounters much higher surface diffusion barriers and is immobilized on the Au(111) surface [96]. Therefore, the partially decomposed diferrocenes with an Fc group at one end and a FeCp complex at the other end are ideal candidates for surface-mounted molecular rotors. The FeCp terminal can serve as an anchor and the Fc group rotates around this anchor (Fig. 11.12c). In their experiment, they also demonstrated molecular rotation at Au(111) surface by using molecules with variable functional terminal groups, for example, thiol group.

The researchers also found that the rotation of molecules is temperature dependent. They have investigated the threshold temperature of molecular rotation by STM observations at different temperatures from 5 to 40 K. By cooling the samples to below 20 K, most partially decomposed diFc-14 molecules are frozen at the surface. The molecular rotors observed are thermally activated with a typical threshold temperature of about 20–40 K.

11.3.6 Rolling Motion of a Single Molecule at the Surface

In the above-mentioned examples, the rotation or translational movement of molecules is in-plane at the surface. In such cases the molecules normally lie flat on the surface and the rotation axis of the motions is therefore oriented perpendicular to the surface. Barrier for the motions is normally low and can be excited by STM tunneling current or activated thermally by controlling the substrate temperature. Except for the in-plane motion, the rolling of molecules vertical to the substrate is a very interesting subject, since it is also a useful motion at the macroscopic scale.

In 2007, Leonhard Grill et al. reported an elegant experiment to detect rolling motion of a wheel–axle–wheel molecule at the surface [68]. They used a STM tip to manipulate laterally a molecule and induce a rolling motion of the two-wheel molecule. This nanowheel molecule consists of two triptycene wheels, connected by an axle of carbon atoms ($C_{44}H_{24}$, see Fig. 11.13a). The size of the molecule is quite small with two 0.7 nm wheels in diameter and a 0.6 nm axle in length. After adsorption on a Cu(110) surface, the molecules exhibit a independent rotation of each wheel around the central axle. The Cu(110) lattice, which exhibits an anisotropic corrugation due to its close-packed rows of copper atoms in the [1–10] direction, holds anisotropic diffusion barrier for the adsorbate. Thus this anisotropic lattice is convenient for the rolling of molecules along its wide lattice direction ([001] direction of the Cu(110) substrate).

The molecules appear in the STM image (3 nm × 3 nm, see Fig. 11.13c) as two bright protrusions that correspond to two wheels of the molecules. After deposition onto a clean Cu(110) surface, the molecules are found in various orientations, in particular with their carbon axle parallel and perpendicular to the close-packed rows, which are the most interesting configurations by means of a rolling motion. The calculated STM image is in good agreement with the experimental ones, consisting of two bright spots at a distance of about 0.87 nm, which correspond to the intramolecular distance between two wheels of a single molecule (Fig. 11.13b).

The characteristics of the molecular movements are recorded in the STM feedback loop manipulation signal in real time. An arrow in Fig. 11.13c marks the pathway of the STM tip during manipulation of a two-wheel molecule. During the tip manipulation on the molecule, the tunneling current curves were taken, which shows sawtooth-shaped signals (see curves I and II in Fig. 11.13c). These signals

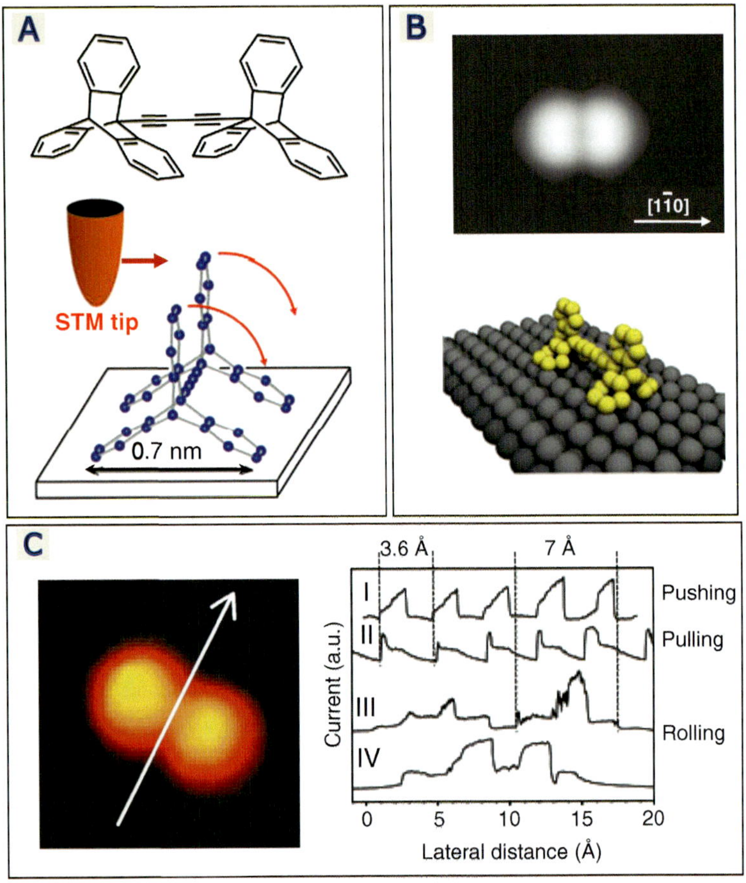

Fig. 11.13 (**a**) Chemical structure of two-wheel rolling molecule ($C_{44}H_{24}$) and scheme of a manipulation using an STM tip to induce a rolling motion (*red arrows* indicate the rotation of the wheels). Hydrogen atoms of the molecule are not plotted. (**b**) Upper: calculated molecular configuration of the wheel-dimer on Cu(110), with an orientation of the molecular shaft parallel to the substrate close-packed copper rows. Down: the corresponding calculated STM image (3.3 nm × 2.2 nm). (**c**) STM images (3 nm × 3 nm) of the wheel molecule on Cu(110) surface before the manipulation (the arrows mark the pathway along [001] direction of the substrate) and the corresponding manipulation signals (periodicities are indicated by *dashed lines*). Reprinted with permission from [68]. Copyright 2007 by Nature Publishing Group

can be assigned to a pushing (ascending edge) and a pulling (descending edge) mode of tip manipulation [97], separately. The 0.36 nm periodicity of these two signals corresponds to the distance between two atomic rows on the Cu(110) lattice. Therefore, the molecule hops from one copper row to the next along the [001] direction during manipulation.

But additionally, a hat-shaped signal (curve III in Fig. 11.13c) is found, which is assigned to the rolling motion of the nanowheel molecule, because its periodicity is about 0.7 nm and corresponds to a lateral motion of a wheel after a 120° rotation (around the molecular axle). The value is over two rows of the Cu(110) lattice along the [001] direction. A standard hopping motion over two rows can be excluded because the signal does not show any pulling or pushing signature. Thus the experiments show that a rolling motion is only possible if the molecules are manipulated perpendicular to the copper rows. The short pathway of the molecular motions can thus be considered as a fingerprint of the rolling mechanism of the molecule itself. A rolling motion occurs only if the molecule is manipulated perpendicular to its axle and along a surface direction of sufficient corrugation. This is the first example of a rolling motion of a single molecule on the solid surface.

Besides the above-reported two-wheel molecules, in another research group headed by Prof. James M. Tour of Rice University, the researchers designed four-wheel molecules and demonstrated a special nanomachine at a Au(111) surface in 2009 [98]. The home-made molecule is much complicated and is called as a nanodragster or nanovehicle. It is a mixed wheeled nanovehicle and has structural similarity to the macroscopic dragster: two small front wheels on a short axle and

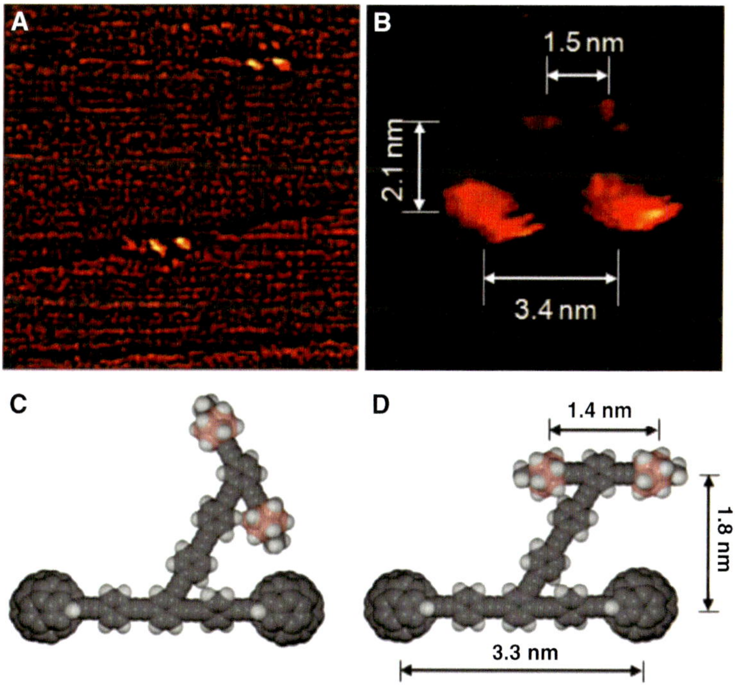

Fig. 11.14 Four-wheel molecular nanodragsters on an Au(111) substrate. (**a**) A filtered, large-area STM image (0.2 nA, –100 mV, 63.8 nm × 63.8 nm) of two nanodragsters in different orientations near step edges. (**b**) A close-up STM image (0.25 nA, –75 mV) of the nanodragster indicating the spacing of the front and rear wheels. (**c, d**) Models of two conformations as they are seen in the STM image in (**a**). (**d**) The calculated distances that roughly match the measured dimensions in (**b**). Reprinted with permission from [98]. Copyright 2009 by the American Chemical Society

two large rear wheels on a longer axle. As for its chemical structure, it is composed of a p-carborane-wheeled front axle and a C_{60}-wheeled rear axle. In the STM observations of this molecule on a clean Au(111) substrate, all four wheels are visible (Fig. 11.14), but the front p-carborane wheels appear much smaller than the rear C_{60} wheels.

The authors have reported that these nanodragster molecules have multiple conformations accessible when adsorbed on the Au(111) surface, and Fig. 11.14c, d shows examples of two such structures. If the front and rear axles are parallel, the rolling of the wheels will provide a linear motion of the whole molecule, whereas if two axles are angled, circular motion should be expected. It means that the rolling motion of the molecular wheels drives a translational movement of the molecule entity, like driving a real car. The STM images in Fig. 11.14a show two molecules with two different conformations positioned near the step edges of the substrate, and in the closeup image the p-carborane wheels can be visible (Fig. 11.14b). The sizes of the molecule in the STM image are in agreement with theoretic optimized models. The authors expected that the p-carborane front wheels would reduce the motion barrier compared to the nanocars with all wheels comprised of C_{60} groups, which were often introduced in the previous reports from the same research group [81, 82, 99, 100]. Thus these multifunctional molecules hold advantages over their old versions, such as the structural stability and the control of directional motion.

11.4 Array of Molecular Motors at Surfaces

The ability to move individual molecules with an STM tip is a powerful first step toward building complex molecular machines at the atomic scale. But for practical applications of such molecular machinery, it must be possible to construct it easily and at low cost, on a large scale. The key satisfying these requirements is finding systems of molecules that assemble themselves into the desired distributions on tailor-made surfaces. Exploring the self-assembly of rotated molecules on metallic surfaces provides application potential for the development of molecular nanodevices.

As the above-mentioned single tetra-*tert*-butyl zinc phthalocyanine (TBZP) molecules on the reconstructed Au(111) surface, it possesses a well-defined rotation axis fixed on the surface. By careful control, the researchers found that these single-molecule rotors form large-scale ordered arrays due to the regular reconstruction of the gold surface [16]. At almost all elbow sites of the reconstructed substrate, an off-center rotation axis is formed by chemical bonding between a nitrogen atom of the molecule and the gold adatom on the surface, which gives a well-defined anchoring point while the molecules have rotation.

Figure 11.15a shows an STM image of a large-scale array of TBZP molecules on Au(111). The molecules adsorb predominantly at the elbow positions of the surface reconstruction of the Au(111) surface. The surface reconstruction in this case acts as an atomically precise template for the selective adsorption of TBZP molecules on the Au(111) surface. High-resolution image (Fig. 11.15b) reveals a feature of

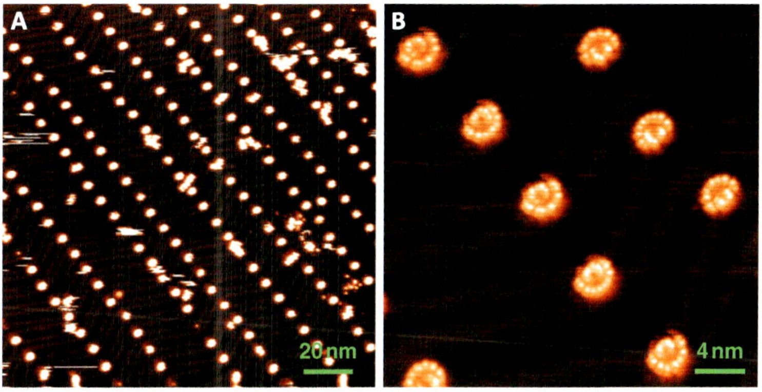

Fig. 11.15 (**a**) STM image of large-scale ordered array of single TBZP molecular rotors on a reconstructed Au(111) surface. (**b**) High-resolution STM image of single molecular rotors showing a "folding-fan" feature at elbow sites. The molecular rotors at two different elbow sites show different features due to the modulation by corrugation ridges (taken at 78 K by 0.07 nA, –1.3 V). Reprinted with permission from [16]. Copyright 2008 by the American Physical Society

the adsorbed molecules that is reminiscent of a folding fan. The "folding-fan" feature is actually the low-frequency image of a high-frequency molecular rotation on Au(111), driven by thermal energy as the above analysis.

The data demonstrate the self-assembled construction of a well-ordered array of single-molecule rotors on a gold surface, each of which is anchored to a fixed point on the surface and is free to rotate around a well-defined axis on the molecule. These results represent a potential step forward in the large-scale construction of arrays of molecular motors and even the integration of molecular nanomachines.

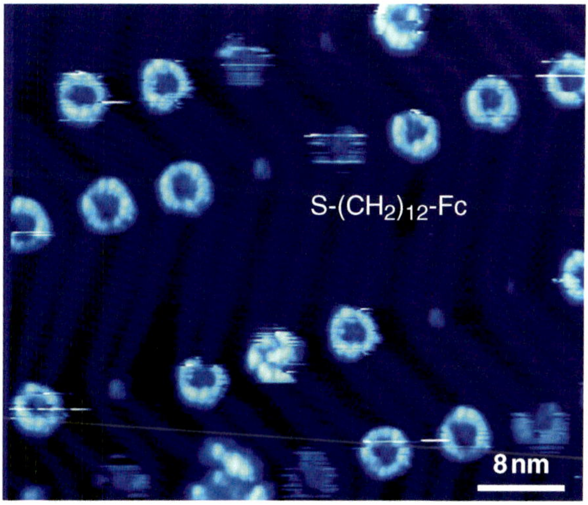

Fig. 11.16 STM image (acquired at 78 K by 5 pA, –1 V) showing a rotor array of S-$(CH_2)_{12}$-Fc molecules at an Au(111) surface. The dehydrogenated thiol groups serve as the anchors of a rotation. Reprinted with permission from [26]. Copyright 2009 by the American Chemical Society

Similar arrays of molecular rotors anchored by dehydrogenated thiol groups on the Au(111) surface are also reported [26]. Because of the easier dehydrogenation and bonding of the thiol groups with gold surfaces, thiol-anchored molecular rotors have formed ordered arrays of molecular rotors at elbow sites on the Au(111) surface (Fig. 11.16). Furthermore, the strong bonding of thiol groups with the gold atoms makes it possible to reduce or eliminate the area covered by the condensed molecules and to make a uniform distribution of the rotors by adjusting the deposition amount and the substrate temperature.

11.5 Outlook

The motion of single molecules plays an important role in nanoscale engineering. In order to realize the control of molecular motions at solid surfaces, a fixed rotation axis is an essential step toward the eventual fabrication of molecular rotors. One practical way is setting up strong interaction (i.e., designing a chemical bond)

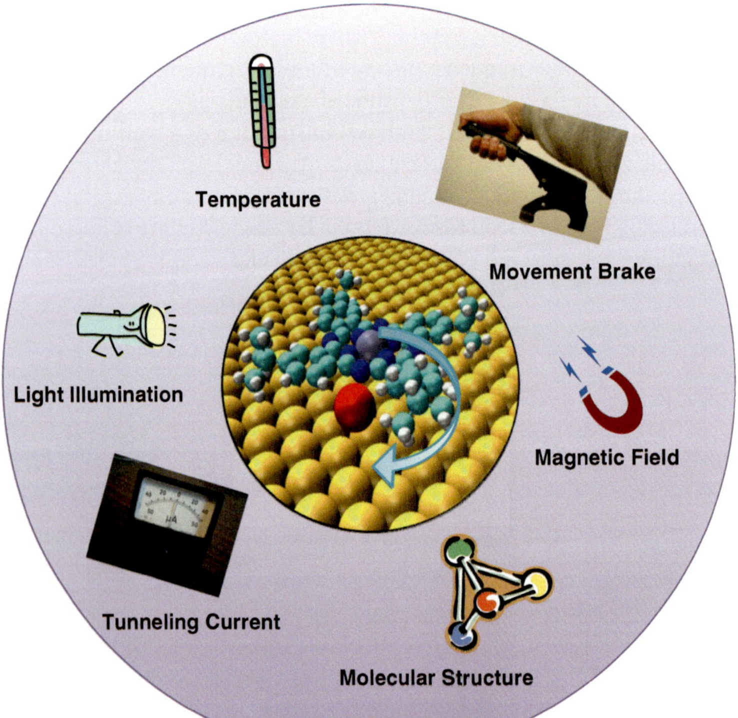

Fig. 11.17 Possible ways for control or manipulation of the movements of molecular rotors, such as changing the surrounding temperature, adding/removing movement brake, increasing/decreasing tunneling current, and modifying molecular structure

between an atom of the molecule and the underneath surface, which can offer not only a rotation axis at the surface but also a strong attachment which limits the excessive freedom of molecules, for example, translational movement. The control of the rotation direction and the adjustment of rotary speed/frequency of molecular rotors are the basic requirements for real applications in nanodevices, however, they still leave immense challenges for scientists. The exploration of patterned surface (like reconstructed metallic surfaces) is also a promising research subject, since they can be used as atomically well-defined templates for formation of large-scale molecular rotor arrays. And for the artificial nanomachines such as molecular motors and generators, multiple approaches of manipulation could be applied to achieve the proposed aims, for example, temperature variation, light illumination, electric/magnetic fields modulation, chemical structure modification and intrinsic/environmental brake mechanism (Fig. 11.17). By combining various methods and techniques, there is a great prospect to find some practical industrial solutions and finally lighten a new revolution of molecular machines.

11.6 Conclusion

The controllability of molecular motion is critical, which may convert external energy into motion at the single-molecule level. The rotational motion of single molecules at solid surfaces has been realized by several ways, for instance, by adjusting the surrounding temperatures or by employing external optical/electrical field. In some cases the motion state and stationary state of the molecular motors have even been switched and manipulated by the technique of STM. For molecular rotors, further efforts would be needed for the control of the direction and speed/frequency of movements. Individual molecular rotors would assemble into large-scale ordered arrays while keeping their original functions, which would be desirable and of essential importance for eventual applications in molecular nanodevices.

Acknowledgments The authors would like to thank S. X. Du, D. X. Shi, L. Gao, X. Lin, Z. H. Cheng, Z. T. Deng, N. Jiang, W. Ji, J. T. Sun, and Y. Y. Zhang for invaluable assistance in experiments and theoretical simulations. Work at IOP was partially supported by the NSFC, MOST, and CAS in China.

References

1. J.V. Barth, G. Costantini, K. Kern, Engineering atomic and molecular nanostructures on surfaces. Nature **437**, 671 (2005)
2. J.V. Barth, Molecular architectonic on metal surfaces. Annu. Rev. Phys. Chem. **58**, 375 (2007)
3. A.R. Pease, J.O. Jeppesen, J.F. Stoddart, Y. Luo, C.P. Collier, J.R. Heath, Switching devices based on interlocked molecules. Acc. Chem. Res. **34**, 433 (2001)
4. A. Facchetti, M.H. Yoon, T.J. Marks, Gate dielectrics for organic field-effect transistors: new opportunities for organic electronics. Adv. Mater. **17**, 1705 (2005)

5. O.M. Yaghi, M, O'Keeffe, N.W. Ockwig, H.K. Chae, M. Eddaoudi, J. Kim, Reticular synthesis and the design of new materials. Nature **423**, 705 (2003)
6. Q. Liu, Y.Y. Zhang, N. Jiang, H.G. Zhang, L. Gao, S.X. Du, H.J. Gao, Identifying multiple configurations of complex molecules in dynamical processes: time resolved tunneling spectroscopy and density functional theory calculation. Phys. Rev. Lett. **104**, 166101 (2010)
7. N. Jiang, Y.Y. Zhang, Q. Liu, Z.H. Cheng, Z.T. Deng, S.X. Du, H.J. Gao, M.J. Beck, S.T. Pantelides, Diffusivity control in molecule-on-metal systems using electric fields. Nano Lett. **10**, 1184 (2010)
8. L. Gao, H.J. Gao, Scanning tunneling microscopy of functional nanostructures on solid surfaces: manipulation, self-assembly, and applications. Prog. Surf. Sci. **85**, 28 (2010)
9. W. Ji, Z.Y. Lu, H. Gao, Electron core-hole interaction and its induced ionic structural relaxation in molecular systems under x-ray irradiation – reply. Phys. Rev. Lett. **99** (2007)
10. L. Gao, W. Ji, Y.B. Hu, Z.H. Cheng, Z.T. Deng, Q. Liu, N. Jiang, X. Lin, W. Guo, S.X. Du, W.A. Hofer, X.C. Xie, H.J. Gao, Site-specific Kondo effect at ambient temperatures in iron-based molecules. Phys. Rev. Lett. **99**, 106402 (2007)
11. M. Feng, L. Gao, S.X. Du, Z.T. Deng, Z.H. Cheng, W. Ji, D.Q. Zhang, X.F. Guo, X. Lin, L.F. Chi, D.B. Zhu, H. Fuchs, H.J. Gao, Observation of structural and conductance transition of rotaxane molecules at a submolecular scale. Adv. Funct. Mater. **17**, 770 (2007)
12. D. Shi, W. Ji, X. Lin, X. He, J. Lian, L. Gao, J. Cai, H. Lin, S. Du, F. Lin, C. Seidel, L. Chi, W. Hofer, H. Fuchs, H.J. Gao, Role of lateral alkyl chains in modulation of molecular structures on metal surfaces. Phys. Rev. Lett. **96**, 226101 (2006)
13. L. Gao, Z.T. Deng, W. Ji, X. Lin, Z.H. Cheng, X.B. He, D.X. Shi, H.J. Gao, Understanding and controlling the weakly interacting interface in perylene/Ag(110). Phys. Rev. B **73**, 075424 (2006)
14. M. Feng, X.F. Guo, X. Lin, X.B. He, W. Ji, S.X. Du, D.Q. Zhang, D.B. Zhu, H.J. Gao, Stable, reproducible nanorecording on rotaxane thin films. J. Am. Chem. Soc. **127**, 15338 (2005)
15. Y.L. Wang, W. Ji, D.X. Shi, S.X. Du, C. Seidel, Y.G. Ma, H.J. Gao, L.F. Chi, H. Fuchs, Structural evolution of pentacene on a Ag(110) surface. Phys. Rev. B **69**, 075408 (2004)
16. L. Gao, Q. Liu, Y.Y. Zhang, N. Jiang, H.G. Zhang, Z.H. Cheng, W.F. Qiu, S.X. Du, Y.Q. Liu, W.A. Hofer, H.J. Gao, Constructing an array of anchored single-molecule rotors on gold surfaces. Phys. Rev. Lett. **101**, 197209 (2008)
17. K. Svoboda, C.F. Schmidt, B.J. Schnapp, S.M. Block, Direct observation of kinesin stepping by optical trapping interferometry. Nature **365**, 721 (1993)
18. M. Schliwa, G. Woehlke, Molecular motors. Nature **422**, 759 (2003)
19. M. Schliwa, *Molecular Motors* (Wiley-VCH Verlag, Weinheim, 2003)
20. M.J. Schnitzer, S.M. Block, Kinesin hydrolyses one ATP per 8-nm step. Nature **388**, 386 (1997)
21. T. Atsumi, L. Mccarter, T. Imae, Polar and lateral flagellar motors of marine vibrio are driven by different ion-motive forces. Nature **355**, 182 (1992)
22. B. Wang, P. Kral, Chemically tunable nanoscale propellers of liquids. Phys. Rev. Lett. **98**, 266102 (2007)
23. A. Cavalcanti, B. Shirinzadeh, R.A. Freitas, Jr., T. Hogg, Nanorobot architecture for medical target identification. Nanotechnology **19**, 015103 (2008)
24. G.S. Kottas, L.I. Clarke, D. Horinek, Artificial molecular rotors. J. Michl, Chem. Rev. **105**, 1281 (2005)
25. J. Vacek, J. Michl, Artificial surface-mounted molecular rotors: molecular dynamics simulations. Adv. Funct. Mater. **17**, 730 (2007)
26. D. Zhong, T. Blomker, K. Wedeking, L. Chi, G. Erker, H. Fuchs, Surface-mounted molecular rotors with variable functional groups and rotation radii. Nano Lett. **9**, 4387 (2009)
27. J. Vacek, J. Michl, Molecular dynamics of a grid-mounted molecular dipolar rotor in a rotating electric field. Proc. Natl. Acad. Sci. USA **98**, 5481 (2001)
28. P. Kral, H.R. Sadeghpour, Laser spinning of nanotubes: a path to fast-rotating microdevices. Phys. Rev. B **65**, 161401 (2002)

29. S. Tan, H.A. Lopez, C.W. Cai, Y. Zhang, Optical trapping of single-walled carbon nanotubes. Nano Lett. **4**, 1415 (2004)
30. J. Berna, D.A. Leigh, M. Lubomska, S.M. Mendoza, E.M. Perez, P. Rudolf, G. Teobaldi, F. Zerbetto, Macroscopic transport by synthetic molecular machines. Nat. Mater. **4**, 704 (2005)
31. K. Petr, S.J. Tamar, Current-induced rotation of helical molecular wires. Chem. Phys. **123**, 184702 (2005)
32. J.E. Green, Wook J. Choi, A. Boukai, Y. Bunimovich, Johnston-E. Halperin, DeE. Ionno, Y. Luo, B.A. Sheriff, K. Xu, Shik Y. Shin, H.-R. Tseng, J.F. Stoddart, J.R. Heath, A 160-kilobit molecular electronic memory patterned at 1011 bits per square centimetre. Nature **445**, 414 (2007)
33. T.R. Kelly, H. De Silva, R.A. Silva, Unidirectional rotary motion in a molecular system. Nature **401**, 150 (1999)
34. R.D. Astumian, Thermodynamics and kinetics of a Brownian motor. Science **276**, 917 (1997)
35. A.P. Davis, Molecular machines: knowledge is power!. Nat. Nanotechnol. **2**, 135 (2007)
36. J.K. Gimzewski, C. Joachim, R.R. Schlittler, V. Langlais, H. Tang, I. Johannsen, Rotation of a single molecule within a supramolecular bearing. Science **281**, 531 (1998)
37. C. Manzano, W.H. Soe, H.S. Wong, F. Ample, A. Gourdon, N. Chandrasekhar, C. Joachim, Step-by-step rotation of a molecule-gear mounted on an atomic-scale axis. Nat. Mater. **8**, 576 (2009)
38. K.V. Mikkelsen, M.A. Ratner, Synthesis and properties of viologen functionalized poly (3-alkylthienylenes). Chem. Rev. **87**, 113 (1987)
39. P. Kral, Nonequilibrium linked cluster expansion for steady-state quantum transport. Phys. Rev. B **56**, 7293 (1997)
40. H. Park, J. Park, A.K.L. Lim, E.H. Anderson, A.P. Alivisatos, McP.L. Euen, Nanomechanical oscillations in a single-C60 transistor. Nature **407**, 57 (2000)
41. C.-C. Kaun, T. Seideman, Current-driven oscillations and time-dependent transport in nanojunctions. Phys. Rev. Lett. **94**, 226801 (2005)
42. B. Wang, L. Vukovic, P. Kral, Nanoscale rotary motors driven by electron tunneling. Phys. Rev. Lett. **101**, 186808 (2008)
43. M.M. Pollard, M. Lubomska, P. Rudolf, B.L. Feringa, Controlled rotary motion in a monolayer of molecular motors. Angew. Chem. Int. Ed. **46**, 1278 (2007)
44. N. Katsonis, M. Lubomska, M.M. Pollard, B.L. Feringa, P. Rudolf, Synthetic light-activated molecular switches and motors on surfaces. Prog. Surf. Sci. **82**, 407 (2007)
45. N. Henningsen, K.J. Franke, I.F. Torrente, G. Sehulze, B. Priewisch, K. Ruck-Braun, J. Dokic, T. Klamroth, P. Saalfrank, J.I. Pascual, Inducing the rotation of a single phenyl ring with tunneling electrons. J. Phys. Chem. C **111**, 14843 (2007)
46. M.D. Alvey, J.T. Yates, K.J. Uram, Electron-stimulated-desorption ion angular distributions of negative ions. J. Chem. Phys. **87**, 7221 (1987)
47. N. Wintjes, D. Bonifazi, F.Y. Cheng, A. Kiebele, M. Stohr, T. Jung, H. Spillmann, F. Diederich, A supramolecular multiposition rotary device. Angew. Chem. Int. Ed. **46**, 4089 (2007)
48. A.E. Baber, H.L. Tierney, E.C.H. Sykes, Quantitative single-molecule study of thioether molecular rotors. ACS Nano **2**, 2385 (2008)
49. O.P.H. Vaughan, F.J. Williams, N. Bampos, R.M. Lambert, A chemical switchable molecular pinwheel. Angew. Chem. Int. Ed. **45**, 3779 (2006)
50. D.O. Bellisario, A.E. Baber, H.L. Tierney, E.C.H Sykes, Engineering dislocation networks for the directed assembly of two-dimensional rotor arrays. J. Phys. Chem. C **113**, 5895 (2009)
51. R.A. Van Delden, M.K.J. ter Wiel, M.M. Pollard, J. Vicario, N. Koumura, B.L. Feringa, Unidirectional molecular motor on a gold surface. Nature **437**, 1337 (2005)
52. G. London, G.T. Carroll, T.F. Landaluce, M.M. Pollard, P. Rudolf, B.L. Feringa, Light-driven altitudinal molecular motors on surfaces. Chem. Commun. **45**, 1712 (2009)

53. B.C. Stipe, M.A. Rezaei, W. Ho, Inducing and viewing the rotational motion of a single molecule. Science **279**, 1907 (1998)
54. G. Binnig, H. Rohrer, Scanning Tunneling Microscopy. Helv. Phys. Acta **55**, 726 (1982)
55. G. Binnig, H. Rohrer, Scanning tunneling microscopy – from birth to adolescence. Rev. Mod. Phys. **59**, 615 (1987)
56. A. Zhao, Q. Li, L. Chen, H. Xiang, W. Wang, S. Pan, B. Wang, X. Xiao, J. Yang, J.G. Hou, Q. Zhu, Controlling the Kondo effect of an adsorbed magnetic ion through its chemical bonding. Science **309**, 1542 (2005)
57. P. Wahl, L. Diekhöner, G. Wittich, L. Vitali, M.A. Schneider, K. Kern, Kondo effect of molecular complexes at surfaces: ligand control of the local spin coupling. Phys. Rev. Lett. **95**, 166601 (2005)
58. N. Tsukahara, K.-I. Noto, M. Ohara, S. Shiraki, N. Takagi, Y. Takata, J. Miyawaki, M. Taguchi, A. Chainani, S. Shin, M. Kawai, Adsorption-induced switching of magnetic anisotropy in a single iron(II) phthalocyanine molecule on an oxidized Cu(110) surface. Phys. Rev. Lett. **102**, 167203 (2009)
59. X. Chen, Y.S. Fu, S.H. Ji, T. Zhang, P. Cheng, X.C. Ma, X.L. Zou, W.H. Duan, J.F. Jia, Q.K. Xue, Probing superexchange interaction in molecular magnets by spin-flip spectroscopy and microscopy. Phys. Rev. Lett. **101**, 197208 (2008)
60. B.C. Stipe, M.A. Rezaei, W. Ho, Single-molecular vibrational spectroscopy and microscopy. Science **280**, 1732 (1998)
61. B.C. Stipe, M.A. Rezaei, W. Ho, Inducing and viewing the rotational motion of a single molecule. Science **279**, 1907 (1998)
62. J.K. Gimzewski, C. Joachim, Nanoscale science of single molecules using local probes. Science **283**, 1683 (1999)
63. J.A. Stroscio, D.M. Eigler, Atomic and molecular manipulation with the scanning tunneling microscope. Science **254**, 1319 (1991)
64. P. Avouris, Manipulation of matter at the atomic and molecular-levels. Acc. Chem. Res. **28**, 95 (1995)
65. F. Rosei, M. Schunack, Y. Naitoh, P. Jiang, A. Gourdon, E. Laegsgaard, I. Stensgaard, C. Joachim, F. Besenbacher, Properties of large organic molecules at surfaces. Prog. Surf. Sci. **71**, 95 (2003)
66. C. Joachim, J.K. Gimzewski, Electronics using hybrid-molecular and mono-molecular devices, A. Aviram. Nature **408**, 541 (2000)
67. D.M. Eigler, C.P. Lutz, W.E. Rudge, An atomic switch realized with the scanning tunnelling microscope. Nature **352**, 600 (1991)
68. L. Grill, K.H. Rieder, F. Moresco, G. Rapenne, S. Stojkovic, X. Bouju, C. Joachim, Rolling a single molecular wheel at the atomic scale. Nat. Nanotechnol. **2**, 95 (2007)
69. N. Koumura, R.W.J. Zijlstra, R.A. van Delden, N. Harada, B.L. Feringa, Light-driven monodirectional molecular rotor. Nature **401**, 152 (1999)
70. K. Tashiro, K. Konishi, T. Aida, Metal bisporphyrinate double-decker complexes as redox-responsive rotating modules. Studies on ligand rotation activities of the reduced and oxidized forms using chirality as a probe. J. Am. Chem. Soc. **122**, 7921 (2000)
71. T.R. Kelly, R.A. Silva, H. De Silva, S. Jasmin, Y. Zhao, A rationally designed prototype of a molecular motor. J. Am. Chem. Soc. **122**, 6935 (2000)
72. M.K.J. ter Wiel, R.A. van Delden, A. Meetsma, B.L. Feringa, Increased speed of rotation for the smallest light-driven molecular motor. J. Am. Chem. Soc. **125**, 15076 (2003)
73. D.A. Leigh, J.K.Y. Wong, F. Dehez, F. Zerbetto, Miniaturized gas ionization sensors using carbon nanotubes. Nature **424**, 174 (2003)
74. D. Horinek, Molecular dynamics simulation of an electric field driven dipolar molecular rotor attached to a quartz glass surface. J. Michl. J. Am. Chem. Soc. **125**, 11900 (2003)
75. H. Jian, J.M. Tour, En route to surface-bound electric field-driven molecular motors. J. Org. Chem. **68**, 5091 (2003)

76. J.V. Hernandez, E.R. Kay, D.A. Leigh, Periodic mesoporous dendrisilicas. Science **306**, 1532 (2004)
77. W.R. Browne, B.L. Feringa, Making molecular machines work. Nat. Nanotechnol. **1**, 25 (2006)
78. E.R. Kay, D.A. Leigh, F. Zerbetto, Synthetic molecular motors and mechanical machines. Angew. Chem. Int. Ed. **46**, 72 (2007)
79. S.J.H. Griessl, M. Lackinger, F. Jamitzky, T. Markert, M. Hietschold, W.M. Heckl, Incorporation and manipulation of coronene in an organic template structure. Langmuir **20**, 9403 (2004)
80. J.K. Gimzewski, T.A. Jung, M.T. Cuberes, R.R. Schlittler, Scanning tunneling microscopy of individual molecules: beyond imaging. Surf. Sci. **386**, 101 (1997)
81. Y. Shirai, A.J. Osgood, Y.M. Zhao, K.F. Kelly, J.M. Tour, Directional control in thermally driven single-molecule nanocars. Nano Lett. **5**, 2330 (2005)
82. Y. Shirai, A.J. Osgood, Y. Zhao, Y. Yao, L. Saudan, H. Yang, Y.-H. Chiu, L.B. Alemany, T. Sasaki, J.-F. Morin, J.M. Guerrero, K.F. Kelly, J.M. Tour, Surface-rolling molecules. J. Am. Chem. Soc. **128**, 4854 (2006)
83. D.L. Keeling, Bond breaking coupled with translation in rolling of covalently bound molecules. Phys. Rev. Lett. **94**, 146104 (2005)
84. F. Chiaravalloti, L. Gross, K.-H. Rieder, S.M. Stojkovic, A. Gourdon, C. Joachim, F. Moresco, A rack-and-pinion device at the molecular scale. Nat. Mater. **6**, 30 (2007)
85. C. Wöll, S. Chiang, R.J. Wilson, P.H. Lippel, Determination of atom positions at stacking-fault dislocations on Au(111) by scanning tunneling microscopy. Phys. Rev. B **39**, 7988 (1989)
86. J.V. Barth, H. Brune, G. Ertl, R.J. Behm, Scanning tunneling microscopy observations on the reconstructed Au(111) surface-atomic-structure, long-range superstructure, rotational domains, and surface-defects. Phys. Rev. B **42**, 9307 (1990)
87. C.C. Perry, S. Haq, B.G. Frederick, N.V. Richardson, Face specificity and the role of metal adatoms in molecular reorientation at surfaces. Surf. Sci. **409**, 512 (1998)
88. H. Röder, E. Hahn, H. Brune, J.P. Bucher, K. Kern, Building one-dimensional and two-dimensional nanostructures by diffusion-controlled aggregation at surfaces. Nature **366**, 141 (1993)
89. A. Wander, M.A. Van Hove, G.A. Somorjai, Molecule-induced displacive reconstruction in a substrate surface: ethylidyne adsorbed on Rh(111) studied by low-energy-electron diffraction. Phys. Rev. Lett. **67**, 626 (1991)
90. S. Stepanow, T. Strunskus, M. Lingenfelder, A. Dmitriev, H. Spillmann, N. Lin, J.V. Barth, C. Woll, K. Kern, Deprotonation-driven phase transformations in terephthalic acid self-assembly on CU(100). J. Phys. Chem. B **108**, 19392 (2004)
91. M. Peter, C.S. Dan, J.T. Yates, Jr., Gold-adatom-mediated bonding in self-assembled short-chain alkanethiolate species on the Au(111) surface. Phys. Rev. Lett. **97**, 146103 (2006)
92. L. Limot, J. Kröger, R. Berndt, A. Garcia-Lekue, W.A. Hofer, Atom transfer and single-adatom contacts. Phys. Rev. Lett. **94**, 126102 (2005)
93. H.-G. Zhang, J.-H. Mao, Q. Liu, N. Jiang, H.-T. Zhou, H.-M. Guo, D.-X. Shi, H.-J. Gao, Manipulation and control of a single molecular rotor on Au(111) surface. Chin. Phys. B **19**, 018105 (2010)
94. T.A. Jung, R.R. Schlittler, J.K. Gimzewski, H. Tang, C. Joachim, Controlled room-temperature positioning of individual molecules: molecular flexure and motion. Science **271**, 181 (1996)
95. W. Katrin, M. Zhongcheng, K. Gerald, S. Jesus Cano, L. Christian Muck, G. Stefan, E. Gerhard, F. Roland, C. Lifeng, W. Wenchong, Z. Dingyong, F. Harald, Oligoethylene chains terminated by ferrocenyl end groups: synthesis, structural properties, and two-dimensional self-assembly on surfaces. Chem. Eur. J. **12**, 1618 (2006)

96. D.Y. Zhong, J. Franke, T. Blomker, G. Erker, L.F. Chi, H. Fuchs, Manipulating surface diffusion ability of single molecules by scanning tunneling microscopy. Nano Lett. **9**, 132 (2008)
97. L. Bartels, G. Meyer, K.-H. Rieder, Basic steps of lateral manipulation of single atoms and diatomic clusters with a scanning tunneling microscope tip. Phys. Rev. Lett. **79**, 697 (1997)
98. G. Vives, J. Kang, K.F. Kelly, J.M. Tour, Molecular machinery: synthesis of a "nanodragster". Org. Lett. **11**, 5602 (2009)
99. Y. Shirai, J.-F. Morin, T. Sasaki, J.M. Guerrero, J.M. Tour, Recent progress on nanovehicles. Chem. Soc. Rev. **35**, 1043 (2006)
100. G. Vives, J.M. Tour, Molecular machinery: synthesis of a "nanodragster". Acc. Chem. Res. **43**, 473 (2009)

Chapter 12
Nanophotonic Devices Based on ZnO Nanowires

Qing Yang, Limin Tong, and Zhong Lin Wang

12.1 Introduction

ZnO nanowires (NWs) have attracted a great deal of interest because of their unique semiconducting, piezoelectric, biocompatibility, and optoelectronic properties, which are the fundamentals for their applications in electronics, photonics, biology, environmental science, and energy [1–3]. The attractive features of ZnO for photonic applications include its wide bandgap (3.37 eV), high exciton binding energy (60 meV), relatively high refractive index ($n > 2$ at visible spectral range), and several other manufacturing advantages of ZnO, including the availability of large area substrates at a relatively low cost, amenability to wet chemical etching, great tolerance to high-energy radiation, and long-term stability. Additionally, ZnO exhibits the most splendid and abundant configurations of nanostructures that one material can form. ZnO nanostructures can be grown by a variety of methods, especially by low-cost and low-temperature methods. They have great potential for a variety of photonic technological applications, such as optical interconnect [4, 5], ultraviolet laser [5–9], photodetector [10–15], dye-sensitized solar cell (DSSC) [16–20], and light-emitting diode (LED) [21–26], as shown in Fig. 12.1. Furthermore, because ZnO is also a piezoelectric material, the coupling of optical, mechanical, and electrical properties of ZnO NW provides new opportunities for fabricating functional devices [3, 27–29], aiming at improving the performance of optoelectronic devices [28, 29] and providing an effective method to integrate optomechanical devices with microelectronic systems [27].

All of these potential advantages motivate intense interest in ZnO NWs, so that several reviews of various aspects of this interesting material have been published in recent years [1–3, 30–32]. In this chapter, we will focus on photonic devices based on ZnO NWs and provide an overview of pure optics, optoelectronics (optic and

Z.L. Wang (✉)
School of Materials Science and Engineering, Georgia Institute of Technology, Atlanta, GA 30332-0245, USA
e-mail: zlwang@gatech.edu; zhong.wang@mse.gatech.edu

W.L. Zhou, Z.L. Wang (eds.), *Three-Dimensional Nanoarchitectures*,
DOI 10.1007/978-1-4419-9822-4_12,

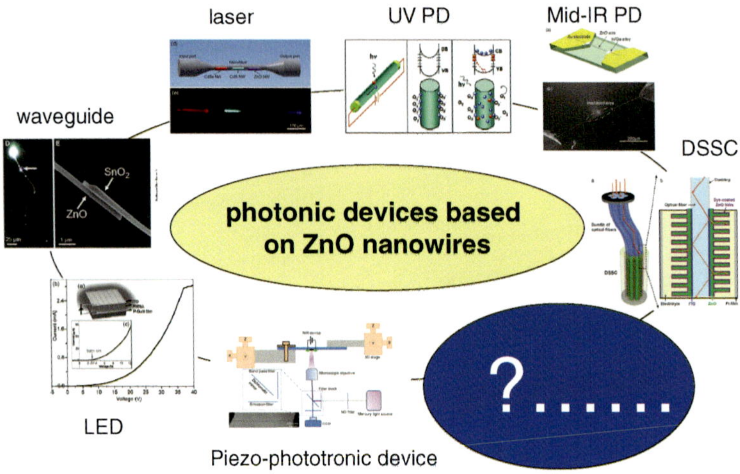

Fig. 12.1 Typical photonic devices based on ZnO NWs

electronic coupling), and piezo-phototronics (optoelectronic–mechanical coupling) devices fabricated by ZnO NWs.

12.2 Pure Optical Devices Based on ZnO NWs

ZnO is a direct wide bandgap semiconductor, with a high exciton binding energy of 60 meV and a large refractive index across the visible spectral range ($n > 2$). Such NWs are transparent throughout the visible spectral region, due to the large room temperature bandgap $E_{gap} = 3.37$ eV of ZnO. These attractive features make ZnO NWs an ideal candidate for optical devices such as excellent subwavelength waveguide [4, 5] and ultraviolet laser [7–9, 30]. ZnO is a highly polar semiconductor that is often used for nonlinear optics, e.g., second-harmonic generation in ZnO NW [33, 34].

12.2.1 ZnO NW Subwavelength Waveguides and Their Applications

Waveguides are very important and fundamental elements in photonic devices. Minimizing the width of the waveguides and the ability to manipulate light within submicrometer volumes are vital for highly integrated light-based devices to be realized.

Conventional single-mode fibers are huge on a nanotechnology scale, with core diameters about five times the wavelength and 125-μm claddings. However, single-mode fiber diameter can be reduced by increasing the core–cladding refractive index

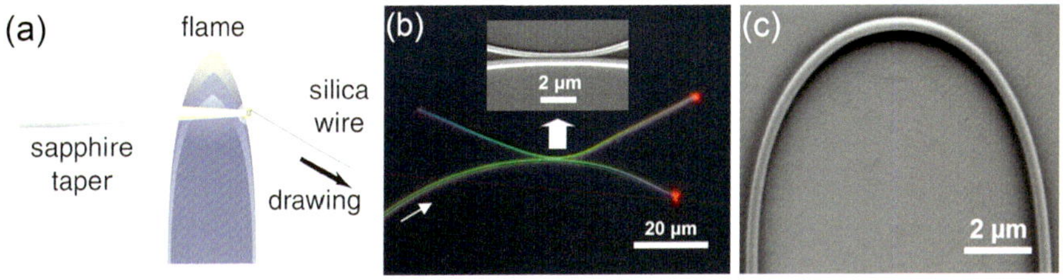

Fig. 12.2 (**a**) Schematic diagram of the drawing of the wire from a coil of micrometer-diameter silica wire wound around the tip of a sapphire taper. The sapphire taper is heated with a CH_3OH torch with a nozzle diameter of about 6 mm. The wire is drawn in a direction perpendicular to the sapphire taper [35]. (**b**) Optical micrograph of an optical coupler assembled using two tellurite glass NWs (350 and 450 nm in diameter, respectively) on the surface of a silicate glass. The coupler splits the 633-nm-wavelength light equally [38]. (**c**) SEM image of an elastically bent 320-nm-diameter silicate glass NW [38]

difference. Harvard University researchers fabricated fibers down to diameters as small as 40 nm, creating bare dielectric waveguides surrounded by air by top-down physical drawing method (Fig. 12.2a) [35]. As the diameter decreases, an increasing fraction propagates along the surface of the waveguide, as for analogous microwave guides. Propagation loss for a typical 400-nm waveguide is below 0.001 dB/mm [36], that is, orders of magnitude higher than standard fibers, but it is promising for nanophotonics in which transmission distances would be very short. Furthermore, because light propagates on the outside of the waveguide, it can be coupled between nanofibers by touching them, without having to align their cores precisely with each other [37, 38] (Fig. 12.2b). Surface propagation makes the nanofibers potentially attractive for sensing applications [39]. And because their diameters are so small, they can be bent very tightly [37, 38] (Fig. 12.2c). The applications of subwavelength optical fiber include optical sensors [39], nonlinear optics [36], fiber couplers [37, 38], interferometer [40], resonators [41], lasers [42], and atom trapping and guiding [43]. Figure 12.3 shows some typical applications of subwavelength optical fiber.

It is becoming more and more important to find materials with several properties to act as links between devices, based on different phenomena and interactions, which can be processed at the nanoscale. This includes heterogeneous functions by coupling properties such as optics, electronics, mechanics, and magnetism. Compared with the insulator dielectric fibers such as silica, glass, and polymer subwavelength optical fibers, chemically synthesized semiconductor NWs have several special features that make them good multifunctional photonic building blocks, including inherent one dimensionality, a diversity of optical and electrical properties, good size control, and low surface roughness. In 2004, M. Law et al. demonstrated that semiconductor nanoribbon (nanowire) could act as subwavelength optical waveguides [4]. They also demonstrated the assembly of ribbon waveguides with NW light sources and detectors as an important step toward building NW photonic circuitry.

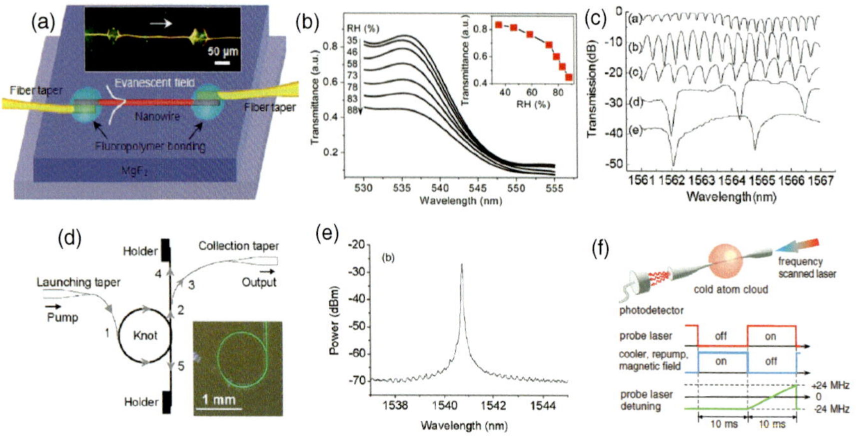

Fig. 12.3 Typical applications of glass and polymer subwavelength optical fiber. (**a**) Schematic illustration of a polyacrylamide (PAM) single NW humidity sensor. *Inset*: optical microscope image of an MgF_2-supported 410-nm-diameter PAM NW with a 532-nm-wavelength light launched from the left side. The *white arrow* indicates the direction of light propagation [39]. (**b**) Transmittance of an MgF_2-supported 410-nm-diameter PAM NW exposed to atmosphere of RH from 35 to 88%. *Inset*: transmittances of the NW at 532-nm wavelength [39]. (**c**) Transmission spectra of a microfiber knot with varied diameter. The diameter of the knot is reduced in steps. The transmission intensities for the different knot sizes are offset for clarity [41]. (**d**) Schematic diagram of the structure of a microfiber knot laser with pump and signal light paths. The pump light is launched into the knot from port 1, and the signal light is collected from port 2. *Inset*: an optical microscope image of an Er:Yb-doped phosphate glass microfiber knot pumped at a wavelength of 975 nm. The green upconverted photoluminescence is clearly seen [42]. (**e**) Single longitudinal mode laser emission with pump power above the threshold [42]. (**f**) Schematic experimental setup for atom trapping and guiding using ultrathin silica fiber [43]

Figure 12.4a shows light injection into a SnO_2 ribbon by an optically pumped ZnO NW. Because of the reflection at the ends of the NW and nanoribbon, the propagation light is modulated by the ribbon cavity (Fig. 12.4c). The opposite configuration can be used as a PD if light is input from one NW locally and the coupling semiconductor NW can absorb the light and provide electrical signals (Fig. 12.4d). The author used an NSOM tip to excite the nanoribbon to provide sufficient spatial resolution to detect waveguided light and eliminate the scattered background laser. ZnO NW can act as a detector in this case because it can weakly absorb subbandgap light. It is quite possible to get much higher sensitivity if CdS or CdSe NW is used, which has a narrower bandgap to detect the propagation light in ZnO or SnO_2 NW.

In order to launch light from commercial laser to subwavelength waveguides, several methods have been used, such as evanescent coupling [35, 44], lens focus [45], and butt coupling [46]. Figure 12.5a, b shows typical schematic diagrams of evanescent coupling and lens focus coupling, respectively. In lens focus method, a laser beam with its wavelength falling in the absorption band of the semiconductor NW is focused by a lens to excite the NW for obtaining luminescent emission. The photoluminescence, usually centered on some specific wavelengths and available

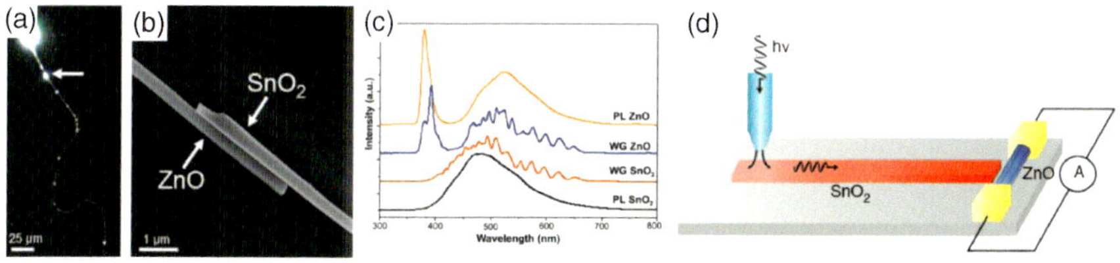

Fig. 12.4 (**a**) A true-color dark-field PL image of a ZnO NW (56-μm long, at *top*, pumped at 3.8 eV) channeling light into a SnO_2 nanoribbon (265-μm long, at *bottom*). The *arrow* denotes the location of the junction. (**b**) An SEM image of the wire–ribbon junction. (**c**) Spectra of the coupled structures taken at different excitation and collection locations. From *top to bottom*: unguided PL of the NW, waveguided (WG) emission from the ZnO wire collected at the bottom terminus of the ribbon, waveguided emission from the SnO_2 ribbon excited just below the junction and collected at its bottom terminus, and unguided PL of the ribbon. The emission from the ZnO NW is modulated during its transit through the nanoribbon cavity. (**d**) Schematic drawing of integrated electrical detection of light at a ribbon/wire junction. Figures adopted from [4]

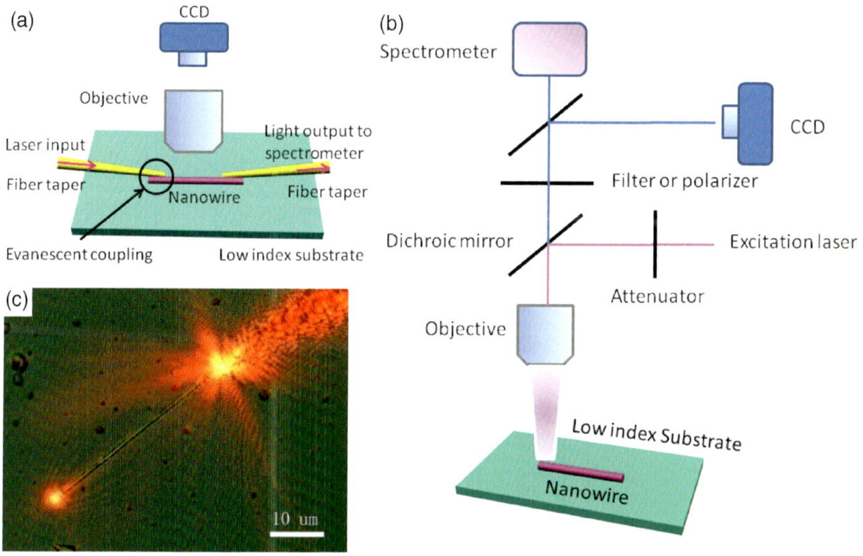

Fig. 12.5 Schematic diagram of evanescent coupling (**a**) and lens focus method (**b**) to launch light from commercial laser to subwavelength waveguides. (**c**) Optical micrograph of launching 633-nm-wavelength light into a ZnO NW by means of a silica nanotaper

within relatively narrow spectral bands, is used for optical characterization of the NW. This method provides several advantages, e.g., easy to control the polarization of excitation laser and more convenient to couple to free space laser. However, with this technique, the wavelength of the light that can be used is limited to the available spectral range of the photoluminescence and the coupling efficiency is relatively low because the detection geometry in Fig. 12.5b only permits collection of a small fraction of the total output power since light in the waveguide is primarily in the direction along the NW axis. In order to improve the collection efficiency of lens focus method, Zimmler introduced a "head-on" geometry to collect the emission along the NW [9]. Schematically, an NW is partially suspended in air and partially resting on a substrate. The NW is then excited uniformly along its entire length, and the emission is collected from one end, at an angle of 90° from the excitation beam.

The evanescent coupling method (Fig. 12.5a, c) is an alternative promising method to launch light into a single semiconductor NW using a silica fiber nanotaper, which is valid for launching light into a single NW from ultraviolet to infrared spectral range with a much higher coupling efficiency. Because light propagates on the outside of the waveguide, it can be coupled between nanofibers by touching them, without having to align their cores precisely with each other [37, 38]. Theoretical investigation showed that the coupling between two subwavelength waveguides is a strong coupling [47]. Compared with weakly coupled waveguides, strongly coupled NWs can provide a much smaller transfer length without sacrificing the high coupling efficiency (maximum coupling efficiency > 90%), providing opportunities for efficiently sending light into or coupling light out of the low-dimensional NWs, as well as for developing highly compact photonic devices with optical NWs.

12.2.2 Optically Pumped Lasers in ZnO NWs

Semiconductor lasers based on cadmium sulfide (CdS) [8], zinc oxide (ZnO) [6–9], and gallium nitride (GaN) NWs [48] and gallium antimonide (GaSb) subwavelength wires [49] have gained considerable attention. Such devices could potentially generate highly localized intense monochromatic light in a geometry ideally suited for efficient coupling into nanophotonic elements such as quantum dots, metallic nanoparticles, plasmonic waveguides, and even biological specimens. Thus, NW lasers could become a critical component in the study and development of novel nanoscale photonic elements. For wide bandgap semiconductor materials, excitonic recombination is a more efficient radiative process and can facilitate lower threshold-stimulated emission than electron–hole plasma process [50, 51]. To achieve efficient excitonic laser action at room temperature, the binding energy of the exciton must be much greater than the thermal energy at room temperature (26 meV). In this regard, ZnO is a good candidate because of its high exciton binding energy (60 meV).

As early as 2001, Huang et al. demonstrated room temperature ultraviolet lasing in ZnO NW arrays grown on sapphire substrate by vapor transport and condensation process [6]. The samples were optically pumped by the fourth harmonic of Nd:yttrium–aluminum–garnet laser (266-nm, 3-ns pulse width). The threshold is about 40 kW/cm^2. The diameter of the NWs varied from 20 to 150 nm and length up to 10 μm. The origin of the laser oscillation in such small NWs remains controversial. The author thought the single-crystalline, well-faceted NWs acted as natural resonance cavities. However, in another systematic study of laser action in ZnO NWs, Zimmler et al. found that NWs with diameters smaller than 150 nm did not

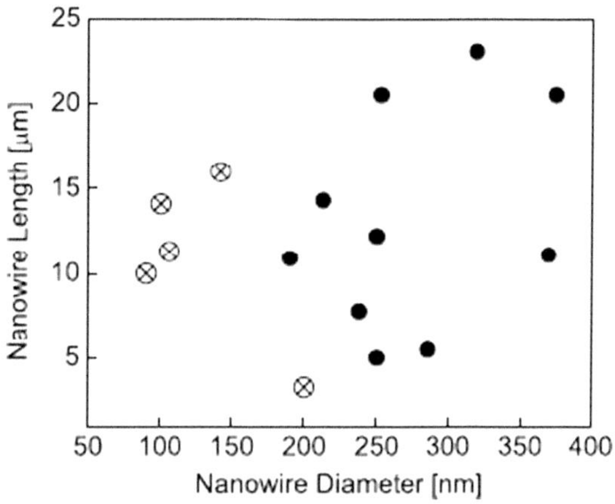

Fig. 12.6 Experimental results on lasing for NWs of different dimensions: *crosses* (×) indicate NWs which did not lase and *circles* (•) indicate those which did. Figures adopted from [9]

reach threshold, independent of the NW length (Fig. 12.6) [9]. This is because the reflection coefficient that is related to the confinement factor is dependent on the diameter of the NWs. For example, for the most confined mode in a ZnO NW on a SiO_2/Si substrate, the fraction of the mode intensity inside the NW decreases from ~85% for $D = 150$ nm to <1% for $D = 100$ nm [9].

Laser oscillation occurs when the round-trip gain matches the round-trip losses; in NWs the latter are typically dominated by the mirror losses (transmission at the end facets). The unambiguous evidence of optically pumped laser oscillation includes several obvious features. First, the laser spectra are dominated by sharp emission lines: their intensity is orders of magnitude greater than the spontaneous emission background (Fig. 12.7a, $I_{ex} > 300$ kW/cm^2), while the spontaneous emission spectra are broad and featureless and light is emitted essentially isotropically along the NW (Fig. 12.7a, b, $I_{ex} < 200$ kW/cm^2). Second, in the pump intensity dependence of the total output power curve, laser threshold is approached when the output power exhibits a superlinear increase with pump intensity, and the spectra consist of a broad emission with the addition of sharp (FWHM < 0.4 nm) emission lines (Fig. 12.7a, 200 kW/cm^2 $< I_{ex} <$ 300 kW/cm^2). Finally, highly directional emission characteristic of laser should be observed, which means the emission from the NW ends dominates (Fig. 12.7b, $I_{ex} > 300$ kW/cm^2).

In most of the semiconductor NW laser configurations, a semiconductor NW not only is a gain medium but also acts as a laser cavity. However, due to the small diameter of the NW and the existence of substrate, significant evanescent field exists outside the NW body that may introduce significant losses and limit the quality factor (Q factor) [52–54]. Both theoretical and experimental works have shown the potential to achieve high Q factor optical cavities and low threshold lasers based on hybrid NW structures [55, 56]. In 2008, Q. Yang et al. reported a hybrid laser combining a semiconductor NW gain section and a microfiber knot cavity. These two components were integrated in a hybrid device to combine high gain of semiconductor NWs and high Q factors of microfiber knot cavities [7].

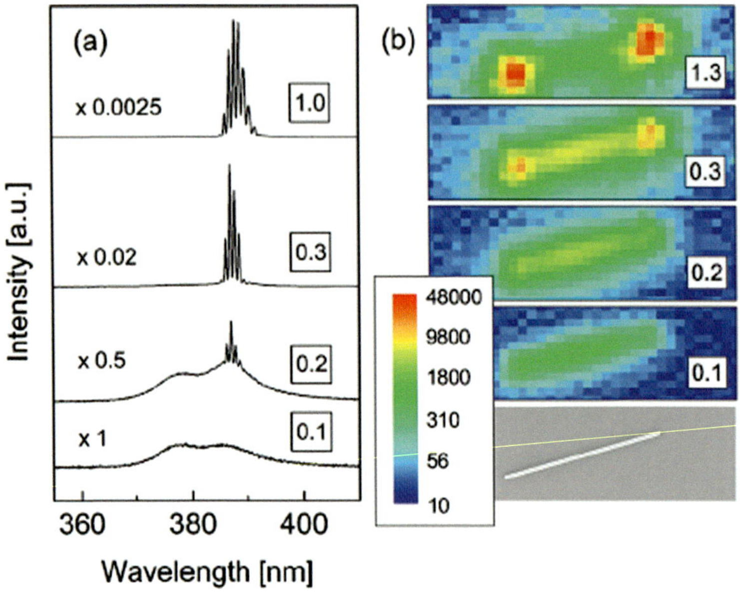

Fig. 12.7 Laser oscillation in ZnO NWs. (**a**) Output spectra vs pump intensity of a 12.2-μm-long, 250-nm-diameter ZnO NW. (**b**) Scanning electron microscopy and CCD images under different pump intensities for the same NW as in **a**. The *labels* indicate the pump intensity in units of MW/cm^2. The *color scale* indicates the number of counts. Figures adopted from [9]

The hybrid structure consists of a single or multiple ZnO NWs attached to a silica microfiber knot cavity (see Fig. 12.8). The pump laser is a frequency-tripled Nd:YAG (yttrium–aluminum–garnet) laser pulse (355 nm, 6 ns, and 10 Hz). Figure 12.9a shows the photoluminescence (PL) spectra of the hybrid structure at different pump intensities. When the pump power exceeds the threshold for laser oscillation, sharp peaks in the PL spectrum appear and the PL peak at the primary lasing wavelength abruptly increases while the spectral width of the PL decreases. The output power is concentrated in a narrow emission range (391 nm $< \lambda <$ 392 nm). With increasing power (pump level higher than 0.27 μJ/pulse), the laser emission range broadens and there is a slight redshift. These may be attributed to heating, bandgap renormalization, carrier-induced refractive index change, or the emergence of electron–hole plasma [5, 57].

Close-up views of the two distinct laser spectra are shown in Fig. 12.9b, c. The mode spacing measured from the lasing spectra in Fig. 12.9b is about 0.04 nm, corresponding to a calculated knot diameter of about 800 μm, which is in good agreement with the measured effective cavity length of 780 μm. The measured linewidth of the lasing mode is about 0.04 nm, corresponding to a Q factor of about 10^4.

Figure 12.10 shows the dependence of spectrally integrated emission intensity on the pump energy. The pump energy is measured at the untapered input port. The

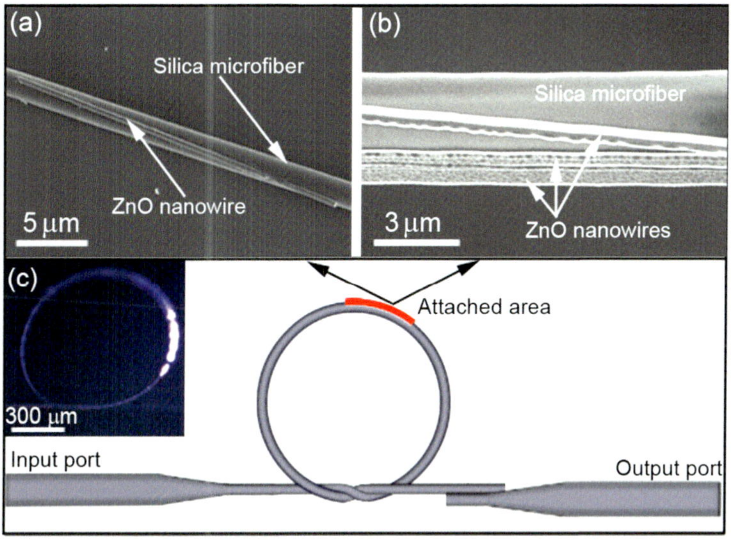

Fig. 12.8 (**a**) Scanning electron microscope (SEM) image of attached area of a 25-μm-long, 350-nm-diameter ZnO NW and a 780-μm-diameter microfiber knot assembled with a 1.8-μm-diameter silica microfiber. (**b**) SEM image of attached area of three ZnO NWs and a 728-μm-diameter silica microfiber knot assembled with a 3.5-μm-diameter silica microfiber; the diameters of the ZnO NWs are 500, 480, and 600 nm, respectively. (**c**) Schematic diagram of the structure of a hybrid laser. *Upper inset*: optical microscope image of the hybrid structure in Fig. 12.1**a** pumped by 355-nm-wavelength laser pulses. Figures adopted from [7]

lasing threshold estimated is about 0.13 μJ/pulse. A slope change and a good linearity of the pump power-dependent output are obviously observed when the pump energy exceeds the threshold.

More than one semiconductor NW can be integrated into the hybrid structure. Figure 12.11 shows the spectrally integrated emission intensity from a hybrid structure combining three ZnO NWs and a 728-μm-diameter microfiber knot cavity (Fig. 12.8b). The lasing threshold estimated is about 0.026 μJ/pulse.

The hybrid laser provides low threshold and narrow linewidth due to the combination of the high gain of semiconductor NWs and high Q factor of microfiber knot cavities. The hybrid structure, when integrated with other semiconductor NWs, should allow similar operation from ultraviolet to near-infrared spectral range.

A single compact multicolor laser system that generates red–green–ultraviolet three-color laser collected from the same end of a commercial fiber was demonstrated, as shown in Fig. 12.12 [8]. The laser system consists of three distinct semiconductor NWs (CdSe, CdS, and ZnO) and a silica microfiber. The pump energy is coupled into NW by the evanescent field existing outside the microfiber, and the PL will be coupled back into the microfiber by end emission and evanescent field of the NW.

CdSe, CdS, and ZnO NWs used in this study are direct bandgap II–VI materials with the bulk bandgaps enabling light emission from ultraviolet to visible

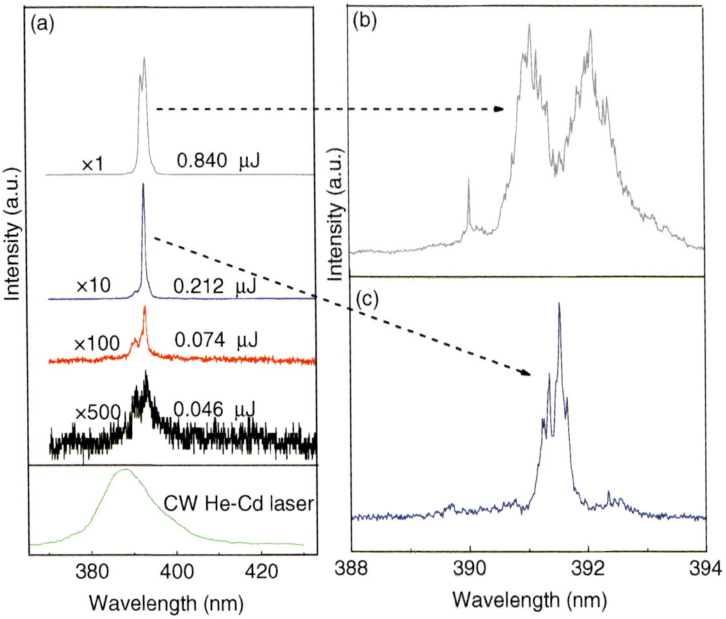

Fig. 12.9 (**a**) Output spectra vs pump energy of the hybrid structure (the same structure shown in Fig. 12.8**a**. (**b, c**) Close-up views of two laser spectra in (**a**). Figures adopted from [7]

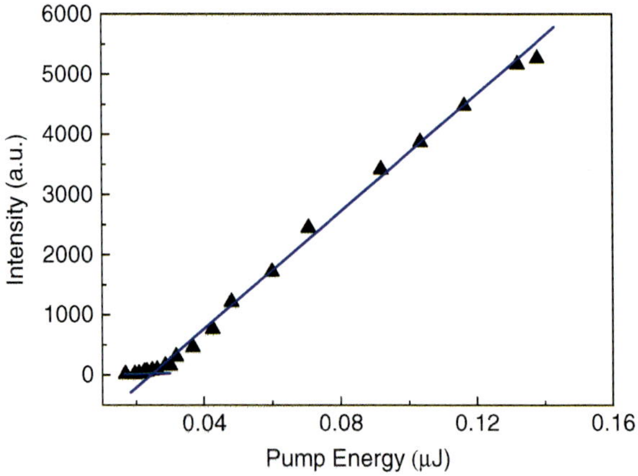

Fig. 12.10 Integrated emission intensity vs pump energy of the hybrid structure (the same structure used in Fig. 12.8**a**. Figures adopted from [7]

region as confirmed for individual NWs using PL measurements [58]. As shown in Fig. 12.12e, red–green–ultraviolet PL from CdSe, CdS, and ZnO NWs are observed obviously along the microfiber. The end facets of the NW will serve as the two mirrors of Fabry–Perot cavity because of the large refractive index contrast between the NW (the refractive indices: ZnO 2.45 at 391 nm, CdS 2.6 at 519 nm, and CdSe 2.78 at 743 nm) and the surrounding air [9, 52, 59]. The laser emission can be

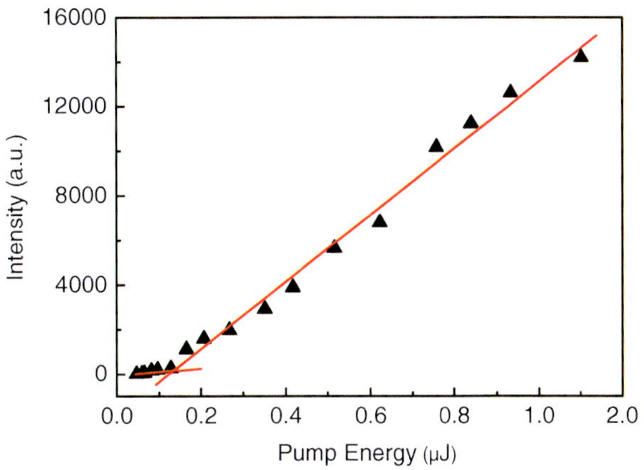

Fig. 12.11 Integrated emission intensity vs pump energy of a hybrid structure with three ZnO NWs (the same structure shown in Fig. 12.8**b**). Figures adopted from [7]

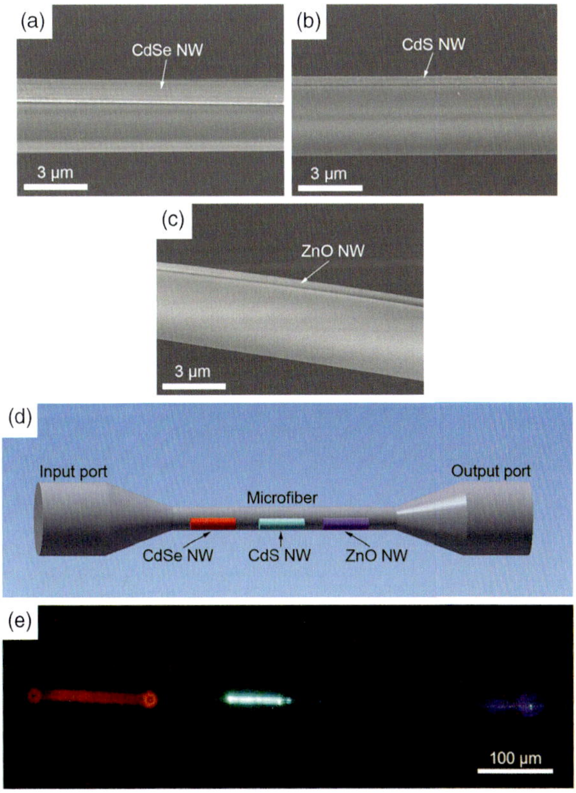

Fig. 12.12 (**a**) Scanning electron microscope (SEM) image of a 153-μm-length, 977-nm-diameter CdSe NW attached to a 3.3-μm-diameter silica microfiber. (**b**) SEM image of a 66.7-μm-length, 370-nm-diameter CdS NW attached to the same silica microfiber. (**c**) SEM image of a 72.6-μm-length, 306-nm-diameter ZnO NW attached to the same silica microfiber. (**d**) Schematic configuration of the red–green–ultraviolet three-color laser. (**e**) CCD image of the hybrid structure pumped by 355-nm-wavelength laser pulses. Figures adopted from [8]

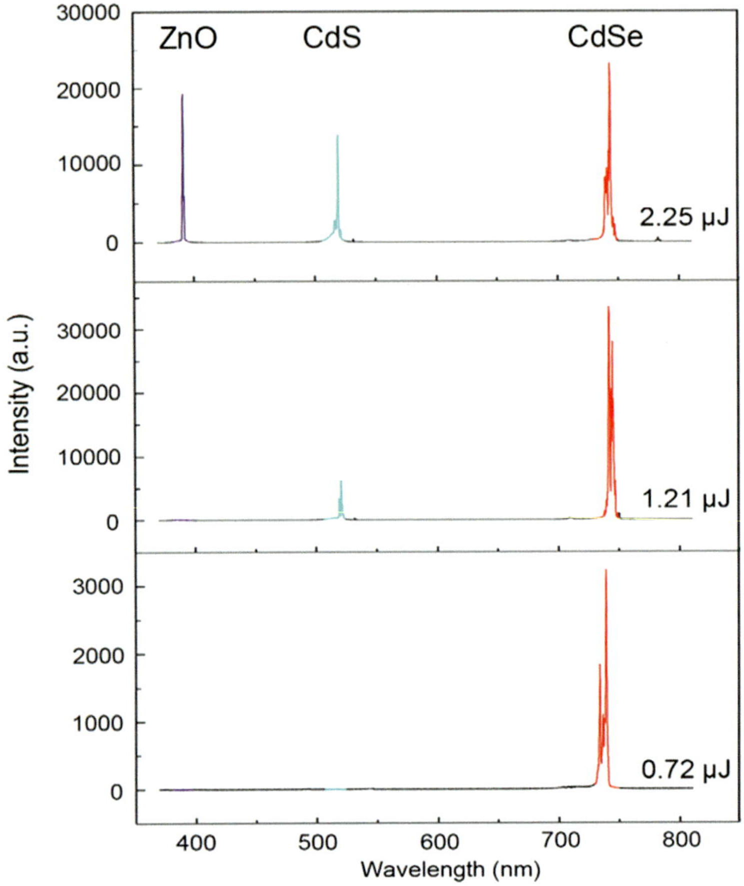

Fig. 12.13 Emission spectra of the three-color laser under different pump energies. The structure used here is the same one as shown in Fig. 12.12**e**. Figures adopted from [8]

observed when the round-trip gain is larger than the round-trip losses. Figure 12.13 shows the optical spectra for the hybrid structure laser (the same structure shown in Fig. 12.12e) as a function of pump intensity. Single-color (CdSe), dual-color (CdSe and CdS), and three-color (CdSe, CdS, and ZnO) lasers are obtained in sequence with increasing pump intensity. According to Fig. 12.13, when pump energy is higher than 1.3 μJ, three spatially and spectrally distinct lasing groups (centered at 391, 519, and 743 nm, respectively) can all be measured at the same output port simultaneously, which is consistent with lasing emission from ZnO, CdS, and CdSe, respectively.

The close-up view spectra for the three distinct NW lasing groups of the hybrid laser (the same structure shown in Fig. 12.12e) are shown in Fig. 12.14a, in which multimode can be observed. The mode spacings measured from the lasing groups originated from ZnO NW, CdS NW, and CdSe NW are 0.17, 0.8, and 0.85 nm, respectively, in good agreement with the length of the three NWs. The linewidths

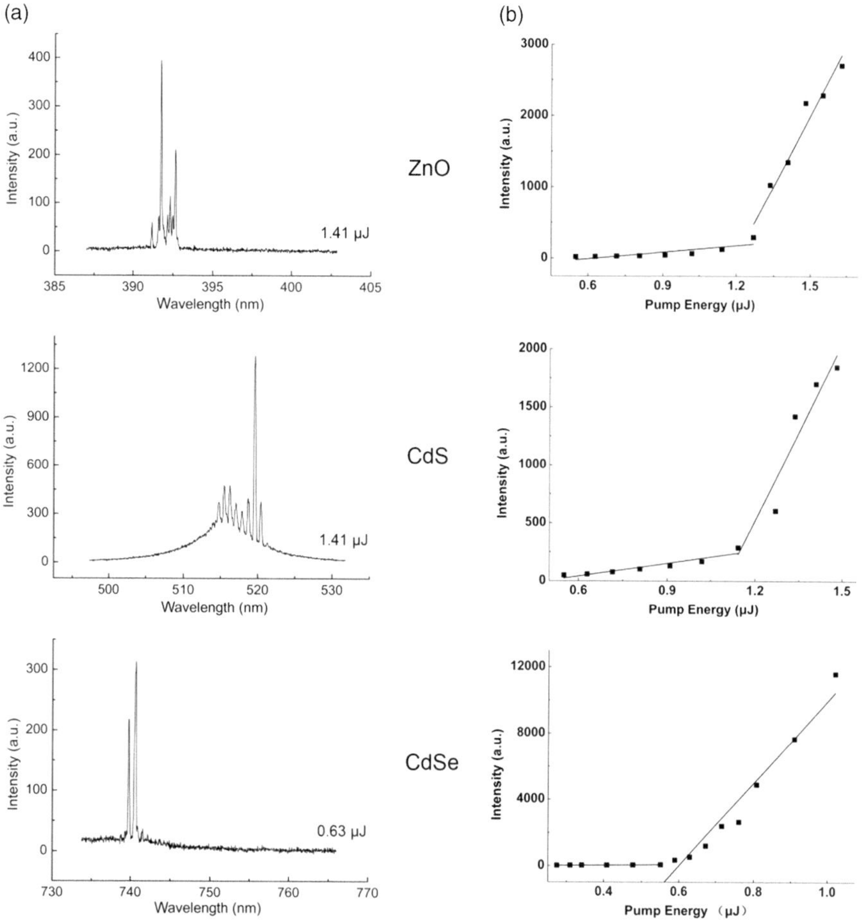

Fig. 12.14 (**a**) Close-up view spectra for the three distinct NW lasing groups of the hybrid laser. (**b**) Peak intensity vs pump energy of the hybrid structure laser for the three lasing groups. The structure used here is the same one as shown in Fig. 12.12**e**. Figures adopted from [8]

of the three lasing groups are 0.7, 0.6, and 0.57 nm, corresponding to the Q factors (quality factor: $Q = \lambda/\Delta\lambda$, where λ is the center of the wavelength and $\Delta\lambda$ is the full width at half-maximum of the cavity mode) of 558, 865, and 1303, respectively. In such a low Q factor optical cavity, laser can also be observed due to the large confinement factor of the high refractive index of NW materials [60].

In order to use the multicolor laser as a practical light source, the threshold of the different lasing groups is crucial. Figure 12.14b shows the peak intensity vs pump energy of the three lasing groups originating from the three distinct NWs. The measured thresholds of the three lasing groups from CdSe, CdS, and ZnO NWs are about 0.6, 1.1, and 1.3 μJ, respectively. Considering the factors that will affect the threshold such as coupling efficiency, taper loss, and scattering loss, the actual pump energy is lower than the threshold given in Fig. 12.14b. In general, the threshold is proportional to $\exp(1/Q\Gamma)$, where Q is the quality factor of the laser cavity and Γ is the confinement factor of a lasing mode, which is almost determined by the fraction-guided power in the fundamental mode η [9, 60, 61]:

$$\eta = 1 - [(2.4e^{-1/v})^2/V^3] \tag{12.1}$$

$$V = \kappa(d/2)(n^2 - n_0^2)^{0.5} \tag{12.2}$$

where $\kappa = 2\pi/\lambda$, d is the diameter of the NW, and n and n_0 are the refractive index of the NW and microfiber, respectively. Q factor can be calculated by Eq. (12.1). It is noted that the threshold of the distinct NWs could be modified by controlling the diameter and the length of the NW. Besides, coupling strength will also affect the lasing threshold in this experiment. Generally, the smaller the diameter of the microfiber and the NW, the higher the coupling efficiency between the microfiber and NW, which means a small diameter is good to reduce the lasing threshold. However, the smaller the diameter of the NW, the lower the end facets reflection, which means a small diameter will reduce the Q factor of the NW cavity and increase the lasing threshold. Thus it is important to select the NW with the proper length and diameter to make their thresholds closely. Otherwise, the NW which has the low threshold will be damaged by over pumped. The compact multicolor laser will lead to many advanced applications in future optoelectronic technology, such as full-color laser display, high-resolution laser printing; medicine; and biology.

12.2.3 Nonlinear Optical Devices Based on ZnO NWs

ZnO is a highly polar semiconductor that is often used for frequency doubling of intense ultrashort laser pulses [62]. Second-harmonic generation (SHG) and two-photon absorption in ZnO NWs have been reported [33, 34, 63, 64].

T. Voss et al. studied the second-harmonic generation from both arrays of ZnO NWs and single NWs under excitation with intense femtosecond pulses [34]. They achieved SHG signal and an efficient two-photon excitation of a broad internal photoluminescence in single NWs (Fig. 12.15). They also found that the excitation of the NW arrays with femtosecond laser pulses induces significant heating of the upper parts of the freestanding NWs (Fig. 12.16). Finite-element simulation of the temperature distribution in a free-standing nanowire placed on a sapphire substrate is shown in Fig. 12.16, where a constant heat source is placed on top of the nanowire to simulate the heat intake due to the laser excitation. The figure demonstrates that the temperature in the heated nanowire is increasing from the substrate to the top, a substantial increase of more than $\Delta T = 200$ K in the upper 1 μm of the wire compared to the substrate being kept at room temperature.

In one-photon absorption (OPA) regime, the excitation photon energy has to be larger than the bandgap of ZnO NWs. ZnO NWs require optical excitation or pumping with deep UV, coherent light to produce stimulated emission. Multi-photon absorption-induced emission process in ZnO NWs provides an alternative approach

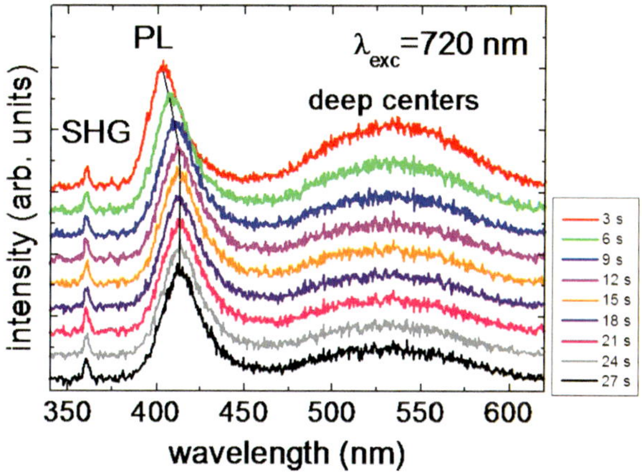

Fig. 12.15 Emission of the NW array at certain time steps 3–27 s after the excitation laser has been unblocked. Figures adopted from [34]

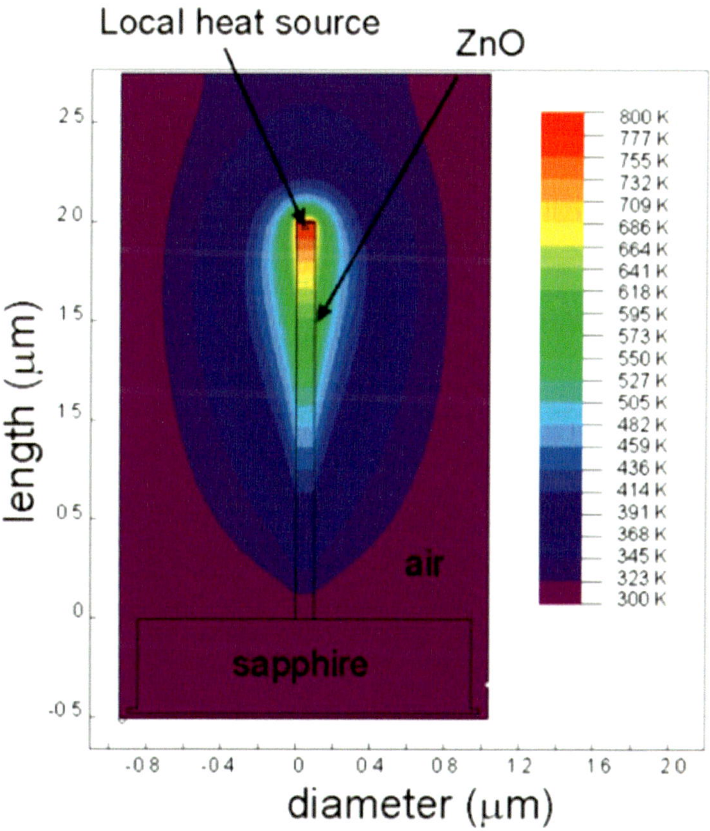

Fig. 12.16 Finite-element simulation of the temperature distribution in a freestanding NW placed on a sapphire substrate. The diameter of the nanowire was set to d = 100 nm, the length to l = 2 μm, and the heat conductivities were assumed to be k_{ZnO} = 10 W/(K cm). Figures adopted from [34]

to fabricating ZnO NW laser excited by near-infrared (IR) radiation with nanosecond or femtosecond pulses. The nonlinear interaction between the applied optical field and ZnO nanostructures leads to the simultaneous absorption of two or more photons of subbandgap energy through a virtual state assisted interband transition, producing electron–hole pairs in the excited states, and subsequently, the band-edge emission via their radiative recombination [33, 63].

UV lasing actions via near-IR excitations at the wavelength of 800 nm have been demonstrated in 2006 [64]. However, this configuration needs a high excitation threshold of 80 mJ/cm^2, because the band-edge transition was induced by the off-resonant two-photon absorption (TPA) with a substantially low efficiency for producing upconverted emission. In a later investigation, C. Zhang et al. reported the resonant TPA-induced lasing performance in ZnO NWs [63]. Room temperature laser operation at a remarkably low threshold of $\sim$160 μJ/cm^2 was demonstrated with femtosecond pulse excitation at 700 nm (Fig. 12.17a). In Fig. 12.17b, it is worth noting that the threshold of TPA-induced lasing is over two orders of magnitude lower than the earlier reported one and differs from the threshold of OPA-induced lasing only by a factor of 3 in the experimental results. In bulk semiconductors, TPA is a third-order nonlinear optical process characterized by a significantly lower efficiency than that of OPA [65–67]. The thresholds for TPA-induced lasing in bulk ZnSe, ZnSSe, and ZnO are consequently two or three orders of magnitude higher than that achieved in OPA-induced lasing processes [68, 69]. The author explained that the intensified modal field in ZnO NWs, in combination with the quartic field dependence of the TPA process, leads to the very low threshold TPA lasing in ZnO NWs.

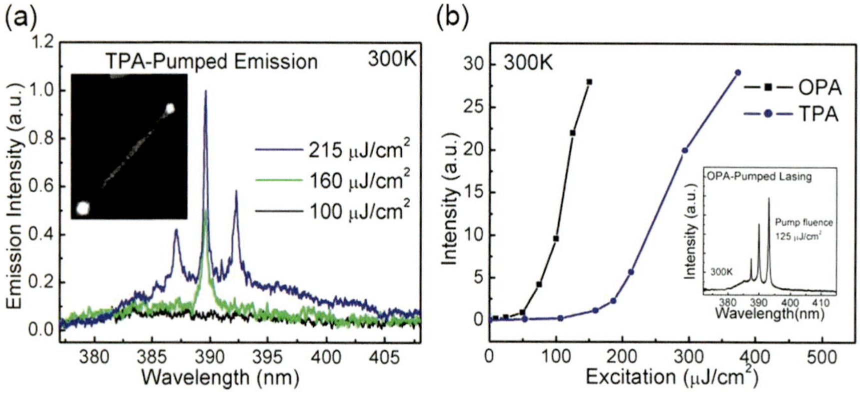

Fig. 12.17 (**a**) Two-photon pumped emission spectra from ZnO NWs for different excitation fluences. The *inset* of (**a**) is the photomicrograph of a lasing ZnO NW pumped in the TPA regime. (**b**) Integrated emission intensity from ZnO nanorods pumped by single-photon and two-photon processes vs excitation fluence. The *inset* of (**b**) shows the one-photon pumped lasing spectrum of ZnO NWs. Figures adopted from [63]

12.3 Optoelectronic Devices Based ZnO NWs

Various optoelectronic devices based on ZnO NWs have been reported, including photodetectors (PDs) [10–14], dye-sensitized solar cells (DSSCs) [16–20], and LEDs [21–26]. Electrically pumped random laser has been achieved on ZnO films [70, 71] and ZnO nanorod arrays [72]. Electrically pumped single-mode lasing emission was also observed based on the self-assembled n-ZnO microcrystalline film/p-GaN heterojunction diode [73]. In this section, we will discuss about some typical optoelectronic devices based on ZnO NWs.

12.3.1 ZnO NW Ultra-sensitive UV and Infrared PDs

As a wide bandgap semiconductor, ZnO is of special interest for application in PDs, in particular visible-blind UV detectors. For the applications of PDs, fast response time, fast reset time, high selectivity, high responsivity, and good signal-to-noise ratio are commonly desired characteristics. Due to the large surface-to-volume ratio and reduced dimensionality of the device area, ZnO nanostructures are expected to have very high photon conductivity gain [10]. In detail, the extremely high photoconductive gain is attributed to the presence of oxygen-related hole-trap states at the NW surface, which prevents charge carrier recombination and prolongs the photocarrier lifetime as shown in Fig. 12.18.

In the dark, oxygen molecules adsorb at the NW surface and capture the free electron present in the n-type semiconductor forming a low-conductivity depletion layer

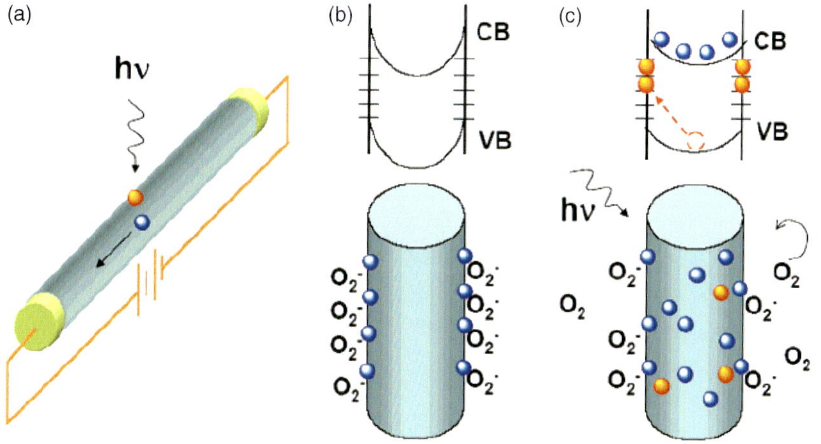

Fig. 12.18 (**a**) Schematic of an NW photoconductor in the dark and upon illumination with photon energy above E_g. (**b, c**) Trapping and photoconduction mechanism in ZnO NWs. Figures adopted from [10]

near the surface, which provides a low dark current for the photodetector fabricated by ZnO NWs (Fig. 12.18b). In the energy band diagrams, there is a band bending near the NW surface. Under UV illumination, electron–hole pairs are photogenerated and holes migrate to the surface and are trapped, leaving behind unpaired electrons in the NW that contribute to the photocurrent. In the energy band diagram, the band bending decreases. In ZnO NWs, the lifetime of the unpaired electrons is further increased by oxygen molecules desorption from the surface when holes neutralize the oxygen ions. Under an applied electric field, the unpaired electrons are collected at the anode, which leads to the increase in conductivity (Fig. 12.18c). Either the unpaired electrons are collected at the anode or they recombine with holes generated when oxygen molecules are readsorbed and ionized at the surface. This hole-trapping mechanism through oxygen adsorption and desorption in ZnO NWs enhances the NW photoresponse and leads to high internal photoconductive gain as shown in Fig. 12.19. The photoconductive gain is defined as $G = (I_{ph}/P)/(h\nu/q)$, where P is the power absorbed in the NW, photocurrent $I_{ph}=I_{light}-I_{dark}$, h is the Planck constant, ν is the frequency of the incident light, and q is the elementary charge. Photoconductive gain as high as $G = 2 \times 10^8$ has been achieved in their experiments due to the extremely long photocarrier lifetime combined with the short carrier transit times. The decrease of the gain at relatively high light intensities is a manifestation of hole-trap saturation.

Much effort has been devoted to enhance the sensitivity of ZnO NW UV PDs. It is also of interest to improve the response and recovery time of ZnO NW PDs. For example, it has been demonstrated that CdTe quantum dots with bandgap energy of 1.5 eV are photosensitizers to enhance the photoresponse of ZnO NWs [11]. Lao et al. have improved the photosensitivity of the ZnO NW UV PDs by functionalizing the surfaces of ZnO using the polymers that have a high absorption at the UV ranges [74]. Doping with appropriate metal atoms has also been shown to dramatically enhance the photosensitivity of ZnO NWs as a result of avalanche photomultiplication [17].

It is well known that Schottky barrier (SB), PIN diodes, and superlattices can increase photocurrent lifetime and result in enhancement of sensitivity [75–79] in

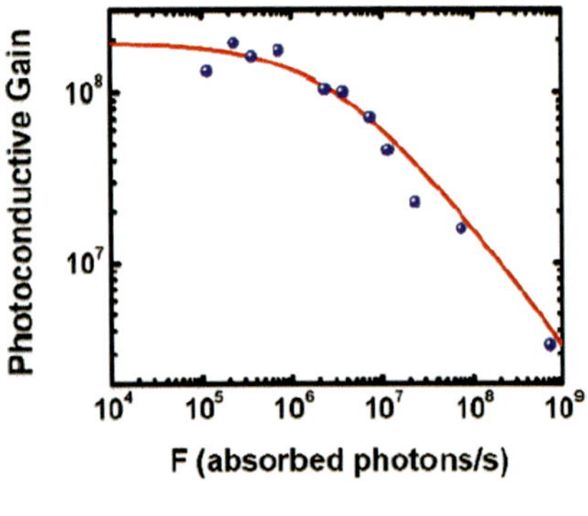

Fig. 12.19 Estimate of the photoconductive gain relative to the photon absorption rate in the ZnO NW. Figures adopted from [10]

bulk or film semiconductor PDs. In ZnO NW PDs, in addition to ZnO material properties, metal contact or the contact between ZnO NW and other materials also significantly affects the device performance [12, 80–82]. Zhou et al. have reported that by utilizing Schottky contacts instead of ohmic contacts, the sensitivity of ZnO NWs for UV light has been improved by four orders of magnitude and the reset time has been drastically reduced from 417 to 0.8 s [12]. By further surface functionalization with function polymers, the reset time has been reduced to ~20 ms even without correcting the electronic response of the measurement system.

As shown in Fig. 12.20, ohmic-contact ZnO NW PDs show high linear *I–V* characteristic curve in the dark and upon UV illumination. By illuminating the device using a 365-nm UV source at a power density of ~30 μW/cm^2, the photon conductance was improved by only ~15%. After ~260-s continuous illumination, the current was still unsaturated. More importantly, the reset time of the sensor was ~417 s, and the current could not recover to its initial state even after ~2500 s. Figure 12.21 shows typical *I–V* characteristics of Schottky-type (ST) ZnO NW PDs both in the dark and upon UV light illumination. The PDs were more sensitive when the Schottky barrier was reversely biased. The current increases from 0.04 to 60 nA. The differences in device performances between the two types of PDs can therefore attribute to the SB at the ZnO/Pt interface. The more rapid photocurrent decay in the ST device is mostly dictated by the electrical transport property of the SB. Upon turning off the UV light, the photon-generated electrons and holes in the interface region decreased dramatically, and the oxygen is only required to be readsorbed close to the interface to modify the SB height.

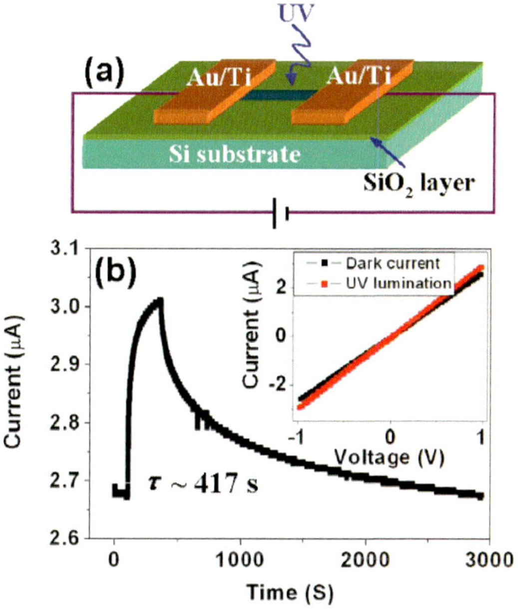

Fig. 12.20 (**a**) Schematic of a ZnO NW UV PD with ohmic contacts. (**b**) Photon response of a ZnO NW UV PD, made using Ti/Au electrodes and at a bias of 1 V, when illuminated by ~30-μW/cm^2, 365-nm UV source. The *inset* shows the corresponding *I–V* characteristics in dark or under UV illumination. Figures adopted from [12]

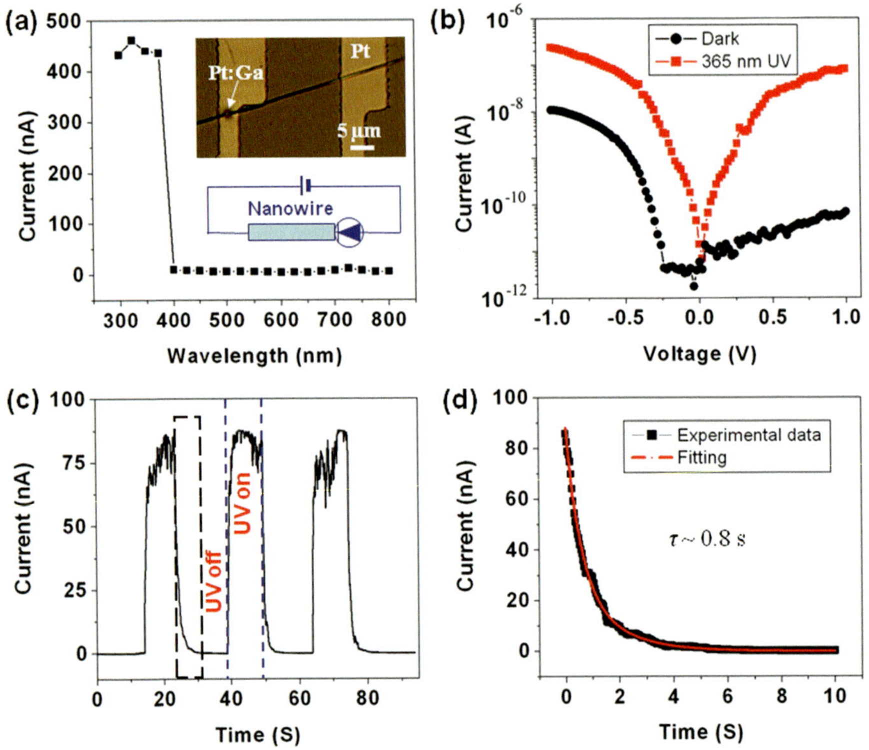

Fig. 12.21 (**a**) Photon response spectrum of the ZnO NW UV PD as a function of wavelength of incident light. *Upper inset* is an optical image of ST ZnO NW PD. *Lower inset* shows the schematic structure of the device. (**b**) *I–V* characteristics of a PD both in the dark (*black circle*) and under 365-nm UV illumination (*red rectangle*). (**c**) Time dependence of the photocurrent growth and decay under periodic illumination of the 365-nm UV light on the device. The bias on the device is 1 V. (**d**) Experimental curve (*black*) and fitted curve (*red*) of the photocurrent decay process. Figures adopted from [12]

Also, a ZnO NW FET work as a UV photodetector was also demonstrated [13]. Devices which could function as both UV and visible PDs based on ZnO/Si heterojunctions were also reported [83, 84].

So far most of the researches on ZnO NW detection are focused on UV photodetection. Except ZnO NW UV PDs, ZnO subwavelength wires can also be used for fast-response mid-infrared detection, considering the relatively strong absorption in mid-IR regime (about 1 mm^{-1} from 8 to 30 μm) [85] and the excellent chemical and thermal stabilities of ZnO microwire. Mid-infrared (IR) detection, typically relying on thermal or photoresponse [86], has wide applications in the fields of medicine [87], remote sensing [88], environmental monitoring [89], and telecommunications [90]. Generally, IR photon detectors, including photoconductors, quantum well PDs, quantum dot PDs, and superconductor detectors [86, 91], offer advantages of fast response and high sensitivity, but usually require low-temperature operation with complex cooling equipments [88]. Thermal detectors, such as thermocouple detectors and resistance thermal detectors, provide the possibility for room temperate operation with broadband response, but usually suffer from slow response times due to relatively large thermal inertia of the sensitive elements [86, 88]. One way to speed up the response of a thermal detector is reducing the thermal inertia or equivalently the size of the sensitive element, through adopting air–bridge microstructure [92]. W. Dai et al. demonstrated room temperature operation mid-IR thermal detection based on ZnO subwavelength wires with response time down to 1.3 ms at 10.6-μm wavelength [15].

The submicrometer wire was placed across two Ti/Au electrodes sputtered on the plate, as schematically illustrated in Fig. 12.22a. A typical as-fabricated detection structure, consisting of a 2.0-μm-diameter, 760-μm-length ZnO wire, is shown in Fig. 12.22b. Light from a coherent K-250 CO_2 laser, centered at the wavelength of 10.6 μm, was used to irradiate the ZnO wire. The laser beam was focused by a ZnSe lens (focus length = 5.0 cm) to a 220-μm-diameter spot on the ZnO wire. Since the dark resistance of the ZnO wire is very large, a constant illumination from a halogen lamp (about 6800 lx) is applied on the ZnO wire for stable and reliable measurement of the response of the ZnO wire.

When a ZnO wire absorbs mid-IR light, its temperature rises, leading to a change in the resistance that can be used to retrieve the intensity of the incident light. Typical *I–V* characteristics of the ZnO wire IR photodetector show that the resistance increases linearly with intensity of the irradiation. For example, for a 2.9-μm-diameter, 520-μm-length ZnO wire under irradiation of 10.6-μm-wavelength

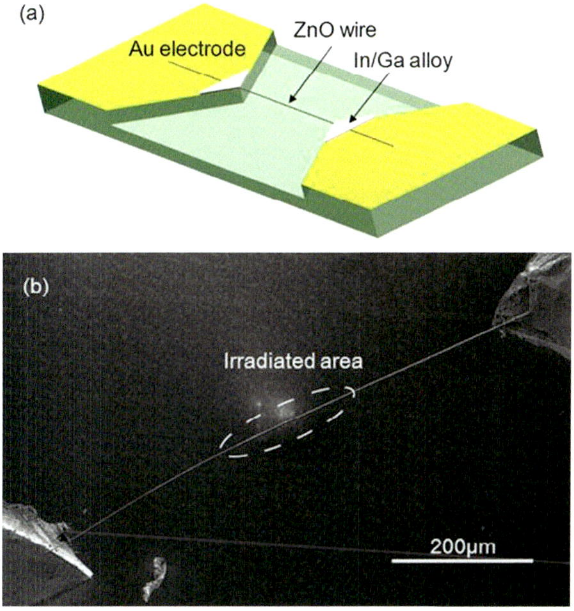

Fig. 12.22 The schematic structure (**a**) and SEM image of a ZnO wire on a grooved glass plate (**b**). Figures adopted from [15]

light, as the light intensity increases to 22.8 mW, the current reduces by 38%. Supposing only the 226-μm-length irradiated part of ZnO wire is elevating temperature, the resistance of this part increases by 144%. By means of frequency domain analysis [93], the author obtained a noise intensity of 700 $\mu V/Hz^{1/2}$ (at 1 kHz), which corresponds to a noise equivalent power (NEP) of 5.8 $\mu W/Hz^{1/2}$.

The investigation on the influence of ambience on ZnO wire for mid-IR detection showed that although the background resistance varies in different atmospheres, the amplitude and response time are insensitive to the ambient gases (Fig. 12.23). However, the response of ZnO NW for UV photodetection is significantly influenced by ambience [94]. The estimated response time of the ZnO wire mid-IR detection is about 1.3 ms when the resistance of ZnO wire falls from 37.2 to 34.5 MΩ in air, which is much faster than other types of room temperature operated microbolometers or thermocouples [95–100] and three orders of magnitude faster than that in ZnO NW UV PDs [10, 101, 102].

In UV and mid-IR spectral ranges, ZnO wires have different response mechanisms. When mid-IR photon is absorbed by ZnO wires, the photon energy is converted to thermal energy, heating up the ZnO wire. When temperature rises, thermal lattice vibration becomes stronger, leading to stronger scattering of carriers in ZnO wire, which in turn reduces the mean free paths of carriers, resulting in the increasing of the resistance. On the other hand, the rising temperature may excite electrons to the conduction band and increase the density of carriers, resulting in the decreasing of the resistance. However, at room temperature and above, with halogen lamp illumination, electrons on the shallow doping levels have already been excited, and the free-carrier density is almost saturated. Therefore, heating of the ZnO wires

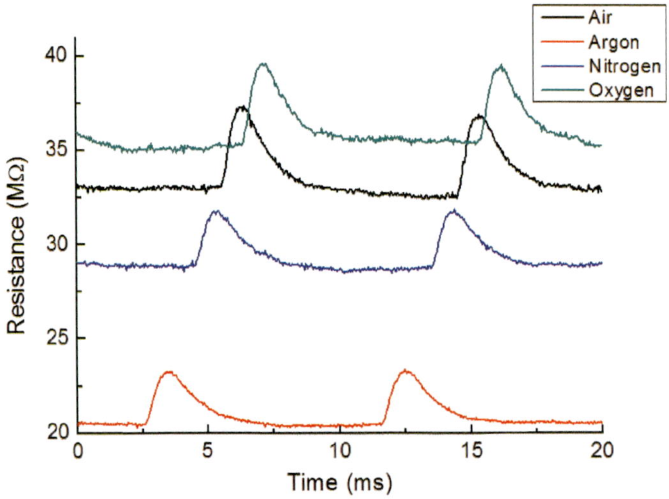

Fig. 12.23 Response of a 2.0-μm-diameter, 760-μm-length ZnO wire to 10.6-μm-wavelength light irradiation measured in different atmospheres. Figures adopted from [15]

would not increase the density of carriers obviously, and the resistance of the ZnO wire would increase when it is irradiated by mid-IR light.

As a kind of thermal detection, the response time of the ZnO wire detection can be theoretically estimated using the time constant [92]

$$\tau = H/G \tag{12.3}$$

where the heat capacity (of the irradiated part of the ZnO wire) H is about 5.5×10^{-9} J/K and the thermal conductivity G is about 5.0×10^{-6} W/K [103, 104]. Calculated τ is about 1.1 ms, which coincides well with the measured value of 1.3 ms. The low thermal inertia of ZnO wire allows the response time down to the order of millisecond. Although the light used in this work is a monochromatic 10.6-μm-wavelength laser, the fast and sensitive response of the ZnO wire can be extended to a wider mid-IR spectrum owing to the broadband absorption of ZnO in the mid-IR spectral range and will be promising for fast-response mid-infrared detection.

12.3.2 Dye-Sensitized Solar Cells Based on ZnO NWs

In our need for highly efficient, low-cost, and CO_2-free sources of energy, solar energy is one of the most promising sustainable energy resources for the future. Among the different emerging photovoltaic options, excitonic solar cells (XSCs) appear to be promising candidates for achieving the basic criteria for large-scale commercialization: they are highly efficient devices that employ low-cost materials and offer the possibility of being fabricated by large-scale and inexpensive (solution processing) techniques [105–107]. The most important examples of XSCs are organic solar cells (OSCs), hybrid solar cells (HSCs), and dye-sensitized solar cells (DSSCs). Nanostructured materials, such as nanoparticles, nanorods, nanosheets, and core–shell, are key constituents of excitonic solar cells. While TiO_2 is commonly used as a DSSC material, there is an interest in DSSC applications of ZnO NWs as an electron transport material. It presents properties closely related to the best semiconductor oxide used up to date, TiO_2 [107], but contrary to the former, it is possible to obtain ZnO in a wide variety of nano-forms by low-cost and scalable synthesis methods [108, 109]. In this chapter we will briefly discuss the application of ZnO nanostructures, particularly vertically aligned ZnO NWs as the electron transport material in DSSCs. We describe the evolution and future potential for the application of ZnO NWs in next-generation excitonic solar cells. We summarize the most applied techniques used for device fabrication and review the advantages and disadvantages observed during its application.

ZnO-based DSSCs with a very wide range of ZnO morphologies and fabrication methods, dyes used, and consequently photovoltaic performances have been reported. Currently, the application of ZnO nanostructured electrodes made of colloidal nanoparticles has achieved up to 6.58% efficiency [110]. The keys to improve the conversion efficiency of DSC include increasing surface area and maintaining good electron transport. Efforts to improve the DSC conversion efficiency by increasing surface area was made by the application of ZnO nanostructures like nanoflower ($\eta = 1.9\%$) [111], nanosheets ($\eta = 3.9\%$) [112], or its application together with other oxides such as mesoporous SnO_2–ZnO (6.34%) [113], core–shell ZnO–TiO_2 (9.8%) [114], or doped ZnO nanoparticles (6%) [115]. Compared with nanoparticles having the large surface area, vertical NW or nanorod arrays could provide high electron transport (Fig. 12.24) [20]. For example, dye-sensitized solar cells (DSSCs) based on hydrothermally grown ZnO nanorods have been reported in 2005 [16]. ZnO NWs exhibited improved electron transport and collection compared to ZnO nanoparticles, as illustrated in Fig. 12.25. The current from the ZnO NWs as a function of roughness factor (a measure of the internal surface area) increased up to the longest wires studied, without showing any signs of saturation or reduction, indicating efficient charge transport and collection. The power conversion efficiency of 1.5% was obtained for NWs with length in the range 18–24 μm [16]. Furthermore, hierarchically structure (3.5%) [116] and porous single crystal structure (5.6%) [117] could provide large surface area and high electron transport simultaneously (Fig. 12.24).

Most of the designs above are based on a 2D planar substrate, which has a relatively low surface area that limits the dye-loading capacity and restricts mobility and adaptability for remote operation. Moreover, the increasing surface area is limited by the requirement that the electron transport distance d remain significantly smaller than the electron diffusion length L_n in order to minimize recombination of electrons with holes or other species. For wire-based SCs, in which light is

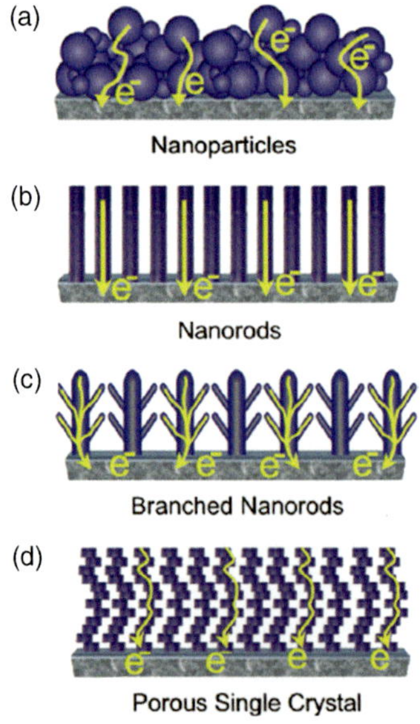

Fig. 12.24 Schematic representation of the possible electron path taking place on different nanostructured electrodes made with (**a**) nanoparticles, (**b**) nanorods, (**c**) branched nanorods, and (**d**) porous single crystal. Figures adopted from [20]

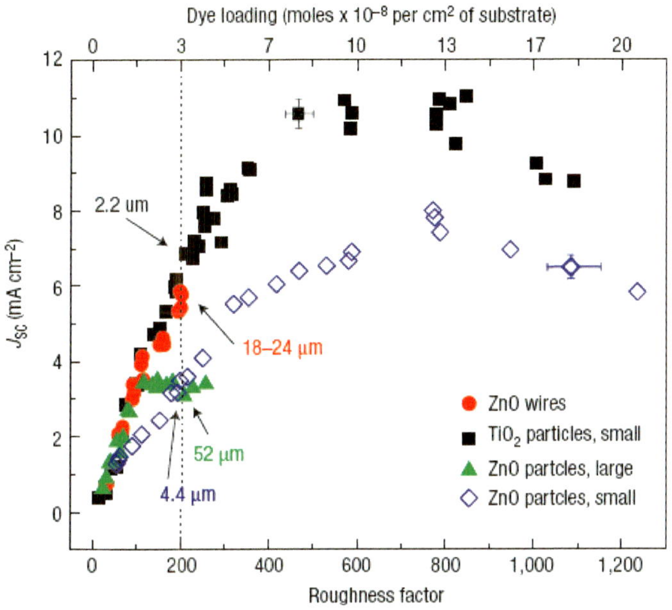

Fig. 12.25 Comparative performance of nanowire and nanoparticle cells. Figures adopted from [16]

illuminated perpendicular to the wire [118, 119], the shadow effect from the entangled wire-shaped electrode may limit the enhancement in power efficiency. Z. L. Wang's group made dye-sensitized solar cells with a higher effective surface area by fabricating the cells around a quartz optical fiber which is called 3D DSSC [18]. B. Weintraub et al. removed the cladding from optical fibers, grew zinc oxide NWs along the surface, treated them with dye molecules, and surrounded the fibers by an electrolyte and a metal film that carries electrons off the fiber [18]. The design and principle of the device are shown in Fig. 12.26. Photons bounce inside the fiber as they travel, so there are more chances to interact with the solar cell and produce more current. The cells are six times more efficient than a zinc oxide cell with the same surface area.

Figure 12.27a demonstrates the 3D DSSC concept of a cell fabricated on a cylindrical optical fiber. Two typical configurations were considered: light illumination normal to the fiber axis (NA) and parallel to the fiber axis (PA), as shown in Fig. 12.27d (insets). The plot of current density against voltage (J–V curve) shows the open circuit voltage V_{OC}, short-circuit current density J_{SC}, fill factor FF, and energy conversion efficiency $h = \mathrm{FF} \times V_{OC} \times J_{SC}/P_{in}$, where P_{in} is the incident light power density. It is apparent that the axial illumination configuration yields an enhanced efficiency. To properly characterize the enhancement in energy conversion efficiency, the efficiency enhancement factor (EEF) is defined as the ratio of power efficiencies for the PA and NA cases, that is, $\mathrm{EEF}=\eta_{PA}/\eta_{NA}$. For a total of five DSSCs, the EEF ranges from 4 to 18 (Fig. 12.27c). The large value is partially

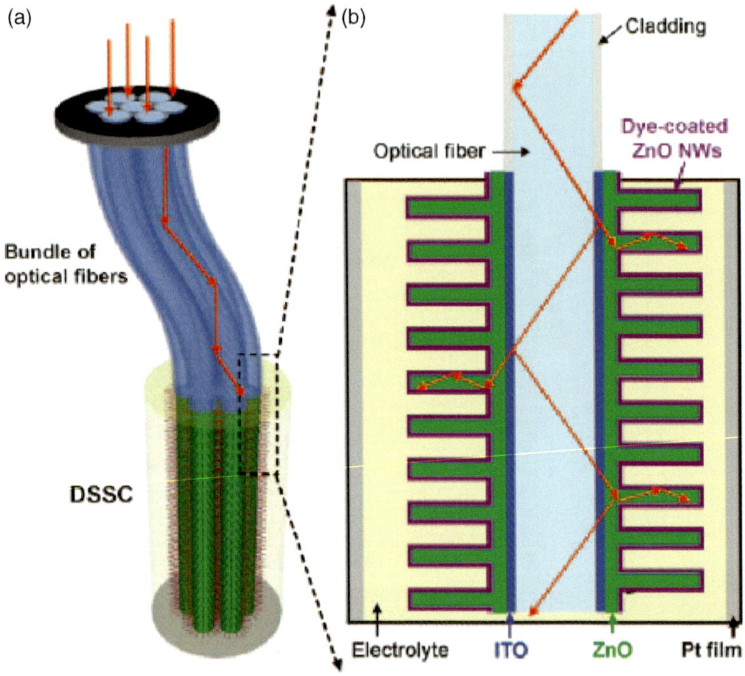

Fig. 12.26 Design and principle of a 3D DSSC. The cross section of the fiber can be cylindrical or rectangular. (**a**) The 3D DSSC is composed of optical fibers, and ZnO NWs are grown vertically on the fiber surface. The *top segment* of the bundled optical fibers utilizes conventional optical fibers and allows for remote transmission of light. The *bottom segment* consists of the 3D DSSC for solar power generation at a remote/concealed location. (**b**) Detailed structure of the 3D DSSC. Figures adopted from [18]

due to the hybrid structure and partially to the geometrical configuration of the Pt film electrode.

The absolute efficiency of the cylindrical fiber in the PA case is still limited by the curved geometry of the fiber and the short mean free path of the generated charges. The highest efficiency with this configuration is 0.45%. An improved design takes advantage of the rectangular optical fiber geometry (Fig. 12.28). ZnO NWs can be grown uniformly on all four sides of a fiber. NA illumination (Fig. 12.28d, case 1) represents the configuration similar to a 2D DSSC arrangement, while PA illumination (Fig. 12.28d, case 2) measurement was conducted as 3D DSSC. The PA case has a significantly enhanced current density. For a total of eight devices, the efficiency of the 3D design for the PA case is enhanced by a factor of up to 6. In addition, a broader photo-action in the red region is seen in the PA orientation, thus suggesting that longer wavelength photons can be more efficiently converted to electrons farther down the optical fiber where the overall light intensity is diminished.

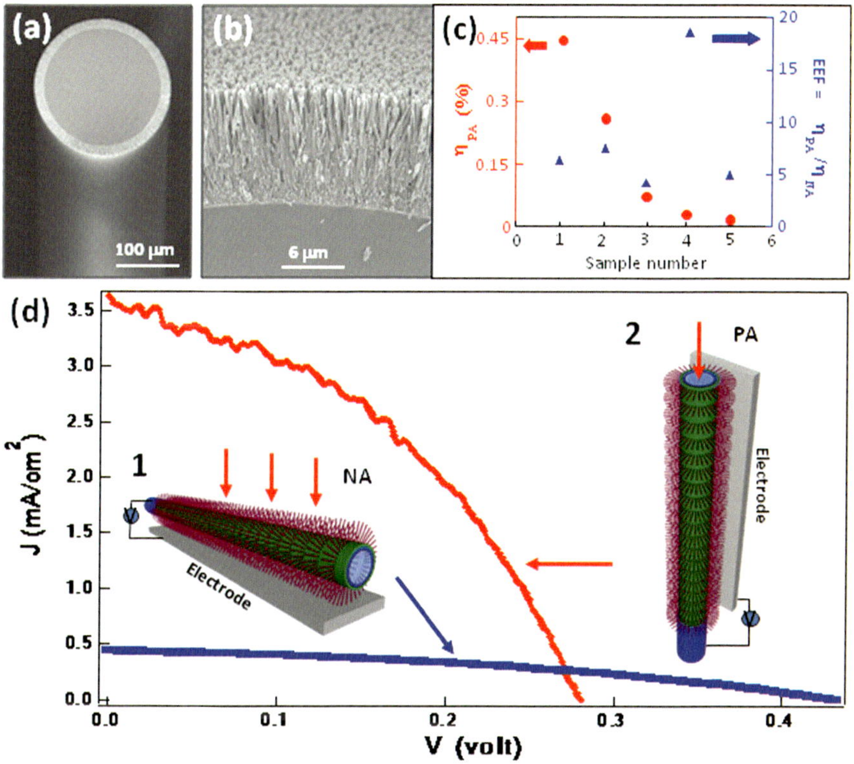

Fig. 12.27 Cylindrical optical fiber based 3D DSSC and its performance. (**a**) Low-magnification SEM image of a quartz fiber with uniformly grown ZnO NWs on its surface. (**b**) High-magnification SEM image showing the densely packed ZnO NWs on the fiber surface. (**c**) Plot of EEF and the corresponding energy conversion efficiencies for five 3D DSSCs. The data variation is mainly attributed to fluctuations in SC packaging. (**d**) *J–V* curves of the DSSC under one full-sun illumination (AM1.5 illumination, 100 mW/cm^2). The illumination is (*1*) normal to the fiber axis (NA; 2D case) and (*2*) parallel to the fiber axis (PA; 3D case). A corresponding efficiency enhancement factor (EEF) = 6.1 has been achieved by converting the 2D DSSC to 3D DSSC. Figures adopted from [18]

The 3D DSSC has several outstanding features from both physical perspective and application view, such as increasing the electron transport distance, the ability to be concealed and located underground or in deep water, and environmental friendly. A sun-tracking system would not be necessary for such cells and would work on cloudy days when light is diffuse [120].

A new approach to fabricate 3D dye-sensitized solar cells (DSSCs) is integrating planar optical waveguide and NWs to the device configuration. The ZnO NWs are grown normally to the quartz slide. The 3D cell is constructed by alternatively stacking a slide and a planar electrode. The slide serves as a planar waveguide for light propagation. Each time when light reaches waveguide–NW interface, photons are coupled into the ZnO NWs and then are absorbed by the dye molecules to generate

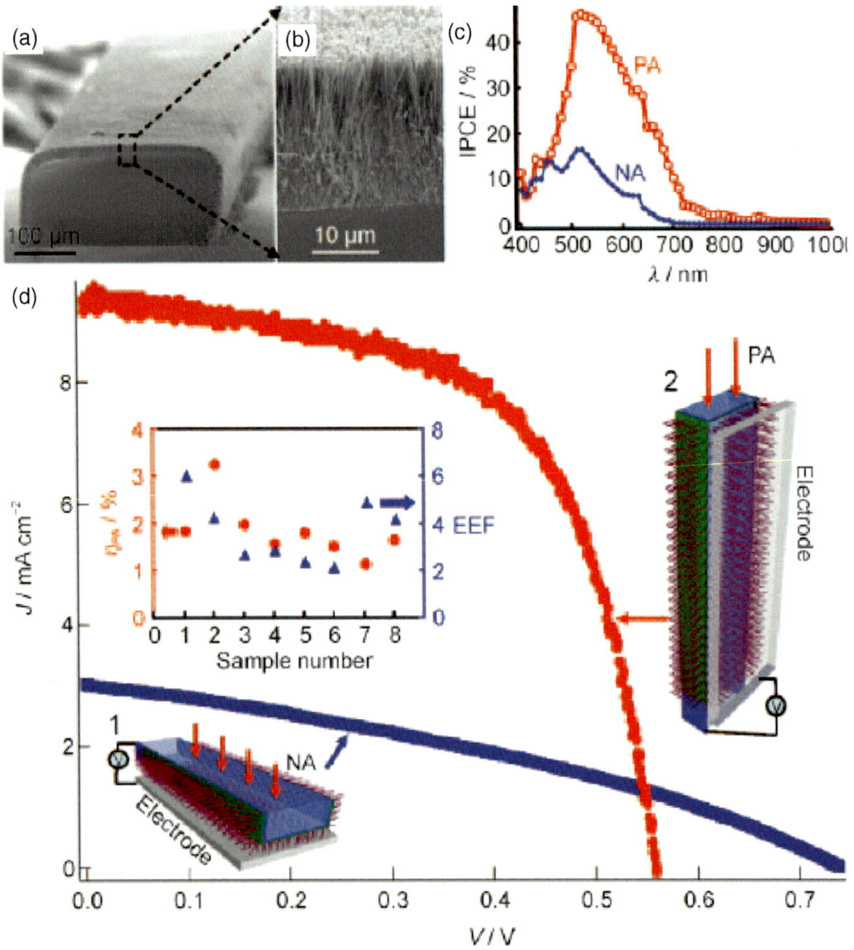

Fig. 12.28 Rectangular optical fiber based 3D DSSC and its performance. (**a**) Low-magnification SEM image of a quartz fiber with uniformly grown ZnO NWs on three sides. (**b**) High-magnification SEM image showing the densely packed ZnO NWs on the fiber surface. (**c**) Typical incident photon to electron conversion efficiency (IPCE) measured for the PA and NA cases from a DSSC. (**d**) Current density *J* and voltage *V* curves of a DSSC under one full-sun illumination oriented (*1*) normal to the fiber axis (NA; 2D case) and (*2*) parallel to the fiber axis (PA; 3D case). A corresponding efficiency enhancement factor (EEF) = 4.34 has been achieved by converting the 2D DSSC to the 3D DSSC. The *inset* shows a plot of EEF and the corresponding energy conversion efficiencies for eight 3D DSSCs. Figures adopted from [18]

electricity. On average, the enhancement of energy conversion efficiency by a factor of 5.8 has been achieved when light propagating inside the slide is compared to the case of light illumination normal to the surface of the slide from outside; and the full sun efficiencies have been achieved up to 2.4% for ZnO NWs. This may be an effective approach for developing large-scale 3D solar cells with high efficiency [17].

Surprisingly, the highest efficiency of XSCs applying ZnO is only about 6% [117] which is less than the 11.3% obtained with the best DSC applying TiO_2 [121, 122]. Thus, it is important to improve the performance of the ZnO NW based DSSCs. Possible methods to improve the performance included improvements in absorption, exciton dissociation, and charge collection (either by improved transport or reduced recombination). In general, the lower efficiency of ZnO-based DSSC compared to TiO_2-based DSSC is partly attributed to surface area differences as well as instability of ZnO in acidic solutions, which results in formation of Zn ion–dye complexes [20]. Lower efficiency of ZnO-based DSSCs compared to TiO_2-based DSSCs is also partly due to the fact that dyes and electrolytes commonly used in DSSCs have been optimized for TiO_2. In DSSCs, all important processes, i.e., dye adsorption, electron injection, and recombination, occur at the surface of nanostructures and as such would be affected by surface defects. While improvements in the efficiency of ZnO-based DSSCs have been achieved, overall efficiency is still lower compared to devices based on titania. A possible solution to this problem would be to explore dyes and electrolytes specifically developed for ZnO rather than simply comparing the performance with materials optimized for TiO_2.

12.3.3 Single ZnO NW and NW Array Light-Emitting Diodes

In light-emitting devices, ZnO can be used for different purposes, such as active semiconductor layer [21–26], electrode [123, 124], current spreading layer [125], as well as buffer layer [126]. Except active semiconductor layer, the other applications mainly rely on ZnO films and hence they are beyond the scope of this chapter. Here we will focus on the devices where ZnO NW is one of the active semiconductor layers in the device.

The LED based on ZnO NW as active layer can be divided into two parts: homojunction LEDs and heterojunction LEDs. In homojunction LEDs, p- and n-doped active layers are both ZnO NWs. In heterojunction LED normally ZnO NWs are grown or deposited on various substrates to form devices.

Reports on homojunction LEDs based on ZnO have been less common, compared to various heterojunction devices. Different dopants and doping methods have been reported for the achievement of p–n junctions. For example, As ion implantation can be used to fabricate p–n junction in ZnO nanorods [127]. The nanorods were grown by vapor deposition on an n-Si substrate with ZnO seed layer. The EL spectra were dependent on As ion fluence: for high fluence ($10^{15}/cm^2$), red emission was obtained, while for low fluence ($10^{14}/cm^2$), dominant emission was UV emission. ZnO homojunction can also be fabricated by annealing the NWs grown on GaAs substrates through the diffusion of As during annealing [128, 129]. The turn-on voltage was above 4.0 V and the corresponding ultraviolet electroluminescence spectra were obtained for the applied forward voltage above 30 V (20 mA). More recently, catalyst-free p–n homojunction ZnO NW arrays in which phosphorus (P) and zinc (Zn) served as p- and n-type dopants, respectively, have also been synthesized by a controlled in situ doping process for fabricating efficient ultraviolet light-emitting devices [130]. The ZnO p–n NW arrays were grown on n-type Si(001) wafers with thin ZnO films on the top by using a thermal vapor deposition method in a three-zone furnace. The doping transition region defined as the width for P atoms gradually occupying Zn sites along the growth direction can be narrowed down to sub-50 nm. The electroluminescence spectra from the p–n ZnO NW arrays distinctively exhibit short-wavelength emission at 342 nm and the blue shift from 342 to 325 nm is observed as the operating voltage further increases (measured at relatively high voltage > 35 V).

While different device architectures and fabrication methods have been reported, comprehensive studies aiming at improving our understanding of ZnO-based LEDs have been scarce. Nevertheless, some factors affecting device performance have been identified. It has been shown that the performance of ZnO-based devices can be very sensitive to the presence of hydrogen and passivation layer [131]. In addition to hydrogen, it is expected that native defects also play a significant role in device properties of ZnO-based LEDs.

While stable and reproducible p-doping of ZnO is still being pursued, heterojunction n-ZnO/p-substrate LEDs have attracted considerable attention. Though a variety of other semiconductors have been used in ZnO heterojunction LEDs, GaN remains one of the most common ones, among the many reasons the main being due to similar crystallographic and electronic properties of ZnO and GaN. A variety of devices have been made by different methods, and different performances and emission colors have been obtained. It was proposed that NW-based devices may exhibit higher efficiency compared to thin film based devices due to the possibility to achieve increased light extraction [132–134]. They are also expected to have improved efficiency due to improved injection through a nanosized junction [21, 22, 135, 136]. Both vapor deposition and solution method were used to grow ZnO NWs on GaN substrates. Figure 12.29a, b shows the EL intensity and spectrum of a high brightness LED dependent on the forward bias voltage by directly growing

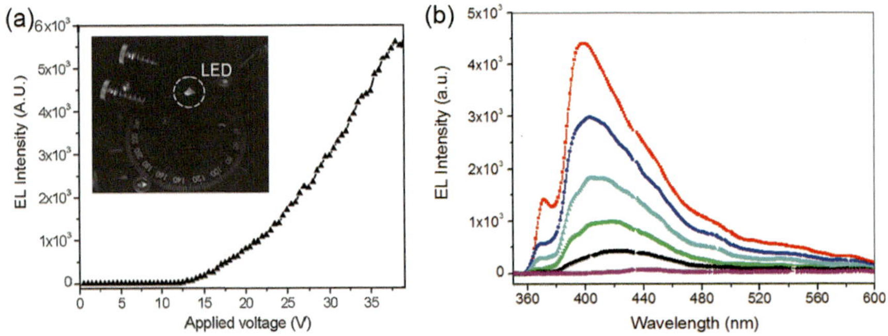

Fig. 12.29 (**a**) A plot of electroluminescence (EL) intensity vs forward bias fixing the emission wavelength at 400 nm. The *inset* is a typical lighting image of the (n-ZnO NWs)/(p-GaN film) LED device, taken using a commercial digital camera Nikon D70, lens: 70 mm, aperture: f/5, exposure time: 3 min. (**b**) The electroluminescence spectrum of the (n-ZnO NWs)/(p-GaN film) LED device under various forward bias voltages (10, 15, 20, 25, 30, 35 V). Figures adopted from [22]

n-type ZnO NW arrays on p-GaN wafers by vapor deposition [22]. The emission light became visible to the naked eye when the bias voltage exceeded ~10 V, and the intensity increased rapidly when the forward bias was larger than the threshold (Fig. 12.29a). The picture taken by commercial digital camera shows the LED's high brightness. UV emissions around 370 nm were observed as well as strong peaks centered at 400–440 nm wavelength (Fig. 12.29b). A blue shift was observed in the EL with the increase of bias voltage, indicating the modification of external voltage to the band profile in the depletion region. The EL intensity was affected by UV illumination due to the excitation of residual charge carriers by UV and the change in p–n junction energy gap. The 370-nm UV emission was first enhanced and then dropped after UV illumination, indicating its stronger dependence on density of charge carriers in ZnO. The 400-nm blue emission was less dependent on the UV excitation.

In most of the ZnO NW/p-GaN LEDs, the n-ZnO NWs are randomly distributed on the substrate, thus light is spotty and nonuniform (Fig. 12.30) [137], which largely limits their applications in high-performance optoelectronic devices. Recently, S. Xu et al. demonstrated the capability of controlling the spatial distribution of the blue/near-UV LEDs composed of position-controlled arrays of n-ZnO NWs on a p-GaN thin film substrate [135]. The device was fabricated by a conjunction of low-temperature wet chemical methods and electron beam lithography (EBL). Under forward bias, each single NW is a light emitter. By Gaussian deconvolution of the emission spectrum, the origins of the blue/near-UV emission are assigned particularly to three distinct electron–hole recombination processes.

Figure 12.31e is the optical image of a lighted-up LED at a biased voltage of 10 V. In the device, all of the NWs are connected in parallel and each single NW is a light emitter. Brightness difference among the individual NWs comes probably from the current crowding effect and different serial contact resistances, and therefore different injection currents through the individual NW. The pitch between each lighting spot shown in Fig. 12.31e is 4 μm and the resolution is 6350 dpi.

For LEDs, it should be noted that the majority of studies only report EL spectra in arbitrary units; for further progress of the research in this field, characterization in

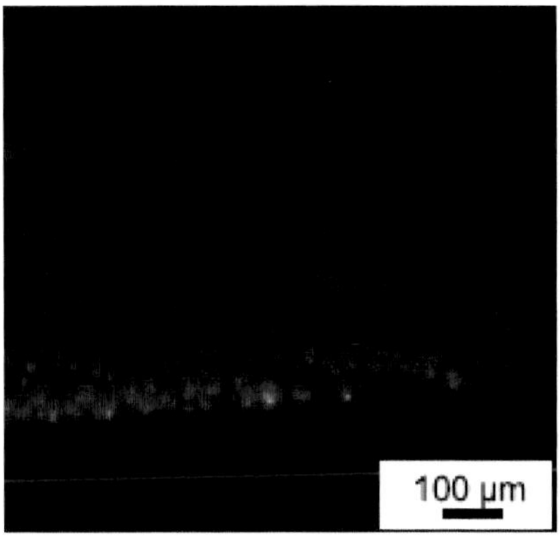

Fig. 12.30 Light emission photograph of the ZnO buffer layer/ZnO nanowire array/p-GaN film heterojunction diode. Figures adopted from [137]

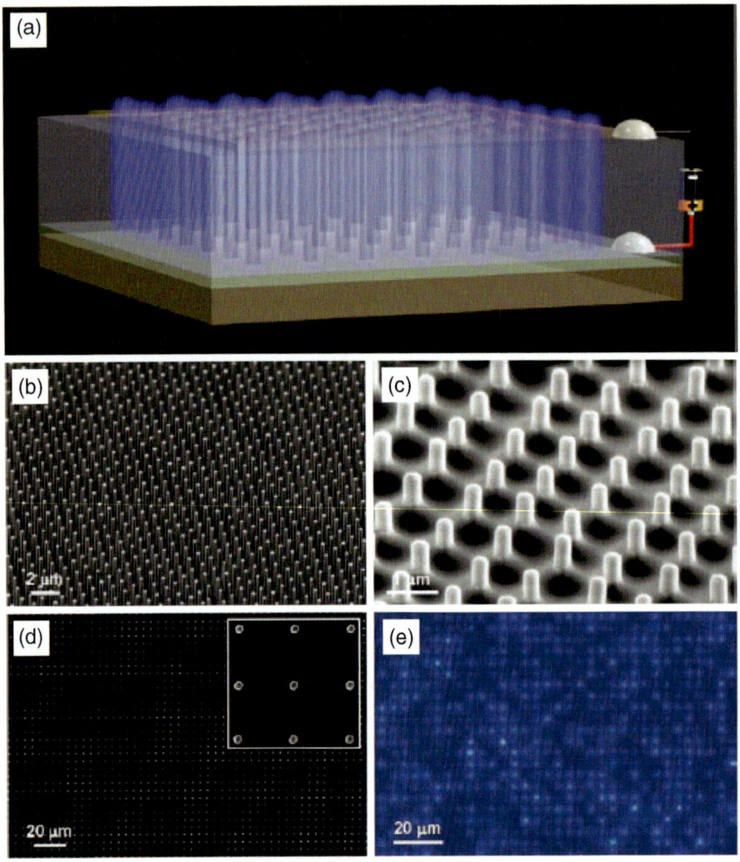

Fig. 12.31 (**a**) Design overview of the LED. (**b**) 60° tilt SEM view of the as-grown patterned vertical ZnO nanowires with a width about 300 nm and (**c**) after they are coated with SiO_2 and wrapped with PMMA, and the tips are exposed. (**d**) Top SEM image of the ZnO nanowire arrays. The pitch and layout of the nanowire arrays are readily controlled by the EBL. (**e**) The optical image of a turned-on LED (*artificial bluish color*). Figures adopted from [135]

terms of brightness, emission power, or emission efficiency is needed. Figure 12.32 shows the room temperature external quantum efficiency vs d.c. injection current characteristics. The external quantum efficiency was calculated by acquiring the ratio of the output light power and the input electrical power. The external quantum efficiency of the LED is about 2.5%, which is considerably high for a single p–n junction-based LED, and such data are reproducible and consistent for several devices. As the biased voltage/injection current is gradually increased, the external quantum efficiency becomes steady (Fig. 12.32), which indicates that the serial resistance or the nonradiative recombination through the defects, e.g., Auger recombination, does not increase in proportion with increasing injection current.

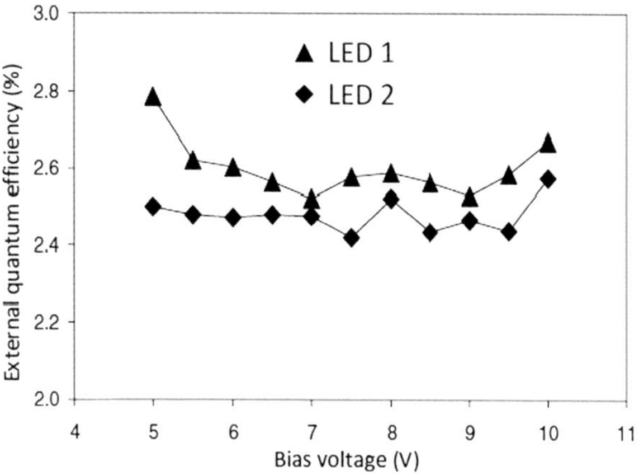

Fig. 12.32 External quantum efficiencies of two heterostructural LEDs as a function of the biased voltage/injection current. The efficiency was determined only when the LED was turned on and the light output power was stably registered with the power meter. Figures adopted from [135]

While most ZnO nanorod/p-GaN devices exhibited light emission under forward bias, a device which lighted up under reverse bias was also reported [21]. Lighting up under reverse bias has been attributed to the presence of large band offset at GaN/ZnO interface, which enables tunneling, and the tunneling probability increases as the reverse bias voltage increases. Light emission under both forward and reverse biases was observed from heterojunctions consisting of p-GaN/nanocomposite SiO_2:ZnO/n-ZnO. Under forward bias, the devices exhibited UV emission from ZnO and blue–violet emission from p-GaN, while under reverse bias, attributed to avalanche breakdown, UV emissions from both GaN and ZnO were observed in addition to a small contribution from p-GaN in the blue–violet region [138].

Besides p-GaN, ZnO NW heterojunction LEDs based on a combination of ZnO NW with a variety of materials have been reported, such as p-type silicon [139], p-$CuAlO_2$ [140], p-polymer [23, 141, 142], p-SiC [70, 143], and p-CrO_3 [144]. Single n-ZnO NW/p-Si heterojunctions were also reported, using NWs with different growth directions [25, 26]. The EL from ZnO NWs grown by vaporizing a mixture of ZnO and graphite powder shows broadband emission, extending from 350 nm to beyond 850 nm [25], while the EL from ZnO NWs grown by vaporizing pure ZnO powder shows UV emission near the bandgap of ZnO [26]. Spin-on-glass (SOG) has also been used as an insulating layer in single ZnO NW/p-Si heterojunction. It has been shown that SOG layer provides thinner coverage at the top of the NW. The single NW LEDs exhibited rectifying *I–V* curves, both with and without thin (7–8 nm) SiO_2 layer, but electroluminescence (sharp UV emission) was observed only in the devices with SiO_2 layer. This result is quite encouraging for application potential of n-ZnO/p-Si heterojunction LEDs, since the formation of SiO_2 during the ZnO NW growth may be difficult to avoid, especially for growth conditions favoring more stoichiometric composition such as higher oxygen partial pressure. Obviously, from the differences in the reported results in the literature in terms of turn-on voltage and emission spectra, the device performance is strongly dependent on the device architecture and ZnO nanomaterial properties, which are in turn affected by native defects and fabrication conditions.

Comparisons of the best devices reported in the literature based on thin films and on nanostructures do not reveal that nanostructure-based devices are substantially better [31]. This is likely due to the fact that ZnO properties are significantly affected by native defects and large surface area nanostructures may be significantly affected by surface/interface defects and dangling bonds. While there is a great potential for further development of these devices, comprehensive study of ZnO nanomaterial properties and their relationship to the device performance (including the roles of native defects and surfaces and interfaces) is still needed.

12.3.4 Electrically Pumped Random Lasing from ZnO Nanorod Arrays

In recent years, there have been several reports of electrically pumped lasing from ZnO. One type of device in which electrically pumped random lasing based on ZnO films and ZnO nanorod arrays was reported [71, 145, 146]. In one case, the device structure was polycrystalline ZnO film or ZnO nanorod arrays on Si, covered by SiO_x layer and Au contact [71]. Threshold current about 70 mA was observed for these devices. The author believed that the recurrent scattering and interference of the enough strong electroluminescent UV light in the in-plane random cavities formed in the ZnO film leads to electrically pumped UV random lasing. Using the same device structure, the same group also demonstrated electrically driven random lasing from ZnO nanorod arrays on Si [72]. Figure 12.33 shows the schematic diagram of an MIS device based on the ZnO nanorod array on Si substrate. Figure 12.34a shows the current–voltage (*I*–*V*) characteristic of a typical device. Herein, forward/reverse bias means that the gate electrode of Au is connected to positive/negative voltage. As can be seen, the device exhibits a rectifying behavior to a great extent. It is found that the MIS devices based on ZnO nanorod arrays are electroluminescent only under forward bias. Figure 12.34b shows the evolution of the EL spectra for a device with the increase of forward bias voltage. Sharp peaks

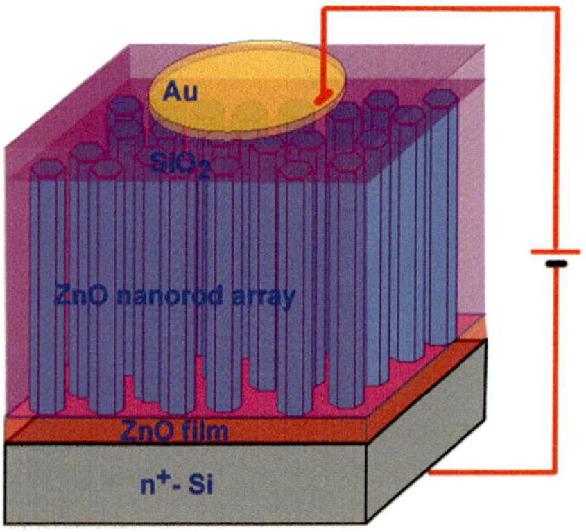

Fig. 12.33 Schematic diagram of the metal (Au)–insulator (SiO_2)–semiconductor (ZnO nanorod array) structure on Si substrate. Figures adopted from [72]

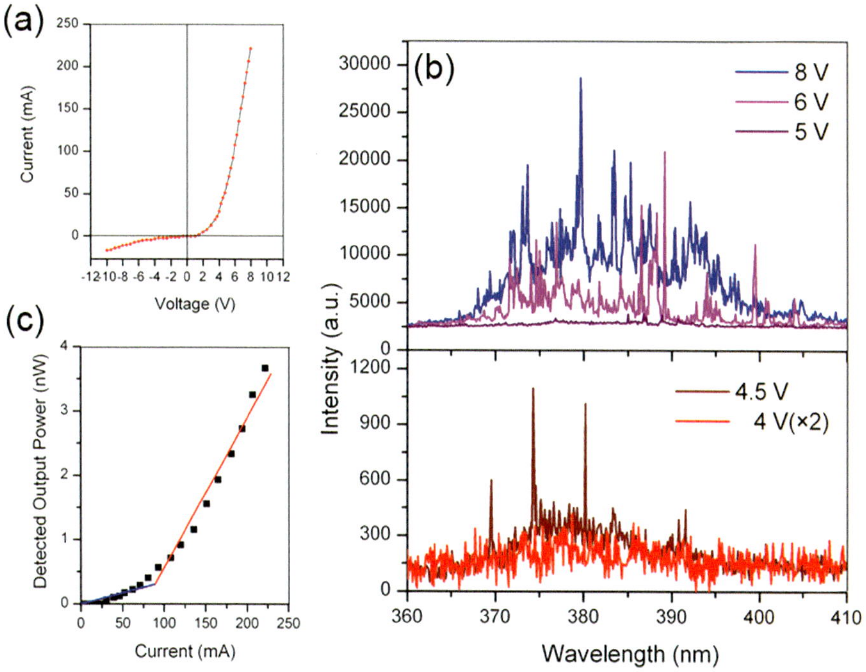

Fig. 12.34 (**a**) Current–voltage characteristic of the MIS device based on ZnO nanorod array. (**b**) EL spectra of the device under different forward bias voltages. (**c**) Detected output power as a function of the injection current. Figures adopted from [72]

in the spectral range from 360 to 400 nm could be observed. The multiple sharp peaks in the spectra represent different lasing modes. The detected output power as a function of injection current is shown in Fig. 12.34c. Above a threshold current, as shown by a solid line plotted to guide the eyes, the output power increases linearly with the injection current. Such a linear dependence is due to the gain saturation that forms an intrinsic aspect of an amplifying system above the threshold [147].

Very low threshold current of 0.8 mA was reported for devices consisting of p-GaN/MgO/n-ZnO heterojunctions [146]. Lasing was also reported in a device consisting of a single ZnO quantum well (1 nm) with MgZnO barriers (1.5 nm), sandwiched between n-ZnO (Ga:ZnO) and p-ZnO (Sb:ZnO) [145]. Devices exhibited threshold current of 25 mA. More recently, the electrically pumped single-mode lasing emission located at 407 nm with a full width at half-maximum (fwhm) of 0.7 nm was observed based on the self-assembled n-ZnO microcrystalline film/p-GaN heterojunction diode [73]. The spectrum recorded from the top surface shows only a weak and broad UV emission, while a lasing action could be observed from the edge, which indicates the laser cavity may be formed directionally parallel to the substrate. While achievement of electrically pumped lasing in ZnO is a great success, several matters need to be clarified, such as the exact role of SiO_2 in MIS random lasing structure and the method to provide high-quality factor cavity in electrically pumped laser. As we can see from the above results, for electrically pumped laser structure, no precise design cavity has been used. Most of the cavity formation is through scattering, thus random lasing is achieved rather than single-mode or multimode lasing in optically pumped laser. High-quality material and high quality factor for electrically driven ZnO nanomaterial laser are needed to develop the laser used in applications.

12.4 Piezo-phototronic Devices Based on ZnO NWs

ZnO is a material that simultaneously has semiconductor, optical, pyroelectric, and piezoelectric properties; besides the well-known coupling of semiconductor with optical properties to form the field of optoelectronics, additional novel effects could be proposed by three-way coupling semiconductor, optical properties, and piezoelectric properties to form a field of piezo-phototronics [148].

Figure 12.35 shows a schematic diagram about the three-way and two-way coupling among piezoelectricity, photoexcitation, and semiconductor, which are the bases of piezotronics (piezoelectricity–semiconductor coupling), piezo-photonics

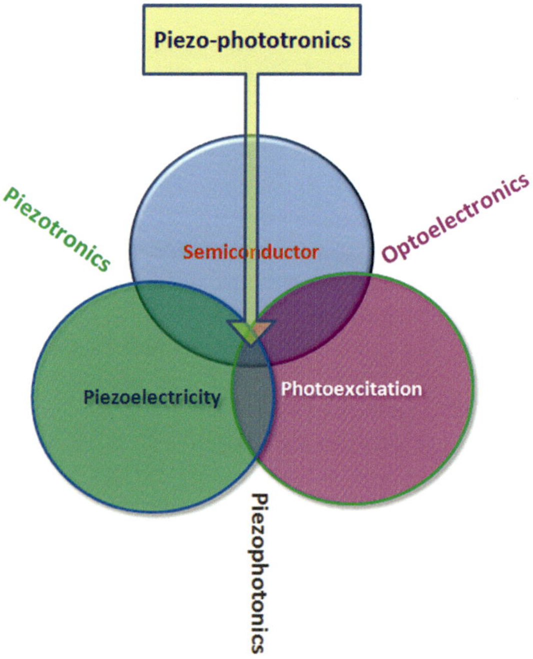

Fig. 12.35 Schematic diagram showing the three-way coupling among piezoelectricity, photoexcitation, and semiconductor, which is the basis of piezotronics (piezoelectricity–semiconductor coupling), piezo-photonics (piezoelectric–photoexcitation coupling), optoelectronics, and piezo-phototronics (piezoelectricity–semiconductor–photoexcitation). The core of these coupling relies on the piezopotential created by the piezoelectric materials. Figures adopted from [148]

(piezoelectric–photoexcitation coupling), optoelectronics, and piezo-phototronics (piezoelectricity–semiconductor–photoexcitation).

The coupling of optical, mechanical, and electrical properties of ZnO NW provides new opportunities for fabricating functional devices [3, 27–29] and an effective method to integrate optomechanical devices with microelectronic systems [27]. In this section, we will discuss in detail two types of optoelectronic devices which can be improved by piezopotential.

12.4.1 Optimizing the Power Output of a ZnO Photocell by Piezopotential

Y. Hu et al. demonstrated that the output of a photocell could be optimized by tuning the strain in the NW [28]. An externally applied strain produces a piezopotential in the microwire, which tunes the effective Schottky barrier (SB) height of the microwire at the local contact, consequently changing the electrical parameter of the device. A back-to-back metal–semiconductor–metal contacted microwire was used to illustrate the effect of the piezopotential on the performance of a photocell. Figure 12.36 shows the measured photon current of a device as a function

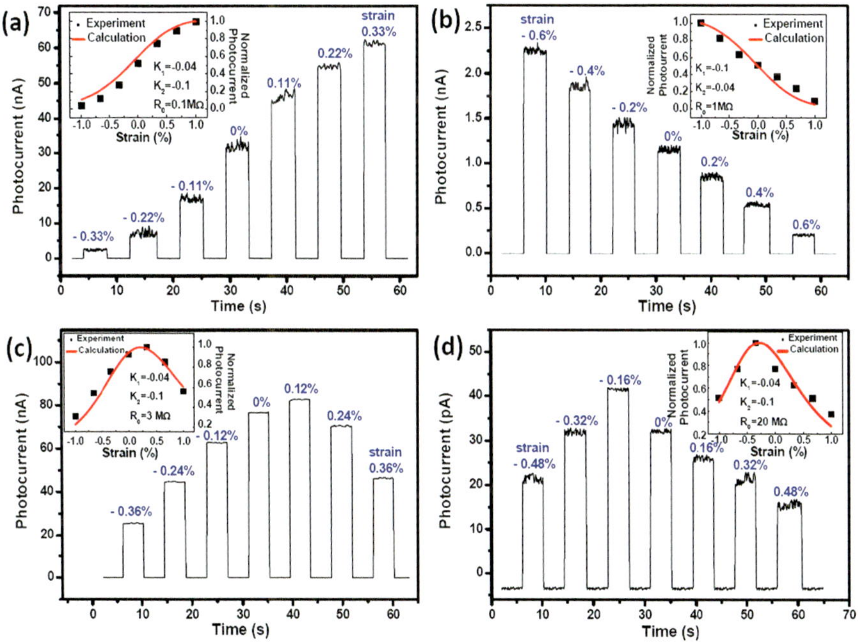

Fig. 12.36 Output current responses to the strain applied on the device, which can be divided into four categories: (**a**) increasing, (**b**) decreasing, and with a maximum under (**c**) applying positive strain or (**d**) negative strain. The *inset* is the calculated results to indicate the output current behavior based on the related parameters, showing a similar changing trend compared with the experimental data after normalization. Figures adopted from [28]

of the applied strain. The applied strain will alter the effective heights of the two SBs and thus the characteristic of the microwire photocell. Four kinds of characteristic relationships between the output current and the applied strain have been observed. The first one is that the output current increases with applied strain, as shown in Fig. 12.36a. The second kind is just opposite to the first one: the output current decreases when the applied strain is increased (Fig. 12.36b). The third and the fourth are similar. They both have a maximum output current responding to the applied strain, but for the former one, the maximum point occurred in the tensile strain range (Fig. 12.36c), while the latter is in the compressive strain range (Fig. 12.36d).

By exciting an SB structure using a laser that has photon energy higher than the bandgap of the semiconductor, electron–hole (e–h) pairs are generated at the interface region. If the height of the SB is too high, the generated e–h pairs cannot be effectively separated, resulting in no photon-induced current. If the SB is too low, the e–h pairs are easily recombined even after a short separation and again there is no photon current. There exists an optimum SB height that gives the maximum output photon current. By using the tuning effect of piezopotential to the SB, the optimum SB could be found for the maximum of the photon current.

12.4.2 Enhancing Sensitivity of a Single ZnO Micro-/NW Photodetector by Piezo-phototronic Effect

The basic principle of a photon detector is based on photoelectric effect, in which the e–h pairs generated by a photon are separated by either a p–n junction or an SB. In such a case, the height of the SB, for example, is important for the detection sensitivity of the photon detector. The sensitivity of the PDs may be improved by applying the proper strain, thus by tuning the SB height in the device [29].

Figure 12.37 is the experimental setup, which is an integrated system that could operate mechanical, optical, and electrical measurements simultaneously. Figure 12.38a shows absolute photocurrent relative to excitation intensity under different strains with a natural logarithmic scale. It can be seen that the photocurrent is largely enhanced for pW-level light detection by using piezoelectric effect. And it is pointed out that the effect of strain is much larger for weak light detection than for strong light detection. Because the dark current did not change under strain, the sensitivity, responsivity, and detectivity of the photodetector increased under compressive strain. The responsivity of the photodetector is, respectively, enhanced by 530, 190, 9, and 15%, respectively, upon 4.1 pW, 120.0 pW, 4.1 nW, and 180.4 nW UV light illumination onto the wire by introducing a –0.36% compressive strain in the wire (Fig. 12.38b), which effectively tuned the SB height at the contact by the produced local piezopotential. The sensitivity for weak light illumination is especially enhanced by introducing strain, although the strain much smaller effect on the sensitivity to stronger light illumination.

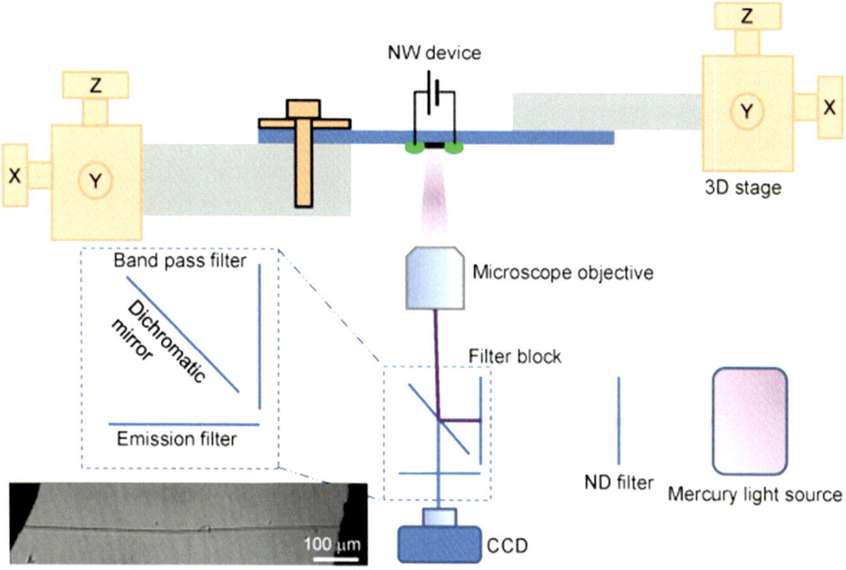

Fig. 12.37 Schematic diagram of the measurement system to characterize the performance of the piezopotential-tuned photodetector. An optical microscopic image of a ZnO wire device is shown. Figures adopted from [29]

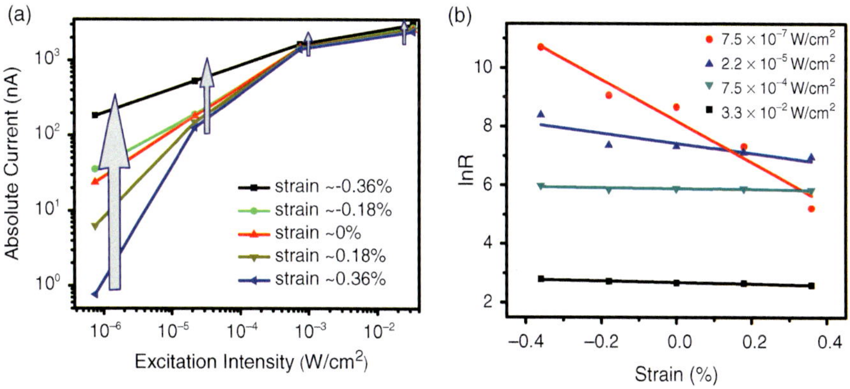

Fig. 12.38 (**a**) Absolute photocurrent relative to excitation intensity under different strains. (**b**) Responsivity (units A/W) as a function of strain under different excitation light intensities on a natural logarithmic scale. Figures adopted from [29]

12.5 Conclusions

In summary, ZnO NWs are important 1D nanostructures that have important applications in photonic field including pure optical devices, optoelectronic devices, and piezo-phototronic devices.

The attractive features of high refractive index, high exciton binding energy, high polarity, and transparence across the visible spectral range make ZnO NWs an ideal candidate for optical devices such as excellent subwavelength waveguide, ultraviolet laser, and nonlinear devices. Until now, in the optically pumped ZnO NW laser and nonlinear devices, pulse laser was used only as the pumped source for the achievement of ZnO NW laser. It is possible to develop techniques to fabricate ZnO NW laser and nonlinear devices by continuous wavelength laser pumping at room temperature due to the high exciton binding energy of ZnO.

The two-way coupling of optical and semiconductor properties of ZnO is the basis of optoelectronic devices based on ZnO NWs. ZnO NWs could be used to fabricate high-sensitive UV and fast-response IR PDs, and high-efficiency LEDs and DSSCs. To realize the great promises of inexpensive, highly efficient light-emitting and photovoltaic devices, better control and understanding of the properties of ZnO material is necessary, which is expected to result in controllable and reproducible achievement of desired properties and device performance. This also includes control and/or passivation of the surfaces and interfaces, since in any nanomaterial-based device, due to high surface-to-volume ratio of the nanomaterial, surface properties will significantly affect the charge transport and recombination processes.

Finally, the three-way coupling of optical, semiconductor, and piezoelectric properties of ZnO is the basis of piezo-phototronic devices. This effect allows tuning and controlling of electro-optical process by strain-induced piezoelectric potential, with potential applications in LED, photocell and solar cell, and photon detector, aiming at improving the performance of optoelectronic devices and providing an effective method to integrate optomechanical devices with microelectronic systems.

Concerning device applications of ZnO NW, comprehensive studies aimed at understanding and controlling fundamental properties of ZnO and optimization of device structures, in particular, surfaces and interfaces, are essential for improving device performance and further applications.

References

1. Y.W. Heo et al., ZnO nanowire growth and devices. Mater. Sci. Eng. R-Rep. **47**(1–2), 1–47 (2004)
2. Z.L. Wang, ZnO nanowire and nanobelt platform for nanotechnology. Mater. Sci. Eng. R-Rep. **64**(3–4), 33–71 (2009)
3. Z.L. Wang et al., Lateral nanowire/nanobelt based nanogenerators, piezotronics and piezo-phototronics. Mater. Sci. Eng. R-Rep. **70**(3–6), 320–329 (2010)
4. M. Law, D.J. Sirbuly, J.C.J. JohnsonGoldberger, R.J. Saykally, P.D. Yang, Nanoribbon waveguides for subwavelength photonics integration. Science **305**, 1269–1273 (2004)

5. J.C. Johnson et al., Optical cavity effects in ZnO nanowire lasers and waveguides. J.Phys. Chem. B **107**(34), 8816–8828 (2003)
6. M.H. Huang et al., Room-temperature ultraviolet nanowire nanolasers. Science **292**, 1897–1899 (2001)
7. Q. Yang et al., Hybrid structure laser based on semiconductor nanowires and a silica microfiber knot cavity. Appl. Phys. Lett. **94**, 101108 (2009)
8. Y. Ding et al., Nanowires/microfiber hybrid structure multicolor laser. Opt. Expr. **17**, 21813–21818 (2009)
9. M.A. Zimmler et al., Laser action in nanowires: Observation of the transition from amplified spontaneous emission to laser oscillation. Appl. Phys. Lett. **93**, 051101 (2008)
10. C. Soci et al., ZnO nanowire UV photodetectors with high internal gain. Nano Lett. **7**(4), 1003–1009 (2007)
11. R.S. Aga et al., Enhanced photoresponse in ZnO nanowires decorated with CdTe quantum dot. Appl. Phys. Lett. **91**(23), 232108 (2007)
12. J. Zhou et al., Gigantic enhancement in response and reset time of ZnO UV nanosensor by utilizing Schottky contact and surface functionalization. Appl. Phys. Lett. **94**(19), 191103 (2009)
13. W. Kim, K.S. Chu, ZnO nanowire field-effect transistor as a UV photodetector; Optimization for maximum sensitivity. Phys. Status Solidi A-Appl. Mater. **206**(1), 179–182 (2009)
14. M.W. Chen et al., Photoconductive enhancement of single ZnO nanowire through localized Schottky effects. Opt. Expr. **18**(14), 14836–14841 (2010)
15. W. Dai et al., ZnO subwavelength wires for fast-response mid-infrared detection. Opt. Expr. **17**(24), 21808–21812 (2009)
16. M. Law et al., Nanowire dye-sensitized solar cells. Nat. Mater. **4**(6), 455–459 (2005)
17. Y.G. Wei et al., Planar waveguide-nanowire integrated three-dimensional dye-sensitized solar cells. Nano Lett. **10**(6), 2092–2096 (2010)
18. B. Weintraub, Y.G. Wei, Z.L. Wang, Optical fiber/nanowire hybrid structures for efficient three-dimensional dye-sensitized solar cells. Angew. Chem. Int. Ed. **48**(47), 8981–8985 (2009)
19. F. Xu et al., Hierarchical ZnO nanowire-nanosheet architectures for high power conversion efficiency in dye-sensitized solar cells. J.Phys. Chem. C **114**(6), 2776–2782 (2010)
20. I. Gonzalez-Valls, M. Lira-Cantu, Vertically-aligned nanostructures of ZnO for excitonic solar cells: A review. Energy Environ. Sci. **2**(1), 19–34 (2009)
21. W.I. Park, G.C. Yi, Electroluminescence in n-ZnO nanorod arrays vertically grown on p-GaN. Adv. Mater. **16**(1), 87–90 (2004)
22. X.M. Zhang et al., Fabrication of a high-brightness blue-light-emitting diode using a ZnO-nanowire array grown on p-GaN thin film. Adv. Mater. **21**(27), 2767–2770 (2009)
23. X.W. Sun et al., A ZnO nanorod inorganic/organic heterostructure light-emitting diode emitting at 342 nm. Nano Lett. **8**(4), 1219–1223 (2008)
24. M.A. Zimmler et al., A two-colour heterojunction unipolar nanowire light-emitting diode by tunnel injection. Nanotechnology **18**(39), 395201 (2007)
25. J.M. Bao et al., Broadband ZnO single-nanowire light-emitting diode. Nano Lett. **6**(8), 1719–1722 (2006)
26. M.A. Zimmler et al., Scalable fabrication of nanowire photonic and electronic circuits using spin-on glass. Nano Lett. **8**(6), 1695–1699 (2008)
27. M. Eichenfield et al., Optomechanical crystals. Nature **462**(7269), 78–82 (2009)
28. Y.F. Hu et al., Optimizing the power output of a ZnO photocell by piezopotential. ACS Nano **4**(7), 4220–4224 (2010)
29. Q. Yang et al., Enhancing sensitivity of a single ZnO micro-/nanowire photodetector by piezo-phototronic effect. ACS Nano **4**(10), 6285–6291 (2010)
30. M.A. Zimmler, F. Capasso, S. Muller, C. Ronning, Optically pumped nanowire lasers: Invited review. Semicond. Sci. Technol. **25**, 024001 (2010)

31. A.B. Djurisic, A.M.C. Ng, X.Y. Chen, ZnO nanostructures for optoelectronics: Material properties and device applications. Prog. Quantum Electron. **34**(4), 191–259 (2010)
32. L. Schmidt-Mende, J.L. MacManus-Driscoll, ZnO-nanostructures, defects, and devices. Mater. Today **10**(5), 40–48 (2007)
33. C.F. Zhang et al., Femtosecond pulse excited two-photon photoluminescence and second harmonic generation in ZnO nanowires. Appl. Phys. Lett. **89**(4), 042117 (2006)
34. T. Voss et al., Nonlinear optics with ZnO nanowires. Phys. Status Solidi B-Basic Solid State Phys. **246**(2), 311–314 (2009)
35. L. Tong et al., Subwavelength-diameter silica wires for low-loss optical wave guiding. Nature **426**, 816–819 (2003)
36. S. Leon-Saval et al., Supercontinuum generation in submicron fibre waveguides. Opt. Expr. **12**, 2864–2869 (2004)
37. L. Tong et al., Assembly of silica nanowires on silica aerogels for microphotonic devices. Nano Lett. **5**, 259 (2005)
38. L. Tong et al., Photonic nanowires directly drawn from bulk glasses. Opti. Expr. **14**, 82–87 (2006)
39. F.X. Gu et al., Polymer single-nanowire optical sensors. Nano Lett. **8**(9), 2757–2761 (2008)
40. Y.H. Li, L.M. Tong, Mach-Zehnder interferometers assembled with optical microfibers or nanofibers. Opt. Lett. **33**(4), 303–305 (2008)
41. X. Jiang et al., Demonstration of optical microfiber knot resonators. Appl. Phys. Lett. **88**, 223501 (2006)
42. X. Jiang et al., Demonstration of microfiber knot laser. Appl. Phys. Lett. **89**, 143513 (2006)
43. G. Sague et al., Cold-atom physics using ultrathin optical fibers: Light-induced dipole forces and surface interactions. Phys. Rev. Lett. **99**(16), 163602 (2007)
44. Q. Yang et al., An evanescent coupling approach for optical characterization of ZnO nanowires. Chin. J.Semicond. **27**(3), 241–244 (2006)
45. D.J. Sirbuly et al., Semiconductor nanowires for subwavelength photonics integration. J.Phys. Chem. B **109**, 15190–15213 (2005)
46. X.W. Chen, V. Sandoghdar, M. Agio, Highly efficient interfacing of guided plasmons and photons in nanowires. Nano Lett. **9**(11), 3756–3761 (2009)
47. K.J. Huang, S.Y. Yang, L.M. Tong, Modeling of evanescent coupling between two parallel optical nanowires. Appl. Opt. **46**(9), 1429–1434 (2007)
48. P.J. Pauzauskie, D.J. Sirbuly, P.D. Yang, Semiconductor nanowire ring resonator laser. Phys. Rev. Lett. **96**(14), 143903 (2006)
49. A.H. Chin et al., Near-infrared semiconductor subwavelength-wire lasers. Appl. Phys. Lett. **88**(16), 163115 (2006)
50. Y. Kayanuma, Quantum-size effects of interacting electrons and holes in semiconductor microcrystals with spherical shape. Phys. Rev. B **38**(14), 9797 (1988)
51. W. Wegscheider et al., Lasing from excitons in quantum wires. Phys. Rev. Lett. **71**(24), 4071 (1993)
52. A.V. Maslov, C.Z. Ning, Reflection of guided modes in a semiconductor nanowire laser. Appl. Phys. Lett. **83**(6), 1237–1239 (2003)
53. Z.-Y. Li, K.-M. Ho, Bloch mode reflection and lasing threshold in semiconductor nanowire laser arrays. Phys. Rev. B **71**(4), 045315 (2005)
54. R. Hauschild, H. Kalt, Guided modes in ZnO nanorods. Appl. Phys. Lett. **89**(12), 123107 (2006)
55. H.G. Park et al., Microstadium single-nanowire laser. Appl. Phys. Lett. **91**(25), 251115 (2007)
56. C.J. Barrelet et al., Hybrid single-nanowire photonic crystal and microresonator structures. Nano Lett. **6**(1), 11–15 (2006)
57. D.M. Bagnall et al., Optically pumped lasing of ZnO at room temperature. Appl. Phys. Lett. **70**(17), 2230–2232 (1997)

58. Y. Huang, X.F. Duan, C.M. Lieber, Nanowires for integrated multicolor nanophotonics. Small **1**(1), 142–147 (2005)
59. E.D. Palik, *Hand Book of Optical Constants of Solids II* (Academic, London, 1985)
60. M.K. Seo et al., Modal characteristics in a single-nanowire cavity with a triangular cross section. Nano Lett. **8**(12), 4534–4538 (2008)
61. D. O'Carroll, I. Lieberwirth, G. Redmond, Microcavity effects and optically pumped lasing in single conjugated polymer nanowires. Nat. Nanotechnol. **2**(3), 180–184 (2007)
62. U. Neumann et al., Second-harmonic efficiency of ZnO nanolayers. Appl. Phys. Lett. **84**(2), 170–172 (2004)
63. C.F. Zhang et al., Low-threshold two-photon pumped ZnO nanowire lasers. Opt. Expr. **17**(10), 7893–7900 (2009)
64. C.F. Zhang et al., Multiphoton route to ZnO nanowire lasers. Opt. Lett. **31**(22), 3345–3347 (2006)
65. Z.W. Dong et al., Multi-photon excitation UV emission by femtosecond pulses and nonlinearity in ZnO single crystal. J.Phys.-Condens. Matter **19**(21), 216202 (2007)
66. X.J. Zhang, Ji W., S.H. Tang, Determination of optical nonlinearities and carrier lifetime in ZnO. J.Opt. Soc. Am. B-Opt. Phys. **14**(8), 1951–1955 (1997)
67. J.A. Bolger et al., Nondegenerate 2-Photon Absorption-Spectra of ZnSe, ZnS and ZnO.Opt. Commun. **97**(3–4), 203–209 (1993)
68. W. Wunstel, C. Klingshirn, Tunable laser-emission from wurtzite-type-II-VI compounds. Opt. Commun. **32**(2), 269–273 (1980)
69. X.H. Yang et al., Two-photon pumped blue lasing in bulk ZnSe and ZnSSe. Appl. Phys. Lett. **62**(10), 1071–1073 (1993)
70. E.S.P. Leong, S.F. Yu, UV random lasing action in p-SiC(4H)/i-ZnO-SiO_2 nanocomposite/n-ZnO: Al heterojunction diodes. Adv. Mater. **18**(13), 1685–1688 (2006)
71. X.Y. Ma et al., Electrically pumped ZnO film ultraviolet random lasers on silicon substrate. Appl. Phys. Lett. **91**(25), 251109 (2007)
72. X.Y. Ma et al., Room temperature electrically pumped ultraviolet random lasing from ZnO nanorod arrays on Si. Opt. Expr. **17**(16), 14426–14433 (2009)
73. Z. Guo et al., Electrically pumped single-mode lasing emission of self-assembled n-ZnO microcrystalline film/p-GaN heterojunction diode. J.Phys. Chem. C **114**(36), 15499–15503 (2010)
74. C.S. Lao et al., Giant enhancement in UV response of ZnO nanobelts by polymer surface-functionalization. J.Am. Chem. Soc. **129**(40), 12096–12097 (2007)
75. J.P. Vilcot et al., Temperature effects on high-gain photoconductive detectors. Electron. Lett. **20**(2), 86–88 (1984)
76. R.R. Mehta, B.S. Sharma, Photoconductive gain greater than unity in CdSe films with Schottky barriers at contacts. J.Appl. Phys. **44**(1), 325–328 (1973)
77. W. Futako et al., Photoconductivity gain over 10 at a large electric field in wide gap a-Si: H. J.Non-Cryst. Solids **227**, 220–224 (1998)
78. G.H. Dohler, Doping superlattices (N-I-P-I crystals). IEEE J.Quantum Electron. **22**(9), 1682–1695 (1986)
79. S.R. Kurtz et al., High photoconductive gain in lateral InAsSb strained-layer superlattice infrared detectors. Appl. Phys. Lett. **53**(20), 1961–1963 (1988)
80. G. Cheng et al., The unsaturated photocurrent controlled by two-dimensional barrier geometry of a single ZnO nanowire Schottky photodiode. Appl. Phys. Lett. **93**(12), 123103 (2008)
81. G. Cheng et al., Erratum: "The unsaturated photocurrent controlled by two-dimensional barrier geometry of a single ZnO nanowire Schottky photodiode" [Appl. Phys. Lett. 93, 123103 (2008)]. Appl. Phys. Lett. **94**(21), 219904 (2009)
82. Z.M. Liao et al., Influence of temperature and illumination on surface barrier of individual ZnO nanowires. J. Chem. Phys. **130**(8), 084708 (2009)

83. H.H. Huang et al., Zero-biased near-ultraviolet and visible photodetector based on ZnO nanorods/n-Si heterojunction. Appl. Phys. Lett. **94**(6), 063512 (2009)
84. Z. Guo et al., Visible and ultraviolet light alternative photodetector based on ZnO nanowire/n-Si heterojunction. Appl. Phys. Lett. **93**(16), 163501 (2008)
85. V. Srikant, D.R. Clarke, On the optical band gap of zinc oxide. J.Appl. Phys. **83**(10), 5447–5451 (1998)
86. A. Rogalski, Infrared detectors: An overview. Infrared Phys. Technol. **43**(3–5), 187–210 (2002)
87. M. Jackson, M.G. Sowa, H.H. Mantsch, Infrared spectroscopy: A new frontier in medicine. Biophys. Chem. **68**(1–3), 109–125 (1997)
88. P.G. Datskos et al., Remote infrared radiation detection using piezoresistive microcantilevers. Appl. Phys. Lett. **69**(20), 2986–2988 (1996)
89. P. Werle et al., Near- and mid-infrared laser-optical sensors for gas analysis. Opt. Lasers Eng. **37**(2–3), 101–114 (2002)
90. K. Karstad et al., Detection of mid-IR radiation by sum frequency generation for free space optical communication. Opt. Lasers Eng. **43**(3–5), 537–544 (2005)
91. B. Cabrera et al., Detection of single infrared, optical, and ultraviolet photons using superconducting transition edge sensors. Appl. Phys. Lett. **73**(6), 735–737 (1998)
92. D.P. Neikirk, W.W. Lam, D.B. Rutledge, Far-infrared microbolometer detectors. Int. J.Infrared Millimeter Waves **5**(3), 245–278 (1984)
93. S.K. Mitra, *Digital Signal Processing: A Computer Based Approach* (McGraw-Hill, New York, NY, 2001)
94. E. Schlenker et al., On the difficulties in characterizing ZnO nanowires. Nanotechnology **19**(36), 365707 (2008)
95. V.R. Mehta et al., Silicon-integrated uncooled infrared detectors: Perspectives on thin films and microstructures. J.Electron. Mater. **34**(5), 484–490 (2005)
96. H.C. Wang et al., IR microbolometer with self-supporting structure operating at room temperature. Infrared Phys. Technol. **45**(1), 53–57 (2004)
97. M. Garcia, et al., IR bolometers based on amorphous silicon germanium alloys. J.Non-Cryst. Solids **338–40**, 744–748 (2004)
98. E. Iborra et al., IR uncooled bolometers based on amorphous $Ge_xSi_{1-x}O_y$ on silicon micromachined structures. J.Microelectromech. Syst. **11**(4), 322–329 (2002)
99. P.G. Datskos, N.V. Lavrik, S. Rajic, Performance of uncooled microcantilever thermal detectors. Rev. Sci. Instrum. **75**(4), 1134–1148 (2004)
100. J. Fonollosa et al., Limits to the integration of filters and lenses on thermoelectric IR detectors by flip-chip techniques. Sens. Actuator A-Phys. **149**(1), 65–73 (2009)
101. H. Kind, et al., Nanowire ultraviolet photodetectors and optical switches. Adv. Mater. **14**(2), 158–160 (2002)
102. Q.H. Li et al., Adsorption and desorption of oxygen probed from ZnO nanowire films by photocurrent measurements. Appl. Phys. Lett. **86**(12), 123117 (2005)
103. R.A. Robie, H.T. Haselton, B.S. Hemingway, Heat capacities and energies at 298.15 K of $MgTiO_3$ (geikielite), ZnO (zincite), and $ZnCO_3$ (smithsonite). J.Chem. Thermodyn. **21**, 743–749 (1989)
104. T. Olorunyolemi et al., Thermal conductivity of zinc oxide: From green to sintered state. J.Am. Ceram. Soc. **85**(5), 1249–1253 (2002)
105. I. Gur et al., Air-stable all-inorganic nanocrystal solar cells processed from solution. Science **310**(5747), 462–465 (2005)
106. G. Yu et al., Polymer photovoltaic cells-enhanced efficiencies via a network of internal donor-acceptor heterojunctions. Science **270**(5243), 1789–1791 (1995)
107. B. Oregan, M. Gratzel, A low-cost, high-efficiency solar-cell based on dye-sensitized colloidal TiO_2 films. Nature **353**(6346), 737–740 (1991)
108. L.E. Greene et al., Low-temperature wafer-scale production of ZnO nanowire arrays. Angew. Chem. Int. Ed. **42**(26), 3031–3034 (2003)

109. Y.G. Wei et al., Wafer-scale high-throughput ordered growth of vertically aligned ZnO nanowire arrays. Nano Lett. **10**(9), 3414–3419 (2010)
110. M.Saito, S. Fujihara, Large photocurrent generation in dye-sensitized ZnO solar cells. Energy Environ. Sci. **1**(2), 280–283 (2008)
111. C.Y. Jiang et al., Improved dye-sensitized solar cells with a ZnO-nanoflower photoanode. Appl. Phys. Lett. **90**(26), 263501 (2007)
112. E. Hosono et al., The fabrication of an upright-standing zinc oxide nanosheet for use in dye-sensitized solar cells. Adv. Mater. **17**(17), 2091–2094 (2005)
113. S. Ito et al., Fabrication and characterization of mesoporous SnO_2/ZnO-composite electrodes for efficient dye solar cells. J.Mater. Chem. **14**(3), 385–390 (2004)
114. Z.S. Wang et al., A highly efficient solar cell made from a dye-modified ZnO-covered TiO_2 nanoporous electrode. Chem. Mat. **13**(2), 678–682 (2001)
115. A.D. Goncalves et al., Synthesis and characterization of ZnO and ZnO:Ga films and their application in dye-sensitized solar cells. Dalton Trans. 2008(11), 1487–1491 (2008)
116. T.P. Chou et al., Hierarchically structured ZnO film for dye-sensitized solar cells with enhanced energy conversion efficiency. Adv. Mater. **19**(18), 2588–2592 (2007)
117. S. Haller et al., Design of a hierarchical structure of ZnO by electrochemistry for ZnO-based dye-sensitized solar cells. Phys. Status Solidi A-Appl. Mater. **207**(10), 2252–2257 (2010)
118. M.R. Lee et al., Solar power wires based on organic photovoltaic materials. Science **324**(5924), 232–235 (2009)
119. X. Fan et al., Wire-shaped flexible dye-sensitized solar cells. Adv. Mater. **20**(3), 592–595 (2008)
120. K. Bourzac, Wrapping solar cells around an optical fiber. http://www.technologyreview.com/energy/23829/. Retrieved 31 Oct 2009. Technology Review, 30 Oct 2009
121. F.F. Gao et al., A new heteroleptic ruthenium sensitizer enhances the absorptivity of mesoporous titania film for a high efficiency dye-sensitized solar cell. Chem. Commun., 2008(23), 2635–2637 (2008)
122. Y. Chiba et al., Dye-sensitized solar cells with conversion efficiency of 11.1%. Jpn. J.Appl. Phys. **45**(24–28), L638–L640 (2006)
123. J.K. Sheu et al., Ga-doped ZnO transparent conductive oxide films applied to GaN-based light-emitting diodes for improving light extraction efficiency. IEEE J.Quantum Electron. **44**(11–12), 1211–1218 (2008)
124. J.J. Berry, D.S. Ginley, P.E. Burrows, Organic light emitting diodes using a Ga: ZnO anode. Appl. Phys. Lett. **92**(19), 193304 (2008)
125. D.B. Thompson et al., Light emitting diodes with ZnO current spreading layers deposited from a low temperature aqueous solution. Appl. Phys. Expr. **2**(4), 042101 (2009)
126. S.F. Chen et al., Improved performances in top-emitting organic light-emitting diodes based on a semiconductor zinc oxide buffer layer. J.Lumin. **128**(7), 1143–1147 (2008)
127. Y. Yang et al., A p-n homojunction ZnO nanorod light-emitting diode formed by As ion implantation. Appl. Phys. Lett. **93**(25), 253107 (2008)
128. J.Y. Zhang et al., Ultraviolet electroluminescence from controlled arsenic-doped ZnO nanowire homojunctions. Appl. Phys. Lett. **93**(2), 021116 (2008)
129. Z.P. Wei et al., Room temperature p-n ZnO blue-violet light-emitting diodes. Appl. Phys. Lett. **90**(4), 042113 (2007)
130. M.T. Chen et al., Near UV LEDs made with in situ doped p-n homojunction ZnO nanowire arrays. Nano Lett. **10**(11), 4387–4393 (2010)
131. Y.L. Wang et al., Dielectric passivation effects on ZnO light emitting diodes. Appl. Phys. Lett. **92**(11), 112101 (2008)
132. S.J. An et al., Enhanced light output of GaN-based light-emitting diodes with ZnO nanorod arrays. Appl. Phys. Lett. **92**(12), 121108 (2008)
133. K.K. Kim et al., Enhanced light extraction efficiency of GaN-based light-emitting diodes with ZnO nanorod arrays grown using aqueous solution. Appl. Phys. Lett. **94**(7), 071118 (2009)

134. L.C. Chen, C.B. Chung, GaN-based light-emitting diode with ZnO nanotexture layer prepared using hydrogen gas. Appl. Surf. Sci. **254**(20), 6586–6589 (2008)
135. S. Xu et al., Ordered nanowire array blue/near-UV light emitting diodes. Adv. Mater. **22**(42), 4749–4753 (2010)
136. E. Lai et al., Vertical nanowire array-based light emitting diodes. Nano Res. **1**, 123–128 (2008)
137. R. Guo et al., Electroluminescence from ZnO nanowire-based p-GaN/n-ZnO heterojunction light-emitting diodes. Appl. Phys. B-Lasers Opt. **94**(1), 33–38 (2009)
138. M.K. Wu et al., Ultraviolet electroluminescence from n-ZnO-SiO_2-ZnO nanocomposite/p-GaN heterojunction light-emitting diodes at forward and reverse bias. IEEE Photonics Technol. Lett. **20**(21–24), 1772–1774 (2008)
139. H. Sun, Q.F. Zhang, J.L. Wu, Electroluminescence from ZnO nanorods with an n-ZnO/p-Si heterojunction structure. Nanotechnology **17**(9), 2271–2274 (2006)
140. B. Ling et al., Electroluminescence from a n-ZnO nanorod/p-$CuAlO_2$ heterojunction light-emitting diode. Physica E **41**(4), 635–639 (2009)
141. C.S. Rout, C.N.R. Rao, Electroluminescence and rectifying properties of heterojunction LEDs based on ZnO nanorods. Nanotechnology **19**(28), 285203 (2008)
142. R. Konenkamp, R.C. Word, M. Godinez, Ultraviolet electroluminescence from ZnO/polymer heterojunction light-emitting diodes. Nano Lett. **5**(10), 2005–2008 (2005)
143. M. Willander et al., Zinc oxide nanorod based photonic devices: Recent progress in growth, light emitting diodes and lasers. Nanotechnology **20**(33), 332001 (2009)
144. Y.Y. Xi et al., Effect of annealing on the performance of CrO_3/ZnO light emitting diodes. Appl. Phys. Lett. **94**(20), 203502 (2009)
145. S. Chu et al., Electrically pumped ultraviolet ZnO diode lasers on Si. Appl. Phys. Lett. **93**(18), 181106 (2008)
146. H. Zhu et al., Ultralow-threshold laser Realized in Zinc Oxide. Adv. Mater. **21**(16), 1613–1617 (2009)
147. D.S. Wiersma, The physics and applications of random lasers. Nat. Phys. **4**(5), 359–367 (2008)
148. Z.L. Wang, Piezopotential gated nanowire devices: Piezotronics and piezo-phototronics. Nano Today **5**(6), 540–552 (2010)

Chapter 13
Nanostructured Light Management for Advanced Photovoltaics

Jia Zhu, Zongfu Yu, Sangmoo Jeong, Ching-Mei Hsu, Shanui Fan, and Yi Cui

13.1 Introduction

The increased concern about detrimental long-term effects of emission of CO_2 and other greenhouse gases and the decreased availability of fossil fuel sources are driving tremendous research efforts for renewable energy technologies. Solar cells, which harvest energy directly from sunlight and convert it into electricity, are widely recognized as an essential component of the future energy portfolio. In spite of a substantial drop in module cost in the past several decades, significant technology improvements in both device performance and manufacturing cost are still necessary for photovoltaics to be economically competitive for terawatt-scale applications.

Currently photovoltaic production is dominated by crystalline silicon modules, which represent about 90% of the market. Even though module cost is recently decreasing substantially as the production rate is increasing, it is estimated that costs for wafer-based Si modules will be in the range of $1–1.5/$W_p$ in the next 10 years, significantly higher than $0.33/$W_p$ target set by US Department of Energy for utility-scale applications. Therefore, there has been a significant effort over the past decade in the development of thin-film solar cells [1] that do not require the use of silicon wafers and therefore can be manufactured at much reduced cost. Unlike wafers with their 200–300 μm thicknesses, thin-film solar cells have thickness typically in the range of 1–2 μm, deposited on cheap substrates such as glass or stainless steel. Currently, there are three leading thin-film technologies: cadmium telluride (CdTe) [2], copper indium diselenide ($CuInSe_2$) [2], and hydrogenated amorphous silicon (a-Si:H) [3, 4]. While thin-film solar cells, notably CdTe modules from First Solar, have demonstrated significant cost reduction, with a compelling low manufacturing cost of around $1/$W_p$, there is still much room for further improvement. For most photovoltaic technologies, efficiencies achieved in the lab are still significantly lower than the theoretical limit, due to both optical and electrical losses. The

Y. Cui (✉)
Department of Materials Science and Engineering, Stanford University, Stanford, CA 94305, USA
e-mail: yicui@stanford.edu

W.L. Zhou, Z.L. Wang (eds.), *Three-Dimensional Nanoarchitectures*,
DOI 10.1007/978-1-4419-9822-4_13,

large capital cost calls for the development of high-throughput processes. Probably one of the biggest concerns regarding thin-film solar cells of CdTe and $CuInSe_2$ is their usage of scarce elements like tellurium and indium, which could fundamentally limit their scales of applications.

For both performance improvement and cost reduction, light trapping is critical to the development of next-generation photovoltaic devices. It is an essential component to boost the efficiency toward the theoretical limit, since it can not only minimize optical losses, including inefficient absorption near the bandgap and reflection at the interfaces but also reduce transport losses due to shortened carrier collection length. By enabling efficient light absorption within much thinner materials, light-trapping design is also critical for manufacturing cost reduction, since it can greatly reduce film thickness, improve the throughput, and expand the range and quality of materials. For CdTe and $CuInSe_2$, the benefit of the implementation of light-trapping design is more significant since their ultimate application scales will be expanded through reduced usage of scarce elements.

However, efficient light-trapping design is rather challenging, primarily due to the broadband nature of the solar spectrum and cost restrictions of module fabrication. An examination of current photovoltaic technologies reveals that the light-trapping designs vary greatly for different kinds of devices. For c-Si solar cells, a representative example is illustrated in Fig. 13.1, known as the "PERL" structure [5]. Even though the process of PERL cells involves lithography, which largely limits their application scales, the PERL structure is a classical demonstration of light trapping in c-Si solar cells. An inverted pyramid surface and double-layer antireflection coatings are the main features for the light-trapping scheme. The inverted pyramids have two main functions. First, they allow light to reflect multiple times before escaping, which minimizes the reflection loss to a large extent. Second, they refract light into large angles, thereby dramatically increasing the optical path length. For these two functions to work properly, the feature size of these pyramids should be typically around tens of micrometers, significantly larger than the wavelength of the utilized solar spectrum.

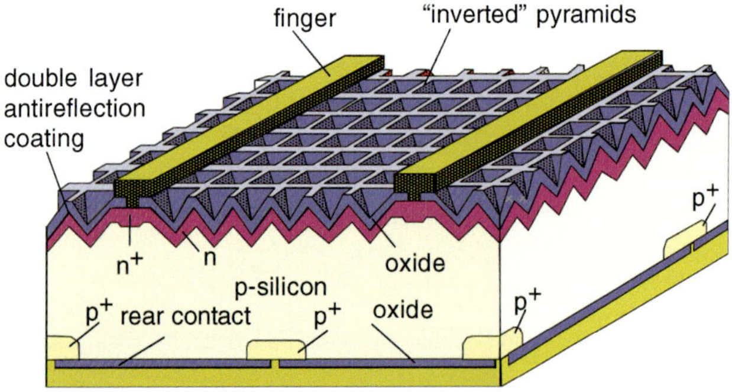

Fig. 13.1 PERL (passivated emitter, rear locally diffused) cell structure [5]

For thin-film solar cells with absorbers of only around 1–2 μm thick, the large size pyramid design is not feasible. In such cells, a transparent conducting oxide (TCO) layer with a quarter wavelength thickness is typically used as an antireflection coating, and a layer of highly reflective metal, such as silver or aluminum, is used as a rear-surface mirror to enhance the light path length by reflection. In some cases, as in a-Si:H solar cells, a layer of randomly textured TCO (aluminum-doped zinc oxide as an example) is used to scatter the light to further enhance the optical path. However, the quarter wavelength antireflection layer works efficiently only within a narrow range of wavelengths and incident angles. The absorption enhancement effects of random textured oxide obtained to date are far from optimal, calling for a better understanding and further technological improvement of light trapping in the sub-wavelength regime (Fig. 13.2).

The ideal light-trapping design for the next-generation solar cells should work for a broad range of the spectrum with a feature size in the sub-wavelength regime, and it must be achievable at very large scale. Nanostructures, with a scale comparable to the wavelengths of visible light, enable an unprecedented manipulation of the flow of photons; therefore, they are widely considered as promising candidates for the advanced light-trapping design. A variety of nanostructure-based photon management designs have been proposed and extensively pursued recently. For example, a significant amount of nanophotonic structures have been reported in the literature for both fundamental understanding and novel applications, including photonic crystals, metamaterials [6], and plasmonics [7, 8]. However, this chapter focuses on nanostructure-based photon management designs specifically for photovoltaic applications, with essentially two functions: suppressing reflection and enhancing absorption across a broad range of the solar spectrum. Plasmonic solar cells, a new type of solar cells which use the scattering from noble metal nanoparticles excited at their surface plasmon resonance, is only briefly mentioned. Interested readers are directed to other sources for further information [9–17].

13.2 Fabrication of Nanowire and Nanocone Arrays

The precise tailoring of morphologies is an essential requirement for any nanostructure to efficiently manipulate the flow of photons. Various parameters such as shapes, diameters, and spacings of these nanostructures must be well controlled across a wide range and be achieved in a large scale. With rapid development of

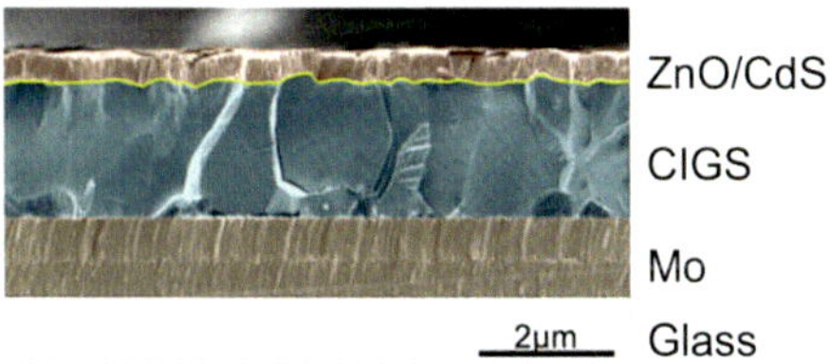

Fig. 13.2 Cross-sectional scanning electron microscopy images of a $Cu(InGa)Se_2$ (CIGS) solar cell [2]

nanotechnology over the past decade, a handful of methods have been developed to fabricate nanostructures in a large scale. For example, vapor–liquid–solid method has been used to synthesize nanowire arrays with high diameter control [18–20]. However, it is very challenging to control spacings between nanowires at the length scale of light wavelength. Solution chemistry is an alternative for synthesis of nanowire arrays [19, 21–23], although the control of both spacing and diameter is limited. Electron beam and photo-lithography have a very good controllability of feature size down to 100 nm or even smaller, but the cost is high and the throughput is low, not feasible for large-scale applications at the current stage. Few processes developed so far can fabricate nanostructures with a fine control of shapes, diameters, and spacings in a large scale. Nanosphere lithography, which combines colloidal nanoparticle synthesis and state-of-the-art fabrication techniques, is one of the most promising ways to achieve this goal [24, 25].

13.2.1 Method

By combining Langmuir–Blodgett (LB) assembly with reactive ion etching (RIE), we have developed a large-scale and low-temperature process to fabricate nanostructures. It provides precise control of diameters, spacings, and shapes across a wide range, from tens of nanometers to several micrometers [26, 27], and can be applied to a large variety of materials. Figure 13.3 shows the general fabrication process, in which silicon is used as an example. Monodisperse SiO_2 nanoparticles, synthesized in-house, are assembled into a close-packed monolayer on top of a silicon wafer using the Langmuir–Blodgett (LB) method. Monodisperse SiO_2 particles, with diameters from 50 to 800 nm, are produced by a modified Stöber synthesis. The particles are modified with aminopropyldiethoxymethylsilane so as to terminate them with positively charged amine groups, preventing aggregation. The diameter and spacing of the nanoparticles can be further tuned by selective and isotropic RIE of SiO_2 (Fig. 13.3b). The RIE etching is based on fluorine chemistry, using a mixture of O_2 and CHF_3. Si nanowires and nanocones can similarly be obtained by using Cl_2-based selective and anisotropic RIE (Fig. 13.3c). The diameter and the spacing of these nanostructures are determined by the initial nanoparticle sizes and both SiO_2 and Si etching times. SiO_2 particles can be removed by hydrofluoric acid (HF), if needed (Fig. 13.3d).

13.2.2 Shape Control: Nanowires and Nanocones

One unique advantage of our process is the control over the shape. Depending on the conditions of RIE, either nanowire or nanocone can be obtained. There are several mechanisms behind the formation of nanocones. First, Cl- and Br-free radicals arrive at the Si surface from all directions during RIE, inducing some isotropic etching of Si or undercutting during the supposedly anisotropic steps. Second, the etching selectivity of Si to SiO_2 is around 26; therefore, the extent of lateral etching

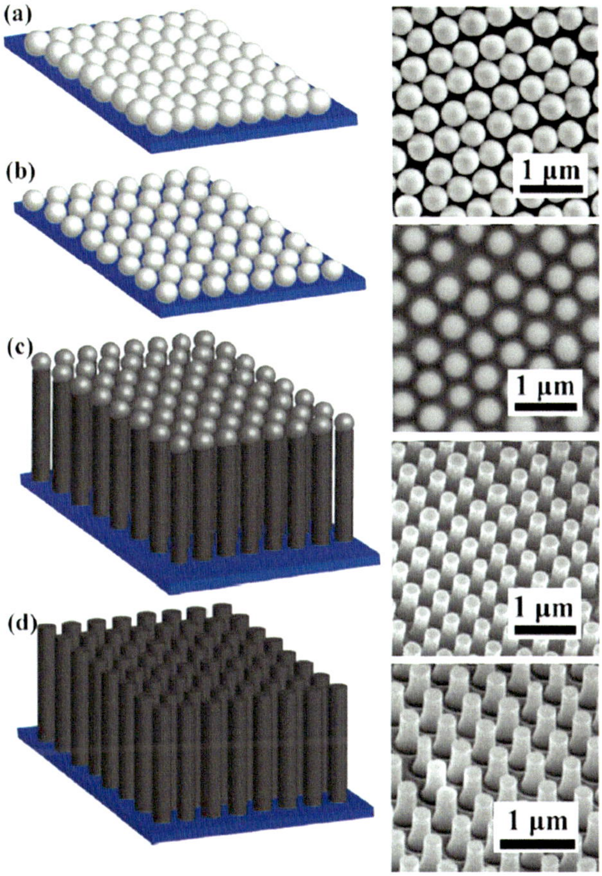

Fig. 13.3 Fabrication process of nanowires

will increase due to mask erosion when using SiO_2 masks. Last, tapered sidewalls can occur when the etched products are redeposited during etching, since the redeposition rate decreases from the bottom to the top of the pillars.

With the understanding of formation mechanisms, the undercutting can be utilized to form uniquely sharp nanocones through control of the etching conditions. The aspect ratio and the tip radius of these nanocones can be precisely controlled. First, Si nanowires are formed by Cl_2-based anisotropic RIE. Second, C_2ClF_5/SF_6 is used for further isotropic etching of preformed nanowires, which creates undercut and sharpens the nanowires (Fig. 13.4a). Figure 13.4b–d shows the SEM images at different stages of the sharpening process. Accompanying the sharpening is the shrinkage of SiO_2 spheres. The combination of anisotropic and isotropic etching can lead to a sharpening of the tips to a radius of curvature of 5 nm, opening up the opportunities for refractive index matching, which will be explained in detail in Section 13.3.

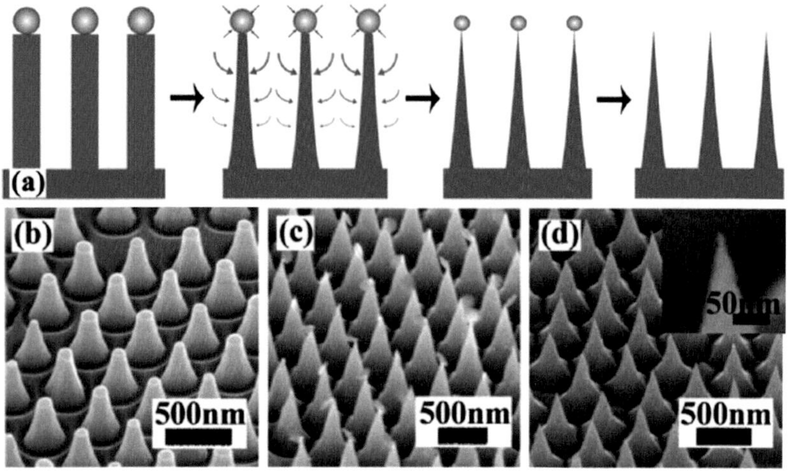

Fig. 13.4 Schematic of sharpening process with *thicker arrows* indicating faster etching rate

13.2.3 Diameter and Spacing Control

Besides the shapes of these nanowire and nanocone arrays, the diameter (D) and spacing (S) can also be rationally designed. Since the center-to-center distance of neighboring nanostructures is D+S, the SiO_2 nanoparticles can be chosen to have an initial diameter of D+S and to be etched by $S/2$. The diameter of the initial SiO_2 nanoparticles can be precisely controlled from 50 nm to 1 μm during the synthesis. RIE etching can be controlled with an accuracy of ~10 nm. Thus, we have precise control of diameter and spacing over a wide range. Figures 13.5 and 13.6 give two examples of fine control of diameter and spacing, respectively. Figure 13.5 shows the nanowires with desired diameters between 60 and 600 nm. Figure 13.6 shows the nanowires with desired spacing between 50 and 400 nm.

13.2.4 Large-Scale Process

Another important characteristic of our process is that it can be applied to a large scale. As explained above, the scale achievable in our process is essentially defined by the area of SiO_2 particles' coverage. We have developed a variety of processes to assemble the nanoparticles in a large area.

Using Langmuir–Blodgett method, we have demonstrated that close-packed monolayers of nanoparticles can be produced on a wafer scale with a reasonable throughput. Figure 13.7 shows a photograph of a 4-in. wafer covered uniformly by a monolayer of 200-nm-diameter SiO_2 particles. Scanning electron microscope (SEM) images at four randomly picked locations far from each other show that a monolayer of particles covers the whole wafer.

Fig. 13.5 SEM images of nanowire arrays with uniform diameter of (**a**) 60 nm, (**b**) 125 nm, (**c**) 300 nm, and (**d**) 600 nm

We have also developed a rod coating method for assembling silica nanoparticles on both rigid and flexible substrates (Fig. 13.8a) [28]. This method can be directly transferred to large-scale roll-to-roll processing. This wire-wound rod coating method has been widely used to deposit an even amount of fluid over a moving surface to manufacture office products and flexible packaging [29]. We extend this technique to nanoparticle colloids to produce controllably two- and three-dimensional arrays of silica nanoparticles.

Silica nanoparticles are synthesized by a modified Stöber process as described above. A silica nanoparticle ink is prepared by mixing the nanoparticles with poly-4-vinylphenol (PVPh; Sigma-Aldrich) in ethanol. The typical concentration of nanoparticle and PVPh is 50 g/L and 0.2 wt%, respectively. The coating assembly consisted of a wire-wound rod, which is a stainless steel rod with stainless steel wire wound around it. The nanoparticle ink is dropped onto the substrate and the rod is pulled across, leaving behind a volume of solution equal to the

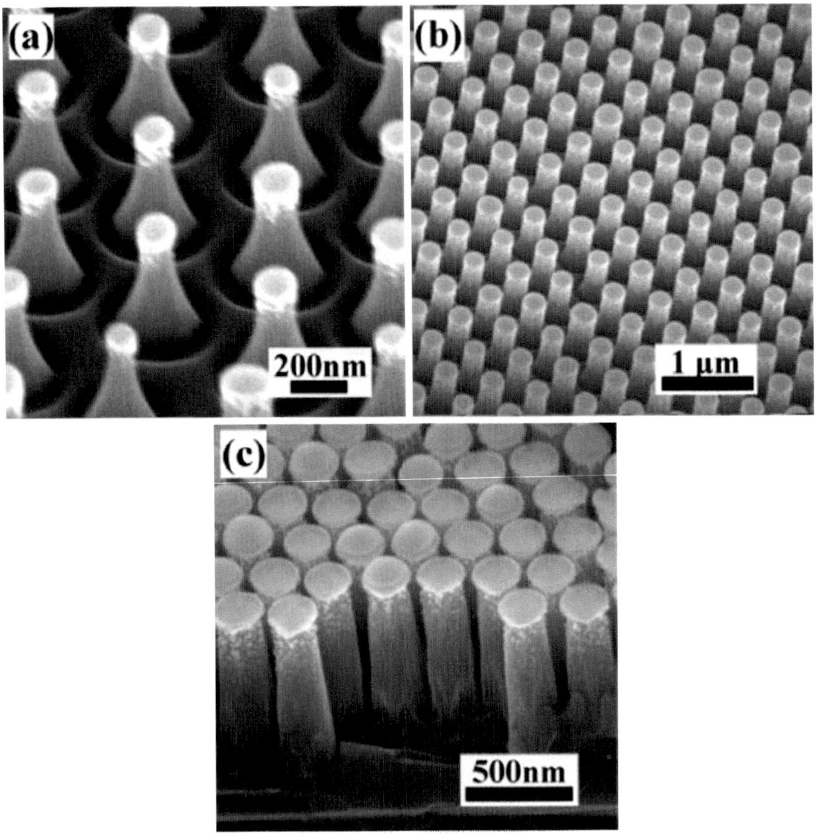

Fig. 13.6 Nanowire arrays with a spacing of (**a**) 400 nm, (**b**) 350 nm, and (**c**) 50 nm

groove space between each wire winding and ultimately leading to a uniform film (Fig. 13.8a). The diameter of the wire on the rod determines the thickness of the wet film. To improve uniformity, the rod can be pulled with an automatic applicator at a constant speed (~2 cm/s). Figure 13.8b shows an optical image of a large-area, close-packed silica nanoparticle monolayer on a flexible plastic, polyethylene terephthalate (PET), substrate fabricated by this process.

Since the assembly process occurs during the drying step, several parameters such as contact angle, evaporation rate, viscosity, and nanoparticle concentration play a critical role in achieving close-packed nanoparticle arrays over large areas. The ink need to completely wet the substrate to form high-quality films. Once the nanoparticle ink is spread evenly by the wire-wound rod and wets the substrate, the solvent starts to evaporate gradually from the part where it was spread first. With the appropriate nanoparticle concentration, the particle-to-particle distance can be controlled such that when the solvent layer becomes as thin as the nanoparticle diameter, the solvent can remain between nanoparticles in the form

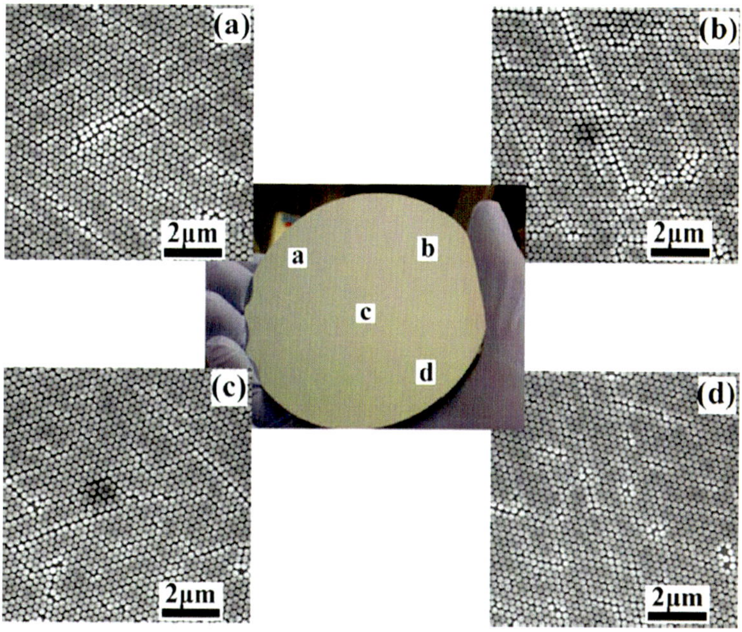

Fig. 13.7 A photograph of a 4-in. wafer with SiO_2 nanoparticle on the surface. (**a**), (**b**), (**c**), and (**d**) SEM images of four random spots with a uniform monolayer of SiO_2 nanoparticles

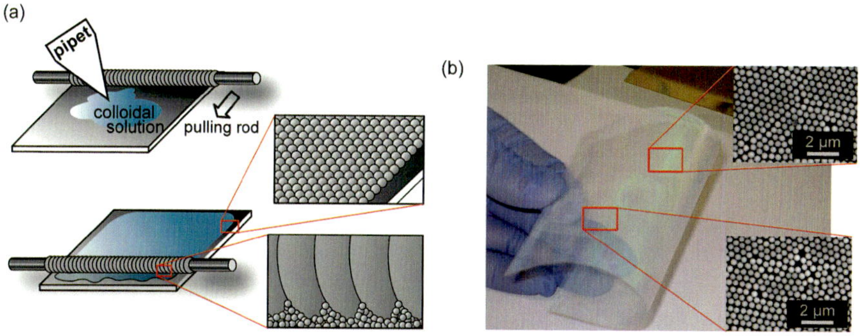

Fig. 13.8 (**a**) Schematic illustration of a wire-wound rod coating method for printing nanoparticle monolayer. (**b**) An optical image of a close-packed silica nanoparticle monolayer printed by the wire-wound rod coating method on a flexible polyethylene terephthalate (PET) substrate. *Two insets* show SEM images of the monolayer

of a meniscus. The meniscus provides a capillary force, which drives the particles together, nucleating a thin-film assembly. This nucleate grows from the convective flux of nanoparticles toward the drying front of the wet film. Figure 13.9a illustrates this assembly mechanism. The ink, a mixture of nanoparticles and PVPh in ethanol, wets silicon and PET substrates completely. The PVPh is added to decrease

Fig. 13.9 (**a**) Schematic illustrations and (**b, c**) SEM images of particle distribution after evaporation of ethanol mixed with poly-4-vinylphenol

the evaporation rate and increase the viscosity. If the solvent is too volatile or its viscosity is too low, instabilities in the array growth may arise and the wet film can easily break into separated droplets. Figure 13.9b, c shows SEM images of resulting good monolayers of nanoparticles formed when using PVPh.

Using the same scalable wire-wound rod coating method, multilayer nanoparticle arrays can also be obtained by increasing the nanoparticle concentration in the ink. Figure 13.10 shows the relationship between the particle concentration and the number of nanoparticle layers formed after the solvent dries. When the nanoparticle concentration increases from 50 to 100 to 200 g/L, the number of nanoparticle layers also proportionally increases from 1 to 2 to 4, respectively, and the uniformity of multiple layers is comparable to that of the monolayer. Considering that a monolayer grows by the convective flux of nanoparticles, we can infer that if the nanoparticle concentration is higher, more particles will flow to the boundary and start to form multiple layers. This concentration dependence was observed for both silicon and PET substrates.

13.3 Photon Management: Antireflection

When light hits the interface between media characterized by different refractive indices, a significant fraction of it is reflected. For example, without any treatment, around 30% of light would be lost due to reflection at the interface between air and Si. Therefore, reflection is a serious problem not only for solar cells but also for many other optoelectronic devices, such as light-emitting diodes (LEDs) [30, 31] and photodetectors. A range of techniques have been developed to reduce reflection for different applications [30, 31]. Now, the industrial standard of antireflection coating for thin-film solar cells is to use a quarter-wavelength transparent layer with destructive interference. However, this technique works only for a narrow range of wavelengths and incident angles.

13.3.1 Nanowires

Over the past few years, nanostructures have been heavily investigated for broadband reflection suppression. Nanowire arrays with moderate filling ratio were found

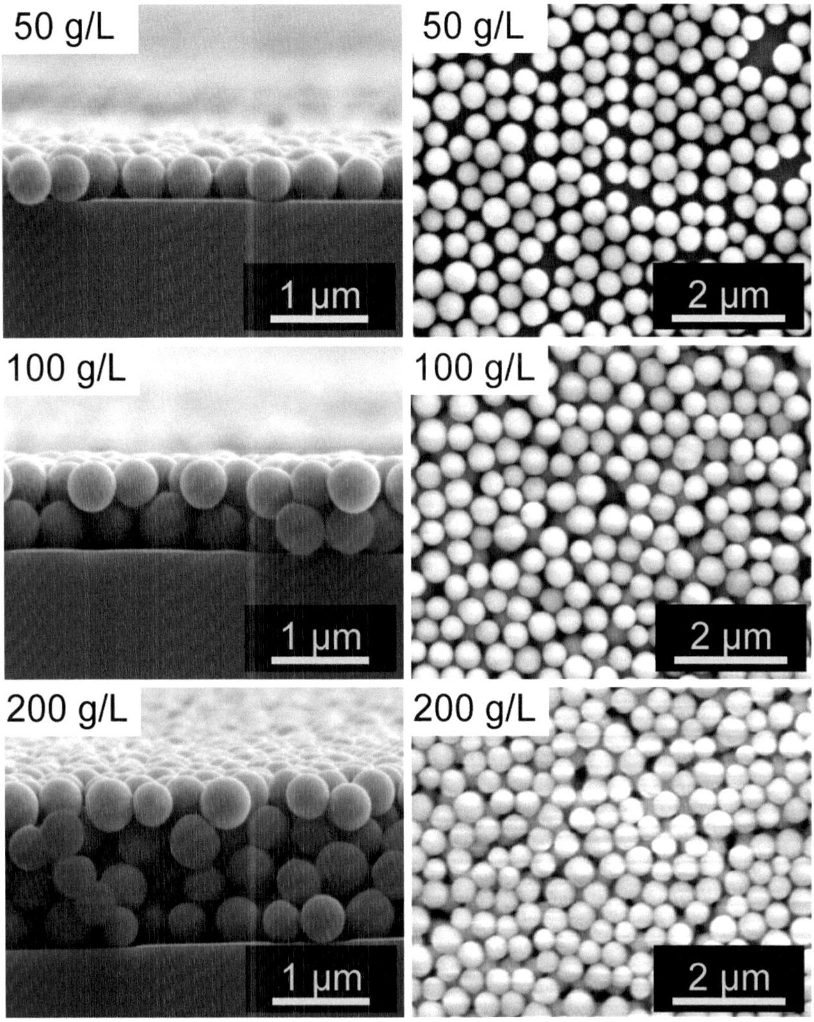

Fig. 13.10 SEM images of silica nanoparticle distribution with different particle concentrations

to be able to greatly reduce reflectance [26, 32–35], since they essentially provide an intermediate refractive index step (Fig. 13.11b). An even more ideal antireflection technique is to provide impedance matching through a gradual reduction of the effective refractive index (Fig. 13.11c). With graded refractive index layers, light experiences only a gradual change of the refractive index instead of hitting a sharp interface (see Fig. 13.11a), and reflection can be efficiently eliminated for a large range of wavelengths and angles of incidence.

One example of nanowire-based, graded refractive index design is shown in Fig. 13.12. Several layers of nanowire arrays of different materials, densities, and tilting angles are carefully designed and deposited in sequence to form a graded

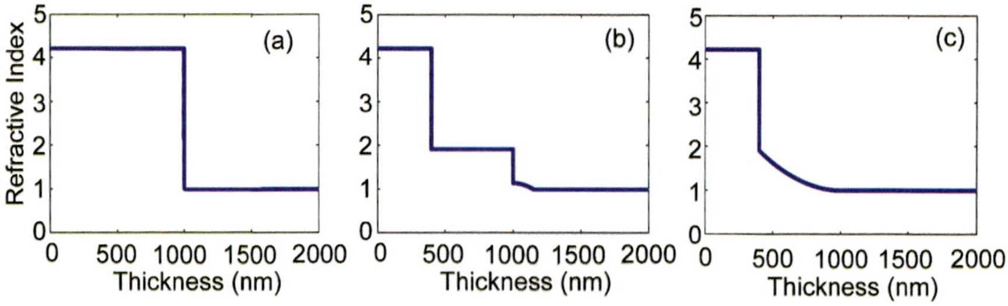

Fig. 13.11 The effective refractive index profiles of the interfaces between air and (**a**) a-Si:H thin film, (**b**) a-Si:H NW arrays, and (**c**) a-Si:H NC arrays

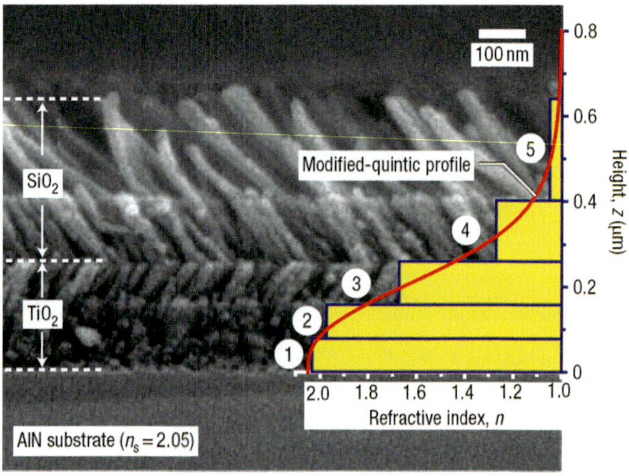

Fig. 13.12 TiO_2–SiO_2-graded index coating. Cross-sectional SEM image of graded index coating with a modified-quintic-index profile. The graded index coating consists of three TiO_2 nanorod layers and two SiO_2 nanorod layers [36]

index coating [36]. It was found that the reflection can be suppressed down to 0.5% for a broad range of wavelengths and angles of incidence.

13.3.2 Nanocones

Nanocone arrays with a gradual change of diameter from the bottom to the top can provide another version of graded index coating [26, 37], which has an efficient broadband antireflection property and great processing advantages. We have performed an experiment to evaluate the antireflection effect of nanowires and nanocones, as compared to thin film. Three samples were fabricated, with a 1-μm-thick a-Si:H thin film deposited onto each substrate. A monolayer of silica nanoparticles was preformed on the second and third samples. After RIE etching, nanowire and nanocone arrays were formed on the second and third samples, respectively, because of different etching conditions as explained in Section 13.2. Figure 13.13 (left) shows an SEM image of an a-Si:H nanowire array after RIE. The dimensions of each nanowire were ~300 nm wide × ~600 nm long. The silica

Fig. 13.13 SEM images of an a-Si:H nanowire array (*left*) and a nanocone array (*right*)

nanoparticles can still clearly be seen on the top of each nanowire. Figure 13.13 (right) shows an SEM image of an a-Si:H nanocone array. Each nanocone was also ~600 nm long. The tip diameter of these nanocones was ~20 nm, while the base diameter was ~300 nm. After RIE, the silica nanoparticles were so small that they were no longer observable on top of the NCs (Fig. 13.14).

Absolute hemispherical measurements, collected with an integrating sphere, were used to quantitatively characterize these three samples (Fig. 13.15, top). The absorption over a wide range of wavelengths (400–800 nm) was measured. With the bandgap of a-Si:H around 1.75 eV, this range covers most of the useful spectral regime for a-Si:H solar cells. Between 400 and 650 nm, nanocone array absorption was maintained above 93%, which was much better than that for both the nanowire arrays (75%) and thin films (64%). The measured total absorption decreased to 88% at 700 nm – corresponding to the a-Si:H bandgap (1.75 eV) – which is also better than that for either nanowires (70%) or thin films (53%).

Their total absorption for different angles of incidence is measured at a wavelength of 488 nm (Fig. 13.15). The sample with a nanocone array demonstrated the highest absorption, i.e., 98.4% around normal incidence, which offers a significant advantage over both nanowires (85%) and thin films (75%). The performance of the nanocone sample also showed a reduced dependence on the angle of incidence and significantly higher absorption at any angle. At angles of incidence up to 60°, the total absorption was maintained above 90%, which compares favorably with 70 and 45% for the nanowire array and thin film, respectively. Since all three samples

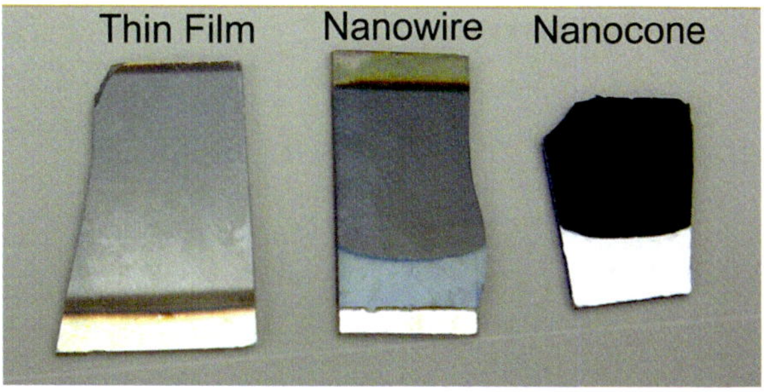

Fig. 13.14 (*left*) a-Si:H thin film, (*middle*) nanowire arrays, and (*right*) nanocone arrays

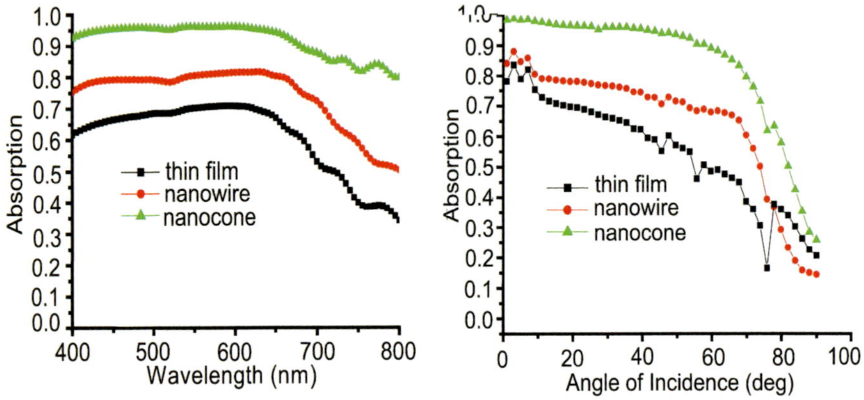

Fig. 13.15 Hemispherical absorption measurement of samples with a-Si:H thin film, a nanowire array, and a nanocone array as top layer: wavelength dependence (at normal incidence) (*top*) and angle of incidence dependence (at wavelength $\lambda = 488$ nm) (*bottom*)

start with 1 μm thickness, which is around the absorption depth of a-Si:H film, the absorption enhancement in the samples of nanowires and nanocones is believed to be mainly due to suppression of reflection from the front surface.

13.4 Photon Management: Absorption Enhancement

Once light is coupled into solar cell devices with much reduced reflection, the next step of photon management is to increase the optical path length within the devices. For crystalline Si solar cells, a pyramid design, with features spanning tens of micrometers, can dramatically increase the optical path length of light with long wavelength. In geometrical optics, where dimensions are generally significantly larger than wavelengths, it is well known that absorption enhancement can be up to $4n^2/\sin^2\theta$, the Yablonovitch limit [38, 39], with n as refractive index of the absorber layer and θ as the angle of the emission cone in the medium surrounding the cell. However, for most of the next-generation solar cell devices, the thickness of the active layer is typically only around 1–2 μm, or less, which is comparable to the wavelengths of visible light. Increasing optical path length at the sub-wavelength regime calls for both a better physical understanding and a novel process development.

13.4.1 Different Mechanisms

Numerous ideas and designs have recently been proposed to use nanostructures for this purpose [40]. One notable example is plasmonic solar cells [9–17]. Nanoparticle arrays of noble metals, such as silver, have been incorporated for absorption

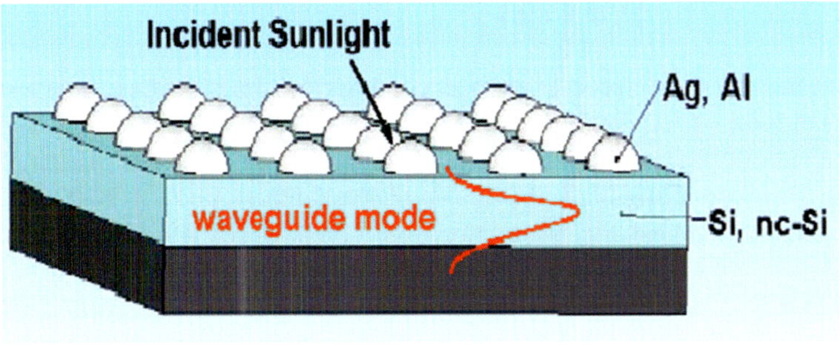

Fig. 13.16 Schematic illustration of plasmonic solar cells [17]

enhancement in the region close to bandgap edge in a variety of devices. In those studies, the main mechanisms were believed to be the large resonant scattering cross section of these particles or plasmonics (Fig. 13.16).

Photonic crystal designs based on well-defined nanostructure arrays are another highly pursued approach [41, 42] (Fig. 13.17). The ultimate scales of practical application of those designs will be determined by the scalability of techniques involved to achieve these photonic crystal structures.

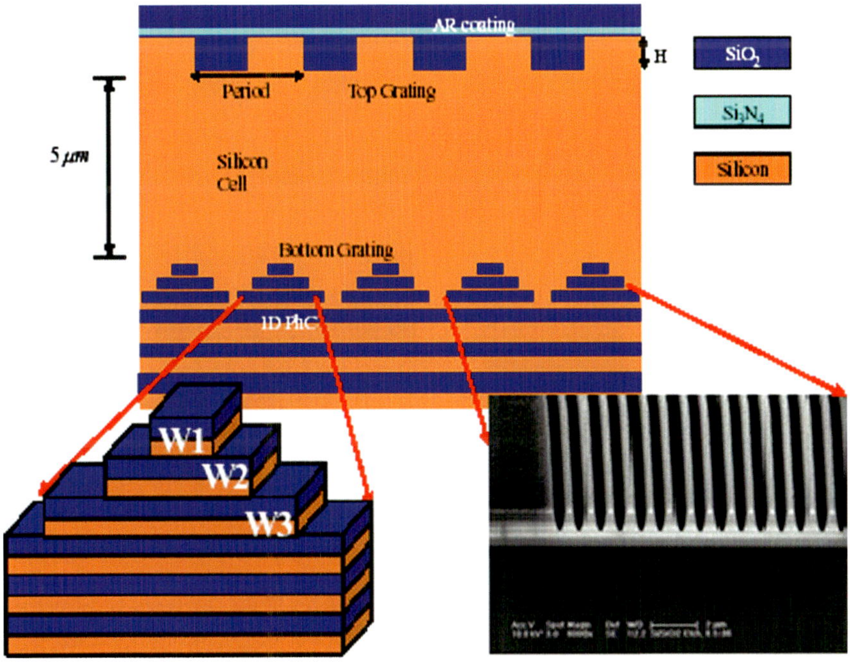

Fig. 13.17 Schematic illustration of photonic crystal design-based solar cells [43]

13.4.2 Nanodome Structures

Based on the fabrication process introduced in Section 13.2, we have demonstrated novel nanodome solar cells. They have periodic nanoscale modulation for all layers, from the bottom substrate through the active absorber to the top transparent contact (Fig. 13.18). These devices combine many nanophotonic effects to both efficiently reduce reflection and enhance absorption over a broad spectral range.

We have chosen a-Si:H solar cells to demonstrate the advantages of the nanodome concept. As the second most produced solar cells, a-Si:H solar cells have several unique advantages. It is based on abundant, non-toxic materials and can be fabricated by low-temperature roll-to-toll processes (around 200°C). More importantly, a-Si:H can absorb light efficiently, with an absorption depth of only 1 μm (at around 1.8 eV), several hundred times thinner than that of crystalline silicon. However, carriers of a-Si:H have poor transport properties, especially a short carrier diffusion length of around 300 nm. In addition, the 10–30% efficiency degradation under light soaking, known as the Stabler–Wronski effect, is found to be less severe with thinner films (below 300 nm). Hence, efficient light harvesting within a much thinner layer (<300 nm) is essential to the device performance of this type of solar cell.

The centerpiece of the nanodome structures is nanocone substrates (Fig. 13.19a). Nanocone glass or quartz substrates are fabricated by Langmuir–Blodgett assembly of close-packed monodisperse SiO_2 nanoparticles followed by reactive ion etching,

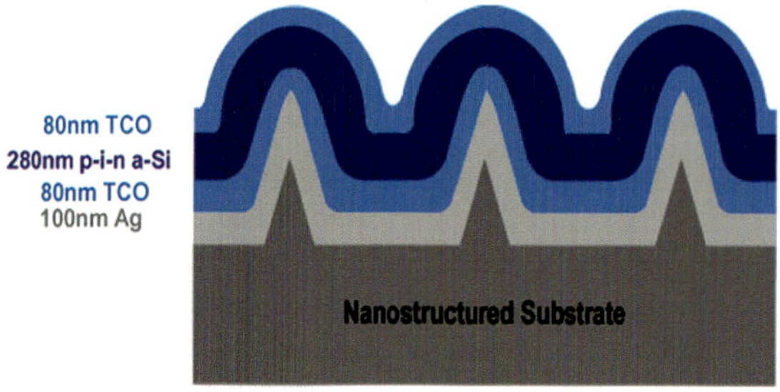

Fig. 13.18 Schematic illustration of nanodome solar cells

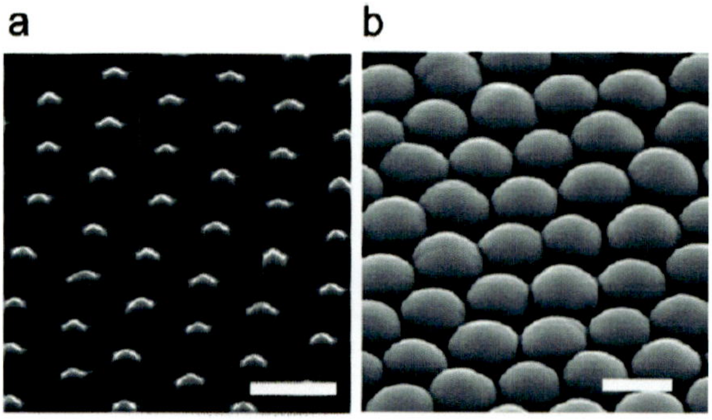

Fig. 13.19 Nanodome a-Si:H solar cell structure. SEM taken at 45° on (**a**) nanocone quartz substrate and (**b**) a-Si:H nanodome solar cells after deposition of multilayers of materials on nanocones. Scale bar, 500 nm

as described in Section 13.2. The base diameters and spacings of nanocones can be controlled in the range of 100–1000 nm, which is relevant to wavelengths in the solar spectrum (Fig. 13.19a). A typical single p–i–n junction nanodome a-Si:H solar cell is conformally deposited on top of the nanocone substrates. The solar cell layers consist of 100-nm-thick Ag as a back reflector, 80-nm-thick transparent conducting oxide (TCO) as both bottom and top electrodes, and a thin a-Si:H active layer of 280 nm (from top to bottom: p–i–n, 10–250–20 nm). After deposition, the nanocone pattern is largely transferred to the top layer, although nanocones become nanodomes, as shown in scanning electron microscopy (SEM) images (Fig. 13.19b).

The unique characteristic of this nanodome structure is that it has periodic nanoscale modulation for all layers, due to the nanostructured bottom layer. Because of this novel geometry, the nanodome solar cells have both antireflection and light-trapping effects. The results of the hemispherical absorption measurement on these nanodome solar cells are a clear indication of both of these two effects. We have measured absorption over a broad wavelength range (400–800 nm), which covers most of the spectrum that is useful for a-Si:H, which has a bandgap of 1.75 eV (or ~710 nm) (note: a-Si:H has a long band tail). For comparison, flat a-Si:H film solar cells with the same device structure and layer thickness were also measured. Since TCO has a lower refractive index (2.2) compared to a-Si:H (~4.23), measurements were carried out with and without the top TCO layer for both types of solar cells. The absorption data in the wavelength range of 400–800 nm under normal incidence are summarized in Fig. 13.20a. Green and black curves are from nanodome devices with and without the top TCO layers, while blue and red ones are from flat film devices with and without the top TCO layers, respectively.

As shown in Fig. 13.20a, over the whole spectrum, nanodome devices show significantly larger absorption than do flat film devices with and without top ITO layers. The weighted absorption, integrated over the whole spectrum under the 1 Sun AM 1.5 illumination condition, is plotted in Fig. 13.21. Nanodomes with a top TCO layer show extremely high total weighted absorption of 94%, significantly higher than 65% achieved in flat films with a top TCO layer (Fig. 13.21).

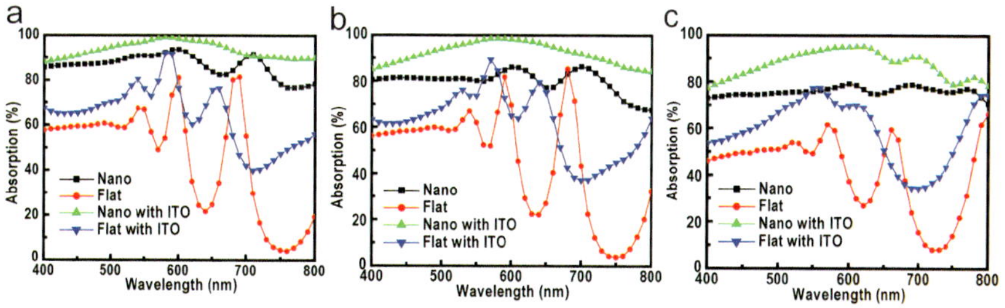

Fig. 13.20 Integrating sphere measurements of absorption under normal incidence (**a**), 30° angle of incidence (**b**), and 60° angle of incidence (**c**). The samples are flat substrates without ITO coating (*red*), flat substrates with ITO coating (*blue*), nanodomes without ITO coating (*black*), and nanodomes with ITO coating (*green*)

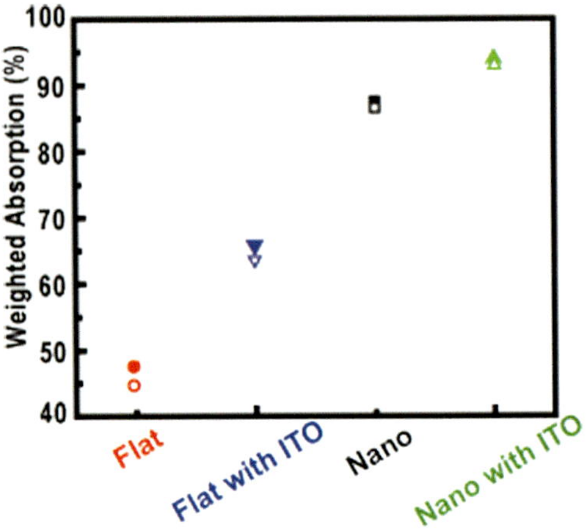

Fig. 13.21 The weighted absorption integrated over the wavelengths of 400–800 nm by experiment (*solid symbols*) and simulation (*hollow symbols*)

As shown in Fig. 13.21, the absorption of nanodome and flat film devices with a top TCO layer is better than that without a top TCO layer, respectively. This is because TCO has a lower refractive index than does a-Si:H so that light reflection is lower for the samples with the top TCO layers. However, this TCO enhancement effect is much less for nanodome devices (from 87 to 94%) than for flat film devices (from 48 to 65%), which indicates that the nanodome geometry without TCO already has very good antireflection property compared to the flat film geometry.

The absorption enhancement comes from both suppressing reflection and increasing optical path length. While it is difficult to separate the two effects, since they are essentially coupled together, the antireflection effect is more dramatic in short-wavelength region. In the case of a-Si:H solar cells, for the wavelength region below 500 nm, all the light loss can be attributed to the light reflection since its absorption depth (~100 nm) is smaller than the a-Si:H layer thickness of 280 nm. As seen in Fig. 13.20a, without the top TCO coating, the absorption of nanodomes is always above 85%, while the flat one is below 60%. Adding the TCO coating improves the absorption above 88 and 65%, respectively. The antireflection effect is due to the tapered shape of nanodome structures, which provides better effective refractive index matching with air, similar to the nanocone case explained in Section 13.3.

The effect of increased optical path length plays a major role in the longer wavelength, where light is less efficiently absorbed. In the case of a-Si:H flat films, for the wavelengths above 550 nm, significant interference oscillations appear in flat film devices. However, nanodome devices still show relatively flat broadband adsorption (Fig. 13.20a). The observed oscillations in flat film devices are Fabry–Perot interference, arising from the long-wavelength light not absorbed by the a-Si:H layer interfering with the reflected light from the top layer of the device. The interference oscillations are greatly reduced for the nanodome devices, suggesting that very little light escapes after reflection by the Ag. Their absence suggests significant light-trapping effects. The most interesting and significant absorption improvement using the nanodome geometry versus the flat film is in the wavelength range of 700–800 nm. a-Si:H has a bandgap of ~710 nm and a long band tail. The absorption coefficient drops quickly when the light wavelength is above ~700 nm. For flat film devices with the top TCO layer, there is a significant reduction of absorption down to 50%, while nanodome devices with the top TCO layer maintain absorption of ~90%. These data suggest that nanodomes can enhance light trapping significantly even for the absorption of photons below the bandgap. However, it is hard to conclude at this moment whether any of the absorption in this wavelength range contributes to short-circuit current.

The dramatically increased optical path length in nanodome structures is due to the underlying nanoscale modulation of the Ag layer, which can couple light along the in-plane dimension, providing a light-trapping mechanism. Compared to the design of plasmonic solar cells, the nanostructured Ag back reflector as bottom contact in our nanodome devices is a better choice: while long-wavelength light is strongly scattered by the modulated Ag back reflector, short-wavelength absorption is not compromised since significant absorption occurs during a single pass through the absorber, before reaching the Ag films. Compared with the Lambertian scattering, which is based on well-understood surface texturing using features much larger than light wavelengths for absorption enhancement, our devices use sub-wavelength nanodome structures, which are more feasible for solar cells with only sub-micrometer-thick absorber layers.

The physical mechanisms of absorption enhancement are further elucidated by solving the Maxwell equations with three-dimensional, finite-difference, time-domain simulation on the experimental device geometry. The cross section of the structure is shown in Fig. 13.22 (left) with incident plane waves polarized in the x-direction. The period of the nanodome array is 450 nm. The dielectric constants of silver and silicon are taken from tabulated experimental data modeled by complex-conjugate, pole-residue method. The simulated nanodome structure can absorb 93%

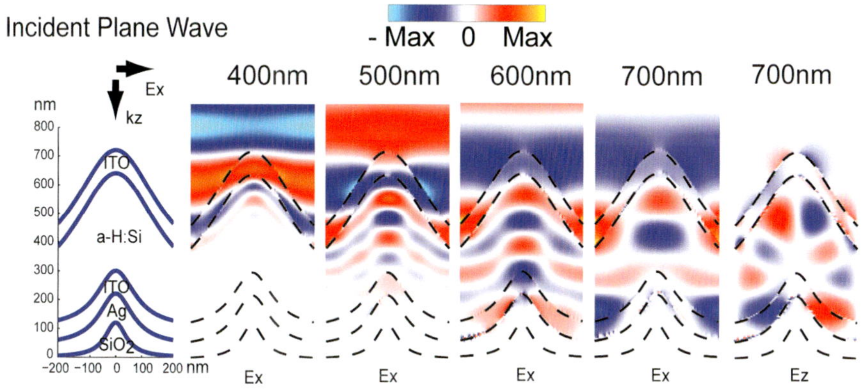

Fig. 13.22 Snapshots of simulated electric fields in the structure for different wavelengths

of normally incident sunlight for the spectral range from 400 to 800 nm (Fig. 13.21, green triangle), which matches well with experimental data. The broadband absorption comes from two contributions. First, the reflection is greatly suppressed by the nanodomes. The dome shape forms a graded refractive index profile, creating a broadband antireflection layer. For short wavelength around 400–500 nm (Fig. 13.22, E_x simulation for 400 and 500 nm), a-Si:H is highly absorptive. Thus, with efficient antireflection effects, all the incident lights are absorbed during a single path through the a-Si:H layer. Second, the nanodome shape can also efficiently couple the incident light into modes that are guided into the a-Si:H layer (note: the dispersion relation of such modes can be strongly influenced by the presence of plasmonic responses of the Ag film). This is particularly important for the long-wavelength regime (Fig. 13.22, E_x simulation for 600 and 700 nm) where a-Si:H is less absorptive and all the incident light cannot be absorbed efficiently during a single path. In Fig. 13.22, the E_z simulation for 700 nm shows the z-component of the electric field, indicating strong guided modes confined inside the nanodome structure. Figure 13.23 shows the absorption spectra with (green line) and without (gray line) metal loss, to evaluate the absorption loss by silver. For the lossless silver case, the weighted absorption for normal incident sunlight is 92% as compared to 93% for the realistic silver case. Therefore, the metal loss contributes only a negligible portion of absorption, which provides another advantage compared to plasmonic solar cells.

It is also very important to evaluate the absorption efficiency over a wide range of incident angles, since sunlight can be quite diffuse in a practical environment. Figure 13.20b, c shows the absorption measurement at incident angles of 30° and 60°, respectively. As the incident angle increases from 0° to 60°, the absorption over the whole spectrum decreases only by 5% for nanodome devices, while the absorption of flat devices decreases by 13%. These data suggest that nanodome devices have an advantage over flat film devices in the real environment. Indeed, nanodome and flat film devices look very different even to the eye. Figure 13.24a shows photographs of nanodomes (left) and flat films (right) without the top TCO layers

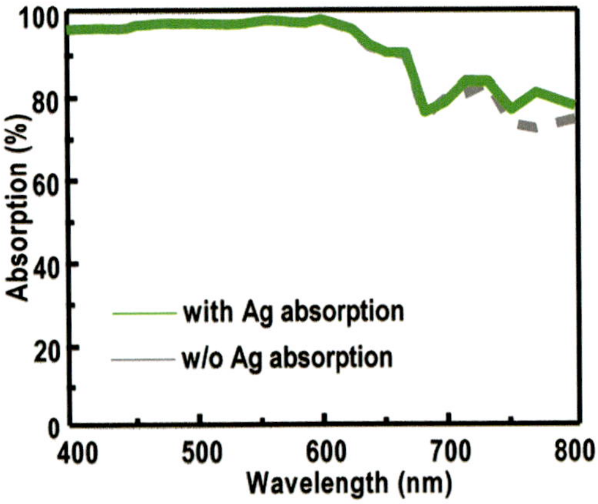

Fig. 13.23 Simulation of absorption spectra for the case with (*green*) and without (*dashed gray*) Ag absorption loss

Fig. 13.24 Photographs of nanodome solar cells (*left*) and flat film solar cells (*right*)

in diffuse light conditions. The flat film devices are mirror-like, highly reflective, and look red because of inefficient light absorption at long wavelengths. However, nanodome devices look black due to efficient antireflection and light trapping.

13.5 Solar Cell Performance

Combining superior antireflection and efficient absorption enhancement, nanodome solar cell devices based on this novel photon management design can achieve near-perfect light absorption with thin active layers. Therefore, significant improvements in short-circuit current and efficiency can be achieved, compared to conventional devices.

To prove how effectively antireflection and light trapping can improve the power conversion efficiency of solar cells, we have tested nanodome and flat film solar cell devices in a solar simulator with 1 sun AM 1.5 G illumination. Excitingly, the nanodome devices show power conversion efficiencies that are 25% higher than those of the flat film devices, made under otherwise identical conditions. An example is shown in Fig. 13.25 (left), in which the nanodome device exhibits a

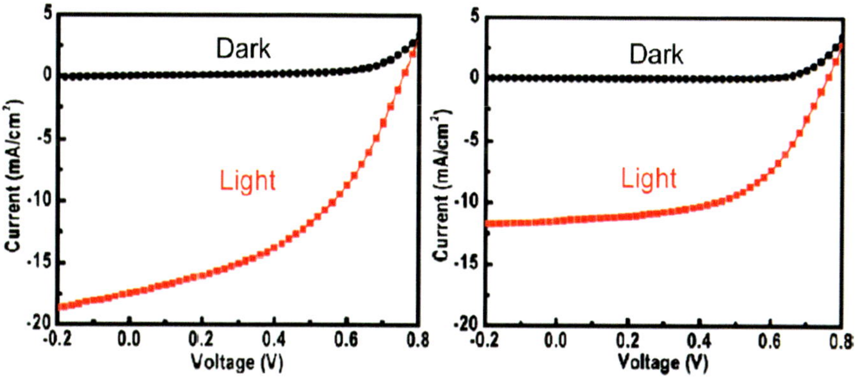

Fig. 13.25 Dark and light I–V curve of solar cell devices for nanodomes (*left*) and flat substrates (*right*)

power efficiency of 5.9% (open-circuit voltage, $V_{oc} = 0.75$ V; short-circuit current, $J_{sc} = 17.5$ mA/cm^2; fill factor, FF $= 0.45$), while the flat device exhibits an efficiency of 4.7% ($V_{oc} = 0.76$ V, $J_{sc} = 11.4$ mA/cm^2, FF $= 0.54$) (Fig. 13.25, right). The significant improvement in power efficiency comes from the large short-circuit current of nanodome devices (17.5 mA/cm^2), which is higher than that (15.6 mA/cm^2) of the world-record, single-junction a-Si:H solar cells with substrate configuration [22], with initial power efficiency of 10.6%. The short-circuit current of nanodome devices is only slightly lower than the theoretical value (20.5 mA/cm^2) limited by the bandgap. It is expected that the efficiency of nanodome devices can be further improved by improving the open-circuit voltage and fill factor, via better material deposition.

13.6 Fundamental Limit of Light Trapping in Nanophotonics

The theory of light trapping was initially developed for conventional cells where the light-absorbing film is typically many wavelengths thick [43, 44]. From a ray optics perspective, conventional light trapping exploits the effect of total internal reflection between the active material and the surrounding medium. By roughening the interface (Fig. 13.26a), one randomizes the light propagation directions inside the material. The effect of total internal reflection then results in a much longer propagation distance inside the material and hence a substantial absorption enhancement. This standard theory, first developed by Yablonovitch [43], shows that the absorption enhancement factor has an upper limit of $4n^2/\sin^2\theta$ [43, 44], where θ is the angle of the emission cone in the medium surrounding the cell.

In a nanostructure film with thickness comparable or even smaller than wavelength scale, such ray optics picture and some of the basic assumptions are no longer applicable. As a result, whether the Yablonovitch limit still holds in the nanophotonic regime remains an open question that is of fundamental importance. Recently, a nanophotonic light-trapping theory based on a rigorous electromagnetic approach

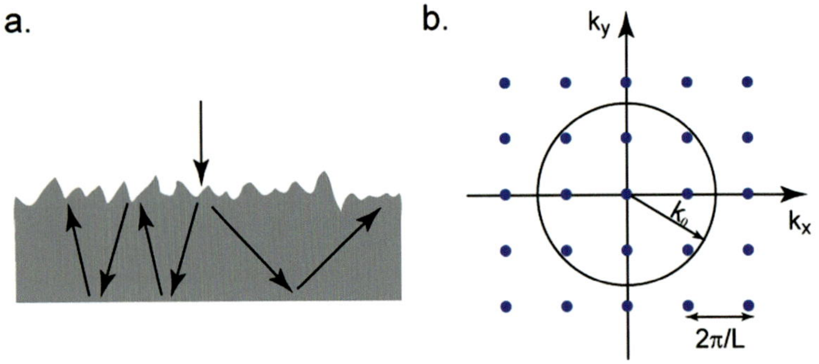

Fig. 13.26 (**a**) Light trapping by randomly textured surface. (**b**) *Blue dots* represent plane waves in the *k* space. Plane waves in the *circle* correspond to free-space propagating modes

was developed [45]. It is shown that the standard limit can be substantially surpassed when optical modes in the active layer are confined to deep sub-wavelength scale, opening new avenues for highly efficient next-generation solar cells.

In a weakly absorptive film where one-path absorption $d\alpha_0 \ll 1$, with d and α_0 being the film thickness and the absorption coefficient, respectively, significant absorption enhancement can only be created by the optical resonances. However, compared to the broad solar spectrum, each individual resonance has very narrow spectral width. Consequently, in order to enhance the absorption over a substantial portion of the solar spectrum, one will have to rely upon a collection of optical resonances. To characterize the contribution of a single resonance to the broadband absorption, we define the spectral cross section:

$$\sigma = \int_{-\infty}^{\infty} A(\omega)\,\mathrm{d}\,\omega \tag{13.1}$$

where $A(\omega)$ is the absorption spectrum of a single resonance. The spectral cross section has units of frequency and has the following physical interpretation: for incident spectrum with a bandwidth $\Delta\,\omega \gg \sigma$, a resonance contributes an additional $\sigma/\Delta\omega$ to the spectrally averaged absorption coefficient.

Using coupled mode theory [46, 47], one can explicitly calculate the spectral cross section of a resonance [45]. The maximum spectral cross section of a single resonance is achieved in the over-coupling regime:

$$\sigma_{\mathrm{max}} = \frac{2\pi\,\gamma_{\mathrm{i}}}{N} \tag{13.2}$$

where γ_{i} is the intrinsic loss rate of the resonance due to material absorption. Equation (13.2) assumes a back reflector behind the medium and that the resonance can equally couple to N plane waves in front of the absorptive medium. N is determined by the periodicity of the structure. For a square lattice of period L, the number of distinct plane waves at frequency ω (Fig. 13.26b) is

$$N = \frac{2\pi\,\omega^2}{c^2}\left(\frac{L}{2\pi}\right)^2 \tag{13.3}$$

In particular, a structure without any periodicity can be viewed as $L \to \infty$.

By summing over the maximal spectral cross section of all individual resonances, we obtain the upper limit for the absorption by a given medium:

$$A_{\mathrm{T}} = \frac{\sum \sigma_{\mathrm{max}}}{\Delta\omega} = \frac{2\pi\,\gamma_{\mathrm{i}}}{\Delta\omega}\,\frac{M}{N} \tag{13.4}$$

where M is number of resonances supported by the medium within the frequency range $\Delta\omega$.

Once the structure of absorptive medium is specified, both M and N can be properly calculated. As an example,

$$M = \frac{8\pi n^3 \omega^2}{c^3} \left(\frac{L}{2\pi}\right)^2 \left(\frac{d}{2\pi}\right) \delta\omega$$

for a bulk medium where the period L and the thickness d are much larger than the wavelength. c is the light speed in vacuum. Combining this result with Eqs. (13.3) and (13.4) and comparing the absorption to single-pass absorption $d\alpha_0$, we can reproduce the well-known Yablonovitch limit.

More importantly, Eq. (13.4) allows us to calculate light-trapping limit in any photonic structure. Of particular interest is a thin film that is only couple of hundreds of nanometers supporting only a single waveguide mode two polarizations. In contrast to the bulk structure, here

$$M = \frac{4\pi n_{\text{wg}}^2 \omega}{c^2} \left(\frac{L}{2\pi}\right)^2 \delta\omega$$

and enhancement factor becomes

$$F = 4n_{wg}^2 \frac{\lambda}{2n_{\text{wg}}d} V \qquad (13.5)$$

where the factor $V = \alpha_{\text{wg}}/\alpha_0$ characterizes the overlapping between the profile of the guided mode and the absorptive active layer and α_{wg} and n_{wg} are the absorption coefficient and the group index of the waveguide mode, respectively. Equation (13.5) in fact becomes $4n^2$ in a dielectric waveguide of $d \approx \lambda/2n$. Therefore, reaching a single-mode regime alone is not sufficient to overcome the Yablonovitch limit. Instead, to achieve the full benefit of nanophotonics, one must either confine the guided mode to a deep sub-wavelength scale or enhance the group index to be substantially larger than the refractive index of the active material, over a substantial wavelength range.

To apply the enhancement mechanisms described above, a toy structure (Fig. 13.27a) is studied to show a light-trapping enhancement factor of $12 \times 4n^2$ with nearly isotropic angular response and over virtually unlimited bandwidth. The cell consists of a 5-nm thin layer of low-index active material ($n_{\text{L}} = \sqrt{2.5}$) on the top of a perfect reflecting mirror. In order to enhance the absorption in the active layer, a transparent cladding layer ($n_{\text{H}} = \sqrt{12.5}$) is introduced on top of the active layer. Such a cladding layer serves two purposes. First, it enhances density of state. The overall structure supports a fundamental mode with group index n_{wg} close to n_{H}, which is much higher than that of the absorbing material. Second, the index contrast between active and cladding layers provides nanoscale field confinement. Figure 13.27b shows the fundamental waveguide mode. The field is highly concentrated in the low-index active layer, due to the well-known, slot-waveguide effect [48]. Thus, the geometry here allows the creation of a broadband high-index

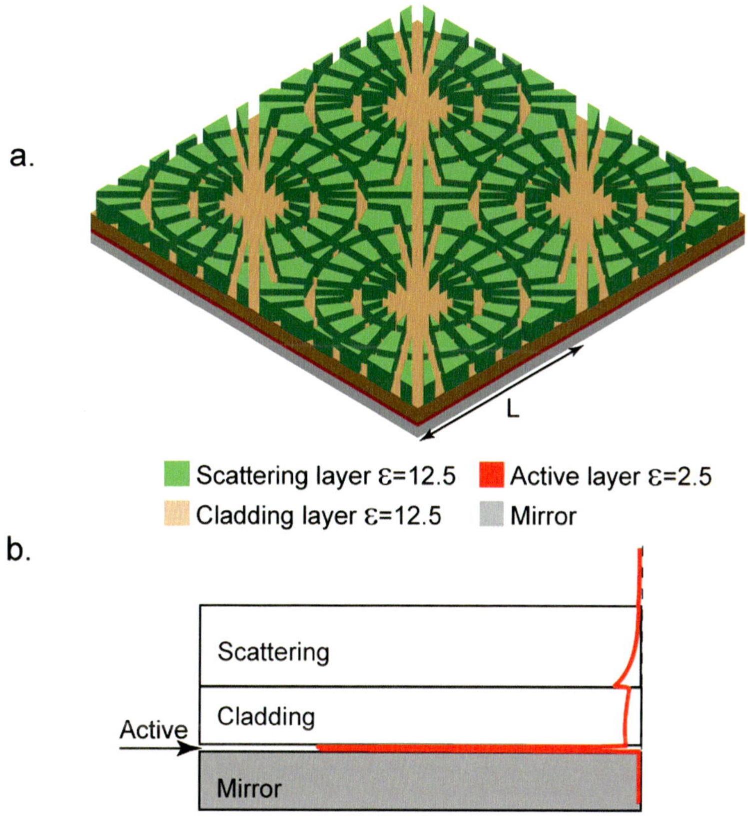

Fig. 13.27 (a) A nanophotonic light-trapping structure. The scattering layer consists of a square lattice of air groove patterns with periodicity $L = 1200$ nm. The thicknesses of the scattering, cladding, and active layers are 80, 60, and 5 nm, respectively. The active layer has a frequency-independent absorption length of 25 μm. The mirror layer is a perfect electric conductor. (**b**) The profile of electric field intensity for the fundamental waveguide mode. Fields are strongly confined in the active layer. To obtain the waveguide mode profile, the scattering layer is modeled by a uniform slab with an average dielectric constant

guided mode, with its energy highly concentrated in the active layer, satisfying the requirement in Eq. (13.3) for high absorption enhancement.

Numerical simulation shows that the device has a spectrally averaged absorption enhancement factor of $F = 119$ (red line) for normally incident light (Fig. 13.28a). This is well above the Yablonovitch limit for both the active material ($4n_{\mathrm{L}}^2 = 10$) and the cladding material ($4n_{\mathrm{H}}^2 = 50$). Moreover, the angular response is nearly isotropic (Fig. 13.28b). Thus such enhancement cannot be attributed to the narrowing of angular range in the emission cone and instead is due entirely to the nanoscale field confinement effect. The theoretical upper limit of light-trapping enhancement calculated from Eq. (13.5) is consistent with the numerical simulation. The example presented here shows that there is substantial opportunity for nanophotonic

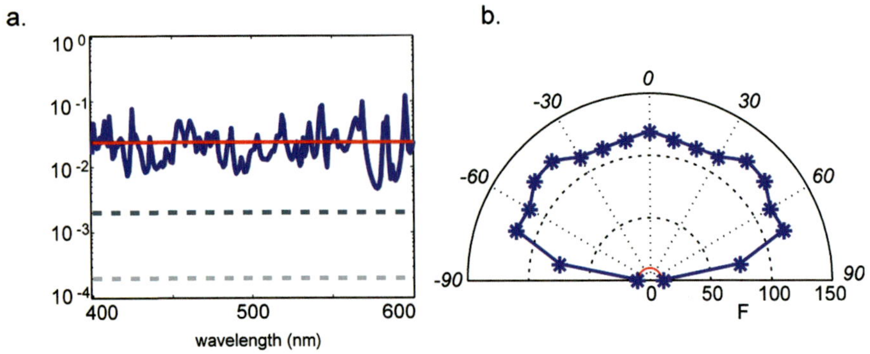

Fig. 13.28 (**a**) Absorption spectrum for normally incident light for the structure shown in Fig. 13.27a. The spectrally averaged absorption (*red solid line*) is much higher than both the single-pass absorption (*light gray dashed line*) and the absorption as predicted by the limit of $4n_L^2$ (*dark gray dashed line*). (**b**) Angular dependence of the spectrally averaged absorption enhancement factor. Incident angles are labeled on top of the semi-circles. The *red circles* represent the $4n_L^2$ limit

light trapping to greatly enhance PV performance using only low-loss dielectric components.

In conclusion, light trapping can greatly enhance absorption, even beyond the conventional limit $4n^2$. For this purpose, one needs to create as many optical resonances as possible in a given volume, for example, by increasing the density of state and confining modes in nanoscale. Therefore, engineering nanostructures present new opportunities for advanced light management to significantly improve solar cell's efficiency.

13.7 Summary and Outlook

Advanced photon management is essential for the development of next-generation photovoltaic devices, to achieve higher efficiency, lower cost, and finally enable terawatt-scale applications. Nanostructures with unique scales and geometries hold great promise to play a critical role in the photon management in the subwavelength regime. Even though the rapidly evolving field of nanotechnology has provided a variety of tools to tailor nanostructures, broadband coverage of the solar spectrum and strict cost limitation of solar cell modules make the task of efficient photon management rather challenging.

We have designed a novel nanodome structure which can efficiently suppress reflection while simultaneously enhancing absorption. The centerpiece of this nanodome structure is the nanocone template, which can be achieved through a low-temperature, large-scale process. The nanodome design is not limited to specific materials, and the shapes, diameters, and spacings of nanocone templates can be tailored precisely across a wide range for different types of solar cells. The a-Si:H nanodome solar cells that we used for this demonstration of concept can achieve efficiencies 25% higher than those of flat film control. We have also proved that in the sub-wavelength regime, light trapping can greatly enhance absorption.

An effective integration of fundamental mechanistic understanding and novel processing development is the key to the efficient photon manipulations for solar cell applications. Every step in each of these two domains will ultimately lead to the success of the development of next-generation solar cells and may well spawn many other unexpected scientific discoveries and technological applications.

References

1. A.V. Shah, H. Schade, M. Vanecek, J. Meier, E. Vallat-Sauvain, N. Wyrsch, U. Kroll, C. Droz, J. Bailat, Prog. Photovoltaics **12**, 113–142 (2004)
2. R. Noufi, K. Zweibel, in *2006 IEEE 4th World Conference on Photovoltaic Energy Conversion (WCPEC-4) Waikoloa,* Hawaii, 7–12 May 2006
3. D.E. Carlson, C.R. Wronski, Appl. Phys. Lett. **28**, 671–673 (1976)
4. R.A. Street, *Hydrogenated Amorphous Silicon* (Cambridge University Press, Cambridge, 1991)
5. J. Zhao, A. Wang, P.P. Altermatt, S.R. Wenham, M.A. Green, Sol. Energy Mater. Sol. Cells **41/42**, 87 (1996)
6. N. Engheta, Science **317**, 1698–1702 (2007)
7. T.W. Ebbesen, H.J. Lezec, H.F. Ghaemi, T. Thio, P.A. Wolff, Nature **391**, 667–669 (1998)
8. E. Ozbay, Science **311**, 189–193 (2006)
9. W.L. Barnes, A. Dereux, T.W. Ebbesen, Nature **424**, 824–830 (2003)
10. P.R. Barry, P. Peter, R.F. Stephen, J. Appl. Phys. **96**, 7519–7526 (2004)
11. K.R. Catchpole, A. Polman, Opt. Expr. **16**, 21793–21800 (2008)
12. K.R. Catchpole, A. Polman, Appl. Phys. Lett. **93**, 191113 (2008)
13. D. Derkacs, S.H. Lim, P. Matheu, W. Mar, E.T. Yu, Appl. Phys. Lett. **89**, 093103 (2006)
14. V.E. Ferry, L.A. Sweatlock, D. Pacifici, H.A. Atwater, Nano Lett. **8**, 4391–4397 (2008)
15. N. Keisuke, T. Katsuaki, A.A. Harry, Appl. Phys. Lett. **93**, 121904 (2008)
16. S. Pillai, K.R. Catchpole, T. Trupke, M.A. Green, J. Appl. Phys. **101**, 093105 (2007)
17. H.A. Atwater, in *GCEP Technical Report* (2009)
18. M.H. Huang, Y. Wu, H. Feick, N. Tran, E. Weber, P. Yang, Adv. Mater. **13**, 113–116 (2001)
19. Y. Xia, P. Yang, Y. Sun, Y. Wu, B. Mayers, B. Gates, Y. Yin, F. Kim, H. Yan, Adv. Mater. **15**, 353–389 (2003)
20. T. Martensson, C.P.T. Svensson, B.A.M. Wacaser, W. Larsson, W. Seifert, K. Deppert, A. Gustafsson, L.R. Wallenberg, L. Samuelson, Nano Lett. **4**, 1987–1990 (2004)
21. K.-S. Cho, D.V. Talapin, W. Gaschler, C.B. Murray, J. Am. Chem. Soc. **127**, 7140–7147 (2005)
22. L.E. Greene, M. Law, J. Goldberger, F. Kim, J.C. Johnson, Y. Zhang, R.J. Saykally, P. Yang, Angew. Chem. Int. Ed. **115**, 3139–3142 (2003)
23. Y. Sun, B. Gates, B. Mayers, Y. Xia, Nano Lett. **2**, 165–168 (2002)
24. C.L. Haynes, R.P. Van Duyne, J. Phys. Chem. B **24**, 5599 (2001)
25. A. Kosiorek, W. Kandulski, P. Chudzinski, K. Kempa, M. Giersig, Nano Lett. **4**, 1359 (2004)
26. J. Zhu, Z. Yu, G.F. Burkhard, C.-M. Hsu, S.T. Connor, Y. Xu, Q. Wang, M. McGehee, S. Fan, Y. Cui, Nano Lett. **9**, 279–282 (2009)
27. C.-M. Hsu, S.T. Connor, M.X. Tang, Y. Cui, Appl. Phys. Lett. **93**, 133109–133113 (2008)
28. S.L.H. Jeong, H.R. Lee, E. Garnett, J.W. Choi, Y. Cui, Nano Lett. submitted
29. A.A. Tracton, *Coatings Technology Handbook*, 3rd edn. (CRC Press, London, 2005)
30. T. Fujii, Y. Gao, R. Sharma, E.L. Hu, S.P. DenBaars, S. Nakamura, Appl. Phys. Lett. **84**, 855–857 (2004)

31. H. Kim, K.K. Choi, K.K. Kim, J. Cho, S.N. Lee, Y. Park, J.S. Kwak, T.Y. Seong, Opt. Lett. **33**, 1273–1275 (2008)
32. L. Hu, G. Chen, Nano Lett. 7, 3249–3252 (2007)
33. B.J.P.J. Wei-Lun Min, Adv. Mater. **9999**, NA (2008)
34. T. Lohmuller, M. Helgert, M. Sundermann, R. Brunner, J.P. Spatz, Nano Lett. **8**, 1429–1433 (2008)
35. Y.J. Lee, D.S. Ruby, D.W. Peters, B.B. McKenzie, J.W.P. Hsu, Nano Lett. **8**, 1501–1505 (2008)
36. J.Q. Xi, M.F. Schubert, J.K. Kim, E.F. Schubert, M. Chen, S.-Y. Lin, W. Liu, J.A. Smart, Nat. Photon. **1**, 176–179 (2007)
37. Y.-F. Huang, S. Chattopadhyay, Y.-J. Jen, C.-Y. Peng, T.-A. Liu, Y.-K. Hsu, C.-L. Pan, H.-C. Lo, C.-H. Hsu, Y.-H. Chang, C.-S. Lee, K.-H. Chen, L.-C. Chen, Nat. Nanotechnol. **2**, 770–774 (2007)
38. T. Tiedje, E. Yablonovitch, G.D. Cody, B.G. Brooks, Electron Devices IEEE Trans. **31**, 711–716 (1984)
39. E. Yablonovitch, J. Opt. Soc. Am. **72**, 899–907 (1982)
40. O.L. Muskens, J.G. Rivas, X.F. Mez, R.E. Algra, E.P.A.M. Bakkers, A. Lagendijk, Nano Lett. **8**, 2638–2642 (2008)
41. J.G. Mutitu, S. Shi, C. Chen, T. Creazzo, A. Barnett, C. Honsberg, D.W. Prather, Opt. Expr. **16**, 15238–15248 (2008)
42. M.J.W. Hampton, S.S, Zhou, Z, Nunes, J. Ko, D.J.L. Templeton, E.T. Samulski, J.M. DeSimone, Adv. Mater. **20**, 2667 (2008)
43. E. Yablonovitch, J. Opt. Soc. Am. A **72**, 899–907 (1982)
44. P. Campbell, M.A. Green, J. Appl. Phys. **62**, 243–249 (1987)
45. A.R. Zongfu, Y.S. Fan, arXiv: 1004.2902v2
46. H.A. Haus, *Waves and Fields in Optoelectronics* (Prentice-Hall, Englewood Cliffs, NJ, 1984)
47. S. Fan, W. Suh, J.D. Joannopoulos, J. Opt. Soc. Am. A **20**, 569–572 (2003)
48. V.R. Almeida, Q. Xu, C.A. Barrios, M. Lipson, Opt. Lett. **29**, 1209–1211 (2004)

Chapter 14
Highly Sensitive and Selective Gas Detection by 3D Metal Oxide Nanoarchitectures

Jiajun Chen, Kai Wang, Baobao Cao, and Weilie Zhou

14.1 Introduction

Gas sensors, as modern electronic components, have obtained enormous development since the 1970s [1]. As an interface between electronic systems and our environment, gas sensors collect information from the atmosphere, i.e., they detect whether a particular chemical vapor is present and of what concentration. Recently, gas sensors have played important roles in various applications, such as environmental monitoring, automobile, industrial safety, manufacture quality control, and public security [2, 3]. For example, oxygen sensors in automobiles can ensure complete and efficient gasoline combustion in the engine, specially designed gas sensors can monitor the quality of perfume, and explosive detectors can be used in train stations and airports. In the last decade, major improvements in the capabilities of mobile devices and wireless networks have generated more and more demand of portable or mobile device-integrated sensor systems. One example is the massively distributed gas sensor that can locate the safety threat when they are integrated with geographical positioning systems [4] that are available in many modern mobile phones. These novel applications impose more requirements such as low power consumption, miniaturized size, high sensitivity, and selectivity of gas-sensing systems.

Though the development of man-made gas sensors has obtained lots of progress, there is no gas sensor that can reach the performance of a mammal's olfactory system in terms of sensitivity, selectivity, response speed, and operation temperature [5]. As an example, a human's nose can identify and discriminate more than 10,000 different chemical vapors and a canine's olfactory system is even more sensitive. The extremely high performance of mammal's olfactory systems can be attributed to the hundreds of millions of nanowire-like olfactory receptors. These receptors consist of tremendous surface area and chemical diversity for chemical capture and

W.L. Zhou (✉)
Advanced Materials Research Institute, University of New Orleans, New Orleans, LA 70148, USA
e-mail: wzhou@uno.edu

W.L. Zhou, Z.L. Wang (eds.), *Three-Dimensional Nanoarchitectures*,
DOI 10.1007/978-1-4419-9822-4_14,

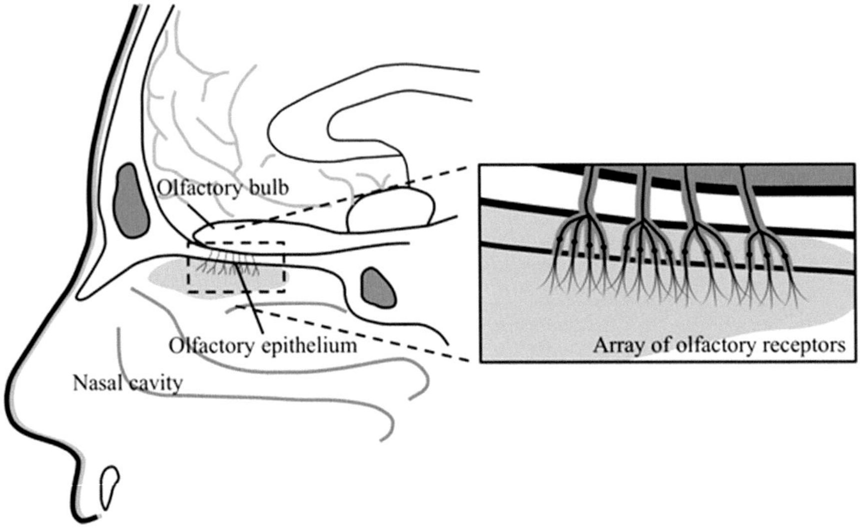

Fig. 14.1 The skull cross section showing the structure of a nasal cavity. The *enlarged inset* is an array of olfactory receptors that are 3D nanostructures. Each receptor has a length of several micrometers and a diameter of several hundred nanometers

recognition. As shown in Fig. 14.1, the inhaled odor molecules interact with specialized receptors located in the olfactory epithelium. Each individual olfactory receptor is several hundred nanometers in diameter and has a length of several micrometers. The large surface area of olfactory receptor arrays provides an efficient way to capture low-concentrated chemical molecules and thus results in high sensitivity. The largely diversified chemical reactivity of these receptors contributes to the strong capability of gas discrimination. Furthermore, it should be pointed out that the signal of chemical–receptor interaction is transmitted and processed in a massively parallel way. This gives the olfactory system fast response speeds.

To pursue sensing performance that is comparable to or surpasses mammal's olfactory systems, many gas sensors based on different detection mechanism have been developed including micromachined cantilever gas sensors [6], surface acoustic wave (SAW) [7, 8], conducting polymer sensors [9–11], metal oxide conductometric sensors [12–14], and optical gas sensors (i.e., optical fiber-based sensors and infrared gas sensors) [3]. In recent years, research groups have also been exploring new phenomena for gas-sensing applications. Wang et al. used piezoelectric oxide to detect the swell in polymers when exposed to some specific organic molecules [15, 16] or the piezoelectric potential at the Schottky contact interface to detect the absorbed gas molecules [17]. The metal–insulator transition (MIT) effect in vanadium oxide is also useful for gas detection [18]. Among all of these detection techniques, the metal oxide conductometric gas sensor is one of the most promising candidates for high-performance gas detection systems due to their low fabrication cost, high sensitivity, and stability [1, 19]. The detection mechanism of metal oxide gas sensors can be described by using the adsorption and desorption of oxygen molecules on the metal oxide surface [20]: Oxygen molecules are adsorbed on the metal oxide surface when the device is exposed to atmosphere. The adsorbed oxygen molecules become negatively charged oxygen adsorbates (O^-, O^{2-}, or O_2^-) when they extract electrons from the material: $O_2 + e \rightarrow nO^-$ (O^{2-}, O_2^-, etc.) Considering the case of n-type metal oxide semiconductor, the adsorbed oxygen molecules cover the surface of the metal oxides forming surface states that can trap electrons from the n-type metal oxides. The surface regions that lose electrons are referred as surface charge depletion layer, the width of which is defined as Debye length. This surface charge depletion layer constrains the effective conduction channel and increases the height of inter-particle energy barriers, as shown in Fig. 14.2a (top). When exposing to reducing molecules, such as H_2S, CO, and NH_3, the oxygen adsorbates react with these gases and release electrons back to the metal oxide semiconductor, increasing the electrical conductance. Depicted in Fig. 14.2a (bottom), the reducing gas molecules consume the oxygen adsorbates and effectively reduce the surface coverage of oxygen adsorbates at equilibrium. Because the width of depletion layer is proportional to the adsorbate coverage, the presence of reducing gases can reduce the width of depletion layer and the height of inter-grain energy barriers, which resulted in increase of overall conductivity of the n-type metal oxides. Contrarily, oxidizing gases like NO_2 and O_3 generally decrease conductance of n-type metal oxides.

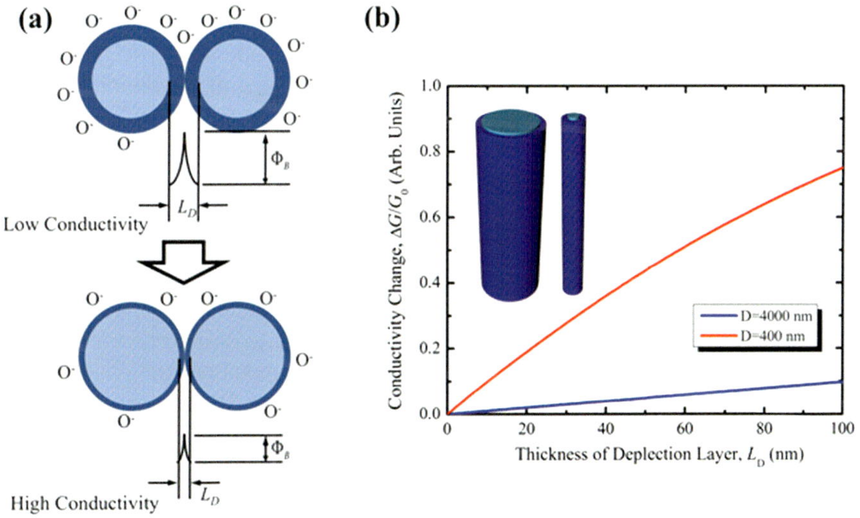

Fig. 14.2 (**a**) Schematic diagrams showing the influence of adsorbed oxygen species on the depletion layer thickness (L_D), boundary energy barriers, and conductivity of n-type materials. *Top*: The n-type materials were exposed to the atmosphere and oxygen species were adsorbed on the surface forming a depletion layer; *bottom*: oxygen adsorbates react with reducing gases, releasing electrons back to the n-type semiconductor and reducing the thickness of the depletion layer. (**b**) Plot of conductivity change ($\Delta G/G_0$) vs. depletion layer thickness (L_D) of nanowires with different diameters

Nanostructured metal oxides generally present much higher sensitivity than their thin or thick film counterparts. Several advantages of nanostructured materials contribute to the high sensitivity, such as large surface-to-volume ratio, feature size close to the characteristic Debye length of semiconductors [21]. The enhancement in sensitivity can be easily explained by comparing two nanowires with different diameters (Fig. 14.2b). In a single nanowire gas sensor, the sensing response is modulated by the width of charge depletion layer. If the size of nanowires is close to the Debye length, a slight change of the depletion layer thickness will result in a significant change in overall nanostructure conductivity. In a special case, the sensitivity is maximized when the nanostructures are completely depleted, namely, the grain size of nanoparticle or the diameter of nanowire is less than the Debye length [22]. Many research groups [23–25] have reported that metal oxide nanostructures can improve the performance of gas sensors. However, there are still lots of factors that should be considered when metal oxide nanostructures are employed for sensing applications. For instance, the nanostructures tend to agglomerate during device processing due to the highly active surfaces, which bring complexity of detection mechanism; the nanostructures may grow into bigger grains if they operate at elevated temperature, which not only cause stability issues but also degrade the sensitivity.

Based on the discussion above, one can summarize that there are three important factors that influence the sensitivity of the sensors based on nanostructures: (1) the intrinsic properties and critical dimensions of nanostructured materials; (2) the structural arrangement of the nanostructures in the gas sensor; and (3) the concentration profiles of gases in the device structures, which are defined by the porosity and the distribution of catalysts. Many efforts have been put into the engineering of individual nanostructures (i.e., nanowires, nanotubes, and nanoparticles), which majorly focus on the control of properties and feature size of the nanomaterials. Recently, 3D nanostructures, such as vertically aligned nanowire arrays, nanotube arrays, opal, and inverted opal structures, attracted more and more interests in the applications of gas detection due to their flexibility of structural control. Three-dimensional nanostructures can make full use of the nanostructure surface and their intrinsic channels allowed gas molecules transport into and out of the materials, which is beneficial for sensitivity, response speed, and operation temperature. The structural configuration of 3D nanostructures is directly related to the sensing performance, which introduce more controllable parameters to optimize the design of nanostructure-based gas sensors. Additionally, catalytic metal decorations [26–29] or surface functionalization [30] can also be applied on the 3D nanostructured gas sensors to further enhance the sensitivity and selectivity.

14.2 Highly Sensitive Gas Detection by Stand-alone 3D Nanosensors

Many metal oxide 3D nanostructures have been fabricated and most of them are suitable for gas-sensing applications. These structures include nanowire arrays [31], branched nanowires [32], flower-like clusters [33], and opal and inverted opal

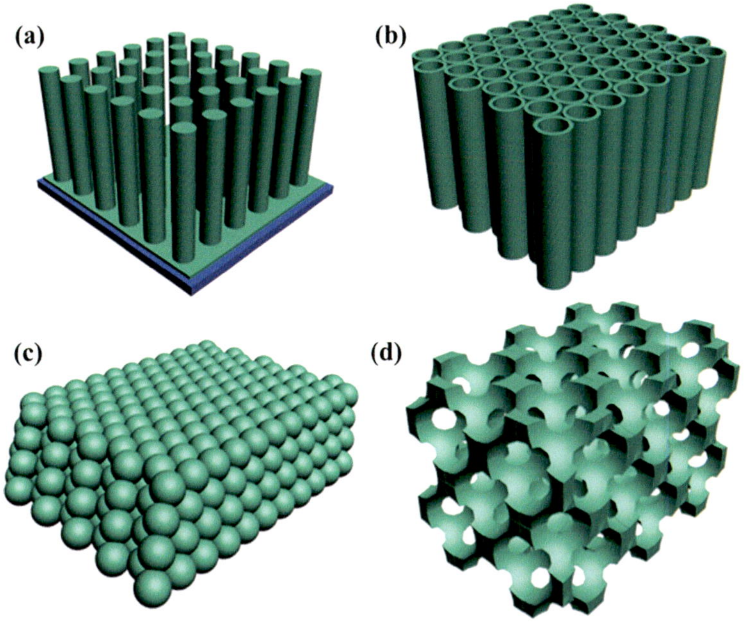

Fig. 14.3 Schematics of four major 3D nanostructures: (**a**) nanowire array, (**b**) nanotube array, (**c**) opal structure, and (**d**) inverted opal structure

structures [34, 35]. In this section, we will focus on two types of nanostructures: nanowire/nanotube arrays and opal/inverted opal structures (shown in Fig. 14.3). These types of nanostructures have been studied extensively and their fabrication techniques (i.e., vapor–liquid–solid or vapor–solid chemical vapor deposition, thermal evaporation, hydrothermal, anodic oxidization/electrochemistry, self-assembly, and colloidal chemistry) are relatively mature.

14.2.1 Metal Oxide Nanowire/Nanotube Array Gas Sensors

14.2.1.1 Nanowire Arrays

Unlike lateral nanodevices which adopt lots of processes from microelectronics, 3D nanostructured sensors require exploration in fabrication methods. This presents the challenge of how to maintain the 3D nanowire arrays during the processing. A microfabrication method has been employed in several reports on 3D nanowire array-based electronic devices [36–38]. This method can be described using a WO_3 nanowire array as an example. The schematic diagrams and corresponding field-emission scanning electron microscopy (FESEM) images are shown in Fig. 14.4. The WO_3 nanowire array has a monoclinic structure, which was prepared by thermal evaporation methods [37]. A pre-deposited conducting layer composed of metal

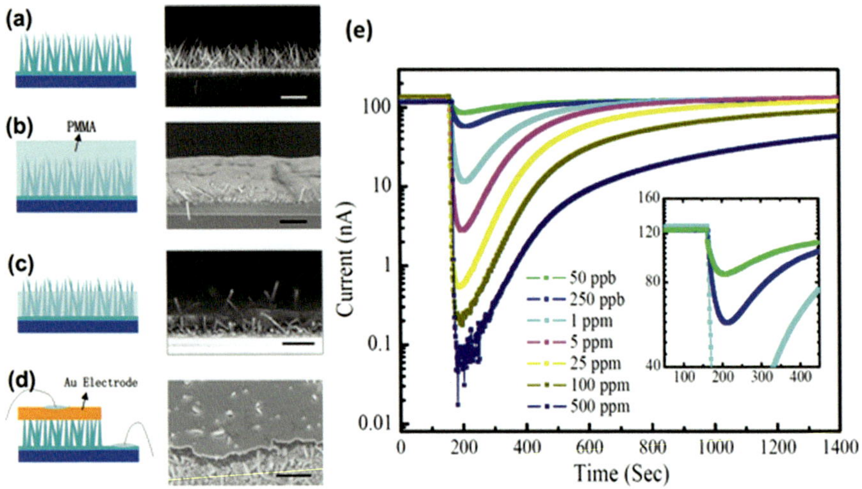

Fig. 14.4 (**a–d**) Schematics and SEM images of the procedures of gas sensor fabrication based on WO_3 nanowire arrays: (**a**) WO_3 nanowire array with a seed layer at the bottom; (**b**) nanowire array after PMMA coating; (**c**) nanowire array after oxygen plasma etching; (**d**) as-prepared sensor after top electrode deposition and PMMA removal. (**e**) Sensing response of the sensor to air-diluted NO_2 with different concentrations. The *inset* is enlarged from the response curve of low concentrations. All scale bars are 4 μm. Reprinted with permission from [37]

or conducting metal oxides on the supporting substrates can work as the bottom electrode of the device (Fig. 14.4a). A nanowire growth seed layer can also be used. The major problem is the preparation of the suspended top electrode. In this method, the whole nanowire array was first covered with poly(methyl methacrylate) (PMMA) by spin coating, shown in Fig. 14.4b. The nanowire tips were then exposed by oxygen plasma etching of the PMMA with a carefully controlled etch rate (Fig. 14.4c). A metal layer was deposited as the top electrode and finally, the PMMA was removed by acetone, as shown in Fig. 14.4d. Depending on the nature of metal oxide and the electrode metals, an annealing process may be needed to reduce the contact resistance and improve the device stability. The sensing responses are given in Fig. 14.4e. Inherited from the intrinsic high sensing performance of WO_3 and the large surface area of 3D nanostructures, the WO_3 nanowire array device shows high sensitivity to NO_2. The sensor can detect NO_2 with a concentration down to 50 ppb. The resistance change can be as high as 3400% even at a relatively low concentration such as 5 ppm. The response speed of the sensor is very fast, giving responses in less than 30 s. From the inset of Fig. 14.4e, one can find that the response signal is much larger than the background noise. Even at a concentration as low as 50 ppb, it can be identified easily. The suppression of electrical conductivity of the sensors when exposed to oxidizing NO_2 gas indicates that the WO_3 materials show n-type properties. The response mechanism can be explained by the conventional oxygen reaction model mentioned in the introduction part. It should be pointed out that the recovery time of these sensors is more than 800 s, which is much longer than the response speed. This indicates that the adsorption process is much easier than the desorption process.

The assembly of 3D nanostructures can also be realized in a finely controlled in situ micromanipulation system installed under FESEM [39]. The gas sensor is fabricated from two CuO nanowire arrays, which are prepared by thermal oxidization of the Cu plate at atmospheric pressure [40]. The schematic of the assembly process is shown in Fig. 14.5a. A 2 mm × 2 mm piece of CuO nanowire array was first attached to a thin copper foil, and another piece of CuO nanowire array of the same size was installed on the tip of the micromanipulator (Kleindiek MM3A-EM). The nanowire arrays were positioned face-to-face, while the micromanipulator pushed one array toward the other. Under the FESEM, the space between the nanowires can be monitored. The micromanipulators have a movement step of 2 nm, which ensures the accurate positioning of the nanowire arrays avoiding short-circuit or breakage. The real-time captured FESEM images of the nanowire arrays approaching each other are shown in Fig. 14.5b. The device conductivity increases since more nanowires get connected when the micromanipulator pushes the nanowire array. The

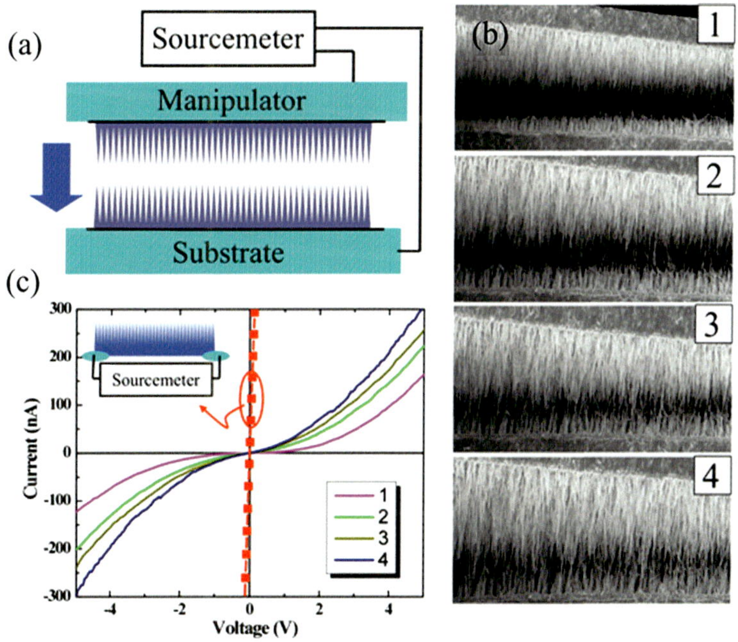

Fig. 14.5 Assembly processes of CuO nanowire arrays. (**a**) Schematic diagram of the setup for the assembly of two vertically aligned CuO nanowire arrays. Two pieces of nanowire array (2 mm × 2 mm) were attached to a copper plate and the micromanipulator tip. By activation of the manipulator, the distance between the two arrays can be adjusted. (**b**) In situ observation of the assembly process when pushing one nanowire array to the other. (**c**) *I*–*V* curves measured at different stages of the assembly process by a sourcemeter. The *I*–*V* curve of a nanowire array connected by silver paste is also plotted (*square symbols*). Reprinted with permission from [39]

nonlinear *I–V* curves indicated energy barriers formed between adjacent nanowires. The normalized response curves are depicted in Fig. 14.6a. The device can detect H_2S with a detection limit of 500 ppb. The CuO nanowires worked as a p-type conduction channel in the device. Exposing the nanowires to reducing H_2S gases at low concentrations results in the decrease of device conductivity. However, highly concentrated H_2S gases increase the device conductivity and the device becomes unrecoverable. This phenomenon was explained by a chemical reaction model at the nanowire-to-nanowire interfaces. Figure 14.6b, c shows FESEM images before and after highly concentrated H_2S exposure, respectively, indicating that an exotic layer was formed on the nanowire surface. The layer was confirmed to be CuS by composition and structural characterization. CuS is a metallic material that will significantly improve the nanowire-to-nanowire contact, resulting in an increase in

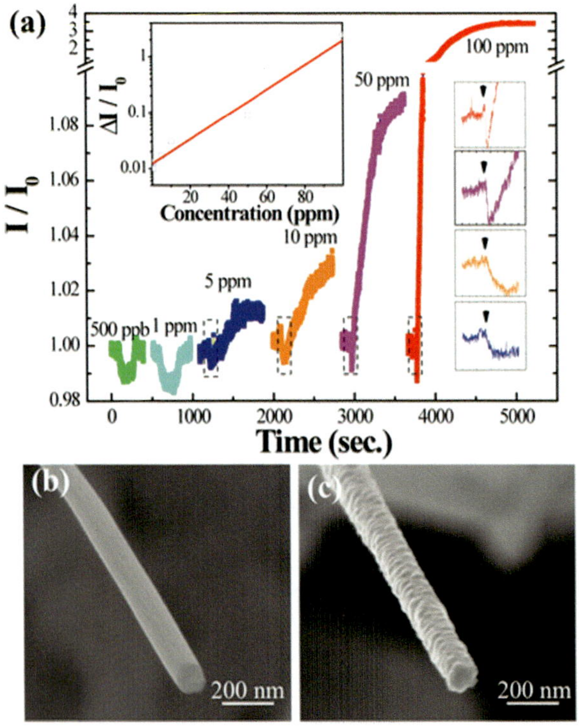

Fig. 14.6 (**a**) Sensing response of a CuO nanowire array sensor to air-diluted H_2S at elevated temperature (160°C). The device can detect H_2S at concentrations higher than 500 ppb. The device conductivity dropped at low concentrations while increased at concentrations higher than 5 ppm. The initial drops of the response curves are also given in *insets*. The concentration dependence of the normalized sensor responses ($\Delta I/I_0$) is also plotted in the *inset*. SEM images of a CuO nanowire taken (**b**) before and (**c**) after exposure to 100 ppm H_2S. Two-stage model was used to explain the sensing response: CuO presented p-type properties at low concentration, which induced conductivity drop when exposed to reducing gas. A surface reaction process forming CuS metallic layer explained the conductivity increase at high concentrations. Reprinted with permission from [39]

conductivity. The device also shows small response to H_2, CO, and NH_3 indicating good selectivity.

Other than conventional microfabrication technique and in situ micromanipulation, there are many other methods under development for 3D nanostructured device fabrication. For ZnO nanowire arrays grown by metalorganic chemical vapor deposition (MOCVD), a thin film can be formed on the top of the nanowire array by adjusting the supersaturation condition [41]. This film is used as the top electrode to connect the ZnO nanowires. This technique simplifies the top electrode fabrication. However, it relies on the growth properties of the materials and is not easy to extend to other nanowire arrays.

14.2.1.2 Nanotube Arrays

The fabrication of metal oxide nanotube arrays by anodic oxidization is a well-developed method, especially the arrays of TiO_2 nanotubes [42–44] and Al_2O_3 [45–47] nanotubes, which have been studied extensively. Each of the aforementioned arrays can be used in 3D gas sensor fabrication. The anodic-oxidized nanotube arrays are generally attached to one another forming continuous films that make the nanotube array freestanding. The electrode preparation for this type of nanotube array is relatively easy compared to nanowire arrays.

Chen et al. [42] have demonstrated that TiO_2 nanotube arrays can be grown through direct anodization of Ti foil in a fluoride-containing ethylene glycol. During the growth process, small pits that form at the inter-pore region induce small tube-to-tube separation. This allows faster gas diffusion, which may improve sensing performance. Figure 14.7a, b shows as-grown TiO_2 nanotube arrays. The as-grown nanotube arrays are amorphous and post-annealing is required to improve the crystallinity. An ultrasonic process in ethanol water solution can be used to separate the nanotube array from the Ti foil. Both the nanotube arrays before and after the separation can be used for sensor fabrication. The electrodes can be easily formed by sputtering on one or both sides of the nanotube array films. The schematic diagram of the device structure and the measurement system is given in Fig. 14.7c. The TiO_2 nanotube array sensors were used for hydrogen detection (Fig. 14.7d). At elevated temperatures, 1000 ppm nitrogen-diluted hydrogen can induce 6800 % increase in conductivity when nanotubes with optimized length are used. In this type of 3D nanostructured sensors, the thickness of nanotube walls, the nanotube length, and the inter-tube distance are the important factors that control the sensitivity. For instance, Paulose et al. demonstrated enormous enhancement (more than seven orders of magnitude) in sensitivity by reducing the thickness of nanotube walls through process parameters [48]. It worths to point out that the geometry and material of metal electrodes have significant effects on the sensitivity. In the case of nanotube array, putting the electrode at the same side will promote the contribution from the tube-to-tube contact and applications of catalytic noble metal as electrode materials can also increase the sensitivity. Additionally, this type of freestanding nanostructures is generally fragile and sensitive to mechanical force; therefore, handling and transfer techniques may be needed for future fabrication of integrated devices.

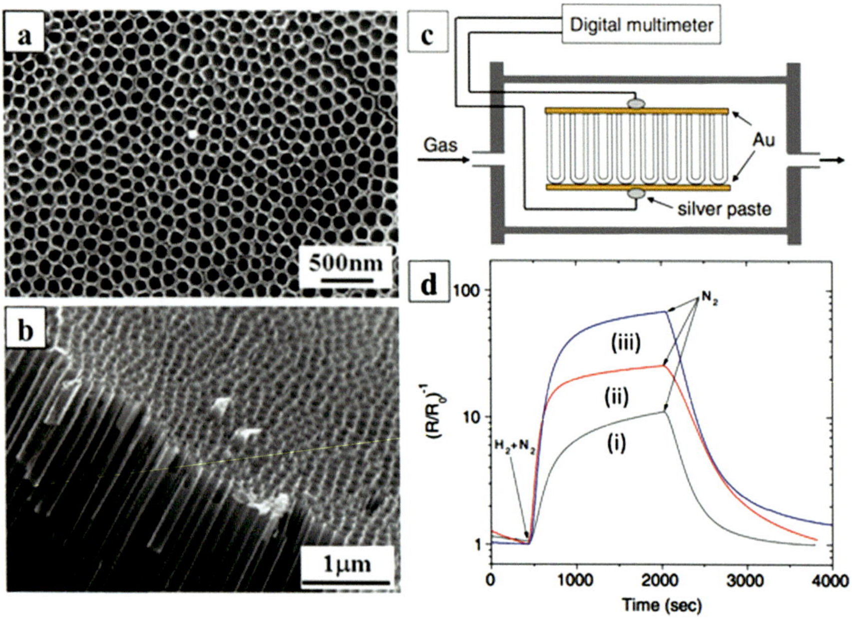

Fig. 14.7 (**a**) Top view and (**b**) cross-sectional view images of as-prepared TiO_2 nanotube arrays anodized in ethylene glycol. (**c**) Schematic diagram of TiO_2 nanotube array sensors loaded in a sensor-testing system. (**d**) The sensing responses of a 25-μm TiO_2 nanotube array on Ti substrate (i), freestanding TiO_2 nanotube array with thickness of 25 μm (ii), and 15 μm (iii). The testing gas is 1000 ppm nitrogen-diluted hydrogen and measurements were carried out at 200°C. Reprinted with permission from [42]

Furthermore, nanotube arrays may not only work as the functioning materials in a gas sensor but also play a role as a structural template to prepare other metal oxide nanotubes that are difficult to grow. For instance, insulating metal oxide nanotube arrays, such as Al_2O_3, prepared by anodic oxidization, can work as structural templates for the preparation of nanotube arrays of gas-sensing metal oxides (SnO_2, ZnO, In_2O_3, WO_3, MoO_3, etc.). Figure 14.8 depicted a general method for preparation of metal oxide nanotube arrays. The precursors for gas-sensitive metal oxides can infiltrate into the nanotube template and cover the entire nanotube surface. After appropriate annealing process, polycrystalline nanoparticle films can be formed on the inner surface of nanotubes. With the electrode metal deposited, the metal oxide coating contributes the major part of the conductivity due to its relative low resistance compared to the insulating template. Artzi-Gerlitz et al. have grown WO_3 nanotube arrays using anodic-oxidized Al_2O_3 nanotube membranes as templates and their sensing performance has been tested [46]. The WO_3 nanotube array sensors presented about three orders of magnitude improvement in sensitivity over their thin-film counterparts. This is an effective route to prepare nanotube arrays with metal oxide sensing materials that are difficult to grow as nanotube arrays.

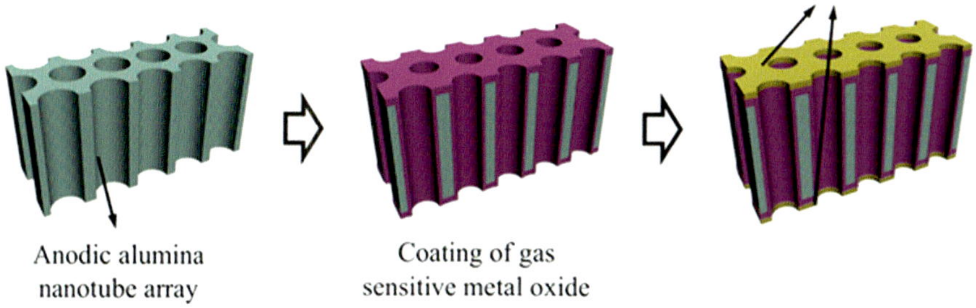

Fig. 14.8 Schematic diagram of fabrication procedures for metal oxide nanotube array based on anodic nanotube template

14.2.2 Gas Sensors Based on Opal and Inverted Opal Nanostructures

The periodic opal and inverted opal structures are generally considered as photonic crystals and their fabrication has become a mature technique in nanotechnology society [34, 35]. Actually, the first several attempts for gas detection by opal or inverted opal structures were based on optical measurements, in which the diffraction or refractive index was monitored when functional polymers were exposed to target vapors [49–52]. Using the opal or inverted opal structures as templates, 3D metal oxide conductometric gas sensors can also be prepared [49, 53–55]. The fabrication of metal oxide inverted opal structures is shown in Fig. 14.9a. Under properly controlled conditions, monodisperse polystyrene (PS) spheres can form a film with face center cubic (fcc) closely packed arrangement on the glass substrate, which is known as opal structure. The space between PS spheres is then filled by metal oxide precursor solution through capillary force. Inverted opal structures of metal oxides are formed after removing the PS spheres by annealing process. The pore sizes in the as-prepared inverted opal structure are controlled by the diameter of PS spheres. Figure 14.9b provides the procedures for the preparation of metal oxide opal structures. Following similar method for PS opal structure, silica opal structure can be formed on the substrate. PS precursor solution then infiltrates into the opal structure forming a PS inverted opal structure after heat treatment and silica removal. The PS inverted opal structures work as structure templates for preparation of metal oxide opal structures. Obviously, both of them are very convenient methods to prepare 3D nanostructures for gas-sensing applications. For instances, the doping and composition control can be easily realized by the precursor solutions, and the wall thickness may also be tuned by processing parameters.

The first high-performance metal oxide conductometric gas sensor based on opal and inverted opal structures was reported by Scott et al. [55]. Figure 14.10a, b shows the as-prepared SnO_2 opal and inverted opal structures. The hollow spheres in inverted opals are connected to each other through empty channels, which originate from the sphere-to-sphere contact points in the PS opal structures. And the

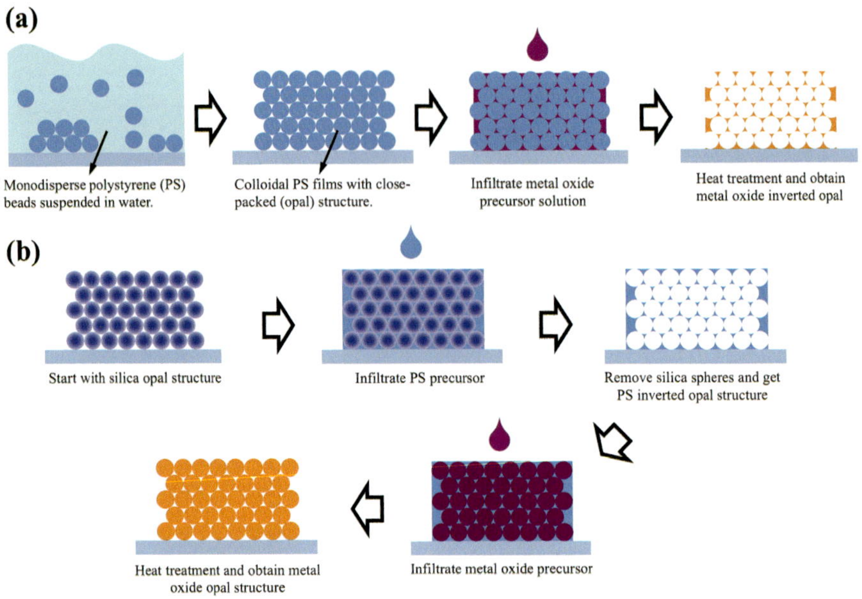

Fig. 14.9 Schematics of fabrication processes of opal and inverted opal structures

spaces between the metal oxide spheres in the opals also formed connected network of channels. These channels allowed gases diffused into the inner part of the structure. The diameters of PS and silica spheres used in the opal structures were about 250 nm, which defined the feature size of the opal and inverted opal structures. The sensing responses to CO in dry or humidity environment have been presented in Fig. 14.10c. The devices showed similar responses in dry and humidity environment, implying resistance to the interference from humidity. As shown in Fig. 10d, the characteristic parameters, prefactor and response order, were extracted from the experimental data following the empirical response formula

$$G = AP^{\beta}$$

where G is the sensing response, P is the partial pressure of testing gas, A is the prefactor, and β is the response order. The inverted opals, which can be treated as 3D network of thin bridges with numerous interconnected junctions and gas diffusion channels, presented nearly "ideal" gas-sensing behavior. Baratto et al. further presented a doping process when creating 3D metal oxide nanostructures with opal templates [53]. The sensors were tested with several environmental gases. The sensing responses to reducing gases (CO and methanol) are negligible when the operating temperature was set to 200°C. The device showed significant responses to NO_2, and the lowest detection limit is less than 1 ppm. By comparing with pure SnO_2 inverted opal structures, it can be concluded that the dopant introduced with the precursors can further increase the device performance.

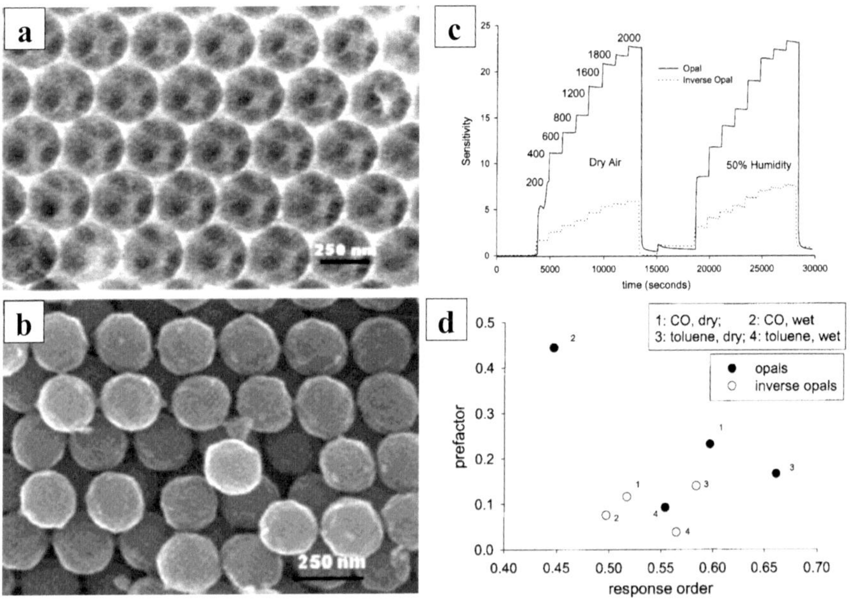

Fig. 14.10 (**a**) SEM image of as-prepared SnO_2 inverted opal structure. (**b**) SEM image of SnO_2 opal structures. (**c**) Sensing responses toward CO at 400°C. (**d**) Prefactor and response order deduced from experimental data for the evaluation of sensor performance. Reprinted with permission from [55]

14.3 Sensor Arrays Based on 3D Nanostructured Gas Sensors

In this section, we introduce sensor arrays based on 3D nanostructures for gas discrimination. Before the invention of the electronic nose, major research of selective detection had been focused on the development of sensing materials, through which the sensor responds specifically to a particular gas of interest. This method is generally referred to as the traditional "lock-and-key" approach [56]. A gas sensor with good selectivity requires the sensor to respond to one specific gas and to be inert to other gases. This is a very strict requirement and difficult to be realized, especially when discrimination between two active gases with similar properties is needed. Unlike the "lock-and-key" approach, the biological olfactory systems use cross-reactive behaviors of individual olfactory receptors. The selective responses of olfactory receptor are not strictly required, while the strong ability of gas discrimination is realized through a massive variety of chemical properties of olfactory receptors [5, 57–59]. Mimicking the cross-reactive detection of olfactory systems, sensor arrays consisting of several gas sensors with different materials were used to create electronic noses for gas discrimination [60–66]. Recently, the research of electronic noses has been extended to nanomaterials utilizing carbon nanotubes [63], metal oxide nanowires [64], etc. However, the major obstacle to an electronic nose based on 3D nanostructures is the availability of 3D nanostructured gas-sensing materials. In the last section, several template methods were introduced, which can be used as a general route to prepare 3D metal oxide nanostructures.

The growth of ZnO nanowire arrays has been studied extensively and well-aligned ZnO nanowire arrays have been fabricated by various methods [67–71] with high reproducibility and excellent alignment. ZnO nanowire arrays can function as structural templates for the growth of gas-sensing metal oxide nanowire arrays. Figure 14.11 shows the schematics of the procedures for the preparation of metal oxide nanowire arrays using ZnO nanowire arrays as structural templates. Following Xu's method [71], the core ZnO nanowire arrays were first grown by combining e-beam lithography patterning and hydrothermal growth. The e-beam lithography defines the growth position of the nanowires, as shown in Fig. 14.11a. After the hydrothermal process, ZnO nanowires grew from the PMMA windows that were patterned by e-beam lithography (Fig. 14.11b). The ZnO nanowire arrays were then loaded into a thin-film deposition system for shell layer coating. Various methods can be used to prepare the shell metal oxide layer, including dc/rf sputtering, pulsed laser deposition, sol–gel method, or other wet chemical methods. Since most of these methods are well developed and have strong capability for stoichiometry control, they can prepare most of the known gas-sensing materials. Similarly, the widely used catalytic metal nanoparticles can also be deposited [26]. After the shell layer coating (Fig. 14.11c), the top electrode was prepared by the methods mentioned before (Fig. 14.11d).

Figure 14.12 provides the structural analysis of ZnO nanowire arrays before and after the SnO_2 and Pd nanoparticles were coated with RF sputtering. Obvious changes in surface roughness of the nanowire arrays indicate successful coating of the nanoparticles. Detailed structural characterizations were observed by transmission electron microscopy. It is important to point out that finely controlled deposition parameters are needed for the vapor-phase deposition methods (sputtering and PLD) to achieve homogeneous coating. The deposition species generally reach the sidewall of nanowires with a glancing angle, which may induce special structures in the coating layers [71, 73].

Three devices with different catalytic metal coatings (Pd, Pt, and Au) were used to form a sensor array for gas discrimination. The configuration of the sensor array system is shown in Fig. 14.13. *I–V* characteristics were first obtained for all three sensors, which confirmed the contact properties between the metal electrodes and the metal oxide nanowires. The conductivity of all the sensors was monitored continuously with a sourcemeter and a switch module when the air-diluted testing gases were introduced into the testing chamber sequentially. For this specific system, sensing responses toward H_2S, NO_2, NH_3, CO, and H_2 with serial concentrations were obtained. The response data form a knowledge base that describes the gas discrimination capability of the sensor array and can also be used as reference data for future gas detection [5]. Many statistical methods, categorized as supervised and unsupervised methods, can be used to analyze the response data. In most electronic nose systems, supervision methods have been used for gas discrimination, including principal component analysis (PCA), partial least squares (PLS), multiple linear regression (MLR), principal component regression (PCR), and discriminant

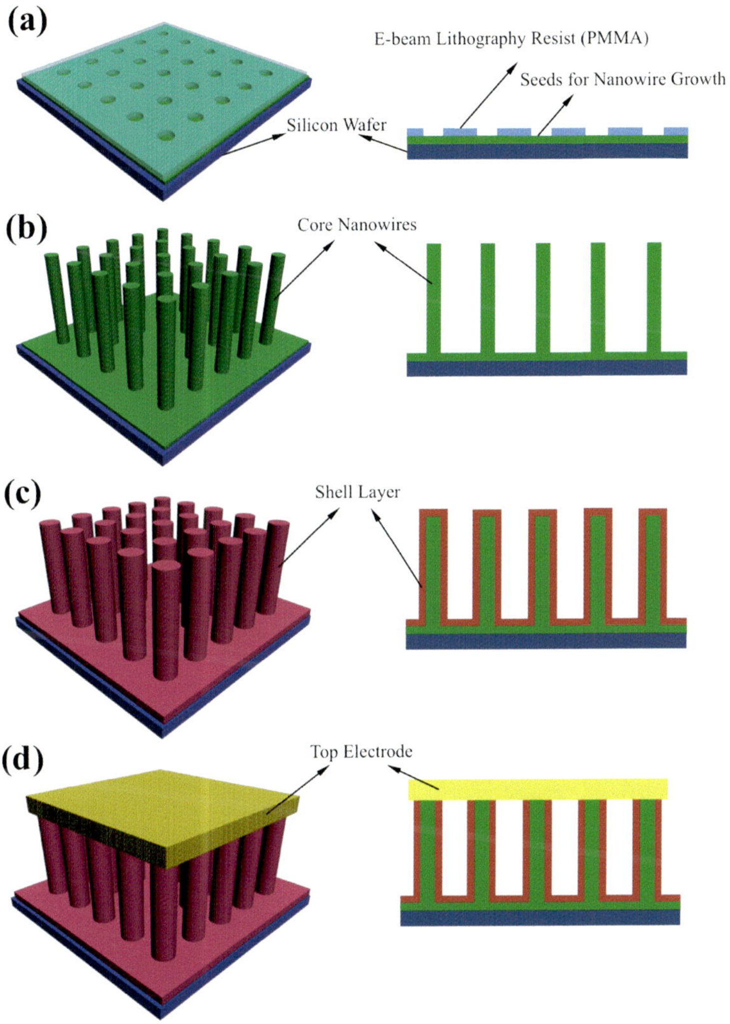

Fig. 14.11 Schematic diagrams showing the fabrication procedures of 3D nanosensors using vertically aligned nanowire array as a template. (**a**) e-Beam-patterned PMMA windows defined the growth position of the nanowires. (**b**) Nanowires grew from the PMMA windows forming the core nanowire array as the template. (**c**) Gas-sensing layer was coated onto the nanowire arrays, which can be realized through various methods, such as sputtering, PLD, and sol–gel. (**d**) Nanowire array sensors can be prepared by microfabrication

function analysis (DFA) [74]. PCA is widely used in electronic noses based on nanostructured materials because it is an unsupervised method that requires no or little prior knowledge. The PCA data for the first two principal components (PC1 and PC2) are plotted in Fig. 14.14a. The cumulative variance of the principal components is 98.80%, indicating that the 2D plot preserves the major part of information

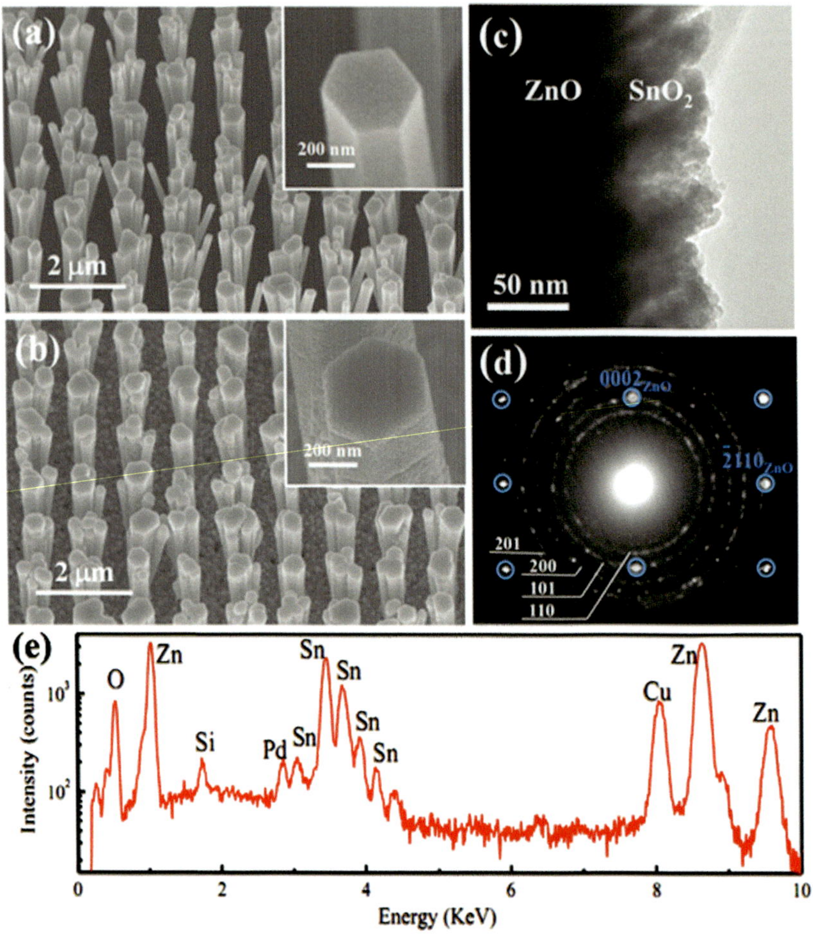

Fig. 14.12 FESEM images of (**a**) as-grown ZnO nanowire array and (**b**) nanowire array after SnO_2 and Pd coatings. (**c**) TEM image on the edge of the nanowires showing the column structure of the SnO_2 coating. (**d**) Selected area electron diffraction pattern from the nanowire along ZnO $[01\bar{1}0]$ zone. The diffraction rings are identified as rutile SnO_2. (**e**) X-ray energy dispersive spectrum of the nanoparticle-coated nanowires (unpublished data)

in the response data. From the plot, one can find that the sensor array shows a strong discrimination capability for these gases. There are some clusters (H_2S and NH_3; CO and H_2) with overlapping regions, bringing ambiguity for discrimination of these gases. This problem may be solved by introducing new response variables such as response speed or recovery time. Figure 14.14b shows the plot after including the response speeds as *z*-axis, indicating improved selectivity between all of the tested gases.

As mentioned before, the sensing material library with 3D nanostructures is still not fully explored. Only a small part of these materials have been presented

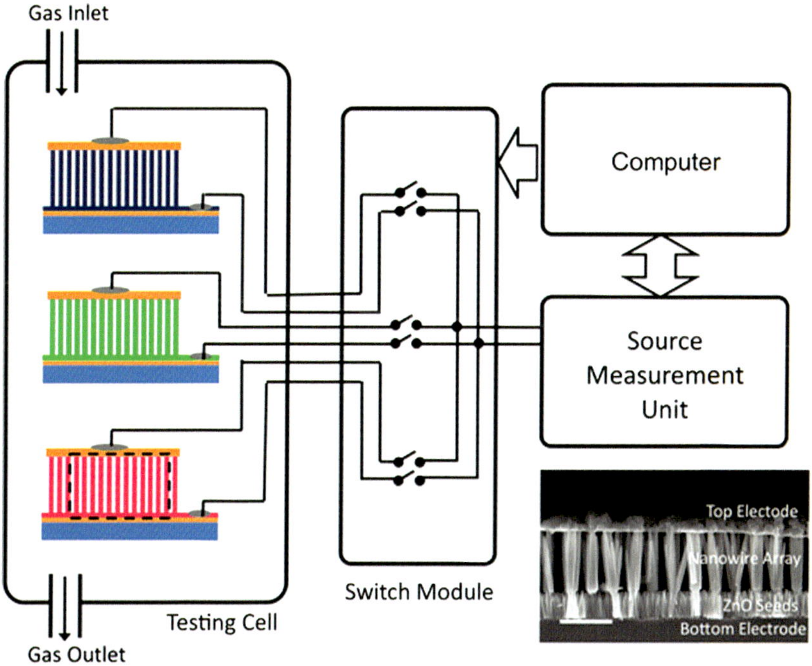

Fig. 14.13 Diagrams of sensor array testing system. The conductivity of three sensors is monitored simultaneously by a sourcemeter and a switch module. The *bottom-right inset* is a cross-sectional FESEM image of a nanowire array sensor. The scale bar is 2 μm

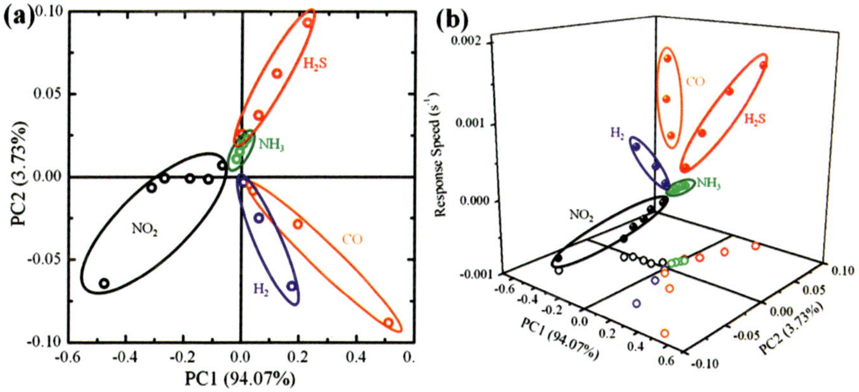

Fig. 14.14 (**a**) PCA analysis of the sensing responses of the sensor array. The data points from same gas are grouped. (**b**) Plot after incorporating response speeds with the PCA data (unpublished data)

and characterized compared to their counterparts of ceramic, thin films, or lateral nanostructures. The template methods that use nanowire arrays or opal structures are feasible ways to access the full material library and will benefit the preparation of cross-reactive electronic noses based on 3D nanostructures.

14.4 Conclusion Remarks

The intrinsic properties of nanostructures, such as large surface-to-volume ratio and feature sizes close to the Debye length, have proved to be beneficial for improving the sensing performance of metal oxide gas sensors. Their 3D structures may further increase the sensor performance due to their tremendous surface area and porous structures. However, there are several demonstrations of gas sensors based on 3D nanostructures due to the difficulty and stability of 3D nanostructure fabrication and device integration. In this review, we introduced several successful fabrications of gas sensors based on 3D nanostructures. The device performance is promising in terms of sensitivity, selectivity, operation temperature, etc. General routes for 3D nanostructure fabrication have also been reviewed, which can make full use of the library of gas-sensing materials. We also mentioned several template-based methods for the fabrication of 3D nanostructures. Because the gas-sensing applications only require the functional materials to be polycrystalline and sometimes the boundaries of polycrystalline materials may also improve the sensing performance, many methods can be used to create the 3D nanostructures based on a pre-defined 3D template. The capabilities of sol–gel, wet chemistry, sputtering, and pulsed laser deposition for this purpose were demonstrated.

In comparison with lateral nanostructured gas sensors, the applications of 3D nanostructures in gas detection are still limited. There are lots of work that can be done in the future:

(1) The structure parameters (nanowire diameters, inter-nanowire distances, spacing between opal structures, etc.) should be finely tuned to obtain optimized factors for gas sensing.
(2) More gas-sensing materials should be included in the future for 3D nanostructure fabrication. For example, incorporating dopant or catalyst into the known materials or preparing 3D nanostructures of complex metal oxides.
(3) The recovery time of 3D nanostructured gas sensors is generally much longer than the response speed. This may be attributed to the larger surface areas of the 3D structures that induce fast adsorption and slow desorption. Some new approaches for device recovery, such as UV exposure, may be investigated.
(4) New techniques may be developed for 3D nanostructured fabrication. A recent example is to prepare gas sensors using several or even individual standing nanowires [75]. These gas sensors have a miniaturized form factor, high sensitivity (ppb level), and room temperature detection capability, which can be applied in portable platform such as mobile phone and handheld detectors.
(5) Using multiple 3D nanostructure gas sensors to form cross-reactive sensor arrays is another trend for future research. This would be based on a fully explored 3D nanostructured material library.

The pursuit of a man-made gas detection system superior to biological olfactory systems is an interdisciplinary topic that not only involves materials and sensor research but also relies on technology and fundamental scientific research on signal transmission, processing, and neuroscience.

Acknowledgments The authors acknowledge the financial support from DARPA Grant No. HR0011-07-1-0032 and research grants from Louisiana Board of Regents Contract Nos. LEQSF (2007-12)-ENH-PKSFI-PRS-04 and LEQSF (2008-11)-RD-B-10.

References

1. P. Grundler, *Chemical Sensors – An Introduction for Scientists and Engineers* (Springer, Leipzig, 2006)
2. J. Hesse, J.W. Gardner, W. Gopel, *Sensors Applications* (Wiley-VCH, Weinheim, 2003)
3. J. Chou, *Hazardous Gas Monitors: A Practical Guide to Selection, Operation, and Applications* (McGraw-Hill Professional, New York, 1999)
4. M. Kuorilehto, M. Kohvakka, J. Suhonen, P. Hamalainen, M. Hannikainen, T.D. Hamalainen, *Ultra-Low Energy Wireless Sensor Networks in Practice* (Wiley, Chichester, England, 2007)
5. T.C. Pearce, S.S. Schiffman, H.T. Nagle, J.W. Gardner, *Handbook of Machine Olfaction* (Wiley-VCH, Weinheim, 2003)
6. D. Lange, O. Brand, H. Baltes, *Cantilever-Based CMOS Sensor Systems* (Springer, New York, NY, 2002)
7. A. D'Amico, E. Verona, SAW sensors. Sens. Actuators **17**, 55–66 (1989)
8. C.M. Harris, Seeing SAW potential. Anal. Chem. **75**, 355A–358A (2003)
9. H. Bai, G. Shi, Gas sensors based on conducting polymers. Sensors **7**, 267–307 (2007)
10. J. Janata, M. Josowicz, Conducting polymers in electronic chemical sensors. Nat. Mater. **2**, 19–24 (2003)
11. D.T. McQuade, A.E. Pullen, T.M. Swager, Conjugated polymer-based chemical sensors. Chem. Rev. **100**, 2537–2574 (2000)
12. G. Korotcenkov, Gas response control through structural and chemical modification of metal oxide films: state of the art and approaches. Sens. Actuators B Chem. **107**, 209–232 (2005)
13. G. Korotcenkov, The role of morphology and crystallographic structure of metal oxides in response of conductometric-type gas sensors. Mater. Sci. Eng. R Rep. **61**, 1–39 (2008)
14. G. Sberveglieri, *Gas Sensors – Principles, Operation and Developments* (Kluwer, Boston, MA, 1992)
15. C. Lao, Y. Li, C.P. Wong, Z.L. Wang, Enhancing the electrical and optoelectronic performance of nanobelt devices by molecular surface functionalization. Nano Lett. **7**, 1323–1328 (2007)
16. C.S. Lao, Q. Kuang, Z.L. Wang, M.-C. Park, Y. Deng, Polymer functionalized piezoelectric-FET as humidity/chemical nanosensors. Appl. Phys. Lett. **90**, 262107 (2007)
17. T.-Y. Wei, P.-H. Yeh, S.-Y. Lu, Z.L. Wang, Gigantic enhancement in sensitivity using schottky contacted nanowire nanosensor. J. Am. Chem. Soc. **131**, 17690–17695 (2009)
18. E. Strelcov, Y. Lilach, A. Kolmakov, Gas sensor based on metal-insulator transition in VO_2 nanowire thermistor. Nano Lett. **9**, 2322–2326 (2009)
19. W. Gopel, J. Hesse, J.N. Zemel, *Sensors, A Comprehensive Survey* (Wiley-VCH, Weinheim, 1989–1995).
20. N. Barsan, U. Weimar, Conduction model of metal oxide gas sensors. J. Electroceram. **7**, 143–167 (2001)
21. A. Kolmakov, M. Moskovits, Chemical sensing and catalysis by one-dimensional metal oxide nanostructures. Ann. Rev. Mater. Res. **34**, 151–180 (2004)
22. Y. Shimizu, M. Egashira, Basic aspects and challenges of semiconductor gas sensors. MRS Bull. **24**, 18–24 (1999)
23. K. Kalantar-zadeh, B. Fry, *Nanotechnology Enabled Sensors* (Springer, New York, NY, 2007)
24. J.G. Lu, P. Chang, Z. Fan, Quasi-one-dimensional metal oxide materials – synthesis, properties and applications. Mater. Sci. Eng. R Rep. **52**, 49–91 (2006)

25. G. Shen, P.-C. Chen, K. Ryu, C. Zhou, Devices and chemical sensing applications of metal oxide nanowires. J Mater. Chem. **19**, 828–839 (2009)
26. A. Kolmakov, D.O. Klenov, Y. Lilach, S. Stemmer, M. Moskovits, Enhanced gas sensing by individual SnO_2 nanowires and nanobelts functionalized with Pd catalyst particles. Nano Lett. **5**, 667–673 (2005)
27. N.S. Ramgir, I.S. Mulla, K.P. Vijayamohanan, A room temperature nitric oxide sensor actualized from Ru-doped SnO_2 nanowires. Sens. Actuators B Chem. **107**, 708–715 (2005)
28. H.T. Wang, B.S. Kang, F. Ren, L.C. Tien, P.W. Sadik, D.P. Norton, S.J. Pearton, J. Lin, Detection of hydrogen at room temperature with catalyst-coated multiple ZnO nanorods. Appl. Phys. A **81**, 1117–1119 (2005)
29. H.T. Wang, B.S. Kang, F. Ren, L.C. Tien, P.W. Sadik, D.P. Norton, S.J. Pearton, J. Lin, Hydrogen-selective sensing at room temperature with ZnO nanorods. Appl. Phys. Lett. **86**, 243503 (2005)
30. L.C. Tien, D.P. Norton, B.P. Gila, S.J. Pearton, H.-T. Wang, B.S. Kang, F. Ren, Detection of hydrogen with SnO_2-coated ZnO nanorods. Appl. Surf. Sci. **253**, 4748–4752 (2007)
31. Z.L. Wang, Zinc oxide nanostructures: growth, properties and applications. J. Phys. Condens. Matter **16**, R829–R858 (2004)
32. C. Cheng, B. Liu, H. Yang, W. Zhou, L. Sun, R. Chen, S.F. Yu, J. Zhang, H. Gong, H. Sun, H.J. Fan, Hierarchical assembly of ZnO nanostructures on SnO_2 backbone nanowires: low-temperature hydrothermal preparation and optical properties. ACS Nano **3**, 3069–3076 (2009)
33. T. Krishnakumar, R. Jayaprakash, N. Pinna, N. Donato, A. Bonavita, G. Micali, G. Neri, CO gas sensing of ZnO nanostructures synthesized by an assisted microwave wet chemical route. Sens. Actuators B Chem. **143**, 198–204 (2009)
34. O.D. Velev, P.M. Tessier, A.M. Lenhoff, E.W. Kaler, A class of porous metallic nanostructures. Nature **401**, 548–548 (1999)
35. J.E.G.J. Wijnhoven, W.L. Vos, Preparation of photonic crystals made of air spheres in titania. Science **281**, 802–804 (1998)
36. J.Y. Park, D.E. Song, S.S. Kim, An approach to fabricating chemical sensors based on ZnO nanorod arrays. Nanotechnology **19**, 105503 (2008)
37. B. Cao, J. Chen, X. Tang, W. Zhou, Growth of monoclinic WO_3 nanowire array for highly sensitive NO_2 detection. J. Mater. Chem. **19**, 2323–2327 (2009)
38. E. Lai, W. Kim, P. Yang, Vertical nanowire array-based light emitting diodes. Nano Res. **1**, 123–128 (2008)
39. J. Chen, K. Wang, L. Hartman, W. Zhou, H_2S detection by vertically aligned CuO nanowire array sensors. J. Phys. Chem. C **112**, 16017–16021 (2008)
40. X. Jiang, T. Herricks, Y. Xia, CuO nanowires can be synthesized by heating copper substrates in air. Nano Lett. **2**, 1333–1338 (2002)
41. M.-C. Jeong, B.-Y. Oh, O.-H. Nam, T. Kim, J.-M. Myoung, Three-dimensional ZnO hybrid nanostructures for oxygen sensing application. Nanotechnology **17**, 526–530 (2006)
42. Q. Chen, D. Xu, Z. Wu, Z. Liu, Free-standing TiO_2 nanotube arrays made by anodic oxidation and ultrasonic splitting. Nanotechnology **19**, 365708 (2008)
43. J.M. Macak, H. Tsuchiya, L. Taveira, S. Aldabergerova, P. Schmuki, Smooth anodic TiO_2 nanotubes. Angew. Chem. Int. Ed. **44**, 7463–7465 (2005)
44. M. Paulose, K. Shankar, S. Yoriya, H.E. Prakasam, O.K. Varghese, G.K. Mor, T.A. Latempa, A. Fitzgerald, C.A. Grimes, Anodic growth of highly ordered TiO_2 nanotube arrays to 134 μm in length. J. Phys. Chem. B **110**, 16179–16184 (2006)
45. H. Masuda, K. Fukuda, Ordered metal nanohole arrays made by a two-step replication of honeycomb structures of anodic alumina. Science **268**, 1466–1468 (1995)
46. R. Artzi-Gerlitz, K.D. Benkstein, D.L. Lahr, J.L. Hertz, C.B. Montgomery, J.E. Bonevich, S. Semancik, M.J. Tarlov, Fabrication and gas sensing performance of parallel assemblies of metal oxide nanotubes supported by porous aluminum oxide membranes. Sens. Actuators B Chem. **136**, 257–264 (2009)

47. W. Lee, R. Scholz, U. Gosele, A continuous process for structurally well-defined Al_2O_3 nanotubes based on pulse anodization of aluminum. Nano Lett. **8**, 2155–2160 (2008)
48. M. Paulose, O.K. Varghese, G.K. Mor, C.A. Grimes, K.G. Ong, Unprecedented ultra-high hydrogen gas sensitivity in undoped titania nanotubes. Nanotechnology **17**, 398–402 (2006)
49. C.-Y. Kuo, S.-Y. Lu, S. Chen, M. Bernards, S. Jiang, Stop band shift based chemical sensing with three-dimensional opal and inverse opal structures. Sens. Actuators B Chem. **124**, 452–458 (2007)
50. J.H. Holtz, S.A. Asher, Polymerized colloidal crystal hydrogel films as intelligent chemical sensing materials. Nature **389**, 829–832 (1997)
51. C.F. Blanford, R.C. Schroden, M. Al-Daous, A. Stein, Tuning solvent-dependent color changes of three-dimensionally ordered macroporous (3DOM) materials through compositional and geometric modifications. Adv. Mater. **13**, 26–29 (2001)
52. V.N. Bogomolov, S.V. Gaponenko, I.N. Germanenko, A.M. Kapitonov, E.P. Petrov, N.V. Gaponenko, A.V. Prokofiev, A.N. Ponyavina, N.I. Silvanovich, S.M. Samoilovich, Photonic band gap phenomenon and optical properties of artificial opals. Phys. Rev. E **55**, 7619–7625 (1997)
53. C. Baratto, G. Faglia, G. Sberveglieri, A. Sutti, G. Calestani, C. Dionigi, Inverse opal structure of SnO_2 and SnO_2: Zn for gas sensing. IEEE International Conference on Sensors 2005, Irvine, California, pp. 1196–1200 (2005)
54. A. Sutti, C. Baratto, G. Calestani, C. Dionigi, M. Ferroni, G. Faglia, G. Sberveglieri, Inverse opal gas sensors: Zn (II)-doped tin dioxide systems for low temperature detection of pollutant gases. Sens. Actuators B Chem. **130**, 567–573 (2008)
55. R.W.J. Scott, S.M. Yang, G. Chabanis, N. Coombs, D.E. Williams, G.A. Ozin, Tin dioxide opals and inverted opals: near-ideal microstructures for gas sensors. Adv. Mater. **13**, 1468–1472 (2001)
56. K.J. Albert, N.S. Lewis, C.L. Schauer, G.A. Sotzing, S.E. Stitzel, T.P. Vaid, D.R. Walt, Cross-reactive chemical sensor arrays. Chem. Rev. **100**, 2595–2626 (2000)
57. M.M. Mozell, M. Jagodowicz, Chromatographic separation of odorants by the nose: retention times measured across in vivo olfactory mucosa. Science **181**, 1247–1249 (1973)
58. K. Persaud, G. Dodd, Analysis of discrimination mechanisms in the mammalian olfactory system using a model nose. Nature 299, 352–355 (1982)
59. J.R. Stetter, W.R. Penrose, Understanding chemical sensors and chemical sensor arrays (electronic noses): past, present, and future. Sensors Update **10**, 189–229 (2002)
60. J.M. Baik, M. Zielke, M.H. Kim, K.L. Turner, A.M. Wodtke, M. Moskovits, Tin-oxide-nanowire-based electronic nose using heterogeneous catalysis as a functionalization strategy. ACS Nano **4**, 3117–3122 (2010)
61. P.-C. Chen, F.N. Ishikawa, H.-K. Chang, K. Ryu, C. Zhou, A nanoelectronic nose: a hybrid nanowire/carbon nanotube sensor array with integrated micromachined hotplates for sensitive gas discrimination. Nanotechnology **20**, 125503 (2009)
62. P.-C. Chen, G. Shen, C. Zhou, Chemical sensors and electronic noses based on 1-D metal oxide nanostructures. IEEE Trans. Nanotechnol. **7**, 668–682 (2008)
63. A. Star, V. Joshi, S. Skarupo, D. Thomas, J.-C.P. Gabriel, Gas sensor array based on metal-decorated carbon nanotubes. J. Phys. Chem. B **110**, 21014–21020 (2006)
64. V.V. Sysoev, B.K. Button, K. Wepsiec, S. Dmitriev, A. Kolmakov, Toward the nanoscopic "electronic nose": hydrogen vs carbon monoxide discrimination with an array of individual metal oxide nano- and mesowire sensors. Nano Lett. **6**, 1584–1588 (2006)
65. V.V. Sysoev, J. Goschnick, T. Schneider, E. Strelcov, A. Kolmakov, A gradient microarray electronic nose based on percolating SnO_2 nanowire sensing elements. Nano Lett. **7**, 3182–3188 (2007)
66. C. Wongchoosuk, A. Wisitsoraat, A. Tuantranont, T. Kerdcharoen, Portable electronic nose based on carbon nanotube-SnO_2 gas sensors and its application for detection of methanol contamination in whiskeys. Sens. Actuators B Chem. **147**, 392–399 (2010)

67. L.E. Greene, M. Law, D.H. Tan, M. Montano, J. Goldberger, G. Somorjai, P. Yang, General route to vertical ZnO nanowire arrays using textured ZnO seeds. Nano Lett. **5**, 1231–1236 (2005)
68. D. Ito, M.L. Jespersen, J.E. Hutchison, Selective growth of vertical ZnO nanowire arrays using chemically anchored gold nanoparticles. ACS Nano **2**, 2001–2006 (2008)
69. L. Vayssieres, Growth of arrayed nanorods and nanowires of ZnO from aqueous solutions. Adv. Mater. **15**, 464–466 (2003)
70. S. Xu, N. Adiga, S. Ba, T. Dasgupta, C.F.J. Wu, Z.L. Wang, Optimizing and improving the growth quality of ZnO nanowire arrays guided by statistical design of experiments. ACS Nano **3**, 1803–1812 (2009)
71. S. Xu, Y. Wei, M. Kirkham, J. Liu, W. Mai, D. Davidovic, R.L. Snyder, Z.L. Wang, Patterned growth of vertically aligned ZnO nanowire arrays on inorganic substrates at low temperature without catalyst. J. Am. Chem. Soc. **130**, 14958–14959 (2008)
72. K.D. Harris, A. Huizinga, M.J. Brett, High-speed porous thin film humidity sensors. Electrochem. Solid-State Lett. **5**, H27–H29 (2002)
73. T. Smy, D. Vick, M.J. Brett, S.K. Dew, A.T. Wu, J.C. Sit, K.D. Harris, Three-dimensional simulation of film microstructure produced by glancing angle deposition. J. Vac. Sci. Technol. A **18**, 2507–2512 (2000)
74. P.C. Jurs, G.A. Bakken, H.E. McClelland, Computational methods for the analysis of chemical sensor array data from volatile analytes. Chem. Rev. **100**, 2649–2678 (2000)
75. P. Offermans, M. Crego-Calama, S.H. Brongersma, Gas detection with vertical InAs nanowire arrays. Nano Lett. **10**, 2412–2415 (2010)

Chapter 15
Quantum Dot-Sensitized, Three-Dimensional Nanostructures for Photovoltaic Applications

Jun Wang, Xukai Xin, Daniel Vennerberg, and Zhiqun Lin

15.1 Introduction

As one of the major renewable energy sources, solar energy has the potential to become an essential component of future global energy production. Commercially available single-crystalline silicon solar cells exhibit a power conversion efficiency (PCE) of ~15%; however, these first generation devices suffer from the high cost of silicon fabrication [1]. Although polycrystalline silicon solar cells offer dramatically reduced fabrication costs than do their single-crystalline counterparts, their *PCE* is markedly lower [1]. Despite having high solar-to-electric energy conversion efficiency, the use of silicon-based solar cells is restricted by their lack of flexibility, high manufacturing and installation cost, and heavy weight. As a result, many research efforts have been made to utilize tailored nanostructured semiconductors, including conjugated polymers, carbon nanotubes, and quantum dots (QDs), for the development of next generation light-weight, low-cost, high-efficiency solar cells. For example, by using ruthenium dye to sensitize titania nanoparticle films, a *PCE* higher than 10% has been achieved [2]; Coakley et al. [3–5] studied organic/inorganic bulk heterojunction (BHJ) photovoltaic cells by infiltrating semiconducting conjugated polymer, poly(3-hexylthiophene) (P3HT), into mesoporous titania film with perpendicular pores throughout the film. Alivisatos et al. [6–11] fabricated hybrid solar cells by spin coating thin films of a mixture of CdSe nanocrystal and conjugated polymer and studied the effects of nanocrystal shape, film morphology, and post-treatment on the device performance. Advincula [12], Xu et al. [13], and Skaff et al. [14] directly grafted conjugated polymer on the QDs, showing enhanced charge transfer at the organic–inorganic interface. Ma et al. [15–17], Xue et al. [18–21], and Mutolo et al. [22] studied the solar cells made by

Z. Lin (✉)
Department of Materials Science and Engineering, Iowa State University, Ames, IA 50011-2300, USA
e-mail: zqlin@iastate.edu

W.L. Zhou, Z.L. Wang (eds.), *Three-Dimensional Nanoarchitectures*,
DOI 10.1007/978-1-4419-9822-4_15,

physically mixing conjugated polymer and fullerene. However, to date, the *PCE* of these solar cells is still low due to the inefficient charge transport to the electrode and significant charge recombination before reaching the electrodes. In this context, improving the solar-to-electric energy conversion efficiency has been one of the biggest issues in solar energy research. Dye-sensitized solar cells (DSSCs), one of the most promising of several alternative cost-effective concepts for solar energy conversion, have received considerable attention over the past decade [2]. To further improve the *PCE* of DSSCs, much research has been performed to optimize three constituents: the photosensitizer, electron-collecting semiconductor, and hole-conducting electrolyte. Owing to their outstanding optical properties, QDs have been utilized for application in solar cells with the possibility of obtaining a *PCE* exceeding the traditional Shockley–Queisser limit of efficiency of 32% [23, 24]. Among several schemes of QD solar cells, QD-sensitized solar cells, which are the focus of this chapter, have the most potential to increase the maximum attainable conversion efficiency of solar energy. There are a number of advantages to utilizing QDs as sensitizers, not least being their capacity for multiple exciton generation (MEG) [25–36]. When incident energy is higher than the bandgap, instead of dissipation in terms of heat loss, as in *p–n* junction solar cells, QDs (e.g., PbSe [36]) have been shown to generate many excitons per photon. This is an efficient way to increase photocurrent, and thus enhance power conversion efficiency. Next, QDs have tunable optical properties as a function of particle size, which facilitate the absorption of sunlight. Finally, QDs exhibit better heterojunction formation with solid hole conductors. To date, various QDs (e.g., CdSe, CdS, CdTe, InAs, InP, and PbSe) have been demonstrated as a substitute for ruthenium dye in conventional DSSCs [23, 37–40]. Another important modification to conventional DSSCs lies in the photoanodes (e.g., TiO_2 or ZnO). Large bandgap semiconductor nanowires and nanotubes have recently been employed as photoanodes and exhibited attractive potential for high-performance solar cells. Unlike the commonly used sintered TiO_2 nanoparticle film, the precise orientation of the crystalline nanowires and nanotubes makes them excellent electron percolation pathways for vectorial charge transfer between interfaces [41], thereby reducing the chance for charge recombination [42–50]. This is in sharp contrast to TiO_2 nanoparticle film, where the structural disorder at the contact between adjacent TiO_2 nanoparticles leads to the increased scattering of electrons (i.e., trapping/de-trapping for charge hopping) and reduced electron mobility.

In this chapter, recent progress in QD-sensitized, three-dimensional nanostructures for use in photovoltaic applications is discussed, including the synthesis and surface functionalization of colloidal QDs, chemical bath deposition of QDs, and QD-sensitized, three-dimensional nanostructures (including semiconductor nanoparticles, nanowires, and nanotubes). This treatment of QD nanostructure fabrication is followed by an investigation of electronic interactions between QDs and large bandgap semiconductors. The chapter concludes with an outlook on the future of QD-sensitized solar cells.

15.2 Quantum Dot-Sensitized Solar Cells

15.2.1 Overview

Semiconductor nanocrystals, known as QDs (e.g., CdSe, CdS. CdTe, InAs, InP, and PbSe), absorb visible to near-IR light and inject electrons into large bandgap semiconductors (e.g., TiO_2 and ZnO). They show the promising potential to substitute for ruthenium-based dyes in conventional DSSCs [23, 37–40]. QDs possess tunable band edge, efficient visible absorption, and MEG capability [23, 25, 28, 30–35] and thus have been studied, both theoretically and experimentally, for use in solar energy conversion. Theoretical calculation shows that as many as seven excitons can be generated upon absorption of one photon by QDs at a certain energy level [34], and an ideal *PCE* higher than 40% can be expected [23, 39].

QDs are usually synthesized via colloidal chemistry and chemical bath deposition. Colloidal chemistry yields QDs with well-defined size, shape, and chemical composition, as well as good solvent solubility due to the fact that the surface is capped mostly by a hydrophobic ligand. Surface functionalization on QDs is necessary to enhance their filling into hydrophilic nanoporous films and to promote the interaction between QDs and electron acceptor surface, thereby increasing the photocurrent in QD-sensitized solar cells. A commonly adopted approach for surface functionalization of QDs is ligand exchange, i.e., replacing the original hydrophobic ligand with bifunctional molecules. QDs made by chemical bath deposition render much better filling of QDs in nanoporous films. However, control over the size and shape of QDs needs to be further investigated and optimized.

To date, in most QD-sensitized solar cells, mesoporous TiO_2 films are used as electron acceptors (i.e., photoanode). QD-sensitized solar cells have been assembled by infiltrating QDs or by chemical bath deposition into these mesoporous electron acceptor films. Hole transport electrolytes are injected between the QD-doped photoanode and the counter electrolyte (e.g., Pt-coated conductive substrate). High-efficiency light absorption by QDs, efficient filling of QDs, good interaction between QDs and the electron acceptor, and high-performance electrolyte are the keys to fabricating highly efficient QD-sensitized solar cells. Large bandgap semiconductor nanowires and nanotubes have recently been developed to replace mesoporous TiO_2 film in order to achieve better electron transport and reduce charge recombination.

15.2.2 Synthesis of Quantum Dots and Surface Functionalization

QDs are highly emissive, spherical nanoparticles that are a few nanometers in diameter [51, 52]. They provide a functional platform for a new class of materials for use in solar cells [7, 8, 11], LEDs [53, 54], tunable lasers [55], optical storage media

[56], nonradiative energy transfer [56–58], biosensors [59–63], and bioimaging [64, 65]. For QDs such as CdSe [66, 67], variation of particle size provides continuous and predictable changes in light absorption and fluorescence emission due to their quantum-confined nature. An appropriate surface passivation with a monolayer of coordinating ligands is crucial to ensuring the solubility and miscibility of QDs with the host environment and to retaining the spectroscopic properties of the materials by preventing QDs from aggregating. Ligand exchange permits derivatization with a broad range of functional groups.

Among all the synthetic approaches, colloidal synthesis is the most widely used technique for preparation of high-quality QDs. These colloidal semiconductor nanocrystals are usually synthesized from precursor compounds dissolved in solutions. The synthesis is based on a three-component system composed of precursors, organic surfactants, and solvents [57]. By tuning the reaction conditions, QDs with controlled size, shape, and chemical composition can be readily obtained [68–71]. Figure 15.1 shows examples of CdSe/CdS core/shell QDs of different shapes [68].

QDs synthesized through conventional organometallic high-temperature growth procedures are functionalized with hydrophobic ligands (e.g., trioctylphosphine oxide, TOPO) [69, 71]. However, on many occasions, it is desirable to prepare water-soluble QDs for use in bioimaging, biosensors, and QD-sensitized TiO_2 nanocrystal solar cells. In the latter context, due to the size-dependent optical properties and MEG capability [25–28, 32], QDs can be exploited as the sensitizer in QD-sensitized TiO_2 solar cells. Placing the water-soluble QDs in intimate direct contact with the hydrophilic TiO_2 nanocrystals is expected to facilitate the efficient charge transfer from QDs to TiO_2 in QD-sensitized TiO_2 solar cells. This can be realized by utilizing both QDs (e.g., dithiocarbamate-functionalized QDs with the carboxyl group on the surface [72]) and TiO_2 (e.g., the hydroxyl group on the TiO_2 surface [50, 73, 74]) with complementary functional groups, which allow them to react with each other under mild conditions.

Bifunctional molecules with the thiol and carboxyl groups at each end (e.g., mercaptopropionic acid, MPA [38, 44, 74]) have been used as effective ligands to prepare water-dispersible QDs. The thiol group provides the chemical affinity

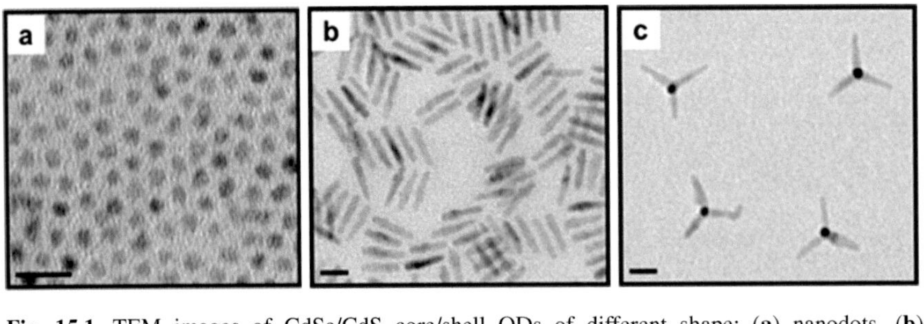

Fig. 15.1 TEM images of CdSe/CdS core/shell QDs of different shape: (**a**) nanodots, (**b**) nanorods, and (**c**) tetrapods; scale bars, 20 nm (reprinted with permission from [68]. Copyright © American Chemical Society)

to QDs, while the carboxyl group imparts the water solubility. Notably, the use of thiols suffers from their instability against oxidation [75–77]. This drawback has recently been overcome by using dithiocarbamate moieties as ligands due to their strong chelate-type binding to metal atoms [78, 79]. The resulting dithiocarbamate-functionalized QDs exhibited improved resistance against photooxidation. However, to obtain the dithiocarbamate ligand, a multi-step synthesis was needed. Later on, a spontaneous assembly of dithiocarbamate on a gold surface was demonstrated by simply exposing gold to carbon disulfide (CS_2) and secondary amine [80]. This approach was then successfully extended to CdSe/ZnS QDs to tune the surface properties of QDs [76]. Basically, TOPO-functionalized CdSe/ZnS QDs were added to the solution containing equivalent amount of CS_2 and glycine (an amino acid-based ligand), followed by stirring at room temperature for several hours to convert CdSe/ZnS–TOPO to dithiocarbamate-functionalized CdSe/ZnS [76]. Glycine, dissolved in the mixture of chloroform and methanol ($CHCl_3$:MeOH 1:1), provides both a source of amine group for the formation of dithiocarbamate via its reaction with CS_2 and a source of carboxyl group for aqueous solubility. It is noteworthy that only very dilute CdSe–TOPO QD concentrations (i.e., 10^{-3} M) were used in the ligand exchange due to the limited solubility of glycine in the $CHCl_3$/MeOH mixture [76]. Thus, the yield of D-CdSe QDs was rather low. However, for the use of water-soluble QDs in bioimaging and solar cells, and also from a cost-effective point of view, a route to large-scale production of QDs is highly desirable. Moreover, these dithiocarbamate-functionalized CdSe/ZnS QDs were formed via the ligand exchange in one phase (i.e., in the $CHCl_3$/MeOH mixture). As a result, additional steps (i.e., precipitation and redissolution) were necessary to obtain the CdSe/ZnS QD aqueous solution [76]. To this end, a simple biphasic ligand exchange approach has recently been developed to prepare water-soluble, dithiocarbamate ligand-capped CdSe QDs on a large scale. This approach takes advantage of the good solubility of glycine in water [72]. Figure 15.2 shows the successful transformation of TOPO-functionalized CdSe in chloroform from water insoluble to water soluble by replacing TOPO with dithiocarbamate moieties. Other approaches, including the use of polyelectrolytes as surface-capping ligands, were also developed to transfer hydrophobic QDs into water [81]. It is worth noting that the terminal functional groups (i.e., carboxyl group) offer the possibility to further control the surface properties of QDs or link to functional molecules.

Chemical bath deposition (CBD) is another way to deposit QDs onto semiconducting metal oxide electron collectors, dispensing with need for the use of surface ligand-capped QDs [38–40, 44, 82]. A much better interface between QDs and an electron-accepting semiconductor and more efficient filling of QDs into mesoporous film can be achieved using CBD. In a CBD process, anionic and cationic precursors react slowly under a given condition. Traditional CBD adds two ionic precursors together and controls the reaction rate, usually by controlling the pH of solution to adjust the ionic concentration [83–86]. A new means to the rate control is to separate cationic and anionic precursors and mix them by immersing the metal oxide alternatively into two precursor solutions [37, 87]. This method is called successive ionic

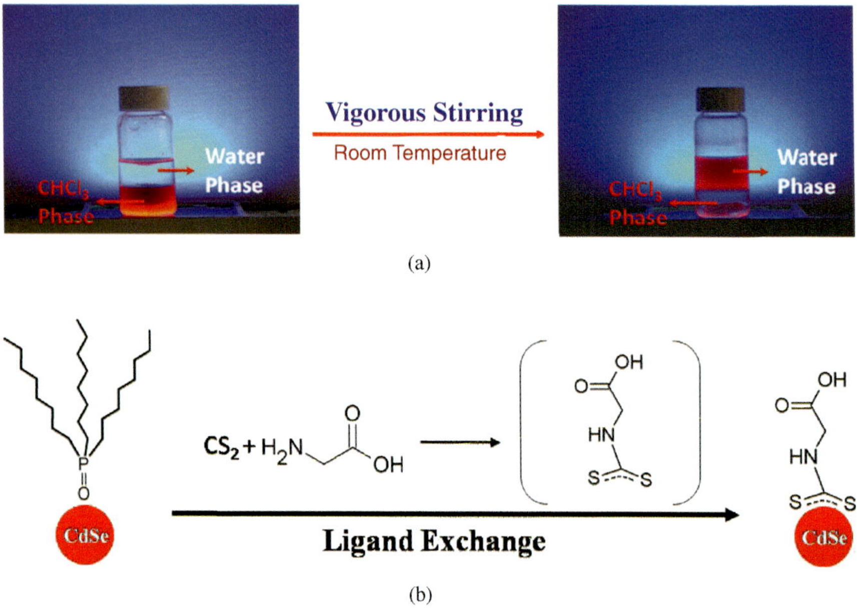

Fig. 15.2 (**a**) Transformation of hydrophobic CdSe QDs in chloroform into hydrophilic QDs in water. (**b**) Schematic illustration of the ligand exchange with dithiocarbamate (reprinted with permission from [72]. Copyright © The Royal Society of Chemistry)

layer adsorption and reaction (SILAR). Compared to the traditional CBD method, the SILAR process is more efficient (less than 1 h as opposed to several hours to overnight process), well controlled in terms of the density and size of the QDs, and selective with respect to only acting on the metal oxides (in comparison to on all surrounding parts, including the container surface), making SILAR the best process to deposit QDs on mesoporous metal oxides [37, 82, 88, 89].

SILAR has been successfully employed for the preparation of various QDs (e.g., PbS, CdSe, and CdTe) of II to VI elementals on semiconductor electron collectors (usually TiO_2) with photocurrent yields up to 70% [37]. Two solutions, one containing anions and the other containing cations, are prepared and the metal oxide electron collector (e.g., TiO_2) is immersed into one of the ionic solutions. Trace amount of ion adsorbs to the surface of the metal oxide. After rinsing with water or ethanol, the metal oxide is soaked into the other oppositely charged ion solution. The residual ion on the metal oxide surface reacts with the oppositely charged ion in the solution to form a thin layer of semiconductor QDs. QDs with desired thickness can be achieved by repeating the deposition cycles several times.

One obstacle for the SILAR process is the preparation of ionic solution, especially metal anionic solution, as it is very easily oxidized in air. To date, most of the SILAR experiments are based on metal sulfide QDs, for example, PbS and CdS [48, 89–91], due to the high stability of sulfide ions. Recently, a breakthrough has been made by Gratzel et al. [88]. Unstable selenide and telluride anion solutions were stabilized by adding a reducing agent, $NaBH_4$, into an ethanol solution containing selenium oxide or tellurium oxide, as described below.

$$SeO_2/TeO_2 + 2NaBH_4 + 6C_2H_5OH \rightarrow Se^{2-}/Te^{2-} + 2Na^+ + 2B(O_2H_5)_3 + 5H_2 + 2H_2O$$

Obvious color change can be observed during the process, indicating the reduction reaction from Se^{4+}/Te^{4+} to Se^{2-}/Te^{2-}. Subsequently, the SILAR process was carried out under an inert gas atmosphere to avoid oxidization by air [88].

15.2.3 Quantum Dot-Sensitized Nanoparticle Films

Mesoporous TiO_2 films, consisting of nanocrystalline TiO_2 particles, have been extensively used as photoanodes and sensitized with ruthenium dyes in conventional dye-sensitized solar cells (DSSC) [2]. Due to the large surface area to volume ratio in these films, efficient dye loading and incident light absorption are achieved. The film is prepared from pre-synthesized TiO_2 colloid solution via the "doctor blade" technique, followed by annealing at high temperature to yield a crystalline anatase phase. The nanoparticles can be synthesized by a variety of methods, including hydrolysis of a Ti organic complex (e.g., Ti tetraisopropoxide [2, 38, 92]), templated synthesis [93, 94], and microemulsion process [95]. In addition to using ruthenium dyes as photosensitizers, QDs have recently been exploited to sensitize TiO_2 for use in solar cells. Because of their unique optical properties and MEG characteristic, QD-sensitized solar cells have exhibited great potential for solar energy conversion. Typically, two approaches have been employed to deposit QDs into mesoporous TiO_2 films, that is, the solution deposition of pre-synthesized colloidal QDs and chemical bath deposition of QDs. As previously discussed, QDs produced from colloidal chemistry possess well-controlled size, shape, chemical composition, and surface properties; however, efficient infiltration of QDs into the mesopores and effective electronic interaction between QDs and TiO_2 remain a challenge. By contrast, chemical bath deposition of QDs provides efficient filling of the mesopores through the atomic-scale add-up process, but the size and shape of deposited QDs cannot be well controlled.

In early studies, pre-synthesized colloidal QDs were deposited on mesoporous TiO_2 films to yield QD-sensitized solar cells. Although the *PCE* of these solar cells is low, considerable knowledge have been gained regarding the optimization of device performance [96–98]. Low bandgap QDs, such as InP and InAs, have been investigated as photosensitizers due to their high absorption coefficient and optimized bandgap for the full spectral absorption of sunlight [39, 40]. The photovoltaic performance of InP and InAs QD-sensitized solar cells is shown in Fig. 15.3.

The low *PCE* can be attributed to the following: (1) insufficient filling of hydrophobic QDs into hydrophilic mesoporous TiO_2 films; (2) less effective electronic interaction between QDs and TiO_2 surfaces; and (3) rapid corrosion of QDs by iodide-based electrolyte. Thereafter, efforts have been made to optimize

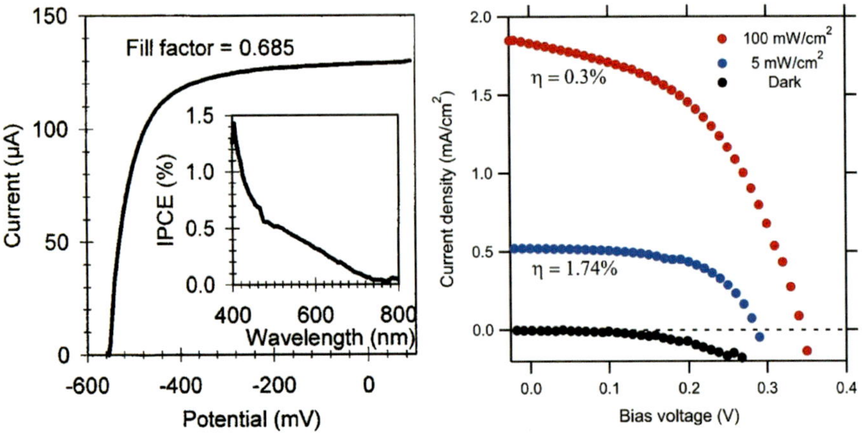

(a) (b)

Fig. 15.3 (**a**) Current–voltage characteristic of InP-sensitized TiO_2 solar cells. *Inset* shows the photocurrent performance. InP QDs of size of 6.5 nm were used (reprinted with permission for [40]. Copyright © American Chemical Society); (**b**) current–voltage characteristic of InAs QD-sensitized TiO_2 solar cells, exhibiting a *PCE* of 1.74% under 5 mW/cm^2 illumination and 0.3% under 100 mW/cm^2 illumination; InAs QDs of size of 3.4 nm were used (reprinted with permission from [39]. Copyright © American Chemical Society)

QD-sensitized solar cells. For example, Co redox electrolyte [39, 82, 88, 89] and polysulfide electrolyte [38] have been used to provide long-term stability for QDs in the devices; in addition, QDs and TiO_2 surfaces have been engineered to promote electronic interaction (i.e., charge injection from QDs to TiO_2) [38, 43, 50, 96, 98]. Chemical bath deposition has also been used for high-efficiency deposition of QDs into mesopores [83, 85, 87, 88, 90, 99–101].

Since pre-synthesized QDs are usually capped with hydrophobic ligand, for example, TOPO, surface functionalization of QDs is a commonly used approach to control surface properties [81]. Surface functionalization not only provides desired surface properties but also influences the intrinsic physical properties of QDs. For example, enhanced quantum yield of CdSe QDs was observed upon ligand exchange with dithiocarbamate ligand [72]. In order to increase loading of QDs and improve electronic interaction between QDs and TiO_2, one widely used method is to use bifunctional molecules (e.g., mercaptopropionic acid, MPA) as the linker to obtain molecular linking between QDs and TiO_2 [38]. As shown in Fig. 15.4, CdSe QDs are chemically linked to mesoporous TiO_2 films. The TiO_2 surface was first functionalized with molecules that have carboxyl and thiol groups at each end. The carboxyl groups were preferentially attached to the TiO_2 surface, while the terminal thiol groups were anchored to the CdSe QD surface via ligand exchange with TOPO [38]. A scheme illustrating charge injection, charge transport, and charge recombination upon the visible light excitation of CdSe QDs that are linked to TiO_2 is given in Fig. 15.5. By using bifunctional molecules as the linker, CdSe QDs can be effectively assembled on the TiO_2 surface, thereby facilitating efficient electron

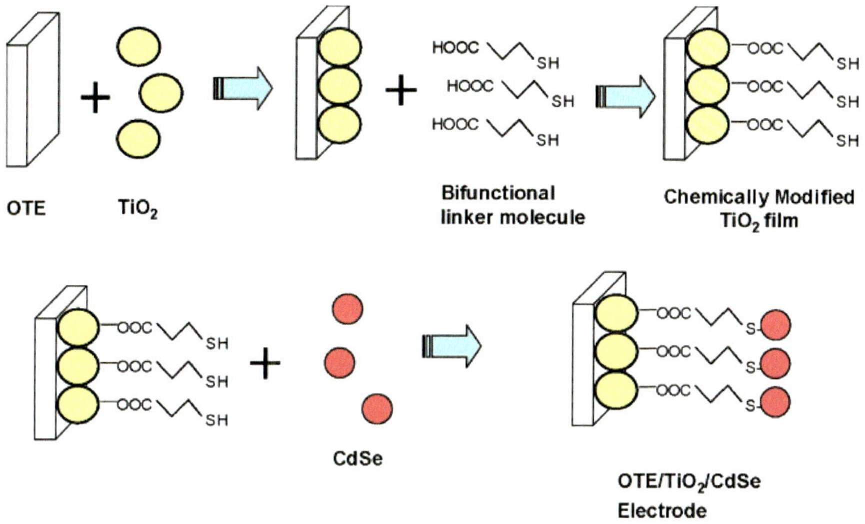

Fig. 15.4 Schematic illustration of chemical linking of CdSe QDs to TiO_2 surface using bifunctional molecules, for example, mercaptopropionic acid (MPA) (reprinted with permission from [38]. Copyright © American Chemical Society)

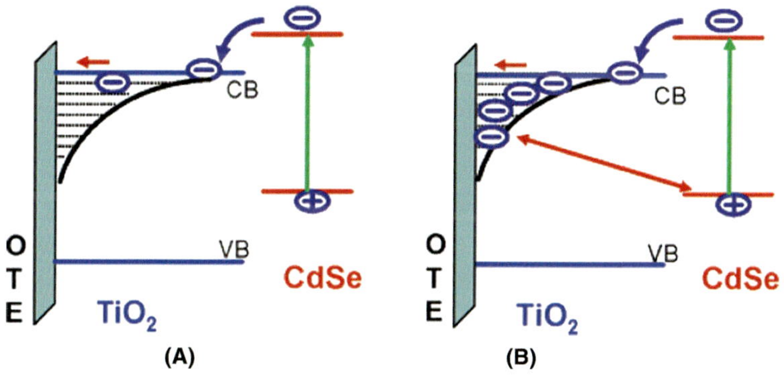

Fig. 15.5 Schematic illustration of the charge injection, charge transport, and charge recombination in CdSe QD-sensitized mesoporous TiO_2 films at (**a**) low and (**b**) high excitation intensities (reprinted with permission from [38]. Copyright © American Chemical Society)

injection from the conduction band of QDs to TiO_2, which was confirmed by femtosecond transient absorption and emission quenching characterizations [38]. It is worth noting that higher incident light intensities can lead to increased scattering as well as charge recombination due to the accumulated carriers in TiO_2 domains, which results in low overall *PCE* [38, 39].

Oleic acid-capped CdSe QDs have been used as a substitute for TOPO-capped CdSe QDs in QD-sensitized solar cells. It was found that surface capping with oleic acid increased the loading of QDs in mesoporous TiO_2 film, extended the absorption range, and possibly suppressed the surface charge recombination [96]; the TiO_2

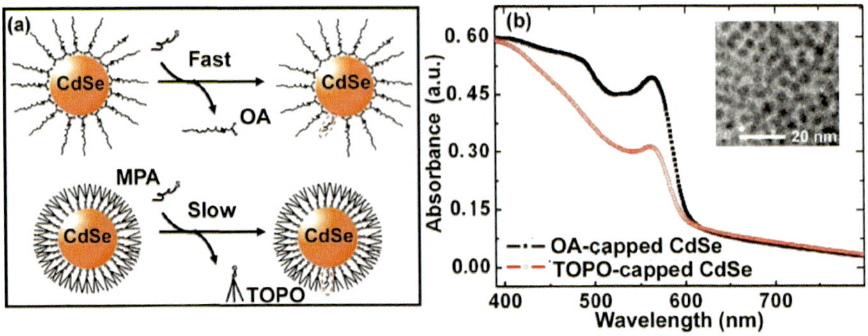

Fig. 15.6 (**a**) Schematic illustration of the ligand exchange between oleic acid and MPA, and TOPO and MPA. (**b**) Absorption spectra of oleic acid-capped and TOPO–capped, CdSe QD-sensitized films over a 5-μm-thick transparent TiO_2 layer. *Inset* shows the TEM of 3-nm oleic acid-capped CdSe QDs (reprinted with permission from [96] Copyright © American Institute of Physics)

film was functionalized with the bifunctional molecule MPA prior to loading the QDs. The ligand exchange between oleic acid and MPA, TOPO and MPA, and the absorption spectra of TiO_2 films infiltrated with oleic acid and TOPO-capped CdSe QDs are shown in Fig. 15.6. The difference between two capping ligands is the bonding strength to CdSe surface. The weak bonding between the carboxyl group and CdSe allowed for the formation of strong bonding between MPA and CdSe upon the ligand exchange, leading to a high rate of ligand exchange between oleic acid and MPA. On the other hand, TOPO molecules interact strongly with CdSe, preventing MPA from approaching the CdSe surface, which results in a low rate of ligand exchange between TOPO and MPA. As a result, improved loading of CdSe QDs and extended absorption were achieved, and the overall *PCE* of QD-sensitized solar cells with oleic acid-capped CdSe showed 64% improvement over those with TOPO-capped CdSe [96].

According to a recent investigation of the charge injection rate from QDs to TiO_2 using transient absorption technique, the size-dependent optical properties of QDs make it possible to engineer the semiconductor bandgap [74]. It was found that as the QD's size decreases (i.e., larger bandgap), the electron transfer rate from QDs to TiO_2 increases, which is due to the increased energy difference between the conduction band of CdSe and that of TiO_2 (Fig. 15.7). As a result, the fastest electron transfer rate was observed from 2.4-nm CdSe QDs. A recent study on the modulation of electron injection in CdSe–TiO_2 system revealed that medium alkalinity affects the energetic band position of TiO_2, and thus influences the charge injection process between CdSe QDs and TiO_2 [102]. Figure 15.8 shows the pH-dependent emission spectra, emission response of CdSe on the TiO_2 film, and emission lifetime response of CdSe on the TiO_2 and SiO_2 films. SiO_2 is an inert substrate and does not directly influence the radiative recombination process in CdSe; therefore it was used as the control experiment. It was clearly shown that a mild acidic medium, if favorable for the electron injection from CdSe QDs to TiO_2, exhibits a nearly

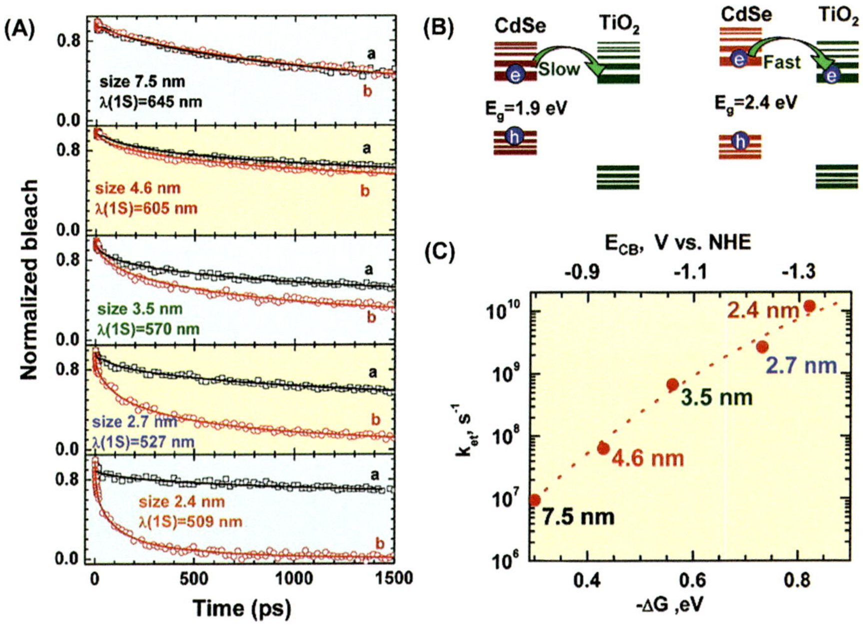

Fig. 15.7 (**a**) The transient recovery recorded at the bleaching maximum following 387 nm laser pulse excitation of CdSe QDs in 1:1 ethanol/THF-containing mercaptopropionic acid (**a**) without and (**b**) with linked TiO_2 particles. (**b**) Schematic illustration of the principle of electron transfer from quantized CdSe into TiO_2 and (**c**) the dependence of electron transfer rate constant on the energy difference between the conduction bands. Top axis represents assumed CdSe conduction band energy positions (reprinted with permission from [74] Copyright © American Chemical Society)

fivefold increase in the apparent electron transfer rate constant upon decreasing the pH from 12 to 5.5 based on the emission lifetime measurement [102]. This result is of great importance for promoting charge injection from QDs by controlling the pH of the medium. It is worth nothing that the change in the emission response is reversible, making it possible to use CdSe–TiO_2 system for sensing the pH of the medium [102].

One challenge in QD-sensitized solar cells is to capture photogenerated electrons as quickly as they are generated and subsequently transport them to the electrode in an efficient manner [103]. Recently, QD–C_{60} composite film was deposited on mesoporous TiO_2 films for the efficient capture of a photogenerated electron [103, 104]. C_{60} and its derivatives have been widely used in organic thin-film solar cells because of their outstanding electron-accepting property [105]. Electrophoretic deposition of CdSe–C_{60} composites provides better interaction between CdSe and C_{60} by encapsulating CdSe QDs with C_{60} clusters. Emission quenching and emission lifetime measurements showed that C_{60} can efficiently capture the electrons from photoexcited CdSe QDs, which was further confirmed by the photoelectrochemical experiment, i.e., the observed photocurrent generation efficiency within

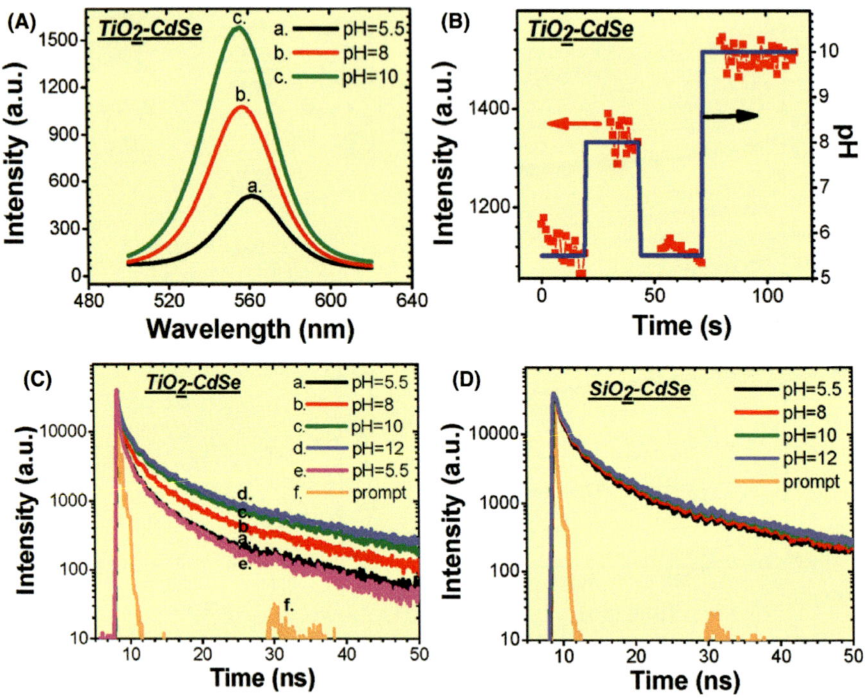

Fig. 15.8 (**a**) Emission spectra and (**b**) emission response of CdSe on TiO_2 film at three different pH values. Distilled water was flowed through the cell before changing pH. (**c** and **d**) Emission lifetime responses of CdSe on TiO_2 and SiO_2 films at varying pH, respectively (reprinted with permission from [102] Copyright © American Chemical Society)

CdSe–C_{60} films is two orders of magnitude greater than that of the CdSe films alone (Fig. 15.9). Other carbon nanostructures, owing to their unique electronic and optical properties, can also be used for efficiently capturing photogenerated electron in QDs, for example, stacked-cup carbon nanotubes (SCCNT) [104]. Electrophoretic deposition can efficiently deposit SCCNT–CdSe on the TiO_2 electrode; ultrafast electron transfer from photoexcited CdSe QDs to SCCNT was achieved and confirmed by emission quenching and transient absorption measurement. Moreover, it was found that decreasing the size of QDs further increased the charge transfer rate [104]. These observations suggested that integrating CdSe QDs with carbon nanostructures may be a promising direction toward the development of high–performance, QD-sensitized solar cells.

Infiltrating QDs into mesoporous TiO_2 films remains a challenge in QD-sensitized solar cells, mainly because the size of QDs is much larger than that of ruthenium dye molecules and non-polar solvent (i.e., toluene) for dispersing QDs cannot wet the hydrophilic surface of TiO_2. Chemical bath deposition (CBD), unlike the pre-synthesized colloidal QDs capped with ligand, involves the successive ionic deposition and reaction of metal ions on the TiO_2 surface. Anionic and cationic

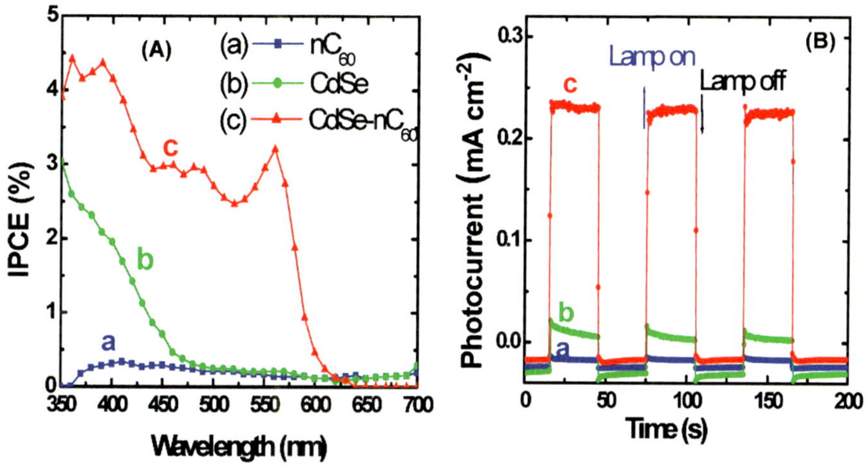

Fig. 15.9 (**a**) Dependence of the incident photon to current generation efficiency (*IPCE*) on the incident wavelength and (**b**) the photocurrent response of electrodes to ON–OFF cycles of illumination: (**a**) C60, (**b**) CdSe QDs, and (**c**) CdSe–C_{60} clusters on SnO_2 film electrodes. The electrolyte was 0.1 M Na_2S in water (reprinted with permission from [103] Copyright © American Chemical Society)

precursors are usually prepared in aqueous or polar solvent, which can wet the hydrophilic TiO_2 surface, thereby achieving a higher degree of filling. In addition, CBD of QDs provides intimate contact between QDs and the TiO_2 surface, resulting in much better charge transfer efficiency. In the conventional CBD process, anionic and cationic precursors react slowly under a given condition. Two ionic precursors are usually put together and the reaction speed can be controlled by the pH value of the solution and the ionic concentration [83–86]. Recently, a new way to control the reaction rate has been developed by simply immersing the electrode alternately into two precursor solutions [37, 87]. This method, as discussed above, is called successive ionic layer adsorption and reaction (SILAR). Various QDs have been successfully prepared by CBD technique [37, 82, 87–89]. The most recent advances in the preparation of stable Se^{2-} and Te^{2-} precursors make it possible to deposit CdSe and CdTe QDs onto mesoporous TiO_2 films [88]. The absorption spectra, the digital image of mesoporous TiO_2 films after SILAR deposition of CdSe QDs, and TEM images on the CdSe–TiO_2 particles are shown in Fig. 15.10. As the deposition cycles increased, the color of the TiO_2 film became deeper, indicating the increased amount of CdSe and increased size of CdSe QDs. The UV–vis absorption spectra also confirmed the successive deposition of QDs onto mesoporous TiO_2. From TEM and HRTEM characterizations, it is clear that SILAR-deposited CdSe QDs had intimate contact with TiO_2 nanoparticles, thereby providing effective electronic interaction between CdSe and TiO_2 without the use of any molecular linkers (e.g., MPA).

It is worth noting that a thin terminating layer of CdTe could dramatically increase charge collection efficiency. Transient photovoltage and photocurrent

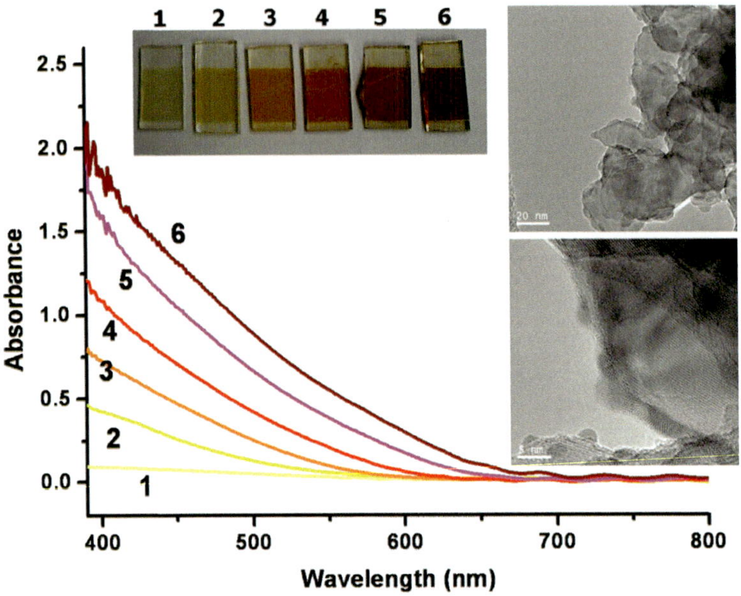

Fig. 15.10 Absorption spectra of 2-μm-thick film made of 20-nm TiO_2 after the SILAR deposition of CdSe QDs (1–6 cycles) and photographs of the corresponding films (*inset*; *left*). TEM images of CdSe QDs/TiO_2 particles after six cycles of the SILAR process (reprinted with permission from [88] Copyright © American Chemical Society)

decay measurements suggested a lower charge recombination with the existence of a thin CdTe terminating layer, which served as a barrier layer and retarded the recombination between photogenerated electrons and the oxidized cobalt electrolyte [88]. As a result, a longer electron diffusion length could be achieved, exhibiting an overall *PCE* over 4% under 100 W/m^2 illumination. Upon optimizing the structure of QD-sensitized solar cells, a cascade electrode of TiO_2/CdS/CdSe has recently been developed for solar cell applications, in which the reorganization of energy levels between CdS and CdSe formed stepwise band-edge levels to facilitate the electron injection and hole recovery of CdS and CdSe QDs; in addition, CdS and CdSe QDs have a complementary effect in the light harvesting. In the end, an overall *PCE* of 4.22% was obtained under AM1.5G illumination, which is the highest reported value for QD-sensitized solar cells [87].

15.2.4 Quantum Dot-Sensitized Nanowire Arrays

As discussed above, the most commonly used semiconductor electrode for sensitized solar cells are mesoporous TiO_2 films composed of crystalline TiO_2 nanoparticles. However, the random dispersion of TiO_2 nanoparticles in thin film leads to increased scattering of free electrons and electron trapping at the interfaces between nanoparticles, thereby reducing electron mobility [106]. In this context, the use of aligned, single-crystalline, wide bandgap semiconductor nanorods and nanowires in DSSCs allows for direct electron transport along the long axis of nanorods and nanowires that are normal to the electrode [42, 45, 106–109]. Recently, ZnO single-crystalline nanowires have been extensively studied in terms of synthesis and application in DSSCs [110]; the vectorial electron transport along the nanowire axis resulted in improved charge collection efficiency. However, due to its poor chemical stability, limited *PCE* has been achieved in ZnO nanowire-based sensitized solar cells. Due to their unique optical properties and MEG capability, QDs show the potential to enhance the performance of ZnO nanowire-based sensitized solar cells [44].

An array of ZnO nanowires were grown vertically from a fluorine-doped tin oxide (FTO)-conducting substrate (Fig. 15.11). Subsequently MPA-functionalized CdSe QDs were chemically anchored to the ZnO nanowire surface via the interaction of a terminal carboxyl group on functionalized CdSe and the hydroxyl group on the ZnO surface. Upon photoexcitation, excitons were generated in CdSe QDs, and then electrons were injected from CdSe QDs to ZnO nanowires. On the initial attempt, the grafting density of MPA-capped CdSe QDs to ZnO nanowires was low. This was increased by modifying the ZnO surface with oxygen plasma prior to the QD adsorption. Oxygen plasma treatment may charge the nanowire surface, create dangling bonds through ion bombardment, remove surface contaminants, and saturate the surface hydroxyl groups [44, 49], which in turn contributed to the increased grafting of QDs onto ZnO nanowires. Photoelectrochemical results showed that both the *PCE* (0.4% under AM1.5G illumination) and short-circuit current density of the quantum dot-sensitized solar cells assembled using the oxygen plasma-treated ZnO nanowires were more than an order of magnitude higher

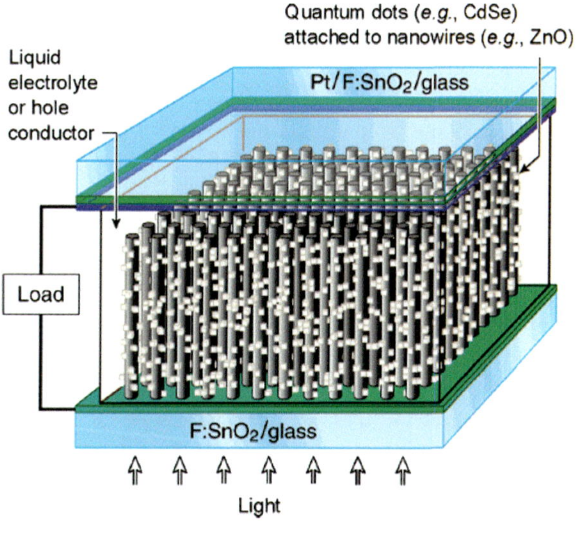

Fig. 15.11 Schematic of the quantum dot-sensitized solar cell. An array of ZnO nanowires, grown vertically from an F-doped SnO_2/glass substrate and decorated with CdSe QDs, serve as the photoanode. A second F-doped SnO_2/glass substrate, coated with a 100-Å layer of Pt, is the photocathode. The space between the two electrodes is filled with a liquid electrolyte (reprinted with permission from [44]. Copyright © American Chemical Society)

than those assembled using non-treated nanowires [44]. Single-crystalline TiO_2 nanowires were directly grown on a transparent conductive substrate via a solvent thermal technique, and their application in DSSCs has recently been demonstrated [42, 45]. These single-crystalline TiO_2 nanowire arrays are also promising photoanodes for quantum dot-sensitized solar cells. Moreover, TiO_2 has much better chemical stability than does ZnO, making it suitable for stable, high-performance solar cells.

Although the performance of quantum dot-sensitized nanowire solar cells is low, further optimization, including the size and chemical composition of QDs, the selection of surface-capping ligand, chemical bath deposition (CBD) of QDs, and the high-performance hole transport electrolyte, is expected to enhance the *PCE* of this type of solar cell.

Recently, photocurrent in silicon nanowires or quantum walls induced by non-radiative resonant energy transfer (NRET) from adjacent layers of lead sulfide QDs was observed using time-resolved photocurrent measurements [108, 111]. This NRET between efficient photo absorbers (e.g., QDs) and adjacent quantum channels with high charge transport mobility may offer a viable alternative to the limitations of carrier transport and collection faced by excitonic solar cells (e.g., quantum dot-sensitized solar cells).

15.2.5 Quantum Dot-Sensitized Nanotube Arrays

Highly ordered, vertically oriented, TiO_2 nanotube arrays of different aspect ratios and surface qualities have recently been fabricated as alternative nanoscale architectures to substitute for sintered TiO_2 nanoparticle films in DSSC [46, 49, 112]. They also exhibited promising performance for use in gas sensors [113–115], water splitting [112, 116–118], photocatalysts [119, 120], and cell separation [121, 122].

In an electrochemical anodization process, TiO_2 nanotube arrays were formed via self-organization of TiO_2 as a result of a delicate balance of electrochemical oxidation of Ti into TiO_2, electrical field-induced dissolution of TiO_2, and the chemical dissolution of TiO_2 by fluorine ions as the fluorine ions quickly react with TiO_2. Therefore, it has been a challenge to grow highly ordered TiO_2 nanotube arrays with high aspect ratio [112, 123–128]. Recently, by adjusting the pH of the electrolyte, using fluorine-containing buffer solution, or applying nonaqueous electrolyte, TiO_2 nanotube arrays with film thickness greater than 1000 μm have been obtained. By tuning the electrochemical anodization conditions (i.e., electrolyte, temperature, and anodization potential, etc.), highly ordered TiO_2 nanotube arrays with pore diameters ranging from 20 to 100 nm, tunable wall thicknesses, and variable film thicknesses can be obtained [41, 129, 130].

It is noteworthy that due to the weak connection between the TiO_2 nanotubes and the Ti substrate through the TiO_2 "barrier layer," a thick freestanding TiO_2 nanotube film can be readily obtained by mechanically detaching it from the Ti substrate via a mild ultrasonication [131, 132]. Alternatively, selective chemical etching of the metallic Ti substrate using a bromine-containing methanol solution also yielded freestanding TiO_2 nanotube arrays; however, the employed etching solution was highly corrosive [119]. After a long anodization time (e.g., 60 h), the top surface of the TiO_2 nanotube arrays was covered by a thin layer of TiO_2 nanowires, which formed due to the electric field-induced chemical etching of nanotubes in a "bamboo-splitting" mode [133]. This thin layer of nanowires can be readily removed by ultrasonication, exposing the nanotubes buried underneath [131].

The ability to produce a freestanding TiO_2 nanotube film makes it possible to transfer the film to a variety of substrates (e.g., Si or transparent, conductive, FTO-coated glass). On the other hand, the poor mechanical strength of the freestanding film makes it difficult to handle, especially in the case of a very thin TiO_2 film. Therefore, it is crucial to grow the TiO_2 nanotubes directly from a desired substrate (e.g., transparent FTO-coated glass). For example, for the use of TiO_2 nanotube arrays in DSSCs, it is more advantageous to exploit TiO_2 nanotubes obtained by anodization on a transparent, conductive substrate than those formed on a metallic Ti substrate. In the so-called frontside-illuminated DSSC, the incident light directly illuminates the dye-sensitized TiO_2 nanotubes through the FTO-coated glass. However, in the "backside-illuminated" mode, the incident photons have to pass through the semi-transparent top electrode and the redox electrolyte (e.g., I_3^-/I^-) before reaching the dye-sensitized TiO_2 nanotube, leading to a significantly less efficient absorption of the incident light.

In comparison to transferring a freestanding TiO_2 nanotube film onto the transparent, conductive substrate for use in DSSCs, an intimate contact between TiO_2 nanotube arrays and the transparent conductive substrate is readily present when anodizing Ti thin film that is predeposited on the substrate by sputtering [e.g., radio frequency (RF)-sputter deposition] [134, 135]. Interestingly, the as-prepared black TiO_2 nanotube film obtained after anodization transforms into a transparent film after high-temperature annealing to induce crystallinity [130, 135]. The optically transparent TiO_2 nanotube arrays have also been used as sensing elements for hydrogen sensing, exhibiting very promising performance compared to the nanotube arrays formed on Ti substrates [136]. It is worth noting that the attainable thickness of TiO_2 nanotube arrays is restricted to only a few hundred nanometers because of limitations on the ability to deposit thicker, high-quality Ti on the substrate under current experimental conditions (e.g., 360-nm-thick TiO_2 nanotubes obtained from anodization of 500-nm-thick pure Ti film deposited on the substrate) [46, 135]. Very recently, an elegant study has demonstrated that a 1000-nm-thick TiO_2 nanotube film can be fabricated by anodizing the deposited Ti film in an ethylene glycol electrolyte [137].

TiO_2 nanotube arrays are amorphous directly after anodization. Many potential applications require the use of crystallized TiO_2 because of its unique electrical and optical properties. Furthermore, specific crystalline phases are ideal for different applications; the anatase phase of TiO_2 is preferred in DSSC and photocatalysis, whereas the rutile form is mostly used in the area of dielectrics and high-temperature oxygen gas sensors [138]. To transform the amorphous TiO_2 into the crystalline phases (anatase, rutile, etc.), thermal annealing is often carried out. Notably, the TiO_2 nanotubular structures are retained after annealing at high temperature, and the phase conversion has been confirmed by X-ray diffraction. Additionally, Raman scattering can also be performed to determine the crystalline phases of TiO_2 nanotubes due to different crystalline phases having specific Raman signatures. For example, the anatase phase has Raman peaks at 145, 198, 399, 516, and 640 cm^{-1}, while the rutile phase has Raman peaks at 143, 240, 447, and 612 cm^{-1}. The amorphous TiO_2 nanotube arrays showed a broad spectrum; by contrast, the TiO_2 nanotubes annealed at 500°C in air exhibited characteristic peaks of the anatase phase [131, 139]. Thermal stability studies on the TiO_2 nanotube arrays showed that the nanotubular structure was stable with annealing temperature up to 600°C; with thermal treatment at higher temperatures, the nanotubular structure collapsed due to the expansion of crystalline domains in the nanotube walls [139]. When the annealing temperature was higher than 700°C, the crystalline phase transformed from anatase to rutile, as confirmed by Raman scattering [139]. According to the systematical studies on crystalline formation in TiO_2 nanotube arrays, an optimal annealing temperature was between 400 and 600°C. This heat treatment allowed the formation of highly crystalline anatase TiO_2 while maintaining the nanotubular structures [139]. Figure 15.12 shows the SEM characterization of highly ordered TiO_2 nanotube arrays and the formation of anatase phase as revealed by Raman scattering [49].

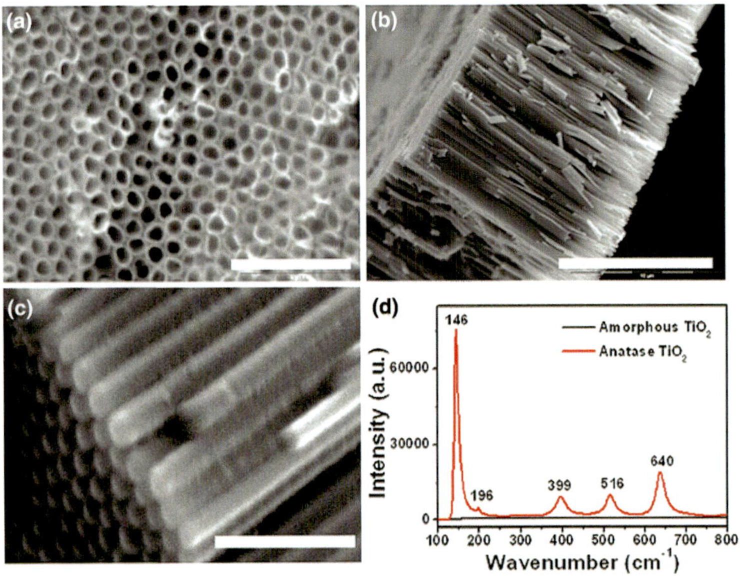

Fig. 15.12 Structural characterization of highly ordered TiO_2 nanotube arrays: (**a**) SEM topology, scale bar, 1 μm; (**b**) SEM cross-sectional view, scale bar, 10 μm; (**c**) SEM cross-sectional view, scale bar, 1 μm; and (**d**) Raman spectra of amorphous (*black curve*) and crystalline (*red curve*) TiO_2 nanotube arrays, where the Raman peaks at 145, 196, 399, 516, and 640 cm^{-1} suggest the formation of anatase TiO_2 (reprinted with permission from [49]. Copyright © American Chemical Society)

As noted above, highly ordered TiO_2 nanotube arrays have been used as a substitute for conventional TiO_2 nanoparticle films in DSSCs [43, 50, 73, 112, 129, 140, 141]. A 2.9% *PCE* was achieved using 360-nm-thick transparent TiO_2 nanotube arrays to assemble DSSCs in a frontside illumination mode [73]. A *PCE* as high as 6.89% has recently been achieved in a backside illumination mode using TiO_2 nanotube arrays that were grown on Ti foil with film thicknesses of 20 μm [112]. Very recently, the efficiency of dye-sensitized TiO_2 nanotube solar cells has been further improved to 7.37% upon TiO_2 surface treatment with $TiCl_4$ and oxygen plasma [49]. In highly ordered TiO_2 nanotube arrays, each nanotube orients perpendicular to the film surface and goes through the entire film, thus providing a separate channel for electron transport. Compared to the TiO_2 nanoparticle film-based DSSCs, enhanced charge collection efficiency and enhanced light scattering in TiO_2 nanotube array-based DSSCs were observed [50]. From the dye loading measurement, it was found that the dye loading in both TiO_2 nanotube arrays and TiO_2 nanoparticle films was similar for identical film thicknesses [50].

When CdSe QDs were used to sensitize TiO_2 [38, 74], higher *PCE* was obtained by using highly ordered TiO_2 nanotube arrays to replace the TiO_2 nanoparticle films [43]. Pre-synthesized CdSe QDs were molecularly linked to TiO_2 nanotube surfaces using the bifunctional molecule, MPA. Compared to mesoporous TiO_2 films, the difference in TiO_2 morphology had little effect on the charge injection rate but influenced the electron transport within the film. The maximum *IPCE* obtained with CdSe/TiO_2(NT) (45%) is greater than that with CdSe/TiO_2(NP) (35%). The nanotubular TiO_2 architecture provides a better scaffold for the construction of QD solar cells, and the photovoltaic performance can be further improved by fabricating rainbow QD-sensitized solar cells, in which QDs of different sizes are successively deposited into highly ordered TiO_2 nanotubes to render a full spectrum absorption of sunlight (Fig. 15.13) [43].

Chemical bath deposition (CBD) of CdS QDs was also carried out on highly ordered anodic TiO_2 nanotube arrays [48]. The CBD approach possesses several advantages, including sufficient filling of QDs in the nanotubes and intimate contact between QDs and the TiO_2 surface, as revealed by TEM imaging of the TiO_2

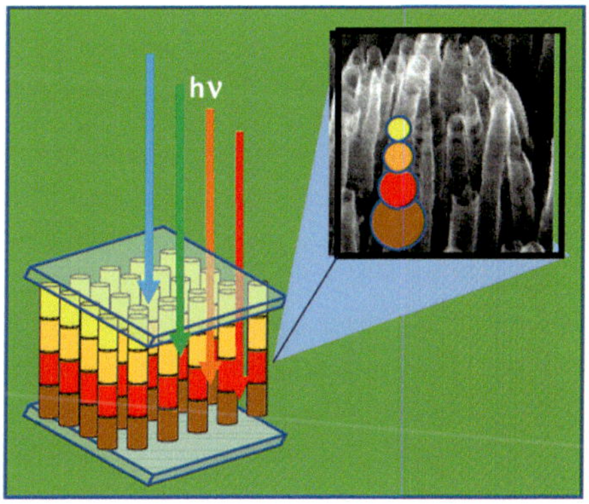

Fig. 15.13 A rainbow solar cell assembled with different-sized CdSe QDs on a TiO_2 nanotube array (reprinted with permission from [43]. Copyright © American Chemical Society)

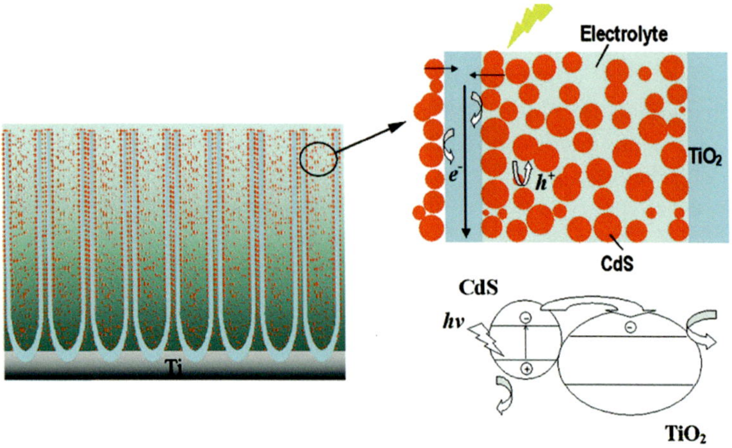

Fig. 15.14 Schematic illustration of the CdS-modified TiO_2 nanotube film electrode and charge transfer process between CdS and TiO_2 (reprinted with permission from [48]. Copyright © American Chemical Society)

nanotube after CBD deposition of CdS QDs [48]. Figure 15.14 shows the scheme of CdS QD-modified TiO_2 nanotube arrays and the charge transfer process between CdS and TiO_2. A cell efficiency of 4.15% was obtained from the CdS-modified TiO_2 nanotube arrays [48].

By taking advantage of recent advances in the preparation of stable Se^{2-} and Te^{2-} precursors, CdSe, PbSe, and CdTe QDs can be deposited into TiO_2 nanotube arrays through the SILAR process. Due to the flexibility of tuning the structural parameters of TiO_2 nanotube arrays, it is possible to obtain optimized filling of QDs as well as a balance between light absorption and charge transport. Thus enhanced solar energy conversion efficiency can be expected from using QDs to sensitize highly ordered TiO_2 nanotube arrays.

15.2.6 Investigation of Charge Injection in Quantum Dot-Sensitized Solar Cells

15.2.6.1 Generation of Excited Electrons

According to the quantum mechanics theory, the energy of an electron outside an atom is not continuous but quantized. There is a finite number of different energy levels available for an electron to reside in. These energy levels correspond to given atomic orbitals. When the atoms are bound together to form a large structure, such as a molecule, these atomic orbitals will interact with each other and split into new orbitals. The total number of the split orbitals is equal to the sum of the original atomic orbitals. When an extremely large number of atoms, for instance, 10^{20}, are bound together, the number of split orbitals is also extremely large. Therefore, the energy difference between two orbitals will be extremely small, and the finite energy levels can be considered as continuous bands. The band filled with valence electrons is called the valence band, while the unoccupied band is termed the conduction band. The Pauli exclusion principle indicates that only two electrons with different spin momentum can occupy one orbital. In the stationary (ground) state, which is the most stable state, the total energy of an atom reaches a minimum. All the electrons will stack from the lowest energy orbital to a higher energy orbital. The highest orbital occupied by an electron in the stationary state is called the highest occupied molecular orbital (HOMO) and the lowest unoccupied orbital is called the lowest unoccupied molecular orbital (LUMO). The energy difference between the HOMO and the LUMO is the bandgap.

For a semiconducting QD, such as CdSe, photons encounter atoms inside the semiconductor and transfer the energy to the atoms. If the energy of an incident photon is higher than the bandgap of atoms inside the QD, one of the electrons which absorbs the photon's energy will be excited from the valence band to conduction band, generating an excited electron and a hole. In order to absorb the visible light, the bandgap of semiconductor should be around 1–2 eV.

In general, the bandgap is intrinsically dictated by the material in question. For example, TiO_2 has a bandgap of 3.4 eV and ZnO has a bandgap of 3.3 eV [92]. Figure 15.15 shows the bandgaps of several semiconductors in contact with aqueous electrolyte at pH 1. Since QDs have small particle sizes, their bandgaps exhibit a strong size dependence. This phenomenon is attributed to the quantum confinement effect, which can be observed when the diameter of a particle is of the same magnitude as the wavelength of the electromagnetic wave incident upon it [142]. Since the number of orbitals is proportional to the number of atoms in the semiconductor nanoparticles, a smaller particle size will induce a smaller number of orbitals, which distances the energy between different orbitals, making the energy band discrete. As a result, the bandgap of QDs will be increased as the particle size decreases. Therefore, the smaller the size of QDs is, the larger energy the photons should have to excite the electrons inside the particle, i.e., a blueshift of the absorption spectrum occurs as the QD size decreases. An empirical sizing curve of CdX (X = S, Se, Te) QDs is shown in Fig. 15.16. The size of ultrasmall CdSe [143] and CdS [144] QDs was obtained by XRD characterization, and others were obtained by TEM measurement. The empirical fitting functions of the curves in Fig. 15.16 are given as follows [145]:

$$\text{CdTe: } 9.8127 \times 10^{-7}\lambda^3 - 1.7147 \times 10^{-3}\lambda^2 + 1.0064\lambda - 194.84$$

$$\text{CdSe: } D = 1.6122\times10^{-9}\lambda^4 - 2.6575\times10^{-6}\lambda^3 + 1.6242\times10^{-3}\lambda^2 - 0.4277\lambda + 41.57$$

$$\text{CdS: } D = -6.6521 \times 10^{-8}\lambda^3 + 1.9557 \times 10^{-4}\lambda^2 - 9.2352 \times 10^{-2}\lambda + 13.29$$

where D (nm) is the size of the QDs and λ (nm) is the wavelength of the first absorption peak of the sample.

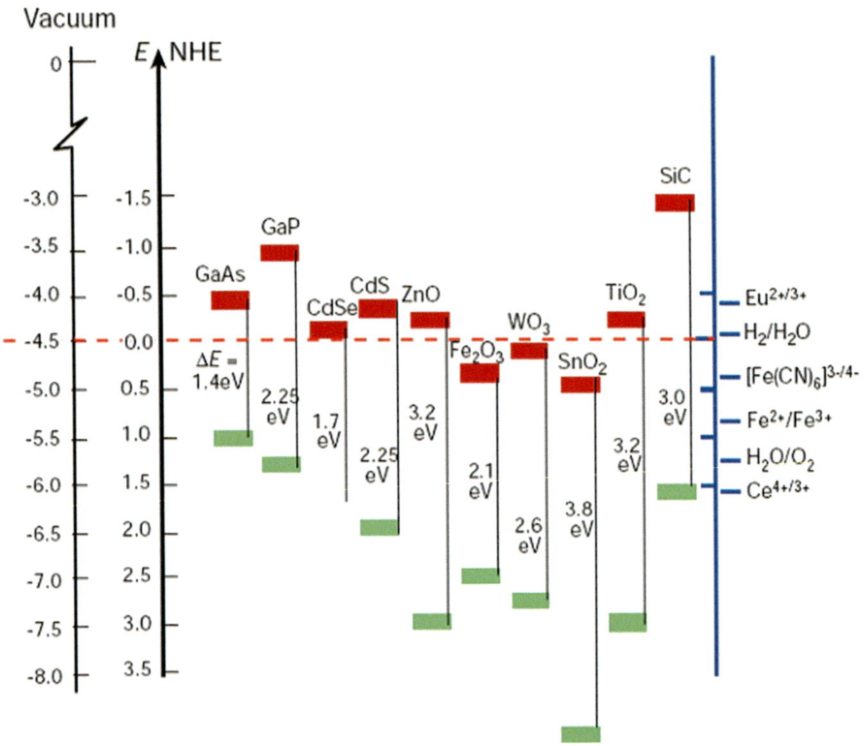

Fig. 15.15 Energy levels of several semiconductors in contact with aqueous electrolyte at pH 1 (reprinted with permission from [92]. Copyright © Nature Publishing Group)

In addition to tunable optical properties, QDs have the ability to utilize hot electrons [146] and generate multiple electrons upon absorption of a single photon of sufficient energy (i.e., MEG characteristic) [25, 30–34, 147]. The electron is multiplied through the impact ionization, which is the inverse process of Auger recombination (AR) [29, 31, 148]. In the AR process, when an excited electron is recombined with a hole, the energy transfers to an adjacent excited electron and pushes the latter electron to a higher excited position; while in the opposite process, one electron at a higher excited position relaxes to a lower excited position and transfers its energy to an electron at stationary state, thus exciting the stationary electron. Therefore, one photon with an energy level high enough to generate excited electrons at higher positions would have the ability to produce two or more excited electrons. MEG has been considered as an efficient way to increase photocurrent in QD-sensitized solar cells.

15.2.6.2 Recombination and Transportation of Excited Electrons

When an electron is excited from the valence band to the conduction band, a hole is generated in the valence band. The hole has a positive charge and strongly attracts

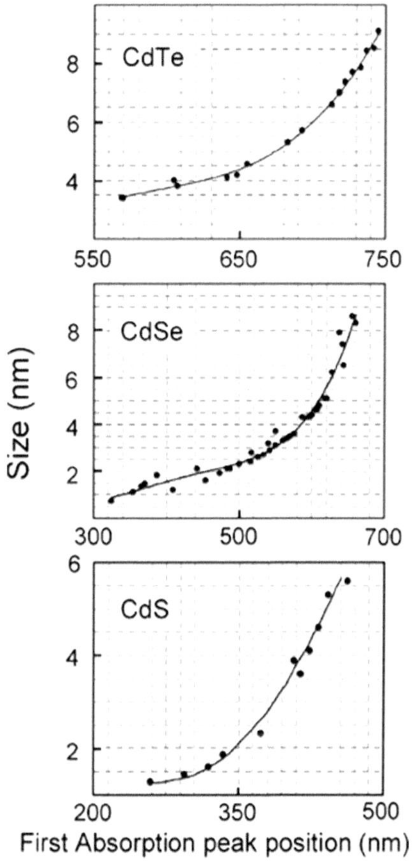

Fig. 15.16 Sizing curve of CdTe, CdSe, and CdS QDs (reprinted with permission from [145]. Copyright © American Chemical Society)

an excited electron, forcing the electron to recombine with it. In this case, the electron would stay inside the QD and not transfer to the electron acceptor. The very high *IPCE* of QDs (up to 80%) and relatively low external power conversion efficiency (*EPCE*) (less than 4%) indicates that only a very small amount of electrons generated in QDs can actually be transferred to the acceptor, and the rest recombine with holes inside the QDs. Since the recombination process needs the presence of both electrons and holes, charge separation is needed to decrease the recombination rate by separating the excited electrons and holes [66, 149–151].

Spontaneous charge separation at the donor–acceptor interface is the electron injection from QDs to the electron acceptor. The transfer rate of electrons from QDs to the acceptor is determined by several parameters, among which the difference of the level of conduction bands between QDs and the acceptor is the most important. The charge transfer rate can be estimated by comparing the lifetime of excited electrons inside QDs with and without the electron acceptor [74]:

$$k = \frac{1}{\tau_{\mathrm{QD+Electron\ Acceptor}}} - \frac{1}{\tau_{\mathrm{QD}}}$$

The kinetics of the electron transfer can be described using Marcus theory [152, 153] in a nonadiabatic reaction in the classical activation limit, which implies the electron transfer rate is exponential to a quadratic function of the driving force, i.e., the difference of the Gibbs free energy between QDs and the acceptor. The charge injection rate from QDs to the electron acceptor exhibits a strong size-dependent characteristic according to the transient absorption measurement [74]. The energy difference between QDs and the acceptor is determined by the difference of their conduction band energies. The larger the difference, the stronger the driving force, and, in turn, the faster the electron transfer rate [74].

Since the mass of an electron is much less than that of a hole, the shift in the conduction band is more significant than the shift in the valence band in QDs [154]. Therefore the conduction band energy can be expected to become more negative with a decrease in the QD size, as shown in Fig. 15.17 [1]. The electron transfer rate increases as the energy difference increases. The change of CdSe QD size from 7.5 to 2.4 nm resulted in a change in the energy difference from 0.3 to 0.8 eV, resulting in as high as three orders of magnitude increase in the transfer rate, i.e., from 10^7 to 10^{10} s^{-1} (Fig. 15.17) [1]. The average electron lifetime in the 2.4-nm CdSe QDs is 83 ps (in the CdSe/TiO_2 system) as compared to 2–50 ps in the CdS/TiO_2 and CdSe/TiO_2 systems [155, 156].

In addition to the efficient electron injection from QDs to the acceptor, the removal of holes in QDs can also promote charge separation. While the electrons transfer from the higher to lower conduction band, the holes transfer from the lower to higher valance band under the same driving force of Gibbs free energy. Although the size variation cannot influence the valance band level as much as the conduction band level, the design toward using different materials with different energy levels to yield a core/shell or multilayer structure has been introduced to facilitate the charge separation by efficient extraction of holes in the valence band [87, 88]. Figure 15.18

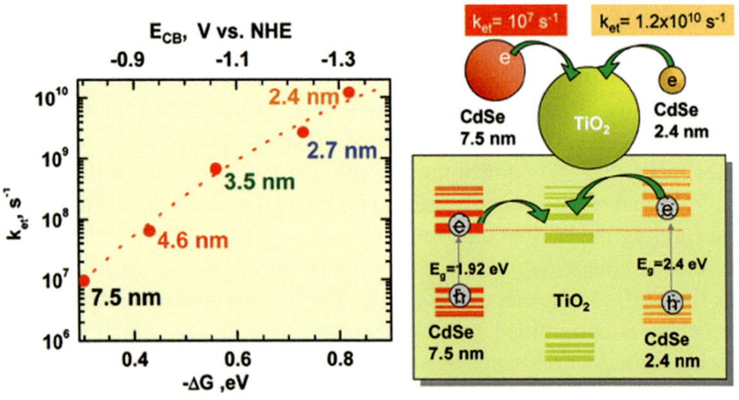

Fig. 15.17 *Left*: The dependence of electron transfer rate constant on the energy difference between the conduction bands of QDs and TiO_2. *Right*: Schematic illustration of the principle of electron transfer from two different-sized CdSe QDs to TiO_2 nanoparticles (reprinted with permission from [1]. Copyright © American Chemical Society)

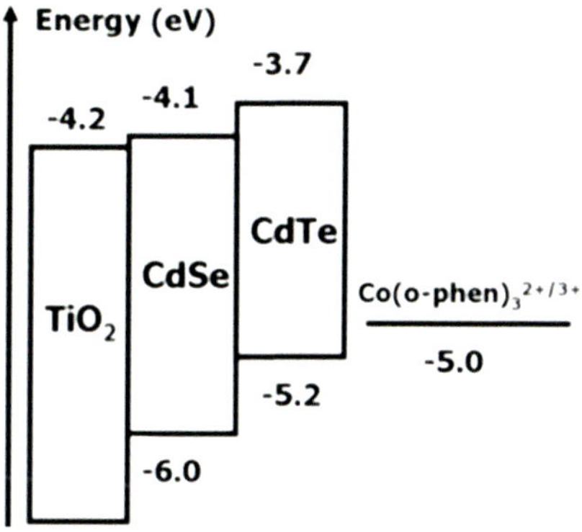

Fig. 15.18 Energy diagram of the bulk band offsets in the TiO_2/CdSe/CdTe system (reprinted with permission from [88]. Copyright © American Chemical Society)

shows an example of utilizing a thin layer of CdTe QDs to effectively extract holes in CdSe QDs, thereby allowing for enough time for the electron injection from CdSe QDs to TiO_2 [88]. The level of conduction bands for CdTe, CdSe, and TiO_2 is –3.7, –4.1, and –4.2 eV, respectively; thus the electrons generated in CdSe transfer to TiO_2. On the other hand, the level of valence bands for CdTe, CdSe, and TiO_2 is –5.0, –5.2, and –6.0 eV, respectively; therefore, the holes generated in CdSe transfer to CdTe. Compared to the CdSe/TiO_2 system with only the electron transfer, the extraction of holes by CdTe in the CdTe/CdSe/TiO_2 system lowered the concentration of holes in CdSe. As a result, the electron and hole recombination rate decreased.

Core/shell QDs have also been introduced for efficient charge separation [157]. In the case of CdSe/ZnS core/shell QDs, the conduction band of ZnS is higher than that of CdSe, while the valance band is lower, suggesting that both the electron and the hole generated from CdSe would have less chance to transfer outside the QDs through the ZnS barrier (Fig. 15.19). They are trapped inside the CdSe core. This trapping effect is not advantageous for charge separation and photocurrent generation. However, the ZnS shell works as a protection layer, increasing the half-life of the CdSe core from 2.4 to 84.1 min; a total of 21.2% initial efficiency is retained even after 20 h of continuous illumination, as compared to 90% loss during 2 h illumination of the CdSe core [157].

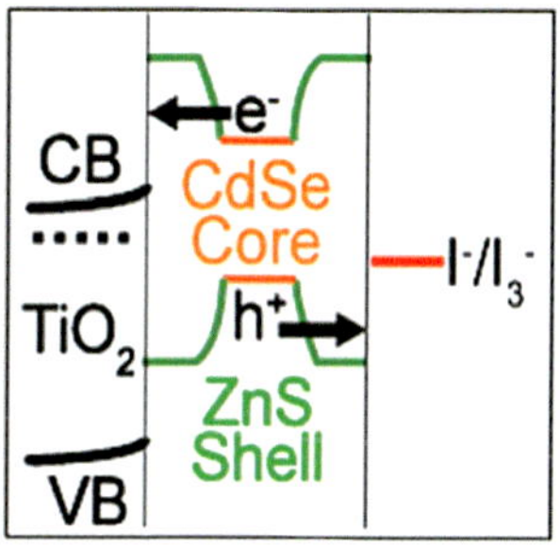

Fig. 15.19 Energy band position of CdSe/ZnS core/shell QDs (reprinted with permission from [157]. Copyright © American Chemical Society)

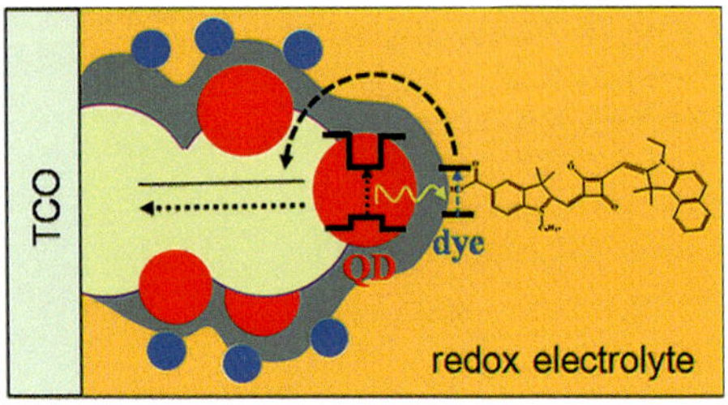

Fig. 15.20 Schematic representation of QD-sensitized solar cells based on nonradiative energy transfer (reprinted with permission from [158]. Copyright © American Chemical Society)

In order to take advantage of the protective ability of ZnS and increase the photocurrent, a novel design of QD-sensitized solar cells based on nonradiative Forster resonance energy transfer (FRET) has been proposed (Fig. 15.20) [158]. The excited electron and the hole within QDs are considered as an oscillating dipole donor, and the energy can be transferred to a nearby dipole acceptor (organic dye in this design) through resonance effects without direct contact of the two dipoles [159]. This transfer requires a short distance between the donor and the acceptor, typically 1–10 nm, to sustain the interaction. The excited donor is treated as an oscillating dipole, which transfers its energy to another dipole with the same resonance frequency. Namely, the fluorescence emission spectrum of the donor overlaps the absorption spectrum of the acceptor. CdSe was used as the donor QD, and unsymmetrical squaraine dye was used as the acceptor. The ZnS barrier prevented direct electron injection from the CdSe core to nanocrystalline TiO_2; thus all excited electrons in the CdSe donor were trapped, recombined, and transferred energy to the acceptor dye molecules. As a result, the dye was excited and transferred excited electrons to nanocrystallized TiO_2 through a thin layer of amorphous TiO_2 [158]. FRET may also take place at a QD film with a single composition but different size [160]. When QDs of different sizes are closely packed together, which may happen due to uncontrollable size under chemical bath deposition, the smaller QDs that have a larger bandgap and therefore higher emission energy would serve as donors to transfer energy to the acceptors (i.e., larger QDs) that have smaller bandgaps and lower absorption energies.

15.3 Outlook

QD-sensitized solar cells have the transformative potential to increase the maximum attainable conversion efficiency of solar energy because of the outstanding optical properties and MEG capability of QDs. However, optimization of QD-sensitized solar cells is needed to achieve higher *PCE* and long-term stability. Developing viable, revolutionary strategies to organize QD assemblies on electrode surfaces is important to improve the performance of QD-sensitized solar cells. These include the use of QD/carbon nanostructures to promote the electron injection from QDs to TiO_2; the deposition of more than one QD onto the TiO_2 surface to not only increase the light absorption but also facilitate the charge separation by efficient extraction of excited electrons and holes; the novel, efficient deposition techniques, such as SILAR, to better control the deposition process and QD quality; the use of highly ordered nanostructures for electron transport to enhance the charge collection efficiency while reducing the chance of charge recombination; the surface modification and activation of TiO_2 surfaces to further enhance the electronic interaction between QDs and electron acceptors; and the novel design (e.g., QD-sensitized solar cells based on FRET) to enhance the performance of the resulting solar cells. Therefore, future efforts should be focused on high-efficiency light absorption by QDs, efficient deposition/impregnation of QDs, improved electronic interaction between QDs and electron acceptors, high-performance hole-transporting electrolytes, etc. All these are the key to yielding ultimate high-efficiency QD-sensitized solar cells. With increased interest in solar-to-electric energy conversion, and recent advances in QD synthesis and fabrication of novel three-dimensional nanostructures, we envision breakthroughs in developing QD-sensitized solar cells with enhanced performance at reduced cost.

References

1. P.V. Kamat, Quantum dot solar cells. Semiconductor nanocrystals as light harvesters. J. Phys. Chem. C **112**(48), 18737–18753 (2008)
2. B. Oregan, M. Gratzel, A low-cost, high-efficiency solar-cell based on dye-sensitized colloidal Tio2 films. Nature **353**(6346), 737–740 (1991)
3. K.M. Coakley, Y.X. Liu, C. Goh, M.D. McGehee, Ordered organic-inorganic bulk heterojunction photovoltaic cells. MRS Bull. **30**(1), 37–40 (2005)
4. K.M. Coakley, Y.X. Liu, M.D. McGehee, K.L. Frindell, G.D. Stucky, Infiltrating semiconducting polymers into self-assembled mesoporous titania films for photovoltaic applications. Adv. Funct. Mater. **13**(4), 301–306 (2003)
5. K.M. Coakley, M.D. McGehee, Photovoltaic cells made from conjugated polymers infiltrated into mesoporous titania. Appl. Phys. Lett. **83**(16), 3380–3382 (2003)
6. I. Gur, N.A. Fromer, A.P. Alivisatos, Controlled assembly of hybrid bulk-heterojunction solar cells by sequential deposition. J. Phys. Chem. B **110**(50), 25543–25546 (2006)
7. I. Gur, N.A. Fromer, M.L. Geier, A.P. Alivisatos, Air-stable all-inorganic nanocrystal solar cells processed from solution. Science **310**(5747), 462–465 (2005)
8. W.U. Huynh, J.J. Dittmer, A.P. Alivisatos, Hybrid nanorod-polymer solar cells. Science **295**(5564), 2425–2427 (2002)
9. W.U. Huynh, J.J. Dittmer, W.C. Libby, G.L. Whiting, A.P. Alivisatos, Controlling the morphology of nanocrystal-polymer composites for solar cells. Adv. Funct. Mater. **13**(1), 73–79 (2003)
10. W.U. Huynh, X.G. Peng, A.P. Alivisatos, CdSe nanocrystal rods poly(3-hexylthiophene) composite photovoltaic devices. Adv. Mater. **11**(11), 923–927 (1999)
11. D.J. Milliron, I. Gur, A.P. Alivisatos, Hybrid organic–nanocrystal solar cells. MRS Bull. **30**(1), 41–44 (2005)
12. R.C. Advincula, Hybrid organic-inorganic nanomaterials based on polythiophene dendronized nanoparticles. Dalton Trans (23), 2778–2784 (2006)
13. J. Xu, J. Wang, M. Mitchell, P. Mukherjee, M. Jeffries-El, J.W. Petrich, Z.Q. Lin, Organic-inorganic nanocomposites via directly grafting conjugated polymers onto quantum dots, J. Am. Chem. Soc. **129**(42), 12828–12833 (2007)

14. H. Skaff, K. Sill, T. Emrick, Quantum dots tailored with poly(para-phenylene vinylene), J. Am. Chem. Soc. **126**(36), 11322–11325 (2004)
15. W.L. Ma, A. Gopinathan, A.J. Heeger, Nanostructure of the interpenetrating networks in poly(3-hexylthiophene)/fullerene bulk heterojunction materials: Implications for charge transport. Adv. Mater. **19**(21), 3656–3659 (2007)
16. W.L. Ma, C.Y. Yang, X. Gong, K. Lee, A.J. Heeger, Thermally stable, efficient polymer solar cells with nanoscale control of the interpenetrating network morphology. Adv. Funct. Mater. **15**(10), 1617–1622 (2005)
17. W.L. Ma, C.Y. Yang, A.J. Heeger, Spatial fourier-transform analysis of the morphology of bulk heterojunction materials used in "plastic" solar cells. Adv. Mater. **19**(10), 1387–1390 (2007)
18. J.G. Xue, S.R. Forrest, Bipolar doping between a molecular organic donor-acceptor couple. Phys. Rev. B **69**(24), 245322 (2004)
19. J.G. Xue, B.P. Rand, S. Uchida, S.R. Forrest, Mixed donor-acceptor molecular heterojunctions for photovoltaic applications. II. Device performance. J. Apply. Phys. **98**(12), 124903–124909 (2005)
20. J.G. Xue, S. Uchida, B.P. Rand, S.R. Forrest, Asymmetric tandem organic photovoltaic cells with hybrid planar-mixed molecular heterojunctions. Appl. Phys. Lett. **85**(23), 5757–5759 (2004)
21. J.G. Xue, S. Uchida, B.P. Rand, S.R. Forrest, 4.2% efficient organic photovoltaic cells with low series resistances. Appl. Phys. Lett. **84**(16), 3013–3015 (2004)
22. K.L. Mutolo, E.I. Mayo, B.P. Rand, S.R. Forrest, M.E. Thompson, Enhanced open-circuit voltage in subphthalocyanine/C-60 organic photovoltaic cells. J. Am. Chem. Soc. **128**(25), 8108–8109 (2006)
23. M.C. Hanna, A.J. Nozik, Solar conversion efficiency of photovoltaic and photoelectrolysis cells with carrier multiplication absorbers. J. Apply. Phys. **100**(7), 074510 (2006)
24. W. Shockley, H.J. Queisser, Detailed balance limit of efficiency of P-N junction solar cells, J. Apply. Phys. **32**(3), 510–519 (1961)
25. R.J. Ellingson, M.C. Beard, J.C. Johnson, P.R. Yu, O.I. Micic, A.J. Nozik, A. Shabaev, A.L. Efros, Highly efficient multiple exciton generation in colloidal PbSe and PbS quantum dots. Nano Lett. **5**(5), 865–871 (2005)
26. A. Luque, A. Martí, A.J. Nozik, Solar cells based on quantum dots: multiple exciton generation and intermediate bands. MRS Bull. **32**, 236 (2007)
27. J.M. Luther, M.C. Beard, Q. Song, M. Law, R.J. Ellingson, A.J. Nozik, Multiple exciton generation in films of electronically coupled PbSe quantum dots. Nano Lett. **7**(6), 1779–1784 (2007)
28. J.E. Murphy, M.C. Beard, A.G. Norman, S.P. Ahrenkiel, J.C. Johnson, P.R. Yu, O.I. Micic, R.J. Ellingson, A.J. Nozik, PbTe colloidal nanocrystals: synthesis, characterization, and multiple exciton generation. J. Am. Chem. Soc. **128**(10), 3241–3247 (2006)
29. A.J. Nozik, Quantum dot solar cells. Physica. E **14**(1–2), 115 (2002)
30. R.D. Schaller, V.M. Agranovich, V.I. Klimov, High-efficiency carrier multiplication through direct photogeneration of multi-excitons via virtual single-exciton states. Nat. Phys. **1**(3), 189–194 (2005)
31. R.D. Schaller, V.I. Klimov, High efficiency carrier multiplication in PbSe nanocrystals: implications for solar energy conversion. Phys. Rev. Lett. **92**(18), 186601 (2004)
32. R.D. Schaller, M.A. Petruska, V.I. Klimov, Effect of electronic structure on carrier multiplication efficiency: comparative study of PbSe and CdSe nanocrystals. Appl. Phys. Lett. **87**(25), 253102 (2005)
33. R.D. Schaller, M. Sykora, S. Jeong, V.I. Klimov, High-efficiency carrier multiplication and ultrafast charge separation in semiconductor nanocrystals studied via time-resolved photoluminescence. J. Phys. Chem. B **110**(50), 25332–25338 (2006)
34. R.D. Schaller, M. Sykora, J.M. Pietryga, V.I. Klimov, Seven excitons at a cost of one: redefining the limits for conversion efficiency of photons into charge carriers. Nano Lett. **6**(3), 424–429 (2006)

35. A. Shabaev, A.L. Efros, A.J. Nozik, Multiexciton generation by a single photon in nanocrystals. Nano Lett. **6**(12), 2856–2863 (2006)
36. M.T. Trinh, A.J. Houtepen, J.M. Schins, T. Hanrath, J. Piris, W. Knulst, A.P.L.M. Goossens, L.D.A. Siebbeles, In spite of recent doubts carrier multiplication does occur in PbSe nanocrystals. Nano Lett. **8**(6), 1713–1718 (2008)
37. R. Plass, S. Pelet, J. Krueger, M. Gratzel, U. Bach, Quantum dot sensitization of organic-inorganic hybrid solar cells. J. Phys. Chem. B **106**(31), 7578–7580 (2002)
38. I. Robel, V. Subramanian, M. Kuno, P.V. Kamat, Quantum dot solar cells. Harvesting light energy with CdSe nanocrystals molecularly linked to mesoscopic TiO2 films. J. Am. Chem. Soc. **128**(7), 2385–2393 (2006)
39. P.R. Yu, K. Zhu, A.G. Norman, S. Ferrere, A.J. Frank, A.J. Nozik, Nanocrystalline TiO2 solar cells sensitized with InAs quantum dots. J. Phys. Chem. B **110**(50), 25451–25454 (2006)
40. A. Zaban, O.I. Micic, B.A. Gregg, A.J. Nozik, Photosensitization of nanoporous TiO2 electrodes with InP quantum dots. Langmuir **14**(12), 3153–3156 (1998)
41. C.A. Grimes, Synthesis and application of highly ordered arrays of TiO2 nanotubes. J. Mater. Chem. **17**(15), 1451–1457 (2007)
42. X.J. Feng, K. Shankar, O.K. Varghese, M. Paulose, T.J. Latempa, C.A. Grimes, Vertically aligned single crystal TiO2 nanowire arrays grown directly on transparent conducting oxide coated glass: synthesis details and applications. Nano Lett. **8**(11), 3781–3786 (2008)
43. A. Kongkanand, K. Tvrdy, K. Takechi, M. Kuno, P.V. Kamat, Quantum dot solar cells. Tuning photoresponse through size and shape control of CdSe-TiO2 architecture. J. Am. Chem. Soc. **130**(12), 4007 (2008)
44. K.S. Leschkies, R. Divakar, J. Basu, E. Enache-Pommer, J.E. Boercker, C.B. Carter, U.R. Kortshagen, D.J. Norris, E.S. Aydil, Photosensitization of ZnO nanowires with CdSe quantum dots for photovoltaic devices. Nano Lett. **7**(6), 1793 (2007)
45. B. Liu, E.S. Aydil, Growth of oriented single-crystalline rutile TiO2 nanorods on transparent conducting substrates for dye-sensitized solar cells. J. Am. Chem. Soc. **131**(11), 3985–3990 (2009)
46. M. Paulose, K. Shankar, O.K. Varghese, G.K. Mor, C.A. Grimes, Application of highly-ordered TiO2 nanotube-arrays in heterojunction dye-sensitized solar cells. J. Phys. D: Appl. Phys. **39**, 6 (2006)
47. M. Paulose, K. Shankar, O.K. Varghese, G.K. Mor, B. Hardin, C.A. Grimes, Backside illuminated dye-sensitized solar cells based on titania nanotube array electrodes. Nanotechnology **17**(5), 1446–1448 (2006)
48. W.T. Sun, Y. Yu, H.Y. Pan, X.F. Gao, Q. Chen, L.M. Peng, CdS quantum dots sensitized TiO2 nanotube-array photoelectrodes. J. Am. Chem. Soc. **130**(4), 1124 (2008)
49. J. Wang, Z.Q. Lin, Dye-sensitized TiO2 nanotube solar cells with markedly enhanced performance via rational surface engineering. Chem. Mater. **22**(2), 579–584 (2010)
50. K. Zhu, N.R. Neale, A. Miedaner, A.J. Frank, Enhanced charge-collection efficiencies and light scattering in dye-sensitized solar cells using oriented TiO2 nanotubes arrays. Nano Lett. **7**(1), 69–74 (2007)
51. A.P. Alivisatos, Semiconductor clusters, nanocrystals, and quantum dots. Science **271**, 933 (1996)
52. Y.D. Yin, A.P. Alivisatos, Colloidal nanocrystal synthesis and the organic–inorganic interface, Nature **437**, 664 (2005)
53. S. Coe, W.K. Woo, M. Bawendi, V. Bulovis, Electroluminescence from single monolayers of nanocrystals in molecular organic devices. Nature **420**, 800 (2002)
54. V.L. Colvin, M.C. Schlamp, A.P. Alivisatos, Light-emitting diodes made from cadmium selenide nanocrystals and a semiconducting polymer. Nature **370**, 354 (1994)
55. V.I. Klimov, A.A. Mikhailovsky, S. Xu, A. Malko, J.A. Hollingsworth, C.A. Leatherdale, H.J. Eisler, M.G. Bawendi, Optical gain and stimulated emission in nanocrystal quantum dots. Science **290**, 314 (2000)

56. J. Kimura, S. Maenosono, Y. Yamaguchi, Near-field optical recording on a CdSe nanocrystal thin film. Nanotechnology **14**, 69 (2003)
57. A.R. Clapp, I.L. Medintz, H.T. Uyeda, B.R. Fisher, E.R. Goldman, M.G. Bawendi, H. Mattoussi, Quantum dot-based multiplexed fluorescence resonance energy transfer. J. Am. Chem. Soc. **127**, 18212 (2005)
58. J. Lee, A.O. Govorov, N.A. Kotov, Bioconjugated superstructures of CdTe nanowires and nanoparticles: multistep cascade Forster resonance energy transfer and energy channeling. Nano Lett. **5**, 2063 (2005)
59. W.C.W. Chan, S. Nie, Quantum dot bioconjugates for ultrasensitive nonisotopic detection. Science **281**, 2016 (1998)
60. H. Mattoussi, J.M. Mauro, E.R. Goldman, G.P. Anderson, V.C. Sundar, F.V. Mikkulec, M.G. Bawendi, Self-assembly of CdSe-ZnS quantum dot bioconjugates using an engineered recombination protein. J. Am. Chem. Soc. **122**, 12142 (2000)
61. I.L. Medintz, H.T. Uyeda, E.R. Goldman, H. Mattoussi, Quantum dot bioconjugates for imaging, labelling and sensing. Nat. Mater. **4**, 435 (2005)
62. L. Shi, V. De Paoli, N. Rosenzweig, Z. Rosenzweig, Synthesis and application of quantum dots FRET-based protease sensors. J. Am. Chem. Soc. **128**, 10378 (2006)
63. C.Y. Zhang, Y. H., M. Kuroki, T. Wang, Single-quantum-dot-based DNA nanosensor. Nat. Mater. **4**, 826 (2005)
64. A.P. Alivisatos, The use of nanocrystals in biological detection. Nat. Biotechnol. **22**, 47 (2004)
65. D.R. Larson, W.R. Zipfel, R.M. Williams, S.W. Clark, M.P. Bruchez, F.W. Wise, W.W. Webb, Water-soluble quantum dots for multiphoton fluorescence imaging in vivo. Science **300**, 1434 (2003)
66. A.P. Alivisatos, Perspectives on the physical chemistry of semiconductor nanocrystals. J. Phys. Chem. **100**(31), 13226–13239 (1996)
67. X.G. Peng, T.E. Wilson, A.P. Alivisatos, P.G. Schultz, Synthesis and isolation of a homodimer of cadmium selenide nanocrystals. Angew. Chem. Int. Ed. **36**, 145 (1997)
68. C.L. Choi, K.J. Koski, S. Sivasankar, A.P. Alivisatos, Strain-dependent photoluminescence behavior of CdSe/CdS nanocrystals with spherical, linear, and branched topologies. Nano Lett. **9**(10), 3544–3549 (2009)
69. C.B. Murray, D.J. Norris, M.G. Bawendi, Synthesis and characterization of nearly monodisperse CdE (E = sulfur, selenium, tellurium) semiconductor nanocrystallites. J. Am. Chem. Soc. **115**, 8706 (1993)
70. X. Peng, L. Manna, W.D. Yang, J. Wickham, E. Scher, C. Kadavanich, A.P. Alivisatos, Shape control of CdSe nanocrystals. Nature **404**, 59 (2000)
71. Z.A. Peng, X.G. Peng, Formation of high-quality CdTe, CdSe, and CdS nanocrystals using CdO as precursor. J. Am. Chem. Soc. **123**(1), 183–184 (2001)
72. J. Wang, J. Xu, M.D. Goodman, Y. Chen, M. Cai, J. Shinar, Z.Q. Lin, A simple biphasic route to water soluble dithiocarbamate functionalized quantum dots. J. Mater. Chem. **18**(27), 3270–3274 (2008)
73. G.K. Mor, K. Shankar, M. Paulose, O.K. Varghese, C.A. Grimes, Use of highly-ordered TiO2 nanotube arrays in dye-sensitized solar cells. Nano Lett. **6**(2), 4 215 (2006)
74. I. Robel, M. Kuno, P.V. Kamat, Size-dependent electron injection from excited CdSe quantum dots into TiO2 nanoparticles. J. Am. Chem. Soc. **129**, 4136 (2007)
75. J. Aldana, Y.A. Wang, X.G. Peng, Photochemical instability of CdSe nanocrystals coated by hydrophilic thiols. J. Am. Chem. Soc. **123**, 8844 (2001)
76. F. Dubois, B. Mahler, B. Dubertret, E. Doris, C. Mioskowski, A versatile strategy for quantum dot ligand exchange. J. Am. Chem. Soc. **129**, 482–483 (2007)
77. W. Jiang, S. Mardyani, H. Fischer, W.C.W. Chan, Design and characterization of lysine cross-linked mercapto-acid biocompatible quantum dots. Chem. Mater. **18**, 872–878 (2006)
78. C. Querner, A. Benedetto, R. Demadrille, P. Rannou, P. Reiss, Carbodithioate-containing oligo- and polythiophenes for nanocrystals' surface functionalization. Chem. Mater. **18**, 4817 (2006)

79. C. Querner, P. Reiss, J. Bleuse, A. Pron, Chelating ligands for nanocrystals' surface functionalization. J. Am. Chem. Soc. **126**, 11574 (2004)
80. Y. Zhao, W. Perez-Segarra, Q. Shi, A. Wei, Dithiocarbamate assembly on gold. J. Am. Chem. Soc. **127**, 7328 (2005)
81. T.R. Zhang, J.P. Ge, Y.P. Hu, Y.D. Yin, A general approach for transferring hydrophobic nanocrystals into water. Nano Lett. **7**(10), 3203–3207 (2007)
82. H.J. Lee, J.H. Yum, H.C. Leventis, S.M. Zakeeruddin, S.A. Haque, P. Chen, S.I. Seok, M. Grazel, M.K. Nazeeruddin, CdSe quantum dot-sensitized solar cells exceeding efficiency 1% at full-sun intensity. J. Phys. Chem. C **112**(30), 11600–11608 (2008)
83. L.J. Diguna, Q. Shen, J. Kobayashi, T. Toyoda, High efficiency of CdSe quantum-dot-sensitized TiO2 inverse opal solar cells. Appl. Phys. Lett. **91**(2), 023116 (2007)
84. R.B. Kale, C.D. Lokhande, Systematic study on structural phase behavior of CdSe thin films. J. Phys. Chem. B **109**(43), 20288–20294 (2005)
85. O. Niitsoo, S.K. Sarkar, C. Pejoux, S. Ruhle, D. Cahen, G. Hodes, Chemical bath deposited CdS/CdSe-sensitized porous TiO2 solar cells. J. Photochem. Photobiol. A **181**(2–3), 306 (2006)
86. K. Okazaki, N. Kojima, Y. Tachibana, S. Kuwabata, T. Torimoto, One-step preparation and photosensitivity of size-quantized cadmium chalcogenide nanoparticles deposited on porous zinc oxide film electrodes. Chem. Lett. **36**(6), 712–713 (2007)
87. Y.L. Lee, Y.S. Lo, Highly efficient quantum-dot-sensitized solar cell based on co-sensitization of CdS/CdSe. Adv. Funct. Mater. **19**(4), 604–609 (2009)
88. H. Lee, M.K. Wang, P. Chen, D.R. Gamelin, S.M. Zakeeruddin, M. Gratzel, M.K. Nazeeruddin, Efficient CdSe quantum dot-sensitized solar cells prepared by an improved successive ionic layer adsorption and reaction process. Nano Lett. **9**(12), 4221–4227 (2009)
89. H.J. Lee, P. Chen, S.J. Moon, F. Sauvage, K. Sivula, T. Bessho, D.R. Gamelin, P. Comte, S.M. Zakeeruddin, S. Il Seok, M. Gratzel, M.K. Nazeeruddin, Regenerative PbS and CdS quantum dot sensitized solar cells with a cobalt complex as hole mediator. Langmuir **25**(13), 7602–7608 (2009)
90. C.H. Chang, Y.L. Lee, Chemical bath deposition of CdS quantum dots onto mesoscopic TiO2 films for application in quantum-dot-sensitized solar cells. Appl. Phys. Lett. **91**(5), 053503 (2007)
91. Y. Tachibana, H.Y. Akiyama, Y. Ohtsuka, T. Torimoto, S. Kuwabata, CdS quantum dots sensitized TiO2 sandwich type photoelectrochemical solar cells. Chem. Lett. **36**(1), 88–89 (2007)
92. M. Gratzel, Photoelectrochemical cells. Nature **414**(6861), 338–344 (2001)
93. H. Chen, K. Dai, T.Y. Peng, H.P. Yang, D. Zhao, Synthesis of thermally stable mesoporous titania nanoparticles via amine surfactant-mediated templating method. Mater. Chem. Phys. **96**(1), 176–181 (2006)
94. L.X. Song, Y.M. Lam, C. Boothroyd, P.W. Teo, One-step synthesis of titania nanoparticles from PS-P4VP diblock copolymer solution. Nanotechnology **18**(13), 135605 (2007)
95. P. Lin, C.Y. Xian, Z.P. Mao, Q. Zhu, X.L. Chen, L.M. Zheng, Synthesis and characterization of titania nanoparticles by microemulsion process. Indian J. Chem. A **45**(9), 2017–2020 (2006)
96. J. Chen, J.L. Song, X.W. Sun, W.Q. Deng, C.Y. Jiang, W. Lei, J.H. Huang, R.S. Liu, An oleic acid-capped CdSe quantum-dot sensitized solar cell. Appl. Phys. Lett. **94**(15), 153115 (2009)
97. Y.L. Lee, B.M. Huang, H.T. Chien, Highly efficient CdSe-sensitized TiO2 photoelectrode for quantum-dot-sensitized solar cell applications. Chem. Mater. **20**(22), 6903–6905 (2008)
98. T. Lopez-Luke, A. Wolcott, L.P. Xu, S.W. Chen, Z.H. Wen, J.H. Li, E. De La Rosa, J.Z. Zhang, Nitrogen-doped and CdSe quantum-dot-sensitized nanocrystalline TiO2 films for solar energy conversion applications. J. Phys. Chem. C **112**(4), 1282 (2008)

99. J.L. Blackburn, D.C. Selmarten, A.J. Nozik, Electron transfer dynamics in quantum dot/titanium dioxide composites formed by in situ chemical bath deposition. J. Phys. Chem. B **107**(51), 14154–14157 (2003)
100. S.Q. Fan, D. Kim, J.J. Kim, D.W. Jung, S.O. Kang, J. Ko, Highly efficient CdSe quantum-dot-sensitized TiO2 photoelectrodes for solar cell applications. Electrochem. Commun. **11**(6), 1337–1339 (2009)
101. S.C. Lin, Y.L. Lee, C.H. Chang, Y.J. Shen, Y.M. Yang, Quantum-dot-sensitized solar cells: assembly of CdS-quantum-dots coupling techniques of self-assembled monolayer and chemical bath deposition. Appl. Phys. Lett. **90**, 143517 (2007)
102. V. Chakrapani, K. Tvrdy, P.V. Kamat, Modulation of electron injection in CdSe-TiO2 system through medium alkalinity. J. Am. Chem. Soc. **132**(4), 1228–1229 (2010)
103. P. Brown, P.V. Kamat, Quantum dot solar cells. Electrophoretic deposition of CdSe-C-60 composite films and capture of photogenerated electrons with nC(60) cluster shell. J. Am. Chem. Soc. **130**(28), 8890–8891 (2008)
104. B. Farrow, P.V. Kamat, CdSe quantum dot sensitized solar cells shuttling electrons through stacked carbon nanocups. J. Am. Chem. Soc. **131**(31), 11124–11131 (2009)
105. G. Yu, J. Gao, J.C. Hummelen, F. Wudl, A.J. Heeger, Polymer photovoltaic cells – enhanced efficiencies via a network of internal donor-acceptor heterojunctions. Science **270**(5243), 1789–1791 (1995)
106. M. Law, L.E. Greene, J.C. Johnson, R. Saykally, P.D. Yang, Nanowire dye-sensitized solar cells. Nat. Mater. **4**(6), 455–459 (2005)
107. J.B. Baxter, E.S. Aydil, Nanowire-based dye-sensitized solar cells. Appl. Phys. Lett. **86**(5), 053114 (2005)
108. S.Y. Lu, Z. Lingley, T. Asano, D. Harris, T. Barwicz, S. Guha, A. Madhukar, Photocurrent induced by nonradiative energy transfer from nanocrystal quantum dots to adjacent silicon nanowire conducting channels: toward a new solar cell paradigm. Nano Lett. **9**(12), 4548–4552 (2009)
109. C.K. Xu, P. Shin, L.L. Cao, D. Gao, Preferential growth of long ZnO nanowire array and its application in dye-sensitized solar cells. J. Phys. Chem. C **114**(1), 125–129 (2010)
110. B. Pradhan, S.K. Batabyal, A.J. Pal, Vertically aligned ZnO nanowire arrays in Rose Bengal-based dye-sensitized solar cells. Sol. Energy Mater. Sol. C **91**(9), 769–773 (2007)
111. S. Lu, A. Madhukar, Nonradiative resonant excitation transfer from nanocrystal quantum dots to adjacent quantum channels. Nano Lett. **7**(11), 3443–3451 (2007)
112. K. Shankar, G.K. Mor, H.E. Prakasam, S. Yoriya, M. Paulose, O.K. Varghese, C.A. Grimes, Highly-ordered TiO2 nanotube arrays up to 220 μm in length: use in water photoelectrolysis and dye-sensitized solar cells. Nanotechnology **18**, 065707 (2007)
113. G.K. Mor, M.A. Carvalho, O.K. Varghese, M.V. Pishko, C.A. Grimes, A room-temperature TiO2-nanotube hydrogen sensor able to self-clean photoactively from environmental contamination. J. Mater. Res. **19**(2), 7 (2004)
114. O.K. Varghese, D. Gong, M. Paulose, K.G. Ong, E.C. Dickey, C.A. Grimes, Extreme changes in the electrical resistance of titania nanotubes with hydrogen exposure. Adv. Mater. **15**(7–8), 624–627 (2003)
115. O.K. Varghese, D. Gong, M. Paulose, K.G. Ong, C.A. Grimes, Hydrogen sensing using titania nanotubes. Sensor. Actuat. B: Chem. **93**, 338 (2003)
116. G.K. Mor, K. Shankar, M. Paulose, O.K. Varghese, C.A. Grimes, Enhanced photocleavage of water using titania nanotube arrays. Nano Lett. **5**(1), 191 (2005)
117. O.K. Varghese, C.A. Grimes, Appropriate strategies for determining the photoconversion efficiency of water photo electrolysis cells: A review with examples using titania nanotube array photoanodes. Sol. Energy Mater. Sol. C **92**(4), 374–384 (2008)
118. O.K. Varghese, M. Paulose, K. Shankar, G.K. Mor, C.A. Grimes, Water-photolysis properties of micron-length highly-ordered titania nanotube-arrays. J. Nanosci. Nanotechnol. **5**(7), 1158–1165 (2005)

119. S.P. Albu, A. Ghicov, J.M. Macak, R. Hahn, P. Schmuki, Self-organized, free-standing TiO2 nanotube membrane for flow-through photocatalytic applications. Nano Lett. **7**(5), 1286 (2007)
120. J.M. Macak, M. Zlamal, J. Krysa, P. Schmuki, Self-organized TiO2 nanotube layers as highly efficient photocatalysts. Small **3**(2), 300–304 (2007)
121. J. Park, S. Bauer, P. Schmuki, K. von der Mark, Narrow window in nanoscale dependent activation of endothelial cell growth and differentiation on TiO2 nanotube surfaces. Nano Lett. **9**(9), 3157–3164 (2009)
122. J. Park, S. Bauer, K. von der Mark, P. Schmuki, Nanosize and vitality: TiO2 nanotube diameter directs cell fate. Nano Lett. **7**(6), 1686 (2007)
123. D. Gong, C.A. Grimes, O.K. Varghese, W. Hu, R.S. Singh, Z. Chen, E.C. Dickey, Titanium oxide nanotube arrays prepared by anodic oxidation. J. Mater. Res. **16**(2), 3331 (2001)
124. J.M. Macak, H. Tsuchiya, P. Schmuki, High-aspect-ratio TiO2 nanotubes by anodization of titanium. Angew. Chem. Int. Ed. **44**, 2100 (2005)
125. J.M. Macak, H. Tsuchiya, L. Taveira, S. Aldabergerova, P. Schmuki, Smooth anodic TiO2 nanotubes. Angew. Chem. Int. Ed. **44**, 7463 (2005)
126. H.E. Prakasam, K. Shankar, M. Paulose, O.K. Varghese, C.A. Grimes, A new benchmark for TiO2 nanotube array growth by anodization. J. Phys. Chem. C **111**(20), 7235–7241 (2007)
127. K. Shankar, G.K. Mor, A. Fitzgerald, C.A. Grimes, Cation effect on the electrochemical formation of very high aspect ratio TiO2 nanotube arrays in formamide – water mixtures. J. Phys. Chem. C **111**(1), 21–26 (2007)
128. V. Zwilling, E. Darque-Ceretti, A. Boutry-Forveille, D. David, M.Y. Perrin, M. Aucouturier, Structure and physicochemistry of anodic oxide films on titanium and TA6V alloy. Surf. Interface. Anal. **27**, 9 (1999)
129. G.K. Mor, O.K. Varghese, M. Paulose, K. Shankar, C.A. Grimes, A review on highly ordered, vertically oriented TiO2 nanotube arrays: fabrication, material properties, and solar energy applications. Sol. Energy Mater. Sol. C **90**(14), 2011–2075 (2006)
130. J. Wang, Z.Q. Lin, Anodic formation of ordered TiO2 nanotube arrays: effects of electrolyte temperature and anodization potential. J. Phys. Chem. C **113**(10), 4026–4030 (2009)
131. J. Wang, Z.Q. Lin, Freestanding TiO2 nanotube arrays with ultrahigh aspect ratio via electrochemical anodization. Chem. Mater. **20**(4), 1257–1261 (2008)
132. M. Paulose, H.E. Prakasam, O.K. Varghese, L. Peng, K.C. Popat, G.K. Mor, T.A. Desai, C.A. Grimes, TiO2 nanotube arrays of 1000 mu m length by anodization of titanium foil: phenol red diffusion. J. Phys. Chem. C **111**(41), 14992–14997 (2007)
133. J.H. Lim, J. Choi, Titanium oxide nanowires originating from anodically grown nanotubes the bamboo-splitting model. Small **3**, 1504 (2007)
134. J.M. Macak, H. Tsuchiya, S. Berger, S. Bauer, S. Fujimoto, P. Schmuki, On wafer TiO2 nanotube-layer formation by anodization of Ti-films on Si. Chem. Phys. lett. **428**(4–6), 421–425 (2006)
135. G.K. Mor, O.K. Varghese, M. Paulose, C.A. Grimes, Transparent highly ordered TiO2 nanotube arrays via anodization of titanium thin films. Adv. Funct. Mater. **15**, 1291 (2005)
136. G.K. Mor, O.K. Varghese, M. Paulose, K.G. Ong, C.A. Grimes, Fabrication of hydrogen sensors with transparent titanium oxide nanotube-array thin films as sensing elements. Thin Solid Films **496**, 7 (2006)
137. K. Shankar, J. Bandara, M. Paulose, H. Wietasch, O.K. Varghese, G.K. Mor, T.A. Latempa, M. Thelakkat, C.A. Grimes, Highly efficient solar cells using TiO2 nanotube arrays sensitized with a donor-antenna dye. Nano Lett. **8**, 1654–1659 (2008)
138. O.K. Varghese, D.W. Gong, M. Paulose, C.A. Grimes, E.C. Dickey, Crystallization and high-temperature structural stability of titanium oxide nanotube arrays. J. Mater. Res. **18**, 156 (2003)
139. J. Wang, L. Zhao, V.S.Y. Lin, Z.Q. Lin, Formation of various TiO2 nanostructures from electrochemically anodized titanium. J. Mater. Chem. **19**(22), 3682–3687 (2009)

140. S.H. Kang, J.Y. Kim, Y. Kim, H.S. Kim, Y.E. Sung, Surface modification of stretched TiO2 nanotubes for solid-state dye-sensitized solar cells. J. Phys. Chem. C **111**(26), 9614–9623 (2007)
141. K. Zhu, T.B. Vinzant, N.R. Neale, A.J. Frank, Removing structural disorder from oriented TiO2 nanotube arrays: reducing the dimensionality of transport and recombination in dye-sensitized solar cells. Nano Lett. **7**(12), 3739–3746 (2007)
142. H. Haug, S.W. Koch, *Quantum theory of the optical and electronic properties of semiconductors* (World Scientific Publishing Company, Hackensack, NJ, 1994)
143. V.N. Soloviev, A. Eichhofer, D. Fenske, U. Banin, Molecular limit of a bulk semiconductor: size dependence of the "band gap" in CdSe cluster molecules. J. Am. Chem. Soc. **122**(11), 2673–2674 (2000)
144. T. Vossmeyer, L. Katsikas, M. Giersig, I.G. Popovic, K. Diesner, A. Chemseddine, A. Eychmuller, H. Weller, Cds nanoclusters – synthesis, characterization, size-dependent oscillator strength, temperature shift of the excitonic-transition energy, and reversible absorbency shift. J. Phys. Chem. **98**(31), 7665–7673 (1994)
145. W.W. Yu, L.H. Qu, W.Z. Guo, X.G. Peng, Experimental determination of the extinction coefficient of CdTe, CdSe, and CdS nanocrystals, Chem. Mater. **15**(14), 2854–2860 (2003)
146. R.T. Ross, A.J. Nozik, Efficiency of hot-carrier solar-energy converters. J. Apply. Phys. **53**(5), 3813–3818 (1982)
147. M. Califano, A. Zunger, A. Franceschetti, Efficient inverse auger recombination at threshold in CdSe nanocrystals. Nano Lett. **4**(3), 525–531 (2004)
148. A.J. Nozik, Spectroscopy and hot electron relaxation dynamics in semiconductor quantum wells and quantum dots. Annu. Rev. Phys. Chem. **52**, 193–231 (2001)
149. A. Henglein, Small-particle research - physicochemical properties of extremely small colloidal metal and semiconductor particles. Chem. Rev. **89**(8), 1861–1873 (1989)
150. P.V. Kamat, Photochemistry on nonreactive and reactive (semiconductor) surfaces. Chem. Rev. **93**(1), 267–300 (1993)
151. P.V. Kamat, D. Meisel, *Semiconductor nanoclusters – physical, chemical, and catalytic aspects – Introduction*, vol. 103 (Elsevier Science Publ B V., Amsterdam, 1997), pp. 1–3
152. R.A. Marcus, On theory of electron-transfer reactions .6. Unified treatment for homogeneous and electrode reactions. J. Chem. Phys. **43**(2), 679–701 (1965)
153. R.A. Marcus, N. Sutin, Electron transfers in chemistry and biology. Biochim. Biophys. Acta **811**(3), 265–322 (1985)
154. D.J. Norris, M.G. Bawendi, Measurement and assignment of the size-dependent optical spectrum in CdSe quantum dots. Phys. Rev. B **53**(24), 16338–16346 (1996)
155. J.L. Blackburn, R.J. Ellingson, O.I. Micic, A.J. Nozik, Electron relaxation in colloidal InP quantum dots with photogenerated excitons or chemically injected electrons. J. Phys. Chem. B **107**(1), 102–109 (2003)
156. Q. Shen, K. Katayama, M. Yamaguchi, T. Sawada, T. Toyoda, Study of ultrafast carrier dynamics of nanostructured TiO2 films with and without CdSe quantum dot deposition using lens-free heterodyne detection transient grating technique. Thin Solid Films **486**(1–2), 15–19 (2005)
157. J.B. Sambur, B.A. Parkinson, CdSe/ZnS core/shell quantum dot sensitization of low index TiO2 single crystal surfaces. J. Am. Chem. Soc. **132**(7), 2130–2131 (2010)
158. S. Buhbut, S. Itzhakov, E. Tauber, M. Shalom, I. Hod, T. Geiger, Y. Garini, D. Oron, A. Zaban, Built-in quantum dot antennas in dye-sensitized solar cells. ACS Nano (2010). doi: 10.1021/nn100021b
159. T. Forster, Zwischenmolekulare Energiewanderung Und Fluoreszenz. Ann. Phys. **2**(1–2), 55–75 (1948)
160. C.R. Kagan, C.B. Murray, M. Nirmal, M.G. Bawendi, Electronic energy transfer in CdSe quantum dot solids (vol 76, pg 1517, 1996). Phys. Rev. Lett. **76**(16), 3043–3043 (1996)

Chapter 16
Three-Dimensional Photovoltaic Devices Based on Vertically Aligned Nanowire Array

Kai Wang, Jiajun Chen, Satish Chandra Rai, and Weilie Zhou

16.1 Introduction

The development and application of nanotechnology in renewable energy has opened up new ways to pursue next-generation solar cells which can deliver high efficiency at an economically viable cost [1–2]. A number of nano-photovoltaic (PV) concepts based on semiconductor nanowires have been developed or proposed in recent years, with either inorganic–organic hybrid [3–6] or all-inorganic approaches [7–11]. Among these concepts, of great importance is the use of quasi-one-dimensional nanowire/nanorod array to construct three-dimensional architectures as building blocks for solar light harvesting. For photogenerated carrier collection, the quasi-one-dimensional system structure is perhaps the optimized choice for optoelectronic devices such as solar cells and photodetectors, because it allows for maximally taking the advantages of reduced dimensionality while retaining the last and only needed conduction channel. Besides the possibility of exploring quantum effects when reaching the nanoscopic scale [9–10], even in the mesoscopic scale where the lateral size falls below the carrier diffusion length, the quasi-one-dimensional system could be superior to the bulk material, for instance, by reducing the non-radiative recombination and carrier scattering loss [12–13], through elimination of the unnecessary lateral transport and the resulting recombination loss [14–15].

More importantly, the efficiency of photon capture in the nanowire array structure, namely the light absorption, could be significantly improved by suppressing the reflection. In principle, when illuminated by the sunlight, the high-density nanowire array with the subwavelength scale in dimension can effectively trap incident light due to multiple scattering [16]. Meanwhile, the geometrical features of the wire array enable to harvest the sunlight over a wide range of spectral region and incidence angles. In this regard, nanowire array can act as efficient antireflection coatings (ARCs), in analogy with the specially designed texturing surfaces in

W.L. Zhou (✉)
Advanced Materials Research Institute, University of New Orleans, New Orleans, LA 70148, USA
e-mail: wzhou@uno.edu

W.L. Zhou, Z.L. Wang (eds.), *Three-Dimensional Nanoarchitectures*,
DOI 10.1007/978-1-4419-9822-4_16,

most thin film PV devices [17–18]. In addition, tailoring the light absorption of the nanowire array can be achieved by deliberately choosing appropriate materials and varying the parameters of nanowire including the wire dimensions (length and diameter), shapes, arrangements, and filling ratios [19–20]. Specifically, by simulating the silicon nanowire with a diameter of 50 nm in light harvesting, longer silicon nanowire with large filling ratio was predicted to enhance absorption of lower energy photons [21].

Additionally, the use of vertically aligned nanowire array as an active layer is predicted to reduce the stringent requirements of material purity, volume, and processing difficulty, holding the promise to produce low-cost freestanding PV devices on various flexible substrates. Novel techniques, especially peeling off the nanowire array from substrates, offer the opportunities to recycle the expensive substrates for nanowire synthesis and improve the mechanical stability of the devices [20, 22].

In the past several years, the nanowire arrays have emerged as the building blocks for fabricating various PV devices with different operation mechanisms in different device prototypes. Most of the work has been focused on dye-sensitized solar cells (DSSCs) and quantum dot-sensitized solar cells (QDSSCs) based on wide bandgap semiconductor nanowire/nanotubes array, such as TiO_2 [23–26] and ZnO [2, 27–28], addressing the functionalities of nanowire array as the ideal channel for effective carrier transport and the large aspect ratio medium for sensitizers/absorbers adsorption. In this chapter, we will primary focus on the development of all-inorganic three-dimensional PV devices based on the vertically aligned nanowire array, with particular emphasis on the synthesis and PV devices of type II core–shell nanowire array conducted by our group. The achievements, challenges, and prospects of DSSCs, QDSSCs, and other electrochemical (PEC) solar cells processed from nanowire array, which could be referred in related reviews [29–32], are not covered. Depending on the device configurations, subtopics will be addressed by four different types of PV devices: (i) nanowire array integrated with the substrate, (ii) nanowire array with axial junctions, (iii) nanowire array embedded in thin film, and (iv) core–shell nanowire array. The four types of devices, as schematically illustrated in Fig. 16.1, can also be classified as the axial junction PV devices (Fig. 16.1a, b) and radial junction PV devices (Fig. 16.1c, d). The challenges and prospects involved in exploration of the three-dimensional PV devices are also discussed.

16.2 Photovoltaic Devices Based on Nanowire Array Integrated with the Substrate

In addition to the enhanced light absorption, the integration of a vertically aligned nanowire array on the heteroepitaxial growth substrate could benefit from the single-crystal nature and aligned geometry of nanowire array, which facilitates a fast and

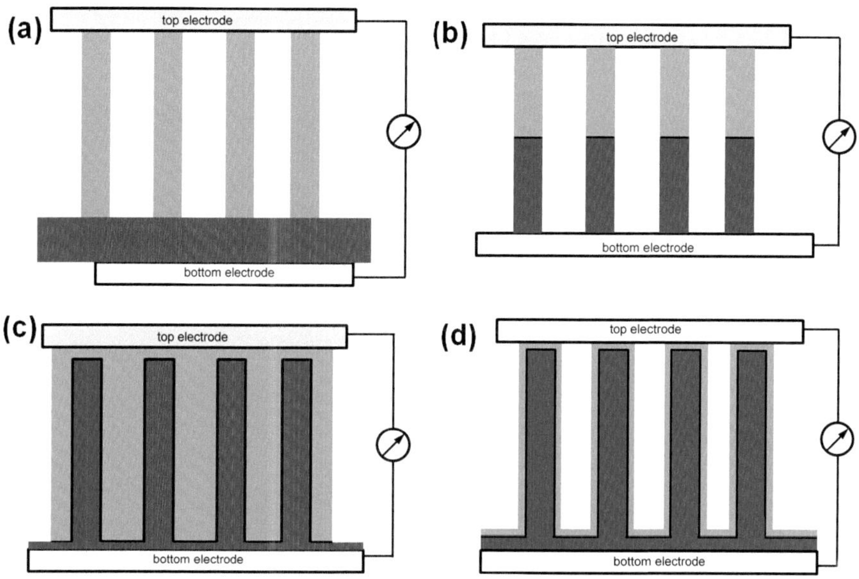

Fig. 16.1 Schematic configuration of three-dimensional PV devices based on (**a**) nanowire array integrated with the substrate, (**b**) nanowire array with axial junctions, (**c**) nanowire array embedded in thin film, and (**d**) core–shell nanowire array

direct conduction of the photogenerated carrier. Despite the challenges remaining in materials synthesis, of special interest is the integration of III–V nanowire arrays on silicon substrate because of the possibility of absorption spectra engineering by material selections. Recently, the realization of direct heteroepitaxial growth of vertically aligned III–V nanowire arrays, such as GaN, InP, GaAs, and InAs [33–35], on silicon substrate paves the way to succeed in integrating the nanowires and the substrate for PV devices fabrication. Tang et al. have fabricated p–n heterojunction PV cells by integrating a vertically aligned Mg-doped GaN nanorods epitaxially grown on n-type Si substrate [36]. Single-crystalline GaN nanorod array was found to act effectively as an antireflection coating to reduce the reflection in visible spectral region and a window layer to enhance spectral response at short-wavelength region (see Fig. 16.2a). The device exhibited well-defined rectifying behavior in the dark with low reverse leakage current and presented a high open-circuit voltage (V_{oc}) of ~0.95 V and a short-circuit current density (J_{sc}) of ~7.6 mA/cm^2. The fill factor (FF) and the conversion efficiency (η) were calculated to be ~0.38 and ~2.73%, respectively, under AM1.5G condition. The conversion efficiency is expected to be further improved by optimizing the electrode thickness because only half of the light would transmit the top electrodes. p–n heterojunction PV devices processed from vertically p-type InAs nanowire array grown heteroepitaxially on n-type silicon substrate were also demonstrated, showing a broad spectral response

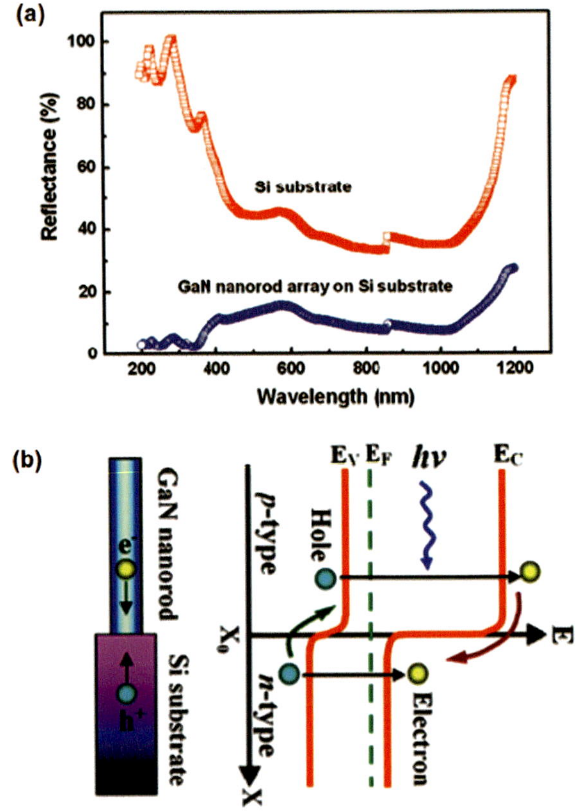

Fig. 16.2 (**a**) Reflectance measurements of the GaN nanorod array on Si substrate and the bare Si substrate, showing the GaN nanowire array suppresses the reflectance over a broad wavelength region. (**b**) Schematic illustration of the separation of photogenerated minority carriers from majority carriers by the built-in electric field across the heterojunction. Adapted with permission from [36] (Copyright 2008 American Chemical Society)

from the visible to the infrared region. The energy conversion efficiency of the typical device was achieved as 2.5% at 110 K and the open-circuit voltage showed temperature-dependent characteristic [37]. In addition, direct integration of p-type GaAs nanowire array on n-type GaAs substrate as PV device was also reported [38]. The device performance was highly dependent on the morphology and doping profile of the nanowire array, which were influenced by the molecular beam epitaxy (MBE) growth temperature. The maximum conversion efficiency was determined as 1.65 % and it is anticipated to be further enhanced by manipulating the beryllium (Be) doping level.

In general, integration of the III–V semiconductor nanowire array with the Si substrate, forming semiconducting heterojunctions, can absorb a different wavelength of the solar spectrum [39]. Significantly, one distinguished feature of this device configuration is that it can easily be incorporated into the existing microelectromechanical systems (MEMS) based on the well-developed silicon technology without complicated procedures. However, before being collected by the electrodes, the photoexcited carrier still travels long distance (see Fig. 16.2b). In other words, the recombination in this device configuration cannot be substantially suppressed, similar to the planar junction solar cells.

16.3 Photovoltaic Devices Based on Nanowire Array with Axial Junctions

Currently, despite the success on PV devices processed from a single nanowire with axial junction [40–41], there are few reports about PV devices based on axial nanowire array, partly because of the limitations of the geometry in charge collections and the challenges in most gas phase routes for materials synthesis. Similar to the device configuration of the nanowire array integrated with the substrate, the charge recombination was not efficiently suppressed during transportation to the corresponding electrodes. The sharp axial p–n junctions and precisely controlled doping level are highly required for efficient charge separation and transportation in this device configuration, which still remains a problem in most of the gas phase routes for nanowire growth. The chemical electrodeposition, which is a promising method in fabrication of the axial junction nanowire array by alternating deposition of the target materials in anodic aluminum oxide (AAO) templates, might suffer from relatively low crystalline quality. Consequently, chemical etching of the initial wafer with predetermined doping level was considered an efficient way to obtain the axial nanowire array. Of particular interest is the silicon nanowire array obtained by metal-assisted chemical etching route with silver, gold, and platinum as catalysts. V. Sivakov et al. reported the optical properties and cell performances based on axial junction silicon nanowires on glass fabricated by wet electroless chemical etching of multicrystalline p^+-n-n^+-doped silicon layers [42]. The procedures for silicon nanowire array fabrication were schematically shown in Fig. 16.3. To form multicrystalline silicon layers, the p^+, n, and n^+ amorphous silicon layers were subsequently deposited by electron beam evaporation (EBE) and crystallized by a laser beam, respectively. Compared with the polished silicon substrates, the etched silicon nanowire array exhibits less reflectance in the range of 300–1000 nm in wavelength, following the same tendency in PV devices based on vertically aligned nanowire arrays. Under AM1.5 illumination, the PV devices showed the best performance of open-circuit voltage, current density, and efficiency as 450 mV, 13 mA/cm^2, and 1.4 %, respectively.

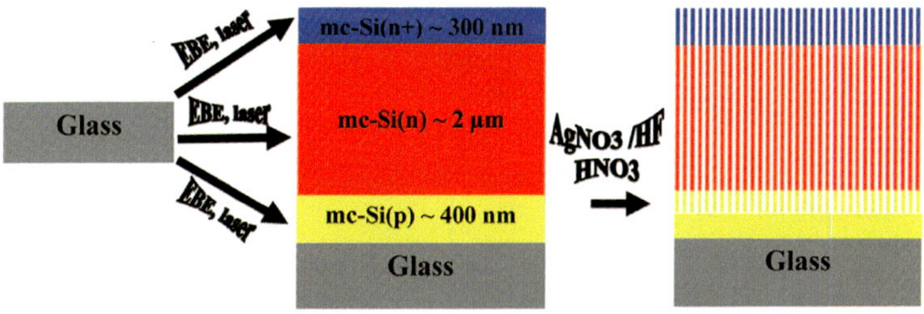

Fig. 16.3 Schematic cross-sectional view of the mc-Si p–n junction layer stack on a glass substrate (*left*) and the Si nanowire array after wet chemical etching (*right*). Reprinted with permission from [42] (Copyright 2009 American Chemical Society)

16.4 Photovoltaic Devices Based on Nanowire Array Embedded in Thin Film

Nanowire array embedded in a thin film of the material having a matching energy band gives rise to an architecture where light absorption and charge carrier separation can be achieved in an orthogonal direction. P. D. Yang's group developed an all-oxide solar cell composed of a vertically aligned n-type ZnO nanowire array covered by p-type Cu_2O nanoparticles, aiming at low-cost, environmentally benign, and stable PV devices [43]. The device performance was found highly dependent on the thickness, morphology, and phase of the nanoparticles, which was originally determined by the process condition. The charge transport through the Cu_2O film could be improved by increasing the grain size in the film. More significantly, it was observed that introducing an immediate insulating layer (i.e., blocking layer) between the absorber and the electrodes directly contacted with the nanowires could greatly improve the overall conversion efficiency. With the atomic layer deposition (ALD) of $\sim$10 nm TiO_2 between Cu_2O and transparent electrode, the device showed an efficiency of 0.053 %, which is 50 times higher than that of the one without the blocking layer. In general, filling the nanowire array by absorbing material to contract the nanowire-based PV device will result in a shunt pathway. The device performance, therefore, will be degraded, especially in the case that the absorber possesses a relatively low resistivity. These observations offer a crucial hint for designing and optimizing the configurations of PV device processed from nanowire directly assembled on the electrode.

Single-crystalline compound semiconductor nanowire arrays have the potential in fabrication of next-generation PV devices, but the requirement of the epitaxial match with substrate is not cost worthy when it comes to practical solar modules. Nonetheless, non-epitaxial catalytic growth of randomly oriented nanowires by the vapor–liquid–solid (*VLS*) method limits the realization of the novel three-dimensional device structures and the consequent improved performances. Template assisted *VLS* growth of highly ordered single-crystalline compound semiconductor nanowires have been proven to be an efficient way to overcome the above-mentioned limitations [22]. Figure 16.4a shows a realization of an n-CdS nanopillar array

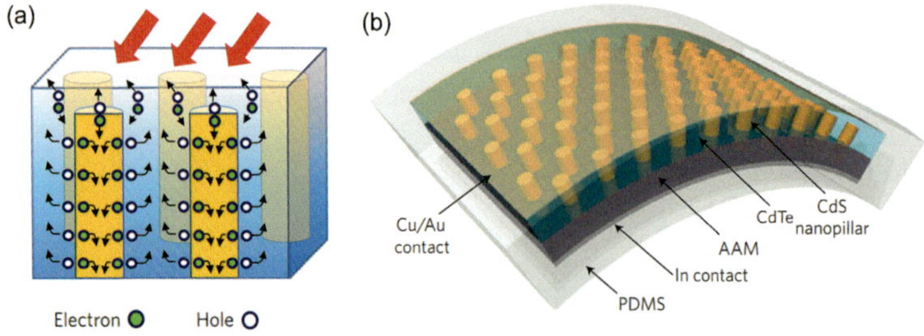

Fig. 16.4 (**a**) Cross-sectional schematics of CdS nanopillars array embedded in CdTe film. (**b**) Schematic diagram of a bendable CdS/CdTe PV device embedded in PDMS. Adapted with permission from [22] (Copyright 2009 Macmillan Publishers Ltd)

embedded in p-CdTe film. Specifically, an efficiency of ~6% is obtained under AM1.5G illumination. To date, this efficiency value is higher than the previously reported values of the devices fabricated to utilize the same concepts of orthogonal architecture for light absorption and charge carrier separation. Moreover, this efficiency value can be, in principle, further improved by optimizing several synthetic and device fabrication processes. This can be achieved by increasing the transparency of the top electrode, decreasing the inter-pillar spacing of the CdS nanopillar array, coating with antireflection layer and reducing the parasitic resistances of the contacts. Additionally, the stable PV devices can be achieved by embedding the semiconductor nanopillar array in a layer of polydimethylsiloxane (PDMS) (see Fig. 16.4b).

In nanowire array embedded in thin film structures, the film quality affects the cell efficiency severely and a proper adjustment of conductivity and thickness is rather crucial to achieve novel PV devices. The parasitic resistances generated by contact electrodes and the increased reflectivity due to contact material may limit the efficiency. Further optimization of these factors remains a challenge. However, a PV device based on nanowire array embedded in thin film has the potency for future practical solar cell modules.

16.5 Photovoltaic Devices Based on Nanowire Array with Core–Shell Structure

Radial p–n/p–i–n junction PV devices processed from individual core–shell nanowire have been intensively studied [44–47]. The core which is generally exposed by dry/wet etching and the shell are bridged across two metal electrodes with different work functions. The key advantage of the radial structure is facilitating the charge collection by suppressing the recombination. The radial p–n junction core–shell nanowire array is, therefore, expected to improve the efficiency by taking additional advantage of the geometry of array for light trapping. Depending on the operation mechanisms, the radial junction PV devices are typically classified as p–n core–shell homojunction and type II core–shell heterojunction. In practical PV devices, combining two operation mechanisms will allow a greater flexibility in optimizing the cell performance.

16.5.1 p–n Core–Shell Homojunction Photovoltaic Devices

Single-crystalline silicon nanowire array has emerged as a promising candidate for solar PV applications [48–49]. The physical model of the device using vertically aligned Si nanowire array with core–shell structure for p–n junction solar cells was first developed in 2005, and the corresponding simulated conversion efficiency was 14.5% [50]. The first Si nanowire array PV device with core–shell structure was demonstrated on stainless steel foil, exhibiting a low efficiency of ~0.1% [49]. The p–n junctions incorporated in the array were formed by depositing a thin amorphous n-type silicon layer on the quasi-aligned p-type silicon nanowire array. The factors affecting the performance, for instance, the impurities and junction qualities, were discussed, which should be taken into account for improving performance of the p–n junction nanowire array PV devices. Garnett and Yang reported a low-temperature wafer-scale etching and thin film deposition method for fabricating silicon n–p core–shell nanowire array PV device [51]. The core–shell Si array was obtained by two-step procedures. Large-area n-type Silicon nanowire array was first obtained by chemically etching the Si substrate, followed by chemical vapor deposition of a p-type Si layer. The devices were fabricated after annealing the core–shell nanowire array at an elevated temperature. The schematics and the cross-sectional SEM image of the core–shell structure are shown in Fig. 16.5a, b. The typical cell had a V_{OC} of 0.29 V, a J_{SC} of 4.28 mA/cm^2, and an FF of 0.33 for an overall efficiency of 0.46%. The reasons for the relatively low efficiency lie in severe interfacial recombination and the higher series resistance, which can be overcome by surface passivation and conductance improvement. In contrast, a novel Si nanohole solar cell incorporating radial p–n junctions was demonstrated with an overall efficiency of 9.51%, proving superiority to those of its counterparts based on Si NWs, planar Si, and pyramid-textured Si [52]. The wafer-scale ordered Si nanohole array was achieved by the combination of deep ultraviolet lithography and metal-assisted etching of p-type Si (Fig. 16.5c). The continuous n-shell layer, schematically shown in Fig. 16.5d, was subsequently formed by phosphorus diffusion on the exposed external and internal surfaces of the Si nanoholes. The final structure of the nanohole array resembles

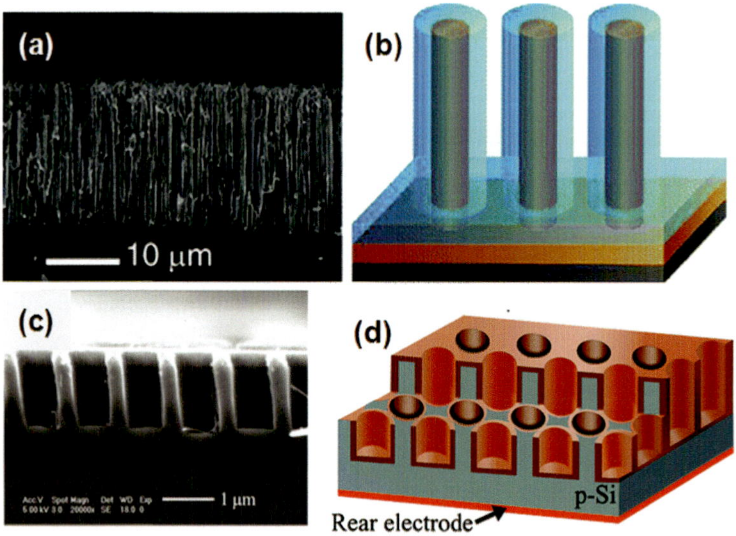

Fig. 16.5 (**a**) Cross-sectional SEM of a completed device silicon nanowire solar cell structure. (**b**) Schematic cell design with the single-crystalline n-Si NW core and the polycrystalline p-Si shell. (**c**) Cross-sectional view of silicon nanoholes. (**d**) Schematics of the Si nanoholes array with radial p–n junctions. Adapted with permission from [51, 52] (Copyright 2009 & 2010 American Chemical Society)

the core–shell nanorod/microrod array with radial p–n junctions. The improved efficiency in Si array solar cells was attributed to the homojunctions via thermal diffusion which facilitate the migration of photoexcited minority carriers across interfaces, and hence reduced interfacial recombination, and the possible optical coupling effect between the nanohole array and the incident light [52–53].

In comparison with silicon, III–V semiconductors in solar applications have more advantages including broader spectra response, higher absorption coefficient, and higher irradiation impedance, which, therefore, could offer more possibilities to achieve higher energy conversion efficiency PV devices. GaAs, with the bandgap of 1.45 eV, is the III–V semiconductor whose absorption is best matched with the solar spectrum. J. Czaban et al. reported the synthesis of core–shell GaAs nanowire array and the influence of the doping process on the operating characteristic of the devices [54]. Vertically aligned GaAs nanowires were grown on GaAs(111)B substrates by the VLS method in a gas source molecular beam epitaxy (MBE) system. Te and Be were used for n and p doping, respectively. Te doping was found to promote radial growth at the expense of axial growth. Cell parameters of the GaAs nanowire device exhibited an I_{sc} of 201 μA, a V_{oc} of 0.2 V, and a fill factor of 0.27. The conversion efficiency of the device was measured as 0.83 % and it degraded with increasing the duration of Te-doping processes, suggesting that the morphologies and dopants distribution impact significantly on the device performances. In addition to the GaAs core–shell nanowire array, a periodic arrangement of InP core–shell nanowire array was also investigated as a p–n homojunction solar cell [55]. The arrangement of the dense nanowire, i.e., the directions and locations, was controlled by catalyst-free selective-area metal organic vapor phase epitaxy method, while Zn and Si were chosen as the dopant to form p-type core and n-type shell. The nanowires were mainly perpendicular to the p-type InP (111)A substrate. The average diameter of the core and the thickness of the shell were 135 nm and 74 nm, respectively. The device exhibited open-circuit voltage, short circuit current, and fill factor levels of 0.43 V, 13.72 mA/cm^2, and 0.57, respectively, under AM1.5G illumination. The solar power conversion efficiency was determined as 3.37 %, which is higher than that of Si core–shell nanowire array possibly because of the uniformity of the array, absence of the impurities (catalysts), as well as the differences in the optical absorption coefficient and bandgap between InP and Si.

In contrast to the heteroepitaxial growth of the nanowire arrays on silicon substrate, direct homoepitaxial growth of III–V semiconductor nanowire arrays on the substrate can minimize lattice-matching strain. In the above cases, however, the substrates also facilitate the light absorption and then carrier transport. To exclude the contribution of the substrate to the overall efficiency in the current device configuration and reuse the expensive substrate, the peel-off technique is highly required [20]. One of the key advantages in III–V semiconductors for solar harvesting is the tunable bandgap realized by varying the compositions. The single nanowire device composed of n-GaN/i-$In_xGa_{1-x}N$/p-GaN coaxial junctions has demonstrated enhanced conversion efficiency [45]. Therefore, exploiting array of such composition-tuned III–V semiconductors for solar application is a promising direction to boost the efficiency further.

16.5.2 Type II Core–Shell Heterojunction Photovoltaic Devices

In nano-architecture PV devices, the charge separation is often facilitated by a type II or staggered energy alignment of a heterojunction, constructed by two materials for which both the valance and conduction bands of one component lie, respectively, lower in energy than those of the other component. Because the lowest energy states for the electrons and holes are in different semiconductors, the energy gradient existing at the interfaces tends to spatially separate electrons and holes on different sides of the heterostructure (Fig. 16.6). Such heterojunctions have been intensively investigated for solar cell applications, including DSSCs [56], QDSSCs [57–59], nanocrystal–polymer hybrid [60], and bilayer nanocrystal films [61]. To improve the carrier transport within the type II scheme, semiconductor nanowires have already been used to serve as the electron transporter in the inorganic–organic hybrid approach [2–4, 6] and a core–shell all-inorganic nanowire architecture has been proposed to improve the carrier transport for both the electron and hole and simultaneously the device stability, using the well-known II–VI and III–V binary semiconductors, such as ZnO/ZnS, ZnO/ZnSe, ZnO/ZnTe, GaN/GaP, and GaN/GaAs [9–10]. Here the type II transverse heterojunction functions similarly as a radial p–n junction [11, 50] but without having to deliberately dope the nanostructure. In addition, the type II core–shell structure may extend the absorption profile to a wavelength longer than that defined by the bandgap of any of the components through a relatively weak interface transition (see Fig. 16.6) [9–10, 62]. When the core–shell nanowire is in the quantum mechanical scale, the "interfacial" absorption occurs between hole states more confined in one component and electron states more confined in the other component; thus, the absorption will be enhanced as compared to the case when both components are in the mesoscopic scale [9–10].

Figure 16.7 depicts the band alignment of the selected II–VI and III–V semiconducting materials, which are commonly employed to construct type II heterojunction for harvesting solar energy. In all of these semiconductors, ZnO is the most attractive materials because it is environmental friendly, composed of abundant and

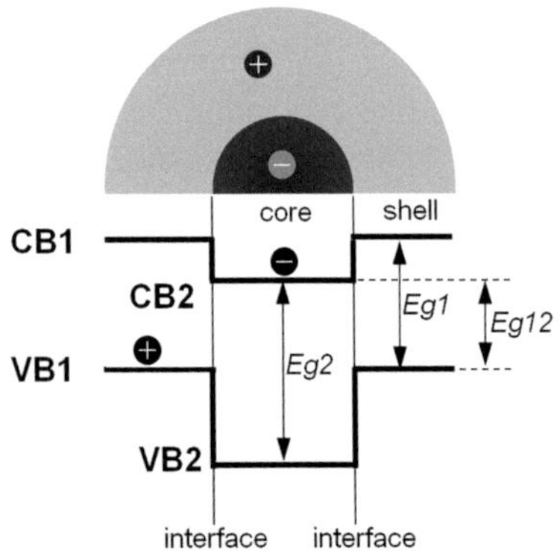

Fig. 16.6 Type II band edge alignment at the heterointerface between two semiconductors tends to spatially separate the electron and holes on different sides of the heterointerface. The emission energy is determined by *Eg12*

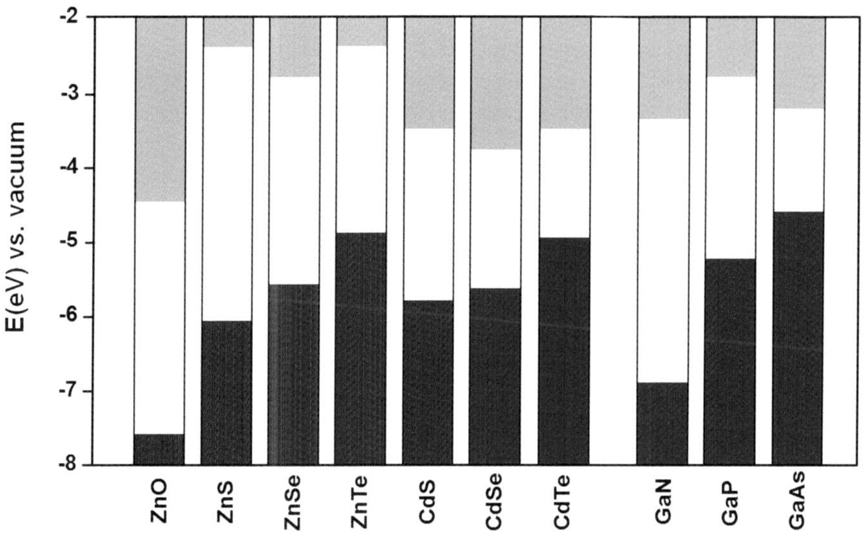

Fig. 16.7 Electronic energy levels of selected II–VI and III–V semiconductors showing the materials selection to construct type II heterojunctions

chemically stable elements. Additionally, intensive work has been devoted to synthesizing various ZnO nanowire arrays on different substrates by hydrothermal, chemical vapor deposition (CVD), pulsed laser deposition (PLD), metal organic chemical vapor deposition (MOCVD), etc. [63–66], providing the opportunities to construct three-dimensional PV device by taking advantage of the unique geometry and the possible quantum size effects. Variation in nanowire length, diameter, seed layer, and tip tapering has a great effect on the antireflection layer performance, which has already been demonstrated theoretically and experimentally [19, 67–68]. In fact, the ZnO nanowire array has been used as an efficient antireflection layer in Si and copper indium gallium selenide (CIGS) thin film solar cell [69–70]. The goal of this section is to discuss recent efforts made to synthesize core–shell structures based on ZnO nanowire arrays for constructing three-dimensional PV device involving type II heterojunctions [15, 71].

16.5.2.1 Synthesis of ZnO/ZnSe and ZnO/ZnS Core–Shell Nanowire Array

Vertically aligned nanowire array with the entire nanowire surface well exposed provides an ideal platform for further engineering. In order to fabricate core–shell nanowire structures, most of the vapor phase thin film fabrication techniques can be used for shell layer deposition. However, some techniques with strong directional particle trajectory, such as evaporation, have a severe shadow effect that will result in uneven coating. PLD is a versatile, effective, and scalable technique in thin film deposition. During PLD, many experimental parameters including wavelength, pulse duration, repetition rate, target-to-substrate distance, substrate temperature, background gas, and pressure can be altered to modify the thin film growth. The core–shell nanowire arrays on indium–tin–oxide (ITO) glass were obtained by two-step synthetic procedures combining chemical vapor deposition for ZnO nanowire array growth and pulsed laser deposition for the shell coating. In this regard, ZnO nanowire array acts as a "substrate" for the shell layer deposition.

To synthesize large-area ZnO nanowire array, a piece of thin zinc foil was used to generate high-pressure zinc vapor in a tube furnace. The Zn foil and ITO substrate were sequentially laid down with a distance of 5 mm in a ceramic boat, which was then transferred to the center of a 2 in. diameter quartz tube furnace. The quartz tube was first evacuated to 10 mtorr, and a 40 sccm argon was introduced into the tube, which was then heated to 600°C at a rate of 30°C/min. When the temperature exceeded 420°C (the melting point of Zn), a 70 sccm air flow was also introduced into the quartz tube. The reaction was maintained for 30 min after the temperature reached 600°C. By controlling the pumping rate, the pressure was kept at 8 torr throughout the nanowire synthesis. The tube furnace was naturally cooled down to room temperature in 3 h. A white-yellowish layer was then harvested on the ITO substrate [64].

The as-synthesized ZnO nanowire array taken as a template was then transferred to hot-wall PLD chamber for ZnSe and ZnS coating. The neodymium-doped yttrium–aluminum–garnet (Nd:YAG) laser was used to ablate the disc target. The laser wavelength, energy density, and pulse frequency were 266 nm, 130 mJ/cm^2, and 2 Hz, respectively. The distance from the target to the ITO substrate with a ZnO nanowire array was 5 cm. Before the deposition, the vacuum system was first evacuated to 1×10^{-3} torr. After the deposition performed at 400°C for 20 min, the sample was annealed at 500°C for 1 h. The room temperature PLD for ZnSe shell coating was also performed in the same system at 27°C for 20 min without any further annealing. In the case of ZnS coating, the parameters were kept the same as that of ZnSe deposition except the deposition temperature was changed to 500°C.

16.5.2.2 Structural and Optical Properties of ZnO/ZnSe Core–Shell Nanowire Array

As shown in Fig. 16.8a, the as-synthesized ZnO nanowire array presented a uniform perpendicular growth of ZnO nanowires on the ITO substrate with an average length of 10 μm and diameters of 80–120 nm. After the PLD followed by annealing, the final nanowires, as shown in Fig. 16.8b, exhibit increased wire diameters and rough surfaces, indicating that ZnSe was successfully deposited on the ZnO nanowires.

A typical low-magnification TEM image of a ZnO/ZnSe core–shell nanowire shows a sharp interface between the core and shell and the ZnSe shell grows directly in the radial direction from the surface of ZnO nanowire with a thickness of 5–8 nm (Fig. 16.8c). At the interface, an epitaxial growth of ZnSe from the ZnO core is observed. The epitaxial growth relationship of the WZ ZnO core and ZB ZnSe shell has been identified as $[0001]_{ZnO}//[100]_{ZnSe}$ and $(2\bar{1}\bar{1}0)_{ZnO}//(011)_{ZnSe}$, respectively, which was further confirmed by the fast Fourier transform (FFT) patterns of ZnO core and ZnSe shell (Fig. 16.8d). Defects were also observed at the interface along

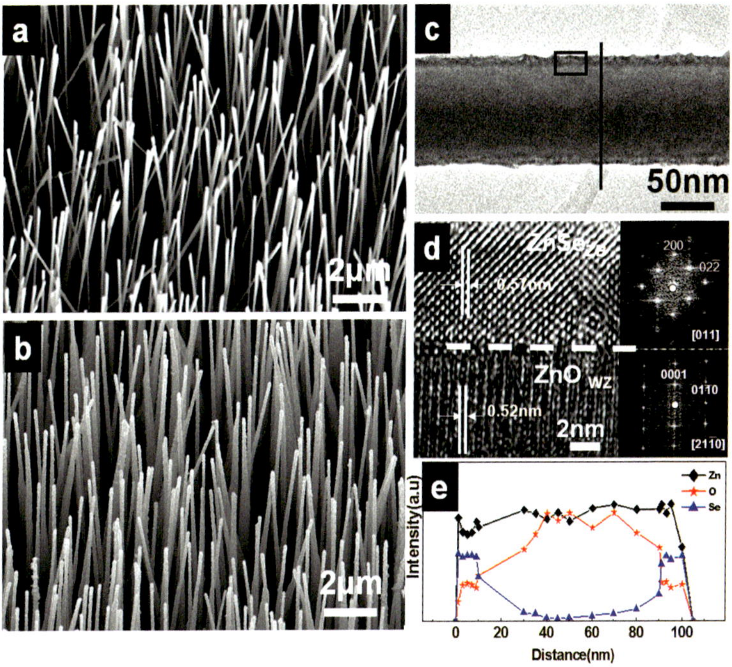

Fig. 16.8 (**a**) SEM image of a well-aligned ZnO nanowire array grown on ITO substrate by CVD. The average length of the nanowires is about 10 μm and the diameters are about 100 nm. (**b**) SEM image of well-aligned ZnO/ZnSe core–shell nanowire array prepared by PLD. (**c**) Low-magnification TEM micrograph of a ZnO/ZnSe core–shell nanowire. A thin layer of ZnSe was coated on the ZnO nanowire. (**d**) High-resolution TEM image of the interface of the core–shell heterostructure, enlarged from the rectangular area in (**c**), showing the epitaxial growth relationship of ZnO wurtzite core and ZnSe zinc blende shell. Fast Fourier transfer (FFT) patterns of rectangular areas in (**d**). (**e**) EDS nanoprobe line-scan on elements Zn, Se, and O, across the ZnO/ZnSe core–shell nanowire, indicated by a *line shown in* (**a**). Reproduced with permission from [15] (Copyright 2008 Wiley-VCH Verlag GmbH & Co. KGaA)

the c-axis of ZnO due to the large lattice mismatch, 8.8% along the nanowire axis between $c = 0.521$ nm for ZnO and $a = 0.567$ nm for ZnSe. The spatial distributions of the atomic composition across the ZnO/ZnSe core–shell nanowire reveal that the ZnO nanowire was homogeneously coated, as shown in Fig. 16.8e. It was also found that epitaxial growth only occurred during high-temperature deposition and no epitaxial growth was found at room temperature deposition, indicating that the epitaxial growth between ZnO and ZnSe demands favorable thermodynamic as well as kinetic conditions. The stability of the epitaxial growth also highly depends on the interplay between the surface energy and mismatch strain. In the case of deposition performed at room temperature, the initial deposited atoms, without gaining enough migrating energy from the system, condense at the very site they arrive at the surface of ZnO and then aggregate with the following deposit atoms, forming a thick layer of polycrystalline ZnSe [15].

The transmission, photoluminescence, Raman, and photoresponse measurements were employed to investigate the optical properties of the ZnO/ZnSe core–shell nanowire. In the typical transmission spectra (Fig. 16.9a), only one absorption peak corresponding to ZnO is observed for bare ZnO nanowires. In contrast, two excitonic absorption peaks are clearly observed at respective excitonic bandgaps (3.31 eV for WZ ZnO and 2.70 eV for ZB ZnSe) in that of ZnO/ZnSe core–shell nanowires, indicating good crystallinity of both core and shell. Figure 16.9b shows the PL spectra of ZnO, ZnO/ZnSe core–shell nanowires. Without any intentional surface passivation effort, for the bare ZnO nanowire array, the PL peak (at 3.235 eV) near the band edge of ZnO is found to be surprisingly very strong. Additionally, the ZnO nanowire array also shows a weak visible emission band (at ~2.49 eV, ~1.5% of the band edge peak). On the one hand, the high PL

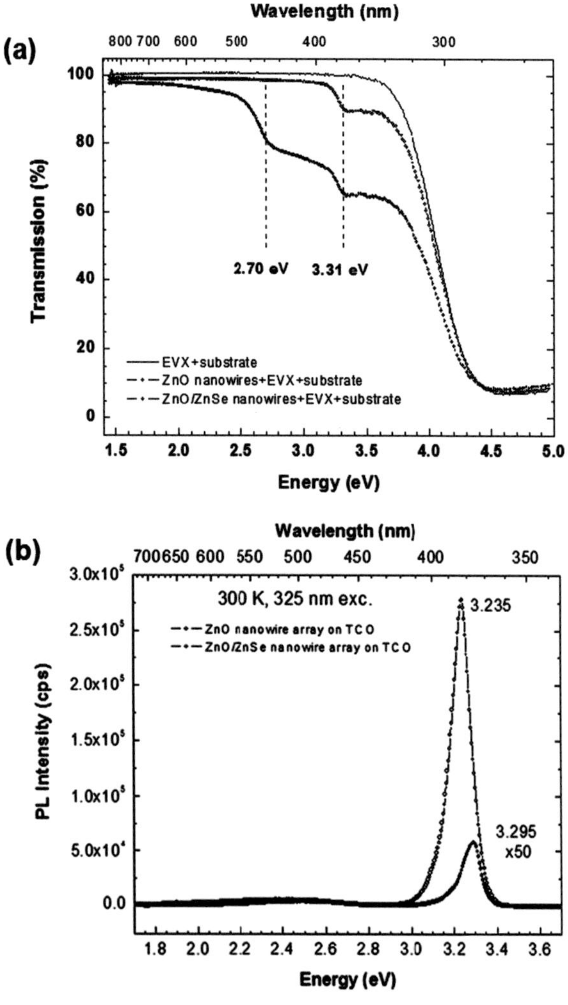

Fig. 16.9 (**a**) Transmission spectra of ZnO nanowires and corresponding ZnO/ZnSe core–shell nanowires. *Two vertical lines* indicate the excitonic bandgap of bulk ZnO and ZnSe, respectively. (**b**) Photoluminescence spectra of ZnO and ZnO/ZnSe nanowire arrays

intensity indicates high interior crystal quality of the nanowires; on the other hand, two photonic crystal-related effects might contribute to the high external quantum efficiency: (1) the waveguide effect of the nanowire eliminates the lateral propagation [72] and (2) a smaller effective dielectric constant of the array, as a result of averaging between the ZnO and air, than the bulk ZnO facilitates the escaping of light from the sample. For the core/shell nanowire array, the band edge emission remains strong but is shifted to 3.295 eV, although reduced by roughly a factor of 300 from that of the bare ZnO nanowire array or ~50 from that of the bulk ZnO reference. There could be several possible reasons for the reduction: (1) the charge separation between the core and shell (namely, the hole relaxes into the shell but the electron remains in the core), (2) non-radiative recombination at the possible defect sites of the core–shell interface, and (3) the absorption of the shell to the excitation and emission photons. Although the ZnSe shell is relatively thin and the absorption of a single path is relatively small (~5% per 10 nm), the multiple scattering in the nanowire array could significantly increase the absorption. Further investigation on the carrier dynamics associated with the core–shell interface and growth optimization is definitely needed, but we could at least conclude that the interface defects are not as detrimental as one might expect for such highly mismatched heterostructure [15].

16.5.2.3 Photoresponse of ZnO/ZnSe Nanowire Array

Photoresponse measurements were performed on a very preliminary device: a single core–shell nanowire sitting on two Au electrodes at the two ends. Since the contacts are symmetric, the device is not expected to function as a single nanowire solar cell [44], and a bias is needed to generate the photocurrent. Figure 16.10 shows the

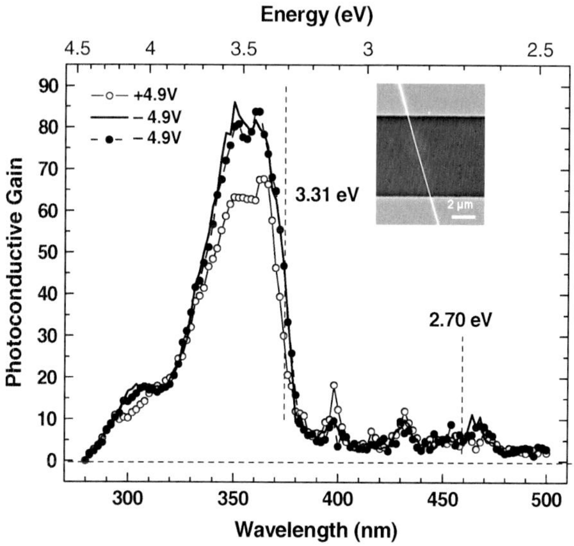

Fig. 16.10 Photoconductivity spectra for a single ZnO/ZnSe nanowire measured under different bias voltages and at different days. The *inset* shows the SEM picture of the single-nanowire device for photoresponse measurements. Copyright 2008 Wiley-VCH Verlag GmbH & Co. KGaA. Reproduced with permission from [15]

photoresponse spectra under positive and negative bias (switching the polarity of the bias voltage). A standard procedure for solar cell quantum efficiency measurement was followed, but with a bias; thus, the result should be considered as photoconductivity gain. Despite the photocurrent being rather small (a few picoamperes maximum) and the gain being significantly smaller than what has been reported previously for pure ZnO nanowires with more robust contacts [73], the result is repeatable, stable, and perhaps more significantly the response shows an extension to the wavelength region below that of ZnO bandgap. The existence of the photoconductivity in the ZnSe spectral region indicates that the carriers are not fully depleted by either defects at the ZnO/ZnSe interface or the ZnSe surface, despite the ZnSe shell being fairly thin [15].

16.5.2.4 Morphologies, Structure and Optical Properties of ZnO/ZnS Nanowire Array

Figure 16.11a shows a typical low-magnification SEM image of an as-synthesized large-area ZnO nanowire array, revealing perpendicular growth of ZnO nanowires on the ITO glass. The ZnO nanowire array, as shown in Fig. 16.11b, has an average length of ~7 μm and diameters in the range of 50–120 nm. SEM image of the nanowire array after the pulsed laser deposition of ZnS is presented in Fig. 16.11c. Compared with the bare ZnO nanowire, a noticeable increase in the diameter and rough surface for the ZnO/ZnS nanowires was also observed. Only the strong hexagonal ZnO(0002) diffractive peak was observed in X-ray diffraction (XRD) patterns (Fig. 16.11d) for the bare ZnO nanowire array, indicating strong preferred orientation along the *c*-axis of wurtzite (WZ) ZnO. No additional diffraction peak was observed in the pattern of ZnO/ZnS nanowires, perhaps because the ZnS layer is fairly thin. However, a slight position shift of the ZnO peak to a lower angle and the increased intensity of the (0002) peak were observed for the ZnO/ZnS core–shell structure, which indicates the possible annealing effect during the deposition process and the presence of lattice expansion caused by the strain [71].

Similar to the ZnO/ZnSe core–shell nanowire, the sharp interface between the core and shell clearly suggests that the ZnO nanowire is fully sheathed along the entire length. The ZnS layer is ~12 nm thick and has a rough surface (Fig. 16.12). Further detailed investigations reveal that the marked inter-planar d spacings of 0.31 and 0.52 nm correspond, respectively, to the $(\bar{1}1\bar{1})$ lattice plane of ZB ZnS with the [011] zone axis and the (0001) lattice plane of WZ ZnO with the $[2\bar{1}\bar{1}0]$ zone axis. Therefore, the core and shell are determined as epitaxial growth with the growth relationship of $[0001]_{ZnO}//[\bar{1}1\bar{1}]_{ZnS}$ and $[01\bar{1}0]_{ZnO}//[21\bar{1}]_{ZnS}$, in contrast to that for the previously studied ZnO/ZnSe system, where $[0001]_{ZnO}//[100]_{ZnSe}$ was observed [15]. The epitaxial growth of WZ ZnS and ZnO has been reported in many nanoheterostructures in the planar form even though a large mismatch (~20%) exists [74–76]. Here, though the lattice mismatch between the WZ ZnO and ZB ZnS is still fairly large (~19%), it appears that various factors, including perhaps the geometry of the "substrate" (the ZnO nanowire), size of the core, growth temperature, and non-thermal equilibrium condition, play roles in successfully growing

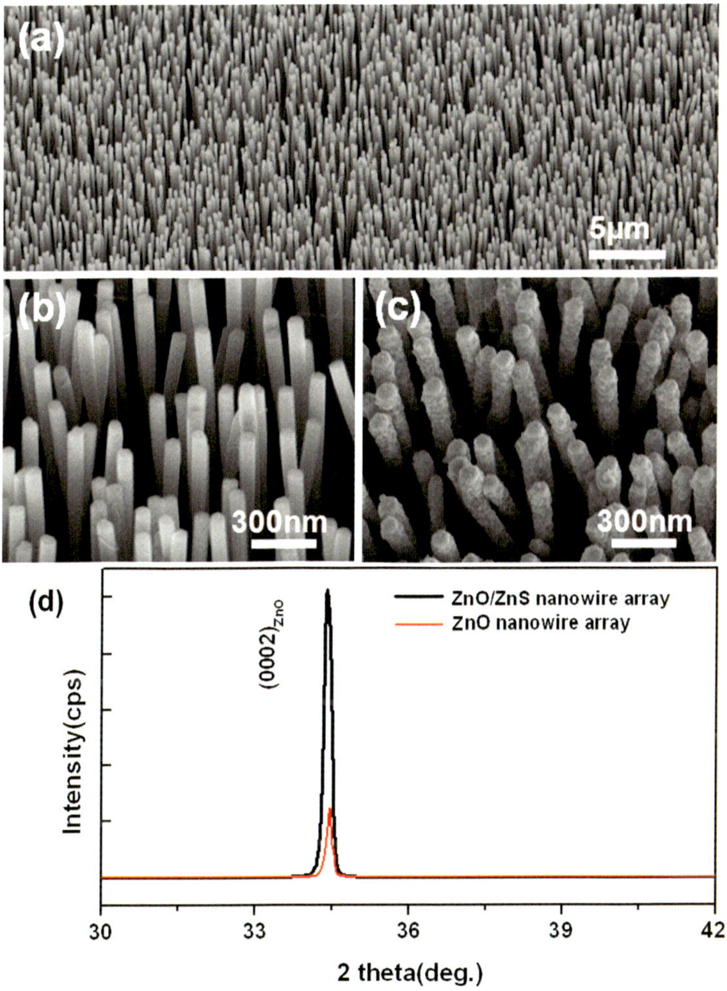

Fig. 16.11 (**a**) Low-magnification SEM image of a well-aligned ZnO nanowire array grown on ITO substrate by chemical vapor deposition. (**b**, **c**) Magnified SEM images of bare ZnO and ZnO/ZnS core–shell nanowire arrays, respectively. (**d**) XRD patterns of ZnO and ZnO/ZnS nanowire arrays. Reprinted with permission from [71] (Copyright 2010 American Institute of Physics)

such a epitaxial core–shell heterostructure that is unlikely to be feasible in the planar form. However, due to the large lattice mismatch, the structure defects, such as stacking faults and lattice distortion, inevitably exist in order to release strain energy when the shell layer grows thicker. Furthermore, the formation of the rough external surface of the ZnS shell could be understood as a transition from the initial cylindrical growth to the island growth in core–shell nanowire system [77].

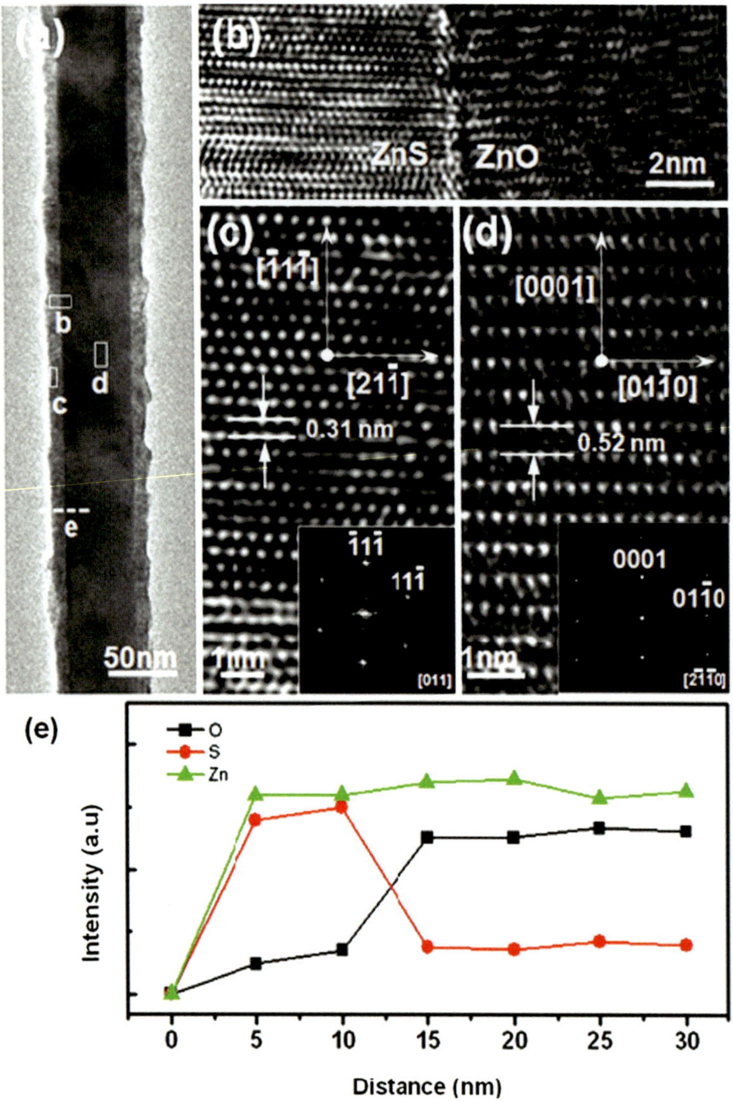

Fig. 16.12 Structural characterization of ZnO/ZnS core–shell nanowire array. (**a**) Low-magnification TEM micrograph of a ZnO/ZnS core–shell nanowire, showing a thin layer of ZnS coated on the ZnO nanowire. (**b**) High-resolution TEM image of the interface of the core–shell heterostructure, enlarged from the *rectangular area in* (***a***). (**c**, **d**) Atomic resolution images of the core and shell areas taken from the *rectangular areas in* (***a***), respectively. The *insets in* (***c***) *and* (***d***) represent the corresponding fast Fourier transfer patterns. (**e**) EDS nanoprobe line scan across the core–shell interface. Reprinted with permission from [71] (Copyright 2010 American Institute of Physics)

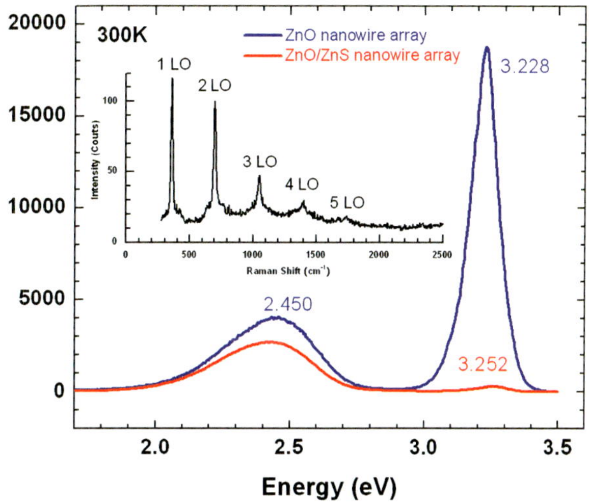

Fig. 16.13 Room temperature photoluminescence spectra of ZnO and ZnO/ZnS core–shell nanowire arrays. *Inset* shows the resonant Raman features of ZnS shell

The optical properties of the ZnO/ZnS core–shell nanowire were investigated by photoluminescence (PL) spectrum and resonant Raman scattering. Figure 16.13 shows the PL spectra of ZnO and ZnO/ZnS nanowire arrays measured at room temperature. The PL spectrum of the bare ZnO nanowire array consists of two emission peaks at 3.228 and 2.450 eV, respectively. Compared with the bare ZnO nanowire array, the peak position of the UV emission in ZnO/ZnS sample shows a small blue shift, and the intensity is reduced by a factor of 63. The intensity of the visible peak is also reduced, but to a much lesser extent (by ~30%). In principle, one should expect that the charge separation effect quenches, instead of enhancing, the PL signal, because the type II core–shell structure should result in depletion of the holes in the ZnO core. That the reduction is relatively small for the visible peak and much more significant for the UV peak may suggest that the visible emission could be from the bulk part of the ZnO nanowire, whereas the UV emission could be surface related and thus more sensitive to the coating that has modified the electronic structure of the bare ZnO nanowire surface, either by introducing non-radiative centers or inducing the transfer of holes to the ZnS shell from the surface-bound excitons [71]. Resonant Raman features with up to fifth-order longitudinal optical (LO) phonons were observed in the same PL measurement in the ZnO/ZnS nanowire array. The spectrum is included in Fig. 16.13. Following the observation of the resonant Raman features in bulk ZnS crystal [78], such features have also been reported for ZnS hollow microspheres [79]. The observation of the multiple orders of resonant Raman features may indicate the good crystalline quality of the material.

16.5.2.5 Photovoltaic Effect of ZnO/ZnS Nanowire Array

The configuration of the PV device, as shown schematically in Fig. 16.14, employs the ZnO/ZnS core–shell nanowire array as the active layer and carrier transport

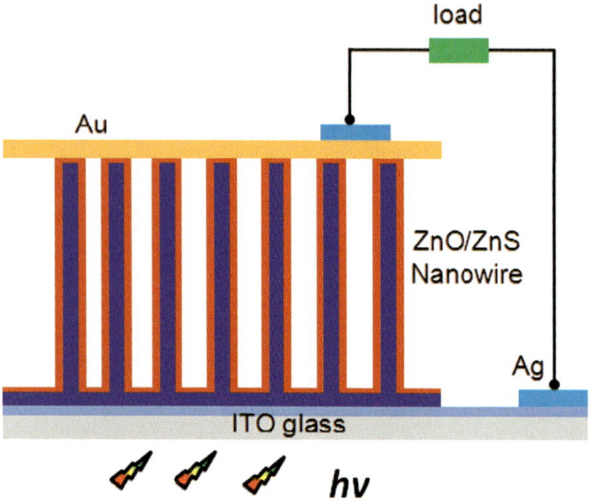

Fig. 16.14 Schematic structure of the PV device based on a type II core–shell nanowire array. Reprinted with permission from [71] (Copyright 2010 American Institute of Physics)

medium, while ITO and gold as the anode and cathode, respectively. In theory, once the incident photon with energy greater than the bandgap of the type II heterojunction is absorbed, the photoexcited electron–hole pair will be generated and immediately separated into the ZnO core and ZnS shell due to the quasi-electric field near the interface. The ZnO nanowire cores and ZnS shells then serve as the transport channels to deliver the carriers to the counterelectrodes. PV devices based on ZnO/ZnS core–shell nanowire array were fabricated using a modified technique of fabricating three-dimensional nanodevices based on vertically aligned nanowire arrays [36, 80]. In brief, the air space of the ZnO/ZnS nanowire array was first filled with photoresist (Shipley 1813) by spin coating, followed by several-second acetone etching to expose the tips of the nanowires (Fig. 16.15a). For electrodes fabrication, a 50 nm thick gold layer was sputtered onto the exposed nanowire tips to form the top electrode, and the bottom (ITO) electrode was exposed by removing the ZnO nanowire near the edge of the substrate via diluted hydrochloride (10 %). Finally, the photoresist was removed by dipping in acetone for several seconds. To maintain the same active area, the top gold electrodes of the devices were sputtered using the same mask. The cross-sectional SEM image of the final device was shown in Fig. 16.15b.

Without applying external bias, the photocurrent generated in the PV device based on ZnO/ZnS core–shell nanowire array increased/decreased faster and represented 30 times stronger than that observed in the device composed of a bare ZnO nanowire array with the same active area, as the incident light (AM1.5G) was turned on/off (Fig. 16.16a). The slow response and the small gain in the ZnO device may be attributed mainly to a heating effect or surface-related adsorbates (O_2^-) [81]. The enhanced photocurrent and faster response observed in ZnO/ZnS, together with the quenching of the UV emission, could indicate the realization of the key feature of the type II heterostructure – the charge separation, although other possibilities (such

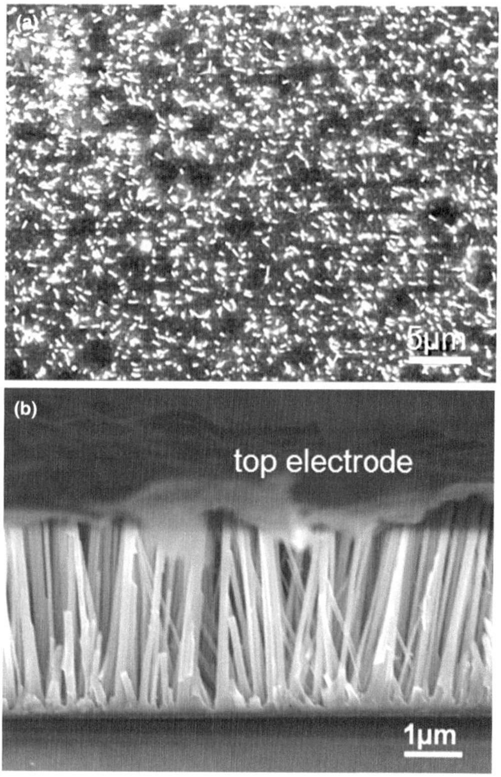

Fig. 16.15 (**a**) A typical top-view SEM image of the ZnO/ZnS nanowire array before electrode deposition, revealing that the tips of nanowires were exposed after the chemical etching. Note that some of the nanowires were still covered by photoresist due to the inhomogeneous photoresist layer and the different etching depth. (**b**) A typical cross-sectional SEM image of PV device based on ZnO/ZnS nanowire array

as the ZnS/Au contact is different from ZnO/Au) cannot be completely excluded. However, because of the large thickness of the nanowire layer, the light absorption occurs mostly in the bulk part of the ZnO nanowire core, the core–shell interface has to serve the function of separating the electrons and holes [68]. Figure 16.16b shows the current–voltage (*I*–*V*) curve of the device under an illumination of 1000 W/m^2. The V_{oc} is 0.09 V, with an FF of 0.28. The energy efficiency could not be given because of the uncertainty of the actual device area. Although the device performance is far from satisfactory, this prototype device demonstrated the feasibility of fabricating a core–shell nanowire solar cell based on the type II heterostructure and a whole array of such devices. Several possible reasons exist for the low performance, including (1) defect formation at the heterostructure interface, (2) quality of the shell material, and (3) quality of the top contact (between the ZnS shell and cathode). Because of the relatively large core nanowire size and the large bandgaps for both materials, this particular device cannot be expected to deliver high-energy efficiency, even in the ideal situation. Further investigation will address these issues and will explore other material combinations that form type II heterojunctions, but with less lattice mismatch and smaller bandgaps.

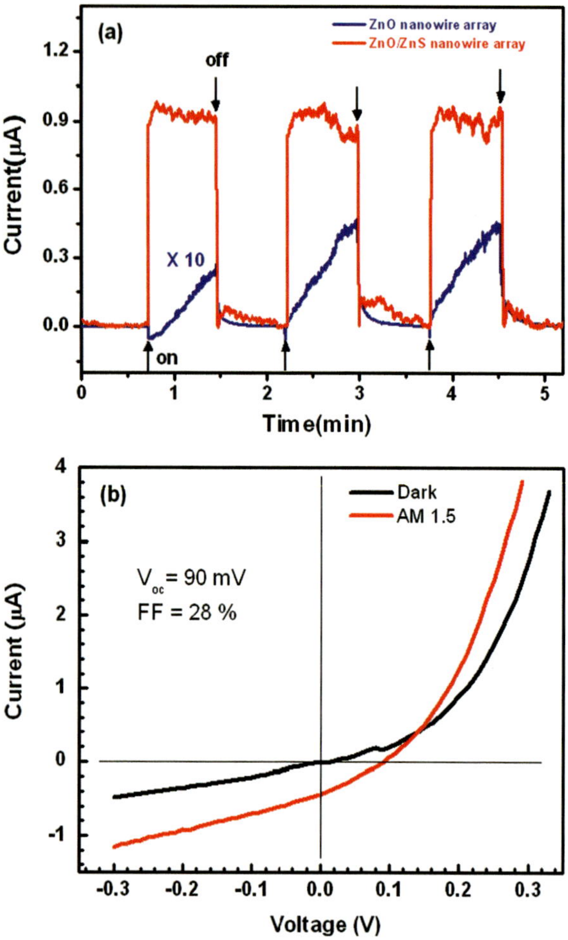

Fig. 16.16 (**a**) Time-dependent photocurrent of bare ZnO and ZnO/ZnS core–shell nanowire arrays without external bias. (**b**) Current (*I*) vs voltage (*V*) for the solar cell in the dark and under simulated AM1.5G illumination with intensity of 1000 W/m^2, showing the PV effect

To conclude, a large-area well-aligned air-stable ZnO/ZnSe and ZnO/ZnS core–shell nanowire array was fabricated directly on ITO substrate by combining CVD and PLD techniques. Morphologies and structure studies showed that the ZnO nanowire array was uniformly and perpendicularly grown on the ITO substrate and the shell with a thickness of 5–10 nm was epitaxially grown on the ZnO nanowire core. The charge separation in the ZnO/ZnS nanowire array was investigated by the PL and photoresponse based on the type II heterojunction-based device. Although the device performance is far from satisfactory at current stage of development, the PV device demonstrated here would expand the opportunities for designing and optimizing nanoscale PV and other optoelectronic devices beyond the conventional p–n junction approach.

16.6 Summary and Perspectives

In this chapter, we have reviewed the recent progress in nanowire array-based PV devices, which were divided into four different PV device configurations: nanowire array integrated with the substrate, nanowire array with axial junctions, nanowire array embedded in thin film, and core–shell nanowire array. Although each configuration has its own advantages and disadvantages in charge collections, all the devices composed of nanowire array have demonstrated enhanced light absorption, which are superior to those of their counterparts based on bulk or thin film materials.

Table 16.1 summarizes the device performances of current nanowire array PV devices measured under AM1.5G condition. Despite tremendous works dedicated to this field, nanowire array PV devices have to date exhibited poor performance mostly due to the severe surface and interface recombination arising from the large aspect ratio of nanowire array. Moreover, poor controllability of morphologies, inefficient device prototype for carrier collection, and challenging manipulation in the electrical properties further deteriorate the expected conversion efficiency improvement. Additionally, the potential quantum effect, especially the size quantization property, in this system was not fully exploited because currently the diameters of the nanowires are far bigger than their corresponding Bohr radii.

To improve the overall conversion efficiency, both theoretical and experimental work on optical engineering, electrical manipulation, and material and device design need to be conducted. For optical engineering, recent work has shown that novel structures of arrays, such as nanohole, tapered nanorod, nanodomes, and nanopillars, could noticeably enhance the light absorption [53, 82]. Future work will need to be centered on optimization of various dimensional parameters including the length, diameter, arrangement, fill ratio, and other unique features, to further tailor the light absorption spectrum and reduce the material consumption. For example, on the one hand, periodic arrangement of nanowire apparently could enhance the overall light absorption, superior to the quasi-aligned or random-distributed nanowires. On the other hand, the periodic arrangement will also create low absorption center owing to the angular absorption profile [20]. Detailed investigations of the nanowire arrangement would clarify the above-mentioned confusions and determine the optimal arrangement. Along with the effort to develop novel architecture, theoretical calculation and modeling on the optical coupling have to be carried out on specific materials in different geometry, which may predict the optimal nanowire diameter to achieve maximum photons absorption. So far, apart from silicon nanowires and nanoholes array, there is little theoretical work focused on other nanostructures, which hinders the development of other nanowire array PV devices beyond silicon. In addition, to push the envelope of the photons collection, traditional concepts, such as antireflective coating, could also be incorporated in the nanowire array PV devices, even though the nanowire array itself could reduce the reflection.

Regarding the efficient carrier collection, first of all, fundamental issues about carrier dynamics in nanoscale need to be fully understood to determine the diameter

Table 16.1 Device performances of current nanowire array PV devices under AM1.5G illumination

Classification	Efficiency (%)	V_{oc} (V)	J_{sc} (mA cm^2)	FF (%)	Description	References
Si	0.10	0.13	1.67	0.28	Quasi-aligned core–shell nanowire array on metal foil	[48]
Si	0.46	0.29	4.28	0.33	Core–shell nanowire array	[50]
Si	9.51	0.57	32.2	0.52	Nanohole array	[51]
mc-Si[a]	4.40	0.45	40.0	0.30	Axial junctions nanowire array on glass	[41]
a-Si:H[b]	5.90	0.75	17.5	0.45	Deposited on quartz nanodome substrate	[81]
GaAs	1.65	0.25	27.4	0.25	Directly integrated with the GaAs substrate	[38]
GaAs	0.83	0.20	0.20	0.27	Core–shell nanowire array	[53]
InP	3.37	0.43	13.72	0.57	Core–shell nanowire array	[54]
GaN/Si	2.73	0.95	7.60	0.38	GaN nanowire array integrated with Si substrate	[36]
InP/Si	0.76	0.19		0.55	InP nanowire array integrated with Si substrate	[37]
CdS/CdTe	0.60	0.62	21.0	0.43	CdS nanowire embedded in CdTe thin film	[22]
ZnO/Cu_2O	0.05	0.15	1.43	0.25	ZnO nanowire embedded in Cu_2O thin film	[42]
ZnO/ZnS		0.09		0.28	Core–shell nanowire array on ITO substrate	[70]

[a]Microcrystalline silicon
[b]Hydrogenated amorphous silicon

or shell thickness in core–shell structure for efficient minority carrier collection. Particularly, reducing the diameter of nanometer for suppressing light scattering or extending light response spectrum will, however, pose challenges to the carrier transportation due to the possible full wire depletion. Second, selection of appropriate semiconductor materials with lower surface recombination velocities, together with the surface passivation of the devices, could greatly suppress the surface recombination. In particular, II–VI semiconductors present lower surface recombination velocities, which are believed to be promising platforms better than silicon and III–V semiconductors [83]. With respect to type II PV device, constructing heterojunction from two semiconductors with less lattice mismatch, for example, ZnTe/CdSe and GaN/GaP, might significantly suppress the interface recombination arising from the surface-induced defects. Furthermore, controllability of doping concentration and homogeneity, which is crucial in tuning the bandgap and electrical properties of materials, has to be fully explored to achieve broader absorption spectra and lower series resistance. Additionally, novel device design is highly expected to reduce the shunt and contact resistance, which could further improve the charge collection. So far, most of the nanowire array PV devices employ the array as efficient carrier channels, posing stringent requirements on the conductivity of the nanowire and hence the synthesis techniques. The nanodome solar cells developed by Yi Cui's group provided an alternative route to collect carriers by subsequently depositing metal and active layers on a nanocone quartz substrate [82].

Currently, the lack of low-cost, large-size, and high-quality nanowire array still remains an obstacle. Thus, more effort is required to exploration of new synthetic routes to obtain large-area nanowire arrays. Given the multiple functionalities of the nanowire and their possible coupling effects, such as piezoelectricity [84] and thermoelectricity [85], the use of nanowire arrays for three-dimensional PV applications is still a burgeoning field with outstanding potential.

Acknowledgments The work was supported by the DARPA Grant No. HR0011-07-1-0032, research grants from Louisiana Board of Regents Contract No. LEQSF(2008-11)-RD-B-10, and American Chemical Society Petroleum Research Fund PRF No. 48796-DN110.

References

1. P.V. Kamat, Meeting the clean energy demand: nanostructure architectures for solar energy conversion. J. Phys. Chem. C **111**, 2834 (2007)
2. M. Law, L.E. Greene, J.C. Johnson, R. Saykally, P. Yang, Nanowire dye-sensitized solar cells. Nat. Mater. **4**, 455 (2005)
3. M. Adachi, Y. Murata, J. Takao, J.T. Jiu, M. Sakamoto, F.M. Wang, Highly efficient dye-sensitized solar cells with a titania thin-film electrode composed of a network structure of single-crystal-like TiO2 nanowires made by the "oriented attachment" mechanism. J. Am. Chem. Soc. **126**, 14943 (2004)
4. J.B. Baxter, E.S. Aydil, Nanowire-based dye-sensitized solar cells. Appl. Phys. Lett. **86** (2005)
5. M. Law, L.E. Greene, J.C. Johnson, R. Saykally, P.D. Yang, Nanowire dye-sensitized solar cells. Nat. Mater. **4**, 455 (2005)
6. Y.M. Kang, N.G. Park, D. Kim, Hybrid solar cells with vertically aligned CdTe nanorods and a conjugated polymer. Appl. Phys. Lett. **86** (2005)

7. Q. Shen, K. Katayama, T. Sawada, M. Yamaguchi, T. Toyoda, Optical absorption, photo-electrochemical, and ultrafast carrier dynamic investigations of TiO2 electrodes composed of nanotubes and nanowires sensitized with CdSe quantum dots. Jpn. J. Appl. Phys. **45**, 5569 (2006)
8. K.S. Leschkies, R. Divakar, J. Basu, E. Enache-Pommer, J.E. Boercker, C.B. Carter, U.R. Kortshagen, D.J. Norris, E.S. Aydil, Photosensitization of ZnO nanowires with CdSe quantum dots for photovoltaic devices. Nano Lett. **7**, 1793 (2007)
9. Y. Zhang, L.-W. Wang, A. Mascarenhas, "Quantum coaxial cables" for solar energy harvesting. Nano Lett. **7**, 1264 (2007)
10. J. Schrier, D.O. Demchenko, L.W. Wang, Optical properties of ZnO/ZnS and ZnO/ZnTe heterostructures for photovoltaic applications. Nano Lett. **7**, 2377 (2007)
11. B.Z. Tian, X.L. Zheng, T.J. Kempa, Y. Fang, N.F. Yu, G.H. Yu, J.L. Huang, C.M. Lieber, Coaxial silicon nanowires as solar cells and nanoelectronic power sources. Nature **449**, 885 (2007)
12. Y. Zhang, M.D. Sturge, K. Kash, B.P. van der Gaag, A.S. Gozdz, L.T. Florez, J.P. Harbison, Temperature dependence of luminescence efficiency, exciton transfer, and exciton localization in GaAs/AlxGa1-xAs quantum wires and quantum dots. Phys. Rev. B **51**, 13303 (1995)
13. Y. Li, F. Qian, J. Xiang, C.M. Lieber, Nanowire electronic and optoelectronic devices. Mater. Today **9**, 18 (2006)
14. Y. Zhang, J. Pern, A. Mascarenhas, W. Zhou, Searching for optimal solar-cell architectures. SPIE Newroom (2008) doi: 10.1117/2.1200811.1388
15. K. Wang, J.J. Chen, W.L. Zhou, Y. Zhang, Y.F. Yan, J. Pern, A. Mascarenhas, Direct growth of highly mismatched type IIZnO/ZnSe core/shell nanowire arrays on transparent conducting oxide substrates for solar cell applications. Adv. Mater. **20**, 3248 (2008)
16. O.L. Muskens, J.G. Rivas, R.E. Algra, Epam Bakkers, A. Lagendijk, Design of light scattering in nanowire materials for photovoltaic applications. Nano Lett. **8**, 2638 (2008)
17. U. Gangopadhyay, S.K. Dhungel, P.K. Basu, S.K. Dutta, H. Saha, J. Yi, Comparative study of different approaches of multicrystalline silicon texturing for solar cell fabrication. Sol. Energy Mater. Sol. Cells **91**, 285 (2007)
18. Y. Inomata, K. Fukui, K. Shirasawa, Surface texturing of large area multicrystalline silicon solar cells using reactive ion etching method. Sol. Energy Mater. Sol. Cells **48**, 237 (1997)
19. Y.J. Lee, D.S. Ruby, D.W. Peters, B.B. McKenzie, J.W. Hsu, ZnO nanostructures as efficient antireflection layers in solar cells. Nano Lett. **8**, 1501 (2008)
20. M.D. Kelzenberg, S.W. Boettcher, J.A. Petykiewicz, D.B. Turner-Evans, M.C. Putnam, E.L. Warren, J.M. Spurgeon, R.M. Briggs, N.S. Lewis, H.A. Atwater, Enhanced absorption and carrier collection in Si wire arrays for photovoltaic applications. Nat. Mater. **9**, 239 (2010)
21. G. Chen, J. Wu, Q.J. Lu, H.R.H. Gutierrez, Q. Xiong, M.E. Pellen, J.S. Petko, D.H. Werner, P.C. Eklund, Optical antenna effect in semiconducting nanowires. Nano Lett. **8**, 1341 (2008)
22. Z.Y. Fan, H. Razavi, J.W. Do, A. Moriwaki, O. Ergen, Y.L. Chueh, P.W. Leu, J.C. Ho, T. Takahashi, L.A. Reichertz, S. Neale, K. Yu, M. Wu, J.W. Ager, A. Javey, Three-dimensional nanopillar-array photovoltaics on low-cost and flexible substrates. Nat. Mater. **8**, 648 (2009)
23. B. Liu E.S. Aydil, Growth of oriented single-crystalline rutile TiO2 nanorods on transparent conducting substrates for dye-sensitized solar cells. J. Am. Chem. Soc. **131**, 3985 (2009)
24. D. Kuang, J. Brillet, P. Chen, M. Takata, S. Uchida, H. Miura, K. Sumioka, S.M. Zakeeruddin, M. Gratzel, Application of highly ordered TiO2 nanotube arrays in flexible dye-sensitized solar cells. ACS Nano **2**, 1113 (2008)
25. J. Wang, Z.Q. Lin, Dye-sensitized TiO2 nanotube solar cells with markedly enhanced performance via rational surface engineering. Chem. Mater. **22**, 579 (2010)
26. W.T. Sun, Y. Yu, H.Y. Pan, X.F. Gao, Q. Chen, L.M. Peng, CdS quantum dots sensitized TiO2 nanotube-array photoelectrodes. J. Am. Chem. Soc.**130**, 1124 (2008)
27. K.S. Leschkies, R. Divakar, J. Basu, E. Enache-Pommer, J.E. Boercker, C.B. Carter, U.R. Kortshagen, D.J. Norris, E.S. Aydil, Photosensitization of ZnO nanowires with CdSe quantum dots for photovoltaic devices. Nano Lett. **7**, 1793 (2007)

28. A.B.F. Martinson, J.W. Elam, J.T. Hupp, M.J. Pellin, ZnO nanotube based dye-sensitized solar cells ZnO nanotube based dye-sensitized solar cells. Nano Lett. **7**, 2183 (2007)
29. P.V. Kamat, Quantum dot solar cells. Semiconductor nanocrystals as light harvesters. J. Phys. Chem. C **112**, 18737 (2008)
30. B. Li, L.D. Wang, B.N. Kang, P. Wang, Y. Qiu, Review of recent progress in solid-state dye-sensitized solar cells. Sol. Energy Mater. Sol. Cells **90**, 549 (2006)
31. M. Toivola, J. Halme, K. Miettunen, K. Aitola, P.D. Lund, Nanostructured dye solar cells on flexible substrates-review. Int. J. Energy Res. **33**, 1145 (2009)
32. S. Anandan, Recent improvements and arising challenges in dye-sensitized solar cells. Sol. Energy Mater. Sol. Cells **91**, 843 (2007)
33. A.L. Roest, M.A. Verheijen, O. Wunnicke, S. Serafin, H. Wondergem, Epam Bakkers, Position-controlled epitaxial III-V nanowires on silicon. Nanotechnology **17**, S271 (2006)
34. B. Mandl, J. Stangl, T. Martensson, A. Mikkelsen, J. Eriksson, L.S. Karlsson, G. Bauer, L. Samuelson, W. Seifert, Au-free epitaxial growth of InAs nanowires. Nano Lett. **6**, 1817 (2006)
35. S.G. Ihn, J.I. Song, T.W. Kim, D.S. Leem, T. Lee, S.G. Lee, E.K. Koh, K. Song, Morphology- and orientation-controlled gallium arsenide nanowires on silicon substrates. Nano Lett.**7**, 39 (2007)
36. Y.B. Tang, Z.H. Chen, H.S. Song, C.S. Lee, H.T. Cong, H.M. Cheng, W.J. Zhang, I. Bello, S.T. Lee, Vertically aligned p-type single-crystalline GaN nanorod arrays on n-type Si for heterojunction photovoltaic cells. Nano Lett. **8**, 4191 (2008)
37. W. Wei, X.Y. Bao, C. Soci, Y. Ding, Z.L. Wang, D. Wang, Direct heteroepitaxy of vertical InAs nanowires on Si substrates for broad band photovoltaics and photodetection. Nano Lett. **9**, 2926 (2009)
38. G.E. Cirlin, A.D. Bouravleuv, I.P. Soshnikov, Y.B. Samsonenko, V.G. Dubrovskii, E.M. Arakcheeva, E.M. Tanklevskaya, P. Werner, Photovoltaic properties of p-doped GaAs nanowire arrays grown on n-type GaAs(111)B substrate. Nano Res. Lett. **5**, 360 (2010)
39. Y.J. Hwang, A. Boukai, P.D. Yang, High density n-Si/n-TiO2 core/shell nanowire arrays with enhanced photoactivity. Nano Lett. **9**, 410 (2009)
40. Y.B. Guo, Y.J. Zhang, H.B. Liu, S.W. Lai, Y.L. Li, Y.J. Li, W.P. Hu, S. Wang, C.M. Che, D.B. Zhu, Assembled organic/inorganic p-n junction interface and photovoltaic cell on a single nanowire. J. Phys. Chem. Lett. **1**, 327 (2010)
41. T.J. Kempa, B.Z. Tian, D.R. Kim, J.S. Hu, X.L. Zheng, C.M. Lieber, Single and tandem axial p-i-n nanowire photovoltaic devices. Nano Lett. **8**, 3456 (2008)
42. V. Sivakov, G. Andra, A. Gawlik, A. Berger, J. Plentz, F. Falk, S.H. Christiansen, Silicon nanowire-based solar cells on glass: synthesis, optical properties, and cell parameters. Nano Lett. **9**, 1549 (2009)
43. B.D. Yuhas, P. Yang, Nanowire-based all-oxide solar cells. J. Am. Chem. Soc. **131**, 3756 (2009)
44. B. Tian, X. Zheng, T.J. Kempa, Y. Fang, N. Yu, G. Yu, J. Huang, C.M. Lieber, Coaxial silicon nanowires as solar cells and nanoelectronic power sources. Nature **449**, 885 (2007)
45. Y.J. Dong, B.Z. Tian, T.J. Kempa, C.M. Lieber, Coaxial group III-nitride nanowire photovoltaics. Nano Lett. **9**, 2183 (2009)
46. Q.L. Bao, C.M. Li, L. Liao, H.B. Yang, W. Wang, C. Ke, Q.L. Song, H.F. Bao, T. Yu, K.P. Loh, J. Guo, Electrical transport and photovoltaic effects of core-shell CuO/C-60 nanowire heterostructure. Nanotechnology **20**, 065203 (2009)
47. C. Colombo, M. Heiss, M. Gratzel, A.F.I. Morral, Gallium arsenide p-i-n radial structures for photovoltaic applications. Appl. Phys. Lett. **94** (2009)
48. T. Stelzner, M. Pietsch, G. Andra, F. Falk, E. Ose, S. Christiansen, Silicon nanowire-based solar cells. Nanotechnology **19**, 295203 (2008)
49. L. Tsakalakos, J. Balch, J. Fronheiser, B.A. Korevaar, O. Sulima, J. Rand, Silicon nanowire solar cells. Appl. Phys. Lett. **91**, 233117 (2007)
50. B.M. Kayes, H.A. Atwater, N.S. Lewis, Comparison of the device physics principles of planar and radial p-n junction nanorod solar cells. J. Appl. Phys. **97** (2005)

51. E.C. Garnett, P. Yang, Silicon nanowire radial p-n junction solar cells. J. Am. Chem. Soc. **130**, 9224 (2008)
52. K.Q. Peng, X. Wang, L. Li, X.L. Wu, S.T. Lee, High-performance silicon nanohole solar cells. J. Am. Chem. Soc. **132**, 6872 (2010)
53. S.E. Han, G. Chen, Optical absorption enhancement in silicon nanohole arrays for solar photovoltaics. Nano Lett. **10**, 1012 (2010)
54. J.A. Czaban, D.A. Thompson, R.R. LaPierre, GaAs core–shell nanowires for photovoltaic applications. Nano Lett. **9**, 148 (2009)
55. H. Goto, K. Nosaki, K. Tomioka, S. Hara, K. Hiruma, J. Motohisa, T. Fukui, Growth of core-shell InP nanowires for photovoltaic application by selective-area metal organic vapor phase epitaxy. Appl. Phys. Exp. **2** (2009)
56. M. Gratzel, Photoelectrochemical cells. Nature **414**, 338 (2001)
57. C.C. Wang, L.C. Chen, T.C. Wang, Nanocrystalline TiO2 solar cells sensitized with chlorophyll and ZnSe quantum dots. J. Optoelectron. Adv. Mater. **11**, 834 (2009)
58. P.R. Yu, K. Zhu, A.G. Norman, S. Ferrere, A.J. Frank, A.J. Nozik, Nanocrystalline TiO2 solar cells sensitized with InAs quantum dots. J. Phys. Chem. B **110**, 25451 (2006)
59. I. Robel, V. Subramanian, M. Kuno, P.V. Kamat, Quantum dot solar cells. Harvesting light energy with CdSe nanocrystals molecularly linked to mesoscopic TiO_2 films. J. Am. Chem. Soc. **128**, 2385 (2006)
60. W.U. Huynh, J.J. Dittmer, A.P. Alivisatos, Hybrid nanorod-polymer solar cells. Science **295**, 2425 (2002)
61. I. Gur, N.A. Fromer, M.L. Geier, A.P. Alivisatos, Air-stable all-inorganic nanocrystal solar cells processed from solution. Science **310**, 462 (2005)
62. S. Kim, B. Fisher, H.J. Eisler, M. Bawendi, Type-II quantum dots: CdTe/CdSe(core/shell) and CdSe/ZinTe(core/shell) heterostructures. J. Am. Chem. Soc. **125**, 11466 (2003)
63. S. Xu, Y. Wei, M. Kirkham, J. Liu, W. Mai, D. Davidovic, R.L. Snyder, Z.L. Wang, Patterned growth of vertically aligned ZnO nanowire arrays on inorganic substrates at low temperature without catalyst. J. Am. Chem. Soc. **130**, 14958 (2008)
64. J.J. Liu, M.H. Yu, W.L. Zhou, Well-aligned Mn-doped ZnO nanowires synthesized by a chemical vapor deposition method. Appl. Phys. Lett. **87**, 172505 (2005)
65. S.S. Lin, J.I. Hong, J.H. Song, Y. Zhu, H.P. He, Z. Xu, Y.G. Wei, Y. Ding, R.L. Snyder, Z.L. Wang, Phosphorus doped Zn1-xMgxO nanowire arrays. Nano Lett. **9**, 3877 (2009)
66. W.N. Lee, M.C. Jeong, J.M. Myoung, Fabrication and application potential of ZnO nanowires grown on GaAs(002) substrates by metal-organic chemical vapour deposition. Nanotechnology **15**, 254 (2004)
67. Y.S. Tian, C.G. Hu, Y.F. Xiong, B.Y. Wan, C.H. Xia, X.S. He, H. Liu, ZnO pyramidal arrays: novel functionality in antireflection. J. Phys. Chem. C **114**, 10265 (2010)
68. R. Tena-Zaera, J. Elias, C. Levy-Clement, ZnO nanowire arrays: optical scattering and sensitization to solar light. Appl. Phys. Lett. **93** (2008)
69. J.Y. Chen, K.W. Sun, Growth of vertically aligned ZnO nanorod arrays as antireflection layer on silicon solar cells. Sol. Energy Mater. Sol. Cells **94**, 930 (2010)
70. L. Ae, D. Kieven, J. Chen, R. Klenk, T. Rissom, Y. Tang, M.C. Lux-Steiner, ZnO nanorod arrays as an antireflective coating for Cu(In,Ga)Se-2 thin film solar cells. Prog. Photovoltaics **18**, 209 (2010)
71. K. Wang, J.J. Chen, Z.M. Zeng, J. Tarr, W.L. Zhou, Y. Zhang, Y.F. Yan, C.S. Jiang, J. Pern, A. Mascarenhas, Synthesis and photovoltaic effect of vertically aligned ZnO/ZnS core/shell nanowire arrays. Appl. Phys. Lett. **96** (2010)
72. M. Law, D.J. Sirbuly, J.C. Johnson, J. Goldberger, R.J. Saykally, P.D. Yang, Nanoribbon waveguides for subwavelength photonics integration. Science **305**, 1269 (2004)
73. C. Soci, A. Zhang, B. Xiang, S.A. Dayeh, D.P.R. Aplin, J. Park, X.Y. Bao, Y.H. Lo, D. Wang, ZnO nanowire UV photodetectors with high internal gain. Nano Lett. **7**, 1003 (2007)
74. M.Y. Lu, J.H. Song, M.P. Lu, C.Y. Lee, L.J. Chen, Z.L. Wang, ZnO-ZnS heterojunction and ZnS nanowire arrays for electricity generation. ACS Nano **3**, 357 (2009)

75. J. Yan, X.S. Fang, L.D. Zhang, Y. Bando, U.K. Gautam, B. Dierre, T. Sekiguchi, D. Golberg, Structure and cathodoluminescence of individual ZnS/ZnO biaxial nanobelt heterostructures. Nano Lett. **8**, 2794 (2008)
76. X. Wu, P. Jiang, Y. Ding, W. Cai, S.S. Xie, Z.L. Wang, Mismatch strain induced formation of ZnO/ZnS heterostructured rings. Adv. Mater. **19**, 2319 (2007)
77. H. Wang, M. Upmanyu, C.V. Ciobanu, Morphology of epitaxial core-shell nanowires. Nano Lett. **8**, 4305 (2008)
78. J.F. Scott, T.C. Damen, W.T. Silfvast, R.C.C. Leite, L.E. Cheesman, Resonant Raman scattering in ZnS and ZnSe with the cadmium laser. Opt. Commun. **1**, 397 (1970)
79. Y.Y. Luo, G.T. Duan, G.H. Li, Resonant Raman scattering and surface phonon modes of hollow ZnS microspheres. Appl. Phys. Lett. **90**, 201911 (2007)
80. B.B. Cao, J.J. Chen, X.J. Tang, W.L. Zhou, Growth of monoclinic WO3 nanowire array for highly sensitive NO2 detection. J. Mater. Chem. **19**, 2323 (2009)
81. C. Soci, A. Zhang, B. Xiang, S.A. Dayeh, D.P.R. Aplin, J. Park, X.Y. Bao, Y.H. Lo, D. Wang, ZnO nanowire UV photodetectors with high internal gain. Nano Lett. **7**, 1003 (2007)
82. J. Zhu, C.M. Hsu, Z.F. Yu, S.H. Fan, Y. Cui, Nanodome solar cells with efficient light management and self-cleaning. Nano Lett. **10**, 1979 (2010)
83. Z.Y. Fan, D.J. Ruebusch, A.A. Rathore, R. Kapadia, O. Ergen, P.W. Leu, A. Javey, Challenges and prospects of nanopillar-based solar cells. Nano Res. **2**, 829 (2009)
84. F. Boxberg, N. Sondergaard, H.Q. Xu, Photovoltaics with piezoelectric core-shell nanowires. Nano Lett. **10**, 1108 (2010)
85. A.I. Hochbaum, R.K. Chen, R.D. Delgado, W.J. Liang, E.C. Garnett, M. Najarian, A. Majumdar, P.D. Yang, Enhanced thermoelectric performance of rough silicon nanowires. Nature **451**, 163 (2008)

Chapter 17
Supercapacitors Based on 3D Nanostructured Electrodes

Hao Zhang, Gaoping Cao, and Yusheng Yang

Climate change, the decreasing availability of fossil fuels vs. the increasing demand for them, and atmospheric pollution caused by combustion engines of automotive systems require society to move toward sustainable and renewable resources [1]. As a result, we observe an increase in renewable energy production from sun and wind, as well as the development of electric vehicles or hybrid electric vehicles with low CO_2 emissions. Because the sun does not shine at night, the wind does not blow on command, and we expect to drive an autonomous car for at least a few hours, energy storage systems are starting to play a larger part in our lives [1, 2]. In response to the needs of modern society and emerging ecological concerns, it is now essential that new, low-cost, and environmentally friendly energy conversion and storage systems are found. At the forefront of these electrochemical energy storage systems are lithium–ion batteries [3, 4], fuel cells [5], solar cells, and supercapacitors (SCs) [6, 7].

High performance of such technologies is mainly achieved through designed sophisticated device structures with multiple materials, for example, carbon nanotube array-based composite electrodes in SCs [8]. As the performance of these devices also depends intimately on the properties of their materials, considerable attention has been paid to the research and development of key materials. However, we still need to improve materials' performance substantially to meet the higher requirements of future systems, ranging from portable electronics to hybrid electric vehicles and large industrial equipment. Now SCs have low energy and moderate power density largely resulting from poor charge and mass transport properties of electrodes. New electrode materials that possess unique nanostructures and/or are chemically modified through molecular or atomic engineering would offer significantly enhanced properties for high-density energy/power storage devices.

In a word, the development of new-generation SCs depends on the development of innovative 3D nanostructured electrodes and materials and on advancing our understanding of electrochemical interfaces at the nanoscale. Further breakthroughs

---

H. Zhang (✉)
Research Institute of Chemical Defense, Beijing 100191, China
e-mail: dr.h.zhang@hotmail.com

W.L. Zhou, Z.L. Wang (eds.), *Three-Dimensional Nanoarchitectures*,
DOI 10.1007/978-1-4419-9822-4_17,

in electrode structures and materials, not incremental changes, hold the key to new generations of energy storage and conversion devices.

This chapter aims to give a concise and useful survey of recent progress on the development and application of nanostructured electrode materials for SCs, starting with a brief overview on SCs and electrode materials.

17.1 Supercapacitors

Lithium-ion batteries were introduced in 1990 by Sony, following pioneering work by Whittingham, Scrosati, and Armand (see [9] for a review). These batteries, although costly, are the best in terms of performance, with energy densities that can reach 180 Wh/kg. Although great efforts have gone into developing high-performance Li-ion and other advanced secondary batteries that use nanomaterials or organic redox couples [10–12], SCs have attracted less attention until very recently. Because Li-ion batteries suffer from a somewhat slow power delivery or uptake, faster and higher power energy storage systems are needed in a number of applications, and this role has been given to the SCs [13]. Also known as electrochemical capacitors or ultracapacitors, SCs are power devices that can be fully charged or discharged in seconds; as a consequence, their energy density (about 5 Wh/kg) is lower than in batteries, but a much higher power delivery or uptake (10 kW/kg) can be achieved for shorter times (a few seconds) [6]. They have had an important role in complementing or replacing batteries in the energy storage field, such as for uninterruptible power supplies (backup supplies used to protect against power disruption) and load leveling. A more recent example is the use of SCs in emergency doors (16 per plane) on an Airbus A380, thus proving that in terms of performance, safety, and reliability SCs are definitely ready for large-scale implementation. A recent report by the US Department of Energy assigns equal importance to supercapacitors and batteries for future energy storage systems [14], and articles on supercapacitors appearing in business and popular magazines show increasing interest by the general public in this topic.

It is undoubted that SCs offer another option to store electrical energy, but with a different balance of performance metrics: energy density is an order of magnitude lower than that of batteries, while power density is several orders of magnitude higher [15]. SCs offer significantly higher energy density than conventional electrostatic capacitors, making them attractive for peak power demands, which require a combination of energy and power densities [6]. Unlike electrostatic capacitors, SCs store energy in a layer of ions that assemble at the interface between the electrolyte and the electrode (with its excess or deficit of surface electronic charge), thereby creating an electrochemical double layer. In addition to the capacitance of the electrochemical double layer, which tracks the surface area of the electrified interface, some materials exhibit pseudo-capacitance arising from surface redox reactions. The magnitude of energy that can be stored per area or volume or mass of the electrode concomitantly increases by the addition of stored electronic charge to yield higher specific capacitance. Although faradic processes occur, the current–voltage characteristics mimic the featureless curves seen for double layer capacitance.

From the introduction above, several types of SCs can be distinguished, depending on the charge storage mechanism as well as the active materials used. Electrochemical double layer capacitors (EDLCs), the most common devices at present, use carbon-based active materials with high surface area. A second group of SCs, known as pseudo-capacitors or redox supercapacitors, uses fast and reversible surface or near-surface reactions for charge storage. Transition metal oxides as well as electrically conducting polymers are examples of pseudo-capacitive active materials. Hybrid capacitors, combining a capacitive or pseudo-capacitive electrode with a battery electrode, are the latest kinds of ECs which benefit from both the capacitor and the battery properties.

Here we give an overview of recent findings as well as an analysis of what the future holds for nanostructured SCs.

17.2 Electrochemical Double Layer Capacitors Based on 3D Nanostructured Electrodes

EDLCs are electrochemical capacitors that store the charge electrostatically using reversible adsorption of ions of the electrolyte onto active materials that are electrochemically stable and have high accessible specific surface area (SSA). Charge separation occurs on polarization at the electrode–electrolyte interface, producing what Helmholtz described in 1853 as the double layer capacitance C:

$$C = \varepsilon_r \varepsilon_0 / Ad \text{ or } C/d = \varepsilon_r \varepsilon_0 / A \tag{17.1}$$

where ε_r is the electrolyte dielectric constant, ε_0 is the dielectric constant in vacuum, d is the effective thickness of the double layer (charge separation distance), and A is the electrode surface area. This capacitance model was later refined by Gouy and Chapman, and Stern and Geary, who suggested the presence of a diffuse layer in the electrolyte due to the accumulation of ions close to the electrode surface. The double layer capacitance is between 5 and 20 $\mu F/cm^2$ depending on the electrolyte used [16]. Specific capacitance achieved with aqueous alkaline or acid solutions is generally higher than in organic electrolytes, but organic electrolytes are more widely used as they can sustain a higher operation voltage (up to 2.7 V in symmetric systems). Because the energy stored is proportional to the square of voltage according to

$$E = 1/2CV^2 \tag{17.2}$$

a three-fold increase in voltage, V, results in about an order of magnitude increase in energy, E, stored at the same capacitance. As a result of the electrostatic charge storage, there is no faradic (redox) reaction at EDLC electrodes. A supercapacitor electrode must be considered as a blocking electrode from an electrochemical point of view. This major difference from batteries means that there is no limitation by the electrochemical kinetics through a polarization resistance. In addition, this surface storage mechanism allows very fast energy uptake and delivery and then better power performance.

The absence of faradic reactions also eliminates the swelling in the active material that batteries show during charge/discharge cycles. EDLCs can sustain millions of cycles whereas batteries survive a few thousand at best. Finally, the solvent of the electrolyte is not involved in the charge storage mechanism, unlike in Li-ion batteries where it contributes to the solid–electrolyte inter-phase when graphite anodes or high-potential cathodes are used. This does not limit the choice of solvents, and electrolytes with high-power performances at low temperatures (down to –40°C) can be designed for EDLCs. However, as a consequence of the electrostatic surface charging mechanism, these devices suffer from a limited energy density. This explains why today's EDLC research is largely focused on increasing their energy performance and widening the temperature limits into the range where batteries cannot operate [7].

The key to reaching high capacitance by charging the double layer is in using high SSA blocking and electronically conducting electrodes. As more SSA leads to larger capacitance, the electrodes in EDLCs are necessarily high surface area objects; they are designed as porous structures and use various high surface area carbons as active materials. Graphitic carbon satisfies all the requirements for EDLC application, including high conductivity, electrochemical stability, and open porosity [17, 18]. Activated, templated, and carbide-derived carbons [19, 20], carbon fabrics, fibers, nanotubes [21], onions [22], and nanohorns [23] have been tested for EDLC applications [17]. In addition, a morphological variety of carbons including powders, nanofoams (such as aerogels), nanofibers, nanotubes, cloths, and nanosheets have been investigated in aqueous and nonaqueous electrolytes: capacitance values as high as 300 F/g have been realized [24–26]. As ions need to approach and be expelled from the electrified interface during charge–discharge, the rate of electrolyte mass transport through the pore network of the electrode structure plays a critical role in determining the magnitude of the normalized capacitance, the frequency of the capacitive response, and thus the power density. Therefore, either the energy density or the power density of EDLCs relies on the nanostructures of carbon materials and the nanoarchitectures of electrodes.

Based on the nanostructure (pore structures) of electroactive carbons and the nanoarchitecture of electrodes, the EDLC electrodes are divided into four categories and expatiated as follows.

17.2.1 Electrodes Based on Activated Carbons and Activated Carbon Fibers: Powdered Carbons with Disordered Pore Structures

The first patent describing the concept of an electrochemical capacitor was filed in 1957 by Becker, who used activated carbon (AC) with a high SSA coated on a metallic current collector in a sulfuric acid solution. Now even after more than a half-century, activated carbons are still the most popular electrode materials for SCs. ACs are the most widely used EDLC materials today, because of their high SSA and moderate cost. ACs are derived from carbon-rich organic precursors by carbonization (heat treatment) in inert atmosphere with subsequent selective oxidation in CO_2, water vapor, or KOH to increase the SSA and pore volume. Natural materials, such as coconut shells, wood, pitch, or coal, or synthetic materials, such as polymers, can be used as precursors. A porous network in the bulk of the carbon particles is produced after activation; micropores (<2 nm in size), mesopores (2–50 nm), and macropores (>50 nm) can be created in carbon grains. Accordingly, the porous structure of carbon is characterized by a broad distribution of pore size. Longer activation time or higher temperature leads to larger mean pore size. The double layer capacitance of AC reaches 100–150 F/g in organic electrolytes; this value can exceed 150–350 F/g in aqueous electrolytes, but at a lower cell voltage because the electrolyte voltage window is limited by water decomposition. A typical cyclic voltammogram of a two-electrode EDLC laboratory cell is presented in Fig. 17.1. Its rectangular shape is characteristic of a pure double layer capacitance mechanism for charge storage according to

$$I = C \times \mathrm{d}V/\mathrm{d}t \tag{17.3}$$

where I is the current, ($\mathrm{d}V/\mathrm{d}t$) is the potential scan rate, and C is the double layer capacitance. Assuming a constant value for C, for a given scan rate the current I is constant, as can be seen from Fig. 17.1, where the cyclic voltammogram has a rectangular shape.

Generally, the larger the specific surface area of carbons, the higher the ability for charge accumulation in the electrode/electrolyte interface. It is well known that micropores (<2 nm in size, as mentioned previously) play an essential adsorption

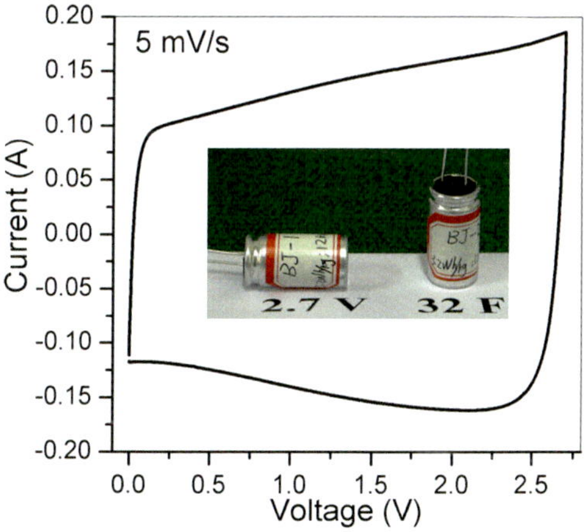

Fig. 17.1 Cyclic voltammetry of a two-electrode laboratory EDLC cell in 1.0 M tetraethylammonium tetrafluoroborate in acetonitrile-based electrolyte, containing activated carbon powders coated on aluminum current collectors. Cyclic voltammetry was recorded at room temperature and potential scan rate of 5 mV/s. *Inset* is a picture of EDLC cells

role for the formation of the electrochemical double layer. However, these micropores must be electrochemically accessible for ions; therefore, the presence of mesopores (2–50 nm) is necessary for efficient charge propagation to the bulk of the electrode material, allowing the so-called frequency response to be fulfilled, i.e., the energy to be extracted at higher frequencies (e.g., 1 Hz). Hence, the availability and wettability of pores, with dimensions adapted to the size of solvated anions and cations which have to be transported from the electrolytic solution, is crucial for high capacitor performance [18]. It can often be found in the literature that the higher the BET specific surface area, the higher the capacitance values. However, for a wide variety of ACs, this trend is not perfectly followed. In fact, the narrow micropores may not contribute to the total double layer capacitance due to a sieving effect whose extent depends on the size of the electrolyte ions [27–29], which explains the absence of proportionality with BET surface area [30]. In addition, the larger values in aqueous electrolyte are essentially justified by a smaller size of solvated ions and a higher dielectric constant than in organic media.

Both ACs and activated carbon fibers (ACFs) are micropore-dominated materials, which is attributed to the activation process. There are also a small quantity of meso- and macropores in these materials [31]. A pore size distribution (PSD) curve of a typical AC is shown in Fig. 17.2. The pore structure of AC and ACF is irregular, i.e., the pores are connected with each other randomly, and some of them are closed pores. These structures will hinder the ion transfer during charge/discharge process and lead to inferior performance, especially under high-power operations.

The microstructure of electrodes based on AC and ACF is also disordered. Because the conductivity of AC and ACF is moderate, conductive additives must be added into the electrode to ensure high electronic conductivity of electrode. It is well known that electrochemical performance is strongly affected by the property of conductive additives. Effects of conductive additive on electrochemical performance of battery have been fully studied [32–34]. SEM image of carbon black clusters, AC particles, and composite electrodes with 5 wt% carbon black is shown in Fig. 17.3. Carbon black clusters are formed by sub-micrometer primary particles and the diameter of carbon black clusters ranges from 2 to 10 μm. The diameter of AC particles ranges from 2 to 20 μm. The morphology and microstructure of

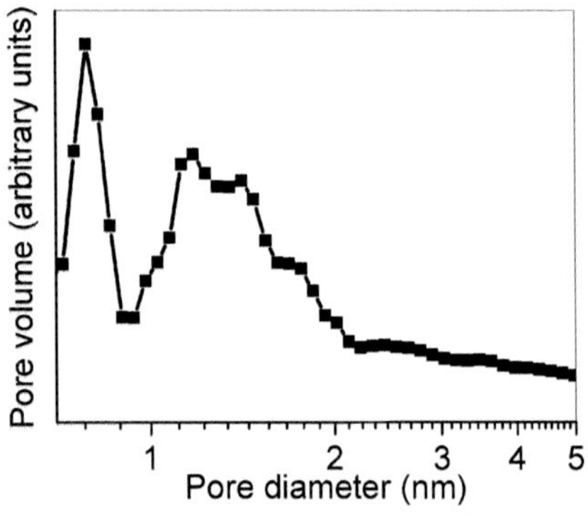

Fig. 17.2 A pore size distribution curve of a typical activated carbon

Fig. 17.3 SEM images of (**a**) carbon black, (**b**) AC, composite electrodes with (**c**) 5 wt% carbon black (from Solid State Ionics). Reprinted with permission from [34]. Copyright 2008 (Elsevier)

the composite electrodes vary with the carbon black contents. Figure 17.3c is a SEM image of an electrode with moderate carbon black (5 wt%). It is obvious that carbon black particles fill the voids among AC particles. The conductive network is formed by both AC and carbon black particles. Carbon black has good electronic conductivity; thus, the composite electrode's electronic conductivity is increased. And there are pores between AC and carbon black particles. However, the conductive paths in these composite electrodes are irregular and zigzag, and the contact points among AC/carbon black particles are overmany, leading to poor conductivity (see Fig. 17.4). The pore structure formed by the accumulation of AC and carbon black particles is also irregular, hindering the ion transfer during charge/discharge process.

17.2.2 Electrodes Based on Carbon Foams, Carbon Aerogels, and Other Monolithic Carbon: Monolithic Carbon with Disordered Micropores

From the viewpoint of many practical applications, monolithic carbons are easier to handle than powdered materials. They also present developed conductive paths at macroscopical scale, leading to better power performance when used as electrodes.

The shaping of carbon materials is often rather difficult, owing to the elastic properties of the grains, and therefore, direct shaping of the materials during their generation can be highly advantageous. In general, carbon monoliths are fabricated

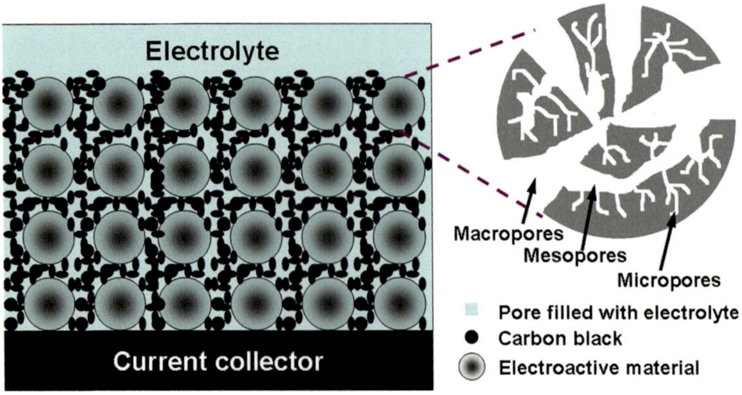

Fig. 17.4 The schematic representation of the microstructure of AC/carbon black composite electrodes

by extrusion or, directly, by wet chemistry (e.g., sol–gel process) [35, 36]. However, fine-tuning of the pore structure of the carbon monoliths is difficult to achieve using these pathways, and they also have the disadvantage of needing binders and additives in the case of extrusion or supercritical drying/freeze-drying during the sol–gel process [37]. In comparison, the nanocasting pathway from shaped precursors provides an opportunity to create monolithic carbons with an ordered or hierarchical structure, and the pore sizes are tailor-made to some extent, if the integrity of the template can be maintained in the cast carbon [38–40]. Monolithic mesoporous carbon with a bicontinuous cubic structure (*Ia*3*d* symmetry) was prepared by using mesoporous silica monoliths as the hard template [41]. This monolithic carbon shows a concentrated pore size of 4.6 nm and a surface area of 1530 m^2/g. The difficulty in preparing such carbon monoliths is, in fact, in the preparation of the noncracked silica monoliths needed as templates. Owing to the stresses involved in the preparation, cracking of the structures can occur at different stages, the most sensitive ones being the drying stage, due to capillary forces, and the calcination stage, due to the temperature-induced stress and/or pressure developing inside the monoliths during the decomposition and combustion of the organic material in the pore system. Thus, the preparation of monolithic mesoporous carbon with an ordered mesostructure is still a great challenge.

Related to the synthesis proposed by Nakanishi [42], silica monoliths with a hierarchical structure containing macropores and mesopores can be prepared by adding poly(ethylene glycol) and/or hexadecyltrimethylammonium bromide as a porogen [43]. Using such silica monoliths as templates and furfuryl alcohol or sucrose as a carbon precursor, carbon monoliths with well-developed porosity are accessible [41–46]. Interestingly, the pore system of the nanocast carbon monoliths can be varied to three- or four-modal porosity by varying the loading amount of furfuryl alcohol in the one-step impregnation. Regardless of loading with the carbon precursor, the obtained carbon monolith is a positive replica of the silica monolith

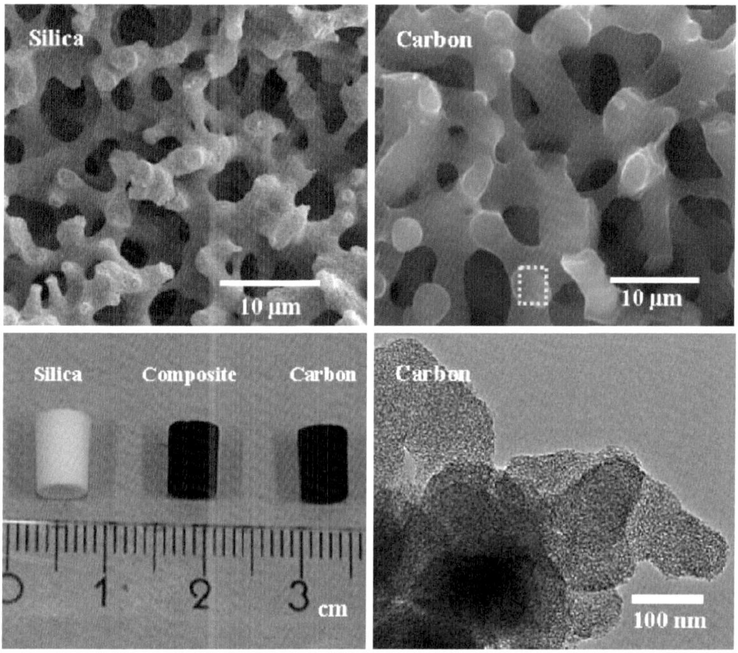

Fig. 17.5 Scanning electron microscopy (SEM) images (*upper*), photograph (*lower, left*), and TEM image (*lower, right*) of silica and carbon monoliths (from Adv. Mater.). Reprinted with permission from [46]. Copyright 2006 (Wiley)

on the micrometer scale, and a negative replica on the nanometer scale, as shown in Fig. 17.5. Combined volume and surface templating, together with the controlled synthesis of the starting silica monoliths used as the scaffold, provides a flexible means of pore size control on several length scales simultaneously [40, 44].

Carbon aerogels, i.e., a monolithic 3D mesoporous network of carbon nanoparticles, are obtained by the pyrolysis of organic aerogels based on resorcinol-formaldehyde (RF) or phenol-furfural (PF) precursors via a sol–gel process. The gel composition (catalyst, precursor, solid ratio) and the pyrolysis temperature determine the microtexture of the final product, especially the particle size and the pore distribution. In order to simplify their production, a supercritical drying of the RF gels is favored with a very low catalyst concentration, with high molar resorcinol to catalyst (*R*/*C*) ratios. The catalyst concentration controls the particle sizes, and the degree of dilution determines the density of the material. The special porosity of aerogels is based on the interconnection of carbon nanoparticles of the same size, leading to a concentrated mesoporous distribution with a specific surface area between 500 and 900 m^2/g and a high pore volume (0.4–2.6 cm^3/g). Carbon aerogels with pore diameters in the range from 3 to 13 nm pointed out the best voltammetry characteristics and the highest capacitance values (70–150 F/g).

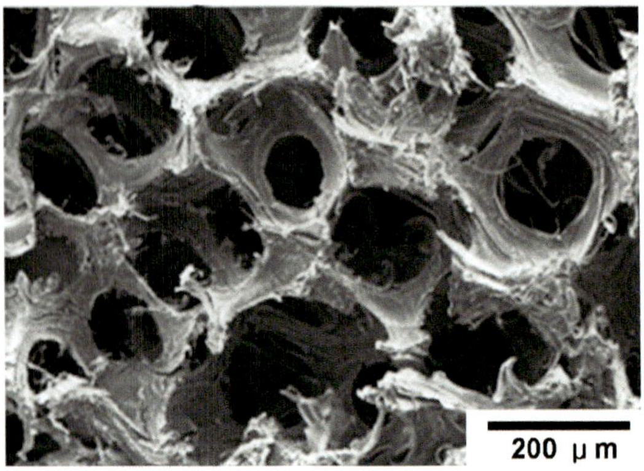

Fig. 17.6 A SEM image of a pitch-derived carbon foam

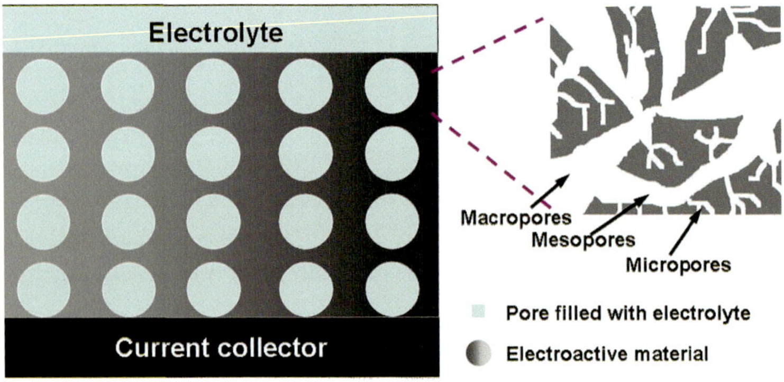

Fig. 17.7 The schematic representation of the microstructure of monolithic carbon electrodes

Carbon foam is another carbon monolithic carbon. Although pitch-based carbon foam (Fig. 17.6) presents developed conductive paths, its SSA is very low, leading to a low specific capacitance. Overall, the conductive network is developed in monolithic carbons. However, their PSD should be optimized to enhance the ionic transfer and the SSA should also be enhanced to obtain high capacitance (see Fig. 17.7).

17.2.3 Electrodes Based on Template Carbons, Graphene, Carbide-Derived Carbons, and Hierarchical Porous Carbons: Powdered Carbons with High Mesopore Ratios or Reasonable PSD

Although monolithic carbons present regular conductive paths, their pore size and pore structure need to be optimized. Template method (also called nanocasting) is a powerful method for creating porous materials that are more difficult to synthesize

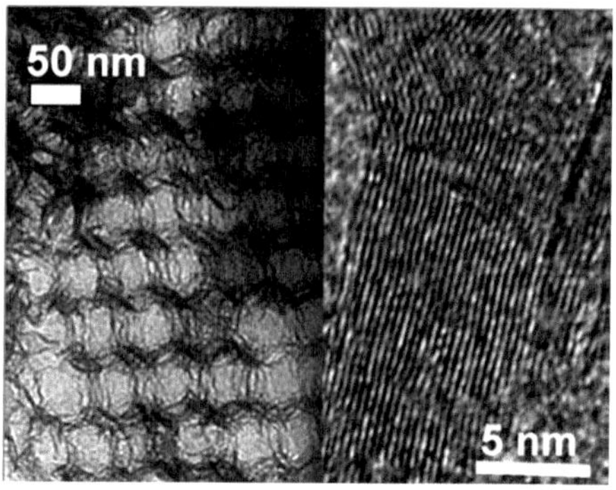

Fig. 17.8 TEM images of ordered nanoporous carbon graphitized under argon at 2500°C (from J. Am. Chem. Soc.). Reprinted with permission from [53]. Copyright 2005 (American Chemical Society)

by conventional processes. Especially, this method is very powerful for the fabrication of mesopore-dominated carbons with regular pore structures. Figure 17.8 shows TEM images of ordered nanoporous carbon prepared by template methods.

The nanocasting pathway to create nanostructured materials involves three main steps: (i) formation of the template; (ii) the casting step with target precursors, including the conversion of the precursor, which is typically molecular, to a solid; and (iii) removal of the template, as shown in Fig. 17.9. Inorganic, ordered porous solids are mostly used as the nanoscale hard template in the first step. For instance, zeolites [47], alumina membranes [48], ordered mesoporous silica [49], ordered mesoporous carbon [50], or, for somewhat larger structure sizes, the assembly structure of colloidal spheres have been employed as a true template to replicate other materials [51]. The pore structures of these parent materials can be transferred to the solid structure of the generated porous materials, while the walls of the parent become the voids of the replica. In order to be able to control the morphology and structural parameters of the replicas, templates with a controllable morphology and structure are highly advantageous. One of the most versatile hard templates is ordered mesoporous silica, since it can be prepared in shapes as diverse as noodle-like, spherical, fibrous, rod-like, and even with chiral morphologies [52–54]. Another requirement for the template is the ability to remove it without affecting the cast. Possibilities are leaching with different agents, melting,

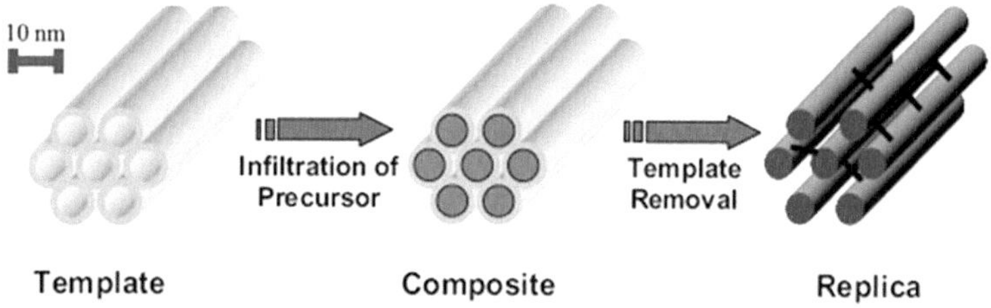

Fig. 17.9 Schematic illustration of the nanocasting pathway (from Adv. Mater.). Reprinted with permission from [46]. Copyright 2006 (Wiley)

or combustion, which is possible with a carbon template. The successful synthesis of carbon with an ordered pore structure was first achieved by Ryoo's group in 1999, whereby MCM-48 was used as a template to create a carbon material (CMK-1) [49]. Previous attempts to replicate the narrower pore systems of zeolites had given indications that a replication should be possible, but the materials lacked the perfection of the CMK-1 structure. Starting with the initial publication from the Ryoo group, many studies were carried out to synthesize mesoporous carbons with ordered structures.

Compared with ACs, template carbons obtain higher rate performance and comparable capacitance, attributed to their higher mesopore ratio and high surface areas [55]. The pore volumes of template carbons are larger than that of ACs, leading to the fact that their packing density is lower than the latter. Therefore, the pore size distribution of template carbons should be further optimized.

Graphene is a novel functional material; its fabrication, microstructure, and application in energy storage have been hot topics these years. Graphene can be prepared by both chemical method and physical method (such as arc discharge method). They present high conductivity and open mesoporous structure (see Fig. 17.10), leading to superior power performance. However, the SSA of graphene is moderate, leading to low specific capacitance. We prepared a graphene material by arc discharge method and studied its pore structures and electrochemical capacitive properties. This graphene presents developed and open mesopore structure (see Fig. 17.10), and its specific surface area and mesopore ratio are 77.8 m^2/g and 74.7%, respectively. When used as an electrochemical capacitor electrode material, graphene gets a capacitance of 12.9 F/g, its cyclic voltammograms show rectangular shape even under a high scan rate of 200 mV/s, and the f_0 on the electrochemical impedance spectroscopy is as high as 18.5 Hz, exhibiting excellent rate capability.

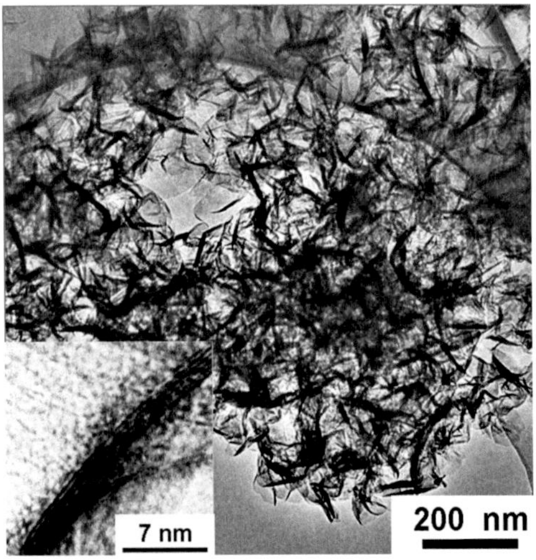

Fig. 17.10 TEM and high-resolution TEM (*inset*) images of graphene

Carbon-derived carbons (CDCs) are another high-performance electrode material with concentrated pore size distributions. These are porous carbons obtained by extraction of metals from carbides (TiC, SiC, etc.) by etching in halogens at elevated temperatures [56]:

$$TiC + 2Cl_2 \rightarrow TiCl_4 + C \qquad (17.4)$$

In this reaction, Ti is leached out from TiC, and carbon atoms self-organize into an amorphous or disordered, mainly sp^2 bonded [57], structure with a pore size that can be fine-tuned by controlling the chlorination temperature and other process parameters. Accordingly, a narrow unimodal pore size distribution can be achieved in the range 0.6–1.1 nm, and the mean pore size can be controlled with sub-ångström accuracy [58]. These materials were used to understand the charge storage in micropores using 1 M solution of NEt_4BF_4 in acetonitrile-based electrolyte [59]. The normalized capacitance (μF/cm^2) decreased with decreasing pore size until a critical value close to 1 nm was reached and then sharply increased when the pore size approached the ion size [60, 61]. The TEM micrographs of TiC-CDC synthesized at various temperatures are shown in Fig. 17.11. High-resolution TEM analysis of TiC-CDC and ZrC-CDC shows a logical progression of increasing order with increasing synthesis temperature occurring roughly in these temperature regimes. At 600°C, both TiC and ZrC-CDCs are amorphous (Fig. 17.11a, d). At 800°C, the structures contain highly curved graphene planes (Fig. 17.11b, e). TiC-CDC produced at 800°C has a higher content of fullerenic shell-type structure than ZrC-CDC synthesized at the same temperature and looked similar to

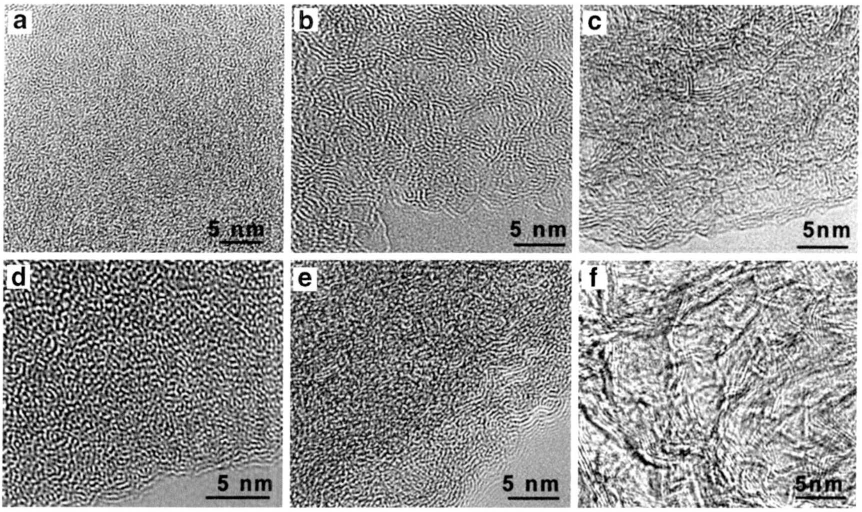

Fig. 17.11 TEM micrographs of TiC-CDC synthesized at (**a**) 600°C, (**b**) 800°C, and (**c**) 1200°C and ZrC-CDC synthesized at (**d**) 600°C, (**e**) 800°C, and (**f**) 1200°C (from J. Power Sources). Reprinted with permission from [61]. Copyright 2006 (Elsevier)

the onion-like structures obtained by Zheng et al. [33]. For a synthesis temperature of 1200°C, straight graphite fringes having interlayer spacing approaching that of pristine graphite (0.335 nm) are clearly visible. It is obvious that CDCs present regular micropore structures. Compared with ACs, CDCs obtain higher packing density because of their optimized PSD. In addition, CDCs present good rate performance, such as a TiC-based CDC with 130 F/g under a current density of 100 mA/cm^2.

As previously mentioned, porous carbon-based SCs are known to suffer from electrode kinetic problems that are related to inner-pore ion transport. The exact mechanism of ion transport within porous textures is very complex, because the tortuosity, connectivity, size distribution, and shape of the pores, as well as the nature of the electrolyte and the solid/liquid interface, all have to be considered [6, 62–64]. Among these factors, the inner-pore ion-transport resistance and the diffusion distance are the most important ones. Large values of these two parameters lead to a significant electrode potential drop (IR drop) and a low ion-accessible surface area at large current values, thus severely reducing the performance of the EDLCs. Microporous activated carbons and CDCs have a long diffusion distance and a high ion-transport resistance, which lead to a large IR drop and a small S_{access} at high current values. On the other hand, ordered mesoporous carbons prepared by template method have mesoporous channels (of about 4–6 nm) with a lower ion-transport resistance and smaller particles (exhibiting a size of 1–2 mm) with a shorter diffusion route. However, the porous texture of mesoporous carbons cannot fulfill high-rate (<2 s) SC applications, where larger mesopores (>10 nm) and nanometer-scale diffusion distances (<100 nm) are required [65].

Recently, Cheng and coworkers reported the electrochemical performance of 3D aperiodic hierarchical porous graphitic carbon (HPGC) as a promising electrode material for high-rate SC applications [65]. The 3D hierarchical porous texture combines macroporous cores, mesoporous walls, and micropores. The physicochemical properties of the electrolyte in the macropores are similar to those of the bulk electrolyte with the lowest resistance [62]. Ion-buffering reservoirs can be formed in the macropores to minimize the diffusion distances to the interior surfaces. Furthermore, the mesoporous walls provide low-resistant pathways for the ions through the porous particles [66–70], and the micropores strengthen the electrochemical double layer capacitance [59, 71]. This 3D architecture composed of macroporous cores and mesoporous walls is self-supported and can overcome nanoparticle aggregation. In this design, three electrochemical processes are involved, namely buffering ions in the macroporous cores, transporting ions through the mesoporous walls, and confining ions in the micropores. Moreover, the localized graphitic structure can enhance the electric conductivity. The experimental results show that both the energy and the power densities at high rate are greatly improved when HPGC replaces AC or mesoporous carbons as the EDLC electrode material.

HPGC can be fabricated by multi-template method or active template method [72]. The microstructure of a typical HPGC is shown in Fig. 17.12. The SEM image (Fig. 17.12a) shows the texture of the macroporous cores, which have a

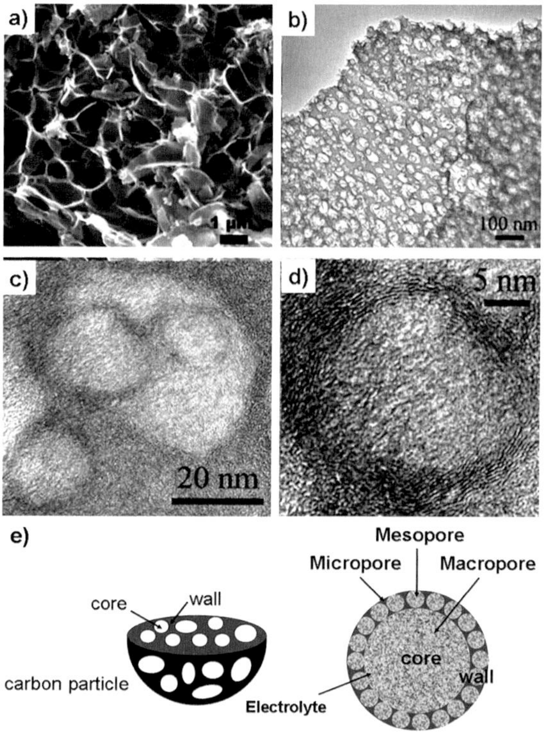

Fig. 17.12 (**a**) SEM image of the macroporous cores of the HPGC material, (**b**) TEM image of the mesoporous walls, (**c**) TEM image showing the micropores, (**d**) high-resolution TEM (HRTEM) image of the localized graphitic mesopore walls, and (**e**) schematic representation of the 3D hierarchical porous texture (from Angew. Chem. Int. Ed.). Reprinted with permission from [65]. Copyright 2008 (Wiley)

diameter of about 1 mm; these cores extend into the particles, thus forming ion-buffering reservoirs. The thickness of the walls around them is less than 100 nm. When the macroporous cores are immersed in the electrolyte, the thin walls around them are covered by it, thus giving a diffusion distance shorter than 50 nm. The TEM image (Fig. 17.12b) reveals the mesoporous texture of the walls. The mesopores have diameters between 10 and 50 nm, and they can provide a short ion-transport pathway through the walls, with a minimized inner-pore resistance. There are micropores around the mesopores (Fig. 17.12c), and the mesopore walls consist of localized graphitic structures, which leads to an enhanced electric conductivity (see Fig. 17.12d). The pore sizes determined from TEM measurements agree well with those derived from nitrogen adsorption analysis. A schematic illustration of a 3D hierarchical porous texture is shown in Fig. 17.12e. The characteristics of a 3D self-supporting macroporous network, including meso–microporous walls and localized graphitic structures, make HPGC materials more convenient than hollow carbon foams or mesoporous carbons (which do not contain 3D macropores) and ordered macroporous carbons (which do not have mesopores or graphitic structures). The new HPGC, which is unique in its structure, shows promise as a good electrode material for advanced EC technology. The new material exhibits high energy and power densities in both aqueous and organic electrolytes.

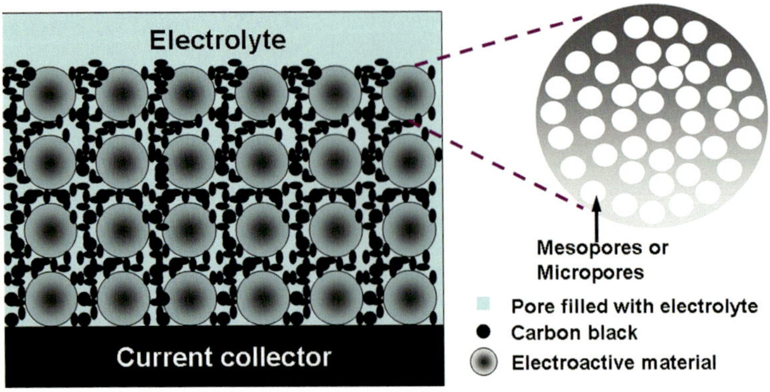

Fig. 17.13 The schematic representation of the microstructure of template carbon electrodes

In a word, template carbons, graphene, carbide-derived carbons, and hierarchical porous carbons present ordered structure at microscopical scale, ensuring to obtain high capacitance or low ion transfer resistance; however, they are powered materials, which are harder to handle than monolithic carbons for the fabrication of the electrodes (see Fig. 17.13). The electrode materials with both controllable pore structure and developed conductive paths at macroscopical scale are highly desired.

17.2.4 Electrodes Based on Carbon Nanotubes: Monolithic Carbons with Developed Mesoporous Structures

Due to their nanometer size and outstanding properties, carbon nanotubes (CNTs) are of great interest for many applications, namely batteries [73], storage of hydrogen [74], flat panel displays [75, 76], chemical sensors [77], and EDLCs [78]. Several researchers [79–82] have explored CNT-based EDLC electrodes, in which most of the CNTs possess entangled structures. Thus, the CNT electrodes have had irregular pore structures that would not facilitate the fast transfer of electrolyte ions and thereby limit any improvement in the electrochemical performance of EDLCs. Compared with entangled CNT (ECNT), carbon nanotube arrays (CNTAs) have more regular pore structures and conductive paths and, therefore, higher effective SSAs [83] and lower ion diffusion resistances. A CNTA can be grown directly or pasted on a current collector to obtain an EDLC electrode [84]. The latter process is also called "transfer technique" [85].

Figure 17.14 illustrates the procedures for the transfer technique. First, a CNTA on silica or silicon substrates, which are the most suitable substrates for CNTA growth, was prepared by chemical vapor deposition (CVD) [86–89]. Second, the CNTA was detached from the original substrate by either cutting or other methods (such as HF etching or surface tension gradient achieved by introducing hot water)

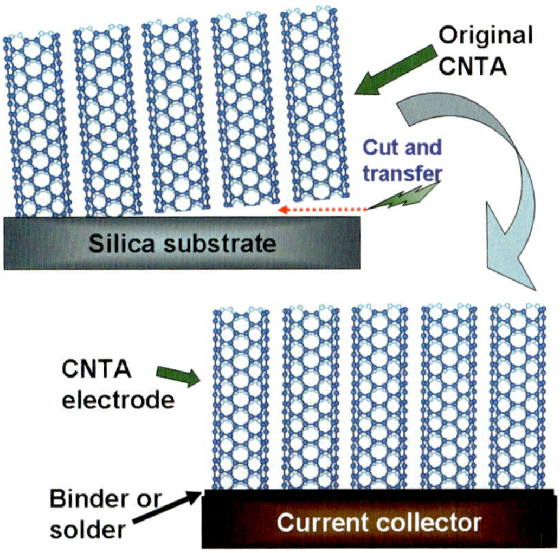

Fig. 17.14 Schematic diagram of the transfer technique for CNTA electrode fabrication (from Energy Environ. Sci.). Reprinted with permission from [87]. Copyright 2009 (The Royal Society of Chemistry)

[90, 91]. Then, the pure CNTA was reattached onto arbitrary surfaces, including Ni foil, Cu foil, Al foil, stainless steel, and other current collectors, by binders (such as graphite milk) [84, 85], solders (such as Sn/Pb, Sn/Au, and Sn/Ag/Cu) [92, 93], or just surface tension [90] to fabricate CNTA electrodes. The SWNT-like CNT presented in Fig. 17.14 can represent either MWNTs or SWNTs. CNTs can also be integrated onto gold surfaces by chemical bonds through a self-assembly process [94, 95]; however, these CNTAs may not be suitable for energy storage applications due to the complicated procedures involved in their fabrication.

CNTA electrode has a lower equivalent series resistance and a better rate capability than activated carbon electrode. This is due to the fact that CNTA possesses a larger pore size and a more regular pore structure. Compared with the transfer technique, growing CNTA directly on current collectors is more attractive and powerful due to its one-step procedure, low electric resistance between CNTA and substrate, and the formation of robust CNT–metal contacts during growth [96, 97]. Usually, CNTA can be grown either by using thin catalyst layers predeposited on substrates or through vapor-phase catalyst delivery. A 10–30 nm Al_2O_3 buffer layer can effectively increase the efficiency of CNT growth by avoiding undesired chemical interaction between catalyst and substrate [86]. This buffer layer also alters the catalyst–support interactions and enhances the growth rate of CNTs [98]. Our results also indicate that the properties of CNTAs are mainly determined by CVD parameters and the properties of Al_2O_3 buffer layers [88, 89, 98], not by the substrates. In addition, the nonconductive Al_2O_3 buffer layer can hardly influence the contact between the CNTA and the substrate, a result of annealing in H_2 which very much affects the Al_2O_3 morphology, creating many holes on the Al_2O_3 surface [88] which ensures the contacts between the substrate and the catalyst particles trapped in these holes, and thus the contacts between the CNT grown from these catalyst particles and the substrate are established.

We carefully investigated the difference between the microstructure, capacitive properties, and energy storage characteristics of a CNTA electrode and an ECNT electrode in ionic liquid electrolytes [99]. The two electrodes were both 1 mm thick and the apparent areas were the same. In addition, the crystalline quality and inherent conductivity of CNTs in ECNT electrode were similar to that in the CNTA electrode. The results obtained from cyclic voltammetry, galvanostatic charge/discharge, and ac impedance showed that the CNTA electrode had higher capacitance, lower resistance, and better rate performance than the ECNT electrode. The mechanism of energy storage in both electrodes comes from the charge separation at the electrode/electrolyte interface, leading to a double layer capacitance. Theoretically, electrode material with high SSA gives high capacitance. Contrary to this, although the SSA of the ECNT electrode (283 m^2/g) is much larger than that of the CNTA electrode (111 m^2/g), the capacitance of the ECNT electrode is lower, which suggests that the SSA utilization of the CNTA electrode is more effective than that of the ECNT electrode. The outer surface of CNTs mainly contributes to the surface area of these two electrodes because the inner space of CNTs is easily blocked by the bamboo-like structures of the inner graphite sheets [100]. These graphite sheets curve inward periodically to form cross walls in the CNT, just like the inner structure of a bamboo; thus, the pores in CNT electrodes are formed by CNT stacking. The N_2 adsorption results indicate that the pore size of the ECNT electrode shows a remarkable decrease compared to ECNT powder [99]. The average pore diameters of ECNT powder and electrode are 36 and 14 nm, respectively. The reason for this pore size decrease is because the nanotubes in the ECNT electrode form bundles easily through the electrode molding process. In addition, polytetrafluoroethylene (PTFE) binder fills up the voids among nanotubes and forms large nanotube/PTFE agglomerates, which result in pore shrinkage and effective surface area loss. Although the pore size distribution (PSD) of the ECNT electrode ranges from 1.1 to 70 nm, the majority of the surface area of the ECNT electrode is contributed by micropores, which are not easily penetrated by big ionic liquid ions, especially at high charge/discharge rates. On the contrary, the surface area of the CNTA electrode is mainly supplied by mesopores. Ion diffusivity plays a key factor to realize EDLCs with good power performance. As shown in Fig. 17.15, the ECNT

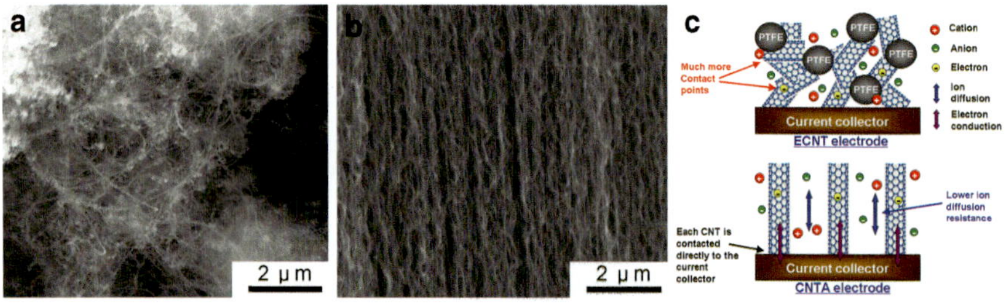

Fig. 17.15 SEM images of an entangled CNT (**a**) and a CNT array (**b**). (**c**) Schematic model comparing the microstructure, ion diffusion, and electron conduction in ECNT and CNTA electrodes (from Energy Environ. Sci.). Reprinted with permission from [87]. Copyright 2009 (The Royal Society of Chemistry)

electrode possesses a large ion diffusion barrier in the inner region of the electrode because of small pore size and irregular pore structure which lead to high internal resistance and inferior power performance. In contrast, CNTs in the CNTA electrode regularly align, giving an electrode with regular pore structure and large pore size. Therefore, compared to the ECNT electrode, the CNTA electrode possesses better ion diffusivity and higher capacitance.

The inherent conductivity of CNTs in CNTA and ECNT is similar; thus, the equivalent series resistance difference is mainly attributed to the electronic resistance difference related to the electrode microstructures and the ion diffusion difference. The CNTA electrode presents lower ion diffusion resistance than the ECNT electrode. Furthermore, each nanotube in the CNTA electrode is contacted directly to the current collector and the nanotubes in CNTA are much longer than that in ECNT; thus, the conductive paths in the CNTA electrode are more regular and shorter than those in the ECNT electrode. The conductive paths in the ECNT electrode form zigzags (by tens) of short CNTs. Every conductive path (from the current collector to the top of the electrode) not only is much longer than 1 mm (electrode thickness) but also contains tens of contact points. This kind of difference in conductive paths has been illustrated by Lira-Cantu and Gonzalez-Valls [101]. In addition, the insulative PTFE binder further increases the resistance of the ECNT electrode. In summary, the CNTA electrode presents lower ion diffusion resistance, higher electronic conductivity, larger pores, and more regular pore structure than the ECNT electrode; thus, the CNTA electrode has higher capacitance, lower resistance, and better rate performance.

CNTA electrodes present good rate performance. We fabricated a button-like device to evaluate the power performance of CNTA-based EC. CNTAs were grown directly on Ta and stainless steel (SS) substrates by chemical vapor deposition [98]. These two electrodes were used as cathode and anode to fabricate a button-like EC (see Fig. 17.16); 1.0 M Et_4NBF_4/PC was the electrolyte. Electrochemical performance measurements indicate that the CNTA-based EC presents high working voltage (3.5 V), long life cycle, and good rate capability. The power and energy densities are 928 kW/kg and 69 Wh/kg, respectively (based on the mass of CNTA), showing that CNTA-based SC is a promising candidate for power applications.

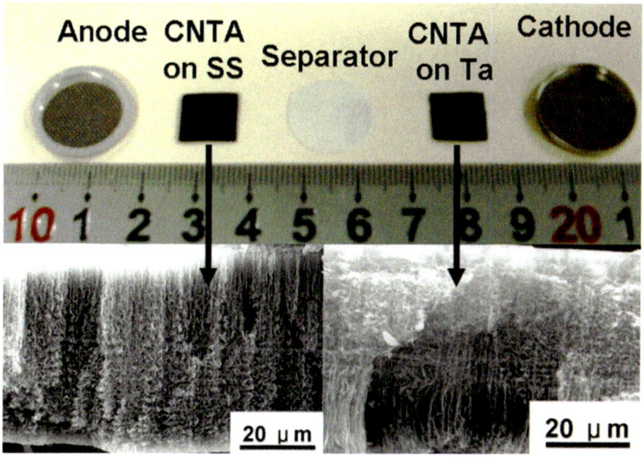

Fig. 17.16 Image of CNTA electrodes used to fabricate the button-like EC. *Insets* are SEM images of CNTAs (from Energy Environ. Sci.). Reprinted with permission from [87]. Copyright 2009 (The Royal Society of Chemistry)

It is obvious that the superior power performance of CNTA electrode is attributed to its ordered 3D nanostructures: developed ionic and electronic conductive paths. Recently Yang and coworkers reported all CNT thin films created by layer-by-layer (LBL) assembly of surface functionalized MWNTs [102]. Negatively and positively charged MWNTs were prepared by surface functionalization, allowing the incorporation of MWNTs into highly tunable thin films via the LBL technique. The pH-dependent surface charge on the MWNTs gives this system the unique characteristics of LBL assembly of weak polyelectrolytes, controlling thickness and morphology with assembly pH conditions. These MWNT thin films have randomly oriented interpenetrating network structure with well-developed nanopores using AFM and SEM, which could be an ideal structure of functional materials for various applications. In particular, electrochemical measurements of these all-MWNT thin film electrodes show high electronic conductivity in comparison with polymer composites with single-wall nanotubes and high capacitive behavior with precise control of capacity.

Figure 17.17 shows the microstructure of CNT electrodes fabricated by (a) normal deposition method and (b) LBL method. The cast film of raw MWNTs shows aggregation and alignment of bundles of MWNTs due to van der Waals interactions and capillary forces. These films exhibit weak mechanical integrity, poor control of thickness, and dense packing, leading to a decrease of the available surface area of the MWNTs. On the other hand, the LBL-assembled MWNT films yield randomly oriented individual MWNTs with well-developed nanoscale pores, making MWNT thin films an ideal electrode structure with mixed ionic/electronic conducting channels. These differences come from the electrostatic repulsion between nanotubes that act against van der Waals interactions that yield close packed aggregates and the electrostatic cross-linking between positively and negatively charged MWNTs during the LBL process that leads to randomly oriented, kinetically driven CNT arrangements in the film. Since the MWNTs have intrinsically high electrical conductivity and higher surface area, these porous network structures can work as fast

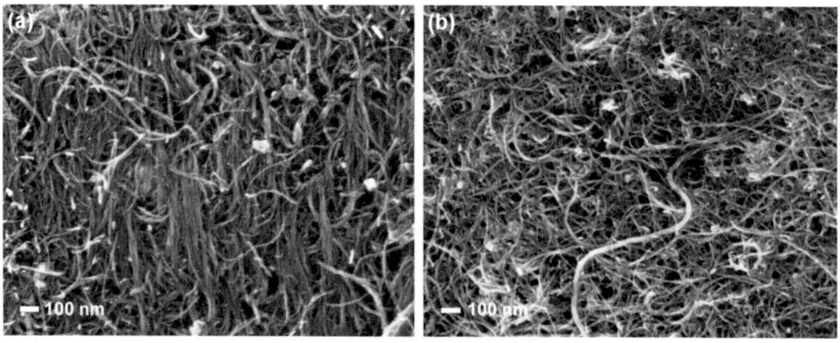

Fig. 17.17 Comparison of the topology of MWNTs using scanning electron microscopy (SEM) images. (**a**) Deposition of raw MWNT solution on silicone wafer after 2 h sonication. (**b**) LBL-assembled MWNT thin film (from J. Am. Chem. Soc.). Reprinted with permission from [102]. Copyright 2009 (American Chemical Society)

electronic and ionic conducting channels, providing the basis for design of the ideal matrix structure for energy conversion as well as energy storage devices.

17.3 Pseudo-capacitors Based on 3D Nanostructured Electrodes

Some SCs use fast, reversible redox reactions at the surface of active materials, thus defining what is called the pseudo-capacitive behavior. Metal oxides such as RuO_2, Fe_3O_4, or MnO_2 [103, 104], as well as electronically conducting polymers [105], have been extensively studied in the past decades. The specific pseudo-capacitance exceeds that of carbon materials using double layer charge storage, justifying interest in these systems. But because redox reactions are used, pseudo-capacitors, like batteries, often suffer from a lack of stability during cycling. Ruthenium oxide, RuO_2, is widely studied because it is conductive and has three distinct oxidation states accessible within 1.2 V. The pseudo-capacitive behavior of RuO_2 in acidic solutions has been the focus of research in the past 30 years [6]. It can be described as a fast, reversible electron transfer together with an electro-adsorption of protons on the surface of RuO_2 particles, according to Eq. (17.5), where Ru oxidation states can change from (ii) up to (iv):

$$RuO_2 + xH^+ + xe^- \leftrightarrow RuO_{2-x}(OH)_x \qquad (17.5)$$

where $0 \leq x \leq 2$. The continuous change of x during proton insertion or de-insertion leads to a capacitive behavior with ion adsorption following a Frumkin-type isotherm [6]. Specific capacitance of more than 600 F/g has been reported [106], but Ru-based aqueous electrochemical capacitors are expensive, and the around 1.0 V voltage window limits their applications to small electronic devices. Less expensive oxides of iron, vanadium, nickel, and cobalt have been tested in aqueous electrolytes, but none has been investigated as much as manganese oxide [107]. Its charge storage mechanism is based on surface adsorption of electrolyte cations C^+ (K^+, Na^+, ...) as well as proton incorporation according to the reaction

$$MnO_2 + xC^+ + yH^+ + (x+y)e^- \leftrightarrow MnOOCxHy \qquad (17.6)$$

MnO_2 micro-powders or micrometer-thick films show a specific capacitance of about 150 F/g in neutral aqueous electrolytes within a voltage window of ≤ 1 V. Accordingly, there is limited interest in MnO_2 electrodes for symmetric devices, because there are no oxidation states available at less than 0 V. However, it is suitable for a pseudo-capacitive positive electrode in hybrid systems. Other transition metal oxides with various oxidation degrees, such as molybdenum oxides, should also be explored as active materials for pseudo-capacitors.

Many kinds of conducting polymers (polyaniline, polypyrrole, polythiophene, and their derivatives) have been tested in SC applications as pseudo-capacitive materials and have shown high gravimetric and volumetric pseudo-capacitance in various nonaqueous electrolytes at operating voltages of about 3 V [105, 108, 109].

When used as bulk materials, conducting polymers suffer from a limited stability during cycling that reduces the initial performance. Research efforts with conducting polymers for supercapacitor applications are nowadays directed toward hybrid systems.

Given that nanomaterials have helped to improve Li-ion batteries [110], it is not surprising that nanostructuring has also affected pseudo-capacitors. Because pseudo-capacitors store charge in the first few nanometers from the surface, decreasing the particle size increases active material usage. Thanks to a thin electrically conducting surface layer of oxide and oxynitride, the charging mechanism of nanocrystalline vanadium nitride (VN) includes a combination of an electric double layer and a faradic reaction (ii/iv) at the surface of the nanoparticles, leading to specific capacitance up to 1200 F/g [111]. A similar approach can be applied to other nano-sized transition metal nitrides or oxides. In another example, the cycling stability and the specific capacitance of RuO_2 nanoparticles were increased by depositing a thin conducting polymer coating that enhanced proton exchange at the surface [112]. The design of specific surface functionalization to improve interfacial exchange could be suggested as a generic approach to other pseudo-redox materials. MnO_2 and RuO_2 films have been synthesized at the nanometer scale. Thin MnO_2 deposits of tens to hundreds of nanometers have been produced on various substrates such as metal collectors, carbon nanotubes, or activated carbons. Specific capacitances as high as 1300 F/g have been reported [113], as reaction kinetics were no longer limited by the electrical conductivity of MnO_2. In the same way, Sugimoto's group has prepared hydrated RuO_2 nanosheets with capacitance exceeding 1300 F/g [114]. The RuO_2 specific capacitance also increased sharply when the film thickness was decreased. The deposition of RuO_2 thin film onto carbon supports both increased the capacitance and decreased the RuO_2 consumption [115, 116]. Thin film synthesis or high SSA capacitive material decoration with nano-sized pseudo-capacitive active material offers an opportunity to increase energy density. Particular attention must be paid to further processing of nano-sized powders into active films because they tend to re-agglomerate into large-sized grains. An alternative way to produce porous films from powders is by growing nanotubes, as has been shown for V_2O_5 [117], or nanorods. These allow easy access to the active material, but can only be produced in thin films so far, and the manufacturing cost will probably limit the use of these sophisticated nanostructures to small electronic devices.

17.3.1 Nanostructured Metal Oxide Electrode Materials

Among the various pseudo-capacitive materials investigated, X-ray amorphous hydrous ruthenium oxide offers the highest capacitance value. The capacitance value of $RuO_2{\cdot}xH_2O$ depends sensitively on the water content and the degree of crystallinity. The excellent performance of hydrous ruthenium oxide as a pseudo-capacitive material arises because it is an innate nanocomposite with a high mobility phase for electron transport, due to metallic conduction along a percolation network of nanocrystalline rutile RuO_2 coupled to high proton mobility along water structured at the rutile network [118]. Operational stability of the device is further enhanced by the chemical stability of RuO_2 in acidic and basic electrolytes. The major issue with RuO_2 is the prohibitively high cost and limited availability of Ru. With an aim to lower cost while increasing the power density relative to carbon-based SCs, a number of transition metal oxides in which the metal ions can support multiple valences (e.g., amorphous or nanostructured MnO_2 and V_2O_5) have been pursued to replace RuO_2 [119, 120]. Manganese oxides have been thoroughly investigated because of their importance in industrial applications, such as catalysis and energy storage [121–123]. Over the years various nanostructured manganese oxides, including dendritic clusters, nanocrystals with different shapes, nanowires, nanotubes, nanobelts, and nanoflowers, have been synthesized [124–131]. A SEM image of manganese oxide dendritic clusters composed of nanotetragonal prisms with square cross-section obtained from hydrothermal reaction is shown in Fig. 17.18. Substitution of less expensive elements such as V and Cr for Ru in RuO_2 as well as composites consisting of RuO_2 on amorphous WO_3 and $Na_{0.37}WO_2{\cdot}xH_2O$ have also been found to exhibit 550–700 F/g at a reduced cost [132, 133].

Many synthesis methods have been reported for the synthesis of nanostructured metal oxides. Among them, solution-based methods are well known for their advantages in tailoring the size and morphology of the nanostructures. It is the uncomplicated sol–gel processing (soft chemistry) method in combination with template synthesis or hydrothermal treatment that produces the most desirable nanostructures with remarkable reliability, efficiency, selectivity, and variety. Template synthesis is a general method for preparing ordered arrays of nanostructures with nanorods/nanotubes/nanocables protruding from the underlying current

Fig. 17.18 SEM images of OMS-2 dendritic clusters composed of nanotetragonal prisms with square cross-section obtained from hydrothermal reaction at 120°C, 12 h. (**a**–**c**) Detailed views of the same cluster with higher magnifications (from J. Am. Chem. Soc.). Reprinted with permission from [124]. Copyright 2005 (American Chemical Society)

collector [134]. Hydrothermal synthesis is another powerful tool to transform transition metal oxides into high-quality nanostructures. Other fabrication methods of nanostructures include the reverse micelle technique, and the size of nanostructures can be tuned easily by keeping the freshly made nanorods in the micellar solution [135].

It is clear that nanostructured metal oxides offer improved energy storage capacity and charge–discharge kinetics, as well as better cyclic stabilities, owing to their huge surface area for redox reaction, short distance for mass and charge diffusion, as well as the added freedom for volume change that accompanies reaction. However, they are powered materials (just like nanostructured template carbons), which are hard to handle for the fabrication of the electrodes (see Fig. 17.13). The electrode materials with both controllable pore structure and developed conductive paths at macroscopical scale are highly desired.

17.3.2 Nanostructured Conducting Polymer Electrode Materials

Conductive polymers exhibit several interesting and important properties, such as metallic conductivity and reversible convertibility between redox states. When the redox states have very different electrochemical and electronic properties, their interconversion gives rise to changes in the polymers' conformations, doping levels, conductivities, and colors, useful attributes if they are to be applied in displays, energy storage devices, actuators, and sensors. Unfortunately, the utilization ratio and the reaction speed are low. One approach to enhancing these properties is decreasing the diffusion distance for the counterions in the polymer. For example, it has been found that nanotubular structures are good candidates for realizing rapid switching between redox states because the counterions can be readily doped throughout the thin nanotube walls.

Over the years much attention has been paid to synthesis of conducting polymers with high electroactive regions by controlling microstructures, and different morphologies of polymers have been obtained [136–140]. Electrochemical deposition and template method are effective for the fabrication of nanostructured polymers. Highly conducting polyaniline (PANI) nanowires have been synthesized in high yield using a very simple electrochemical technique. The morphology of typical PANI nanowires prepared directly on current collectors by deposition method is shown in Fig. 17.19. The formation of such nanowires can be explained invoking seedling growth process. Polyaniline nanowires of different morphologies can be synthesized using appropriate experimental conditions.

A possible growth process of such nanowires is proposed as shown in Fig. 17.20. The polyaniline nanowires are expected to grow via seedling growth process, in which further polyaniline is deposited on the initially deposited nano-sized granules (Fig. 17.20a). As the deposition progresses, an aligned nanowire network is formed. A further deposition results in extended length and in the misalignment of nanowires and cross-links are formed. From there, the secondary growth of polyaniline begins as the incoming polyaniline is distributed along the elongated

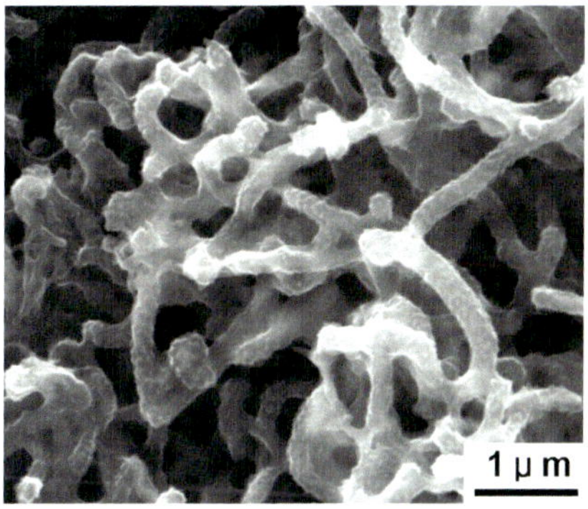

Fig. 17.19 A SEM image of the polyaniline nanowires prepared by electrodeposition

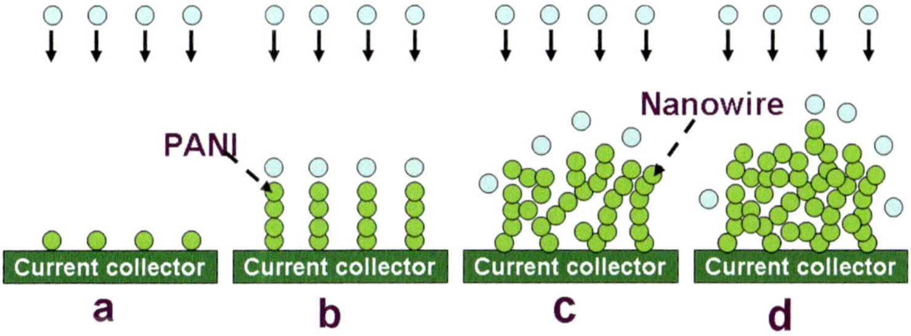

Fig. 17.20 (**a**)–(**d**) Schematics of the possible growth process of the polyaniline nanowires in the order of increasing polyaniline content

surface of the nanowires which results in their corn-like structure as shown in Fig. 17.20c. Polyaniline continues to deposit on the network of the nanowires such that all porous area is filled, as shown in Fig. 17.20d. The final situation is similar to that shown in Fig. 17.19.

The combination of electrochemical deposition and template method is more effective for the preparation of nanostructured polymers. Although the synthesis of conductive polymer nanotubes can be performed using electrochemical template synthesis, the synthetic techniques and underlying mechanisms controlling the nanotube morphologies are currently not well established. The applied potential, monomer concentration, and base electrode shape all play important roles in determining the nano-polymers' morphologies. A mechanism based on the rates of monomer diffusion and reaction allows the synthesis of nanotubes at high oxidation potentials; a mechanism dictated by the base electrode shape dominates at very low oxidation potentials. The SEM and corresponding TEM images of polypyrrole nanostructures obtained at various potentials are shown in Fig. 17.21 [141]. The morphology of nanostructures electrochemically grown in the template pores was predicted by considering the diffusion and reaction kinetics as well as the interaction between pore wall and polymers at high oxidation potentials. However,

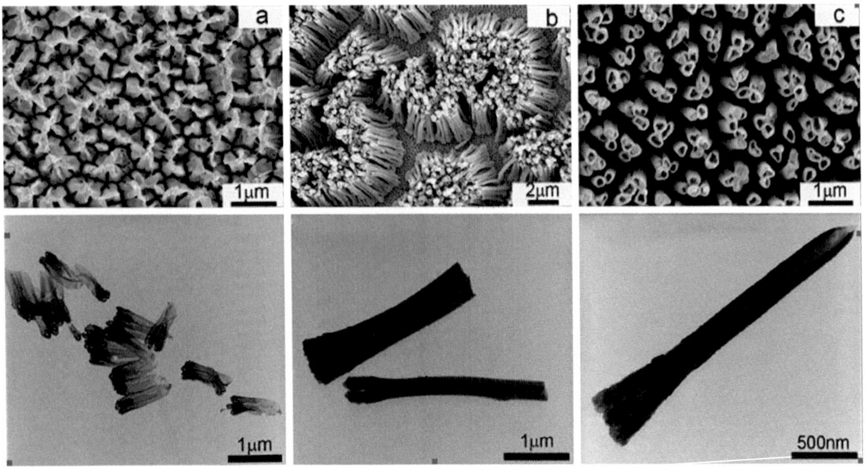

Fig. 17.21 SEM and corresponding TEM images of polypyrrole nanostructures synthesized at (**a**) low, (**b**) middle, and (**c**) high potentials (0.7, 1.4, and 1.8 V, respectively) (from Acc. Chem. Res.). Reprinted with permission from [141]. Copyright 2008 (American Chemical Society)

the base electrode shape plays an important role in determining nanotube morphology at very low oxidation potentials. This elaborated mechanism is applicable to other conductive polymers and inorganic materials and helpful for fabrication of composite structures such as nanotube/nanotube, nanotube/nanowire, or nanotube/nanoparticles. The controlled synthetic technique for nanotubular structures made it feasible to investigate the performances of nanotube-based electrochemical devices. The supercapacitor of poly(3,4-ethylenedioxythiophene) (PEDOT) nanotubes provided high energy density at high power demand.

Just like nanostructured metal oxides, nanostructured polymers offer improved energy storage capacity and charge–discharge kinetics, as well as better cyclic stabilities, owing to their huge surface area for redox reaction, short distance for mass and charge diffusion, as well as the added freedom for volume change that accompanies reaction. However, they are powered materials. The electrode materials with both controllable pore structure and developed conductive paths through electrodes can further enhance the electrochemical performance.

17.4 Hybrid Capacitors Based on 3D Nanostructured Electrodes

Various nanostructured carbon materials present large surface area and developed ionic and electronic conductive paths and some of these carbons have even monolithic structures; therefore, they show high rate capability than any other electrode materials. However, their capacitances are moderate, i.e., less than 300 F/g, which is lower than a lot of pseudo-capacitive materials, because of the double layer mechanism. Pseudo-capacitive materials, such as conducting polymers and metal oxides, have high theoretical capacitance. However, their utilization ratios are low leading to moderate capacitance. Although nanostructured pseudo-capacitive materials can get high utilization due to their high SSA and small particle size, the power performance of pseudo-capacitive materials is poor due to their moderate conductivity. In addition, the life cycle of both conducting polymers and metal oxides is much shorter than carbon-based materials because structure or volume change during redox reactions leads to the loss of active materials. In order to greatly increase the electrochemical performance of conducting polymers and metal oxides, a hierarchical porous structure with high electronic conductivity must be considered [65, 142].

How to fabricate advanced materials obtaining the advantages of both carbons and pseudo-capacitive materials? Doping the latter into the pores of highly conductive and stable carbons is effective to improve their power and cycle performance [143–146]. Considering that some carbons (such as CNTA) have regular pore structure, high SSA, homogeneous property (binder free), and superior conductivity [86, 147], depositing pseudo-capacitive materials on these carbons' (also take CNTA as an example, see Fig. 17.22) electrodes is a promising method for the fabrication of novel composite electrodes with superior capacitive properties. Recently, researchers reported the preparation, microstructure, and capacitive properties of these carbon/pseudo-capacitive material based composite materials. These novel materials can be divided into two categories: metal oxide/carbon electrodes and conducting polymer/carbon electrodes.

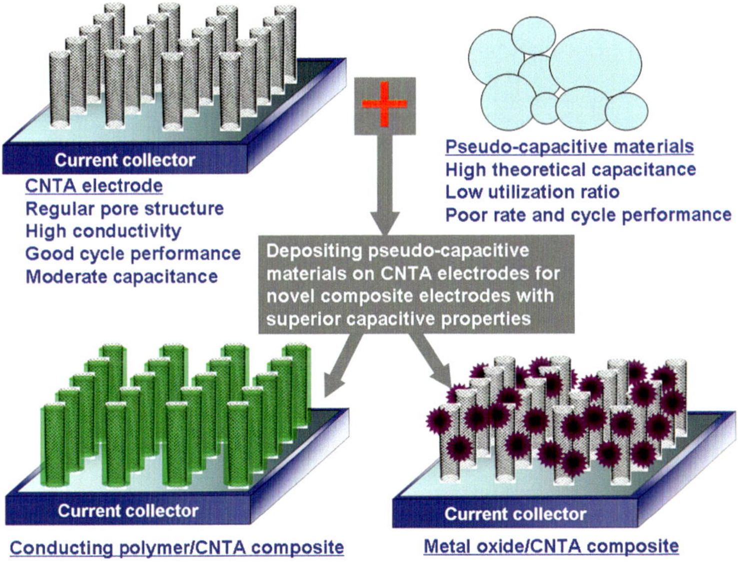

Fig. 17.22 A schematic diagram illustrating how pseudo-capacitive materials may be deposited on CNTA electrodes for novel CNTA-based composite electrodes (from Energy Environ. Sci.). Reprinted with permission from [87]. Copyright 2009 (The Royal Society of Chemistry)

17.4.1 Nanostructured Electrodes Based on Metal Oxides/Carbon Composite

Compared with the powdered carbons, it is obvious that monolithic carbons are more suitable for the deposition of pseudo-capacitive materials. Carbon aerogel, carbon nanotube array, and carbon foam are all suitable to be the substrates of these composite materials.

As mentioned previously, the electrochemical performance of metal oxides is largely affected by its morphology. In the composite materials, not only the distribution of metal oxide but also the structure of the carbon substrate is an important consideration. For example, physical properties of the substrate such as morphology and microstructure, surface area, and pore size affect the nature of the deposited metal oxides as well as the electrochemical properties of the final hybrid device.

Carbon aerogels and related sol–gel-derived nanoarchitectures are particularly attractive for electrochemical capacitor applications due to their inherent structural characteristics that include high specific surface areas, through-connected networks of mesopores and/or macropores, tunable pore sizes ranging from nanometers to micrometers, durable monolithic, moldable forms, and reasonable electronic conductivity (10–40 S/cm) [148]. These same characteristics make carbon aerogels a useful platform for modification with metal oxides. For example, the surface area, porosity, and pore sizes of carbon aerogels can be tuned in order to maximize the weight loadings of incorporated metal oxides. In this way, the hybrid architecture combines the desirable structural and electronic characteristics of the native aerogel with the built-in electrochemical capacitance of metal oxide.

Long and coworkers describe the electroless deposition of manganese oxide under controlled pH conditions within and onto commercially available carbon aerogel substrates [149]. In this general method, the carbon substrate serves as a sacrificial reductant and converts aqueous permanganate (MnO_4^-) to insoluble MnO_2, as described previously for MnO_2 deposition on other powdered carbons. Although this approach offers a straightforward processing protocol, careful control of the reaction is required in order to achieve nanoscale MnO_2 deposits throughout a 3D monolithic carbon architecture. Achieving such non-line-of-sight control is particularly challenging when modifying macroscopically thick porous carbon substrates because the supply and flux of permanganate is likely to be inhomogeneous between the electrode exterior and interior, thereby favoring deposition at the exterior boundary. Moreover, preservation of the native carbon template pore structure is vital for electrochemical performance in order to facilitate electrolyte infiltration and ion transport throughout the internal volume of the hybrid electrode. Such hybrid electrode structures can be produced by using deposition methods that are inherently self-limiting, as recently demonstrated for the electrodeposition of conformal, ultrathin poly(*o*-methoxyaniline) coatings on carbon nanofoam templates [150]. The TEM image of a neutral-deposited MnO_2–carbon aerogel composite is shown in Fig. 17.23. The nanoscale structure of the neutral-deposited composite architecture reveals the presence of layered birnessite MnO_2 ribbons and rods that permeate the carbon nanofoam substrate. Such a 3D design exploits the benefits of a nanoscopic

Fig. 17.23 Transmission electron micrographs of a neutral-deposited MnO_2–carbon (from Nano Lett.). Reprinted with permission from [149]. Copyright 2007 (American Chemical Society)

MnO_2–carbon interface to produce an exceptionally high area-normalized capacitance (1.5 F/cm^2), as well as high volumetric capacitance (90 F/cm^3). Varying the carbon template pore structure, particularly targeting larger pore sizes (100–200 nm) and higher overall porosity, could result in higher mass loadings of MnO_2 and utilization of more internal pore volume.

Considering that CNTA presents larger pores and more regular pore structures than carbon aerogels, we tried to combine the manganese oxide active material with CNTA frameworks and obtained a novel nanostructured electrode with good capacitive properties [151]. Our strategy was to first grow a vertically aligned CNTA on a Ta foil directly by chemical vapor deposition at 800°C and then to electrodeposit manganese oxide on a CNTA "scaffold" by a potentiodynamic method. The SEM image (left inset of Fig. 17.24) shows manganese oxide particles, around 150 nm in diameter, well dispersed on CNTA. The TEM image (right inset of Fig. 17.24) reveals that nanostructured manganese oxide particles are composed of hundreds of surfboard-shaped nanosheets, and the nanosheets of each individual particle originate from the same core, forming a dandelion-like flower. The length and thickness of each surfboard-shaped "petal" are about 50 and 3 nm, respectively. The manganese oxide particle size, distribution, and microstructure can be controlled by electrodeposition parameters, such as CV cycle number and potential range. In brief, more CV cycles and higher upper limit of the potential range result in larger manganese oxide particles and denser particle distribution, respectively, which has a profound effect on the morphology and the electrochemical properties of MnO_x/CNTA composites. The SSA of the manganese oxide nanoflower is as high as 236 m^2/g. Compared with pure CNTA, the MnO_x/CNTA composite exhibits more developed micropores, which can mainly be attributed to the numerous gaps between manganese oxide nanosheets. Besides the micropores in manganese oxide, there are a great many macropores among CNTs. The micropores in manganese oxide and the macropores between CNTs compose a hierarchically porous structure.

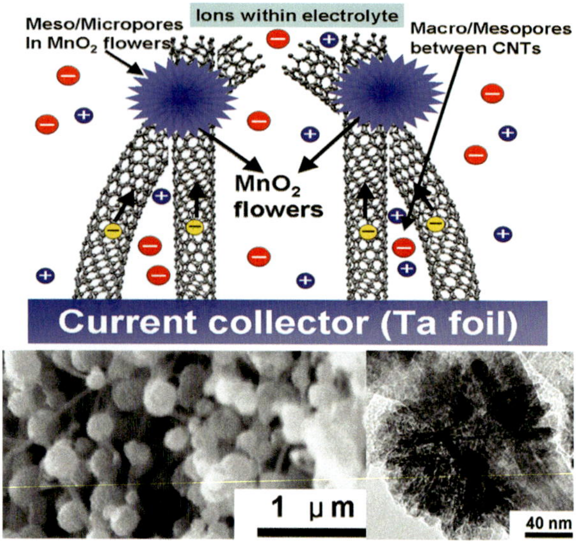

Fig. 17.24 Schematic representation of the microstructure and energy storage characteristics of MnO_x/CNTA composite. The left and right *insets* show a SEM image of MnO_x/CNTA composite and TEM image of MnO_x nanoflower on CNTs, respectively (from Energy Environ. Sci.). Reprinted with permission from [87]. Copyright 2009 (The Royal Society of Chemistry)

In addition, manganese oxide nanoflowers are apt to form at the junctions of CNTs, attributed to manganese oxide tending to nucleate at the junctions of CNTs, rather than at the curved surface of CNTs during electrodeposition. Generally, manganese oxide deposits readily form planar nanosheets on flat substrates [152, 153]. The dandelion-like nanostructure may be due to the manganese oxide deposit nucleating at the junctions of CNTs and then growing radically from the junctions. On this basis, the growth process of the MnO_x/CNTA composite is different from that of the manganese oxide/nanocarbon composite, which is prepared by coating the surface of a carbon aerogel with a thin layer of manganese oxide through a self-limiting reaction. Electrochemical impedance spectroscopy and CV results show that the MnO_x/CNTA electrode has low equivalent series resistance, low charge-transfer resistance, and highly capacitive nature with good ion response [151]. In addition, this electrode retains 97% of its capacitance after 20,000 consecutive cycles, indicative of good electrochemical stability. Good electrochemical stability is an advantage of MnO_x/CNTA composites compared to conducting polymers, whose life cycles are less than 5000 cycles [140, 154].

We electrodeposited manganese oxides on an ECNT electrode and an activated carbon electrode for comparison. Compared with the MnO_x/ECNT composite, the MnO_x/CNTA composite showed not only higher capacitance but also better rate capability [151], attributed to the fact that the MnO_x deposited on ECNT is not uniform and readily forms MnO_x/CNT agglomerates (see Fig. 17.25), resulting in worse electrochemical accessibility and lower ionic conductivity. Furthermore, the conductive paths in the ECNT electrode are inferior to that in the CNTA electrode [99], leading to lower electronic conductivity. Although the capacitance of MnO_x/activated carbon is 201 F/g, the rate performance of this composite is poor,

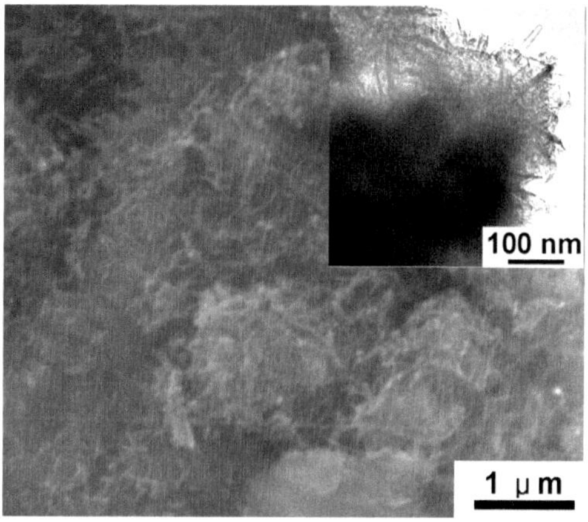

Fig. 17.25 SEM and TEM images (*inset*) of the manganese oxide/ECNT composite (from Nano Lett.). Reprinted with permission from [151]. Copyright 2008 (American Chemical Society)

that is, 24% capacitance retention at 10 A/g, attributed to the low conductivity of the activated carbon substrate [155, 156]. The energy storage characteristics of MnO_x/CNTA composite are illustrated in Fig. 17.24. Manganese oxide nanoflowers are grown directly on nanostructured current collector (CNTA). This geometry has several advantages. First, each manganese oxide nanoflower is connected directly with the current collector (Ta foil) by two or more electron "superhighways" (CNTs); thus, this superior conducting network allows for efficient charge transport and enhances significantly the electronic conductivity of the composite. Second, the high SSA and nanometer size, which reduces the diffusion length of ions within the manganese oxide phase during charge/discharge [140], ensures a high utilization of electrode materials and a high specific capacitance. Third, a hierarchically porous structure enhances greatly ionic conductivity of the composite. Fourth, the use of CNTs with exceptional mechanical properties as a support and the geometry of the manganese oxide nanoflower can remove the cycle degradation problems caused by mechanical problems or volume changes and can overcome nanoparticle aggregation [157, 158]. In addition, as every manganese oxide particle is connected to the conducting framework, the need for binders or conducting additives, which add extra contact resistance or weight, is eliminated. Thus, the MnO_x/CNTA composite electrode presents the best electrochemical capacitive performance.

Broughton reported that adding acetate into a $MnSO_4$ precursor solution can decrease the scale of MnO_x deposits [159]. We added 0.1 M acetate (manganese acetate) into the precursor solution for the preparation of MnO_x/CNTA. The new MnO_x deposits showed different morphology, i.e., flake-like deposits covering the CNTs. These MnO_x flakes present higher capacitance (302 F/g) than nanoflower-like deposits, attributed to higher SSA (322 m^2/g). These results indicate that the microstructure and capacitive properties of MnO_x/CNTA can be modified and improved by additives, such as acetates and cobalt and nickel salts. Considering that nickel oxides, cobalt oxides, magnetite, and other metal oxides can also be prepared by electrodeposition, other metal oxide/CNTA composite electrodes with novel nanostructure and good capacitive performance can be prepared by electrodeposition with careful control of the preparation parameters. Metal oxide/CNTA composites can also be fabricated by sputtering deposition and colloidal methods [160, 161].

17.4.2 Nanostructured Electrodes Based on Polymers/Carbon Composites

Conducting polymers, such as PANI, polypyrrole, and polythiophene, have the advantages of low cost, high conductivity, high doping/dedoping rate during charge/discharge process as well as facile synthesis through chemical and electrochemical methods [140, 162]. Over the years much attention has been paid to synthesis of conducting polymers with high electroactive regions by controlling microstructures, and different morphologies of conducting polymers have been obtained [136–139]. However, in order to further increase capacitive performance of conducting polymers, a smaller scale and hierarchically porous structure must be considered [163]. Like metal oxides, much smaller size can greatly reduce the diffusion length and ensure higher utilization of electrode materials.

Xia and coworkers reported the growth of ordered whisker-like polyaniline on the surface of mesoporous carbon by a chemical synthesis process. The preparation process mainly involves (1) immersing mesoporous carbon in 20% ethanol solution (containing aniline and H_2SO_4) while stirring under vacuum for 1 h; (2) adding an equal volume of 20% ethanol solution (which does not contain aniline or H_2SO_4) to the above-mentioned solution quickly with intensive stirring; and (3) then adding ammonium persulfate drop by drop to the solution mentioned in step 2 while stirring at 0°C. The first step of the preparation process can introduce guest precursor (20% ethanol solution containing some aniline and H_2SO_4) to mesoporous carbon (Fig. 17.26, after process a). During the second step of the preparation process, after an equal volume of 20% ethanol solution has been added to the solution mentioned in the first step quickly with intensive stirring, the concentration of aniline or H_2SO_4 outside of the mesopores is reduced quickly. However, the concentration decrease of solution inside the mesopores is not as quick as outside. The reason is that the narrow and long mesopores (the average diameter of these channels is 3–4 nm and their length is about 1 μm) limit the diffusion process. Furthermore, intensive stirring does not affect the solution inside these mesopores. Accordingly, a concentration gradient is produced between the solution inside the mesopores and the solution outside of the mesopores (Fig. 17.26, after process b). When ammonium persulfate is added to this solution, PANI precipitates will first be formed close to the external surfaces of mesoporous carbon, where the concentration of aniline or H_2SO_4 is higher (Fig. 17.26, after process c). These initial precipitates provide nucleation centers that result in the extending growth of PANI (Fig. 17.26, after process d).

The nanometer-sized PANI thorns and thus formed "V-type" nanopores yield a high electrochemical capacitance performance. Why such a structured electrode

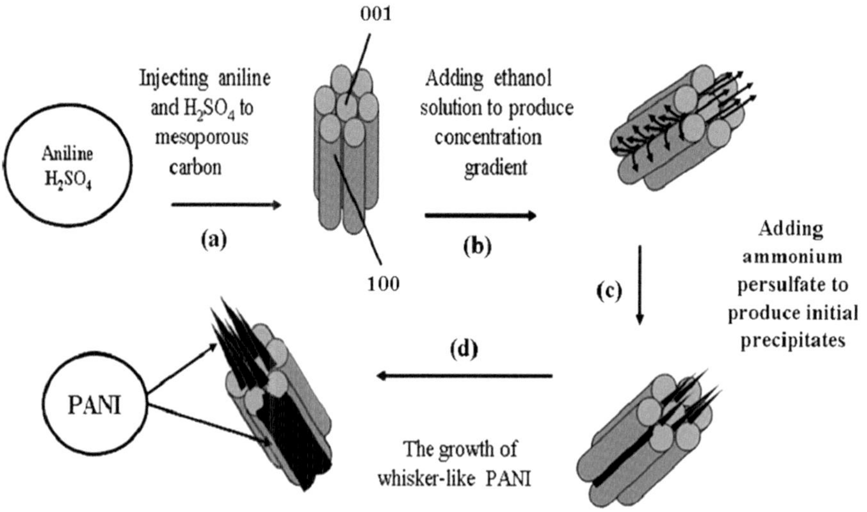

Fig. 17.26 Scheme showing the preparation of whisker-like PANI on the surface of mesoporous carbon. Reprinted with permission from [140]. Copyright 2006 (Wiley)

material shows the best capacitance performance can be summarized as follows (Fig. 17.27). First, the nanopores provide numerous "V-type" channels inside the active material, which facilitates the fast penetration of the electrolyte. In other words, these channels ensure that enough ions contact the surface of the active material in a short time. Second, the diffusion length *L* of ions within the electrode during the charge–discharge process can be estimated as (*Dt*)1/2, where *D* and *t* are the diffusion coefficient and time, respectively. The value of *t* decreases rapidly at

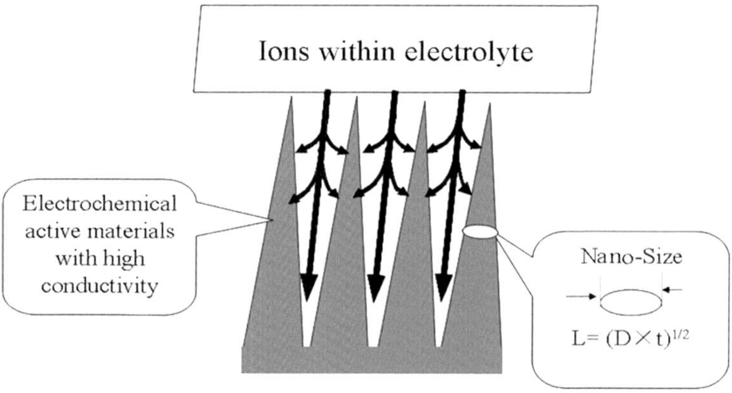

Fig. 17.27 Schematic representation of how the required diffusion length is reduced in supercapacitor electrodes with numerous V-like channels (from Adv. Mater.). Reprinted with permission from [140]. Copyright 2006 (Wiley)

high charge–discharge current density. Therefore the nanometer-size, which greatly reduce the diffusion length (*L*), will ensure the high utilization of electrode materials. Third, the high conductivity of active material and support greatly reduces the energy loss and power loss due to *IR* loss (where *I* and *R* are charge–discharge current and resistance, respectively) at high charge–discharge current density. The specific capacitance of the PANI/mesoporous carbon composite is as high as 900 F/g at a charge–discharge current density of 0.5 A/g (or 1221 F/g for PANI based on the pure PANI in the composite). This value is even higher than that of amorphous hydrated RuO_2 (840 F/g). Furthermore, the capacitance retention of this composite is higher than 85% when the charge–discharge current density increases from 0.5 to 5 A/g, indicating its high-power performance.

Compared with powdered mesoporous carbon, carbon monolith is a more attractive substrate for the fabrication of composite electrode materials. Fan and coworkers adopted a strategy to enhance the capacitive performance of active materials by electrodeposition of PANI onto a hierarchically porous carbon monolith (hereafter abbreviated as HPCM) [164]. HPCM can be used as both a current collector and a high-surface support for conducting polymers (the inset of Fig. 17.28a shows a photograph of a typical HPCM); thus, it enables an intimate contact between the active material and the current collector. Carbon monoliths with large sizes and porosities tailored by parent templates and precursors are much easier to handle than powdered carbon (e.g., activated carbon and carbon nanotubes). Unlike other powdered carbon materials, carbon monoliths do not require any binder or conductive agent (e.g., carbon black) during preparation of the electrode. Also, the limited pore space (e.g., mesopore diameter is ca. 7 nm) in a HPCM can limit the grain growth of the active materials, thus, leading to a high specific capacitance. As a porous support, the network of comparatively large pore sizes can render the electrolyte diffusion into the bulk of the electrode material facile and hence provides fast transport channels for the conductive ions. The advantages of this material are (i) easy handling compared with powdered carbon; (ii) binder-free and conductive agent-free electrode preparation; (iii) facile and fast synthesis; (iv) controlled growth of active materials by limited pore spaces; and (v) excellent performance (specific capacitance, power and energy densities, excellent cycling stability). All these characteristics demonstrate that HPCM can be used as a versatile support for electroactive materials. Of course, there is still much room to further improve the electrode performances by tuning the porosity and composition of the porous carbon monoliths.

In order to further increase capacitive performance of conducting polymers, a smaller scale and hierarchically porous structure must be considered. Much smaller size can greatly reduce the diffusion length and ensure higher utilization of electrode materials. A hierarchically porous structure can facilitate ionic motion and enhance the rate performance of electrode [65, 143]. We realize a desired structure, i.e., a nanobrush-like PANI/CNTA composite electrode, by the combination of a CNTA framework and electrodeposition techniques [142]. The preparation process involves (i) growing a 35 μm high vertically aligned CNTA on a Ta foil directly by chemical vapor deposition (CVD) at 800°C [98] and (ii) electrodepositing a 7 nm

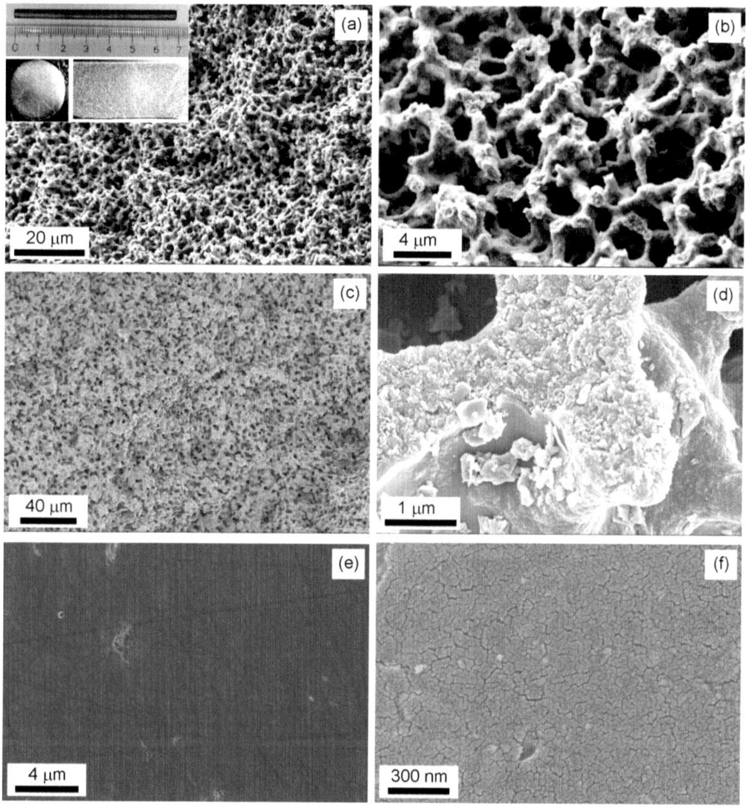

Fig. 17.28 Scanning electron microscopy (SEM) images of (**a**, **b**) HPCM-1 at low and high magnification, respectively; (**c**, **d**) PANI+HPCM-1 at low and high magnification, respectively; (**e**) NPCM; (**f**) PANI+NPCM. *Insets* of (**a**) show a photograph of a typical HPCM rod and SEM images of the surface and cross-section of the whole piece of HPCM after cutting (from Adv. Funct. Mater.). Reprinted with permission from [164]. Copyright 2007 (Wiley).

PANI nanolayer on a CNTA "template" prepared in step one by the CV method. SEM images of the PANI/CNTA composite (left inset of Fig. 17.29) illustrate that there are no PANI shells or agglomerates formed on the top or in the composite, indicating that the PANI deposits are well distributed. A high-resolution TEM image (right inset of Fig. 17.29) clearly illustrates that the graphite structural CNT is covered by a uniform and compact PANI nanolayer, forming a tube-covering-tube unique microstructure. The thickness of this PANI nanolayer is only 7 nm, which is very desirable for SC applications. Moreover, the thickness of the PANI nanolayer can be controlled by electrodeposition parameters, e.g., CV sweeping cycle number (more CV cycles result in thicker PANI layers). The PANI nanolayer is composed of numerous nanoclusters with diameters of about 0.5 nm [142], which implies that this nanolayer is composed of PANI units ($0.59 \times 0.44 \times 0.43$ nm^3) [165]. Compared with CNTA, the PANI/CNTA composite exhibits more developed

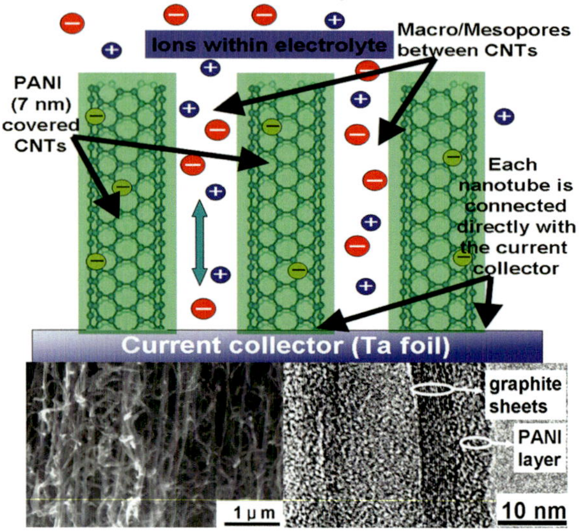

Fig. 17.29 Schematic representation of the microstructure and energy storage characteristics of the PANI/CNTA composite. *Left* and *right insets* are a SEM image of a PANI/CNTA composite and a TEM image of a PANI-covered CNT in the composite, respectively (from Energy Environ. Sci.). Reprinted with permission from [87]. Copyright 2009 (The Royal Society of Chemistry)

micropores (<2 nm), which can be mainly attributed to the gaps between PANI units (see right inset of Fig. 17.29). A developed microporous structure results in the SSA of the composite (352 m^2/g) being higher than the SSA of CNTA (201 m^2/g). Besides micropores in the PANI nanolayer, there are a great many macropores, larger than 50 nm, in the composite nanotubes (see left inset of Fig. 17.29). The micropores in the PANI nanolayer and the macropores between CNTs compose a hierarchical porous structure. The capacitance of the PANI/CNTA composite at low current density is 1030 F/g, which is contributed by both double layer capacitance and pseudo-capacitance, and is higher than other PANI materials [166]. The PANI/CNTA composite still presents 978 and 789 F/g at current densities as high as 118 and 294 A/g, respectively, and the corresponding capacitance retention ratios are 95 and 77%, indicating that high capacitance can be maintained under high-power operations. The capacitance loss of the composite after 5000 consecutive cycles was only 5.5%, which is the best of the conducting polymers and can be attributed to good electrochemical stability of the CNTA/Ta foil substrate in acidic solution and good mechanical properties of the CNTA support. Furthermore, the 7 nm PANI nanolayer allows for better accommodation of the large volume changes during charge/discharge without the initiation of fracture that can occur in bulk- or micron-sized materials [157]. Besides acid resistance and robust property of the CNTA/Ta substrate and the 1D nanostructure of the PANI phase, PANI may form chemical bonds or other strong affinity (such as p–p stacking) with the CNT surface, which needs further investigation [167].

Why the PANI/CNTA composite presents good capacitive performance is illustrated in Fig. 17.29. The nanometer-sized PANI layer coats every CNT in the CNTA uniformly. This geometry has several advantages. First, the PANI nanolayer is connected directly with the current collector (Ta foil) by electron “superhighways”

(CNTs); thus, this superior conducting network allows for efficient charge transport and enhances the electronic conductivity of the composite significantly. Second, the high SSA and nanometer size, which reduce the diffusion length of ions in the PANI phase during charge/discharge, ensure a high utilization of electrode materials and a high specific capacity. Third, the hierarchically porous structure enhances the ionic conductivity of the PANI/CNTA composite greatly. Physicochemical properties of the electrolyte in macropores are similar to those of the bulk electrolyte with lowest resistance [62]. Ion-buffering reservoirs can be formed in macropores between nanotubes to minimize the diffusion distances to the interior surfaces of the PANI phase. Fourth, the use of CNTs with exceptional mechanical properties as a support and the geometry of the nanometer-sized PANI layer can release the cycle degradation problems caused by mechanical problems or volume changes and can overcome nanoparticle aggregation [168, 169]. In addition, as the PANI phase is connected to the conducting framework, the need for binders or conducting additives, which add extra contact resistance or weight, is eliminated. Considering that polypyrrole, polythiophene, and other conducting polymers can also be prepared by electrodeposition, it is expected that these conducting polymer/CNTA composite electrodes with novel nanostructure and good capacitive performance can be prepared by electrodeposition by carefully controlling the preparation parameters. Some pioneering work in this area is reported [141, 170].

17.5 Conclusions and Perspectives

In conclusion, the nanoarchitectures of electrodes, pore structures, surface area, and morphology of materials control the properties and performance of supercapacitors. A notable improvement in performance has been achieved through recent advances in the development of nanostructured materials. Especially, the CNT-based electrodes, hierarchically porous carbons, nanostructured manganese oxides and conducting polymers, and the combination of nanostructured metal oxides and polymers with monolith carbons show superior capacitive properties. Through the development of these novel electrodes, it is concluded that electrodes with 3D ordered/aperiodic architectures and suitable pore size distribution, high surface area, and developed conductive paths are desirable for next-generation supercapacitors.

In the near future, redesign of materials on the nanoscale and novel chemical synthesis and processing approaches to control the nanostructure will have a significant impact in advancing the commercial feasibility of SC technology; incorporating a massively parallel current collector in 3D and on the nanoscale to maximize electron and ion mobility affords an immediate improvement. In addition, development of new composite material/structure combinations should be pursued. Many transition metal oxides exhibit high pseudo-capacitance, in combination with electrochemically active polymers, and can be combined with the novel high SSA micro-/nanostructured carbon frameworks providing high electrical conductivity. Optimization of the electrode nanostructure and interfacial chemistry should be pursued to allow the recording of higher capacitance and fast ion transport in the system for high energy density and fast charge/discharge times. Finally, robustness of the system design and materials (stability to multiple charge/discharge cycles and long-term performance degradation) needs to be addressed, especially for the novel pseudo-capacitor-based systems. One goal of these electrode materials will be to reach capacitance values that approach those for hydrous ruthenium oxide while decreasing the material cost relative to ruthenium-based SC by orders of magnitude.

In the long term (>10 years), controlled 3D integration of opposing electrified interfaces (i.e., cathode and anode), structured ionic layers, and mobile ions within redesigned nanoarchitectures and pore structures will create advanced SCs based on novel composite electrodes that should achieve power densities typical of carbon-based SCs and energy densities comparable to the lithium-ion batteries.

References

1. P. Simon, Y. Gogotsi, Materials for electrochemical capacitors. Nat. Mater. **7**, 845 (2008)
2. J. Tollefson, Car industry: charging up the future. Nature **456**, 436 (2008)
3. M. Armand, J.M. Tarascon, Building better batteries. Nature **451**, 652 (2008)
4. Y. Wang, G. Cao, Developments in nanostructured cathode materials for high-performance lithium-ion batteries. Adv. Mater. **20**, 2251 (2008)
5. B. Sorensen, *Hydrogen and Fuel Cells* (Elsevier Academic Press, London 2005)
6. B.E. Conway, *Electrochemical Supercapacitors: Scientific Fundamentals and Technological Applications* (Kluwer/Plenum Press, New York, NY 1999)
7. R. Kötz, M. Carlen, Principles and applications of electrochemical capacitors. Electrochim. Acta **45**, 2483 (2000)
8. G. Che, B.B. Lakshmi, E.R. Fisher, C.R. Martin, Carbon nanotubule for electrochemical energy storage and production. Nature **393**, 346 (1998)
9. J.M. Tarascon, M. Armand, Issues and challenges facing rechargeable lithium batteries. Nature **414**, 359 (2001)
10. R.J. Brodd, K.R. Bullock, R.A. Leising, R.L. Middaugh, J.R. Miller, E.S. Takeuchi, Batteries, 1977 to 2002. J. Electrochem. Soc. **151**, K1 (2004)
11. M. Armand, J.M. Tarascon, Building better batteries. Nature **451**, 652 (2008)
12. M. Armand, P. Johansson, Novel weakly coordinating heterocyclic anions for use in lithium batteries. J. Power Sources **178**, 821 (2008)
13. J.R. Miller, P. Simon, Electrochemical capacitors for energy management. Science **321**, 651 (2008)
14. US Department of Energy. Basic Research Needs for Electrical Energy Storage (2007) <www.sc.doe.gov/bes/reports/abstracts.html#EES2007>
15. M. Winter, J.O. Besenhard, M.E. Spahr, P. Novak, Insertion electrode materials for rechargeable lithium batteries. Adv. Mater. **10**, 725 (1998)
16. A.G. Pandolfo, A.F. Hollenkamp, Cabron properties and their role in supercapacitors. J. Power Sources **157**, 11 (2006)
17. Y. Gogotsi (ed.), *Carbon Nanomaterials* (CRC Press, Boca Raton, FL, 2006)
18. E. Frackowiak, Carbon materials for supercapacitor application. Phys. Chem. Chem. Phys. **9**, 1774 (2007)
19. A. Janes, E. Lust, Electrochemical characteristics of nanoporous carbide-derived carbon materials in various nonaqueous electrolyte solutions. J. Electrochem. Soc. **153**, A113 (2006)

20. B.D. Shanina, A.A. Konchits, S.P. Kolesnik, A.I. Veynger, A.M. Danishevskii, V.V. Popov, S.K. Gordeev, A.V. Grechinskaya, A study of nanoporous carbon obtained from ZC powders. Carbon **41**, 3027 (2003)
21. D.N. Futaba, K. Hata, T. Yamada, T. Hiraoka, Y. Hayamizu, Y. Kakudate, O. Tanaike, H. Hatori, M. Yumura, S. Iijima, Shape-engineerable and highly densely packed single-walled carbon nanotubes and their application as super-capacitor electrodes. Nat. Mater. **5**, 987 (2006)
22. C. Portet, J. Chmiola, Y. Gogotsi, S. Park, K. Lian, Electrochemical characterizations of carbon nanomaterials by the cavity microelectrode technique. Electrochim. Acta **53**, 7675 (2008)
23. C.M. Yang, Y.J. Kim, M. Endo, H. Kanoh, M. Yudasaka, S. Iijima, K. Kaneko, Nanowindow-regulated specific capacitance of supercapacitor electrodes of single-wall carbon nanohorns. J. Am. Chem. Soc. **129**, 20 (2007)
24. A.G. Pandolfo, A.F. Hollenkamp, J. Power Sources **11**, 157 (2006)
25. A.V. Murugan, K. Vijayamohanan, *Nanomaterials Chemistry* (Wiley-VCH GmbH & Co., Weinheim 2007)
26. P. Simon, A. Burke, Nanostructured carbons: double-layer capacitance and more. Interface **17**, 38 (2008)
27. G. Salitra, A. Soffer, L. Eliad, Y. Cohen, D. Aurbach, Characterization of electrodes for super capacitors: surface area and pore size at the solid-gas and electrode solution interfaces of activated carbons. J. Electrochem. Soc. **147**, 2486 (2000)
28. L. Eliad, G. Salitra, A. Soffer, D. Aurbach, Ion sieving effects in the electrical double layer of porous carbon electrodes: estimating effective ion size in electrolytic solutions. J. Phys. Chem. B **105**, 6880 (2001)
29. L. Eliad, G. Salitra, A. Soffer, D. Aurbach, Proton-selective environment in the pores of activated molecular sieving carbon electrodes. J. Phys. Chem. B **106**, 10128 (2002)
30. Y. Guo, J. Qi, Y. Jiang, S. Yang, Z. Wang, H. Xu, Performance of electrical double layer capacitors with porous carbons derived from rice husk. Mater. Chem. Phys. **80**, 704 (2003)
31. H.F. Stoeckh, Microporous carbons and their characterization: the present state of the art. Carbon **28**, 1 (1990)
32. J.K. Hong, J.H. Lee, S.M. Oh, Effect of carbon additive on electrochemical performance of $LiCoO_2$ composite cathodes. J. Power Sources **111**, 90 (2002)
33. S. Kuroda, N. Tobori, M. Sakuraba, Y. Sato, Charge-discharge properties of a cathode prepared with ketjen black as the electro-conductive additive in lithium ion batteries. J. Power Sources **119–121**, 924 (2003)
34. K. Tatsumi, K. Zaghib, H. Abe, S. Higuchi, T. Ohsaki, Y. Sawada, A modification in the preparation process of a carbon whisker for the anode performance of lithium rechargeable batteries. J. Power Sources **54**, 425 (1995)
35. G. Ertl, H. Knözinger, J. Weitkamp (eds.), *Handbook of Heterogeneous Catalysis*, vol. 1. (Wiley-VCH, Weinheim 1997)
36. H.D. Gesser, P.C. Goswami, Aerogels and related porous materials. Chem. Rev. **89**, 765 (1989)
37. R.W. Pekala, Organic aerogels from the polycondensation of resorcinol with formaldehyde. J. Mater. Sci. **24**, 3221 (1989)
38. B.H. Han, W.Z. Zhou, A. Sayari, Direct preparation of nanoporous carbon by nanocasting. J. Am. Chem. Soc. **125**, 3444 (2003)
39. A. Taguchi, J.H. Smått, M. Lindén, Carbon monoliths possessing a hierarchical, fully interconnected porosity. Adv. Mater. **15**, 1209 (2003)
40. A.H. Lu, J.H. Smått, S. Backlund, M. Lindén, Easy and flexible preparation of nanocasted carbon monoliths exhibiting a multimodal hierarchical porosity. Microporous Mesoporous Mater. **72**, 59 (2004)
41. H. Yang, Q. Shi, X. Liu, S. Xie, D. Jiang, F. Zhang, C. Yu, B. Tu, D. Zhao, Synthesis of ordered mesoporous carbon monoliths with bicontinuous cubic pore structure of la3d symmetry. Chem. Commun. **38**, 2842 (2002)

42. K. Nakanishi, Pore structure control of silica gels based on phase separation. J. Porous Mater. **4**, 67 (1997)
43. J.H. Smått, S.A. Schunk, M. Lindén, Versatile double-templating synthesis route to silica monoliths exhibiting a multimodal hierarchical porosity. Chem. Mater. **15**, 2354 (2003)
44. A.H. Lu, J.H. Smått, M. Lindén, Combined surface and volume templating of highly porous nanocast carbon monoliths. Adv. Funct. Mater. **15**, 865 (2005)
45. Z.G. Shi, Y.Q. Feng, L. Xu, S.L. Da, Preparation of porous carbon-silica composite monoliths. Carbon **41**, 2668 (2003)
46. Z.G. Shi, Y.Q. Feng, L. Xu, S.L. Da, M. Zhang, Synthesis of a carbon monolith with trimodal pores. Carbon **41**, 2677 (2003)
47. T. Kyotani, T. Nagai, S. Inoue, A. Tomita, Formation of new type of porous carbon by carbonization in zeolite nanochannels. Chem. Mater. **9**, 609 (1997)
48. C.R. Martin, Membrane-based synthesis of nanomaterials. Chem. Mater. **8**, 1739 (1996)
49. R. Ryoo, S.H. Joo, S. Jun, Synthesis of highly ordered carbon molecular sieves via template-mediated structural transformation. J. Phys. Chem. B **103**, 7743 (1999)
50. F. Schüth, Endo-and exotemplating to create high-surface-area inorganic materials. Angew. Chem. Int. Ed. **42**, 3604 (2003)
51. O.D. Velev, T.A. Jede, R.F. Lobo, A.M. Lenhoff, Porous silica via colloidal crystallization. Nature **389**, 447 (1997)
52. D. Zhao, J. Sun, Q. Li, G.D. Stucky, Morphological control of highly ordered mesoporous silica SBA-15. Chem. Mater. **12**, 275 (2000)
53. C. Yu, J. Fan, B. Tian, D. Zhao, G.D. Stucky, High-yield synthesis of periodic mesoporous silica rods and their replication to mesoporous carbon rods. Adv. Mater. **14**, 1742 (2002)
54. S. Che, Z. Liu, T. Ohsuna, K. Sakamoto, O. Terasaki, T. Tatsumi, Synthesis and characterization of chiral mesoporous silica. Nature **429**, 281 (2004)
55. W. Li, D. Chen, Z. Li, Y. Shi, Y. Wan, G. Wang, Z. Jiang, D. Zhao, Nitrogen-containing carbon spheres with very large uniform mesopores: the superior electrode materials for EDLC in organic electrolyte. Carbon **45**, 1757 (2007)
56. R. Dash, J. Chmiola, G. Yushin, Y. Gogotsi, G. Laudisio, J. Singer, J. Fischer, S. Kucheyev, Titanium carbide derived nanoporous carbon for energy-related applications. Carbon **44**, 2489 (2006)
57. S. Urbonaitea, S. Wachtmeisterb, C. Mirguetc, E. Coroneld, W.Y. Zoub, S. Csillagb, G. Svenssona, EELS studies of carbide derived carbons. Carbon **45**, 2047 (2007)
58. Y. Gogotsi, A. Nikitin, H. Ye, W. Zhou, J.E. Fischer, B. Yi, H.C. Foley, M.W. Barsoum, Nanoporous carbide-derived carbon with tunable pore size. Nat. Mater. **2**, 591 (2003)
59. J. Chmiola, G. Yushin, Y. Gogotsi, C. Portet, P. Simon, P.L. Taberna, Anomalous increase in carbon capacitance at pore sizes less than 1 nanometer. Science **313**, 1760 (2006)
60. J. Gamby, P.L. Taberna, P. Simon, J.F. Fauvarque, M. Chesneau, Studies and characterisations of various activated carbons used for carbon/carbon supercapacitors. J. Power Sources **101**, 109 (2001)
61. D. Qu, H. Shi, Studies of activated carbons used in double-layer capacitors. J. Power Sources **74**, 99 (1998)
62. D.R. Rolison, Catalytic nanoarchitectures–the importance of nothing and the unimportance of periodicity. Science **299**, 1698 (2003)
63. G. Lee, S. Pyun, Theoretical approach to ion penetration into pores with pore fractal characteristics during double-layer charging/discharging on a porous carbon electrode. Langmuir **22**, 10659 (2006)
64. J.W. Long, B. Dunn, D.R. Rolison, H.S. White, Three-dimensional battery architectures. Chem. Rev. **104**, 4463 (2004)
65. D.W. Wang, F. Li, M. Liu, G.Q. Lu, H.M. Cheng, 3D aperiodic hierarchical porous graphitic carbon material for high-rate electrochemical capacitive energy storage. Angew. Chem. Int. Ed. **47**, 373 (2008)

66. D.W. Wang, F. Li, H.T. Fang, M. Liu, G.Q. Lu, H.M. Cheng, Effect of pore packing defects in 2-D ordered mesoporous carbons on ionic transport. J. Phys. Chem. B **110**, 8570 (2006)
67. W. Xing, S.Z. Qiao, R.G. Ding, F. Li, G.Q. Lu, Z.F. Yan, H.M. Cheng, Superior electric double layer capacitors using ordered mesoporous carbons. Carbon **44**, 216 (2006)
68. A.B. Fuertes, F. Pico, J.M. Rojo, Influence of pore structure on electric double-layer capacitance of template mesoporous carbons. J. Power Sources **133**, 329 (2004)
69. H. Yamada, H. Nakamura, F. Nakahara, I. Moriguchi, T. Kudo, Electrochemical study of high electrochemical double layer capacitance of ordered porous carbons with both meso/macropores and micropores. J. Phys. Chem. C **111**, 227 (2007)
70. T. Morishita, Y. Soneda, T. Tsumura, M. Inagaki, Preparation of porous carbons from thermoplastic precursors and their performance for electric double layer capacitors. Carbon **44**, 2360 (2006)
71. E. Raymundo-PiMero, K. Kierzek, J. Machnikowski, F. Béguin, Relationship between the nanoporous texture of activated carbons and their capacitance properties in different electrolytes. Carbon 44, 2498 (2006)
72. C. Zhao, W. Wang, Z. Yu, H. Zhang, A. Wang, Y. Yang, Nano-$CaCO_3$ as template for preparation of disordered large mesoporous carbon with hierarchical porosities electronic supplementary information (ESI) available: nitrogen adsorption/desorption isotherms, textural parameters and carbon yield data for LMC materials. J. Mater. Chem. **20**, 976 (2010)
73. J. Shu, H. Li, R. Yang, Y. Shi, X. Huang, Cage-like carbon nanotubes/Si composite as anode material for lithium ion batteries. Electrochem. Commun. **8**, 51 (2006)
74. A. Dillon, K. Jones, T. Bekkedahl, C. Kiang, D. Bethune, M. Heben, Storage of hydrogen in single-walled carbon nanotubes. Nature 386 377 (1997)
75. J. Wildoer, L. Venema, A. Rinzler, R. Smalley, C. Dekker, Electronic structure of atomically resolved carbon nanotubes. Nature **391**, 59 (1998)
76. Z. Ren, Z. Huang, J. Xu, J. Wang, P. Bush, M. Siegal, P. Provencio, Synthesis of large arrays of well-aligned carbon nanotubes on glass. Science **282**, 1105 (1998)
77. J. Kong, N. Franklin, C. Zhou, M. Chapline, S. Peng, K. Cho, H. Dai, Nanotube molecular wires as chemical sensors. Science **287**, 622 (2000)
78. R. Baughman, A. Zakhidov, W. de Heer, Carbon nanotubes–the route toward applications. Science **297**, 787 (2002)
79. C. Niu, E. Sichel, R. Hoch, D. Moy, H. Tennent, High power electrochemical capacitors based on carbon nanotube electrodes. Appl. Phys. Lett. 70 1480 (1997)
80. C. Liu, A. Bard, F. Wudl, I. Weitz, J. Heath, Electrochemical characterization of films of single-walled carbon nanotubes and their possible application in supercapacitors. Electrochem. Solid-State Lett. **2**, 577 (1999)
81. C. Emmenegger, P. Mauron, A. Zuttel, C. Nutzenadel, A. Schneuwly, R. Gallay, L. Schlapbach, Carbon nanotube synthesized on metallic substrates. Appl. Surf. Sci. **162–163**, 452 (2000)
82. C. Emmenegger, P. Mauron, P. Sudan, P. Wenger, V. Hermann, R. Gallay, A. Zuttel, Investigation of electrochemical double-layer (ECDL) capacitors electrodes based on carbon nanotubes and activated carbon materials. J. Power Sources **124**, 321 (2003)
83. D. Zilli, P. Bonelli, A. Cukierman, Effect of alignment on adsorption characteristics of self-oriented multi-walled carbon nanotube arrays. Nanotechnology **17**, 5136 (2006)
84. H. Zhang, G. Cao, Y. Yang, Effect of alignment on adsorption characteristics of self-oriented multi-walled carbon nanotube arrays. Nanotechnology **18**, 195607 (2007)
85. Y. Honda, T. Haramoto, M. Takeshige, H. Shiozaki, T. Kitamura, M. Ishikawa, Aligned MWCNT sheet electrodes prepared by transfer methodology providing high-power capacitor performance. Electrochem. Solid-State Lett. **10**, A106 (2007)
86. K. Hata, D.N. Futaba, K. Mizuno, T. Namai, M. Yumura, S. Iijima, Water-assisted highly efficient synthesis of impurity-free single-walled carbon nanotubes. Science **306**, 1362 (2004)

87. G.Y. Xiong, D.Z. Wang, Z.F. Ren, Aligned millimeter-long carbon nanotube arrays grown on single crystal magnesia. Carbon **44**, 969 (2006)
88. H. Zhang, G.P. Cao, Z.Y. Wang, Y.S. Yang, Z.J. Shi, Z.N. Gu, Influence of hydrogen pretreatment condition on the morphology of Fe/Al_2O_3 catalyst film and growth of millimeter-long carbon nanotube array. J. Phys. Chem. C **112**, 4524 (2008)
89. H. Zhang, G.P. Cao, Z.Y. Wang, Y.S. Yang, Z.J. Shi, Z.N. Gu, Influence of ethylene and hydrogen flow rates on the wall number, crystallinity, and length of millimeter-long carbon nanotube array. J. Phys. Chem. C **112**, 12706 (2008)
90. G.Y. Zhang, D. Mann, L. Zhang, A. Javey, Y.M. Li, E. Yenilmez, Q. Wang, J.P. McVittie, Y. Nishi, J. Gibbons, H.J. Dai, Ultra-high-yield growth of vertical single-walled. Carbon nanotubes: hidden roles of hydrogen and oxygen. Proc. Natl. Acad. Sci. USA **102**, 16141 (2005)
91. Y. Murakami, S. Maruyama, Detachment of vertically aligned single-walled carbon nanotube films from substrates and their re-attachment to arbitrary surfaces. Chem. Phys. Lett. **422**, 575 (2006)
92. L.B. Zhu, Y.Y. Sun, D.W. Hess, C.P. Wong, Well-aligned open-ended carbon nanotube architectures: an approach for device assembly. Nano Lett. **6**, 243 (2006)
93. A. Kumar, V.L. Pushparaj, S. Kar, O. Nalamasu, P.M. Ajayan, Contact transfer of aligned carbon nanotube arrays onto conducting substrates. Appl. Phys. Lett. **89**, 163120 (2006)
94. J. Yu, J.G. Shapter, M.R. Johnston, J.S. Quinton, J.J. Gooding, Electron-transfer characteristics of ferrocene attached to single-walled carbon nanotubes (SWCNT) arrays directly anchored to silicon(100). Electrochim. Acta **52**, 6206 (2007)
95. D. Nkosi, K.I. Ozoemena, self-assembled nano-arrays of single-walled carbon nanotube–octa(hydroxyethylthio)phthalocyaninatoiron(ii) on gold surfaces: impacts of SWCNT and solution pH on electron transfer kinetics. Electrochim. Acta **53**, 2782 (2008)
96. T. Hiraoka, T. Yamada, K. Hata, D.N. Futaba, H. Kurachi, S. Uemura, M. Yumura, S. Iijima, Synthesis of single- and double-walled carbon nanotube forests on conducting metal foils. J. Am. Chem. Soc. **128**, 13338 (2006)
97. S. Talapatra, S. Kar, S.K. Pal, R. Vajtai, L. Ci, P. Victor, M.M. Shaijumon, S. Kaur, O. Nalamasu, P.M. Ajayan, Direct growth of aligned carbon nanotubes on bulk metals. Nat. Nanotechnol. **1**, 112 (2006)
98. H. Zhang, G.P. Cao, Z.Y. Wang, Y.S. Yang, Z.N. Gu, Electrochemical capacitive properties of carbon nanotube arrays directly grown on glassy carbon and tantalum foils. Carbon **46**, 822 (2008)
99. H. Zhang, G.P. Cao, Y.S. Yang, Z.N. Gu, Comparision between electrochemical properties of aligned carbon nanotube array and entangled carbon nanotube electrodes. J. Electrochem. Soc. **155**, K19 (2008)
100. P. Ayala, A. Grüneis, T. Gemming, B. Büchner, M.H. Rümmeli, D. Grimm, J. Schumann, R. Kaltofen, F.L. Freire Jr., H.D. Fonseca Filho, T. Pichler, Influence of the catalyst hydrogen pretreatment on the growth of vertically aligned nitrogen-doped carbon nanotubes. Chem. Mater. **19**, 6131 (2007)
101. I. Gonzalez-Valls, M. Lira-Cantu, Vertically-aligned nanostructures of ZnO for excitonic solar cells: a review. Energy Environ. Sci. **2**, 19 (2009)
102. S.W. Lee, B.S. Kim, S. Chen, S.H. Yang, P.T. Hammond, Layer-by-layer assembly of all carbon nanotube ultrathin films for electrochemical applications. J. Am. Chem. Soc. **131**, 671 (2009)
103. N.L. Wu, Nanocrystalline oxide supercapacitors. Mater. Chem. Phys. **75**, 6 (2002)
104. K.R. Prasad, N. Miura, Potentiodynamically deposited nanostructured manganese dioxide as electrode material for electrochemical redox supercapacitors. J. Power Sources **135**, 354 (2004)
105. A udge, I. Raistrick, S. Gottesfeld, J.P. Ferraris, Conducting polymers as active materials in electrochemical capacitors. J. Power Sources **47**, 89 (1994)
106. J.P. Zheng, T.R. Jow, High energy and high power density electrochemical capacitors. J. Power Sources **62**, 155 (1996)

107. H.Y. Lee, J.B. Goodenough, Supercapacitor behavior with KCl electrolyte. J. Solid State Chem. **144**, 220 (1999)
108. A. Laforgue, P. Simon, J.F. Fauvarque, Chemical synthesis and characterization of fluorinated polyphenylthiophenes: application to energy storage. Synthetic Met. **123**, 311 (2001)
109. K. Naoi, S. Suematsu, A. Manago, Electrochemistry of poly(1,5-diaminoanthraquinone) and its application in electrochemical capacitor materials. J. Electrochem. Soc. **147**, 420 (2000)
110. A.S. Arico, P. Bruce, B. Scrosati, J.M. Tarascon, W.V. Schalkwijk, Nanostructured materials for advanced energy conversion and storage devices. Nat. Mater. **4**, 366 (2005)
111. D. Choi, G.E. Blomgren, P.N. Kumta, Fast and reversible surface redox reaction in nanocrystalline vanadium nitride supercapacitors. Adv. Mater. **18**, 1178 (2006)
112. K. Machida, K. Furuuchi, M. Min, K. Naoi, Mixed proton-electron conducting nanocomposite based on hydrous RuO2 and polyaniline derivatives for supercapacitors. Electrochemistry **72**, 402 (2004)
113. M. Toupin, T. Brousse, D. Belanger, Charge storage mechanism of MnO_2 electrode used in aqueous electrochemical capacitor. Chem. Mater. **16**, 3184 (2004)
114. W. Sugimoto, H. Iwata, Y. Yasunaga, Y. Murakami, Y. Takasu, Preparation of ruthenic acid nanosheets and utilization of its interlayer surface for electrochemical energy storage. Angew. Chem. Int. Ed. **42**, 4092 (2003)
115. J.M. Miller, B. Dunn, T.D. Tran, R.W. Pekala, Deposition of ruthenium nanoparticles on carbon aerogels for high energy density supercapacitor electrodes. J. Electrochem. Soc. **144**, L309 (1997)
116. M. Min, K. Machida, J.H. Jang, K. Naoi, Hydrous RuO_2/carbon black nanocomposites with 3D porous structure by novel incipient wetness method for supercapacitors. J. Electrochem. Soc. **153**, A334 (2006)
117. Y. Wang, K. Takahashi, K.H. Lee, G.Z. Cao, Nanostructured vanadium oxide electrodes for enhanced lithium-ion intercalation. Adv. Funct. Mater. **16**, 1133 (2006)
118. W. Dmowski, T. Egami, K.E. Swider-Lyons, C.T. Love, D.R. Rolison, Local atomic structure and conduction mechanism of nanocrystalline hydrous ruo2 from x-ray scattering. J. Phys. Chem. B **106**, 12677 (2002)
119. T. Shinomiya, V. Gupta, N. Miura, Effects of electrochemical-deposition method and microstructure on the capacitive characteristics of nano-sized manganese oxide. Electrochim. Acta **51**, 4412 (2005)
120. Y.U. Jeong, A. Manthiram, Nanocrystalline manganese oxides for electrochemical capacitors with neutral electrolytes. J. Electrochem. Soc. **149**, A1419 (2002)
121. J. Kim, A. Manthiram, A manganese oxyiodide cathode for rechargeable lithium batteries. Nature **390**, 265 (1997)
122. J.W. Long, C.P. Rhodes, A.L. Young, D.R. Rolison, Ultrathin, protective coatings of poly(o-phenylenediamine) as electrochemical proton gates: making mesoporous MnO_2 nanoarchitectures stable in acid electrolytes. Nano Lett. **3**, 1155 (2003)
123. Y. Murakami, K. Konishi, Remarkable co-catalyst effect of gold nanoclusters on olefin oxidation catalyzed by a manganese–porphyrin complex. J. Am. Chem. Soc. **129**, 14401 (2007)
124. J.K. Yuan, W.N. Li, S. Gomez, S.L. Suib, Shape-controlled synthesis of manganese oxide octahedral molecular sieve three-dimensional nanostructures. J. Am. Chem. Soc. **127**, 14184 (2005)
125. M. Yin, S. O'Brien, Synthesis of monodisperse nanocrystals of manganese oxides. J. Am. Chem. Soc. **125**, 10180 (2003)
126. X.H. Zhong, R.G. Xie, L.T. Sun, I. Lieberwirth, W.J. Knoll, Synthesis of dumbbell-shaped manganese oxide nanocrystals. Phys. Chem. B **110**, 2 (2006)
127. L.C. Zhang, Z.H. Liu, H. Lv, X.H. Tang, K. Ooi, Shape-controllable synthesis and electrochemical properties of nanostructured manganese oxides. J. Phys. Chem. C **111**, 8418 (2007)

128. M.S. Wu, P.J. Chiang, J.T. Lee, J.C. Lin, Synthesis of manganese oxide electrodes with interconnected nanowire structure as an anode material for rechargeable lithium ion batteries. J. Phys. Chem. B **109**, 23279 (2005)
129. F.Y. Cheng, J. Chen, X.L. Gou, P.W. Shen, High-power alkaline Zn-MnO_2 batteries using y-MnO_2 nanowires/nanotubes and electrolytic zinc powder. Adv. Mater. **17**, 2753 (2005)
130. V. Subramanian, H.W. Zhu, R. Vajtai, P.M. Ajayan, B.Q. Wei, Hydrothermal synthesis and pseudocapacitance properties of MnO_2 nanostructures. J. Phys. Chem. B **109**, 20207 (2005)
131. Y. Oaki, H. Imai, One-pot synthesis of manganese oxide nanosheets in aqueous solution: chelation-mediated parallel control of reaction and morphology. Angew. Chem. Int. Ed. **46**, 4951 (2007)
132. Y.U. Jeong, A. Manthiram, Amorphous ruthenium-chromium oxides for electrochemical capacitors. Electrochem. Solid-State Lett. **3**, 205 (2000)
133. Y.U. Jeong, A. Manthiram, Amorphous tungsten oxide/ruthenium oxide composites for electrochemical capacitors. J. Electrochem. Soc. **148**, A189 (2001)
134. J.C. Hulteen, C.R. Martin, A general template-based method for the preparation of nanomaterials. J. Mater. Chem. **7**, 1075 (1997)
135. N. Pinna, U. Wild, J. Urban, R. Schlögl, Divanadium pentoxide nanorods. Adv. Mater. **15**, 329 (2003)
136. H.J. Qiu, J. Zhai, S.H. Li, M.X. Wan, Oriented growth of self-assembled polyaniline nanowire arrays using a novel method. Adv. Funct. Mater. **13**, 925 (2003)
137. J.X. Huang, R.B. Kaner, A general chemical route to polyaniline nanofibers. J. Am. Chem. Soc. **126**, 851 (2004)
138. V. Gupta, N. Miura, Large-area network of polyaniline nanowires prepared by potentiostatic deposition process. Electrochem. Commun. **7**, 995 (2005)
139. Z.X. Wei, M.X. Wan, Hollow microspheres of polyaniline synthesized with an aniline emulsion template. Adv. Mater. **14**, 1314 (2002)
140. Y.G. Wang, H.Q. Li, Y.Y. Xia, Ordered whiskerlike polyaniline grown on the surface of mesoporous carbon and its electrochemical capacitance performance. Adv. Mater. **18**, 2619 (2006)
141. S.I. Cho, S.B. Lee, Fast electrochemistry of conductive polymer nanotubes: synthesis, mechanism, and application. Acc. Chem. Res. **41**, 699 (2008)
142. H. Zhang, G.P. Cao, Z.Y. Wang, Y.S. Yang, Z.J. Shi, Z.N. Gu, Tube-covering-tube nanostructured polyaniline/carbon nanotube array composite electrode with high capacitance and superior rate performance as well as good cycling stability. Electrochem. Commun. **10**, 1056 (2008)
143. E.S. Toberer, T.D. Schladt, R. Seshadri, Macroporous manganese oxides with regenerative mesopores. J. Am. Chem. Soc. **128**, 1462 (2006)
144. T. Fukushima, A. Kosaka, Y. Yamamoto, T. Aimiya, S. Notazawa, T. Takigawa, T. Inabe, T. Aida, Dramatic effect of dispersed carbon nanotubes on the mechanical and electroconductive properties of polymers derived from ionic liquids. Small 2, 554 (2006)
145. V. Subramanian, H. Zhu, B. Wei, Synthesis and electrochemical characterizations of amorphous manganese oxide and single walled carbon nanotube composites as supercapacitor electrode materials. Electrochem. Commun. **8**, 827 (2006)
146. E. Frackowiak, F. Béguin, Carbon materials for the electrochemical storage of energy in capacitors. Carbon **39**, 937 (2001)
147. D. Zilli, P.R. Bonelli, A.L. Cukierman, Effect of alignment on adsorption characteristics of self-oriented multi-walled carbon nanotube arrays. Nanotechnology **17**, 5136 (2006)
148. R.W. Pekala, J.C. Farmer, C.T. Alivaso, T.D. Tran, S.T. Mayer, J.M. Miller, B. Dunn, Carbon aerogels for electrochemical applications. J. Non-Cryst. Solids **225**, 74 (1998)
149. A.E. Fischer, K. Pettigrew, D.R. Rolison, R.M. Stroud, J.W. Long, Incorporation of homogeneous, nanoscale MnO_2 within ultraporous carbon structures via self-limiting electroless deposition: implications for electrochemical capacitors. Nano Lett. 7 281 (2007)

150. J.W. Long, B.M. Dening, T.M. McEvoy, D.R. Rolison, Charge insertion into hybrid nanoarchitectures: mesoporous manganese oxide coated with ultrathin poly(phenylene oxide). J. Non-Cryst. Solids **350**, 97 (2004)
151. H. Zhang, G.P. Cao, Z.Y. Wang, Y.S. Yang, Z.J. Shi, Z.N. Gu, Growth of manganese oxide nanoflowers on vertically-aligned carbon nanotube arrays for high-rate electrochemical capacitive energy storage. Nano Lett. **8**, 2664 (2008)
152. J.H. Jiang, A. Kucernak, Electrochemical supercapacitor material based on manganese oxide: preparation and characterization. Electrochim. Acta **47**, 2381 (2002)
153. N. Nagarajan, H. Humadi, I. Zhitomirsky, Cathodic electrodeposition of MnOx films for electrochemical supercapacitors. Electrochim. Acta **51**, 3039 (2006)
154. L.Z. Fan, J. Maier, High-performance polypyrrole electrode materials for redox supercapacitors. Electrochem. Commun. **8**, 937 (2008)
155. Y.J. Kim, Y. Abe, T. Yanagiura, K.C. Park, M. Shimizu, T. Iwazaki, S. Nakagawa, M. Endo, M.S. Dresselhaus, Easy preparation of nitrogen-enriched carbon materials from peptides of silk fibroins and their use to produce a high volumetric energy density in supercapacitors. Carbon **45**, 2116 (2007)
156. A. Jänes, H. Kurig, E. Lust, Characterisation of activated nanoporous carbon for supercapacitor electrode materials. Carbon **45**, 1226 (2007)
157. C.K. Chan, H.L. Peng, G. Liu, K. McIlwrath, X.F. Zhang, R.A. Huggins, Y. Cui, High-performance lithium battery anodes using silicon nanowires. Nat. Nanotechnol. **3**, 31 (2008)
158. R. Liu, S.B. Lee, MnO_2/poly(3,4-ethylenedioxythiophene) coaxial nanowires by one-step coelectrodeposition for electrochemical energy storage. J. Am. Chem. Soc. **130**, 2942 (2008)
159. J.N. Broughton, M.J. Brett, Variations in MnO_2 electrodeposition for electrochemical capacitors. Electrochim. Acta **50**, 4814 (2005)
160. J.S. Ye, H.F. Cui, X. Liu, T.M. Lim, W.D. Zhang, F.S. Sheu, Preparation and characterization of aligned carbon nanotube-ruthenium oxide nanocomposites for supercapacitors. Small 1, 560 (2005)
161. H.J. Ahn, W.B. Kim, T.Y. Seong, $Co(OH)_2$-combined carbon-nanotube array electrodes for high-performance micro-electrochemical capacitors. Electrochem. Commun. **10**, 1284 (2008)
162. P. Novák, K. Müller, K.S.V. Santhanam, O. Haas, Electrochemically active polymers for rechargeable batteries. Chem. Rev. **97**, 207 (1997)
163. L. Liang, J. Liu, C.F. Windisch, G.J. Exarhos, Y. Li, Direct assembly of large arrays of oriented conducting polymer nanowires. Angew. Chem. Int. Ed. **41**, 3665 (2002)
164. L.Z. Fan, Y.S. Hu, J. Maier, P. Adelhelm, B. Smarsly, M. Antonietti, High electroactivity of polyaniline in supercapacitors by using a hierarchically porous carbon monolith as a support. Adv. Funct. Mater. **17**, 3083 (2007)
165. M.J. Winokur, B.R. Mattes, Structural studies of halogen acid doped polyaniline and the role of water hydration. Macromolecules **13**, 8183 (1998)
166. C.C. Hu, J.Y. Lin, Effects of the loading and polymerization temperature on the capacitive performance of polyaniline in $NaNO_3$. Electrochim. Acta **47**, 4055 (2002)
167. D. Tasis, N. Tagmatarchis, A. Bianco, M. Prato, Chemistry of carbon nanotubes. Chem. Rev. **106**, 1105 (2006)
168. M. Hughes, M. Shaffer, N.C. Renouf, C. Singh, G.Z. Chen, D.J. Fray, A.H. Windle, Electrochemical capacitance of nanocomposite films formed by coating aligned arrays of carbon nanotubes with polypyrrole. Adv. Mater. **14**, 382 (2002)
169. E. Frackowiak, V. Khomenko, K. Jurewicz, K. Lota, F. Béguin, Supercapacitors based on conducting polymers/nanotubes composites. J. Power Sources **153**, 413 (2006)
170. J.H. Chen, Z.P. Huang, D.Z. Wang, S.X. Yang, W.Z. Li, J.G. Wen, Z.F. Ren, Electrochemical synthesis of polypyrrole films over each of well-aligned carbon nanotubes. Synthetic Met. **125**, 289 (2001)

Chapter 18
Aligned Ni-Coated Single-Walled Carbon Nanotubes Under Magnetic Field for Coolant Applications

Haiping Hong, Mark Horton, and G.P. Peterson

18.1 Introduction

The discovery of carbon nanotubes (CNTs) has inspired significant interest in recent years, due to the tremendous promise related to their thermal, electrical, mechanical, and functional properties. For example, it has been reported that fluids containing single-walled carbon nanotubes exhibit a thermal conductivity ranging from 2,000 to 6,000 W/mK [1] under ideal circumstances. By contrast, typical heat transfer fluids such as water and oil have thermal conductivity values of only 0.6 and 0.2 W/mK, respectively.

These carbon nanotube fluids, referred to as nanofluids, exhibit substantially improved thermal conductivity values [2–3]. However, a simple composite structure turned out to be ineffective in enhancing the thermal conductivity of the nanofluids. At low nanotube concentrations, no significant improvements in thermal conductivity were reported. At a nanotube loading of 1 vol.%, CNTs ($\sim$1.4 wt%), there is an $\sim$10–20% increase in the thermal conductivity as reported by different groups [4–6]. However, at these high concentrations, the fluid has already lost most of its fluidity and becomes "mud like," which makes the fluid much less useful for coolant and lubricant applications.

A possible explanation for this lack of appreciable thermal conductivity enhancement through compositing with carbon nanotubes may be attributed to the lack of particle alignment and orientation in the fluids. When carbon nanotubes are dispersed in a fluid, there are only random and infrequent chances for them to contact each other to form a percolation network. As a result, only very high concentrations of CNTs produce noticeable thermal conductivity improvements. Therefore, there is a need to introduce nanotube–nanotube physical contacts in the fluids in order to increase the thermal (and electrical) conductivity significantly while keeping the nanotube concentration relatively low to ensure a desirably low fluid viscosity.

H. Hong (✉)
Department of Material and Metallurgical Engineering, South Dakota School of Mines and Technology, Rapid City, SD 57701, USA
e-mail: haiping.hong@sdsmt.edu

W.L. Zhou, Z.L. Wang (eds.), *Three-Dimensional Nanoarchitectures*,
DOI 10.1007/978-1-4419-9822-4_18,

A composite fluid containing high thermal conductivity CNTs and magnetic field-responsive magnetic nanoparticles was prepared with the aim to increase the thermal conductivity of the fluid. Such fluids with enhanced thermal conductivity are useful for a variety of applications, such as heat transfer coolants and lubricants [7, 8]. Magnetic particles in a liquid medium can have a variety of configurations depending on the nature of magnetic particles and the magnetic field strength [9, 10]. Under a relatively strong magnetic field, small magnetic particles form connected networks and also tend to be oriented toward the field direction, which also draws the carbon nanotubes closer and induces more physical contacts. Thus, improved thermal conductivity is anticipated. In a previous investigation, carbon nanotubes and magnetic field-sensitive nanoparticle Fe_2O_3 were combined in water under a magnetic field, and the results confirmed this assumption [11]. Also, enhanced thermal conductivity was observed in the magnetically polarizable nanofluid that consisted of a colloidal suspension of magnetic nanoparticles with an average diameter of 6.7 nm [12].

Based on the above results, it was estimated that magnetically sensitive, metal-encapsulated carbon nanotubes [13] dispersed in water, water/ethylene glycol, or oil, with the help of appropriate chemical surfactant, could also significantly increase the thermal conductivity under an applied magnetic field. The coated layer configuration of the magnetic material could be more beneficial than the presence of isolated magnetic nanoparticles as small magnetic particles tend to be superparamagnetic with a weak response to an applied magnetic field.

Investigation of nanofluid systems containing 0.01, 0.02, and 0.05 wt% of nickel-coated (Ni-coated) magnetically sensitive nanotubes in water under different magnetic field strengths is currently planned, in addition to observation of real-time images of Ni-coated single-walled carbon nanotubes in water and poly-α-olefin and polyester oils under magnetic field by high-speed microscopy [14]. Understanding the behavior of the thermal conductivity of such a composite material system in an applied magnetic field would be valuable in the analysis and synthesis of nanofluids.

18.2 Experiment

Single-walled carbon nanotubes (SWNTs) were obtained from Helix Material Solutions, Inc. (Helix, Richardson, TX). A chemical surfactant, sodium dodecylbenzene sulfonate (SDBS), was purchased from Sigma Aldrich.

Ni-coated SWNTs were prepared using the following methodology:

1. Oxidation: HNO_3:H_2SO_4 (3:1) reflux at 110°C for 6 h.
2. Sensitization and activation: Immersion in 0.1 M $SnCl_2$–0.1 M HCl for 30 min and immersion in 0.0014 M $PdCl_2$–0.25 M HCl for 30 min.
3. Electroless plating: Introduction into electroless bath for 25 min with composition as listed in Table 18.1.

Table 18.1 Electroless bath conditions

Chemical	Concentration (g/L)
$NiSO_4 \cdot 6H_2O$	25
$Na_3C_6H_5O_7 \cdot 2H_2O$	5
$NaH_2PO_2 \cdot H_2O$	15
NH_4Cl	60
$Pb(NO_3)_2$	2.5
Bath temperature	20°C
pH (adjusted with NH_4OH)	~8.0

The Ni-coated SWNTs were placed into the water together with an appropriate amount of chemical surfactant, sodium dodecylbenzene sulfonate (SDBS). The poly-α-olefin-based nanofluid was made with appropriate surfactant. Sonication of the samples was performed using a Branson Digital Sonifier, Model 450. The magnetic field was provided by a pair of Ba ferrite magnet plates (10.2 cm × 15.2 cm × 2.5 cm), and the magnetic field intensity was recorded by a FW Bell Gauss Meter, Model 5060. The thermal conductivity of the samples was obtained by a Hot Disk^{TM} thermal constant analyzer (for detailed information, see www.hotdisk.se). All thermal conductivity measurements were conducted at room temperature with a kapton-covered, double spiral nickel sensor with a radius of 3.189 mm. The measurements were taken at a power of 0.012 W, which was applied for 15 s. The temperature coefficient of resistivity (TCR) was 0.0471 K^{-1} for the nickel sensor.

Microscope images were recorded by a Motion Scope (Redlake MASD Inc., San Diego, CA), Model PCI 2000S. The parameters were set as follows: record rate 250, shutter 1/250, and trigger 70%. The lenses used were WHB 10×/20 and MPlan 10×/0.25. Scanning electron microscopy (SEM) images were acquired using the backscattered electron detector on a Zeiss Supra 40 VP variable pressure system. Transmission electron microscopy (TEM) images were acquired with a Hitachi H-7000 FA.

18.3 Results and Discussion

18.3.1 Thermal Conductivity of Nanofluids Containing Ni-Coated Nanotubes

The thermal conductivity versus time plot for various concentrations of Ni-coated nanotubes (0.01 and 0.02 wt%) in applied magnetic fields (two different intensities) is shown in Fig. 18.1. Without a magnetic field, the thermal conductivity value of Ni-coated, SWNT-loaded nanofluids is around 0.63–0.64 W/mK and is mostly independent of time. Because of very few contacts between Ni-coated carbon nanotubes, the thermal conductivity value of the nanofluid is essentially the value of the DI water itself. Figure 18.1a shows the thermal conductivity versus time plot with

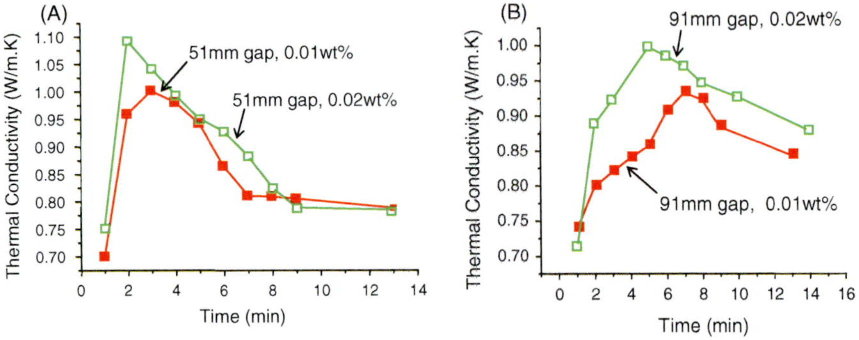

Fig. 18.1 Thermal conductivity versus time plot in different concentrations at two different magnetic field intensities. (**a**) 51-mm gap (0.62 kG) and (**b**) 91-mm gap (0.38 kG)

a 51-mm gap between the two magnets (0.62 kG magnetic field) with two concentrations of Ni-coated SWNT (0.01 wt% and 0.02 wt%) and Fig. 18.1b shows the thermal conductivity versus time plot with a 91-mm gap (0.38 kG) with the same particle concentrations. Under the applied magnetic field, the thermal conductivity of the nanofluid shows very interesting behavior. The thermal conductivity initially increases with time but eventually reaches a peak. When holding the sample in the magnetic field for longer times, the Ni-coated nanotubes gradually move and form bigger clumps of CNTs, and a decrease in the thermal conductivity is seen. The maximum thermal conductivity is approximately 1.10 W/mK for the 51-mm gap (0.62 kG) experiment at 0.02 wt% and 1.01 W/mK for the 91-mm gap (0.38 kG) experiment at 0.02 wt% or about 75 and 60% higher than the thermal conductivity of DI water, respectively. For both the 51-mm and the 91-mm gaps, the thermal conductivity of 0.02 wt% nanofluids is obviously higher than that of 0.01 wt% sample, and the peak of maximum thermal conductivity is slightly left shifted versus time.

These results strongly support our hypothesis that good alignment and orientation of the conducting material in a fluid is critical to the thermal conductivity enhancement of the composite fluid. However, in the case of nanoparticle-containing composite fluids, the thermal conductivity enhancement ratio is not as significant due to the thermal contact resistance in the nanofluid [15].

The next step was to determine what happens to the thermal conductivity of nanofluids containing Ni-coated single-walled carbon nanotubes if the external magnetic field is reduced and maintained at a much smaller field once the thermal conductivity reaches a maximum. Figure 18.2 is a plot of thermal conductivity versus time in a nanofluid containing 0.01 wt% Ni-coated SWNTs; (A) after placing the nanofluid in a 0.62 kG magnetic field for 2–3 min to reach the maximum thermal conductivity, the magnetic field was reduced to 0.18 kG and maintained; (B) after placing the nanofluid in a 0.62 kG magnetic field for 2–3 min to reach the maximum thermal conductivity, the magnetic field was removed completely; and (C) after placing the nanofluid in a constant 0.18 kG magnetic field. It is clearly

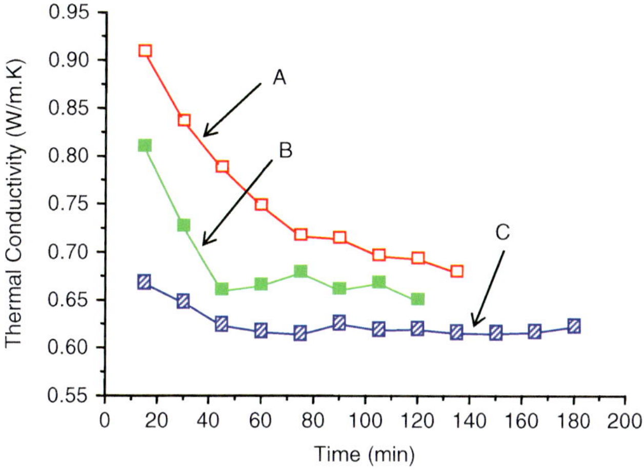

Fig. 18.2 Thermal conductivity versus time plot in 0.01 wt% Ni-coated, SWNT-loading nanofluids. (*A*) After putting fluids in 0.62 kG magnetic field for 2–3 min to reach the maximum thermal conductivity and then reducing the magnetic field to 0.18 kG; (*B*) after putting fluids in 0.62 kG magnetic field for 2–3 min to reach the maximum thermal conductivity and removing the magnetic field; and (*C*) after putting the fluids in a constant 0.18 kG magnetic field

seen from Fig. 18.2 that reducing the applied magnetic field strength after the thermal conductivity of the nanofluid reaches its maximum, rather than removing the magnetic field completely, results in a higher thermal conductivity over time. These results indicate that the gradual magnetic clumping was the cause of the decrease in thermal conductivity. To better understand this, measurements of the thermal conductivity versus time in a constant weak magnetic field (0.18 kG) were conducted, (see sample C in Fig. 18.2). This resulted in a relatively low thermal conductivity value. Further optimization or time-dependent programming of the applied magnetic field is likely to yield more desirable thermal conductivity as a function of time.

It is interesting to note that if one extrapolates the curves A and B to zero time, the two curves would come to a similar starting point, around 1.00 W/mK. Actually, this more or less coincides with the maximum thermal conductivity value of the nanofluid under 0.62 kG magnetic field.

Also, it is interesting to note that different magnetic field strengths influence the maximum thermal conductivity of the nanofluid, the time to reach the maximum, and the shape of the thermal conductivity versus time curve. Figure 18.3 shows the thermal conductivity versus time at different magnetic field strengths (as adjusted by changing the distance between the pair of magnets as shown in Table 18.2). It is seen that the shorter the gap distance (the higher the applied magnetic field), the shorter the time to reach the maximum thermal conductivity. At a 35-mm gap, it is impossible to observe the maximum thermal conductivity value. This is as anticipated since the shorter distance means a stronger magnetic field, which causes the nanotubes to

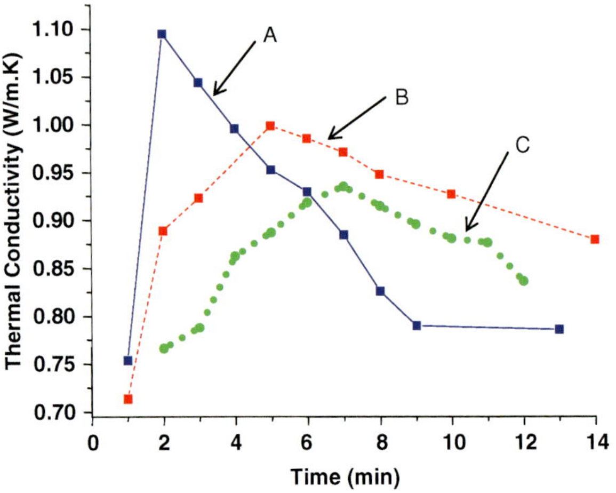

Fig. 18.3 Thermal conductivity versus time plot at different gap distances between two magnets (different magnetic field intensities) at 0.02 wt% concentration. (*A*) 51-mm gap, (*B*) 91-mm gap, and (*C*) 141-mm gap

Table 18.2 Magnetic field dependence of the time to reach peak thermal conductivity and the value of thermal conductivity maximum at the concentration of 0.02 wt% Ni-coated SWNT

Magnet gap (mm)	Magnetic field (kG)	Time to reach maximum TC (min)	TC maximum value (W/mK)
35	0.86	<2	N/A
51	0.62	2	1.10
91	0.38	5	1.01
141	0.18	7	0.95

quickly assemble and align at a rate quicker than we are able to measure the thermal conductivity with our current equipment.

The quantitative data correlating the magnetic field intensity, the time to reach thermal conductivity maximum, and the thermal conductivity value at different gap distances between the two magnets are listed in Table 18.2.

In order to make sure that alignment process dominates thermal conductivity enhancement rather than micro-convection, the magnet height and positions were altered (and hence the magnetic field intensity and orientation) during the time-dependent thermal conductivity measurements to increase particle motion in the fluids. However, no significant differences were observed for the thermal conductivity data. Magnetic field direction was also reversed manually by switching the magnets during the experimental period to see if enhanced tangling and contacts among the magnetically coated nanotubes could be obtained. Again, the thermal conductivity data did not show a significant difference. This illustrates that the thermal conductivity enhancement could be observed along the applied magnetic field direction, but not along the perpendicular direction (data not shown). If the micro-convection assumption is true, then thermal conductivity enhancement in all directions should be comparable. Normally, micro-convection effect only lasts several minutes. The longer timescale thermal conductivity enhanced phenomenon is also strong evidence that thermal conductivity enhancement is not due to micro-convection [16].

18.3.2 Evidence of Magnetic Alignment of Ni-Coated Nanotubes

Evidence of the effective magnetic alignment of Ni-coated nanotubes was seen in a variety of methods. Microscope images of 0.05 wt% Ni-coated SWNT in DI water are shown in Fig. 18.4a before magnetic field was applied and in Fig. 18.4b after magnetic field was applied for 30 min. It is clearly seen in the high-speed microscope video that before the magnetic field, Ni-coated nanotubes are randomly dispersed in water. Like pristine nanotubes, Ni-coated nanotubes also entangle and look like scattered dots (most of these dots are much less than 30 μm in diameter) in the microscope image. With the addition of an external magnetic field, the Ni-coated nanotubes start to stretch, vibrate, and enlarge. Eventually, these randomly dispersed dots form short lines, indicating that aggregated and entangled nanotubes form aligned chains and clusters. It is intuitive that the particles would move toward the direction of magnetic field. However, the Ni-coated nanotubes do not show significant movement under the magnetic field. Instead, they start to precipitate after some time. The lines of black particles along the magnetic field at the bottom of the vessel are clearly observed.

This observed phenomenon coincides very well with the time-dependent thermal conductivity results of Ni-coated nanotubes in water, shown in Figs. 18.1, 18.2, and 18.3. Without a magnetic field, the thermal conductivity value of the nanofluids containing Ni-coated SWNT is around 0.63–0.64 W/mK, which is essentially the thermal conductivity value of DI water. A reasonable explanation could come from the microscope image; too few contacts among these randomly dispersed Ni-coated carbon nanotubes exist to enhance the heat transfer of the fluid.

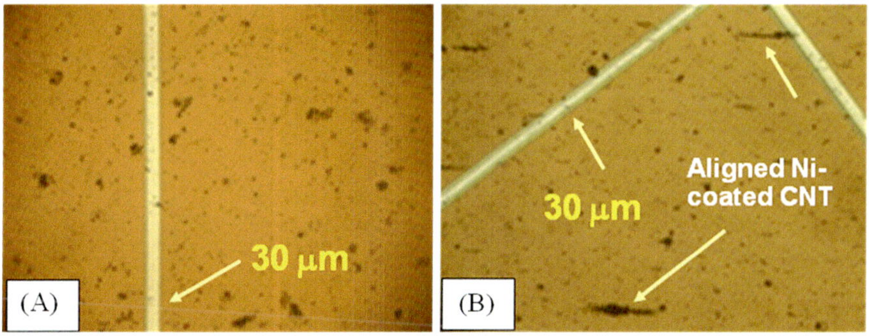

Fig. 18.4 Microscope image of 0.05 wt% Ni-coated SWNT in DI water. (**a**) Before magnetic field is applied and (**b**) after magnetic field (0.62 kG) is applied

Under the magnetic field, the thermal conductivity of the nanofluid shows very interesting behavior. The thermal conductivity initially increases with time but eventually reaches a peak (1.10 W/mK) after 2–4 min of magnetic field exposure, as shown in Table 18.2. These results can be explained by the nanotube alignment process. Microscopic video shows that it takes some time for the nanotubes to be completely oriented. As time in the magnetic field increases, the thermal conductivity decreases. This is due to the excessive agglomeration of nanotubes into larger particles which then start to precipitate or sediment. This was confirmed by the microscopic examinations.

Evidence of Ni-coated nanotube alignment in water-based nanofluids under an applied magnetic field was also obtained through backscattered electron SEM imaging. From Fig. 18.5, it is clearly seen that, under the magnetic field, the nanotubes align well in the magnetic field direction due to the magnetic moment of the Ni coating. The red arrow represents the magnetic field direction. The black material represents the carbon nanotubes. The small bright particles are images from some portion of the coated Ni metal. Since Ni is much heavier than carbon, Ni appears brighter in the backscattered electron image. Energy-dispersive X-ray analysis showed that the bright particles contain nickel. To prepare the SEM sample, drops of nanofluids containing Ni-coated nanotubes were placed on a SEM sample holder and left to dry under the magnetic field. Since it was shown that longtime exposure to a magnetic field will reduce the thermal conductivity value, the samples were placed under a magnetic field only for a short period of time (2–4 min).

Microscope images of 0.05 wt% Ni-coated SWNT in poly-α-olefin (PAO) oil are shown in Fig. 18.6a before magnetic field is applied and Fig. 18.6b after magnetic field is applied for 30 min. Similar Ni-coated SWNT images before and after external magnetic field exposure have been observed in DI water (Fig. 18.4). The alignment of Ni-coated CNTs in PAO oil appears to be less pronounced or slower

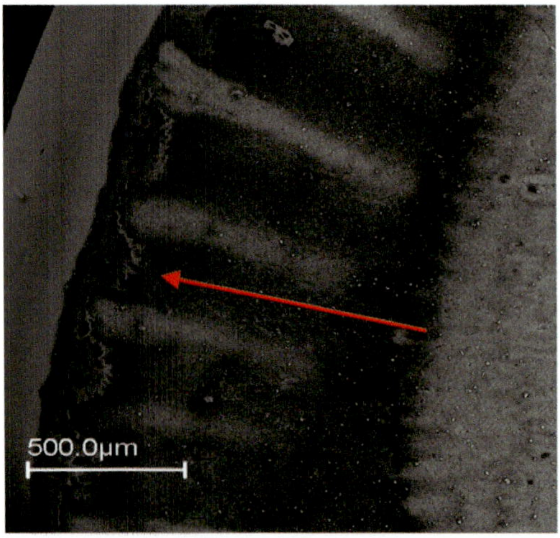

Fig. 18.5 Backscattered electron SEM image of 0.01 wt% Ni-coated single-walled carbon nanotubes aligned with magnetic field. Scale bar is 500 μm. *Red arrow* is magnetic field direction

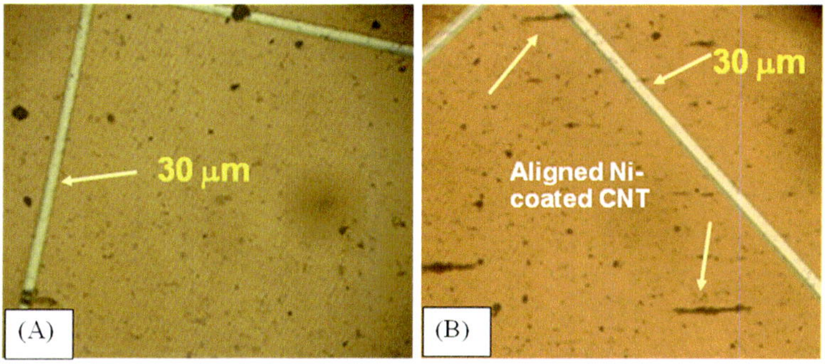

Fig. 18.6 Microscope image of 0.05 wt% Ni-coated SWNT in poly-α-olefin (PAO) oil. (**a**) Before magnetic field is applied and (**b**) after magnetic field (0.62 kG) is applied for 30 min

as compared with that of less viscous water. To better understand how the viscosity could influence the thermal conductivity and alignment process, time-dependent thermal conductivities of Ni-coated nanotubes in various oils with different viscosities were measured after 2–6 min exposure to the magnetic field. The data are listed in Table 18.3. It is clearly seen that, in all samples, the thermal conductivity values increase under the applied magnetic field. However, the viscosity of the fluid does influence the thermal conductivity enhancement ratio. It was seen that the more viscous the fluid, the less the thermal conductivity ratio was enhanced. Water has the lowest viscosity (1 cP) and the highest thermal conductivity enhancement (~80%), while the PAO oil has the highest viscosity (40 cP) and the lowest thermal conductivity enhancement (~11%). The reasonable explanation is that a highly viscous fluid would make it difficult for the Ni-coated nanotube to stretch and move toward alignment.

The macro-geometrical effect of the applied magnetic field on the nanofluid containing 0.05 wt% Ni-coated SWNT was evaluated by taking digital camera images, as shown in Fig. 18.7a in DI water and in Fig. 18.7b in poly-α-olefin (PAO). It is clearly observed that black particles form the lines along the magnetic field at the bottom of the vessel as shown in Fig. 18.7a. While the trend of alignment is the same for the PAO-based nanofluid, it takes more time for Ni-coated nanotubes to align or precipitate to the bottom of the vessel, due to the relatively high viscosity of

Table 18.3 Thermal conductivity maximum values and enhancement ratios for 0.02 wt% Ni-coated SWNT in water and oils under magnetic field strength of 0.38 kG after 2–6 min

Fluid brand names	Component	Viscosity (cP)	TC maximum value (W/mK)	TC increase ratio
De-ionized water	H_2O	1	1.10	0.80
Durasyn 166	Poly-α-olefin	40	0.21	0.11
Royco 500	Polyester	39	0.22	0.16
Royco 800	Polyolester	16	0.24	0.26

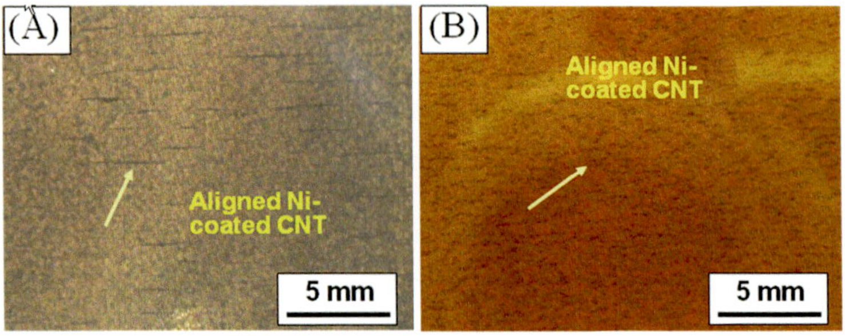

Fig. 18.7 Macroeffect of magnetically aligned, 0.05 wt% Ni-coated SWNT taken by digital camera. (**a**) In DI water and (**b**) in poly-α-olefin (PAO)

the PAO. The effect of base fluid viscosity is also reflected in the respective thermal conductivity enhancements shown in Table 18.3.

In order to make sure that all of these interesting results are solely due to the Ni-coated SWNT and not by Ni particles detached from the nanotubes, several transmission electron microscopy (TEM) pictures were taken. Figure 18.8 shows

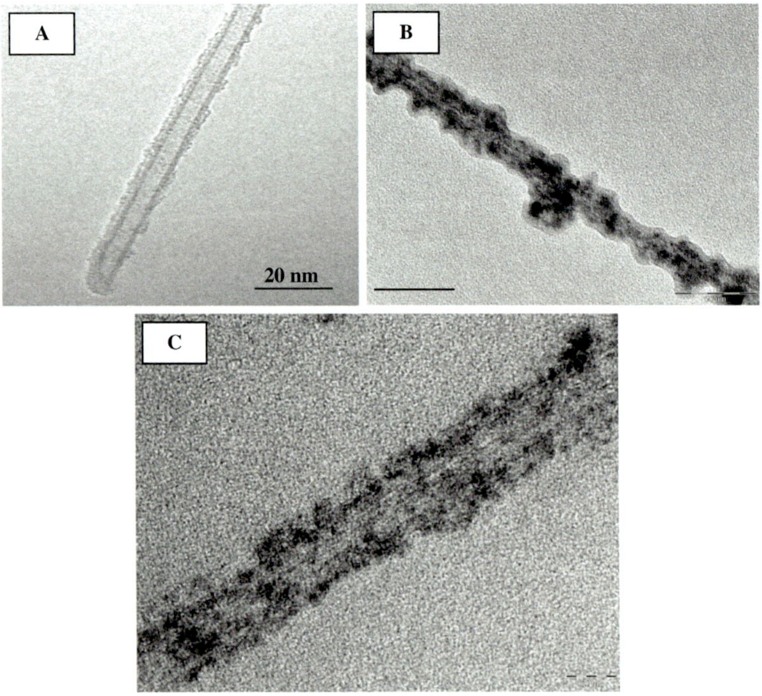

Fig. 18.8 TEM images of carbon nanotube. (**a**) As-received, uncoated, (**b**) Ni-coated nanotube sample before the magnetic field experiment, and (**c**) Ni-coated nanotube sample after the experiment. The scale bar is 20 nm

TEM images of (a) an as-received, uncoated carbon nanotube, (b) a Ni-coated carbon nanotube before the experiment, and (c) a Ni-coated carbon nanotube after the experiment. It is clearly seen that the Ni particles are still attached to the nanotube surface after the magnetic field experiment. This confirms that the results reflect the magnetically guidable nature of the Ni-coated SWNT.

18.4 Conclusion

In summary, we have successfully demonstrated that the thermal conductivity of heat transfer nanofluids containing Ni-coated single-walled carbon nanotubes (SWNTs) can be significantly enhanced in the presence of an applied magnetic field. The explanation for these interesting results is that Ni-coated nanotubes form aligned chains under an applied magnetic field, resulting in improved contacts and improved thermal conductivity. When holding the sample in the magnetic field for longer times, the nanotubes gradually form large clumps of nanotubes and precipitate, thus decreasing the thermal conductivity. The experiment that magnetic field strength was reduced and maintained at a much smaller field right after the maximum thermal conductivity was reached showed enhanced thermal conductivity values for longer times, compared to the case of complete magnetic field removal. Also, time-dependent thermal conductivity experiments indicate that the alignment process dominates the thermal conductivity enhancement rather than micro-convection.

Meanwhile, we have observed the real images of Ni-coated single-walled carbon nanotube in water and oils under magnetic field by high-speed microscopy and confirmed the magnetic alignment process. Initially, the nanotubes are randomly dispersed in the fluid. Upon longer exposure to the magnetic field, the nanotubes gradually stretch out linearly and finally align. The chain length in the images is around 30–150 μm, which is much longer than the real length of an individual nanotube, 5–40 μm. This indicates that nanotubes are aligned but form chains and clusters. Because of the semi-continuous nature of magnetic nickel nanoparticles, as well as the viscosity resistance of the fluid itself, it takes some time for the Ni-coated nanotubes to respond to the applied magnetic field and align. SEM images show that the Ni-coated nanotubes are aligned well under the influence of a magnetic field. TEM images also indicate that nickel remains attached onto the nanotubes after the magnetic field exposure and nanotube movement.

Further research is underway to improve the morphology, the configuration, and the volume of magnetic nanoparticles to make nanofluids with enhanced thermal properties, low viscosity, and longtime stability for coolant applications.

Acknowledgments The financial support of Army Research Lab (Cooperative agreement W911NF-08-2-0022) and NASA EPSCoR (award No. NNX09AU83A) is acknowledged. The author would also like to recognize the following people for their contributions to this chapter: Dustin Thomas and Brian Wright, SDSM&T Department of Materials and Metallurgical Engineering, for their help with thermal conductivity measurements; Lori Groven and Jan Puszynski, SDSM&T Department of Chemical and Biological Engineering, for their help with the preparation of the Ni-coated nanotubes; Edward Duke, SDSM&T Department of Geology and Geological Engineering, for his help with SEM imaging; Chen Li, Department of Mechanical Engineering, University of South Carolina; and Bo Shi, Department of Mechanical Engineering, University of Colorado Boulder, for their help on high-speed microscope image recording. Finally, the author would like to thank Sungho Jin, Department of Mechanical & Aerospace Engineering, University of California San Diego, for his foresight, initial magnetically sensitive nanotube samples and many technical comments.

References

1. S. Berber, Y.K. Kwon, D. Tománek, Unusually high thermal conductivity of carbon nanotubes. Phys. Rev. Lett. **84**, 4613 (2000)
2. X. Wang, X. Xu, S. Choi, Thermal conductivity of nanoparticle-fluid mixture. J. Thermophys, Heat Trans. **13**(4), 474 (1999)
3. S. Choi, Z. Zhang, W. Yu, F.E. Lockwood, E.A. Grulke, Anomalous thermal conductivity enhancement in nanotube suspensions. Appl. Phys. Lett. **79**(14), 2252 (2001)
4. B.H. Kim, G.P. Peterson, Effect of morphology of carbon nanotubes on thermal conductivity enhancement of nanofluids. J. Thermophys, Heat Trans. **21**, 451 (2007)
5. H. Xie, H. Lee, W. Youn, M. Choi, Nanofluids containing multiwalled carbon nanotubes and their enhanced thermal conductivities. J. Appl. Phys. **94**(8), 4967 (2003)
6. H. Hong, J. Wensel, S. Peterson, W. Roy, Disperse the carbon nanotube into polyalphaolefine oils (DURASYN 166) for heat transfer application. Polymeric Mater. Sci. Eng. **95**, 1076 (2006)
7. H. Hong, J. Wensel, F. Liang, W.E. Billups, W. Roy, Heat transfer nanofluids based on carbon nanotube. J. Thermophys, Heat Trans. **21**, 234 (2007)
8. P. Keblinski, J.A. Eastman, D.G. Cahill, Nanofluids for enhanced thermal transport. Mater. Today **8**(6), 36 (2005)
9. S. Jin, M. McCormack, Dispersoid additions to a Pb-free solder for suppression of microstructural coarsening. J. Electron. Mater. **23**, 8 (1994)
10. X.R. Ye, C. Caraio, C. Wang, J.B. Talbot, S. Jin, Room temperature solvent-free synthesis of monodisperse magnetite nanocrystals. J. Nanosci. Nanotechnol. **6**, 852 (2006)
11. H. Hong, B. Wright, J. Wensel, S. Jin, X. Ye, W. Roy, Enhanced thermal conductivity by the magnetic field in heat transfer nanofluids containing carbon nanotube. Synthetic Met. **157**, 437 (2007)
12. J. Philip, P.D. Shima, B. Raj, Nanofluid with tunable thermal properties. Appl. Phys. Lett. **92**, 043108 (2008)
13. Y. Zhang, N.W. Franklin, R.J. Chen, H. Dai, A study of metal coating on suspended carbon nanotubes: towards elucidating metal-tube interactions. Chem. Phys. Lett. **331**, 35 (2000)
14. M. Horton, H. Hong, C. Li, B. Shi, G.P. Peterson, S. Jin, Magnetic alignment of Ni-coated single wall carbon nanotubes in heat transfer nanofluids. J. Appl. Phys. **107**, 104320 (2010)
15. B. Wright, D. Thomas, H. Hong, L. Groven, J. Puszynski, E. Duke, X. Ye, S. Jin, Magnetic field enhanced thermal conductivity in heat transfer nanofluids containing Ni coated single wall carbon nanotubes. Appl. Phys. Lett. **91**, 173116 (2007)
16. P.D. Shima, J. Philip, B. Raj, Magnetically controllable nanofluid with tunable thermal conductivity and viscosity. Appl. Phys. Lett. **95**, 133112 (2009)

Index

W.L. Zhou, Z.L. Wang (eds.), *Three-Dimensional Nanoarchitectures*,
DOI 10.1007/978-1-4419-9822-4,